W9-AOV-290

Ice Sheets

Ice Sheets

TERENCE J. HUGHES

New York Oxford

OXFORD UNIVERSITY PRESS

1998

Oxford University Press

Oxford New York
Athens Auckland Bangkok Bogotá Buenos Aires Calcutta
Cape Town Chennai Dar es Salaam Delhi Florence Hong Kong Istanbul
Karachi Kuala Lumpur Madrid Melbourne Mexico City Mumbai
Nairobi Paris São Paulo Singapore Taipei Tokyo Toronto Warsaw

and associated companies in
Berlin Ibadan

Published by Oxford University Press, Inc.
198 Madison Avenue, New York, New York 10016

Library of Congress Cataloging-in-Publication Data
Hughes, Terence J., 1938–
Ice sheets / Terence J. Hughes.
p. cm.
Includes bibliographical references and index.
ISBN 0-19-506964-1
1. Ice sheets. I. Title.
GB2403.2.H84 1997
551.3'1—dc20 96-28246

1 3 5 7 9 8 6 4 2

Printed in the United States of America
on acid-free paper

To Bev.

My gratitude is tucked
between the pages.

PREFACE

The purpose of this book is to elaborate on three propositions:

1. Quaternary ice sheets trigger abrupt global climate change by advancing and retreating rapidly over large areas (Hughes, 1992a).
2. The triggering mechanism associated with advance of ice sheets is the surge mechanism in ice streams that drain these ice sheets (Hughes, 1992b).
3. The triggering mechanism associated with retreat of ice sheets is the calving mechanism when ice streams stagnate in water (Hughes, 1992c).

Chapter 1 presents various mechanisms, some quite tentative for now, by which Quaternary ice sheets interact with global climate. Chapter 2 presents a research strategy for studying dynamic systems that can be used to determine which of the proposed mechanisms are viable. Chapter 3 applies this strategy to the mechanisms associated with the largest Quaternary ice sheet, the Antarctic Ice Sheet. Chapter 4 presents the basic glaciological theory needed to study ice sheets, including creep and thermal instabilities, which are now only theoretical concepts. Chapter 5 analyzes sheet flow, which occurs over about 90 percent of landscapes covered by ice sheets. Chapter 6 analyzes stream flow, which occurs over only about 10 percent of these landscapes but which discharges some 90 percent of the ice. Chapter 7 analyzes shelf flow, which occurs when ice sheets and their ice streams become afloat. Chapter 8 analyzes calving of ice from floating ice shelves and from ice walls grounded in water of variable depth. Chapter 9 shows how Quaternary ice sheets flipped from interstadial to stadial modes to produce the last glacial maximum. Chapter 10 shows how all of the instabilities presented in Chapter 1, and quantified in subsequent chapters, might interact during the next Quaternary glaciation cycle.

The most speculative concepts presented in these chapters are the Jakobshavns Effect, the White Hole, an Arctic Ice Sheet, ice, dirt, dust, and cloud "machines" that turn on and off with "life cycles" of ice streams, and the location of "water gates" that open to terminate Quaternary glaciation cycles. These concepts are introduced in Chapter 1 and are used to model the next glaciation cycle in Chapter 10. The treatment of glacioisostasy is purely empirical, and the interpretation of glacial geology is purely subjective. The analysis of calving dynamics is purely theoretical. Deformation of wet sediments or till beneath ice sheets and the associated subglacial hydrology are not presented in detail. Field observations show that bed deformation is of minor significance beneath Ice Stream B in Antarctica today (Barclay Kamb, personal communication, 1996) and along the southern lobate margin of the former Laurentide Ice Sheet (George Denton, personal communication, 1996), sites where a deforming bed was thought to control ice motion (e.g., Iverson, 1985; MacAyeal, 1989; Alley, 1990, 1991; Clark, 1992, 1994, 1995; Clark and Walder, 1994; Clark and others, 1996). Iverson and others (1995) and MacAyeal and others (1995)show that most basal shear is at or near the ice-till interface during fast glacial flow, in which case resistance to flow must be bedrock "sticky spots" that poke through the till. Their view is adopted in this book.

A comprehensive treatment of time-dependent three-dimensional thermomechanical models of ice-sheet dynamics, and the numerical methods used in these models, are beyond the scope of this book. State-of-the-art models of this kind and their capabilities were presented in 1995 at the International Symposium on Ice Sheet Modeling (Annals of Glaciology, vol. 23, 1996). A thorough presentation of numerical methods employed in these models, *Lessons in Ice Sheet Modeling*, by Douglas R. MacAyeal, is available on Web site www2.uchicago.edu/pds-macayeal/.

This book challenges the view that ice sheets have only a passive response to Quaternary climate change (e.g., Broecker and Denton, 1989; Lowell and others, 1995). In that view, computer reconstructions of Quaternary ice sheets serve only to provide boundary conditions for computer models of atmosphere and ocean circulation (Anonymous, 1994). That view guided reconstructions of ice sheets at the last glacial maximum for CLIMAP (Denton and Hughes, 1981). It provided boundary conditions for most climate modeling over the subsequent thirteen years. It is the view that justifies the geophysical approach to reconstructing ice sheets during the last deglaciation that is currently represented by the ICE-4G model (Peltier, 1994). In this model, an array of Earth's radii viscoelastically shorten or lengthen in response to changing loads of ice and water on Earth's surface, such that the ocean surface is always a gravitational equipotential surface. By using spherical harmonics, known sea-level curves through time at the ends of some radii can be used to calculate changing ice thicknesses at the ends of other radii if the areal extent of glaciation is known.

This approach is presently supplanting recent innovative ice-sheet modeling that reveals inherent instabilities in ice sheets that can trigger abrupt climate change (e.g., Alley, 1990, 1991; MacAyeal, 1992, 1993a, 1993b; Alley and MacAyeal, 1994; Verbitsky and Saltzman, 1994, 1995).

These instabilities cannot be simulated by the ICE-4G model. Much of the innovative ice-sheet modeling that allows abrupt changes loses acceptance as the ICE-4G model gains acceptance.

Seven drawbacks in the ICE-4G model justfy modeling abrupt changes in ice sheets:

1. Ice-sheet elevations are computed indirectly from mantle dynamics constrained by sea-level data, instead of directly from ice dynamics constrained by surface and basal conditions. In science, direct approaches are preferable to indirect approaches, and it was the direct approach that produced abrupt changes.
2. Mantle dynamics are poorly known compared to ice-sheet dynamics. Research proceeds best from the more known to the less known, not vice versa.
3. Mantle dynamics are too simplistic in the ICE-4G model compared to ice dynamics in the best ice-sheet models. In the ICE-4G model, properties vary only vertically and are represented by only two layers in the mantle, whereas properties vary in all directions and are represented by over twenty layers in the best ice-sheet models.
4. The ICE-4G model ignores the dominant mantle deformation caused by thermal convection and descending lithosphere plates, and allows only elastic lithosphere deformation. In contrast, the best ice-sheet models treat all known deformations associated with ice sheets and their beds.
5. Mantle and lithosphere deformation in the ICE-4G model are constrained mainly by sea-level data that are sparse, nonuniform, often near major tectonic boundaries, and of uneven quality. Ice and bed deformation in ice-sheet models are constrained by glacial geology, which is often better dated and better understood.
6. The ICE-4G model responds only to glacioisostatic crustal loading and unloading by ice and water, the slowest response in Earth's climatic system, so it cannot produce known abrupt changes that are produced by ice-sheet models.
7. Initial states of dynamic and glacioisostatic equilibrium must be assumed at the last glacial maximum in order to use the ICE-4G model, whereas the best ice-sheet models allow nonequilibrium conditions throughout the glaciation cycle.

In addition to the seven general drawbacks inherent in the ICE-4G model, its application by Peltier (1994) to the last deglaciation has five specific drawbacks:

1. The ICE-4G model produced only 105 m of lowered eustatic sea level at the last glacial maximum, whereas 125 m of lowering are obtained when sea-level data are corrected for known rates of tectonic uplift (Edwards, 1995).
2. Ice-sheet elevations are obtained for a mantle three times more dense than ice, so that 200 m of ice thickness results in 133 m of ice elevation for glacioisostatic equilibrium (Peltier, 1995), whereas a mantle four times more dense than ice is more realistic and would give 150 m of ice elevation.
3. Tidal data along the American Atlantic coast require a lower mantle viscosity double that used in the ICE-4G model (Davis and Mitrovica, 1996). This alone invalidates ice-sheet thicknesses calculated by Peltier (1994).
4. Ice-sheet volumes computed as lowerings of eustatic sea level at the last glacial maximum include 22 m from Antarctica, where glacial geology indicates less than 10 m (George Denton, personal communication, 1996), and 22 m from Eurasia, including Siberia, where field studies indicate no ice sheet existed at the last glacial maximum (e.g., Sher, 1991, 1992; Rutter, 1995). The 105 m in Peltier (1994) is therefore reduced by up to 20 m unless the Laurentide Ice Sheet is thickened.
5. The ICE-4G model was unable to even approximate a fit to the excellent sea-level curve in the Gulf of Maine, which was under the Laurentide Ice Sheet at the last glacial maximum (Barnhardt and others, 1995).

These drawbacks justify a return to ice-sheet modeling based on ice dynamics constrained by surface and basal conditions that allow major rapid changes in the size and shape of Quaternary ice sheets, as presented in this book.

I wish to thank Wallace Walters for providing a draftsman for many of the figures; Ellen Wilch, Mauri Pelto, James Fastook, Kirk Maasch, Paul Mayewski, Michael Prentice, and Paul Prescott for allowing me to present figures showing their unpublished work; and my glaciology students for spotting many errors that are not in this book; James Fastook, Roger Hooke, Claire Parkinson, and anonymous referees for spotting many more errors that I missed; and Beverly Hughes for typing endless drafts of the manuscript and producing camera-ready copy. Beyond these individuals, I thank John Brittain for sustaining me as a graduate student in materials science, Johannes Weertman for introducing me to glaciology and for sustaining my glaciological career, Colin Bull for giving me my first job as a glaciologist, Gerald Holdsworth for my first glaciological field work, Harold Borns, Jr., for exposing me to glacial geology, and George Denton for encouraging me to view glaciology in the larger context of Quaternary climatic change. Others too numerous to name have also helped me to reach the point where this book seems like a useful next step. Five, however, must be named: Henry Brecher, Mikhail Grosswald, Johan Kleman, John Mercer, and Charles Swithinbank.

CONTENTS

Ice Sheets

1

A NEW PARADIGM FOR THE ICE AGE

Rhythms within Quaternary Glaciations

A new paradigm is emerging to explain the structure of glaciation cycles during the Quaternary Ice Age. The new paradigm is being imposed by rhythms in the history of Quaternary climatic change recorded in ocean-floor stratigraphy, particularly from sediment cores in the North Atlantic (Heinrich, 1988; Broecker and others, 1992; Andrews and Tedesco, 1992; Lehman and Keigwin, 1992; Eglinton and others, 1992; Bond and others, 1993; Bond and Lotti, 1995), and in ice sheet stratigraphy, particularly from ice cores through the summit of the Greenland Ice Sheet (GRIP Members, 1993; Dansgaard and others, 1993; White, 1993; Taylor and others, 1993; Grootes and others, 1993; Mayewski and others, 1994; 1997). These data show that a general correlation between global ice volume, as recorded by raised or submerged marine shorelines (Richards and others, 1994), and insolation variations caused by known variations in the tilt and precession of Earth's rotation axis, modulated by eccentricity variations in Earth's orbit around the Sun (Imbrie and Imbrie, 1980; Imbrie, 1985), coexists with abrupt changes in ocean-surface temperatures that begin and end within decades or less, last from one to a few centuries or millenia, and accompany abrupt atmospheric temperature changes. The new paradigm aspires to explain these abrupt changes by linking external forcing to the internal dynamics of ice sheets. Forging these links is the central theme of this book. The main glaciological processes in the new paradigm (Hughes, 1992a) are advance of ice sheets by fast currents of ice called ice streams (Hughes, 1992b) and retreat of ice sheets by calving bays that migrate up stagnating ice streams (Hughes, 1992c).

Records of climate change, such as those shown in Figure 1.1, have a definite rhythm. Abrupt changes beginning and ending within 100 years and lasting a few thousand years seem to appear in bundles and are superimposed on gradual changes that have been correlated with insolation changes caused by the 19,000 and 23,000-year cycles of Earth's axial precession and the 41,000-year cycle of Earth's axial tilt (Dansgaard and others, 1993; Bond and others, 1993; Heinrich, 1988; Hays and others, 1976; Imbrie and others, 1984). Abrupt changes may signal reorganizations in major parts of Earth's climate machine, such as the oceans (Broecker and others, 1985) or the ice sheets (Hughes, 1992a). Terminations of the 100,000-year cycles of major Quaternary glaciations may be triggered when abrupt and gradual warming mechanisms coincide (Denton and Hughes, 1983; Broecker, 1989). The emerging new paradigm explores these rhythms. This book focuses attention on the role of ice sheets in the new paradigm. The larger goal of synthesizing all aspects of Quaternary climatic change can be summarized with reference to Figure 1.1.

To succeed, the emerging paradigm must provide a plausible and testable explanation for the structure of the last Quaternary glaciation cycle indicated by the climate proxy data in Figure 1.1 and other climate records. Six observations are particularly noteworthy.

Insolation variations

The last glaciation cycle began when insolation at both mid (50°N) and high (75°N) northern latitudes was at a minimum, around 115 ka BP (thousand ^{14}C years before present), and ended when insolation at these latitudes was at a maximum, around 10 ka BP, as seen in Figure 1.1. All of the climatic changes within the glaciation cycle, as recorded by atmospheric dust, oxygen isotope stratigraphy, ice-rafted detritus, atmospheric carbon dioxide, and sea level, however irregular or abrupt, have a crude correlation with insolation. Therefore, as noted by Hays and others (1976), insolation variations are the "pacemaker" for Quaternary glaciation cycles. These variations at 50°N are caused primarily by a minor cycle of 19,000 years and a major cycle of 23,000 years in precession of Earth's rotation axis, and at 75°N are caused primarily by cycles of 41,000 years in tilt of Earth's rotation axis. Axial precession changes mid-latitude insolation during the equinoxes, which is out of phase across the equator, so when precession increases Northern Hemisphere insolation, it decreases Southern Hemisphere insolation. The

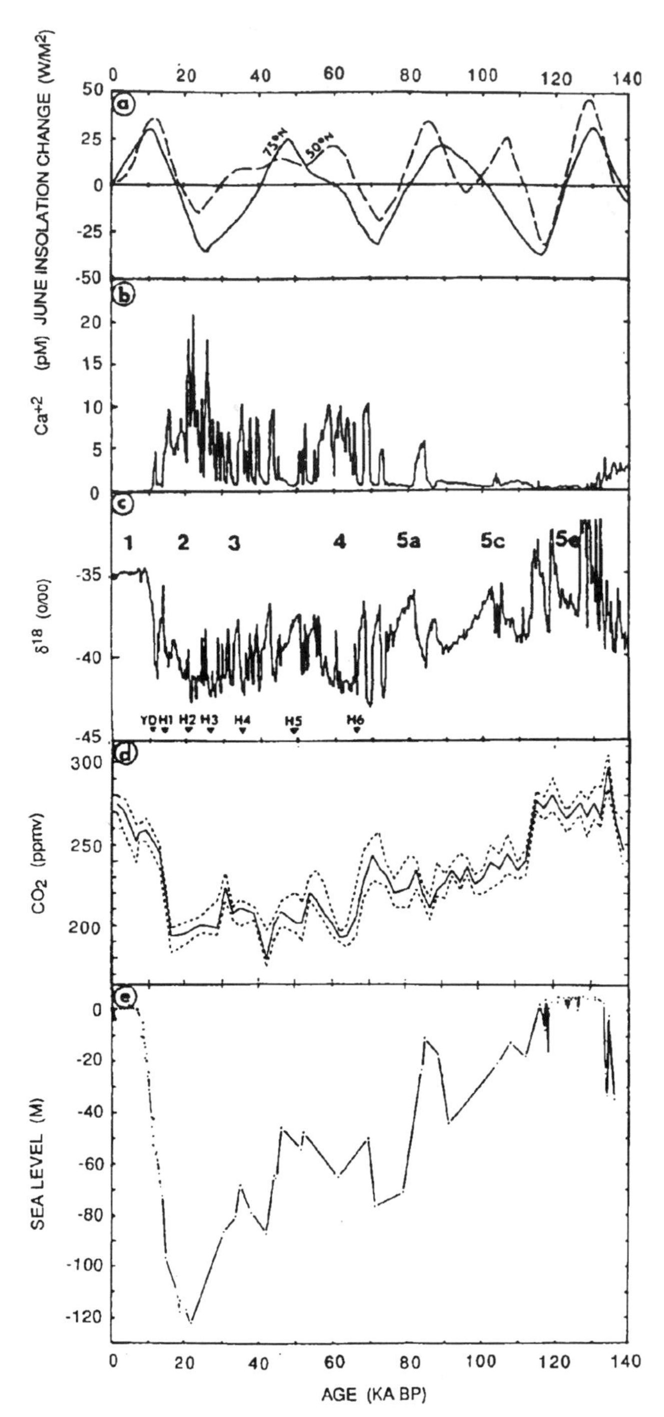

Figure 1.1: Rhythms of the last glaciation cycle. Shown are (a) departures from present-day insolation (Vernekar, 1972), (b) atmospheric dust as Ca^{+2} concentrations in Greenland ice cores (Greenland Ice-core Project Members (GRIP), 1993), (c) $d^{18}O$ in Greenland ice cores linked to marine $d^{18}O$ stages 1–5e and Younger Dryas and Heinrich ice-rafted sediments layers YD-H6 in the North Atlantic (GRIP Members, 1993; Broecker and others, 1992), (d) atmospheric CO_2 concentrations in Antarctic ice cores (Jouzel and others, 1993), and (e) departures from present-day sea level (Fairbanks, 1989; Ota, 1994).

The eccentricity of Earth's orbit around the sun changes with a cycle of about 100,000 years that serves primarily to amplify or reduce insolation variations caused by Earth's axial precession and tilt. Other insolation cycles, especially higher frequency cycles that may trigger abrupt climatic change, may be caused by direct variations in solar energy output.

In the new paradigm, Imbrie and others (1992, 1993) show with Talmudic clarity that Quaternary glaciation cycles are responses to Milankovitch (1930) insolation variations (Talmudic clarity: Careful reading is rewarded, cursory reading is not). They analyzed twenty-six time series of climate proxy records from ocean-floor sediments around the world, fifteen of which span the last 400,000 years, to identify and track patterns of Quaternary limatic change. Variables include summer insolation at 65°N, volume of ice sheets, concentration of atmospheric dust, and records of surface and deep ocean circulation. Analyses of the time series in a first-order dynamic model revealed four degrees of freedom at each Milankovitch insolation frequency: interglacial, preglacial, glacial, and deglacial. The Milankovitch frequencies are 23,000 years for Earth's axial precession, 41,000 years for Earth's axial tilt, and 100,000 years for Earth's orbital eccentricity, in descending order of insolation variations over these cycles. These degrees of freedom are end-member states that describe evolution of the climate system through a glaciation cycle. System responses were distributed in two clusters, one that leads and one that lags maximum ice-sheet volume. The data did not resolve climate change on time-scales of millennia or less. Within these constraints, Imbrie and others (1992, 1993) concluded that Quaternary glaciation cycles were continuous linear responses to insolation cycles of 23,000 and 41,000 years.

During an interglacial stage, such as today, westerly winds pass through Drake Passage and drive the Antarctic Circumpolar Current, which is unobstructed by continents, making it the only surface current strong enough to extend to the ocean floor. Earth's rotation creates an equatorial drift of the top layer of this current (the Ekman drift), which causes a strong upwelling of deeper water. This upwelling literally pulls North Atlantic Deep Water into the Antarctic. Warm westerly winds in the North Atlantic evaporate surface water, increasing its salinity so that it sinks to become North Atlantic Deep Water.

The preglacial stage ends this interglacial stage when reduced summer insolation at 65°N cools westerlies enough to slow or stop production of warm North Atlantic Deep Water. It no longer upwells in the Antarctic Circumpolar Current. This allows Antarctic sea ice to extend further north, which displaces westerly winds into latitudes north of Drake Passage. The Antarctic Circumpolar Current is then weakened because it abuts South America, which diverts part of it northward as a stronger Humbolt Current. The reduced upwelling slows flushing into the atmosphere of carbon dioxide stored in the deep ocean, causing general climatic cooling. General cooling and the shift of westerly winds toward the equator intensifies the thermal gradient to polar latitudes, thereby increasing the flow of warm moisture-bearing winds into high latitudes. Snow precipitated from these winds accumulates to become the Northern Hemisphere ice sheets.

This ushers in the glacial stage. Unable to upwell and recycle in the Antarctic, any remaining North Atlantic Deep Water flows into the Indian and Pacific Oceans, carrying dissolved carbon dioxide into these deep ocean reservoirs, where it largely remains and sustains global climatic cooling. Late Quaternary glaciations having the 100,000-year frequencies of Earth's orbital eccentricity seem to be triggered when the overall reduction of atmospheric carbon dioxide during the Cenozoic reaches a critical threshold that bifurcates Earth's climate into two modes, glacial and interglacial. In the full glacial model, Northern Hemisphere ice sheets advance from high latitudes into mid-latitudes, so their formerly regional control on climate becomes global. The ice sheets also advance onto polar continental shelves, thereby advancing into deeper water. These overextended ice-sheet margins are inherently unstable.

The deglacial stage begins when the inertial effect that large Northern Hemisphere ice sheets has on stabilizing the glacial climate is overcome by the instabilities inherent in these large ice sheets. Their advance was from hard beds, notably the crystalline shields of Canada and Scandinavia, onto water-soaked lacustrine and marine sediments. This advance from a high-traction bed onto a low-traction bed took place primarily as fast ice streams. The overextended ice-sheet margins were therefore thin, so they were exposed to rapid ablation by melting and calving. Basal heat conducted rapidly upward through thin ice to the surface eventually caused the bed to freeze, so stream flow halted and ablation rates exceeded forward ice velocities. Ice-sheet margins then retreated rapidly over glacioisostatically depressed beds which were filling with meltwater along landward margins and being inundated by the sea along seaward margins. The calving rate intensified in deepening water and ice-sheet margins began to float, with grounding lines retreating downslope into subglacial basins beneath the central domes of ice sheets. The resulting gravitational collapse of these ice sheets was irreversible when summer insolation at 65°N was maximized. This shifted westerly winds northward and renewed production of North Atlantic Deep Water, which ended the deglacial stage by releasing carbon dioxide from the ocean into the atmosphere. Interglacial conditions were then restored worldwide.

Atmospheric dust

Atmospheric dust, as recorded in the Greenland Ice Sheet Project (GRIP) corehole through the Greenland Ice Sheet in Figure 1.1 (GRIP Members, 1993), increases as atmospheric circulation becomes more intense. Southward advance of Northern Hemisphere ice sheets during Quaternary glaciation cycles steepens the thermal gradient that transports equatorial heat toward polar latitudes. This intensifies atmospheric circulation, as recorded by the polar circulation index (Mayewski and others, 1994). Intensified circulation promotes desertification in many parts of the world. This is the primary reason for increased atmospheric dust. In Figure 1.1, atmospheric dust peaked at the last glacial maximum, when Northern Hemisphere ice sheets attained their most southerly extent, but dust production was very irregular.

In the new paradigm, Northern Hemisphere ice sheets advance furtherest southward when they become thick enough for a frozen bed to thaw beneath their major ice domes. Melting of basal ice reduces bed traction, causing partial gravitational collapse and spreading of the ice domes. Spreading takes place primarily as ice streams that move over water-soaked sediments in either marine channels or proglacial lakes, since wet sediments provide much less traction than does wet bedrock. Glaciological and glacial geological studies indicate that stream flow has a life cycle lasting from a few years for surging mountain glaciers to a few millennia for major ice streams (Denton and Hughes, 1981). For a major ice stream, a life cycle would end when the ice dome became too low to provide the gravitational potential energy needed to sustain the kinetic energy of stream flow, or when the ice stream had flushed the proglacial lake and advanced onto permafrost or bedrock. Either way, its terminal ice lobe would stagnate, leaving moraines and outwash fans as it retreated. The fine fraction in these deposits is called "glacial flour" and it could be entrained by katabatic winds flowing down ice streams. Some of this dust would be recycled by higher inflowing winds that replaced the katabatic outflow, and some could be carried by westerly winds that tracked south of the ice sheet margins. Since katabatic winds are strong but intermittent, this may account for some of the "spikiness" in the dust record shown in Figure 1.1. These dust spikes have been resolved on a scale of days, which is compatible with the duration of desert dust storms or of katabatic winds flowing off of the Antarctic Ice Sheet today (Bromwich and others, 1990).

A goal of the new paradigm is to model katabatic winds for Northern Hemisphere ice sheets, to see if they would be strong enough to entrain glacial flour in moraines and outwash fans. If so, katabatic winds should then be coupled to regional circulation models for each ice sheet, with the regional models nested in a general circulation model of Earth's atmosphere. This would determine what, if any, atmospheric dust in Figure 1.1 and loess deposits south of Northern Hemisphere ice sheets could be attributed to glacial flour entrained by these katabatic winds.

Oxygen-isotope stratigraphy

Oxygen-isotope variations in Figure 1.1 are part of ice stratigraphy from the GRIP corehole and are therefore a proxy for air temperature at the high Summit Dome of the Greenland Ice Sheet (GRIP Members, 1993). Oxygen-isotope variations from microfossils in ocean-floor sediments reflect ice-sheet volume and, to some lesser degree, ocean temperature at the location where the organisms once lived. Oxygen-isotope stratigraphy from Quaternary ocean-floor sediments reveal irregular "sawtooth" ice volume increases and climatic cooling during Quaternary glaciations, followed by rapid decreases in ice volume and rapid warming that terminated the glaciations (Broecker and van Donk, 1970). Over the last 900,000 years, glaciations lasted about 90,000 years and interglaciations lasted about 10,000 years. Within each glaciation, sawtooth cooling lasted about 80,000 years and terminations lasted about 10,000 years. Better resolution of oxygen-isotope stratigraphy revealed that the sawtooth and termination structure is repeated on a smaller scale as stadials and interstadials within the last glaciation cycle, as shown in Figure 1.1 (Bond and others, 1993). The larger sawteeth are now called Bond cycles and they average about 10,000 years in duration. They consist of smaller sawteeth that last 1000 to 3000 years and are called Dansgaard-Oeschger events (Dansgaard and others, 1993).

In the new paradigm, stadials are caused by outbursts of large tabular icebergs from ice streams that drain Quaternary ice sheets and that have life cycles lasting from 1000 to 3000 years. Iceberg outbursts during these life cycles cool and freshen ocean surfaces, notably in the North Atlantic, for the duration of the life cycle. Since many ice streams drain an ice sheet, both along its landward and seaward margins, they will lower and extend the ice sheet. Landward extension will flush proglacial lakes, which also cools and freshens ocean surfaces. This sequence of life cycles begins when ice sheets become thick enough for the bed to thaw beneath their major ice domes, and the life cycle sequence ends when the ice streams have downdrawn these ice domes enough for the bed to refreeze beneath them, as first demonstrated by MacAyeal (1993a, 1993b). This takes place over the 7000 yr to 15,000 yr (years) time span of Bond cycles, and constitutes a stadial of the glaciation cycle. In this view, Dansgaard-Oeschger events are not warm interruptions in a generally cooling climate, but record iceberg outbursts during life cycles of ice streams that cause cooling interruptions in a warm climate such that their cumulative effect is climatic cooling. Cooling ends only when ice domes have lowered too much to supply ice sufficient to sustain ice streams, so ice-stream life cycles end and bring the stadial to an end. As ice slows when the bed freezes beneath the ice dome and its ice streams, ablation rates by melting or calving of ice then exceed the ice velocity, so the ice margin must retreat. Since basal freezing can be abrupt, the end of a stadial has the abrupt character of a termination.

Ice retreats over a glacioisostatically depressed bed during the following interstadial, and glacial meltwater fills the depression to produce proglacial lakes along the landward margin of ice sheets. Evaporation from adjacent water bodies can double precipitation rates, so that lowered ice domes can be restored to their higher elevations in half the time. Precipitation doubled over the Greenland Ice Sheet when it became more surrounded by seasonally ice-free water at the end of the Younger Dryas (Alley and others, 1993). In the new paradigm, katabatic winds flowing down stagnating ice streams sweep floating ice from lacustrine and marine ice-sheet margins and cause rapid evaporation rates from the windswept water. Much of this water vapor is carried by inflowing air above the katabatic outflowing winds and precipitates as snow over the ice sheet. This precipitation mechanism augments precipitation from convective storm systems generated by the circumpolar vortex and allows the retreating ice sheets to thicken more rapidly than otherwise. It is a mechanism that intensifies as ice sheets thicken because their steepening surface slope creates stronger katabatic winds. Eventually the ice becomes thick enough to begin thawing the bed, causing another stadial during which ice domes lower and ice streams advance.

Ice-rafted detritus

During the last glaciation, ice-rafted detritus called Heinrich layers were deposited in the North Atlantic at the end of Bond cycles (Heinrich, 1988; Bond and others, 1992; Broecker and others, 1992). These carbonate-rich sediments originated from large marine ice streams that drained the Laurentide Ice Sheet, an ice stream in Hudson Strait being the leading candidate (Andrews and Tedesco, 1992; Andrews and others, 1993, 1994; Dowdeswell and others, 1995). The appearance of Heinrich layers and a similar layer at the end of the Younger Dryas are shown in Figure 1.1 and have spacings from 7000 to 10,000 years. Other layers of ice-rafted detritus having spacings from 2000 to 3000 years occur at the end of Dansgaard-Oeschger events, and have been traced to other Quaternary ice sheets in the North Atlantic (Bond and Lotti, 1995) and the North Pacific (Kotilainen and Shackleton, 1995). The iceberg outbursts that transported the ice-rafted detritus seem to have caused regional, and perhaps global, climatic cooling followed by rapid warming that marked the end of Bond cycles and Dansgaard-Oeschger events within Bond cycles (e.g., Keigwin, 1995; Thunell and Mortyn, 1995; Kennett and Ingram, 1995; Lowell and others, 1995).

In the new paradigm, ice-rafted detritus is produced at the end of life cycles for marine ice streams, when basal freezing incorporates the detritus into regelation ice (MacAyeal, 1993a, 1993b; Alley and MacAyeal, 1994; Dowdeswell and others, 1995). After the bed becomes frozen, so that stream flow stops, the detritus-charged basal ice is released when calving bays migrate up the submarine channels occupied by the stagnating ice streams. For example, a calving bay migrated 1000 km up Hudson Strait in 800 years during the last deglaciation (Prest, 1969; Paterson, 1994). Estuarine circulation in Hudson Strait would evacuate the detritus-charged icebergs into the Labrador Sea (Hughes, 1987, 1992c), where the North Atlantic Current would transport them across the North Atlantic as they melted (Andrews and Tedesco, 1992; Dowdeswell and others, 1995).

Owing to the extreme instability of marine ice-stream grounding lines on sills at the seaward end of marine channels like Hudson Strait (Thomas, 1977; Hughes, 1987, 1992a), the rise in global sea level during the life cycles of major ice streams could trigger rapid grounding-line retreat of other large ice streams, regardless of whether they were active or stagnating at the time. For example, ice discharged by Laurentide ice streams might have raised sea level enough to cause the grounding lines of major Antarctic ice streams to retreat, and some of these icebergs may have been transported northward along the Pacific coast of South America by the Humbolt Current. Regional cooling as these icebergs melted may have lowered snowlines and advanced termini of Andean glaciers. This may account for the interhemispheric correlations of late Pleistocene glacial events that Lowell and others (1995) linked to their dates for the of piedmont glacier lobes in the Chilean Andes, although they invoked other explanations. There was certainly no shortage of major ice streams with the capacity to abruptly regulate the volume of ice discharged by Northern Hemisphere and Antarctic ice sheets during the last glaciation and deglaciation, as required by the new paradigm.

Atmospheric carbon dioxide

Atmospheric carbon dioxide can be produced directly from variations in solar energy output or by changes in the reservoir capacity of Earth's biosphere and hydrosphere, including long-term tectonic activity and weathering rates linked to the carbon cycle (Saltzman, 1987; Saltzman and Maasch, 1988, 1991; Verbitsky and Oglesby, 1992; Saltzman and Verbitsky, 1993, 1994, 1995). Along with water vapor and methane, carbon dioxide is a major "greenhouse" gas that can regulate climatic warming and cooling, and its variations during the last glaciation cycle are shown in Figure 1.1 for the Vostok ice core in Antarctica (Jouzel and others, 1993). Although the resolution is not as great, atmospheric carbon dioxide shows thc same "sawtooth" decrease, followed by an abrupt increase at the termination of the last glaciation, that is recorded by oxygen-isotope ratios. However, except at the beginning and the end of the last glaciation cycle, there is no clear correlation with June insolation.

In the new paradigm, forcing by atmospheric carbon dioxide in Figure 1.1 is capable of producing Northern Hemisphere ice sheets and associated ice shelves, notably on Eurasian continental shelves and over the Arctic Ocean at the last glacial maximum, and their subsequent collapse and disintegration during the Holocene (Lindstrom and MacAyeal, 1989, 1993; Lindstrom, 1990; Verbitsky and Oglesby, 1992). The role of the carbon cycle in the new paradigm was known over a century ago (Maasch, 1992), but it has been incorporated into a time-dependent climate model only recently by Saltzman and Verbitsky (1994). In a zero-dimensional model that combines proxy records for ice volume, atmospheric carbon dioxide, and mean ocean temperature as boundary conditions over time for a fast-response equilibrium general circulation model, they computed equilibrated global climates at sequential "snapshots" of time. They found that Cenozoic tectonic outgassing and weathering processes decreased atmospheric carbon dioxide to a level that permitted Quaternary glaciation cycles to begin about 2.5 million years ago and that excited a free oscillator that bifurcated climate with a periodicity of about 100,000 years during the Pleistocene. Imbedded within this cycle were cycles of about 40,000 years at high latitudes and 20,000 years at mid-latitudes that lagged by about 1000 years the respective tilt and precession cycles of Earth's axial rotation, due to inertia inherent in ocean circulation. In the Atlantic Ocean, circulation oscillates between a warm interglacial mode having a deep and broad thermocline, thermohaline circulation dominated by salinity, and weak tropical upwelling of cold carbon-rich water, and a cold glacial mode having a shallower and narrower thermocline, thermohaline circulation dominated by temperature, and strong tropical upwelling of cold carbon-rich water. The cold glacial mode allows expansion of ice sheets and sea ice, which intensifies surface winds, but the upwelling increases atmospheric carbon dioxide that eventually reduces the ice extent by "greenhouse" climatic warming. This initiates other processes that also increase atmospheric carbon dioxide and accelerates termination of the glaciation cycle.

Refinements of the model by Saltzman and Verbitsky (1994) allow simulations of stadials and interstadials within a glaciation cycle that are linked to the life cycles of ice streams that drained Quaternary ice sheets. The model then predicts life cycles for Laurentide ice streams that compare reasonably with the age of several Heinrich layers of ice-rafted detritus that appear at the end of Bond cycles (Verbitsky and Saltzman, 1994, 1995). Further refinements can generate climatic changes on the millennial time scale that produces Dansgaard-Oeschger events (Dupont, 1996).

Sea level

Sea-level data in Figure 1.1 are a composite from several sites, notably Barbados and the Huon Peninsula of New Guinea (Fairbanks, 1989; Ota, 1994). The other climate proxy records are at single sites, but no continuous single-site record of sea level exists over the past 140,000 years. This is a serious problem with sea-level data because problems with dating and tectonic stability are difficult to reconcile. For example, multisite sea-level data dated at about 80 ka BP indicate that sea level was close to present-day level at rapidly rising tectonic sites in New Guinea (Bloom and Yonekura, 1990), slowly rising tectonic sites on the Pacific coast of North America (Muhs and others, 1994), the tectonically stable Atlantic Coastal Plain of Virginia and North Carolina (Szabo, 1985), and sites on Bermuda and the Florida Keys, which lie on tectonically stable platforms (Ludwig and others, 1996). According to other data, sea level was at least 15 m to 18 m lower at tectonically rising Barbados (Bender and others, 1979; Gallup and others, 1994), 19±5 m lower in New Guinea (Chappell and Shackleton, 1986), or even 60 m to 70 m lower according to analysis of deep sea sediments (Dwyer and others, 1995). However, data on the Atlantic Coastal Plain of the United States and on Bermuda are contaminated by continuing glacioisostatic rebound beneath the former Laurentide Ice Sheet, making sea-level anomalously low as the peripheral forebulge migrates northward (Davis and Mitrovica, 1996). Without this, 80 ka

BP (thousand ^{14}C years before present) sea-level data at these sites would be below present-day sea level.

Barbados has the longest record of sea level, giving + 7 m at about 125 ka BP, –19 m at 100 ka BP and 88 ka BP, –65 m at 79 ka BP and 78 ka BP, –74 m at 71 ka BP, –84 m at 30 ka BP, and –118 m at 19 ka BP, before rising in three pulses from –93 m at 14.5 ka BP, from - 53 m at 11.5 ka BP, and from –17 m at 7.5 ka BP (Bard and others, 1990). These three pulses in rising sea level during termination of the last glaciation have been dated in detail on Barbados by Blanchon and Shaw (1995).

In the new paradigm, pulses of rising sea level occur during life cycles of major ice streams that drain Quaternary ice sheets. During their life cycles, marine ice streams discharge flotillas of large icebergs and terrestrial ice streams flush proglacial lakes. Icebergs displace sea water and flushed lakes add fresh water, so both raise sea level. Blanchon and Shaw (1995) attributed the three pulses of rising sea level during the last deglaciation to climatic changes that depend on the extent and dynamics of ice sheets. The pulse at 14.5 ka BP began when widespread thawing of a frozen bed beneath the Laurentide Ice Sheet initiated gravitational collapse of interior domes and life cycles of ice streams which flooded the North Atlantic with icebergs. Lower ice elevations allowed westerly winds to pass over the ice sheet, enhancing precipitation and drawing warm equatorial water into the North Atlantic. Nearly full glacial conditions returned when a refreezing bed shut down the ice streams, so calving bays migrated up marine ice streams and ice-rafted detritus was transported across the North Atlantic. After enhanced precipitation restored interior ice elevations enough to again thaw the bed, ice streams again passed through their life cycles, causing the Younger Dryas pulse at 11.5 ka BP. At that time however, July insolation was nearing a maximum at 60°N, so the southern Laurentide margin continued to retreat after the life cycles ended. Therefore, proglacial lakes that formed when meltwater filled the glacioisostatically depressed landscape were drained catastrophically when the retreating ice opened spillways into the sea. The final pulse began at 7.5 ka BP when rising sea level from collapsing Northern Hemisphere ice sheets initiated grounding-line retreat of Antarctic ice streams, thereby accelerating gravitational collapse of the Antarctic Ice Sheet, especially marine portions grounded below sea level in West Antarctica.

The sea-level curve in Figure 1.1 shows the same "sawtooth" structure seen in the oxygen isotope and carbon dioxide curves during the last glaciation. The fall in sea level associated with stadials is as rapid as the rise in sea level during termination of the glaciation. In the new paradigm, a rapid fall takes place when marine ice sheets grow on Arctic continental shelves and thereby create a huge precipitation sink that also dams rivers flowing toward the Arctic (Hughes, 1992a).

Stable and Unstable Climatic Change

Gradual predictable long-term cycles of climate change overprinted by sudden unpredictable short-term changes having irregular cycles or no cycles at all raise a fundamental question. Is Earth's climate machine stable or unstable during an ice age of the late Quaternary type (Budyko, 1974)? A climate curve showing how global climate changes through time has zero slopes during glacial maxima and minima and has maximum positive or negative slopes during glacial retreat or advance, as shown schematically in Figure 1.2. If the climate machine is fundamentally stable, being controlled by regular insolation changes, then abrupt climate changes are mere perturbations and research should focus on climate dynamics during glacial maxima and minima when global climate approached steady-state equilibria. If the climate machine is fundamentally unstable, however, then abrupt changes are the most important manifestations of instability and research should focus on climate dynamics during glacial advance and retreat, when global climate was cooling and warming most rapidly. Perhaps Earth's climate machine is metastable, with stability controlled by regular cycles that impose rigid external boundary conditions, and instability controlled by irregular inter-actions among components of the climate machine that produce flexible internal boundary conditions.

Earth's climate machine is fueled externally by the regular beat of insolation cycles, but its internal components in the lithosphere, biosphere, cryosphere, hydrosphere, and atmosphere have many interactions which march to their own drummers. This book examines stable and unstable behavior in the most important cryosphere component, continental ice sheets, and attempts to show how their march through time contributes to the rhythms of global climate change in the Quaternary Ice Age.

In ice sheets, sheet flow diverging from interior ice domes converges to become stream flow toward grounded margins, and becomes shelf flow when margins become

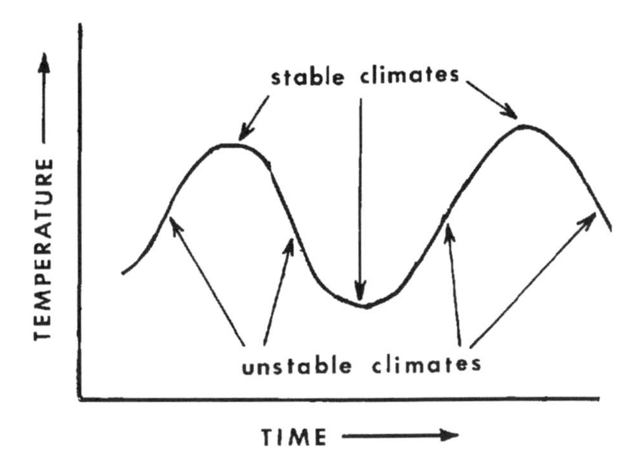

Figure 1.2: Earth's climate stability represented by a curve of mean global surface temperature versus time. A fundamentally stable climate is represented by its maximum perturbations, which are points of zero slope on the curve. A fundamentally unstable climate is represented by its fastest rates of change, which are points of maximum positive and negative slope on the curve.

afloat. Sheet flow is relatively stable and responds to external insolation forcing. Stream flow is fundamentally unstable and responds to internal forcing. Shelf flow is metastable and responds to both external and internal forcing. The effect of stable, unstable, and metastable behavior in dynamic systems is illustrated schematically in Figure 1.3. Applying this behavior to the effect of ice sheets on global climatic stability, slow insolation variations allow ice sheets to fluctuate in a narrow range that produces stadial and interstadial climates. Fast internal mechanisms widen the range of fluctuations to produce unstable advances and retreats that begin and end a global glaciation cycle if the mechanisms reinforce each other and external insolation variations. If the fast internal mechanisms are out of harmony with each other and with insolation variations, they will produce only metastable "noise" on a harmonious climate curve showing changes in ice volume over time, with the noise being local or regional instead of global. Mankind began as hostages to these forcingmechanisms, but if we discover that some mechanisms are finely tuned, perhaps by manipulating these mechanisms we can orchestrate the rhythms of climate change.

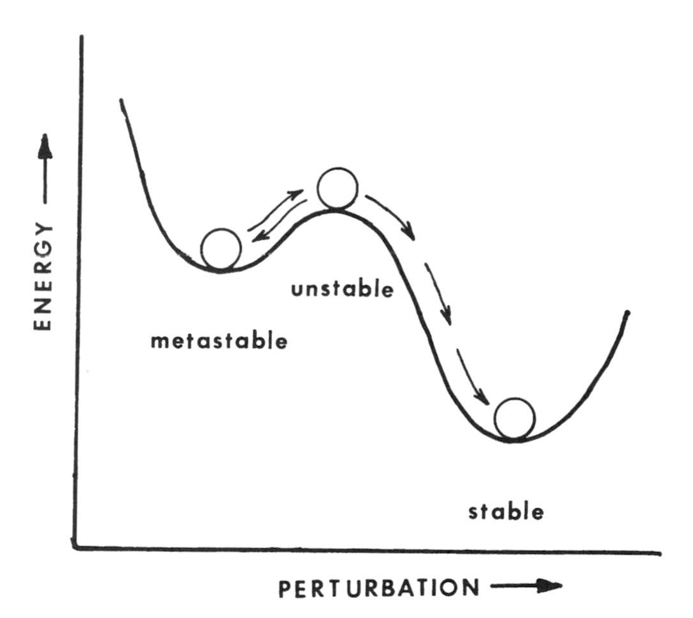

Figure 1.3: A cartoon illustrating metastable, unstable, and stable climate equilibria. Climate is represented by a ball and climate change is represented by perturbations that can roll the ball from a metastable energy dip over an unstable energy hump into a stable energy well. Metastable perturbations are reversible if small and irreversible if large. Unstable perturbations are always irreversible. Stable perturbations are always reversible. If Earth's climate is fundamentally stable, the ball typically rolls within a dip or a well, and rarely rolls over the hump. If Earth's climate is fundamentally unstable, the ball typically rolls over the hump into dips or wells. Wells represent glaciation cycles. Dips represent stadials and interstadials within a glaciation cycle.

The interaction between glaciation and climate change commands our attention more forcefully the more we realize that Earth's rather mild climate today may be finely tuned. Future ice sheets undoubtedly loom over our northern horizon. If their certain advance is temporarily halted by the "greenhouse" global warming we are manufacturing, our manmade Indian Summer will usher in a glacial winter that will be even more cataclysmic because it was forestalled. If Earth's climate is fundamentally unstable, could we nudge it irreversibly toward climates like those on Earth's sister planets, Venus and Mars, the one a torrid hothouse and the other a frigid desert? Fluvial erosion features on Mars show that it once had a climate more like ours. Could a runaway glaciation make Earth an icy tomb? Before Earth's biosphere was established, the Palaeoproterozoic world fluctuated from virtually ice-free environments to environments in which glaciation extended virtually to the equator (Evans and others, 1997). Does Earth's biosphere, which includes us, maintain stable rhythms that prevent a glaciological Götterdämmerung?

the Jakobshavns Effect

Perhaps the most dramatic glaciological manifestation of the equilibrium states represented in Figure 1.3 is the competition between positive and negative feedback mechanisms that are observed on Jakobshavns Isbræ in Greenland. The successive reinforcement of positive feedback, therefore, is called the Jakobshavns Effect (Hughes, 1986a).

The crest of the north-south ice divide of the Greenland Ice Sheet is bowed sharply eastward between 69°N and 72°N. In this latitude band, sheet flow extends from the ice divide to the eastern ice margin, but on the western flank, sheet flow becomes mostly stream flow toward the western ice margin. There, twenty outlet glaciers discharge some 22 percent of Greenland ice into coastal fjords having a combined width of only 80 km (Bader, 1961; Carbonnell and Bauer, 1968). Six of the twenty outlet glaciers discharge 21 percent of the ice and one, Jakobshavns Isbræ (69°N, 50°W), discharges up to 7.6 percent (Bindschadler, 1984). Jakobshavns Isbræ is shown in Figure 1.4. It is the world's fastest-known glacier, with a midsummer velocity of 23 m/d at its terminus (Lingle and others, 1981). In their computer model of ice dynamics, Radok and others (1982) predicted a largely thawed bed on the west flank of the central ice divide and a largely frozen bed on the east flank of the Greenland Ice Sheet. Sharp eastward bowing of the ice divide seems to be caused primarily by a loss of basal traction on the west flank caused by lubrication of the ice-rock interface. Secondary factors are a somewhat higher bed topography on the east that would tend to displace the ice divide eastward, and much heavier snow accumulation on the west that would tend to displace the ice divide westward (Weertman, 1973).

Loss of basal traction generates ice streams. As seen in Figure 1.4, two ice streams merge to form Jakobshavns Isbræ, a major ice stream from the east and a minor ice stream from the north. They bend around an ice fall at the

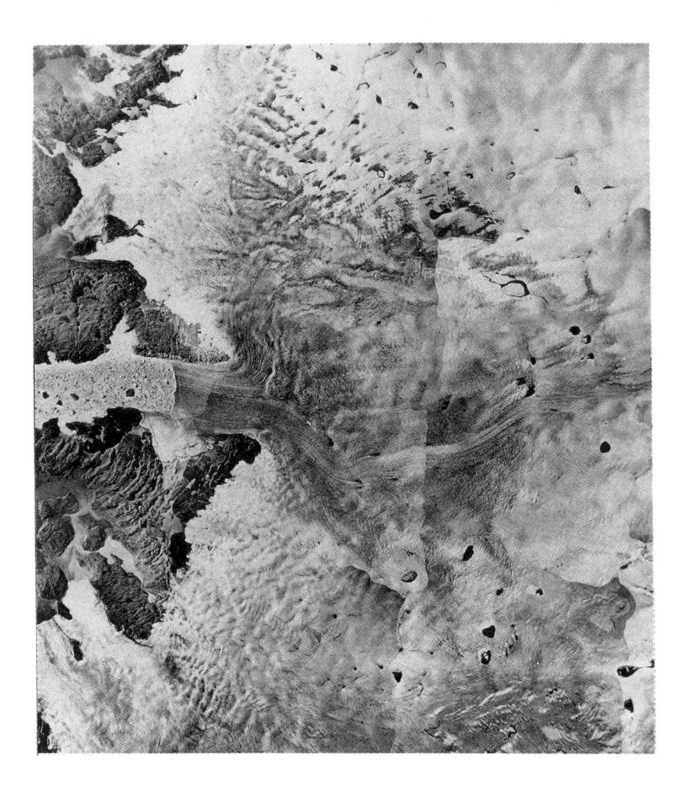

Figure 1.4: An areal photomosaic of Jakobshavns Isbræ (69.2°N, 49.9°W) in Greenland during July. Ice streams from the east and the north join at the head of Jakobshavns Isfjord to become Jakobshavns Isbræ. Photomosaic by H.H. Brecher.

head of Jakobshavns Isfjord and become afloat in water over 800 m deep (Carbonnell and Bauer, 1968). Jakobshavns Isbræ is therefore a floating ice shelf within Jakobshavns Isfjord and, except where it has bedrock pinning points, it has no basal traction at all. Instead, Jakobshavns Isbræ acquires the enormous pulling power of a thick unpinned ice shelf, as analyzed by Hughes (1992b). Loss of basal traction allows Jakobshavns Isbræ and the other west-coast outlet glaciers between 69°N and 72°N to literally pull ice out of the Greenland Ice Sheet. In turn, this pulling power seems to depend on high rates of summer melting over the heavily crevassed surfaces of the ice streams, with much of the meltwater passing through crevasses and lubricating the basal ice-rock interface. This relationship between surface melting and pulling power in a heavily crevassed ice stream is the Jakobshavns Effect (Hughes, 1986a, 1992a, 1992c). The Jakobshavns Effect seems to be a consequence of several positive feedback processes.

Surface roughness

Over all its length, the surface of Jakobshavns Isbræ is a chaos of crevasses and seracs (Echelmeyer and others, 1991). This is the most distinguishing feature of a surging glacier (Paterson, 1981, p. 275). It approximately triples the surface area of the glacier and allows the glacier to absorb more than twice as much solar energy as it otherwise would by presenting greater surface area to direct solar radiation and by permitting multiple reflections of indirect radiation between crevasse walls (Pfeffer and Bretherton, 1987).

Surface melting

July surface melting rates average 0.1 m/d on serac faces in the floating part of Jakobshavns Isbræ (Lingle and others, 1981). Melting rates on all faces, sunlit or in shadow, north or south facing, windward or leeward, vertical or horizontal, were substantial (Echelmeyer and others, 1992). All of this meltwater drains into crevasses to either refreeze or find its way to the bed. Latent heat released by refreezing enhances creep spreading rates because of thermal softening. Meltwater reaching the bed allows greater basal sliding rates in grounded ice because it lubricates the ice-bed interface and increases basal water pressure. An enormous volume of water is involved. A 0.1 m/d midsummer melting rate on a rough surface that triples the surface area of the 8 km long by 6 km wide portion of Jakobshavns Isbræ floating in thickness. Averaging 0.3 m/d in midsummer with 0 m/d in Jakobshavns Isfjord removes 0.3 m/d of vertical ice midwinter gives 55 m/yr of ice thinning over 48 km^2 of floating ice. This releases 2×10^{18} cal/d if all the meltwater refreezes internally; otherwise it delivers 20 km^3/yr of water to the bed. Figures for the grounded part of Jakobshavns Isbræ would be less, but probably close to the same order of magnitude.

Extending flow

If all surface meltwater refreezes internally, the 55 m/yr of vertical thinning over 48 km^2 of floating ice requires a velocity increase of 550 m/yr at the 6 km wide and 800 m thick calving front of Jakobshavns Isbræ, because of volume conservation (provided that water freezing in a crevasse allows the crevasse to reopen). Ice velocity from the fjord headwall to the calving front increases from 15 m/d to 23 m/d (Lingle and others, 1981), giving 2920 m/yr caused by a combination of internal refreezing and creep spreading in the floating ice. Averaging this velocity increase over the 8 km floating length gives a longitudinal strain rate of more than 0.3/yr in extending flow. The mean ice temperature would have to be –5°C in order to maintain this extending strain rate in a glacier of constant thickness floating between fjord sidewalls that were essentially frictionless and parallel (Weertman, 1957b; Paterson, 1981, Table 3.3).

Basal uncoupling

Since lateral shear zones alongside the fjord sidewalls are nearly parallel, a 0.3/yr extending strain rate is also the thinning strain rate. Applying the buoyancy requirement of a floating glacier to the 90 m ice-surface elevation at the headwall grounding line gives a grounding-line ice thickness of about 900 m. Uniform creep therefore thins the floating ice about 270 m/yr, which raises basal ice some 240 m/yr at the grounding line. Surface meltwater reaching the bed, instead of refreezing internally, is a real loss of ice mass. Melting 55 m/yr vertically on the rough surface raises ice at the grounding line about 50 m/yr if all surface meltwater reaches the bed. Raising basal ice at the grounding line 240 m/yr by creep thinning and up to an additional 50 m/yr by surface melting is a powerful means for basal uncoupling, which consists of bedrock lubrication in the grounded glacier, eliminating bedrock pinning points in the floating glacier and causing the grounding line to retreat. Grounding-line retreat of Jakobshavns Isbræ since 1850 has apparently been halted at the head of Jakobshavns Isfjord, where ice pours over an icefall between the east and the north ice streams. The bedrock sill beneath the icefall apparently continues across the ice streams in lowered form, and stream flow continually brings thick ice to the grounding line.

Lateral uncoupling

Stream flow emerges from sheet flow 100 km east of the calving front of Jakobshavns Isbræ. This is evidence that lateral uncoupling alongside an ice stream can begin long before the ice stream becomes an outlet glacier confined between fjord walls. Lateral uncoupling is most complete inside Jakobshavns Isfjord, however, as demonstrated by the nearly constant transverse profile of the ice velocity between the lateral shear zones (Bauer and others, 1967; Carbonnell and Bauer, 1968; Fastook and others, 1995) and tidal bending cracks within the shear zones (Lingle and others, 1981). A cushion of relatively stagnant ice lies between lateral shear zones and the rock sidewalls of the fjord in most places at the surface. This suggests that lateral uncoupling is facilitated by thermal and strain softening within the shear zones, at least near the glacier surface. Glacial sliding at the ice-rock interface may account for true lateral uncoupling at deeper levels if the fjord sidewalls, which curve inward with depth, intersect the shear zone before the glacier becomes afloat. Lateral uncoupling begins in the shear zones alongside the two ice streams that merge at the foot of the icefall to become Jakobshavns Isbræ. Median longitudinal crevasses known as "the Zipper" extends from thin ice at the base of the icefall to the calving front.

Rapid calving

The 23 m/d midsummer velocity of Jakobshavns Isbræ is largely maintained in midwinter (Echelmeyer and Harrison, 1990), so the annual velocity is 8 km/yr and the iceberg calving rate must also be 8 km/yr if the calving front fluctuates about a stable position (both are 7 km/yr averaged along the calving front). Calving is intermittent and produces tabular icebergs that often roll over (Epprecht, 1987). Calving mechanisms have been examined by Reeh (1968), Holdsworth (1977), Robin (1979), and Fastook and Schmidt (1982), among others. It would seem that some bottom crevasses are necessary to allow tabular icebergs to calve from Jakobshavns Isbræ. Bottom crevasses can open from extending flow in the floating glacier (Weertman, 1980) and from tidal flexure along grounding lines (Lingle and others, 1981). Bottom crevasses, being filled with water, can extend up to sea level and meet surface crevasses. Primarily longitudinal and transverse top and bottom crevasses are opened when basal traction effectively vanishes in Jakobshavns Isbræ. These crevasses are intersecting lines of weakness that are transported toward the floating terminus, where they become reactivated by tidal flexure alongside grounding lines (Lingle and others, 1981) and down-bending along the calving front (Reeh, 1968). Tabular icebergs calve along these lines of weakness (Hughes, 1992c).

Positive feedback

All feedbacks seem to be positive in the Jakobshavns Effect. Surface crevasses are a consequence of extending flow, and crevassing is ubiquitous at surge velocities. Surface melting is enhanced greatly by extensive surface crevassing. Surface meltwater that refreezes internally helps to increase the creep rate to over 0.3/yr by releasing huge amounts of latent heat. Surface meltwater that reaches basal ice increases the basal sliding rate of grounded ice by lubricating the ice-rock interface. Increasing both creep and sliding rates increases ice velocity, surface crevassing, and meltwater production, in that order. Ice thinning by creep and by melting raises the floating ice from bedrock pinning points, thereby allowing a faster ice-stream velocity by reducing ice-shelf buttressing. An unpinned ice shelf allows a faster iceberg calving rate because it can thin more rapidly by creep. Enhanced creep thinning shortens the vertical ice thickness that must be fractured by crevasses and allows crevasses to penetrate farther. Rapid iceberg calving reduces the length of floating ice and therefore the potential number of ice-shelf pinning points. An unpinned ice shelf exerts a potentially huge pulling power on the ice stream, because its pulling force increases as the square of the floating ice thickness (Weertman, 1957a) and the maximum pulling force is at the grounding line, because there the floating ice is thickest (Sanderson, 1979).

Negative feedback

The only feature tending to stabilize Jakobshavns Isbræ is Jakobshavns Isfjord, where bedrock basal pinning points, sidewalls, and the headwall prevent the Jakobshavns Effect from spreading laterally and inland (Warren and Hulton, 1990). Without these bedrock constraints, it is difficult to imagine what could prevent the Jakobshavns Effect, as observed in Jakobshavns Isbræ and other outlet glaciers between 69°N and 72°N, from collapsing most of the Greenland Ice Sheet. Indeed, as shown in Figure 1.5, the calving front of Jakobshavns Isbræ retreated 30 km up Jakobshavns Isfjord from 1850 to 1964 (Carbonnell and Bauer, 1968; Prescott, 1995), perhaps as a consequence of climatic warming following the Little Ice Age. The ground-

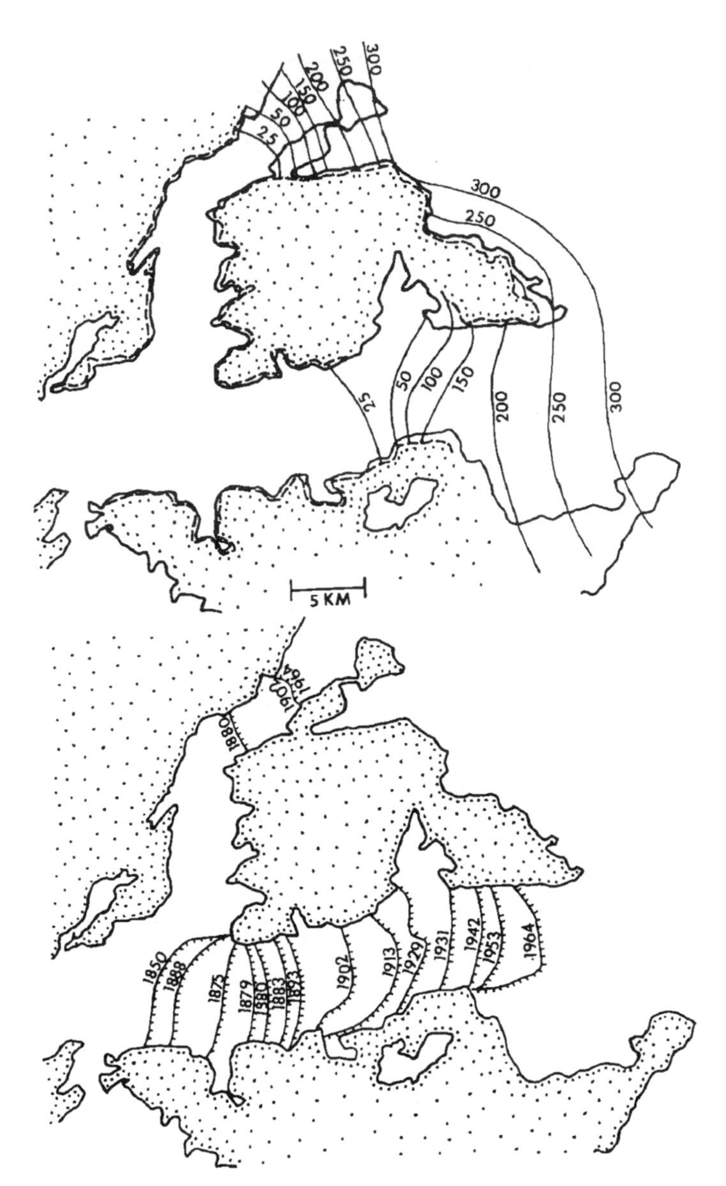

Figure 1.5: Retreat of Jakobshavns Isbræ from 1850 to 1964. *Top*: Glacial erosion "trimlines" showing former ice elevations (m) were mapped photogrammetrically by Prescott (1995). *Bottom*: Positions of the calving front are from Carbonnell and Bauer (1968).

ing line of Jakobshavns Isbræ probably retreated a similar amount, because the sidewalls of Jakobshavns Isfjord are polished and scraped clean of lichens from a height that increases from near sea level to some 200 m above sea level over the 30 km of calving-front retreat. This former ice elevation would require floating ice to have been almost 2000 m thick at the present-day calving front, where floating ice is now 800 m thick or less. So the former ice was probably grounded, unless the fjord is at least 1800 m deep.

Time transgression

At the last glacial maximum, the Greenland Ice Sheet extended far out onto the continental shelf. Today, the Greenland Ice Sheet ends mostly on land in the south, at fjord headwalls in the center, and as a continuous tidewater ice wall along much of the northern coastline. This implies that the Jakobshavns Effect may have migrated northward with climatic warming since the last glacial maximum ended 14,000 years ago. Anthropogenic "greenhouse" warming may telescope this northward migration in the decades ahead and initiate the Jakobshavns Effect in ice streams reaching tidewater in the northern third of Greenland. In southeast Greenland, ice streams converge on Sermilik (62.0°N, 37.7°W) and Kungerdlugssuaq (68.5°N, 32.0°W), which empty into large fore-deepened troughs across the continental shelf that may have been occupied by extensions of these ice streams at the last glacial maximum. In northern Greenland, similar troughs cross the continental shelf where ice streams enter Dove Bugt, Jøkel Bugten, Nioghalvfjerds-fjorden, Hazen Fjord, and Independence Fjord between 76°N and 82°N on the east coast, and enter Melville Bugt, Kane Basin, Hall Basin, and Victoria Fjord between 74°N and 82°N on the west coast. Four huge ice streams are particularly noteworthy: Zachariae Ice Stream (79°N, 22°W), the largest ice stream on Earth (Fahnestock and others, 1993), Humboldt Gletscher (79°N, 62°W), Petermanns Gletscher (81°N, 60°W), and C. H. Ostenfeld Gletscher (82°N, 44°W), see Rignot and others (1997). If the Jakobshavns Effect rejuvenates life cycles in these ice streams, a major deglaciation of the southern third of Greenland will take place in Sermilik and Kungerdlugssuaq on the southeast coast, and will be quickly followed by a major deglaciation of the northern third of Greenland, while the Jakobshavns Effect continues for ice streams in the central third of Greenland. The Jakobshavns Effect, which has been time transgressive from south to north across Greenland since the last glacial maximum, could then conceivably repeat as a time-transgressive rejuvenated deglaciation from south to north that is telescoped into just a few centuries, beginning with maximum "greenhouse" warming anticipated near AD 2050.

Rejuvenated deglaciation

Figure 1.6 shows the redistribution of Greenland ice that is possible if life cycles of major ice streams are rejuvenated by the Jakobshavns Effect, beginning in AD 2050. The ice-stream model proposed by Hughes (1992b) was applied to Greenland ice streams. Submarine troughs crossing the continental shelf were assumed to continue beneath the ice streams. Otherwise, the surface mass balance and the basal topography are the same as those used by Huybrechts and others (1991), but the results are quite different. Their sheet-flow model reduced the area-elevation ratio, whereas rejuvenated stream flow increases this ratio. The ice sheet lowers and advances onto the continental shelf, especially in the fore-deepened troughs of southeastern and northern Greenland. Ice volume decreases by 30 percent, causing sea level to rise about 2 m. Ablation will be dominated by iceberg calving, rather than surface melting, because most surface meltwater enters surface crevasses and refreezes in the Jakcobshavns Effect. Icebergs calving from the rejuvenated

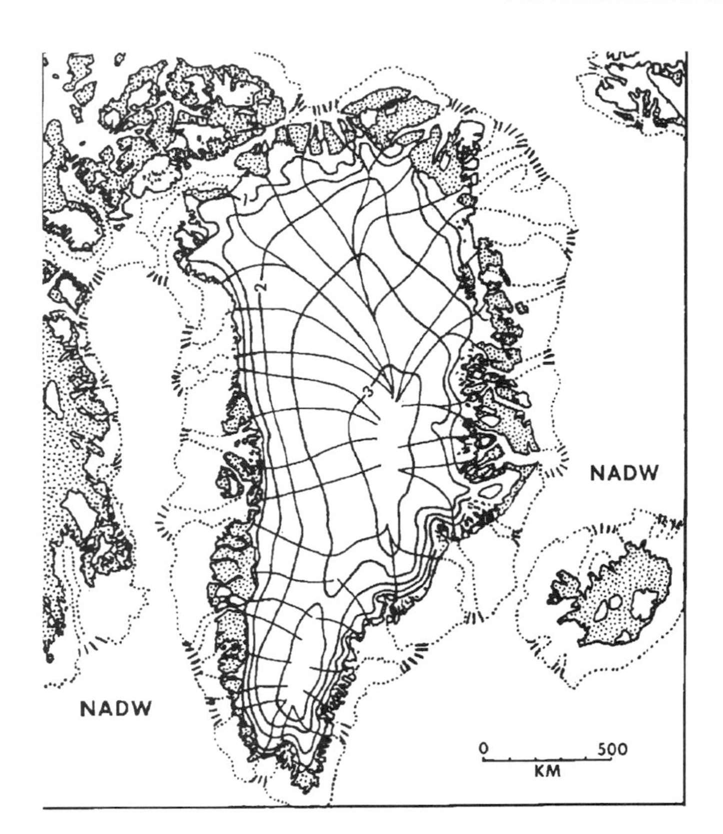

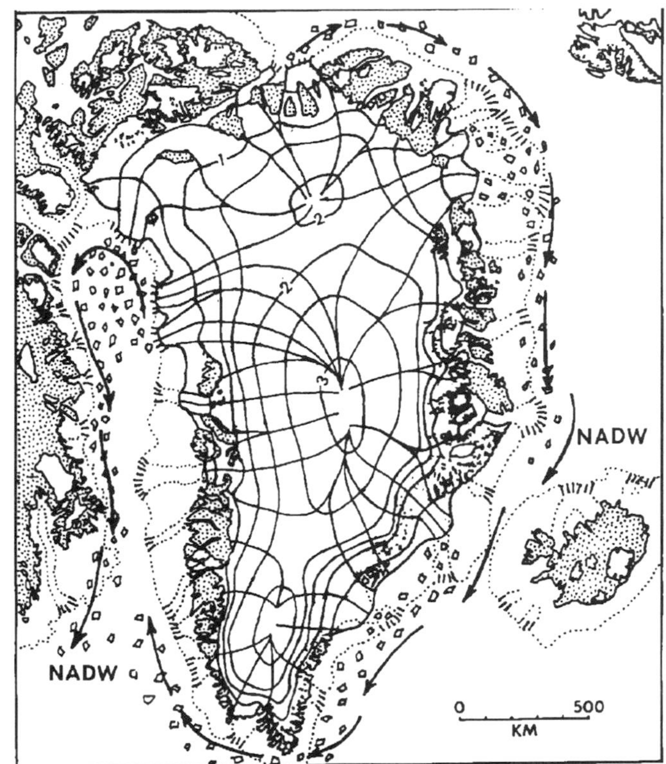

Figure 1.6: Inward downdraw and outward advance of the Greenland Ice Sheet caused by the future initiation of the Jakobshavns Effect in presently stable or dormant ice streams. Shown are ice elevation contours (0.5-km intervals), unglaciated land (dotted areas), the edge of the continental shelf and submarine troughs crossing the continental shelf (dotted lines), depositional fans at the foot of troughs (short bars), sites of North Atlantic Deep Water production (NADW), prevailing ocean-surface currents (arrows), and icebergs transported by currents (small polygons). The Greenland Ice Sheet as it is today (left) and after the Jakocbshavns Effect activates ice streams at the heads of submarine troughs (right).

ice streams will be carried into the North Atlantic by the East Greenland Current for East Greenland ice streams, especially Zachariae Ice Stream, and by the Baffin-Labrador Current for West Greenland ice streams. This iceberg influx will reduce the density of surface water by reducing salinity and by reducing temperature below the temperature of maximum density. If this density reduction prevents surface water from sinking and becoming North Atlantic Deep Water, the oceanic conveyor belt discussed by Broecker and Denton (1989) may shut down and plunge the world into its glacial climate mode. This would be a continental-scale application of the instantaneous glacierization hypothesis proposed by Ives and others (1975) for highland plateaus, because snowline lowering during the glacial mode would also include lowlands as grounded sea ice became marine ice sheets that transgressed onto Arctic coasts (Hughes, 1992a). The overextended Greenland ice margins in Figure 1.6 will retreat rapidly if the conveyer belt does not slow or stop.

The Heartbeat of Ice Sheets

The Jakobshavns Effect can be applied to former ice sheets, specifically to their Innuitian, Laurentide, Scandinavian, Barents, Kara, East Siberian, and Chukchi ice domes which were grounded on ice-cemented permafrost in marine embayments or on Arctic continental shelves. As these ice domes thickened, therefore, the underlying permafrost would eventually thaw and ice streams would develop in interisland channels, straits, submarine troughs, linear gulfs, estuaries, and river valleys that surrounded the ice domes. the Jakobshavns Effect would then take place in these ice streams, giving them "life cycles" consisting of inception, growth, mature, declining, and terminal stages, as depicted in Figure 1.7. Thawing of ice-cemented permafrost along the various kinds of linear depressions would initiate stream flow and inception of ice-stream life cycles. To the extent that these life cycles were largely simultaneous, the ice domes would lower during growth stages, stabilize during the mature stages, and rise during declining stages. After terminal stages, when ice streams shut down because the bed refroze under thinning ice, ice accumulation would begin to restore the elevation of ice domes. If concurrent life cycles repeat during a Quaternary glaciation, expansion and contraction of Northern Hemisphere ice sheets would give the ice sheets a "heartbeat" in computer models that simulate these life cycles, with the rise and fall of ice domes pumping ice into ice streams that, like arteries, transport ice to the extremities of the ice sheet. Therefore, the "heartbeat" exists in ice domes and coincides with "life cycles" of ice streams. Within life cycles, growth stages produce interstadial-stadial transitions, mature stages produce stadials, declining stages produce stadial-interstadial transitions, and terminal stages produce interstadials within the glaciation cycle. These are similar to the preglacial, glacial, deglacial, and interglacial stages of Imbrie and others (1992). A glaciation cycle consists of successive "heartbeats" of ice domes and "life cycles" of ice streams.

Field evidence from Quaternary ice sheets indicates that marine ice streams produced iceberg outbursts and terrestrial ice streams produced lobate moraines during their life cycles. Iceberg outbursts into the North Atlantic constitute an "ice machine." Glacial erosion that produces lobate moraines constitutes a "dirt machine." Entrainment of the dirt fine fraction by katabatic winds that flow down ice streams constitutes a "dust machine." Entrained ice crystals and glacial flour that become cloud–condensation nuclei carried back over the ice sheet by upper inflowing air replacing katabatic outflow constitute a "cloud machine." These four "machines" are depicted in Figure 1.8. The new paradigm postulates that all four "machines" are turned on during ice-stream life cycles and turned off between life cycles. During life cycles, rapid ice flow and downdraw of ice extends ice streams forward beyond the ice sheet and backward into the ice sheet, causing the thawed bed beneath ice streams to merge with the thawed bed beneath interior ice. This enhances downdraw, causing the ice sheet to expand. Between life cycles, basal heat is conducted rapidly upward through stagnating downdrawn ice to the ice surface, causing the bed to freeze and the ice sheet to contract as the extended ice lobes melt back. This "heartbeat" of Quaternary ice sheets is linked to life cycles of their ice streams by changes in the internal temperature field in response to downdraw and buildup of ice during and between life cycles. The "heartbeat" is manifested by abrupt changes in iceberg outbursts,

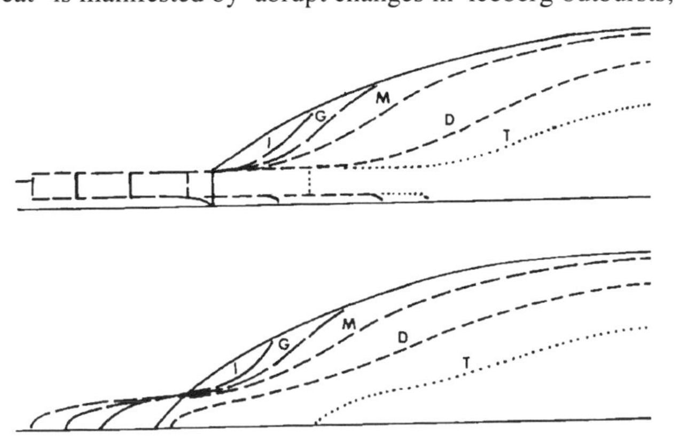

Figure 1.7: A cartoon of the hypothetical life cycle of ice streams. Stages in the life cycle are inception (I), growth (G), mature (M), declining (D), and terminal (T). During the life cycle, a marine ice stream ends in a floating ice shelf (top) and a terrestrial ice stream ends in a grounded ice lobe (bottom).

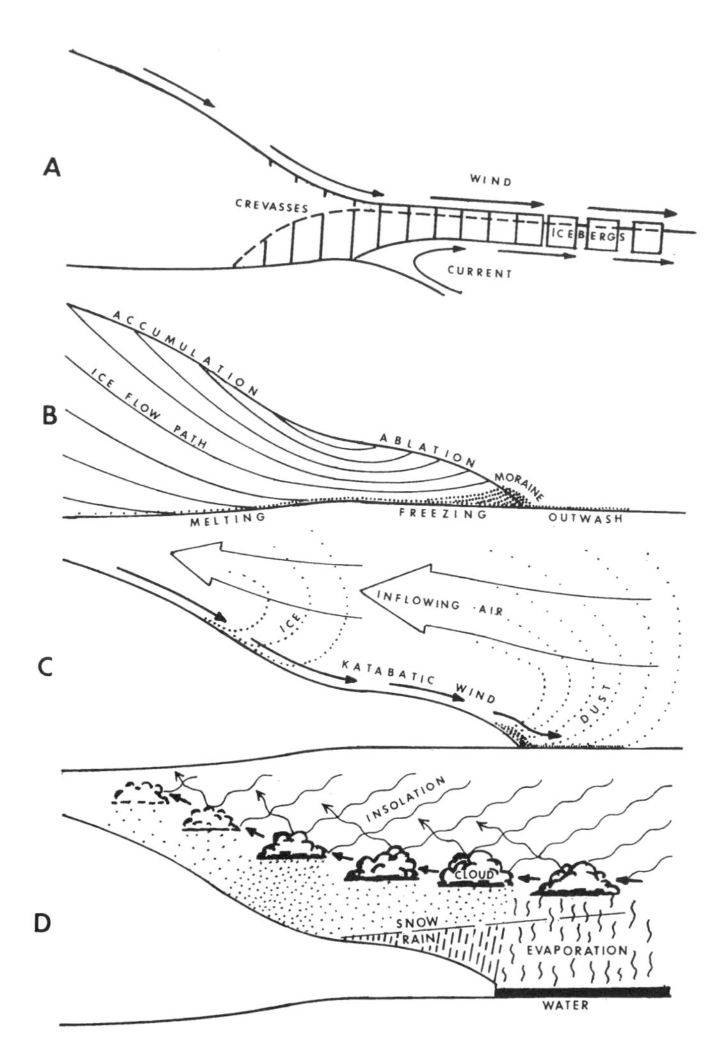

Figure 1.8: A cartoon illustrating hypothetical ice, dirt, dust, and cloud machines that turn on and off with ice-stream life cycles. In the ice machine (A), top and bottom crevasses open at the head of a marine ice stream and meet at the calving front of its floating ice shelf. In the dirt machine (B), eroded basal material is transported to the ice-stream terminus by freezing into basal ice, by suspension in basal meltwater, or by simple shear of the water-soaked material itself. In the dust machine (C), katabatic winds flowing down ice streams entrain ice crystals in the accumulation zone and glacial flour melted out in the ablation zone or washed out in the outwash fan. In the cloud machine (D), ice crystals and glacial flour entrained by katabatic winds become cloud condensation nuclei, with clouds forming from water vapor evaporated from windswept marine and lacustrine ice margins, being transported back over the ice sheet by upper inflowing air, precipitating snow over the ice sheet, and reflecting incoming solar radiation.

glacial erosion, dust production and cloud condensation that accompany life cycles.

All four "machines" have life cycles that depend on the proposed life cycle of ice streams, a life cycle that begins and ends as abruptly and that lasts as long as the events of abrupt climatic changes recorded in cores through the Greenland Ice Sheet and through ocean-floor sediments. A feature of the proposed life cycle of ice streams is that stream flow begins when ice-cemented till or sediments become thawed in linear channels (river valleys, ice-dammed lakes, fjords, interisland channels, etc.), allowing fast flow that migrates into the ice sheet along these channels as the thawing front retreats (Hughes, 1992b). Basal heat is conducted more rapidly to the lowered ice surface and the bed refreezes in the downdrawn zone. This ends the life cycles of the ice streams. Expansion and contraction of the thawed bed beneath interior ice domes and ice streams along its margin can be modeled as the "heartbeat" of an ice sheet that sustains its vitality by regulating the life cycles of its ice streams. The ice sheet expands when the inner and outer basal thawed zones expand and meet, and it contracts when these zones contract and separate. Modeling this "heartbeat" of Quaternary ice sheets during the last glaciation can be constrained by the dated record of stadials and interstadials during the glaciation cycle. Referring to the equilibrium states in Figure 1.3, an ice sheet retreats to its stable core and advances to its unstable limit with each "heartbeat."

Evidence for Late Quaternary abrupt climatic changes is found in the oxygen isotope and atmospheric dust stratig raphy in cores through the Greenland Ice Sheet, first reported by Hammer and others (1985), and in horizons of ice-rafted sediments in ocean-floor cores from the North Atlantic, first reported by Heinrich (1988). Climate changes inferred from these data begin and end over time spans ranging from a few years to a few decades, and last over time spans ranging from a century to a millennium or more. Both time spans encompass the inferred fast-flow time spans of present-day marine ice streams draining the Antarctic Ice Sheet (Bentley, 1987; Clarke, 1987; Shabtaie and others, 1987, 1988; Whillans and others, 1987; Bindschadler and others, 1987; Alley and others, 1987a, 1987b) and of former terrestrial ice streams that produced the lobate southern margins of the Scandinavian Ice Sheet (Andersen, 1981) and the Laurentide Ice Sheet (Mayewski and others, 1981), as well as former Laurentide marine ice streams (Miller and Kaufman, 1990), during the last glaciation. Although reorganizations of the oceanic and atmospheric components of Earth's climatic machine have been cited as causing abrupt climatic changes (Broecker and Denton, 1989; Broecker and others, 1990; Birchfield and Broecker, 1990), ice streams also may trigger these reorganizations (Hughes, 1992a). This possibility should be investigated for present-day ice streams, particularly for Greenland ice streams now experiencing. The Jakobshavns Effect, because fast flow in ice streams apparently can be turned on and off in substantial independence of external climatic forcing (Fastook, 1987a; Lingle and Brown, 1987; MacAyeal, 1989, 1992; Budd and others, 1984; Alley, 1989a, 1989b, 1990, 1991; Hughes 1992b). It may be our last opportunity to understand the role of ice streams in abrupt change before Earth is plunged abruptly into a new cycle of Quaternary glaciation.

The White Hole

Inception of the coming glaciation might begin rapidly, after anthropogenic "greenhouse" warming has run its course, if the subsequent cooling creates a White Hole in the Northern Hemisphere. Caused by thickening sea ice that grounds on shallow Arctic continental shelves and impounds rivers flowing into the Arctic Ocean, the White Hole would be a vast region into which precipitation could enter but not leave, making it analogous to the Black Hole in outer space in that nothing entering can escape. The White Hole might form almost instantaneously and enclose an area comparable in size to that covered by ice sheets during a glacial maximum, causing sea level to drop at the inception of a new glaciation cycle almost as fast as it rose during termination of the last cycle. Little or no glacial geology is mapped and dated to specify boundary conditions needed to numerically model the inception stage of the last glaciation. However, three mechanisms have been proposed for initiating a glaciation cycle. They are illustrated in Figure 1.9. Precipitation is emphasized in the "highland-origin, windward-growth hypothesis" (Flint, 1971, pp. 481–484, 596–601). Convective storm systems precipitate snow as they climb to higher elevations, so ice sheets originate as highland snowfields and grow fastest in the directions that bring in most of the precipitation. Snowline elevation is emphasized in the "instantaneous glacierization hypothesis" (Ives, 1975). Climate cooling lowers the regional snowline so that plateau snowfields rapidly increase in number and size, almost instantaneously glacierizing polar plateaus that remain above the lowered snowline. Sea ice is emphasized in the "marine ice transgression hypothesis" (Hughes, 1986b). Increasing snowfall and basal freezing thicken sea ice, which then grounds on shallow polar continental shelves, with grounding lines migrating into deeper water, allowing marine ice domes to grow behind and ice shelves to grow ahead of the advancing grounding lines. These three hypotheses complement one another. They combine to maximize the area of high albedo that reflects incoming solar radiation, thereby promoting inception of Northern Hemisphere ice sheets during cooling hemicycles of Milankovitch insolation. Figure 1.1 shows that Earth is presently halfway into a cooling hemicycle at mid and high Northern Hemisphere latitudes, with a maximum rate of decreasing insolation in these latitudes.

As seen in Figure 1.10, inception of the last glaciation cycle was as abrupt as its termination. Miller and deVernal (1992) described three conditions that made this possible.

1. Compared to the present day, mid-to-high-latitude Northern Hemisphere insolation decreased in the summer and increased in the winter during inception of the last glaciation cycle. The insolation changes in Figure 1.10 were accompanied by a negative shift in $\delta^{18}O$ that corresponded to a 7 mm/yr fall in eustatic sea level, which is equivalent to an ice accumulation rate of 3000 km^3/yr. In the Arctic, warmer winters kept the Barents, Norwegian, Greenland,

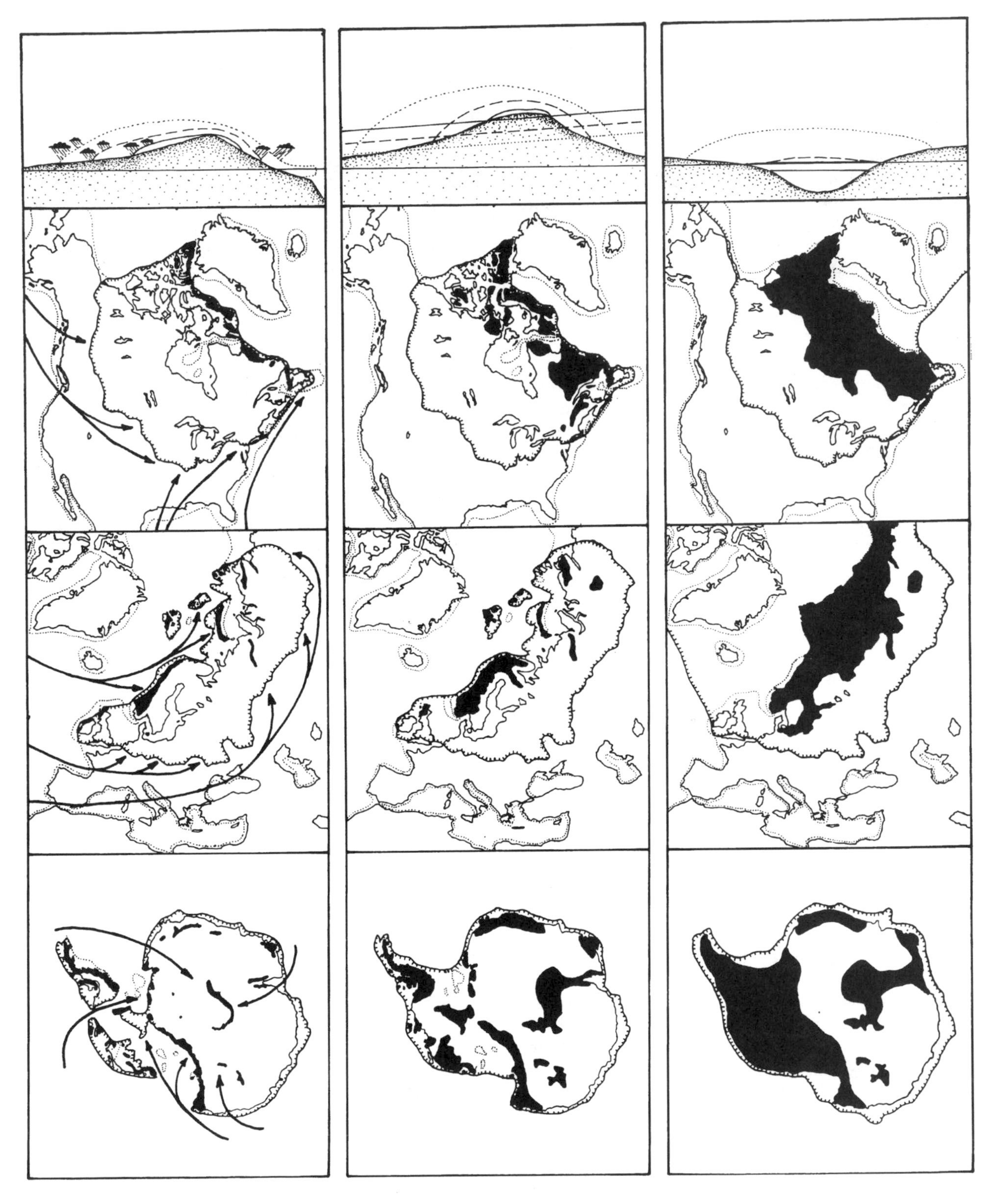

Figure 1.9: Hypotheses for forming Cenozoic ice sheets. The "highland origin, windward growth" (left), "instantaneous glacierization" (center), and "marine ice transgression" (right) hypotheses are compared (top) and applied to the initiation and growth of ice sheets in North America, Eurasia, and Antarctica. Areas of initiation are shown in black, stages in growth are shown as the solid, dashed and dotted profiles at the top, and the maximum Cenozoic extent is shown by the hatched line. Arrows denote the major storm tracks that supplied snow to the growing ice sheets, as inferred from present-day storm tracks (Flint, 1971, p. 74–75; Weyant, 1967; however, see Manabe and Broccoli, 1985).

and Labrador Seas ice free, so they provided the evaporation source for increased winter snowfall over the ice-covered Arctic Ocean. Coupled with decreased summer melting that resulted from cooler summer air temperatures, this in-creased snow precipitation gave the ice cover a positive mass balance. In the coming glaciation, increased Arctic cloudiness caused by "greenhouse" warming could also produce warmer winters and cooler summers by reducing winter black-box radiation into space and by reflecting summer insolation back into space.

2. Compared to lower latitudes, Arctic summers must warm more to melt a given increase in winter snow because total annual snowfall is light in the Arctic. Figure 1.10 quantifies this phenomenon as a nonlinear relationship between snow accumulation rates and mean summer air temperatures, both measured at the equilibrium line of glaciers. The summer temperature increase needed to melt an additional 0.1 m/yr of snow decreases from 0.4°C to 0.1°C when annual snowfall increases from 0.5 m/yr to 4.0 m/yr. Arctic summers cooled, not warmed, as winter snowfall increased during inception of the last glaciation cycle. Therefore, Arctic snowlines would have fallen sharply, lowering the equilibrium line and causing glaciers to advance. This will happen again in the coming glaciation.

3. Both annual temperature and annual precipitation in the Arctic are lowest over the deep Arctic Ocean basins and over the broad Arctic continental shelves and adjacent mainlands of Greenland, North America, and Siberia. Figure 1.10 shows this region which is most sensitive to increased snowfall, with a mean annual temperature less than –7°C and an average precipitation rate under 0.3 m/yr of water equivalent. Inception of the last glaci-ation probably took place within this area, and so may inception of the coming glaciation.

The "fast track" for the coming glaciation proposed by Miller and deVernal (1992) must be hedged in several important respects (Schneider, 1992). However, since a "fast track" commands the most concern, and the last glaciation apparently began abruptly, an imminent coming glaciation will be assumed in this book. Conditions described by Miller and deVernal (1992) for inception of the last glaciation cycle make use of increased snowfall, emphasized in the highland-origin, windward-growth hypothesis; lowered snowlines, emphasized in the instantaneous glacierization hypothesis; and thickening sea ice, emphasized in the marine-ice-transgression hypothesis, as illustrated by Figures 1.9 and 1.10. In the coming glaciation, highland glaciers facing the Barents, Norwegian, Greenland, and Labrador Seas should advance because warmer winter surface water in these seas will make them sources of increased precipitation. Terrestrial ice domes should develop when lowered snowlines instantaneously glacierize polar plateaus on the Arctic islands and on the mainlands of Canada and Siberia. Marine ice domes should develop when sea ice thickens and grounds on the broad, shallow continental shelves of Arctic Canada and Siberia. Floating ice shelves should develop over the deep Arctic Ocean basins where thickening sea ice remains afloat. The highland glaciers, terrestrial ice domes, marine ice domes, and floating ice shelves should coalesce within the area shown in Figure 1.10 to produce an Arctic Ice Sheet that behaves as a single dynamic system. Formation of the Arctic Ice Sheet will bring the inception stage of the coming glaciation cycle to a close.

Figure 1.11 compares the areal extent of the White Hole and the Water Hole impounded by it, which together initiate a Quaternary glaciation cycle, with the areal extent of an Arctic Ice Sheet at the glacial maximum. The two areas are similar in extent. If mean annual precipitation of 250 mm/yr of ice equivalent were to fall over this 24×10^6 km^2 area, then 6000 km^3/yr of ice would accumulate, causing sea level to drop 14 mm/yr.

An Arctic Ice Sheet

Weertman (1976) posed, as glaciology's grand unsolved problem, the existence and the future behavior of the West Antarctic Ice Sheet. It is a "marine" ice sheet grounded an average of 500 m below sea level on the broad Antarctic continental shelf in the Western Hemisphere and it is drained by ice streams, which are fast currents of ice similar to the great rivers that drain the other continents. Weertman asked, "How then did this ice sheet form? Why does it remain in existence? Is it growing or disintegrating at the present time?" The West Antarctic Ice Sheet is grudgingly providing answers to these questions, primarily through large scale glaciological research projects funded by the U.S. National Science Foundation.

Grosswald and Hughes (1995) noted that paleoglaciology also has a grand unsolved problem that is remarkably parallel to glaciology's grand unsolved problem; namely, did an Arctic Ice Sheet develop during Quaternary glaciation cycles? Figure 1.12 shows the extent of a possible Arctic Ice Sheet during a Quaternary glacial maximum. As with the Antarctic Ice Sheet, an Arctic Ice Sheet would have consisted of terrestrial ice domes grounded on land above sea level and marine ice domes grounded on land below sea level, interconnected by floating ice shelves and drained by fast ice streams (Hughes and others, 1977). The idea of an Arctic Ice Sheet is not new (Thomson, 1888; Mercer, 1970). Testing the idea is difficult because evidence for an Arctic Ice Sheet is often obscured by active cryogenic processes on land or submerged on continental shelves. Quaternary terrestrial ice domes in the Northern Hemisphere are well documented. They include domes of the Laurentide and Cordilleran Ice Sheets of North America and of the Scandinavian Ice Sheet of Europe, and the ice caps on the Taimyr Peninsula and the Putorana Plateau of Siberia. For an Arctic Ice Sheet to have existed, these terrestrial ice domes would have to be connected with marine ice domes on the Arctic continental shelves of North America and Eurasia, and the marine ice domes would have to be connected to an ice shelf floating over the deep basins of the Arctic Ocean.

Clues that these marine ice domes existed include raised beaches on islands surrounding the Barents Sea north of

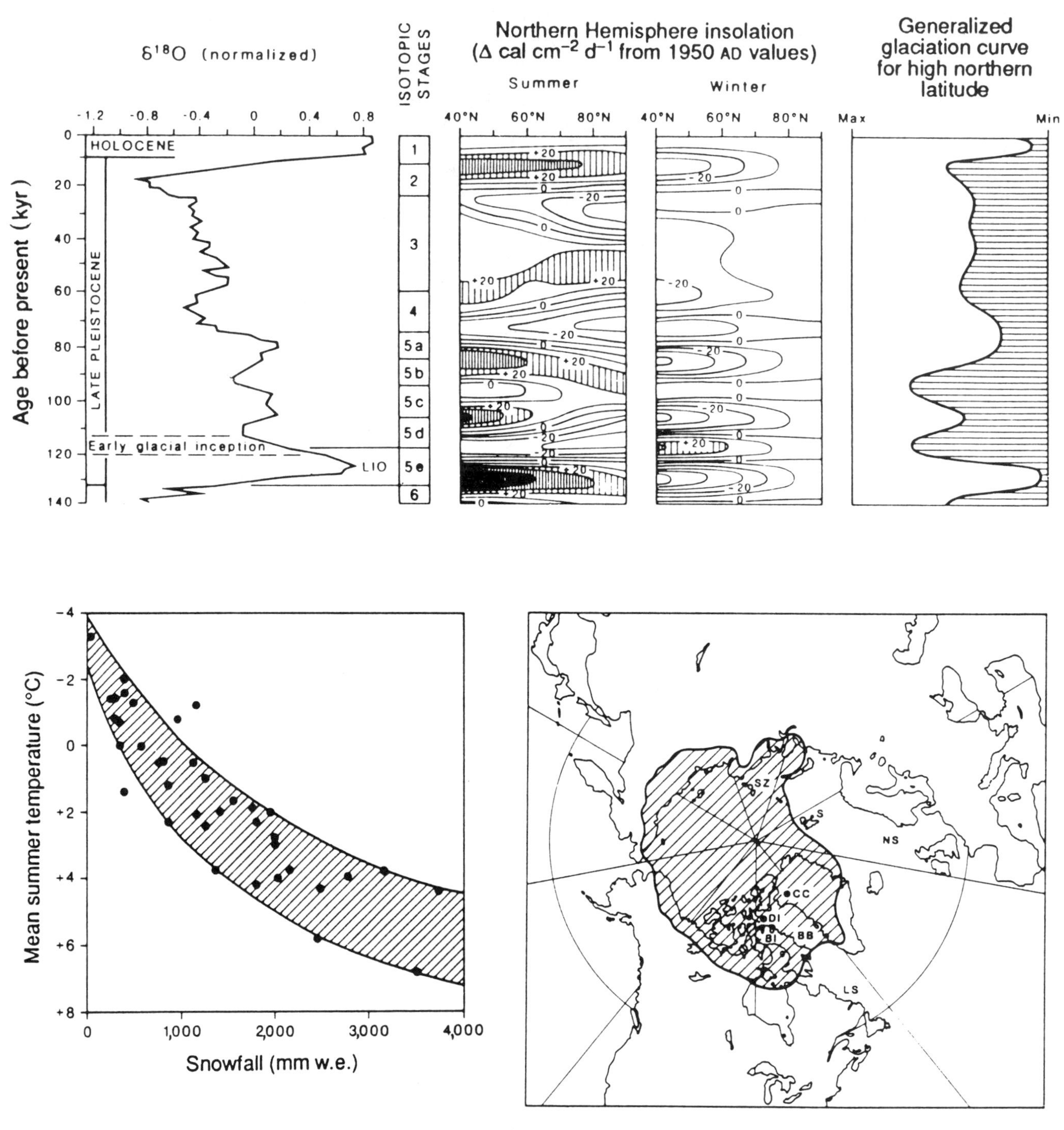

Figure 1.10: Vulnerability of the Arctic to rapid glaciation. *Top left*: Glaciation during the last 140,000 years represented by a tuned oxygen-isotope record using benthic foraminifera as a global ice-volume proxy. *Top center*: Seasonal insolation changes at the top of Earth's atmosphere between latitudes 40°N and 90°N over the last 140,000 years. *Top right*: Generalized land glaciation in northeast Canada, northwest Greenland, and Spitzbergen during the last 140,000 years. *Bottom left*: A plot of the mean summer temperature needed to melt winter snow at the equilibrium line of glaciers (dots), showing that the increase in summer temperature needed to melt a given increase in winter snowfall is greater in low-precipitation cold regions because the envelope (shaded area) enclosing most dots is not linear. *Bottom right*: The Arctic region (shaded area) that is most sensitive to increased precipitation because its mean annual precipitation is less than 0.3 m of water equivalent, making it both cold and dry. Reprinted with permission from Nature, G.H. Miller, and A. deVernal, Vol. 355, p. 245, Figs. 1, 2, and 3, 1992.

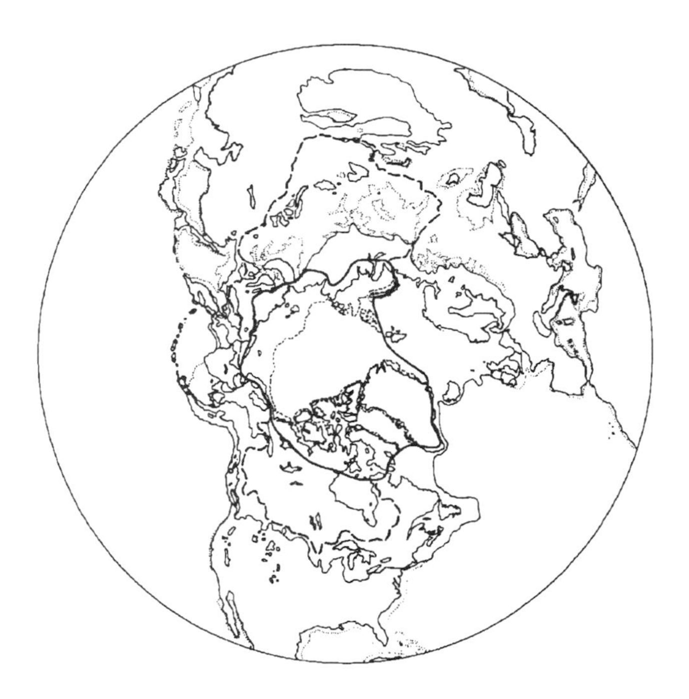

Figure 1.11: A White Hole and an Arctic Ice Sheet (Hughes, 1996a). The White Hole consists of the area within the thick solid line, where the mass balance is positive, which impounds a Water Hole within the watersheds enclosed by the thick dashed lines. Precipitation within the White Hole and the Water Hole is impounded before it returns to the sea, so sea level lowers accordingly until marine ice margins calve and spillways drain lacustrine ice marings. The White Hole initially within the thick solid line is also an Arctic Ice Sheet at inception of a glaciation cycle, and it expands into the area of the Water Hole until it reaches the thin solid lines at the glacial maximum, when light dotted lines denote marine and lacustrine shorelines. Thin dashed lines are ice-shelf grounding lines. Highland glaciation is also shown, with an ice sheet on Tibet (Kuhle, 1995).

Figure 1.12: A hypothetical Arctic Ice Sheet at a Quaternary glacial maximum (Hughes, 1996a). Shown are the limits of glaciation and the marine shoreline (solid lines), the grounding lines of floating ice shelves (dashed lines), the present-day marine and lacustrine shorelines (dotted lines), and the glacial lakes (black areas).

Scandinavia (Schytt and others, 1968) and on the Queen Elizabeth Islands of Arctic Canada (Blake, 1970). Therefore, a marine Barents Sea Ice Sheet joined the Scandinavian Ice Sheet and a marine Innuitian Ice Sheet joined the Laurentide Ice Sheet at the last glacial maximum. Scandinavian Quaternary geologists have now substantiated from the sedimentary record on the Barents Sea floor that a BarentsSea Ice Sheet existed (Elverhøi and Solheim, 1983; Vorren and Kristoffersen, 1986). However, the Innuitian Ice Sheet produced little or no glacial geology on the Queen Elizabeth Islands (England, 1976). If its bed were frozen over these islands, the usual glaciological processes that produce glacial geology would not be active, and glacial geology would be confined to the interisland channels occupied by ice streams sliding on thawed deforming marine sediments, as is indicated by the relatively thick permafrost on the islands and the relatively thin permafrost in the interisland channels today (Taylor, 1988). Clark (1980) showed that an Innuitian Ice Sheet was necessary in order for an inverse computer model of glacioisostatic rebound to account for the dated raised beaches. Ice-rafted dropstones on the Arctic Ocean floor originated from glacial erosion in the northwestern Queen Elizabeth Islands (Bischof and Clark, 1992).

Grosswald (1980, 1993a) showed that the terrestrial glacial geological record on the North Russian Plain can be explained only if marine ice sheets on the Barents Sea and the Kara Sea continental shelves had transgressed onto the Eurasian mainland. Grosswald (1988) and Grosswald and Lasca (1993) subsequently found glacial geological evidence, especially glacial tectonics, for marine ice sheets on the continental shelves of the Laptev, East Siberian, and Chukchi Seas. These data are scattered, perhaps owing to widespread frozen bed conditions, so they are not conclusive evidence for the existence of marine ice sheets at this time. Still, climate modelers are increasingly placing marine ice domes on Arctic continental shelves at the last glacial maximum (Tushingham and Peltier, 1991; deBlonde and others, 1992; Verbitsky and Oglesby, 1992; Peltier, 1994).

With marine ice domes widespread on Arctic continental shelves, only an ice shelf floating over deep Arctic Ocean basins is required to produce an Arctic Ice Sheet at the last glacial maximum. Oxygen isotope records from a Greenland ice core and from a Pacific sediment core led Broecker (1975) to postulate a floating ice cover between 12 m and 2000 m thick at the last glacial maximum. With reasonable mass-balance assumptions, Lindstrom and MacAyeal (1986) found that a dynamic finite-element ice-shelf model produced floating ice 500 m to 1000 m thick as far south as Iceland. An ice-shelf floating in the Norwegian and Greenland Seas may have delayed deglaciation in northeastern Greenland (Spielhagen, 1992). The entire Arctic Ocean, from the Chukchi Foreland to Fram Strait, was abiotic from 33 ka BP to 13 ka BP, with minimal biotic activity from 13 ka BP to 8 ka BP (Jones, 1994). Only a thick sea-ice cover with no leads or a floating ice shelf could suppress all biological activity. Numerous iceberg keel marks trend southward through Fram Strait in water from 450 m to 850 m deep (Vogt and others, 1994). Arctic ice streams would initially release icebergs that became imbedded in the perennial Arctic pack ice. But as outlets for the pack ice became constricted over time, cohesion between pack ice and icebergs would produce a floating ice shelf supplied by the floating tongues of ice streams, as observed for Antarctic ice shelves. A record of this transformation may exist in the Arctic. Mapping the distribution and provenance of Quaternary ice-rafted dropstones in Arctic Ocean basins, as Bischof and Clark (1992) are doing, should reveal a dispersal A pattern that is characteristic of either ice-shelf flow or pack-ice drift. Ice-shelf models (Thomas and MacAyeal, 1982; Lindstrom and MacAyeal, 1986) and sea-ice models (Parkinson and Washington, 1979; Hibler, 1979, 1985) have fundamentally different dynamics and flow regimes.

Future research related to an Arctic Ice Sheet will involve mapping the maximum extent of marine ice sheets on Arctic continental shelves, dating their deglaciation history, establishing whether they were connected across the Arctic Ocean by a floating ice shelf, determining how changing climatic boundary conditions in one sector of a unified Arctic Ice Sheet affected ice dynamics in other sectors, understanding how surge-like life cycles of ice streams may have triggered abrupt climatic change, and investigating whether existence of an Arctic Ice Sheet compels us to recast our knowledge and understanding of the Quaternary. The payoff may complete the paradigm shift in understanding global climatic change.

Like the White Hole, an Arctic Ice Sheet impounds all rivers flowing into the Canadian and the Siberian Arctic. Initial landward advance of the Arctic Ice Sheet would be over permafrost in these areas, but permafrost would eventually thaw beneath the ice-dammed lakes formed by impounded rivers. These lakes would grow until they spilled over into the Great Lakes and Mississippi River watersheds of North America and the Black Sea, Caspian Sea, and Aral Sea watersheds of Eurasia. Soft lacustrine sediments would accumulate in these lakes. Ice sheets thickening over the permafrost would be an insulating blanket that would raise the melting point isotherm at the base of the permafrost, thawing the permafrost from the bottom up until the ice-bed interface became thawed. Thawing the bed would substantially reduce basal traction, allowing partial gravitational collapse of the overlying ice, which would then rush into the ice-dammed lakes. The soft, water-saturated lacustrine sediments would provide virtually no basal traction, so the ice sheet could advance across the area covered by the ice-dammed lakes and beyond, until restoration of basal traction by dry land brought the advance to a halt. Along marine margins of the Arctic Ice Sheet, the forward rush of ice would either create a floating ice shelf or thicken one already in place. According to the new paradigm, the forward rush of ice along both lacustrine and marine margins would be in ice streams, and the resulting expansion would produce the full size of the Arctic Ice Sheet that existed at Quaternary glacial maxima and is depicted in Figure 1.12.

A component of a hypothetical Arctic Ice Sheet that is already close to being a full-sized ice sheet is the Greenland Ice Sheet. Using a mass-balance model for the Greenland Ice Sheet and a maximum "greenhouse" warming of 8.4°C, Huybrechts and others (1991) predicted that sea level would rise 0.4 m by AD 2100 due to increased ablation of Greenland ice, with most of the increased ablation being in southwest Greenland. When they included ice dynamics in the model, ablation was reduced because sheet flow narrowed the ablation zone by steepening the slope of the ice margin and making flowline profiles more convex. Their model did not take ice streams into account, however. Ice streams drain 90 percent of the ice from the present-day Greenland Ice Sheet and have concave flowline profiles that lengthen as ice streams begin their life cycles. Including ice stream dynamics would therefore make the mass balance even more negative during an ice-stream life cycle, because the ablation zone widens and the calving rate increases. Moreover, ice-stream life cycles would propagate northward in Greenland during "greenhouse" warming, affecting the whole ice sheet, not just one sector. This is a consequence of the Jakobshavns Effect, and it is presented in Figure 1.6.

The Life Cycle of Ice Streams

The proposed "ice machine," "dirt machine," "dust machine," and "cloud machine" shown in Figure 1.8 are active during the ice-stream life cycles shown in Figure 1.7. The proposed life cycles of marine and terrestrial ice streams consist of inception, growth, mature, declining, and terminal stages, with a marine ice shelf or a terrestrial ice lobe forming during the growth stage, attaining maximum size during the mature stage, and disintegrating during the declining stage (Hughes, 1992a, 1992b). These are the fast-flow stages of the life cycle, when the ice machine, the dirt machine, and the dust machine are most active. Between life cycles, stream flow reverts to sheet flow. Alley (1990) has outlined a model that allows switches between these two steady-state flow regimes by including the physics and the continuity of glacial ice, basal water, and basal till. In his sheet-flow model, the bed can be thawed but basal meltwater is insufficient to saturate the till enough, or drown bedrock projections enough, to produce stream flow by till deformation or by basal sliding. Basal meltwater is sufficient to mobilize till or increase sliding in his stream-flow model. Generation of basal water and basal till appears to be critical in any model that turns ice streams on and off, and thereby turns on and off the ice, dirt, dust , and cloud machines that may trigger abrupt climatic change. These machines are illustrated in Figure 1.8 and are applied to a hypothetical Arctic Ice Sheet in Figure 1.13.

The ice machine

The ice machine (Figures 1.8a and 1.13a) caused outbursts of icebergs into the North Atlantic during fast-flow stages of marine ice streams discharging from the Laurentide Ice Sheet into Baffin Bay, the Labrador Sea, the Gulf of Saint Lawrence, and the Gulf of Maine (Broecker and others, 1992; Andrews and Tedesco, 1992; MacAyeal, 1992), and discharging from the Eurasian Ice Sheet into the Norwegian Sea and the North Sea (Mercer, 1969; Grosswald, 1980; Sættem and others, 1992; Lehman and Keigwin, 1992). In the new paradigm, tabular icebergs were released along intersecting longitudinal and transverse crevasses that formed at the heads of marine ice streams and enlarged as they were carried to the calving front. Extending flow here along an ice stream was greatest at the peak of its life cycle, so crevassing and iceberg production were also greatest. Iceberg production may have diminished if the calving front retreated into an embayment later in the life cycle, because this reduced the calving perimeter.

Figure 1.13a illustrates the ice machine for Northern Hemisphere ice sheets during the peak of a Quaternary glaciation. Longitudinal and transverse crevasses forming at the heads of marine ice streams become lines of weakness that section these ice streams into blocks that either become imbedded in sea ice or coalesce into floating ice shelves. Eventually the entrained blocks of ice reach warmer water in the North Atlantic, where they separate along the lines of weakness to become tabular icebergs. Mercer (1969) postulated that the Younger Dryas stadial resulted from an iceberg outburst in the Norwegian and Greenland Seas. In Figure 1.13a, transverse tensile crevasses and longitudinal shear crevasses form at the head of marine ice streams, where fast stream flow begins, and are carried passively to the calving front, where icebergs are released. Top crevasses can extend downward for only a few tens of meters (Weertman, 1973), but bottom crevasses can extend upward to a height determined by the basal water pressure and the internal ice temperature (Weertman, 1980). Icebergs are released when top and bottom crevasses meet along intersections of longitudinal and transverse crevasses. In the North Atlantic, these icebergs circulate in a gyre north of the North Atlantic Current, where they cool and freshen surface water, and suppress production of North Atlantic Deep Water.

A major goal of the new paradigm is modeling the stress field and the temperature field for major Quaternary marine ice streams. This can begin by modeling these fields for an idealized ice stream on a flat bed in two dimensions to investigate opening of top and bottom transverse crevasses along the ice-stream center line. Then the solution can be expanded to three dimensions so that shear crevasses along the sides can be modeled. For both kinds of crevasses, depth of crevasse penetration will be determined by fracture criteria for opening crevasses, and on freezing rates for closing crevasses when surface meltwater enters top crevasses and basal meltwater or sea water enters bottom crevasses.

The dirt machine

The dirt machine (Figures 1.8b and 1.13b), originally proposed by Kotef (1974) and Kotef and Pessl (1981), delivered nearly all material eroded beneath the Laurentide and Eurasian Ice Sheets to ice stream termini, because 90 percent or more of interior ice was discharged by ice streams, if ice-stream discharge from the present-day Greenland (Bader, 1961) and Antarctic (Drewry, 1983) Ice Sheets also applied to former ice sheets (Hughes and others, 1985). Rock clasts were comminuted into glacial flour by ice shearing a regolith layer or sliding over bedrock. Ice then converged on ice streams, thereby concentrating and continually resupplying wet glacial flour, which became the slippery substrate that sustained stream flow (Alley, 1991). In the new paradigm, stream flow began when Laurentide ice advanced over water-saturated lacustrine sediments in linear ice-dammed lakes along river valleys or entered coastal fjords and inter-island channels floored by water-saturated marine sediments. Glacial erosion of these sediments provided the initial dirt.

Figure 1.13b illustrates the dirt machine for Northern Hemisphere ice sheets during the peak of a Quaternary glaciation. Erosion of bedrock and its unconsolidated regolith cover, and comminution of this eroded material into "dirt" while it is transported to ice streams, is maximized when continuity exists between a thawed bed beneath the interior ice sheet, where depression of the melting point is greatest and thick ice insulates the bed, and beneath ice streams, where basal heat is generated by rapid ice motion. When an intervening frozen bed disrupts this continuity, eroded and comminuted material from the interior can be delivered to ice streams only by basal freezing that incorporates this material into basal ice. Two other mechanisms are suppressed, bulk transport of the material by laminar shear deformation and transport of its fine fraction in subglacial water channels. These two transport mechanisms are probably substantially more efficient than transport by basal freezing, since eroded material typically constitutes well under half the volume of debris-charged basal ice, this ice is rarely more than a few meters thick, and the debris can slow its creep rate.

A major goal of the new paradigm is to combine the equations of heat and mass transport so ice-sheet models will simulate expansion and contraction of the thawed bed beneath interior ice domes in response to interior downdraw by ice streams as they pass through their life cycles. Downdraw allows ice streams to lengthen at both ends, so the thawed bed beneath these ice streams joins the expanding thawed bed beneath ice domes. After downdraw, basal heat is conducted more rapidly to the ice surface, and these connections are broken by basal freezing. This process produces the "heartbeat" of ice sheets: pulses of rapid expansion and contraction triggered by basal thawing and refreezing. This is equivalent to the binge-purge bulimia cycle modeled by MacAyeal (1993a, 1993b) for the Hudson Bay ice dome and

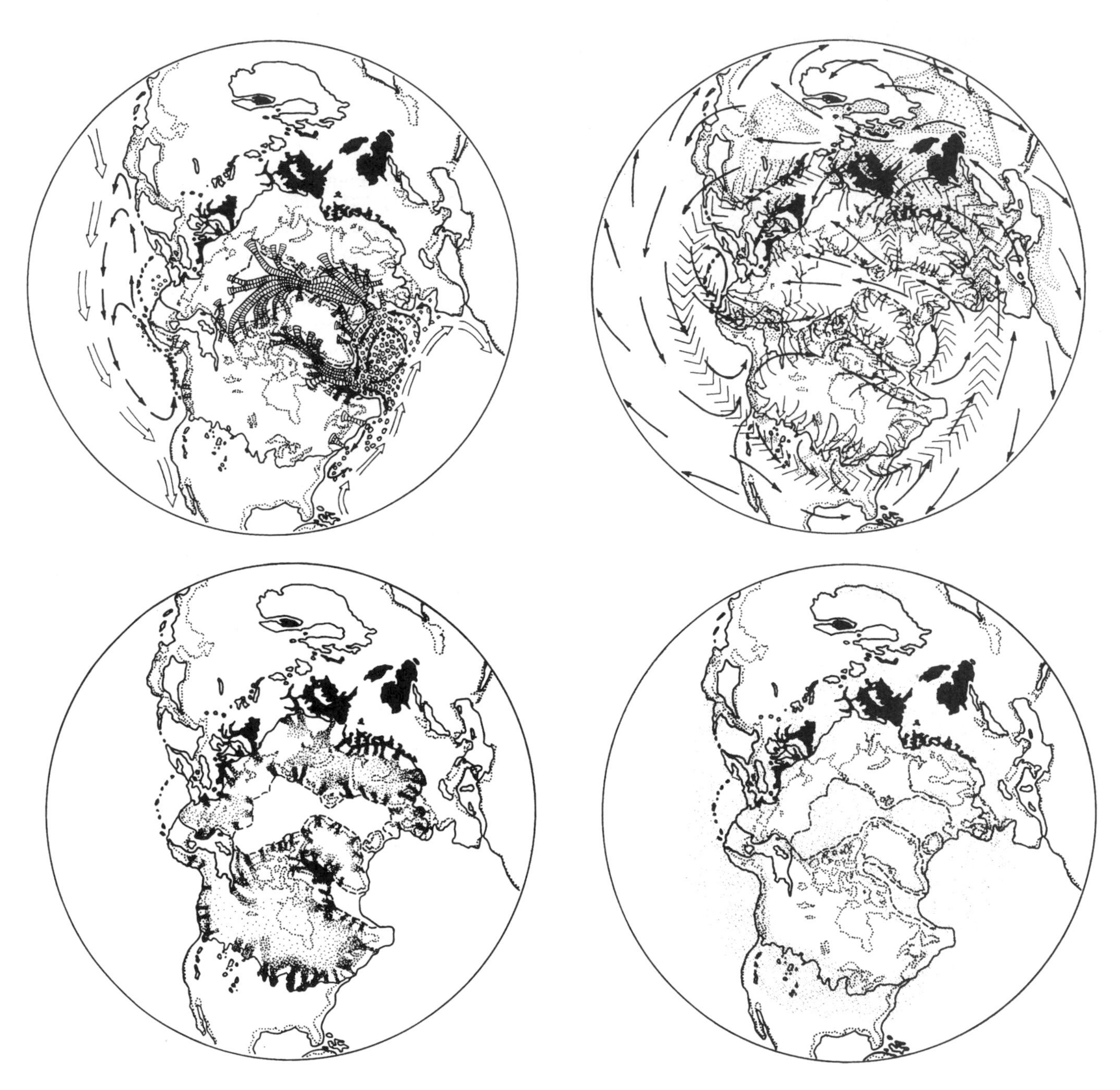

Figure 1.13: Hypothetical ice, dirt, dust, and cloud "machines" operating in a hypothetical Arctic Ice Sheet during a Quaternary glacial maximum (Hughes, 1996a). See caption of Figure 1.12 for ice and water limits. *Top left*: (a) The ice machine. Longitudinal and transverse crevasses (heavy lines) form at the heads of ice streams and are transported passively to the calving fronts of ice streams and ice shelves in the North Atlantic, where they are fracture lines of weakness for releasing icebergs. The icebergs (small polygons) circulate in gyres (narrow arrows) that lie north of the glacial North Atlantic Current and North Pacific Current (wide arrows). *Bottom left*: (b) The dirt machine. Glacial erosion under the ice sheet produces "dirt" (dots) that is transported by glacial flow toward the ice-sheet margin, where it becomes concentrated beneath ice streams and is deposited as end moraines and outwash fans beyond ice streams. The density of dots under the ice sheet denotes the concentration of glacial flour. *Top right*: (c) The dust machine. Ice crystals and glacial flour entrained by katabatic winds (wide arrows) flowing down ice streams become dust particles carried by surface winds (narrow arrows) and the jet stream (herringbone pattern), back over the ice sheet as cloud-condensation nuclei or deposited as loess (dotted areas) south of the ice sheet. *Bottom right*: (d) The cloud machine. Dry and adiabatically-warmed katabatic winds flowing off the ice sheet and westerly winds cause evaporation from ice-free lacustrine and marine ice-sheet margins, the water vapor condenses on entrained ice and dust particles, and becomes a cloud belt (shaded area) along the margin.

the Hudson Strait ice stream of the Laurentide Ice Sheet.

Another major goal of the new paradigm is to include the physics of glacial erosion and deposition in ice-sheet models. This should follow the pioneering work by Alley (1989a, 1989b, 1990, 1991), which includes the transport mechanisms already mentioned, and comminution during transport, since only the fine fraction of "dirt" deposited at ice-stream termini by the dirt machine contributes "dust" to the dust machine. It is necessary to include the longitudinal strain rate in the stress analysis, since the longitudinal deviator stress is the "pulling stress" of ice streams that facilitates downdraw of interior ice (Hughes, 1992b). It is also be necessary to include advective heat transport associated with downdraw, because ice advection is the primary control on basal thawing and freezing in the critical intermediate zone, allowing the ice dome "heartbeat" to pump ice into the ice stream "arteries" of the ice sheet (Fastook, 1993). These effects were not considered by Alley (1990). Ice and heat advection are therefore the primary control on rates of erosion, transport, and deposition of regolith or bedrock beneath the ice sheet. Comminution, which produces the dust for the dust machine, occurs when xregolith clasts striate bedrock and grind against each other as regolith is sheared by the overlying moving ice.

The dust machine

The dust machine (Figures 1.8c and 1.13c) was active along terrestrial margins of former North American and Eurasian ice sheets when katabatic winds flowing down terrestrial ice streams were intense enough to entrain glacial flour deposited on basal debris-charged ice exposed in the ablation zone of terminal ice lobes or deposited from suspension in subglacial streams feeding outwash fans beyond these ice lobes (Hughes, 1992a). Convergence of surface winds on ice streams draining the Antarctic Ice Sheet is recorded by satellite imagery and the orientation of sastrugi (Bromwich, 1989) and has been successfully modeled by Parish and Bromwich (1987), with particular attention to West Antarctic ice streams (Parish and Bromwich, 1986) and David Glacier, a major ice stream draining East Antarctic ice (Bromwich and others, 1990). In the new paradigm, dust generation was greatest during the declining and terminal stages of an ice-stream life cycle, because melt-back of terminal ice lobes exposed more glacial flour and steepened the ice-surface slope, so katabatic winds were stronger and entrained glacial flour more easily. These dust episodes are recorded in Greenland ice cores.

Figure 1.13c illustrates the dust machine for Northern Hemisphere ice sheets during the peak of a Quaternary glaciation. Surface winds flowing off ice sheets are influenced primarily by Earth's rotation (the Coriolis force), especially in high latitudes, and by the surface slope (Parish and Bromwich, 1986). During downdraw of interior ice by lengthening ice streams during their life cycles, the steep surface slope at the heads and along the sides of ice streams draws surface winds into ice streams, where they become gravity-driven katabatic winds of great intensity (Bromwich and others, 1990). A major goal of the new paradigm is to model the ability of these katabatic winds to entrain the fine fraction of end moraines and outwash fans at the termini of terrestrial ice streams along the southern margins of the Quaternary ice sheets. This includes desiccation of these deposits by katabatic winds before they can be entrained, so melting of surface ice and discharge of subglacial water are important considerations.

Figure 1.13c implies that dust produced by the dust machine may get incorporated into strong westerly winds south of the ice-sheet margin, or even into the jet stream above the southern ice-sheet margin (Manabe and Broccoli, 1985; Kutzbach, 1987), and this dust contributes to the loess sheets deposited south of Northern Hemisphere ice sheets. Modeling such processes is beyond the competence of glaciologists as yet, but it is a goal of the new paradigm nonetheless.

The cloud machine

The cloud machine (Figures 1.8d and 1.13d) was active above proglacial lakes and when the dust machine was active. If katabatic winds were strong enough to dislodge and entrain ice crystals from firn on the upper slopes of ice streams, and to dislodge and entrain glacial flour from till deposited on their lower slopes and in outwash fans, these entrained ice and dust particles would then become cloud-condensation nuclei for water vapor that was evaporated from proglacial lakes and then transported back over the ice sheet by higher inflowing winds that replaced air drained by the lower outflowing katabatic winds. In the new paradigm, the dust machine produced a cloud machine that both enhanced snow precipitation along ice-sheet margins and suppressed ablation along these margins by reflecting insolation at the top of the high-albedo clouds. Hence, North American and Eurasian ice sheets were able to regulate their own mass balance in a way that kept their southern margins in advanced positions to 14 ka BP, long after Milankovitch insolation peaked at 23 ka BP. Precipitation from clouds allowed rapid ice-sheet thickening.

Figure 1.13d illustrates the cloud machine for Northern Hemisphere ice sheets during the peak of a Quaternary glaciation. Clouds form from water vapor evaporated from ice-dammed lakes, where evaporation rates are high because katabatic winds are dry, adiabatically warmed and keep the lakes free of ice. The distribution and intensity of katabatic winds flowing down ice streams of the Quaternary ice sheets can be modeled, following Parish and Bromwich (1986) and Bromwich and others (1990), because ice-sheet models will simulate ice streams and compute their surface slopes during ice-stream life cycles. The dynamics of evaporation, cloud condensation, and precipitation, linked to katabatic winds, the entrained dust volume, and the area of ice-free proglacial lakes, can be modeled by atmospheric scientists. If dust and water vapor are lifted to elevations high enough to be transported back over the ice sheet by inflowing winds that replace outflowing katabatic winds, dust particles can be cloud-condensation nuclei that enhance cloud albedo and snow precipitation. Both processes should cause a positive mass-balance shift along the southern ice-sheet margins.

Modeling the "cloud machine" is a major goal of the new paradigm because it allows ice sheets to regulate their own mass balance, substantially independent from insolation variations or other external forcing, while influencing these external variables. Figure 1.13d shows a cloudy belt generated by the cloud machine in the Northern Hemisphere.

Terminations

Quaternary glaciation cycles over the past 700,000 years have ended in rather abrupt terminations about every 100,000 years (Broecker and van Donk, 1970). Within those glaciation cycles, smaller terminations at the ends of "Bond cycles" have taken place every 10,000 to 15,000 years (Bond and others, 1992, 1993), and Bond cycles are clusters of climate fluctuations called "Dansgaard-Oeschger events" that last from 1000 to 3000 years (Dansgaard and others, 1971; Oeschger and others, 1976; Broecker and Denton, 1989). Bond cycles and Dansgaard-Oeschger events are recorded by oxygen-isotope stratigraphy in ocean-sediment and ice-sheet cores, which is sensitive to both ice volume and atmospheric temperature. Bond cycles seem to correlate with changes of sea level (Figure 1.1) and Dansgaard-Oeschger events seem to correlate with temperature variations that affect Alpine glaciers (Denton and Karlén, 1973). In the new paradigm, Bond cycles correspond to clusters of ice-stream life cycles that overlap enough to change sea level significantly and cause an interstadial, whereas Dansgaard-Oeschger events correspond to individual life cycles of major ice streams that have little effect on sea level but have a substantial effect on atmospheric temperature. Many stadials recorded as Bond cycles culminate as "Heinrich events," which produced layers of carbonate sediments, poor in foraminifera, from ice-rafted lithic material that crossed the North Atlantic and can be traced back to major Laurentide ice streams that eroded carbonate source areas (Heinrich, 1988; Broecker and others, 1992; Andrews and Tedesco, 1992). The new paradigm ascribes Heinrich layers to life cycles of major Laurentide marine ice streams. Stadial terminations followed Heinrich events.

Nested life cycles of major ice streams have the capability of drawing down saddles along interior ice divides of ice sheets, especially when life cycles overlap for ice streams draining both flanks of a saddle. Ice-dammed lakes impounded on the landward sides of ice divides would discharge rapidly into the sea if the saddles were lowered nearly to sea level, so that calving bays could carve through the saddles. For large ice-dammed lakes, the volume of fresh water discharged through calving bays of the Arctic Ice Sheet might be sufficient to suppress formation of North Atlantic Deep Water, by flushing out icebergs that then became trapped in the large gyre shown in Figure 1.13a. The gyre formed north of the glacial North Atlantic Current, which followed the Polar Front from Newfoundland to Spain during stadials. If a more northerly interstadial Polar Front was in place, the outburst of fresh water and icebergs could restore the stadial Polar Front. Once the impounded lakes were drained, the Polar Front would return to its interstadial position. Saddles along the ice divide of an Arctic Ice Sheet that would be most vulnerable to downdraw by flanking ice streams, followed by discharge of large proglacial lakes through calving bays, existed over Hudson Bay, the Saint Lawrence estuary, the North Sea, the Barents Sea, and the Laptev Sea and between the East Siberian Sea and the Chukchi Sea. These are the sites of water gates that opened during terminations of Quaternary glaciation cycles, according to the new paradigm, and which contributed to making the terminations irreversible.

Water Gates

Ice streams that drain ice sheets are like rivers that drain continents in one important respect. They both discharge precipitation over vast interior areas at discrete sites along the outer perimeter. Discharge sites for both ice streams and rivers may change abruptly during a glaciation cycle. Abrupt changes of discharge rates and of discharge sites for major ice streams and major rivers may cause abrupt changes in the global climate machine. Several of these abrupt changes have been recorded during the last deglaciation by glacial geology, lacustrine sediments, and marine sediments. These changes record opening and closing of water gates during deglaciation. The same water gates can be expected to open and close during the final collapse of an Arctic Ice Sheet during the coming glaciation. Major ice streams that drain this ice sheet hold the keys that open and close the major water gates. These keys are the life cycles of ice streams. The same keys that turn the ice machine, the dirt machine, the dust machine, and the cloud machine on and off also open and close the water gates.

The southern water gates

In the new paradigm, major ice domes of the Arctic Ice Sheet partially collapse during stadials, with ice streams transporting advances from its stable core on a hard bed to its unstable margin on a soft bed. During the stadials of Quaternary glaciations, the Arctic Ice Sheet controls two major water gates shown in Figure 1.14. Both are well south of the ice sheet. Water draining from North America is discharged primarily into the Gulf of Mexico through the Mississippi River. Water draining from Eurasia is discharged primarily into the Mediterranean Sea through the Dardanelles. Both water gates regulate the flow of fresh water into the North Atlantic.

In North America, opening and closing of the Mississippi water gate is controlled by the life cycles of major Laurentide ice stream occupying the elongated channels of the Great Lakes. During life cycles, the terminal lobes of

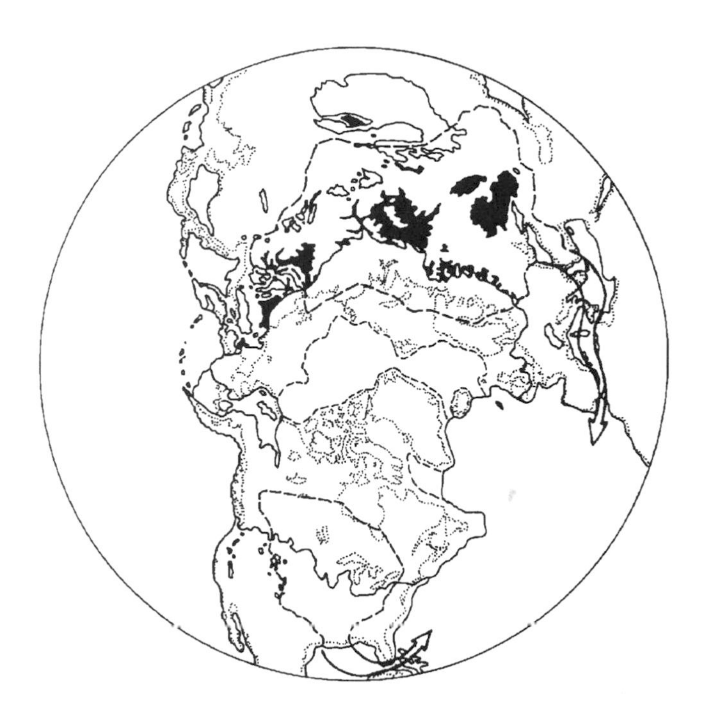

Figure 1.14: Northern Hemisphere glaciation during a glacial stadial when ice sheets thin and advance (Hughes, 1996a). Thin solid lines show ice-sheet margins, the extent of highland glaciation, and shorelines. Thin dashed lines are ice-shelf grounding lies, thick dashed lines enclose watersheds. Thick solid arrows show where watersheds drain into the North Atlantic. Black areas denote proglacial lakes and other glacial lakes within the watersheds. Dotted lines are present-day shorelines.

these ice streams extend southward far enough to prevent most meltwater along the perimeter of the ice sheet from being discharged by the Hudson River, so this meltwater overflows into the Mississippi River watershed, thereby opening the Mississippi water gate. Between life cycles, these ice lobes are nearly stagnant, so they melt back far enough to allow meltwater to be discharged by the Hudson River, thereby closing the Mississippi water gate.

In Eurasia, opening and closing of the Dardanelles water gate is controlled primarily by ice streams in the Ob River and Pechora River valleys east and west of the Ural Mountains, respectively, and by the Baltic Ice Stream. During their life cycles, the Ob and Pechora ice streams have terminal ice lobes that prevent Siberian meltwater from draining westward around the northern Urals, so the meltwater overflows southward into the watershed of the Aral, Caspian, and Black Seas, and through the Dardanelles water gate into the Mediterranean Sea. Between their life cycles, the Ob and Pechora ice streams are nearly stagnant, so their ice lobes melt back far enough to allow Siberian meltwater to drain around the northern Urals and into European Russia. This meltwater still passes through the Dardanelles water gate during life cycles of the Baltic Ice Stream, together with meltwater along the southern ice-sheet margin in Europe. Between life cycles of the Baltic Ice Stream, its terminal ice lobe melts back enough to allow European meltwater, and Asian meltwater not blocked by the Ob and Pechora ice streams, to be discharged through a water gate in the North Sea, probably the English Channel.

Several openings and closing of these southern water gates can occur as ice streams pass through life cycles along the southern margin of the Arctic Ice Sheet during a Quaternary glaciation cycle. Each opening and closing changes the major entry ports of fresh water into the North Atlantic. Ice-stream life cycles end when the bed freezes at reduced ice thicknesses, and the water gates open when the terminal ice lobes retreat.

The central water gates

In the new paradigm, major ice domes of the Arctic Ice Sheet are restored during interstadials, when calving bays return ice from its unstable margin on a soft bed to its stable core on a hard bed. During the interstadials of Quaternary glaciations, early deglaciation favors the central water gates shown in Figure 1.15 over the southern water gates in Figure 1.14, although stadials may temporarily open the southern water gates. The central water gates feed fresh water directly into the North Atlantic where North Atlantic Deep Water is produced during interstadials. Therefore, it is possible that ice streams which open and close the central water gates also turn production of North Atlantic Deep Water off and on, and thereby control climate regionally, if not globally, during stadials and interstadials.

In North America, the principal central water gate is through Cabot Strait in the Gulf of Saint Lawrence. It is open when the life cycle of a major ice stream in the Saint Lawrence estuary draws down interior Laurentide ice enough to collapse the saddle on the ice divide across the Saint Lawrence River valley. Meltwater along the entire western and southern Laurentide ice margin, from Great Bear Lake onward, is then discharged through the water gate in the Gulf of Saint Lawrence. Since this is a major ice stream, its life cycle is long. Therefore, it probably has only one life cycle during early deglaciation, and collapse of the ice saddle prevents rejuvenation of the ice stream. Late in the life cycle, a calving bay migrates up the Saint Lawrence River valley and this opens the water gate (Thomas, 1977). A life cycle of about 2000 years is followed by migration of a calving bay lasting 350 years. Over this time, a calving ice wall retreats 1700 km from the edge of the continental shelf to Lake Ontario, after which some 4×10^5 km^3 of Laurentide meltwater passes through the water gate and into the North Atlantic. This could shut down production of North Atlantic Deep Water by capping the region with a less dense freshwater layer, and thereby cause a stadial early in the deglaciation cycle, perhaps like the Younger Dryas early in the last deglaciation cycle (Fairbanks, 1989). It could also produce a Heinrich layer in the North Atlantic.

In Eurasia, the principal water gates are through the English Channel or the Norwegian Trough in the North Sea. Collapse of the saddle between the British and Scandinavian

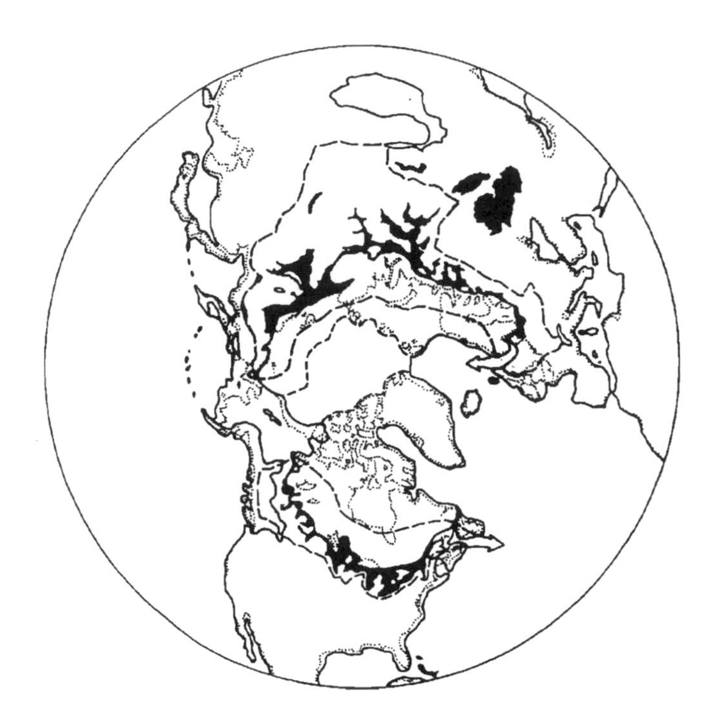

Figure 1.15: Northern Hemisphere glaciation during a glacial interstadial when ice sheets thicken and retreat (Hughes, 1996a). Thin solid lines show ice-sheet margins, the extent of highland glaciation, and shorelines. Thin dashed lines are ice-shelf grounding lies, thick dashed lines enclose watersheds. Thick solid arrows show where watersheds drain into the North Atlantic. Black areas denote proglacial lakes and other glacial lakes within the watersheds. Dotted lines are present-day shorelines.

Figure 1.16: Northern Hemisphere glaciation during a glacial termination when ice sheets collapse (Hughes, 1996a). Thin solid lines show ice-sheet margins, the extent of highland glaciation, and shorelines. Thin dashed lines are ice-shelf grounding lies, thick dashed lines enclose watersheds. Thick solid arrows show where watersheds drain into the North Atlantic. Black areas denote proglacial lakes and other glacial lakes within the watersheds. Dotted lines are present-day shorelines.

domes of the Arctic Ice Sheet closes the English Channel and opens the Norwegian Trough because the Norwegian Trough is deeper. The ice saddle collapses after life cycles of about 300 yr for ice streams draining the north and south flanks of this saddle. Both water gates are closed during life cycles of the Baltic Ice Stream, but one is opened between these life cycles. Probably only one life cycle is possible during early deglaciation, if a life cycle for the Baltic Ice Stream is 1600 years, after which a calving ice wall retreats 1000 km from Kiel Bay to the Gulf of Bothnia in only 20 years. This allows 6×10^5 km^3 of Eurasian meltwater west of the Verkhoyansk Mountains in northeastern Siberia to pass through the water gate. This can cause a shutdown of North Atlantic Deep Water production lasting perhaps 100 years. However, these are order-of-magnitude estimates so the shutdown could be substantially longer and may trigger a major stadial early in the deglacation. An ice-rafting event in the North Atlantic is also possible, one that would not deposit a carbonate layer.

The northern water gates

In the new paradigm, the Arctic Ice Sheet "expires" after its "heartbeat" stops when retreat does not halt at its stable core. Instead, the core itself collapses instead of thickening, and this terminates the glaciation cycle. At the end of Quaternary glaciations, final deglaciation will involve the northern water gates shown in Figure 1.16. By then, retreating southern margins of the Arctic Ice Sheet will have eliminated ice streams that could regulate opening and closing of the central water gates.

In North America, the principal northern water gate is the eastern entrance to Hudson Strait. It opens when a saddle on the ice divide over Hudson Bay is collapsed by ice streams draining its northern and southern flanks. The northern marine ice stream turns eastward in Hudson Strait, where it becomes a floating ice tongue. The southern terrestrial ice stream occupies James Bay, where it becomes a terminal ice lobe. A calving rate of 10 km/yr from the ice tongue, and a calving rate of 6 km/yr for an ice wall calving into a proglacial lake at the end of the ice lobe, open the water gate after these ice streams downdraw the ice saddle from a height of 1500 m to sea level in 800 yr. Collapse of the ice saddle discharges 16×10^4 km^3 of meltwater through the water gate. This is enough to shut down production of North Atlantic Deep Water in the Labrador Sea, and perhaps elsewhere in the North Atlantic. A North Atlantic Heinrich layer will form as carbonate-bearing icebergs are flushed out of

Hudson Strait. Another northern water gate in North America opens when an ice lobe retreats from the Mackenzie River delta (Teller, 1995).

In Eurasia, northern water gates exist in the Barents Sea and the Laptev Sea, and perhaps between the East Siberian and Chukchi Seas. The Barents Sea water gate opens with collapse of an ice saddle over the Barents Sea. The ice saddle is downdrawn by Bjørnøyrenna Ice Stream to the northwest and by an ice stream entering the White Sea to the southwest. This saddle lowers 1000 m to sea level in 600 yr. Calving bays migrating up these ice streams open the water gate in 150 yr, discharging 9 x 10^4 km^3 of meltwater from Europe. The Laptev Sea water gate opens with collapse of an ice saddle over the Laptev Sea. This saddle is downdrawn by a north-flowing marine ice stream ending at a sediment pile in a reentrant angle on the continental slope, and by a south-flowing terrestrial ice stream ending on the Lena River delta. The ice saddle collapses from a height of 600 m in 350 yr. The calving rates for the marine ice stream and for the terrestrial ice stream require an additional 80 yr to open the water gate, after which 3×10^4 km^3 of meltwater is discharged. Water gates may open after an ice saddle collapses between marine ice domes in the East Siberian and Chukchi Seas, or after the ice margin retreats in Long Strait, between Wrangel Island and the Siberian mainland. An ice-dammed lake in the Indigirka River and the Kolyma River lowland discharges through these water gates. About 12×10^3 km of meltwater would be discharged. Discharge through these water gates may temporarily slow or halt production of North Atlantic Deep Water, causing sharp cold spikes in the Quaternary climate. Icebergs flushed out of the Barents Sea may produce layers of ice-rafted asediments in the Norwegian Sea, but without carbonate deposition.

Pleistocene Peopling of the Americas

In the standard paradigm, glaciation of Siberia was limited to mountain glaciers and to ice caps on Arctic islands and on the Putorana and Anadyr plateaus, all of restricted extent. The great bulk of Siberia was a vast, dry, windswept prairie that extended across the steppes of Central Asia into Europe and across the Beringian land bridge into Alaska. Because the wooly mammoth was prominent among the large grazing animals, and the ecology that sustained these herbivores and their predators has no widespread counterpart today, Guthrie (1989, 1990) called this former grassland stretching from Europe to Alaska, the Mammoth Steppe. Faunal and floral dispersions across the Mammoth Steppe were continuous during Pleistocene glaciations. At the termination of the last glaciation, these dispersions included migration of Early Man into Alaska, from which the Americas were first peopled after an ice-free corridor opened between the Laurentide and Cordilleran ice sheets during deglaciation. The standard paradigm is sustained by three lines of evidence.

1. The earliest archeological sites in Alaska are about 12,000 years old, and artifacts have some similarities with the Clovis culture that produced the earliest Paleoindian assemblages to become widespread in North America shortly thereafter (Bonnichsen and Turnmire, 1991). By then, the Beringian land bridge had become submerged by rising sea level, preventing sizable migrations from Siberia to Alaska.
2. Dental traits that link prehistoric and modern people native to northeastern Siberia with their counterparts in the Americas show evolutionary divergences that are compatible with an essentially permanent separation of these populations 12,000 years ago (Turner, 1985, 1986, 1987). Similar dentochronological sepa-ration estimates have been made successfully for other circum-Pacific populations.
3. The time needed to develop the great diversity of Indian languages that survive in the Americas is compatible with proliferation from parent languages associated with three migration waves beginning 12,000 years ago (Greenberg, 1987). Classification of these languages led to discovery of common linguistic elements and development of a model for language dispersal.

A paradigm shift was already under way at the First World Summit Conference on the Peopling of the Americas, held in 1989 at the University of Maine (Morell, 1990). Since then, the three legs sustaining the standard paradigm have come under increasing attack and are now quite wobbly.

1. Archeological sites older than Clovis are gaining more credibility. In South America, these sites include Monte Verde in central Chile, which was inhabited 13,000 years ago and perhaps as early as 33,000 years ago (Dillehay, 1989), and hundreds of rockshelters in northeastern Brazil (Bahn, 1993), including Pedra Furada, which has nearly a continuous record of human habitation from 50,000 to 5000 years ago (Parenti, 1993), and pre-Clovis assemblages about 13,000 years old from Taima-taima in Venezuela (Gruhn and Bryan, 1984). In North America, sites include the Meadowcroft rockshelter in western Pennsylvania, dated between 14,000 yrnd 14,500 years old (Adovasio and others, 1990), Bluefish Cave in the Yukon Territory dated at 24,000 years (Morland and Cinq-Mars, 1989), and the Bow Valley site in western Alberta (Chlachula, 1994), which is between 25,000 and 21,000 years old.
2. Genetic diversity among native American popu-lations does not square with the "three-wave" model based on dental traits (Szathmary, 1993). Mitochrondrion DNA in blood samples for both North American and South American Indians points to four genetically distinct populations in a single "wave" of migration between 21,000 and 14,000 years ago (Horai and others, 1993; Gibbons, 1993; Baer, 1993).
3. A linguistic survival and dispersal model for Paleoindian migration through the Americas allows a single entry across the Beringian land bridge 50,000 years ago or ten en tries beginning 35,000 years ago, as end members in a spectrum of entries that fit the linguistic data (Nichols, 1990). The model of three waves beginning 12,000 years ago falls outside of this spectrum (Simpson, 1992).

The new paradigm is compatible with these new data and models. It provides a reassessment of the land bridge as a highway open to continuous migrations during Pleistocene glaciations. Instead, the land bridge was closed by an "ice gate" during each glacial maximum and perhaps during each stadial and closed by a "sea gate" during interglaciations and perhaps during each interstadial. Migrations were possible only during the brief intervals when both the ice gate and the sea gate were open. During the last glaciation, human migration was restricted to these intervals.

The land bridge

In the new paradigm, marine ice sheets grounded on the Arctic continental shelves of North America and Eurasia were connected by an ice shelf floating over Arctic Ocean basins, and transgressed onto Arctic coasts to merge with highland glaciers and plateau ice caps, to become an Arctic Ice Sheet that behaved as a single dynamic system. Marine ice sheets advanced furthest onto the mainland where coastal lowlands were lowest and most extensive. For this reason, Laurentide ice advanced across the Hudson Bay lowlands to south of the Great Lakes, Scandinavian ice advanced onto the North European Plain, Kara Sea ice advanced into the Central Siberian Lowland, East Siberian Sea ice advanced to the foothills of mountain ranges in western Beringia, and Chukchi Sea ice spread across the Bering Sea land bridge in central Beringia, where the circum-Arctic landmass was lowest and narrowest, and calved directly into the Pacific Ocean, as shown in Figure 1.12. Bering Strait would have been the site of a major ice stream from the Chukchi Sea ice dome into the Bering Sea.

This view is compatible with the existence of a submarine trough in Bering Strait, north-to-south glacial through valleys across Chukchi Peninsula and possibly across Seward Peninsula on opposite sides of Bering Strait, raised marine shorelines that are continuous from Seward Peninsula northward along the Alaskan coast, and glacial tectonics associated with glacial erratics on the north side of Saint Lawrence Island in the Bering Sea (Hughes and Hughes, 1994; Grosswald and Hughes, 1995). Submarine troughs are eroded beneath marine ice streams, glacial through valleys imply that a marine ice dome in the Chukchi Sea overrode ice caps on Chukchi Peninsula and perhaps on Seward Peninsula. Raised shorelines imply that the marine ice dome was extensive enough to transgress and isostically depress the adjacent coastline of Alaska. Glacial geology on Saint Lawrence Island shows that an ice lobe from the north entered the Bering Sea. An ice lobe from the marine ice dome in the Chukchi Sea that crossed the Bering Sea and calved into the Pacific Ocean would have been an "ice gate" that blocked Pleistocene migrations across the land bridge of central Beringia. This ice gate is shown in Figure 1.12.

The ice gate

In the new paradigm, the ice gate across central Beringia would have been closed during the maximum of each Pleistocene glaciation cycle, and perhaps during stadials within glaciation cycles. If an ice lobe from a marine ice dome in the Chukchi Sea was too thin and too transient to isostatically depress the land bridge sufficiently, then rising sea levels during interstadials may not have flooded central Beringia when the ice lobe retreated back into the Chukchi Sea. The land bridge across Bering Strait would then have been open for migration. However, a thicker and more persistent ice lobe could have depressed central Beringia enough to allow flooding of the land bridge during interstadials. In this case, Bering Strait would have been closed to migration during both stadials and interstadials, and migration would have been possible only during brief intervals between the advancing ice lobe and falling sea level, and between the retreating ice lobe and rising sea level, when both the ice gate and the sea gate would have been opened across the land bridge (Hughes and Hughes, 1994).

The sea gate

In the new paradigm, the sea gate across central Beringia is closed to migration during Quaternary interglaciations, and perhaps during interstadials during Quaternary glaciations. Opening and closing of the sea gate depends not only on whether or not a lobe of ice from a marine ice dome in the Chukchi advanced and retreated across the central Beringian land bridge, but also primarily on global ice volume changes during glaciation cycles of the Quaternary Ice Age. The traditional concept is that stadials and interstadials within a glaciation are probably a gradual response to Milankovitch insolation variations controlled by cycles of 41,000 years in the tilt of Earth's rotation axis and by cycles of 19,000 years and 23,000 years in the precession of Earth's rotation axis. This concept is rapidly being recast in terms of Dansgaard-Oeschger events lasting from 1000 to 3000 years that reflect substantial global or regional temperature fluctuations, and that are nested within Bond cycles lasting from 7000 to 15,000 years that also reflect abrupt changes in global sea level. When the relative contributions of atmospheric temperature and sea level become known, times for opening and closing the sea gate across central Beringia will also be established.

A Siberian Cradle for Mankind?

In the new paradigm, the Pleistocene Mammoth Steppe proposed by Guthrie (1989; 1990) may not have existed during glacial maxima and during stadials within each glaciation. There is evidence that neither the characteristic flora nor the megafauna existed at these times in Siberia and Alaska (Colinvaux, 1980; Cwynar and Ritchie, 1982; Ritchie and Cwynar, 1982; Ritchie, 1984). At these times, according to the new paradigm, the cold, dry Mammoth Steppe was transformed into a landscape of huge proglacial lakes impounded by the Arctic Ice Sheet, as shown in Figure 1.12. Environmental transformations restored the Mammoth Steppe when these lakes drained through the various water gates. The Pleistocene climate of Siberia would therefore have been quite variable and subject to abrupt changes. This would have placed a premium on adaptability for survival of Early Man in Siberia, even more so than exists today. It is therefore highly significant that Chlachula and others (1994) ascribes a Middle Pleistocene age to archeological sites on the upper Yenisei River in south central Siberia (55°N, 91°W). Even more significant, Mochanov (1977) ascribes an age of 35,000 years to the Diuktai Cave site and an astonishing age of 1.8 to 3.4 million years to the Diring site, both on the upper Lena River in western Beringia, as reported by Hall (1992). This implies long-term human habitation in eastern Siberia. The Diring site predates the Oldowan sites in East Africa, dated between 1.7 and 2.7 million years in age, and led Mochanov to propose that, unlike the lazy latitudes of Africa, Siberia provided an environment where "stress influenced mutation" to produce "adaptive peaks" for early hominids living "on the edge of existence" and this resulted in Homo sapiens. Siberia, if not the cradle of mankind, may have been its kindergarten.

The possible antiquity of early man in Siberia underlines a persistent question. If people from southeast Asia crossed a stormy shark-infested strait in the deep ocean and occupied Australia 80,000 years ago to become the aboriginal population encountered by European explorers, why were people from northeast Asia unable to cross central Beringia, a broad grassland 1000 km wide teeming with game, with the westerly winds at their backs, and become the aboriginal population in North America until only 12,000 years ago? And, if they crossed earlier, why was their survival so problematic that only a few scattered controversial sites give uncertain testimony to their passage? Perhaps the answer is that this broad highway was blocked by a marine ice sheet during most of the Pleistocene, including the critical last glaciation, after hominids had crossed the threshold into humanity.

2

LIVING IN AN ICE AGE

The Holocene Interglacial

Planet Earth has been in the grip of an ice age for over four million years, long before humans emerged as the dominant species. We are an Ice Age People. Even in the present "interglacial" time, much of Earth remains glaciated. Figure 2.1 compares the present-day coastline of North America with two other coastlines, one that existed at the last glacial maximum 18,000 years ago, when sea level was aproximately 120 m lower, and one that would exist if present-day ice sheets on Antarctica and Greenland collapsed, causing sea level to be over 60 m higher. We realize that one third of the Pleistocene glacial ice is still with us. Without that ice, rising sea level would displace 70 percent of our population. In North America, all of Florida would be under water, Memphis would be an ocean port, New England would be an island, and San Francisco Bay would be 600 km long and 200 km wide. Worldwide, whole countries would vanish, from Denmark to Bangladesh. Global climate and weather would be drastically altered, but exactly how is unknown. All we know for sure is that drastic global climatic changes accompanied the collapse of most Northern Hemisphere ice sheets from 18,000 to 8000 BP (years before present).

Figure 2.1: North American sea-level changes linked to Quaternary glaciations. Sea level during the maximum glaciation is shown by the heavy line, sea level for the present glaciation is shown by the light line, and sea level for complete deglaciation is shown by the black border. The Antarctic ice sheet constitutes ninety percent of the present glaciation. The sea level increase resulting from Antarctic deglaciation would probably initiate disintegration of the Greenland Ice Sheet, which constitutes most of the remaining ten percent. (Taken from Dott and Batten, 1971, Figure 1.6).

Research Strategy for an Ice Age

While he was the director of the Office of Polar Programs at the U.S. National Science Foundation, J. O.Fletcher (1972) proposed a six-point research strategy that can be applied to understanding any dynamic system. Each point addresses a fundamental question about the system:

1. What are its most energetic parts?
2. What factors force motion in these parts?
3. Which of these factors vary over time?
4. What physical processes cause the time variations?
5. Can these processes be quantified theoretically?
6. What experiments will test the theories?

The dynamic system of interest to an Ice Age people is Earth's climatic machine, particularly the interaction between glaciation and climate change. That is the theme of this book.

The dominant manifestations of glaciation are continental ice sheets, such as those now covering Antarctica and Greenland. Ice sheets interface with the atmosphere above them, with the oceans on their seaward perimeter, with land and lakes on their landward perimeter, and with Earth's crust beneath them. Gravity makes ice sheets flow. It pulls the ice downward, causing horizontal spreading during gravitational collapse. Horizontal spreading displaces air, a medium much less dense than ice. Transverse crevasses near the ice margin indicate that ice is pulled forward by gravity, not pushed forward, to convert horizontal segregation of ice and air into vertical segregation, with heavy ice overlain by light air. The horizontal pulling force is therefore a manifestation of the vertical pulling force of gravity. The pulling force gives rise to several instabilities along the margins of ice sheets.

Instabilities along Seaward Margins

The pulling force acting where ice sheets spread into the ocean is illustrated in Figure 2.2. The ice margin becomes afloat when it advances into water that is sufficiently deep, and the longitudinal forward lithostatic gravitational force in ice exceeds the opposing hydrostatic gravitational force in water, leaving a net forward pulling force that is proportional to the square of the floating ice thickness. The floating perimeter of an ice sheet is called an ice shelf, and this net pulling force draws out the ice shelf, causing it to neck

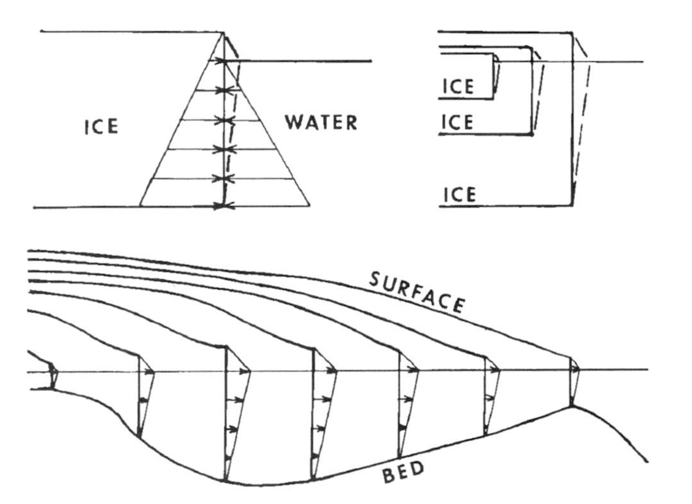

Figure 2.2: The pulling force where a marine ice sheet becomes afloat (Hughes, 1987a). *Top left*: The pulling force is the difference (dashed line) between the lithostatic and hydrostatic forces in ice and water (arrows). *Top right*: The pulling force (the triangle areas) increases with the square of floating ice thickness. Bottom: As the grounding line of a marine ice sheet retreats, the pulling force increases on a downsloping bed and decreases on an upsloping bed.

down with distance from its grounding line, much like the necking of a specimen in a tensile test. Hence, the greatest floating ice thickness, and therefore the strongest pulling force, is at the grounding line of an ice shelf. If ice crossing the grounding line is too slow to overcome the ice thinning rate caused by necking, the grounding line will retreat. Retreat on a downsloping bed will accelerate because the grounding-line ice thickness increases. Conversely, retreat on an upsloping bed decelerates because the grounding-line ice thickness decreases. As shown in Figure 2.2, grounding-line retreat for an ice sheet that has isostatically depressed the floor of a marine embayment will accelerate on the downsloping seaward flank and decelerate on the upsloping landward flank, because the pulling force increases downslope and decreases upslope.

Figure 2.3 illustrates mechanisms that can trigger grounding-line retreat into a marine embayment. In addition to isostatic depression, glacial erosion within the embayment and glacial deposition at the grounding line typically produce a basal sill where the ice sheet becomes afloat. If the grounding line retreats over this sill, it will be on a downward slope and can retreat further irreversibly until it rests on the upsloping landward side at a bed elevation comparable to the sill elevation. Initial retreat over the sill can be caused by (1) a glacial surge that thins grounded ice and extends floating ice, (2) continued erosion and isostatic depression of the sill until it attains the depth of the grounding line, (3) rising sea level that floats the grounding line over the sill, or (4) net melting at the top and bottom ice surfaces that thins the ice at the grounding line. Also shown in Figure 2.3 is iceberg calving as the grounding line retreats, and ferrying of icebergs over the sill, so they do not pile up against the sill and force readvance of the grounding line. This is important because icebergs formed in the marine embayment will be thicker than those formed at the sill, unless necking of the floating ice shelf reduces ice thickness sufficiently along its calving front. Otherwise, icebergs must roll over or disintegrate before they reach the sill, which often happens.

When a ice sheet retreats onto land, calving mechanisms along its seaward margin change, as shown in Figure 2.4. When the pulling force at the grounding line becomes weakened on the upsloping marine bed, a rising snowline due to climatic warming can cause continued retreat. Precipitation falls as snow above the snowline and as rain below the snowline, so ice melting begins to replace ice calving as the dominant ablation mechanism. Although the snowline rises continuously during climatic warming, the ice margin retreats stepwise as one calving mechanism replaces another. In (a) calving occurs along transverse crevasses opened by the longitudinal pulling force seaward of the ice-shelf grounding line. In (b) the calving front has temporarily stabilized along ice rises, where the ice shelf is locally pinned to the sea floor and behind which transverse crevasses do not form because the ice shelf is pushing against the pinning points. In (c) the calving front has migrated around pinning points so that calving occurs along tidal flexure crevasses at the grounding line. In (d) the calving front reaches the

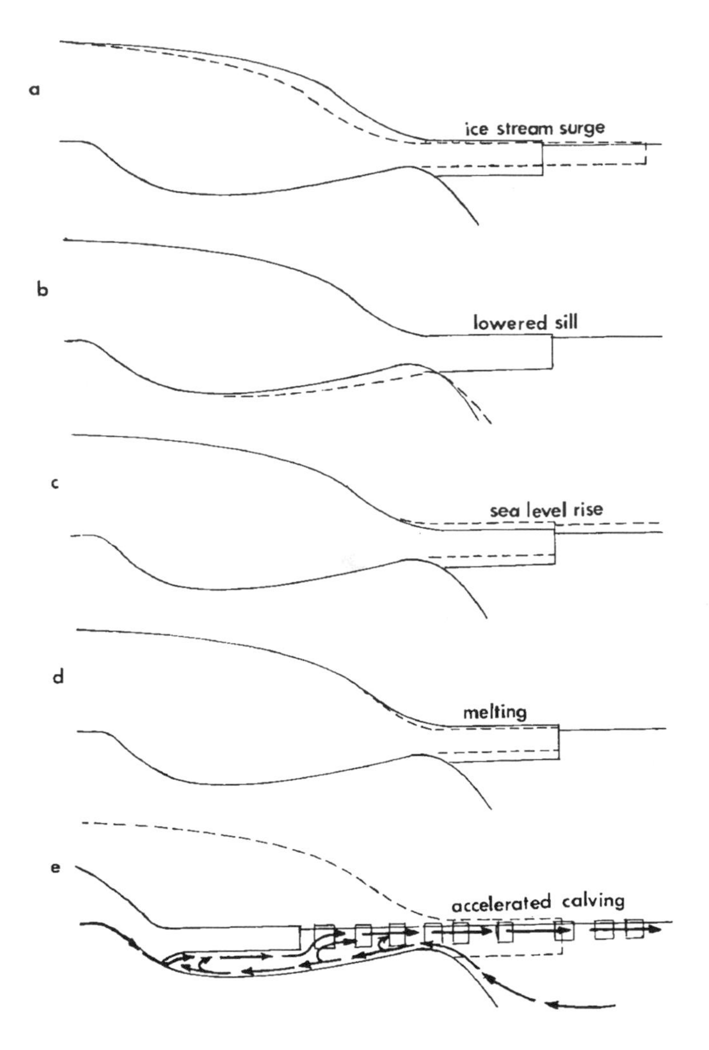

Figure 2.3: Mechanisms triggering collapse of a marine ice sheet. Irreversible collapse occurs when the grounding line migrates over its basal sill as a result of (a) ice thinning during an ice-stream surge, (b) lowering of the sill by glacial erosion and glacioisostatic depression of the bed, (c) rising sea level that lifts the ice shelf, (d) climatic warming that melts upper or lower surfaces of the ice shelf, and (e) accelerated calving and evacuation of icebergs by water currents (arrows).

grounding line so tidal flexure stops and calving occurs above and below an ice groove eroded by wave action at sea level. In (e) the calving front has retreated to the shoreline of a beach, so calving occurs only above the wave-washed groove. In (f) the calving front has retreated from the beach, so the basal groove is gone and the calving ice wall has become a melting ice ramp.

Instabilities along Landward Margins

The snowline shown in longitudinal cross-section in Figure 2.4 lies on an atmospheric surface that generally dips from the equator to the poles, and from continental to maritime

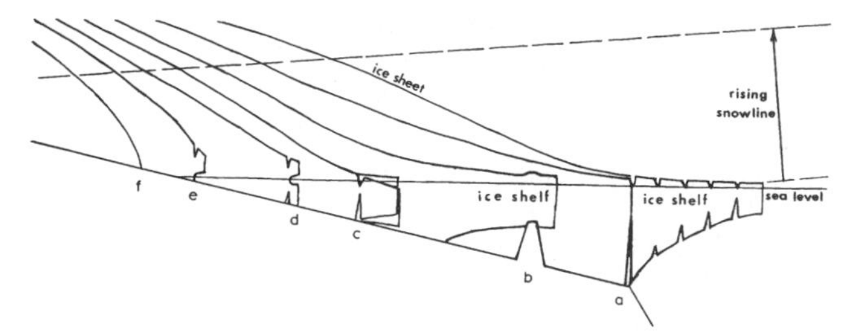

Figure 2.4: Pauses in upslope retreat of a marine ice sheet as the snowline rises. Pauses occur when calving mechanisms change (a) at the outer edge of the continental shelf, (b) along ice-self pinning points on the continental shelf, (c) along the ice-shelf grounding line, (d) along an ice wall grounded in deep water, (e) along an ice wall grounded at the shoreline, and (f) when calving stops on land.

environments, with net snowfall above the surface and net rainfall below the surface. Therefore the surface intersects an ice sheet along an equilibrium line that separates upper (inner) regions of net ice accumulation from lower (outer) regions of net ice ablation, as shown in Figure 2.5. In general, snowlines on this surface will dip toward equatorward margins of the ice sheet, away from poleward margins, and will be nearly horizontal along east-west margins. Surface flowlines for an ice sheet lie perpendicular to ice elevation contour lines, and the intersection of the snowline, flowline, and equilibrium line at the surface of an ice sheet can cause both stable and unstable responses of the ice sheet, depending on the slope of the snowline and the direction of the flowline. Since ice sheets have flowlines in all directions, some sectors can respond reversibly and others can respond irreversibly to migrations of the snowline-flowline-equilibrium-line triple point. Mass-balance analyses have

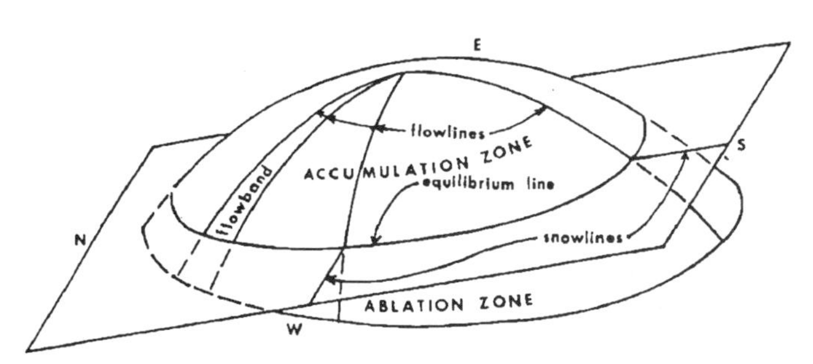

Figure 2.5: Mass-balance conditions in an idealized Northern Hemisphere ice sheet. A snow surface in the atmosphere separates mean annual precipitation as snowfall (above) from rainfall (below), and intersects the ice sheet along an equilibrium line that separates an annual net accumulation zone (above) from an annual net ablation zone (below). A snowline on the snow surface intersects a flowline on the ice sheet at a point on the equilibrium line. A flowband lies between adjacent flowlines. Note that north-south snowlines dip from the equator to the pole, but east-west snowlines are horizontal. Ice sheets are unstable for horizontal snowlines, as seen in Figure 2.6.

have been developed for two extreme cases of the two-dimensional snowline-flowline intersection along a longitudinal cross-section of an ice sheet. Mass balance perturbations causing stable and unstable responses in the ice sheet were first analyzed by Nye (1960) and by Weertman (1961), respectively. Granting their simplifying assumptions, these models remain valid. Accumulation and ablation rates above and below the equilibrium line were assumed to be constant, and isostatic compensation beneath the ice sheet was assumed to be constant. Figure 2.6 illustrates both extreme models, as well as a metastable model.

In the stable response model, the snowline is assumed to depend only on latitude, so it is vertical for north-south transects on an ice sheet and exists only near the equatorward margin. If a positive mass balance strengthens the pulling force, the ice margin advances, thereby expanding the ablation zone without a corresponding expansion of the accumulation zone. Hence, ablation exceeds accumulation, so mass balance turns negative and the ice margin retreats to its original position. If a negative mass balance weakens the pulling force, the ice margin retreats, thereby shrinking the ablation zone without a corresponding shrinkage of the accumulation zone. Hence, accumulation exceeds ablation, so the mass balance turns positive and the ice margin returns to its original position. The mass balance perturbation is reversible in both cases.

In the unstable response model, the snowline is assumed to depend only on altitude, so it is horizontal for east-west transects on an ice sheet. A positive mass balance that strengthens the pulling force and advances the ice margin now expands the accumulation zone without changing the ablation zone, thereby making the mass balance even more positive and making the ice sheet advance even more. A negative mass balance that weakens the pulling force and retreats the ice margin now shrinks the accumulation zone without changing the ablation zone, thereby making the mass balance even more negative and making the ice sheet retreat even more. The mass balance perturbation is irreversible in both cases.

Results from the stable and unstable response models imply that the snowline has a critical slope, with more steep slopes producing a stable response and less steep slopes producing an unstable response. As seen in Figure 2.5, a sloping snowline is the usual case and it allows the metastable response to mass balance perturbations shown in Figure 2.6. Weertman (1961) analyzed this metastable response and Figure 2.7 displays his results. When plotted against distance from the center of ice sheets of various sizes, elevation changes linearly for the snowline but nonlinearly for the equilibrium line. At short distances, corresponding to small ice sheets in the first stages of advance or the last stages of retreat, the snowline elevation passes under the elevation of the equilibrium line so a mass balance perturbation is irreversible. At long distances, corresponding to the steady-state maximum of an ice sheet, the snowline elevation passes over the elevation of the equilibrium line so a mass-balance perturbation is reversible. When the snowline

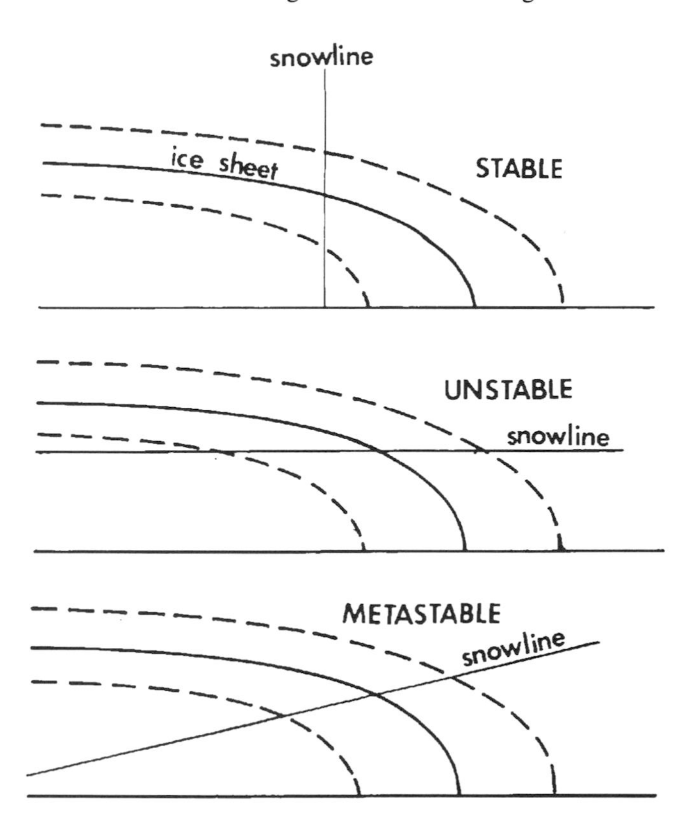

Figure 2.6: Stable, unstable, and metastable responses of ice sheets and local ice domes along ice-sheet margins to perturbations of their margins in relation to the snowline slope (Hughes, 1987a). *Top*: Perturbations change the ablation zone and are reversible when the snowline depends on latitude only. *Middle*: Perturbations change the accumulation zone and are irreversible when the snowline depends on altitude only. *Bottom*: Perturbations change both the ablation and accumulation zones and may be either reversible or irreversible when the snowline depends on both latitude and altitude.

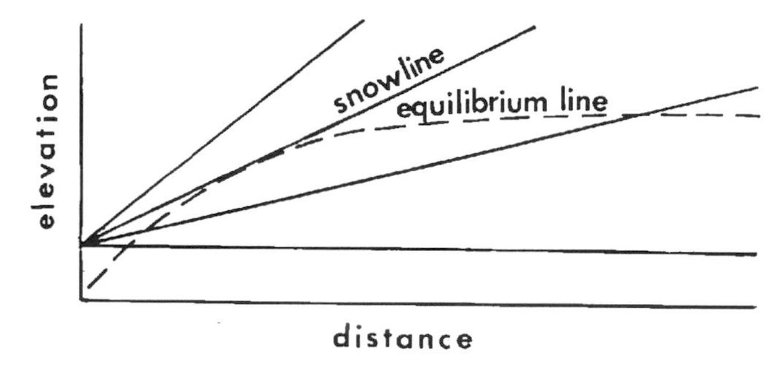

Figure 2.7: The theoretical dependence of the stability of an ice sheet and local ice domes along its margin upon flowline length (Weertman, 1961; Hughes, 1987a). The snowline intersection with the equilibrium line produces a stable response in major interior ice domes having long flowlines and an unstable response in local ice domes along the ice sheet margin because they have short flowlines.

elevation is tangent to the equilibrium-line elevation, a positive perturbation is reversible but a negative perturbation is irreversible.

Applied to continental ice sheets, a reversible ice-sheet response is most likely along equatorward margins, because the snowline has a maximum dip toward the ice sheet, and irreversible responses are most likely along east-west and poleward margins, because the snowline is either horizontal or dips away from the ice sheet. Hence, an ice sheet will advance into the ocean along its eastern, western, and poleward margins until ice calving replaces ice melting as the dominant ablation mechanism and will advance over land along its equatorward margin until the melting rate offsets the advance rate, assuming ice calving into lakes or seas is not involved.

Snowline slope in north-south directions changes during 100,000 year cycles of Pleistocene glaciation, because of changing insolation with latitute. The 23,000 year cycle of Earth's axial precession changes insolation most in mid latitudes, but the 41,000 year cycle of Earth's axial tilt changes insolation most in high latitudes, as shown in Figure 2.8. These slope changes are illustrated in Figure 2.9, and they

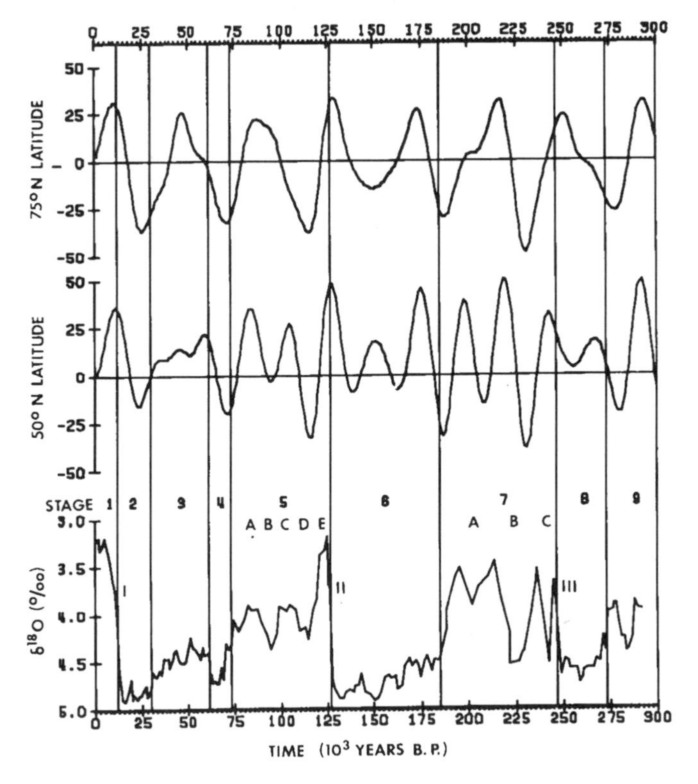

Figure 2.8: Comparison of Northern Hemisphere summer half-year insolation at 75°N and 50°N (Vernekar, 1972) and the oxygen-isotope record from V19-29 (Ninkovitch and Shackleton, 1975). Paleoclimatologists generally assume tight linkage between oxygen-isotope records and global ice volume on a time scale by assigning ages to important peaks (Imbrie and Imbrie, 1980) and stage boundaries (Kominz and others, 1979; Pisias and others, 1980) and then by assuming constant sedimentation rates between chronologic control points.

may be great enough to flip the response to a mass balance perturbation from reversible to irreversible, and vice versa, even along equatorward margins of Pleistocene ice sheets. These flip-flops may be abrupt, even though the snowline slope changes gradually. Hence, abrupt equatorward advances and retreats in response to gradual insolation changes must be considered.

Most intriguing of all is the possibility of instability in local ice domes around the perimeter of ice sheets. In Antarctica, these ice domes often lie between ice streams that form near the ice margin. Interior ice is downdrawn into ice streams, with downdraw strongest at the heads of ice streams so saddles form on the ice ridges between ice streams. These saddles convert ice ridges into local ice domes. If climatic warming, perhaps from increasing atmospheric carbon dioxide causing a "greenhouse effect," produces an equilibrium line around these local ice domes, they may shrink irreversibly until they vanish, according to the instability mechanism depicted in Figure 2.7. This would cause the ice sheet to retreat rapidly over distances equal to the diameters of these local ice domes.

Local ice domes probably existed between ice streams that drained former Pleistocene ice sheets in the Northern Hemisphere. Any ice domes existing between the terminal lobes of terrestrial ice streams along the equatorward margins would probably have equilibrium lines. Therefore, any perturbations of these ice streams would cause irreversible growth or shrinkage of the intervening local ice domes. It is therefore important to consider ice-stream instabilities.

Instabilities in Ice Streams

Ice streams are fast currents of ice that develop near the margins of ice sheets, and they drain more than 90 percent of the ice from present-day ice sheets in Antarctica and Greenland. The glacial geological record is consistent with ice streams controlling ice drained from former Pleistocene ice sheets as well. Figure 2.10 shows present Antarctic ice streams, and Figure 2.11 shows former Northern Hemisphere ice streams. As Bader (1961) described them, "An ice stream is something akin to a mountain glacier; but a mountain glacier is hemmed in by rock slopes, while the ice stream is contained by slower moving surrounding ice." The major instability in certain mountain glaciers is the glacial surge, during which ice moves 10 to 1000 times faster than normal, causing the accumulation zone to lower and the ablation zone to advance with little or no change in overall ice volume (Meier and Post, 1962). Surges last from one to three years, and seem to repeat every ten to thirty years. This implies a similar surge-like cycle for at least some ice streams, with the durations up to thousands of times longer because ice streams drain accumulation areas up to thousands of times larger than those drained by surging mountain glaciers. The drainage basins of major ice streams in present-day ice sheets are as large as the huge watersheds for the major rivers that drain continents. If ice streams that drain marine ice sheets grounded below sea level on continental shelves undergo simultaneous surges, the ice shelf ground-

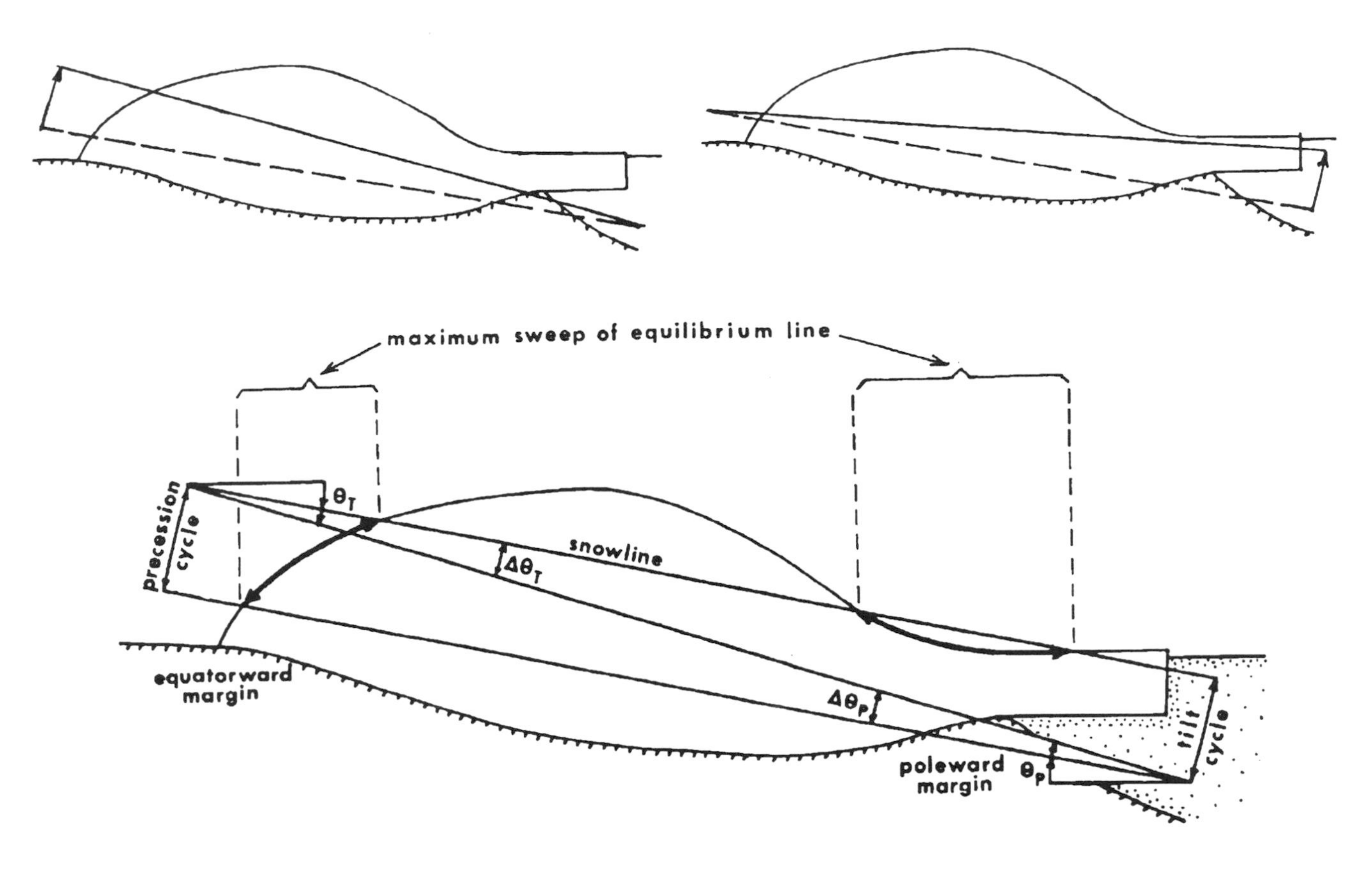

Figure 2.9: A north-south transect across a Northern Hemi-sphere ice sheet during the last glaciation (Hughes, 1987a). The snowline slope is determined by changes $\Delta\theta_T$ of angle θ_T during Earth's 41,000 yr cycle of axial tilt and changes $\Delta\theta_P$ of an angle θ_P during Earth's 23,000 yr cycle of axial precession. *Upper left inset:* Maximum snowline slope ($\Delta\theta_T = 0$). *Upper right inset*: Minimum snowline slope ($\Delta\theta_P = 0$).

ing line will retreat as shown in Figures 2.2 and 2.3, and cause interior ice to lower drastically while the floating ice shelf lengthens rapidly, all without changing ice volume if calved icebergs are included. Figure 2.12 compares the surge of a mountain glacier with the surge of a marine ice sheet caused by simultaneous surges of its ice stream. For comparison with Figure 2.12, Figure 2.13 shows an east-west cross-section of the West Antarctic Ice Sheet (the Antarctic Ice Sheet in the Western Hemisphere and north of the Transantarctic Mountains). The short (left) flowline has a largely convex pre-surge profile above a depressed bed. The long (right) flowline has a largely concave post-surge profile above a rebounding bed. This shifts the ice divide to the left.

Surges of mountain glaciers have been linked to production of basal meltwater that greatly reduces bed traction (Kamb, 1986). Ice streams draining a marine ice sheet may surge if marine sediments in the permafrost condition beneath the ice sheet become thawed, beginning at the grounding line and proceeding along interisland channels and straits into the marine embayment beneath the ice sheet. As interstitial ice in the marine sediments turns to water, allowing the sediments to become mobile by dilation and shear deformation, basal traction reduces greatly, and the surging front retreats above the thawing front. The surge stops when ice has thinned enough for basal heat to be conducted to the surface faster than it is generated by basal shear deformation. The bed will then freeze, halting the surge. Since this process may last for 1000 years or more, it is appropriate to speak of a life cycle for ice streams, instead of a surge cycle. Hypothetical life cycles are illustrated in Figure 1.7. Gravitational collapse of the West Antarctic Ice Sheet, illustrated in Figure 2.13, can propagate into the much larger East Antarctic Ice Sheet through the surge-like behavior of Ice Stream A and Foundation Ice Stream, which drain both East and West Antarctic ice, as shown in Figure 2.14. East Antarctic ice enters these ice stream in a narrow region called the Bottleneck, where it drains through a submarine gap in the Transantarctic Mountain (see Figure 2.15).

Only marine ice streams can be observed on present-day ice sheets. Figure 2.15 identifies the major Antarctic ice streams, which may be in various stages of a life cycle, ranging from inception for Thwaites Glacier, growth for Pine Island Glacier, mature for Ice Stream B, declining for Byrd Glacier, and terminal for Ice Stream C (Hughes, 1992b). Many ice streams drain marine subglacial basins: Byrd basins, Byrd Subglacial Basin in West Antarctica and

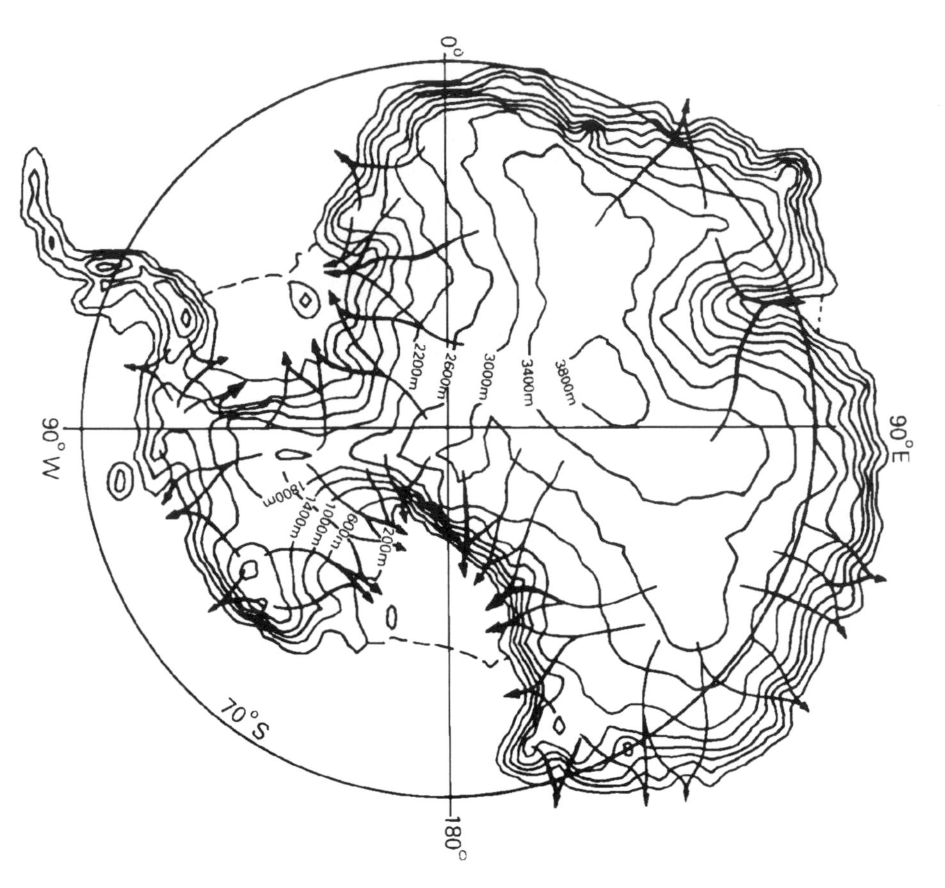

Figure 2.10: Ice-surface topography and major ice streams in present-day Antarctica. Elevations in meters are from Drewry and others (1982). Ice streams and their approximate drainage areas are shown by arrows with forked tails.

Polar, Wilkes, and Aurora Subglacial Basins in East Antarctica. Figures 2.13 and 2.14 suggest that West Antarctic ice may be in an advanced stage of collapse, with the Ross and Filchner-Ronne Ice Shelves being collapsed and floating portions akin to the surging ice sheet in Figure 2.12.

East Antarctic ice has collapsed in the Lambert Glacier ice drainage basin, producing the Amery Ice Shelf, and the area may be poised for additional collapse. If ice streams draining Wilkes Subglacial Basin and Aurora Subglacial Basin are at the inception stage of surge-like life cycles, they could collapse the entire Wilkes Land sector of the East Antarctic Ice Sheet. If gravitational collapse in this sector by surging ice streams were to be on the scale of gravitational collapse in the sector drained by Lambert Glacier, as seen by the inwardly bowing ice elevation contour lines in Figure 2.10, a substantial fraction of the 60 m of sea level now locked in the East Antarctic Ice Sheet would be released. In fact, ancient beaches up to 60 m high are widespread around the world (Brigham-Grette and Carter, 1992; Anderson and Borns, 1994). These beaches are though to be of Pliocene age, when global climate was warmer than today. If much of the East Antarctic Ice Sheet collapsed during warm Pliocene climates, then perhaps ongoing "greenhouse" climate warming today could cause a substantial future collapse of the East Antarctic Ice Sheet. Rapid collapse could be as traumatic for Earth's population as the biblical deluge (Hughes, 1986b).

Legends of a great flood are widespread among the earliest cultures all around the world, notably in the eastern Mediterranean, India, China, and Meso America, where people lived close to sea level. This suggests the possibility that the legends are a collective folk memory of an actual event, dimmed by time, but so traumatic that it lingers still. Such an event may have occurred during the final gravitational collapse of Northern Hemisphere ice sheets at the termination of the last Pleistocene glaciation cycle, when sea level rose in several distinct jumps (Blanchon and Shaw, 1995). The last one, a 6.5-m rise in global sea level, took place 7600 years ago. This would coincide with emergence of the earliest cultures that preserve legends of a great flood. Most notable of all, apart from the biblical deluge, is the account of the flooding of Atlantis that Plato learned from the early Egyptians (Donnelly, 1882, 1949). In addition, Hindu flooding legends appear in the Rig Veda and the Upanges. In Meso America, the Toltecs called the age before the great flood "The First World." Among North American Plains Indians, the Mandans had the most advanced culture and their folklore traced their origins to the Atlantic seaboard, "toward the rising sun," from whence they were driven westward by a Great Flood.

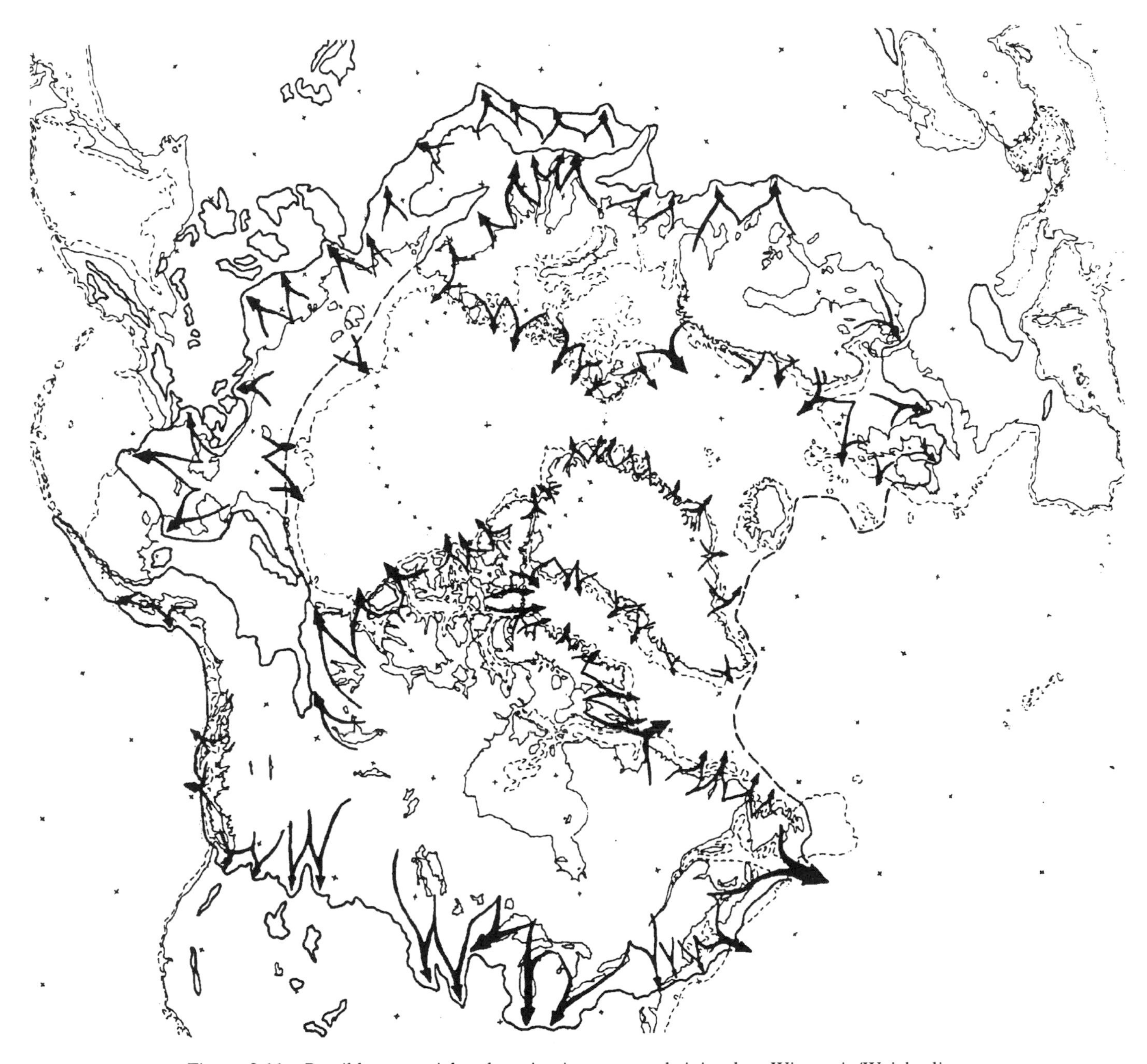

Figure 2.11: Possible terrestrial and marine ice streams draining late Wisconsin/Weichselian ice sheets. Arrows are ice streams postulated to occupy crustal troughs, with the branching tails of arrows enclosing the approximate ice-stream drainage basins. Solid heavy lines are the maximum landward ice-sheet margins between 18,000 BP and 14,000 BP. Broken heavy lines are either ice-shelf or perennial sea-ice margins. Two possibilities are shown for the maximum southern extent of ice sheets in Siberia.

MacAyeal (1992a, 1993a, 1993b) has shown theoretically that a single ice stream, one occupying Hudson Strait and draining the heart of the former Laurentide Ice Sheet over Hudson Bay, could raise sea level by several meters worldwide in only a few centuries if it surged. It apparently did, repeatedly, and produced the Heinrich events in Figure 1.1 (Andrews and Tedesco, 1992). MacAyeal (1992b) has shown that Antarctic ice streams could also behave in this manner.

Gravitational collapse of a marine ice stream is resisted by ice-bed coupling beneath it, lateral traction in shear zones alongside it, and buttressing by an ice shelf beyond it that is confined in a marine embayment and locally pinned to the sea floor at ice rises. Byrd Glacier has all three constraints and has a maximum velocity of 960 m/a across its ground-

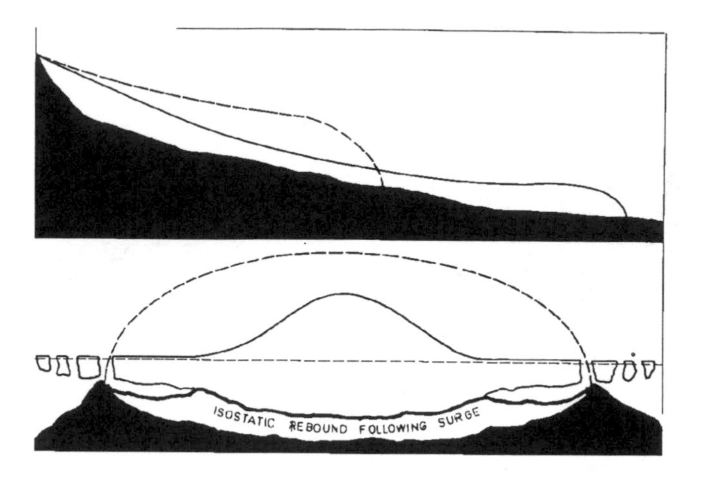

Figure 2.12: Surges as an internal instability phenomenon. A mountain glacier surge (top) begins as an internal instability nears the terminus, causing vertical collapse and horizontal advance. A concave-convex inflection on the top surface migrates upstream while the glacier terminus moves downstream, both beginning at the original equilibrium terminus. The zone between the retreating inflection point and the advancing terminus is the surging portion of the glacier. Conservation of volume is maintained. An ice sheet surge (bottom) causes isostatic uplift and horizontal advance creates a floating ice shelf, assuming the equilibrium ice sheet extended to the sea. If the equilibrium ice sheet has depressed bedrock far enough below sea level and if isostatic rebound cannot keep pace with the surge, then the grounding line will retreat in the surging portion of the ice sheet, which may become completely ungrounded. Conservation of volume is not maintained because the ice shelf continually breaks up to form tabular icebergs.

ing line (Scofield and others, 1991). Pine Island Glacier has only lateral constraint, and moves up to 2800 m/a at its calving front (Sanderson, 1979; Ferrigno and others, 1993). Thwaites Glacier may have little constraint at all, and moves up to 3400 m/a at its calving front (Lindstrom and Tyler, 1984; Lucchitta and others, 1995). The locations of these three ice streams are shown in Figure 2.15, and their ice drainage areas are indicated in Figure 2.10. None of these ice streams has appreciable surface melting in the summer. When large amounts of surface meltwater drain into moulins and crevasses, to eventually lubricate the bed, velocities can reach 7 km/a at the calving front, as is the case for Jakobshavns Isbræ in Greenland, shown in Figure 1.4 (Echelmeyer and Zhongxiang, 1987). These velocities compare with surge velocities for mountain glaciers. The successive relaxation of constraints on ice streams has been called the Jakobshavns Effect, because the most unconstrained ice stream is Jakobshavns Isbræ (Hughes, 1986a). The Jakobshavns Effect may be propagating northward along the east and west margins of the Greenland Ice Sheet, in response to climatic warming after the Little Ice Age ended in the middle of the last century.

Propagation of the Jakobshavns Effect is highly sensitive to boundary conditions. Byrd Glacier is contained along its sides by the rock walls of Byrd Glacier fjord and across its front by the Ross Ice Shelf. It resists these constraints by producing shear rupture crevasses within the fjord and lateral rifts beyond the fjord. As Byrd Glacier leaves the fjord, its velocity is therefore nearly constant across its width. Byrd Glacier has a huge catchment area, over which ice precipitation is light. In contrast, the catchment areas of Thwaites Glacier and Pine Island Glacier are much smaller but ice precipitation is much heavier. Both Thwaites Glacier and Pine Island Glacier calve into the open water of Pine Island Bay polynya, but the floating terminus of Pine Island Glacier is constrained by lateral shear zones along its entire length, whereas the floating terminus of Thwaites Glacier is unconstrained along its sides. There is little summer melting on the surface of these Antarctic ice streams, so the bed is not lubricated by surface meltwater descending into crevasses and moulins, as is the case for ice converging on Jakobshavns Isbræ in Greenland. Therefore, reducing constraints beyond, alongside, and beneath these ice streams allows their maximum velocity to increase by an order of magnitude. In order to propagate the Jakobshavns Effect in ice streams draining the Antarctic Ice Sheet, climate warming may have to cause substantial summer melting around the perimeter of the ice sheet.

Instabilities in Climate

Figure 1.1 shows the climatic record during the last Pleistocene glaciation cycle. Superimposed on the gradual stadial-interstadial cycles that have been correlated with the 23,000-year and 41,000-year cycles of Earth's axial precession and tilt, which cause gradual insolation variations, are abrupt changes lasting from 1000 to 3000 years but beginning and ending in less than 100 years. These abrupt changes are especially evident in changes in atmospheric dust trapped in the ice during firnification of precipitation (Hammer, 1985). This implies abrupt changes in erosion and deposition of glacial flour beneath ice sheets, with deposition concentrated at the termini of ice streams discharging some 90 percent of the ice, having life cycles from 1000 to 3000 years long which begin and end abruptly with abrupt thawing and refreezing of subglacial sediments or till, and being conduits for katabatic winds that entrain glacial flour, see Figures 1.8 and 1.13 (Hughes, 1992a, 1992b, 1996a). Figure 2.16 shows how katabatic winds flowing off the Antarctic Ice Sheet are funneled into the major ice streams (Parish, 1987).

If abrupt changes along the marine margins, terrestrial margins, and in ice streams of ice sheets can cause abrupt climatic changes, then strong links must exist between ice sheets and the ocean and atmosphere components of Earth's climatic machine. Fletcher (1972) noted the dominance of the atmosphere for short-term climatic changes occurring in a few years or less, and the dominance of the oceans for sustaining these changes over centuries or millennia. The reason for this is that Earth's atmosphere and oceans have ratios of 25:1 for kinetic energy, 1:4 for momentum, 1:400 for mass, and 1:1600 for heat capacity. When ice sheets are

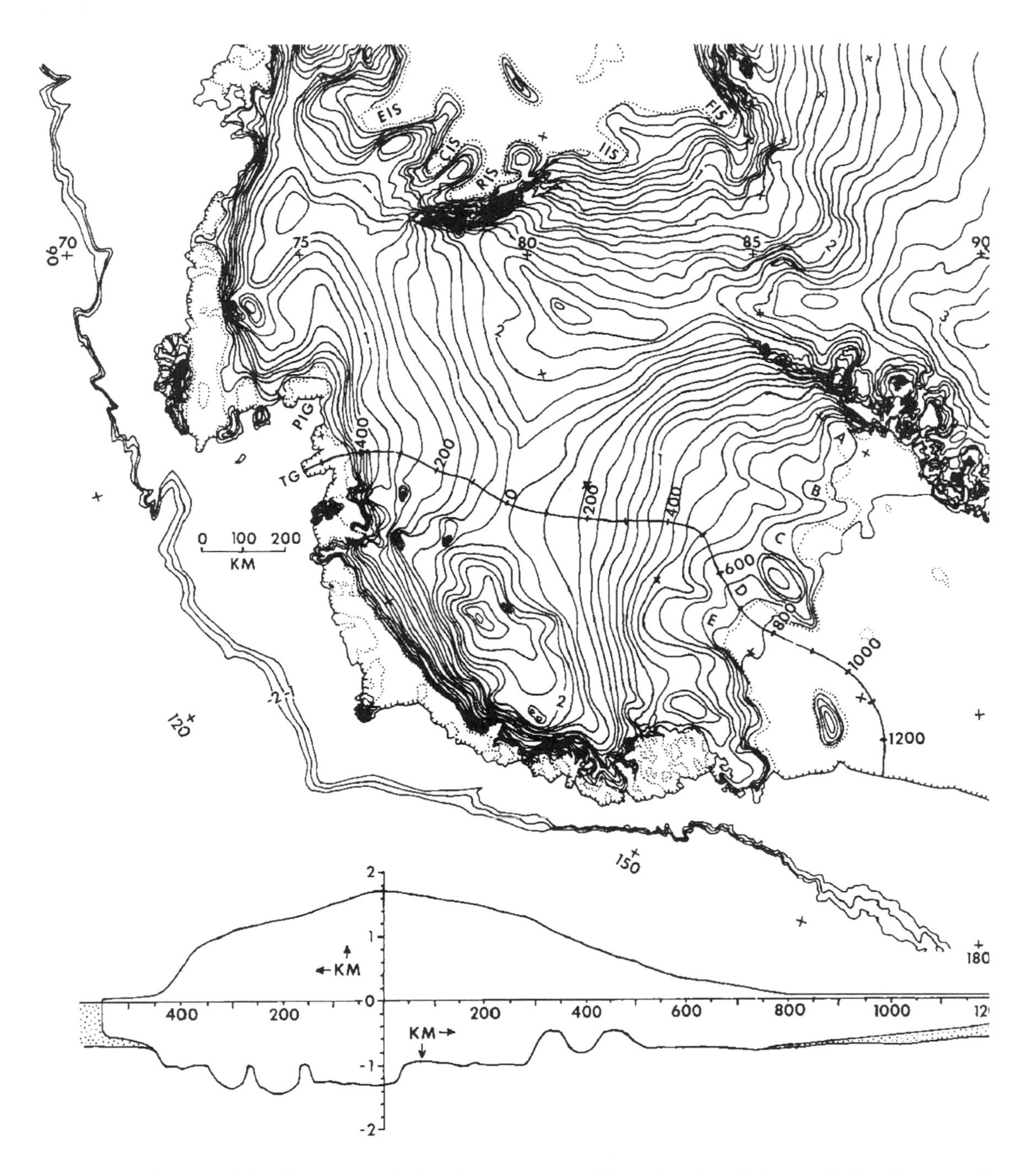

Figure 2.13: Sectors of the West Antarctic Ice Sheet that resemble an ice sheet before and after the surge-like gravitational collapse illustrated in Figure 2.12. *Top*: Downdraw is primarily through the Ross ice streams (A, B, C, D, E), Thwaites Glacier (TG), Pine Island Glacier (PIG), Evans Ice Stream (EIS), Carlson Ice Stream (CIS), Rutford Ice Stream (RIS), Institute Ice Stream (IIS), and Foundation Ice Stream (FIS). Ice elevation contour lines are 0.1 km apart. Ice shelf boundaries are dotted along their grounding lines and hatchured along their calving fronts. *Bottom*: Surface and bed profiles along the flowline in the top view. Note the advanced stage of collapse through Ice Stream D, with its largely post-surge concave flowline profile (see Figure 2.12) and apparently rebounded bed, and the great potential for rapid collapse through Thwaites Glacier, with its largely pre-surge convex flowline profile (see Figure 2.12) and high ice elevations above a steeply dipping bed, conditions that maximize the pulling force in Figure 2.2 (Adapted from Drewry, 1981).

included, the circulation time is 10,000 years for ice sheets, 1000 years for the deep ocean, 10 years for the surface oceans, and 0.1 year for the atmosphere, with a resident time for ice or water of 100,000 years for ice sheets, 5000 years for the deep ocean, 100 years for the surface ocean, and 0.02 years for the atmosphere. This implies that ice sheets have

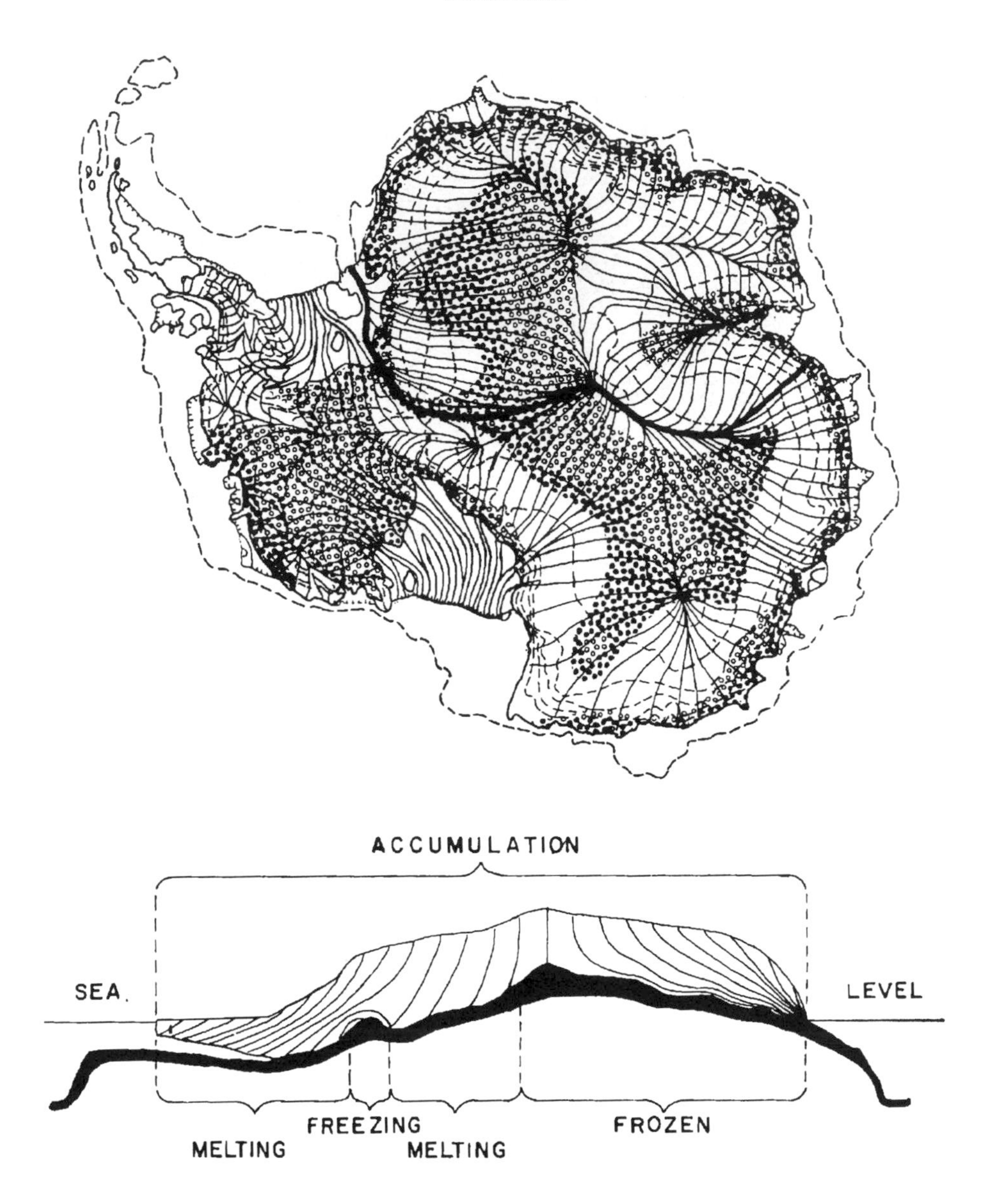

Figure 2.14: Generalized Antarctic ice-sheet circulation. *Top*: The plane view shows the sur-face ice flowlines (thin lines) drawn normal to 500 m ice-elevation contour intervals (dashed lines), the continental shelf (broken line), and the ice-bedrock wet melting zone (white dots), wet freezing zone (black dots), and frozen zone (undotted). *Bottom*: The profile view is along the thick ice flowline in the plan view and shows particle paths (thin lines) and the surface and basal mass balance. The left ice flowline develops a concave top surface profile in Foundation Ice Stream because high basal friction from fast-moving ice keeps the bed wet. The right ice flowline follows an ice ridge which maintains a convex profile because low basal friction from slow-moving ice keeps the bed frozen. (Obtained from the model by Budd and others,1970, as modified by Sugden, 1976).

long term interior dynamics that are stable and can override the more rapid dynamics of the oceans and atmosphere, but the shorter-term unstable dynamics along marine and terrestrial ice-sheet margins and in ice streams may be able to link with ocean dynamics, which have similar time frames.

In applying the research strategy proposed by Fletcher (1972) to the interaction between glaciation and climatic change, answers to the six questions he posed begin to merge for the ice-sheet component of Earth's climatic machine. Referring to components of the climatic machine that are most directly linked to the ice-sheet dynamics in Figures 1.8 and 1.13, tentative answers to Fletcher's questions are:

1. Ice streams are the most energetic parts.
2. Motion in ice streams is driven by a horizontal gravity force that can extend up ice streams to pull ice out of ice sheets, with a reach and strength that grow as ice-bed coupling, side traction, and forward buttressing weaken.
3. The pulling power of ice streams varies over time, apparently waxing and waning through a life cycle that is

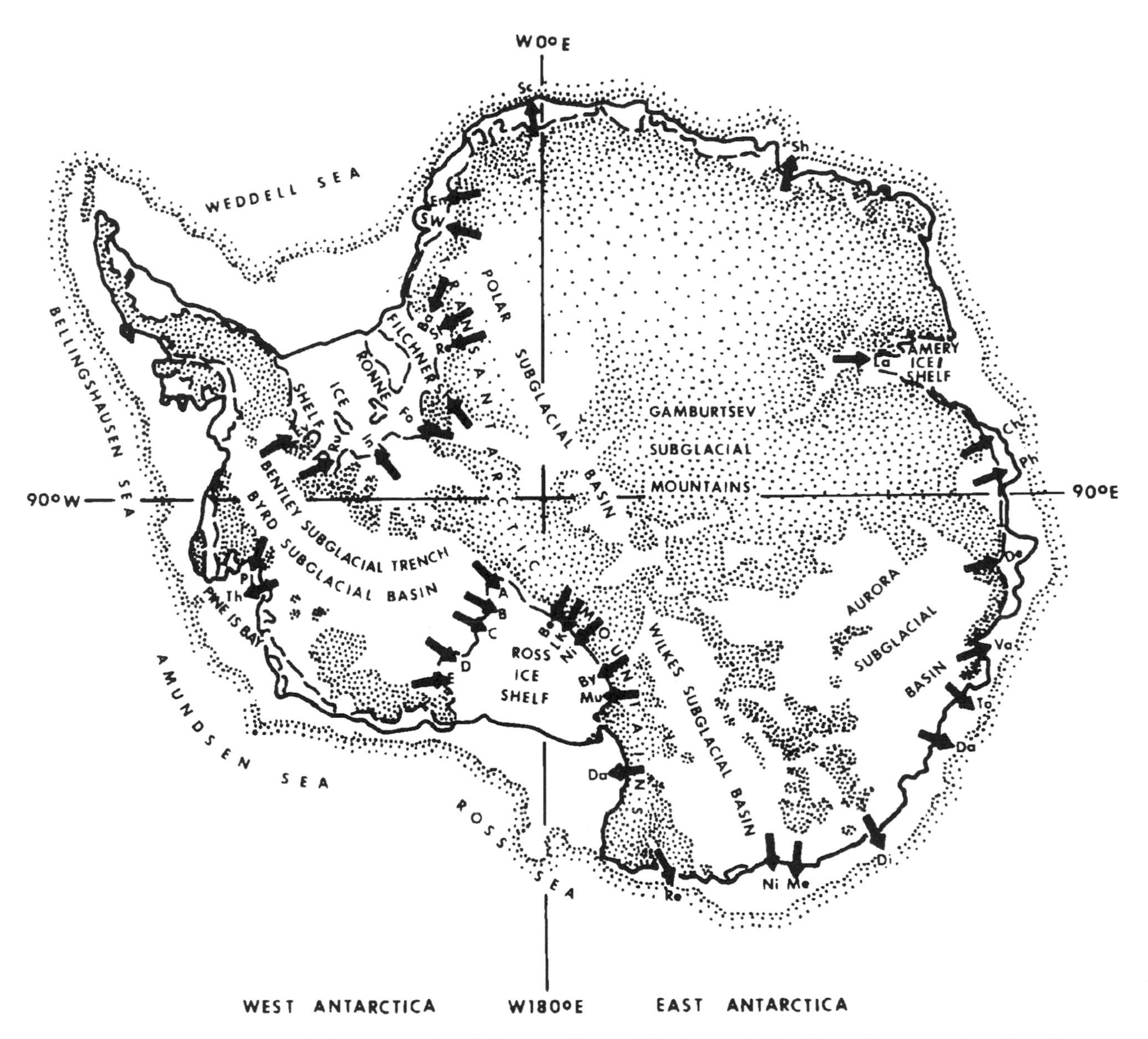

Figure 2.15: Major Antarctic ice streams in relation to collapse of the Antarctic Ice Sheet. Shown are the edge of the Antarctic continental shelf (dotted outer border), the edge of the Antarctic Ice Sheet (heavy solid line), grounding lines of ice shelves (heavy dashed lines), bedrock above sea level (dotted areas), and major ice streams (arrows). The ice shelves are collapsed remnants of an ice sheet that was grounded to the edge of the continental shelf 17,000 years ago. Further collapse is most likely in ice over the four subglacial basins, which lie below sea level and are drained by major Antarctic ice streams (the full extent of Polar Subglacial Basin is unknown). Clockwise from W 0° E, the names of ice streams shown are Schytt (Sc), Shirase (Sh), Lambert (La), Chelyuskintsy (Ch), Philippi (Ph), Denman (De), Vandeford (Va), Totten (To), Dalton (Da), Dibble (Di), Mertz (Me), Ninnis (Ni), Rennick (Re), David (Da), Mulock (Mu), Byrd (By), Nimrod (Ni), Lennox-King (LK), Beardmore (Be), unnamed ice streams A, B, C, D, E, Thwaites (Th), Pine Island (PI), Evans (Ev), Rutford (Ru), Institute (In), Foundation (Fo), Support Force (SF), Recovery (Re), Slessor (Sl), Bailey (Ba), Stancomb-Wills (SW), and Endurance (En).

longer for larger ice streams.

4. Physical processes triggering the time variations are glacial surges for all ice streams, changing sea level for marine ice streams, changing snowline slope for terrestrial eastward or westward directions where snowline slope is ice streams flowing equatorward, and both sea level and snowline changes for ice streams reaching the sea from minimal.

5. Theories quantifying ice-stream dynamics in relation to the dynamics of ice sheets are a major theme of this book.

6. The basic experimental tests of the theories will be numerical modeling experiments showing how well the theo-

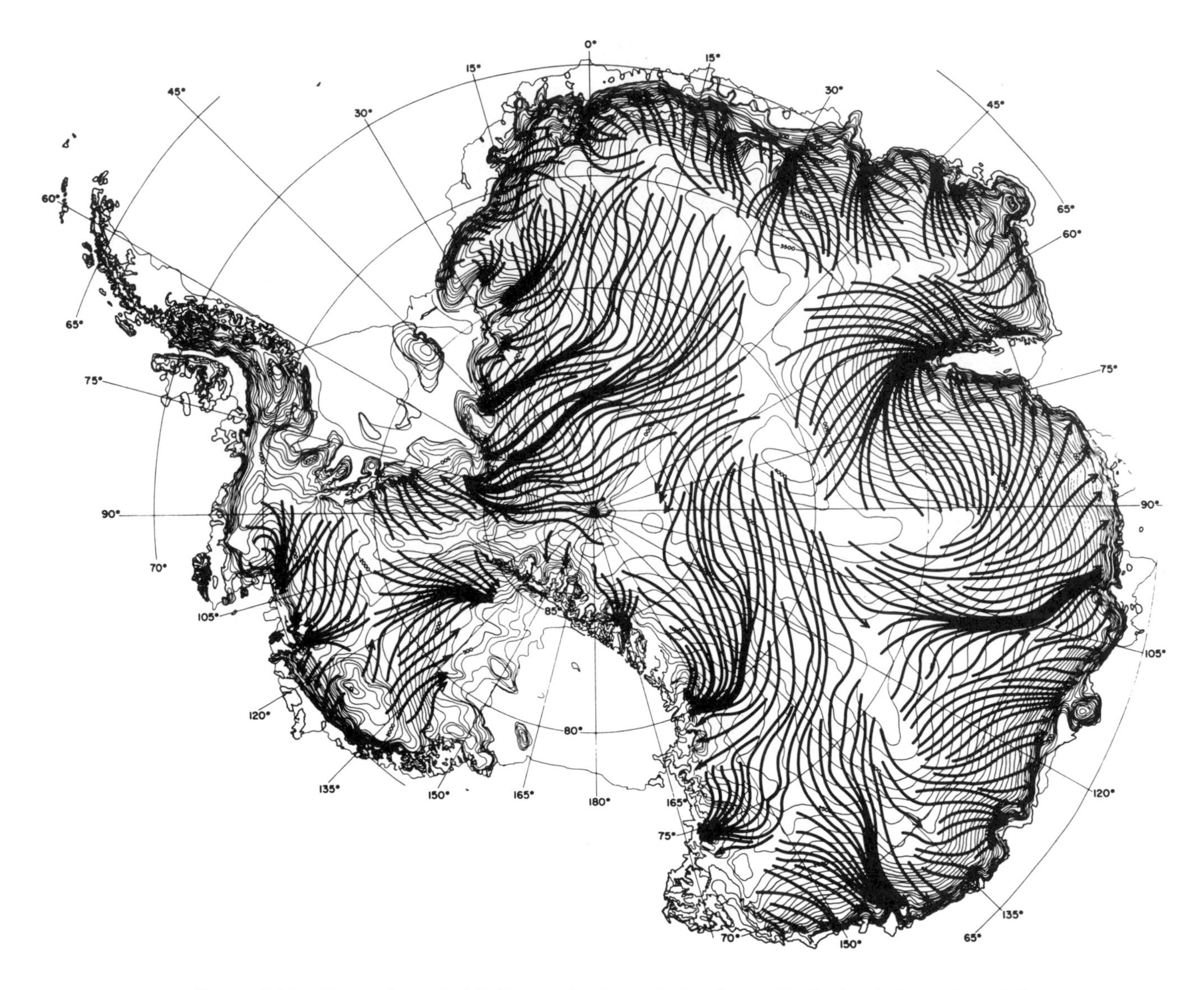

Figure 2.16: The surface wind field over the Antarctic Ice Sheet. Katabatic winds are funneled down ice streams. Reprinted from Parish and Bromwich (1987)with permission from *Nature*.

retical models are able to simulate the last glaciation cycle and the behavior of present-day ice streams.

Applying the Fletcher (1972) research strategy to the larger problem of Earth's climatic machine during an ice age, and to how its cryosphere, hydrosphere, and atmosphere components interact with one another over time, is beyond the scope of this book. However, clues to these interactions today can be found in the Southern Hemisphere, which holds 60 percent of the atmosphere because the climatic equator lies north of the geographic equator, holds 80 percent of the hydrosphere in the Southern Ocean, and holds 90 percent of the cryosphere in the Antarctic Ice Sheet. Earth's climatic equator is north of Earth's geographic Equator because the Southern Hemisphere is the watery hemisphere and hence dominates heat transfer from the ocean to the atmosphere and momentum transfer from the atmosphere to the ocean. The Southern Hemisphere also controls the major seasonal variations in heat transport from the Equator to the poles, because seasonal variations of sea ice around Antarctica double and then halve the area of the Antarctic heat sink. These features make the kinetic energy of atmospheric heat flow much greater in the Southern Hemisphere than in the Northern Hemisphere on an annual average, but with a seasonal variation ranging from several times larger in the southern winter when the extent of Antarctic sea ice is greatest, to about the same in the southern summer when the extent of Antarctic sea ice is least.

When considerations of the dominant role and great variability of the Southern Hemisphere in Earth's climatic machine are added to the suspicion that the marine West Antarctic Ice Sheet is now disintegrating (Hughes, 1973), it is no wonder that Weertman (1974) has called the West Antarctic Ice Sheet "glaciology's grand unsolved problem." This problem, in its global climatic context, will be examined next.

3

CLUES FROM ANTARCTICA

The Southern Hemisphere Climate System

In the cryosphere-hydrosphere-atmosphere climatic system of the Southern Hemisphere, flow in the ice sheet is intrinsically different from flow in the ocean and atmosphere because the former is crystalline and the latter are fluids. The crystalline-fluid distinction between the cryosphere and the hydrosphere-atmosphere components of the Southern Hemisphere heat engine is fundamental. It allows ice-sheet flow to be controlled in large part by internal variables which can be treated independently from the external influence which climatic regime, with minor perturbations (weather) related the ocean and atmosphere exert on the ice sheet. It also explains the low glacial velocity which gives the ice sheet very low dynamic properties compared with the ocean or the atmosphere, as shown in Table 3.1. Here circulation time is the time needed for motion in each component to stop after the driving force is removed, and residence time is the average time water spends in each component in solid, liquid, or vapor form.

Of the three components, the liquid oceanic phase has by far the most mass and momentum, so its energy and mobility are the steady-state standards for the system as a whole. The gaseous atmospheric phase is the high energy-high mobility component whose fluctuations produce weather, and the solid ice-sheet phase is the low energy-low mobility component whose fluctuations produce ice ages. Hence, oceanic circulation determines the steady-state to atmospheric circulation and major perturbations (ice ages) related to glacial circulation.

All three circulation patterns might ultimately be controlled by perturbations in the amount of solar radiation reaching Earth's surface either from Milankovitch cycles in Earth's orbit (Hays and others, 1976) or from fluctuations in atmospheric shielding, primarily by carbon dioxide and dust variations (Hammer and others, 1985).

Dynamics of the Atmospheric Component

Southern Hemisphere atmospheric dynamics are of global importance because that hemisphere's atmospheric flow has more kinetic energy, and seasonal sea-ice there regulates the ocean-atmosphere heat exchange more than in the Northern Hemisphere. As a result, "The meteorological equator is usually well north of the geographic Equator, so the wind field over the Equatorial ocean is mostly a feature of the Southern Hemisphere circulation. Thus, the Equatorial heat inputs to the atmosphere, which force the circulation in both hemispheres, are more influenced by the southern circulation than by the northern. Most of the momentum transfer from the atmosphere to the ocean also occurs in the Southern Hemisphere" (Fletcher, 1972, pp. 5–6).

The atmosphere is divided by temperature maxima and minima into the discrete layers shown in Figure 3.1. Temperature maxima result from photoionization, photodissocia-

Table 3.1: Selected physical characteristics of the Southern Hemisphere components of the cryosphere-hydrosphere-atmosphere dynamic system

	Cryosphere		Hydrosphere		Atmosphere
	Ice Sheets	Sea Ice	Deep	Surface	Troposphere
Circulation time (yr)	1×10^{4}	2×10^{0}	1×10^{3}	1×10^{1}	5×10^{-2}
H_2O residence time (yr)	1×10^{5}	5×10^{-1}	5×10^{3}	1×10^{2}	1×10^{-2}
Heat capacity (cal/°C)	1×10^{22}	1×10^{19}	—	1×10^{24}	1×10^{21}
Mass (kg)	2×10^{19}	2×10^{16}	—	1×10^{21}	2×10^{18}
Momentum (kg m/yr)	6×10^{20}	1×10^{23}	—	5×10^{27}	1×10^{27}
Kinetic energy(kg $m^2 yr^{-2}$)	2×10^{22}	5×10^{29}	—	2×10^{34}	5×10^{35}

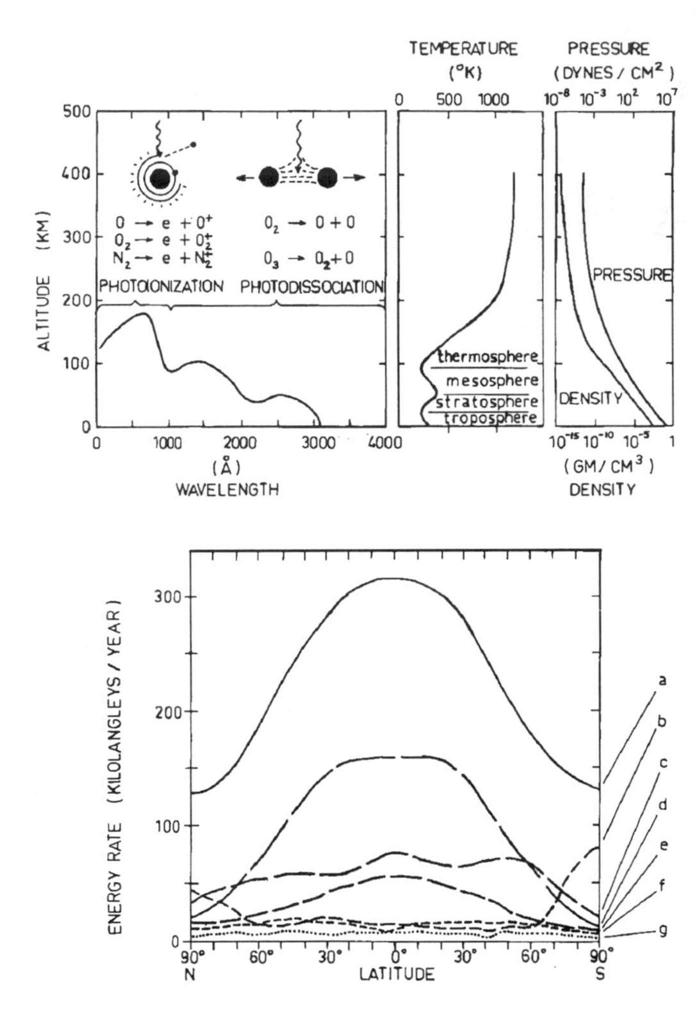

Figure 3.1: Mean vertical and latitudinal disposition of solar radiation. *Top left*: Endothermic reactions heat the atmosphere above 150 km, where photoionization occurs; between 50 and 100 km where photodissociation occurs, and at the Earth's surface where photoabsorption occurs. Exothermic reactions cool the atmosphere at about 80 km where airglow radiation occurs and at about 30 km where thermal emission occurs. *Top right*: The temperature reversals and monochromatic variation of density and pressure with altitude. *Bottom*: Curves show solar radiation energy fluxes; (a) the 100% entering the top of the atmosphere, (b) the 7% reflected by the Earth's surface, (c) the 23% reflected by clouds, (d) the 47% absorbed by the earth, (e) the 14% absorbed by the air, (f) the 6% reflected by air, dust, and water vapor, and, (g) the 3% absorbed by clouds. The top figures are taken from Walker (1967). The bottom figure is from Barry and Chorley, (1971).

tion, and photoabsorption, at absorption altitudes of 180 km, 50 km, and Earth's surface, respectively. The ozone photodissociation temperature maximum is the stratopause, and the minimum between it and the ultraviolet and infrared absorption maxima are the mesopause and the tropopause, respectively. Hence, as altitude increases, the atmosphere cools in the troposphere, heats in the stratosphere, cools in the mesosphere, and heats in the thermosphere. More than 80 percent of atmospheric gases are concentrated in the troposphere, and the temperature decrease with elevation makes this region of the atmosphere unstable so that thermal convection is necessary. Thermal convection controls weather and climate.

The troposphere is the most energetic part of the atmosphere because it is the most massive and interacts directly with Earth's surface in the exchange of heat, momentum, moisture, and gases, which determines the global climatic patterns. It is therefore the atmospheric layer most essential to the cryosphere-hydrosphere-atmosphere climate machine (Miller and Thompson, 1970). In this climate machine, the oceans drive the troposphere primarily by heat exchange and the troposphere drives the oceans primarily by momentum exchange. Troposphere and ocean circulation are perturbed most strongly by Earth's rotation and geography (Strahler, 1973). Both flow patterns are also strongly influenced by the effect of the cryosphere, especially in relation to the inclination of Earth's rotational axis to the ecliptic plane (Fletcher, 1969). This is the main cause of seasonal variations in atmosphere and ocean circulation. The driving force for atmospheric circulation is the thermal gradient between the Equator and the poles.

In modeling atmospheric circulation, a first approximation considers a water-covered, nonrotating, noninclined Earth. Hot equatorial air rises toward the tropopause but is diverted toward the poles in order to conserve its angular momentum by keeping its radius of gyration rather constant. As the hot air rises, it cools adiabatically, and water vapor condenses; as the air converges toward the poles, its molecules become more crowded together. Cooling, condensation, and convergence all combine to increase air density, and as the air currents converge near the poles the increasingly heavy air begins to fall back toward Earth's surface. Downward flow at the poles diverges along Earth's surface, where it absorbs heat and water vapor as it approaches the Equator. This decreases its density, and near the Equator it rises again to complete the convective circuit. Hence, air circulates in a single closed convection cell for each hemisphere in the first approximation.

A second approximation to global atmospheric circulation permits rotation of the water-covered, noninclined Earth. When viewed from the rotating surface, Northern Hemisphere flow is diverted to its right and Southern Hemisphere flow is diverted to its left by the theoretical Coriolis force. This deflection greatly increases the path length of air circulating between the Equator and the poles. Hence, air moving toward the poles becomes too heavy to remain aloft by the time it nears the 30° parallels, and air moving toward the Equator becomes too light to remain at the surface by the time it nears the 60° parallels. Consequently, the single convection cell is short-circuited by a descending flow belt near 30° and an ascending flow belt near 60°. The single cell breaks down into an equatorial Hadley cell and two polar cells, loosely coupled across temperate latitudes by Ferrel cells, as shown in Figure 3.2.

In the days of sailing ships, mariners called the equatori-

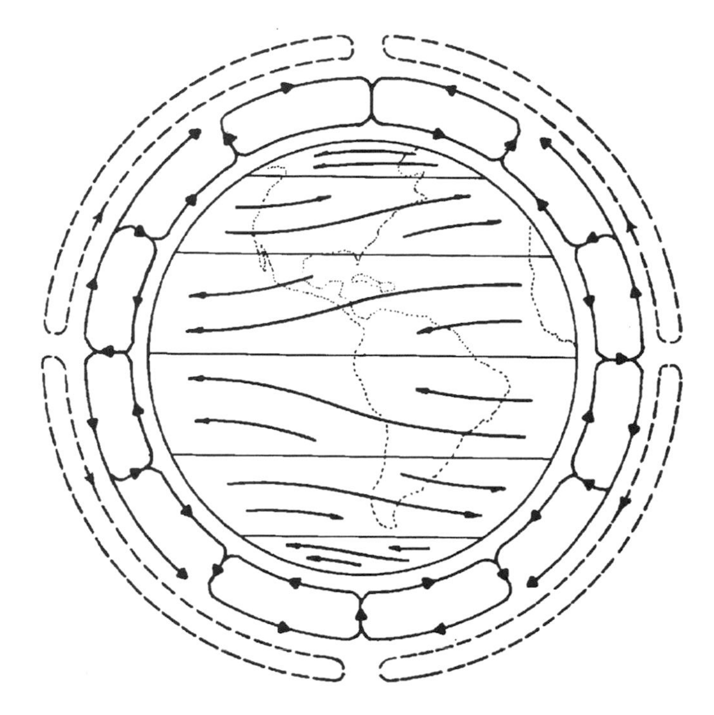

Figure 3.2: Generalized atmospheric circulation. Tropospheric flow (solid arrows), consists of an equatorial convection cell, polar convection cells, and connecting flow, all having a strong zonal component increasing toward the poles due to the Earth's rotation. Zonal flow in the equatorial cell causes the northeast and southeast trade winds north and south of the equator, respectively. Zonal flow in the polar cells causes the polar easterlies. Westerlies result from zonal flow in the connecting flow between the equator and the poles. Stratospheric flow (broken arrows), consists of a single convection cell with descending flow over the equator and ascending flow over the poles (taken from Duxbury, 1971, Figure 11.4).

al latitudes of rising Hadley air "the doldrums" because seas were calm, and sinking Hadley air created choppy water in the "horse latitudes" because whitecapped waves looked like the manes and tails of running horses. The Ferrel cells are driven passively by coupling to active thermal convection in the Hadley and polar cells. Since the Coriolis force increases from zero at the Equator to a maximum at the poles, surface flow in the equatorial Hadley cell is the gentle northeast and southeast trade winds in the respective Northern and Southern Hemispheres, surface flow in the mid-latitude Ferrel cell is the brisk to roaring westerlies in both hemispheres, and surface flow in the polar cells is the easterlies, especially in the Southern Hemisphere, because it is mostly water, so topographic deflections of winds are minimal. A single weak convection cell may exist in the stratosphere circulation, as shown in Figure 3.2. Hence, the second approximation succeeds in reproducing several basic features of actual atmospheric circulation, particularly in the watery Southern Hemisphere.

Further approximations to actual atmospheric circulation involve the wave theory of global atmospheric circulation, which best relates flow in temperate latitudes to the flow in the convection cells depicted in Figure 3.2 (Pfeffer, 1967) Incorporating axial inclination, surface albedo, and geographic features into this theory require numerical General Circulation Models (GCMs) run by high-speed computers. These considerations strengthen westerly winds and weaken easterly winds in high latitudes, for example.

Atmospheric circulation is thermal circulation. Hence, it depends first of all on the amount of solar energy and how this energy is reflected, absorbed, and transmitted in and through Earth's atmosphere. This in turn depends on the composition of the atmosphere. For example, dust and carbon dioxide concentrations are increasing. Absorption and reflection of shortwave solar radiation by stratospheric dust tends to cool Earth's surface. Absorption and reflection of long-wave terrestrial radiation by tropospheric carbon dioxide tends to warm Earth's surface (the "greenhouse" effect). Absorption of solar radiation at Earth's surface depends on the surface albedo. Southern Hemisphere surface albedo seasonal variations are very large because the high-albedo summer ice cover is half the winter ice cover owing to sea ice melting around Antarctica (Schwerdtfeger, 1970a). Fletcher (1969) has summarized the resulting heat balance and exchange.

Ascending air around the circumference of the Antarctic polar convection cell, as shown in Figure 3.2, is immediately diverted toward the left with respect to the eastward rotating Earth's surface. However, the air also moves south along the temperature-pressure gradient between the warm Southern Ocean and the cold Antarctic Ice Sheet. The former flow is zonal, and the latter flow is meridional. The combined flow creates the circumpolar vortex, with air spiraling in toward the cold center of the polar ice plateau of Antarctica. This cold center lies in East Antarctica and is about 1000 km from the South Pole. Hence, the circumpolar vortex is not symmetrical about Earth's rotation axis, causing an inflow of cold sinking air over East Antarc-tica and warm rising air over West Antarctica (Schwerdtfeger, 1970a, pp. 265–280).

Strong Coriolis effects near the South Pole suppress themeridional heat transfer until meridional temperature-pressure gradients become strong enough to force some meridional flow to cut across the dominant zonal flow, thereby transporting heat toward Antarctica. This short-circuiting of zonal flow spawns cyclonic systems north of the Ross Sea, the Bellingshausen Sea, Queen Maud Land, and Wilkes Land. These cyclones drift east and south in the circumpolar vortex. The meridional temperature-pressure gradient is greatest near the edge of the pack ice, where the ocean-to-atmosphere heat transfer rate has a sharp change, and after sunset, when the air cooling rate is greatest. Hence, this gradient has two annual maxima between 40°S and 60°S, one during the vernal equinox, when the pack ice boundary lies in these latitudes, and one after the autumnal equinox, when night predominates in the same latitudes. These are the times of the equinoctial storms in the Southern Ocean, resulting from intense cyclonic activity caused by meridional flow short-circuiting zonal flow. The meridional temperature-pressure gradient maxima move south with the shrinking summer pack ice boundary and the shortening summer season (Schwerdtfeger, 1970a, pp. 261, 264, 277, 283–284).

The four dominant cyclonic systems sustained by baroclinic instability in the circumpolar vortex ventilate Antarctica with warm, moist air from the ice-free parts of the Southern Ocean. This process continues through the Antarctic winter so that a relatively warm air layer in the lower troposphere blankets the continent above a thin layer of cold air at the surface of the ice sheet. Heating is via both sensible heat from advection flow and latent heat from precipitation. Warm air just above the cold surface air layer is heated by the intense radiative heat loss from the high-albedo snow surface (Schwerdtfeger, 1970a, pp. 276–278). Hence, a temperature inversion exists near the surface of the ice sheet, with the temperature maximum in the lower troposphere, as illustrated schematically in Figure 3.3. This phenomenon causes a surface inversion wind to blow continuously down the gentle slope of the interior ice sheet. The inversion wind is a consequence of the horizontal-temperature gradient through the inversion layer over sloping terrain, as seen in Figure 3.3. Hence, the heavy upslope cold surface air rushes downslope. Coriolis effects deflect this surface wind to the left, so that it spirals with anticlockwise rotation down the polar plateau. Surface frictional effects damp out with distance above the snow, allowing diverging clockwise flow in the circumpolar vortex to descend and become the diverging anticlockwise inversion flow (Mahrt and Schwerdtfeger, 1970). Upon reaching the sea, inversion flow becomes the polar easterlies.

Inversion winds do not disrupt the inversion layer and are therefore an equilibrium phenomenon. They are mild and continuous, cause no significant snow drift, and predominate along the gentle slopes of the ice sheet interior. Katabatic winds are the nonequilibrium counterparts. They are violent and discontinuous, can cause massive drifting, and predominate on steep slopes of the ice-sheet periphery, where they may create polynyas (ice-free areas) in the sea offshore (Schwerdtfeger, 1970a, pp. 292–294). In some respects, katabatic winds are the atmospheric counterpart of glacial surges in ice streams. They arise because of rapid adiabatic warming of surface air as elevation and latitude decrease near the coast. This causes reservoirs of heavy cold air to accumulate at the head of ice streams. Eventually an instability develops, and the cold air rushes downward until its reservoir is exhausted. Figure 2.19 shows how ice streams control surface wind flow off the Antarctic Ice Sheet by converting it into funnelled katabatic winds (Parish, 1987).

Precipitation over the Antarctic Ice Sheet originates mainly from water vapor in the updraft over the Southern Ocean between the surface subpolar westerlies and polar easterlies in Figure 3.3. This water vapor condenses as clouds when it adiabatically cools and enters the circumpolar vortex. Hence, a thick cloud cover typically covers this part of the Southern Ocean and is carried toward the continent by the great cyclonic systems offshore (Zilman and Dingler, 1973). These swirling cloud masses gather at altitudes mainly below the polar ice plateau, so they must ascend as

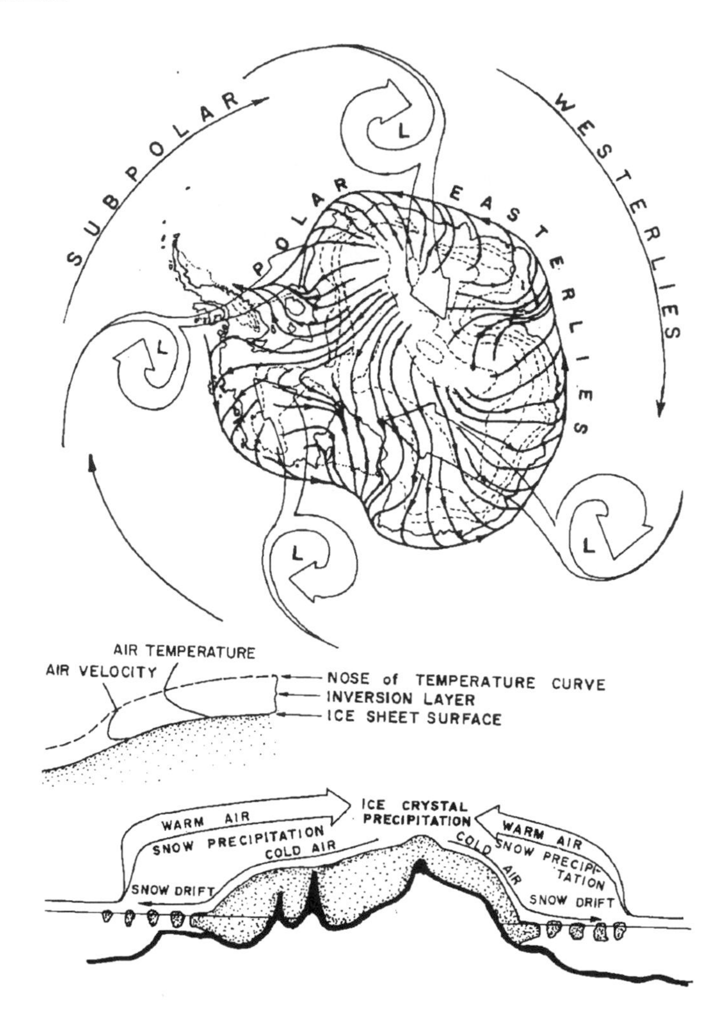

Figure 3.3: Generalized atmospheric circulation (top) and precipitation (bottom) patterns in the Antarctic. Baroclinal instability in the subpolar westerlies creates cyclonic storms in which surface winds spiral inward and upward until they become trapped in the circumpolar vortex of the upper arm of the polar convection cell. The circumpolar vortex begins as converging clockwise flow in the subpolar westerlies, then spirals upward in the cyclones along the polar front and becomes clockwise diverging westerly flow which brings precipitation to the polar plateau of Antarctica. As the diverging winds cool, first snow precipitates to blanket the margin and then ice crystals blanket the interior of the Antarctic ice sheet. The cold air sinking over the continent spirals anticlockwise down the ice sheet as inversion winds, to become the polar easterlies in the lower arm of the polar convection cell. Lower (narrow arrows) and upper (wide arrows) air circulation and ascending flow (widening arrows) are shown.

they move onto the continent. Coastal precipitation is often combined with fog, becomes snowfall when the cloud banks cool while ascending the polar plateau, and culminates in almost continuously falling ice crystals over the central polar plateau where the water vapor descends into the cold surface air below the temperature inversion (Schwerdtfeger, 1970a, pp. 300–302). The accumulation pattern can be greatly changed by drifting and blowing snow. The volume of such snow transport increases sharply, perhaps exponentially, with wind speed (Budd, Dingle, and Radok, 1966). Drifting and blowing are not critical over the polar plateau where gentle inversion winds prevail, but can transport great volumes of snow out to sea in coastal regions where violent katabatic winds occur (Schwerdtfeger, 1970a, pp. 303–307).

To summarize, the most energetic part of the atmosphere is the troposphere, especially the troposphere-ocean interface. The most energetic parts of the Antarctic troposphere are the circumpolar vortex, the equinoctial cyclonic storms associated with baroclinic instability in the circumpolar vortex, and katabatic winds associated with baroclinic instability around the margin of Antarctica. Figure 3.3 illustrates the generalized air circulation in the most energetic parts of the atmosphere above the Southern Ocean and the Antarctic Ice Sheet. Motion in these parts is driven by heat exchange between the ice-free Southern Ocean and the Antarctic heat sink, modified by the solar radiation balance, Earth's rotation and inclination, and the size and shape of the ice cover. These relate to Earth's orbital cycles, atmospheric constituents, sea-ice fluctuations, and ice-sheet stability.

Dynamics of the Oceanic Component

Surface ocean circulation is driven mainly by surface winds modified by Earth's rotation, and deep ocean circulation is driven mainly by surface ocean circulation modified by thermohaline circulation. Thermohaline circulation is density driven, because water gets more dense as its temperature decreases (to 4°C for fresh water) and salinity increases (the density maximum of saline sea water is at 0°C.)

The ocean is divided by thermohaline horizons into distinct zones that also define the major regions of horizontal and vertical circulation. Figure 3.4 shows these zones in plan view for Earth's ocean surface and in profile for a north-south section through the Atlantic Ocean, the only ocean having broad mass exchange with both Arctic and Antarctic seas. The thin layer of surface water is driven by the surface winds shown in Figure 3.2, but modified by Coriolis effects and the continental borders. The Antarctic Divergence occupies the low-pressure belt of ascending air around the circumference of the south polar convection cell. Water to the south is driven by the polar easterlies derived from inversion and katabatic winds spiraling anticlockwise off the Antarctic Ice Sheet. Water to the north is driven by surface westerlies in the Ferrel cell, which connects atmospheric circulation in the equatorial and polar convection cells of Figure 3.2. Although surface air converges in the updraft, the strong Coriolis effect causes flow to be almost completely zonal. Hence, convergence of surface air is too weak to converge the zonal flow of surface water, which diverges due to the much stronger Ekman drift (Duxbury, 1971, pp. 216–217). The divergent Ekman drift is about 500 m thick and results from the leftward displacement of Southern Hemisphere sur-

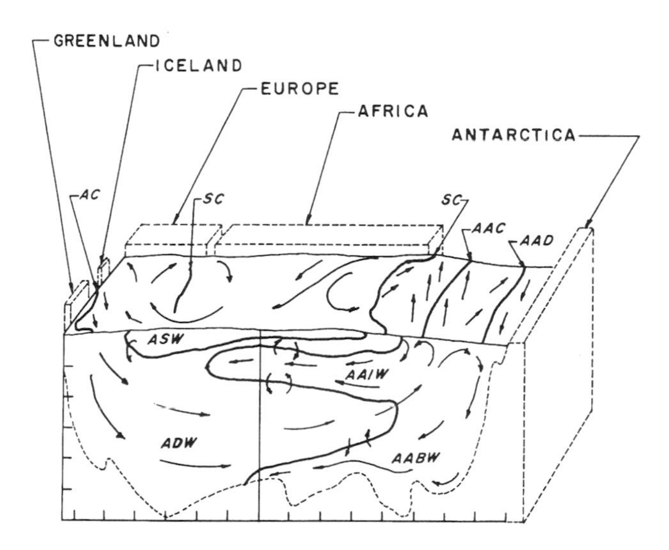

Figure 3.4: Generalized ocean circulation. Horizontal global surface circulation (top) and vertical Atlantic north-south circulation (bottom) are shown. Legend: AAD = Antarctic Divergence, AAC = Antarctic Convergence, SC = Subtropical Convergence, EQC = Equatorial Convergence, EQD = Equatorial Divergence, AC = Arctic Convergence, AABW = Antarctic Bottom Water, AAIW = Antarctic Intermediate Water, ADW = Atlantic Deep Water, ASW = Atlantic Surface Water (From Duxbury, 1971).

face-water currents due to Earth's rotation, giving the water layer influenced by wind friction a net horizontal flow normal to and left of the surface wind. Hence, surface water driven by the polar easterlies has a net southern flow and surface water driven by the westerlies has a net northern flow. Consequently, surface water must diverge where these wind systems meet; this is the Antarctic Divergence.

The Antarctic Divergence causes an upwelling of Circumpolar Deep Water which feeds the diverging Antarctic Surface Water. The southern arm of the diverging Antarctic Surface Water is diverted to the left by the polar easterlies as it moves south, until it sinks from densification by salt enrichment beneath freezing sea ice to become the Antarctic Bottom Water formed off the continental shelf of Antarctica. The northern arm of the diverging Antarctic Surface Water is diverted to the left by the westerlies, and as the Ekman drift transports it north, it meets and underthrusts warmer Atlantic Surface Water to become Antarctic Intermediate Water. The underthrust zone is called the Antarctic Convergence.

Further north, in the horse latitudes of descending Hadley circulation, a Subtropical Convergence occurs as a result of northward Ekman drift of surface water driven by the westerlies and southward Ekman drift of surface water driven by the southwest trade winds; compare Figures 3.2 and 3.4. Another Subtropical Convergence occurs in the Northern Hemisphere horse latitudes between the northeast trade winds and the westerlies and is also a result of Ekman drift, which now diverts surface water to the right of the wind.

Descending flow along the Subtropical Convergence in the North Atlantic is the main source of Atlantic Deep Water in Figure 3.4. This is heavy water, because surface evaporation has increased its salinity. Less robust subtropical deep water also forms in the Pacific and Indian Oceans by the same mechanisms. The north-south orientation of the Americas, Eurafrica, and Asia-Australasia diverts zonal wind-driven surface flow. As shown in Figure 3.4, Coriolis-/Ekman effects transform this flow into large gyres, which rotate clockwise in the North Pacific and North Atlantic, and anticlockwise in the South Pacific, South Atlantic, and Indian Oceans. In all cases, Ekman drift drives high-salinity surface water toward the centers of these gyres, where the water sinks owing to its greater density. The sinking column of water possesses two kinds of momentum, one due to its rotation around the axis of Earth (planetary vorticity) and one due to its rotation about the axis of the gyre (relative vorticity). It preserves the former by keeping that radius of gyration constant, which means moving to lower latitudes as it sinks. Hence, descending flow under the subtropical gyres is also flow toward the Equator.

Water moving downward and toward the Equator must be balanced by an upward flow of Equatorial waters drifting toward the poles. The upward flow occurs under the Equatorial Divergence at the western end of each subtropical gyre shown in Figure 3.4. This ascending flow also brings cold deep water upward to mix with the warm surface water layer. The source of the cold deep water must be the polar regions, and this water must move in a bottom current along the western margins of the Pacific, Atlantic, and Indian Oceans (Stewart, 1969). All three oceans have broad boundaries with the Southern Ocean, but only the Atlantic Ocean has a broad boundary with the Arctic Ocean. Hence, the Southern Ocean is the major source of the cold deep water, and its western bottom current is primarily Antarctic Bottom Water.

Antarctic Bottom Water production is intimately associated with the extent of sea ice in the Southern Ocean, and probably also with the large floating ice shelves of Antarctica. Since the sea ice cover decreases by 82 percent from a maximum of 19×10^6 km^2 at the autumnal equinox to a minimum of 3.5×10^6 km^2 at the vernal equinox, it is clear that Antarctic Bottom Water production is largely seasonal. The greatest centers of production are under the pack ice, especially during the austral winter, when salt is being rejected at the freezing ice-water interface. The cold temperature and salt enrichment of sea water at the interface increases its density so that it sinks to the floor of the Antarctic continental shelf, being replaced by Antarctic Surface Water flowing south from the Antarctic Divergence. When the buildup of cold, high salinity water on the continental shelf becomes great enough, it flows down the continental slopes to the deep ocean basins, where it is then called Antarctic Bottom Water. Like katabatic winds, this intermittent water flow may be akin to a glacial surge.

The Antarctic Peninsula causes large clockwise gyres to develop in the Bellingshausen and Weddell Seas, and these are the two main production centers for Antarctic Bottom Water. Of these two, the Weddell Sea is the most important center because it has much greater areas of pack ice, shelf ice, and continental shelf. Clockwise flow in the Bellingshausen and Weddell gyres is a result of the clockwise Antarctic Circumpolar Current (also called the West Wind Drift) driven by the westerlies north of the Antarctic Divergence and a weaker anticlockwise current driven by polar easterlies south of the Antarctic Divergence. The Antarctic Peninsula separates these two seas and diverts the zonal currents to create the two gyres. Ekman drift causes surface water to diverge from the centers of these gyres, and Circumpolar Deep Water must rise into these centers.

In addition to the Bellingshausen and Weddell Seas, other important sites of Antarctic Bottom Water are the ocean floors north of the Ross Sea, the Adélie Coast, the Amery Ice Shelf, the Shackleton Ice Shelf, and probably to some extent all along the continental slopes of Antarctica. Antarctic Bottom Water north of the Ross Sea is unusually salty and suggests salt enrichment due to the freezing of sea water onto the underside of the cold Ross Ice Shelf (Zumberge, 1964; Jacobs and others, 1970; Gordon, 1973).

Antarctic Bottom Water flowing north from the continental slopes of Antarctica has an anticlockwise westward rotation due to the Coriolis/Ekman effect. The westward flow is blocked at several places by the circum-Antarctic midocean ridge and other submarine features. Flow is diverted north through transform faults which breach the midocean ridge and is eventually driven clockwise by the Antarctic Circumpolar Current to become the deepest level of Circumpolar Deep Water. As such, it has a northward Ekman drift and eventually becomes the cold deep water currents along the deep western boundaries of the Pacific, Atlantic, and Indian Oceans described earlier. Earth's rotation piles this water against the western ocean boundaries.

Cold deep water rising near the Equator must move toward the poles in order to preserve its planetary vorticity. In so doing, it mixes with salty water and returns in modified form. Salty North Atlantic Deep Water blends with return flow in the Atlantic, and the increasing Ekman drift turns this flow east to form a zone of Circumpolar Deep Water warmer and saltier than the lower Antarctic Bottom

Water zone. The entire Circumpolar Deep Water layer spirals clockwise toward Antarctica and must therefore rise in order to preserve its planetary vorticity. It becomes the feeder current for upwelling water in the Antarctic Divergence and in the centers of the Bellingshausen and Weddell Sea gyres. Hence, it brings salty northern water into the Southern Ocean to mix with fresh Antarctic water created by precipitation and melting glacial ice.

To summarize, the most energetic part of the Southern Ocean is the surface layer, where horizontal wind-driven motion is strongest and vertical thermohaline convective motion begins. The Antarctic Circumpolar Current, extending to the ocean floor with relatively minor attenuation, is the Southern Ocean's most energetic current; it probably transports more water than any other ocean current. The most energetic Southern Ocean gyres are those in the Bellingshausen and Weddell Seas, particularly the latter. Figure 3.5 illustrates these general features of water circulation in the Southern Ocean. Note the similarity with the most energetic parts of the atmosphere summarized earlier: first, the air-sea interface; second, the circumpolar air and sea current north of the Antarctic Divergence; and third, the cyclonic storms and gyres south of the Antarctic Divergence. Fletcher (1969) emphasizes the dominant influence of pack ice on the energetics of the sea-air interface, and Gordon (1973) emphasizes the dominant influence of bottom topography on water flowing in the major currents and gyres. Over the Southern Ocean north of West Antarctica, sea-ice cover is greatest, the Antarctic Circumpolar Current is strongest, and cyclonic gyres are largest. Disintegration of the West Antarctic Ice Sheet would therefore have a profound influence on circulation in the Southern Ocean and, by extension, general ocean circulation. It would greatly alter the absolute extent of sea ice and its seasonal variations. It would drastically modify flow in the Antarctic Circumpolar Current. It would redistribute and perhaps destroy the great Bellingshausen and Weddell gyres. The effect on the production and distribution of Antarctic Bottom Water would beprofound, with far-reaching ecological consequences. If the West Antarctic Ice Sheet has disintegrated in the past, as Mercer (1968a) proposed, these consequences can be studied using the techniques of marine geology (Watkins, 1973), marine paleontology (Bandy, 1973; Kellogg and others 1979), marine biology (El-Sayed, 1973; McWhinnie, 1973; Hedgpeth, 1973), and marine sedimentation (Anderson and others, 1979, 1980; Domack, 1982). These disciplines can be combined with physical oceanography (Gordon, 1973) in order to understand how the Southern Ocean responds to an inherently unstable West Antarctic Ice Sheet.

Dynamics of the Cryospheric Component

The cryosphere has four components: permafrost, sea ice, land ice, and the seasonal snow cover on unfrozen ground. Flow in permafrost and snow are nil, except on steep slopes, and sea-ice flow is synonymous with surface water circulation in the polar oceans. Land-ice flow is creep of

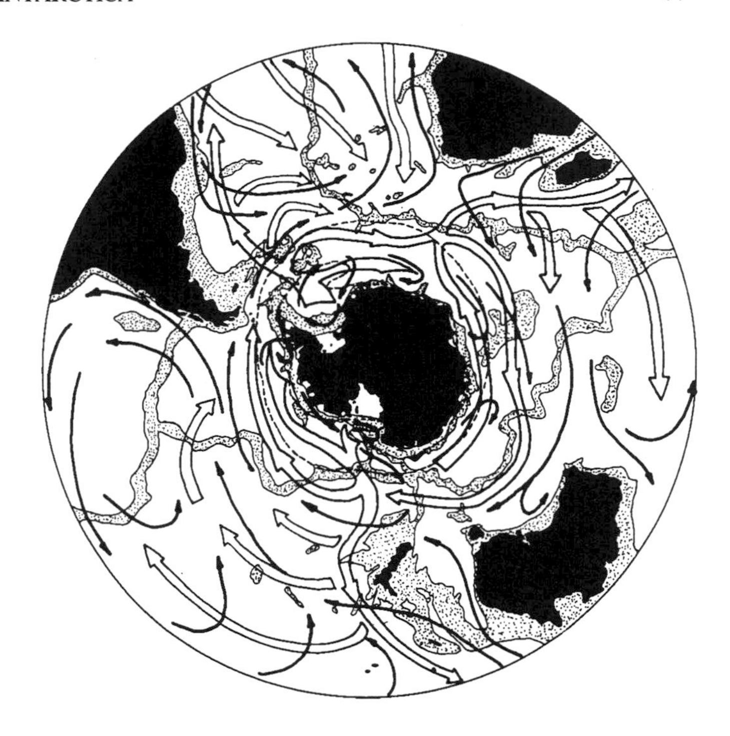

Figure 3.5: Generalized Southern Ocean circulation. Top surface circulation (thin black arrows) is driven mainly by momentum exchange from atmospheric surface winds (see Figures 3.2 and 3.3), and bottom surface circulation (wide white arrows) is driven mainly by thermohaline gradients at the seasonal limits of pack ice (dashed lines). Both patterns of circulation are influenced by the earth's rotation and by topography; top currents by terrain above sea level (black areas), and bottom currents by terrain above 1500 m below sea level (dotted areas). Top and bottom currents tend to complement each other in the clockwise Antarctic Circumpolar Current. (From Stewart, 1969, Figure 13; Schwerdtfeger, 1970a, and Duxbury, 1971).

freshwater ice, both in the grounded and the floating portions of glaciers and ice sheets. The Antarctic Ice Sheet is Earth's dominant cryospheric component, especially in the Southern Hemisphere.

Flow in the Antarctic Ice Sheet, like flow in the ocean and the atmosphere, has different properties in various layers, as shown in Figure 3.6. The top firn layer extends from 60 to 80 m depth, depending upon the initial snow density and is the layer in which continuous air channels con-stitute a permeable matrix. Longitudinal strains and cold temperatures predominate, with elastic deformation followed by minimal plastic deformation leading to brittle fracture and large transverse crevasses in zones of extending flow. Lithostatic pressure seals air channels below the firn layer so that ice now constitutes an impermeable matrix in which the air channels become trapped air bubbles. This is a thick ice layer with slowly varying low temperatures and high densities, with longitudinal strain rates still predominating but causing steady-state creep instead of brittle fracture. Below this layer is a basal ice layer in which traction at the bed increasingly

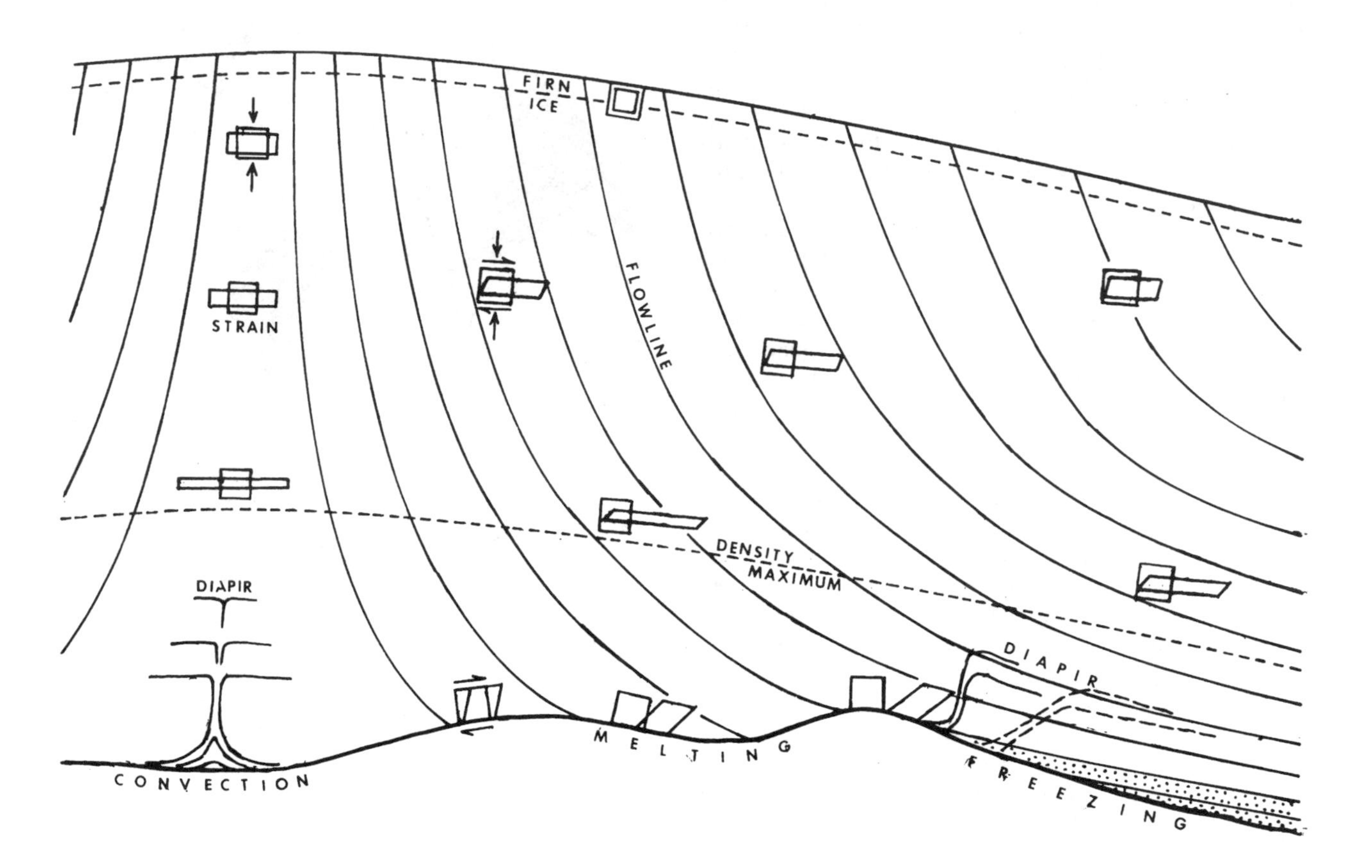

Figure 3.6: Flow in the Antarctic Ice Sheet. Flow beneath the ice divide is by pure shear from vertical compression. Flow at the bed is by simple shear from bed traction and by sliding if the bed is thawed. Flow elsewhere combines simple shear and pure shear, shown by the deforming squares in plane strain. Flowlines advect surface ice downslope toward the bed. Regelation ice is dotted. The two thermal convection diapirs, beneath and beyond the ice divide, are conjectural.

transforms the pure shear produced by longitudinal strain into simple shear produced by laminar strain. Trapped air bubbles get smeared out by the shear strain and the air is ultimately dissolved into the ice-crystal structure as a result of increasing lithostatic pressure and thermomechanical energy with depth. The frictional heat generated by shear raises the temperature, and density falls when thermal expansion and dissolving air expand the ice-crystal lattice. Transient creep replaces steady-state creep near the bed, because shear stress continually changes at an ice-rock interface having significant roughness.

Flow in the Antarctic Ice Sheet is always a result of the conversion of potential gravitational energy into kinetic energy of motion. General ice-sheet flow, as summarized by Budd (1969), results when the effective viscosity of ice is unable to prevent gravity from drawing ice toward the center of the earth. When this occurs an ice mass collapses; surface lowering causes peripheral spreading for sheet flow, and peripheral spreading causes surface lowering for stream flow (Hughes, 1992b).

In addition to mechanical instability resulting in general ice-sheet flow, thermal instability may lead to convective glacial flow. This is a consequence of cold higher-density, ice overlying hot lower-density ice, with the resulting potential gravitational energy being converted into kinetic energy of motion by means of transient diapiric activity which mixes the two layers (Hughes, 1985). The energetically most efficient means of mixing is for the cold, heavy upper ice to move downward en masse, as shown in Figure 3.6, with the hot, light lower ice being squeezed (almost squirted) upward in a transient array of narrow ribbons or pipes, each encased in a shear band sheath. Ocean-floor spreading ridges and crustal hot spots are manifestations of this kind of thermal convection in Earth's mantle. Advective flow of ice from the center to the edge of the ice sheet strings out convective flow, so isolated transient pipes in the interior become ribbons aligned with advective flow toward the margin. These ribbons of intense shear can exist as the sides of ice streams, and they may have a role in converting sheet flow into stream flow, especially where bed topography channels ice (Hughes, 1992b; Raymond, 1996). The possibility of thermal convection in polar ice sheets is discounted by glaciologists, but it may be the Achilles heel of ice-sheet models that ignore it.

General ice-sheet flow is modified around the margins of the Antarctic Ice Sheet, notably in gaps through bordering mountain ranges, where sheet flow is converted into stream flow as ice discharges into coastal fjords (Weertman, 1963). Conversion to stream flow occurs even if coastal mountains are buried by the ice sheet or are absent, because flow is concentrated in troughs and fjords where ice is thickest (Weertman and Birchfield, 1982). Frictional heat is generated by the concentration of flow in ice streams, and bed erosion due to glacial sliding begins when frictional heat thaws a frozen bed (Weertman, 1964). Erosion widens, deepens, and lengthens the ice streams, and as they cut further inland they eventually form a network of deep ice-stream beds which control drainage of the entire ice sheet. In this way, ice streams modify the original subglacial topography by erosion until they control the topography and therefore ice-sheet drainage. Because of its concave surface and its fast flow, an ice stream at the peak of its life cycle is a potential surge basin of the ice sheet .

This description of sheet flow, diapiric flow, and stream flow show that, to a considerable extent, an ice sheet can control its own dynamics by internal processes inherent in glacial flow. Flow is also controlled by mass accumulation and ablation at the surface and the bed of the ice sheet, and these are external processes responding to climatic conditions (although a major ice sheet can to a large degree mold the surrounding climate). Basal accumulation and ablation zones are zones where basal water freezes and basal ice melts. These zones control glacial erosion and deposition at the bed, with the basal mass balance being strongly influenced by the surface mass balance (Weertman, 1966; Budd and others, 1970, 1971; Hughes, 1981).

In general, a terrestrial ice sheet will have a central accumulation zone and a peripheral ablation zone at its surface and frozen central and peripheral zones bordering a thawed intermediate zone at its base, if it is centered on highlands, as is the East Antarctic Ice Sheet. Bordering the thawed intermediate zone are an inner zone of basal melting and an outer zone of basal freezing. A marine ice sheet can have a thawed central basal zone if it lies over a deep marine basin, as does the West Antarctic Ice Sheet. Ice streams have thawed beds that may cut across the outer frozen zone. Equilibrium lines separate all these zones on both top and bottom surfaces, and equilibrium lines migrate to satisfy the dynamic requirements of the ice sheet. At one extreme, net accumulation occurs over the entire top surface, and the bottom surface is everywhere frozen to the bed. This causes maximum growth of the ice sheet. At the other extreme, net ablation occurs over the entire top surface, and the bottom surface is everywhere a thawed melting bed. This causes maximum shrinkage of the ice sheet. The actual positions of these equilibrium lines are determined by rates of surface and basal accumulation and ablation, surface temperature, basal geothermal heat flux and topography, ice thickness, the distribution of ice flow between internal creep and basal sliding, and climatic warming or cooling rates. If thawed beds under peripheral ice streams extend across the marginal frozen basal zone to join the intermediate thawed basal zone, then stream flow controls ice drainage over the interior thawed bed. This enhances the possibility of an ice-stream surge spreading inland to consume the interior heart of an ice sheet when its thawed bed breaches the outer frozen zone (Weertman, 1969; Clarke and others, 1977).

As shown in Figure 3.7, the present mass-balance regime of the Antarctic Ice Sheet has no significant top surface ablation zone. Top surface accumulation generally decreases toward the center of East Antarctica (Giovinetto and others, 1989), where cyclonic storms originating in the Southern Ocean seldom reach. No-cloud precipitation predominates over the polar plateau and is compressed into ice at about 80 m depth. Flow diverges toward the coast from four major ice domes, three in East Antarctica and one in West Antarctica. The highest East Antarctic ice dome is over the subglacial Gamburtsev Mountains, near the pole of inaccessibility (the furthest point from sea in Antarctica), with lower ice domes centered over Queen Maud Land and Wilkes Land. The West Antarctic ice dome is centered around the nunataks of the Executive Committee Range in Marie Byrd Land. Minor ice domes are found over subglacial highlands, around nunataks, at various sites in the Transantarctic Mountains, on the Antarctic Peninsula, and along the coast, especially between major ice streams.

The most stable ice dome is the one over the Gamburtsev Mountains, since other ice domes vanish when the ice sheet grows and more ice domes appear when the ice sheet shrinks. Ice flowing north from the Gamburtsev dome is channeled into the huge Lambert Glacier, which terminates in the Amery Ice Shelf. A thawed bed is expected under the Lambert Glacier; otherwise, ice in this drainage basin is probably frozen to the bed. Ice flowing south from the Gamburtsev ice dome breaches the Transantarctic Mountains, with a major ice divide extending across Marie Byrd Land to the Antarctic Peninsula. Flow east of this ice divide forms the Weddell Sea ice drainage basin, which feeds the Filchner-Ronne Ice Shelf, and flow west of this ice divide forms the Ross Sea ice drainage basin, which feeds the Ross Ice Shelf. These three drainage basins constitute nearly two thirds of the Antarctic Ice Sheet, as shown in Figure 3.7. Major ice streams downdraw interior ice, thereby producing a sequence of saddles that separate minor ice domes along the ice divide.

To summarize, the most energetic parts of the Antarctic Ice Sheet are basal ice, any interior convective diapirs, and the peripheral ice streams, with activity over a thawed bed being more energetic than activity over a frozen bed because of basal sliding. Note that these three regions of high energy, a basal boundary layer, a convective plume, and currents of flow, all have their counterparts in atmosphere and ocean circulation. Motion is gravity-driven in each case, but controlled by the basal mass-balance regime at the bed, a density inversion in any interior convecting regions, and subglacial bed topography under the peripheral ice streams. Figure 3.7 illustrates the generalized pattern of ice flow in the Antarctic Ice Sheet at present. The bed is frozen over the subglacial

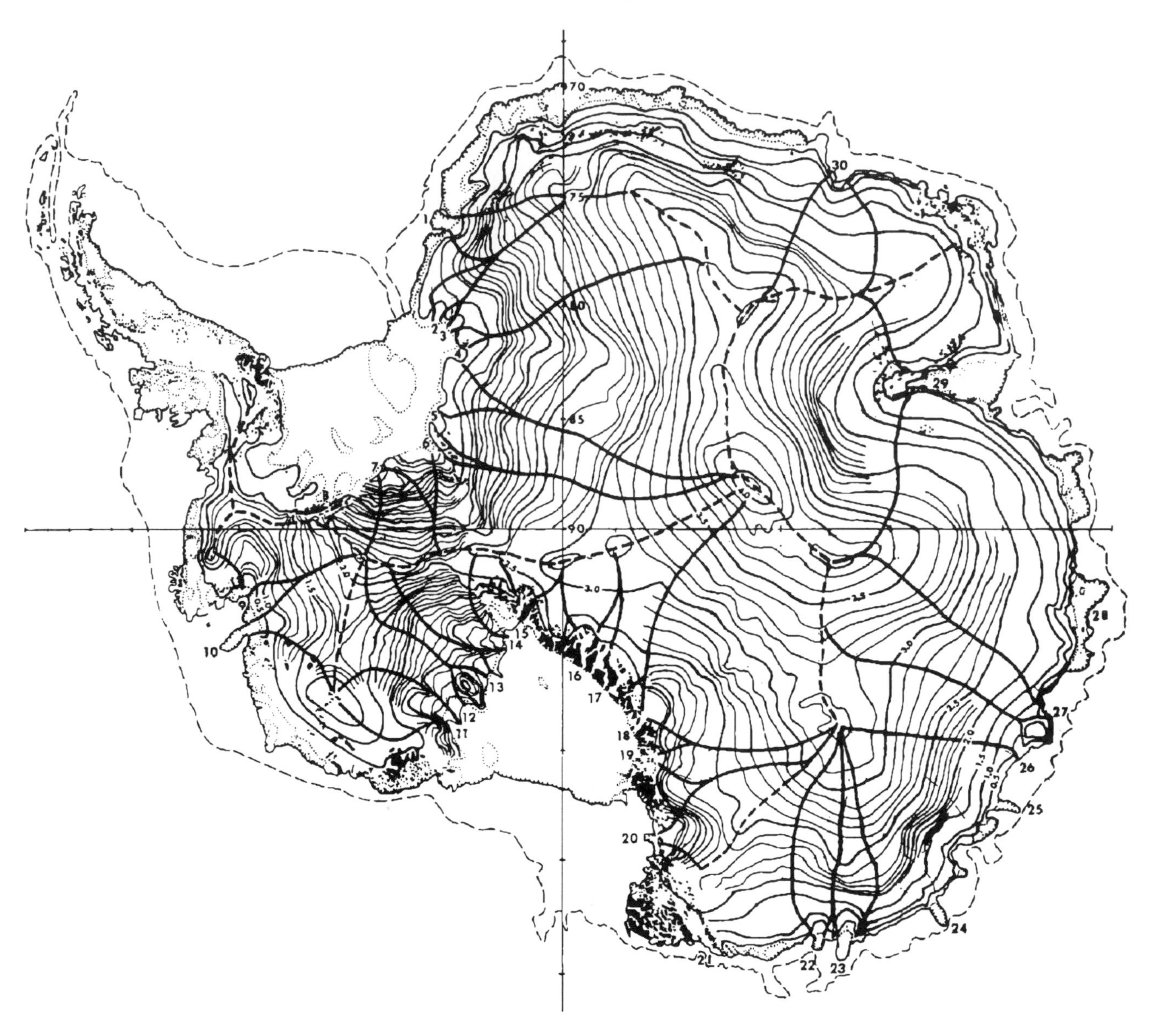

Figure 3.7: Major Antarctic ice streams and their catchment areas. Catchment areas extend from ice divides (*dashed*) and lie within flowbands normal to ice elevation contours (km). Bed topography is mapped for all but the first five. Ice streams are numbered counterclockwise from left: 1. Stancomb-Wills Glacier, 2. Baily Ice Stream, 3. Slessor Glacier, 4. Recovery Glacier, 5. Support Force Glacier, 6. Foundation Ice Stream, 7. Institute Ice Stream, 8. Rutford Ice Stream, 9. Pine Island Glacier, 10. Thwaites Glacier, 11. Ice Stream E, 12. Ice Stream D, 13. Ice Stream C, 14. Ice Stream B, 15. Ice stream A, 16. Beardmore Glacier, 17. Nimrod Glacier, 18. Byrd Glacier, 19. Mulock Glacier, 20. David Glacier, 21. Rennick Glacier, 22. Ninnis Glacier, 23. Mertz Glacier, 24. Dibble Glacier, 25. Dalton Glacier, 26. Totten Glacier, 27. Vanderford Glacier, 28. Denman Glacier, 29. Lambert Glacier, 30. Shirasebreen

Gamburtsev Mountains of central East Antarctica and over the peripheral subglacial highlands of East and West Antarctica. In between, the bed is thawed due to the basal geothermal heat flux and frictional heat from internal shear deformation in ice (especially basal shear, shear bordering any diapiric activity, and shear alongside ice streams). A thawed bed also exists under the large outlet glaciers, especially Lambert Glacier and the huge ice streams feeding the Ross and Filchner Ice Shelves. In West Antarctica, the thawed bed under major ice streams joins the thawed bed under the inland subglacial basin, and this portion of the ice sheet has the nonequilibrium concave surface profile characteristic of a surge basin. Basal meltwater is forced toward the edge of the ice sheet by the pressure gradient from the overburden ice and loses heat by conduction to the cold upper ice layer that moves closer to the bed by advective flow. Eventually the basal meltwater begins to freeze, and eroded debris being washed down the pressure gradient is frozen onto the sole of the ice sheet. The transition from basal melting to basal freezing therefore marks the boundary between clean ice and debris-charged ice, and this boundary intersects the ice surface if a surface ablation zone exists. In the Antarctic Ice Sheet, net accumulation occurs over nearly all of the ice surface, and net ablation is almost entirely by calving of ice slabs and icebergs, although basal meltwater losses may be important under ice streams and basal melting or freezing can be substantial under ice shelves (Jacobs and others, 1996). It is evident from Figure 3.7 that ice streams control drainage of the Antarctic Ice Sheet and that the West Antarctic Ice Sheet appears to have lowered in a way that suggests unstable surge conditions. If so, West Antarctic ice streams are the obvious focus for studying ice-sheet gravitational collapse. Also noteworthy are the giant East Antarctic ice streams feeding the Filchner, Ross, and Amery Ice Shelves, and other large East Antarctic ice streams draining the Polar, Wilkes, and Aurora Subglacial Basins. The thawed beds of these ice streams may pass through the outer basal frozen zone and produce concave "surge basins" that might collapse the East Antarctica Ice Sheet (Hughes, 1987b). The surge potential for Antarctic ice streams has been studied by McInnes and others (1985), Jenssen and others (1985), McInnes and others (1986), and Radok and others (1986).

Relating Ice-Sheet Stability to Ice-Surface Slopes

Stability of the Antarctic Ice Sheet seems to be related to its surface slope, with the relatively stable East Antarctic Ice Sheet having a largely convex surface and the relatively unstable West Antarctic Ice Sheet having a largely concave surface (Hughes, 1973, 1975, 1977). A concave ice surface is typical of stream flow, and it is transitional between the convex surface of sheet flow and the flat surface of shelf flow. The transition occurs when basal sliding becomes dominant (Weertman, 1957a, 1957b), and it is inherently unstable (Weertman, 1974). A major theme of this book is to present the physical variables and constraints that determine the surface slope, and therefore the stability, of an ice sheet. The result is an expression for the ice-surface slope that applies for (1) sheet flow, stream flow, and shelf flow, (2) diverging, converging, and curving flow, (3) variable surface accumulation and ablation rates, (4) frozen, thawed, melting, and freezing conditions at the bed, (5) variable basal water and ice overburden pressures, (6) beds ranging from rigid bedrock and ice-cemented permafrost to wet or dry soft sediments and till, (7) variable bed topography and bathymetry, (8) variable rheological properties for ice and bed deformation, (9) variable glacial isostatic adjustments of the bed to the changing ice load, and (10) the major stress and stress gradients induced by flow.

Let the horizontal x direction be positive along flowlines from the grounded ice margin to the ice divide, the horizontal y direction be along ice-surface contour lines, and the vertical z direction be positive upward from sea level, as shown in Figure 3.8 for grounded ice-sheet margins that end as floating ice shelves, calving ice walls, or terminal ice lobes. In finite-difference flowline models, the expression for a surface elevation change Δh in horizontal distance Δx along a flowline, which is the centerline of a flowband that experiences the ten influences listed above and along which flow instabilities occur that affect surface slope $\Delta h/\Delta x$, is derived from the force balance and the mass balance along the flowband.

The force balance along an ice-sheet flowband

The longitudinal gravitational pulling force, shown at the calving fronts and grounding lines of ice shelves in Figure 2.2 and shown for sheet flow, stream flow, and shelf flow portions of an ice-sheet flowline profile in Figure 3.8, is merely a reduction in lithostatic pressure in the direction of ice flow. Along the flowline, this reduction arises from the surface slope for sheet flow, from surface slope and basal hydrostatic pressure for stream flow, and from basal hydrostatic pressure for shelf flow. In all three cases, reduced lithostatic pressure is demonstrated by the primary transverse crevasses that arc across the heads of ice streams and extend along the calving fronts of ice shelves. These crevasses are tensile features. They open because lithostatic pressure is reduced in the forward direction when ice moves longitudinally to displace air, in an attempt to satisfy the first-order gravitational tendency for planetary mass to occupy strata in which density decreases outward from Earth's center. In this strata, ice would occupy a spherical shell between either land or sea and the atmosphere. That ice sheets are pulled, not pushed, longitudinally to occupy this density niche is proven by the transverse crevasses, which would not open if ice were being pushed longitudinally as a passive response to downward gravitational pulling (Hughes, 1992b).

Longitudinal gravitational pulling is resisted by a longitudinal deviator tensile stress, which reduces lithostatic pressure in the ice sheet in the direction of longitudinal motion. For grounded ice, pulling is also resisted by a basal shear stress caused by longitudinal ice motion. In ice streams, this ice motion is also resisted by a side shear stress. In ice shelves, motion is resisted by shear along the sides of a confining embayment and by shear over or around basal "pinning points" that produce ice rumples or ice rises on the ice shelf. These are all kinematic deviator stresses because they resist ice motion that attempts to remove deviations from lithostatic pressure in longitudinal directions.

In Figure 3.8, vertical gravitational pulling force $(F_z)_G$ in a flowband of width w is mass m of a column having density ρ and volume V times gravity acceleration g_z in the vertical z direction, such that $(F_z)_G = m\ g_z = \rho\ V\ g_z$. Ice columns of height h_I are shown for sheet flow, stream flow, and shelf flow, where the flowline surface is convex, concave, and flat, respectively. Basal water pressure would support water columns of height h_W in these ice columns. Let P_I and P_W be the respective lithostatic and hydrostatic pres-

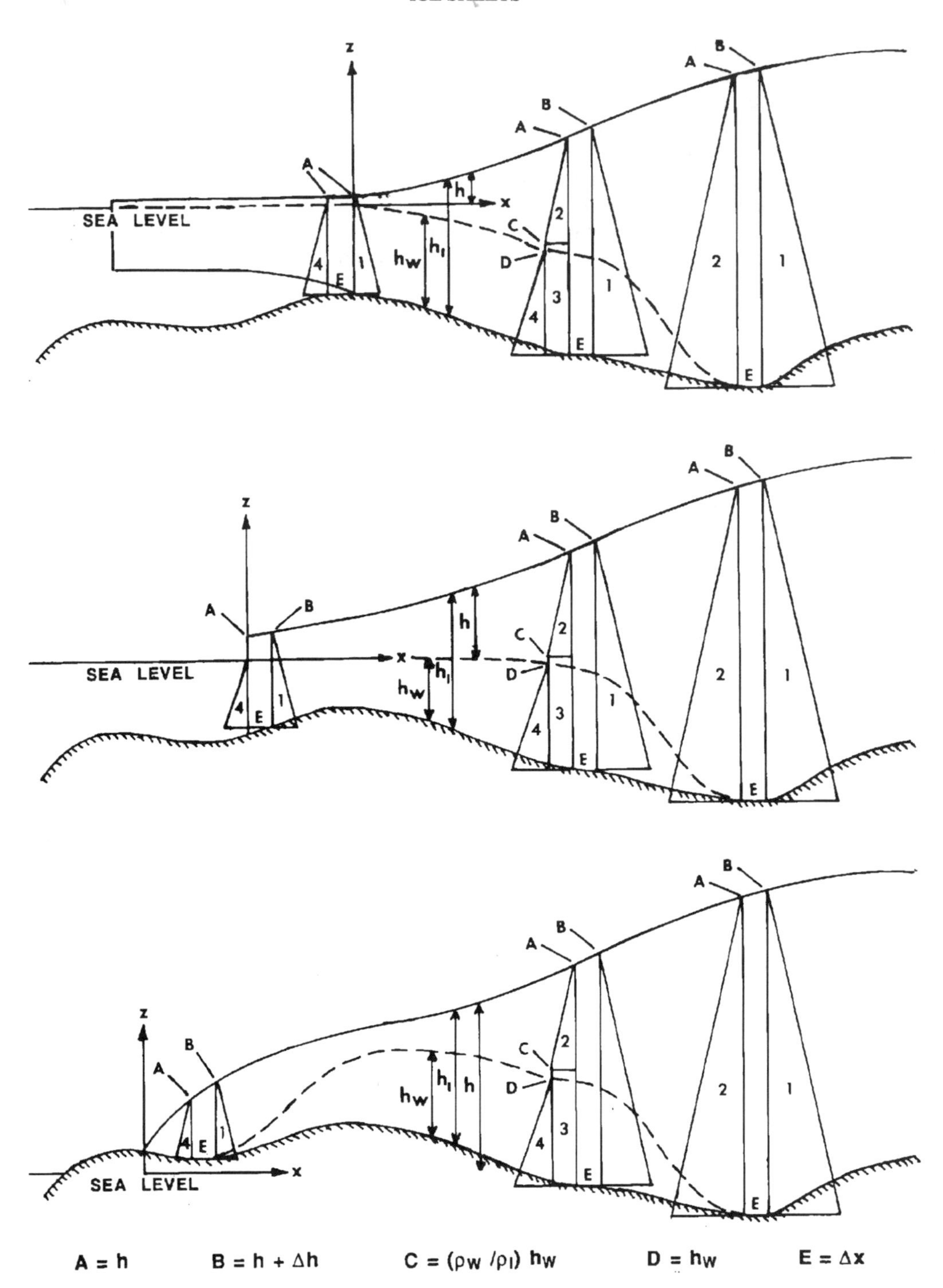

Figure 3.8: A geometrical representation of the net horizontal gravitational force along an ice-sheet flowline. Horizontal axis x is at sea level and begins at the grounding line of a floating ice shelf (*top*), at a calving ice wall grounded in water (*center*), or at the terminus of an ice lobe grounded on land (*bottom*). Basal water pressure would force basal water to height h_W above the bed in ice of height h_I above the bed and height h above sea level, all measured in the vertical z direction. The net horizontal gravitational force pulls ice forward, thereby thinning ice and causing a change Δh in ice surface elevation in horizontal distance Δx along the flowline. It is obtained by subtracting from the area of triangle 1, the combined areas of triangle 2, rectangle 3, and triangle 4. The slopes of triangles 1 and 2 show the increase of ice lithostatic pressure toward the bed. The slope of triangle 4 shows the increase of water hydrostatic pressure toward the bed.

sures P at the base of these respective ice and water columns having basal area A_z = w Δx normal to z for width w and longitudinal length Δx. Then $(F_z)_G = P A_z$. At the base of an ice column, where $P_I = \rho_I g_z h_I$:

$$(F_z)_G = \rho_I V_I g_z = \rho_I (h_I w \Delta x) g_z = P_I A_z = P_I w \Delta x \quad (3.1)$$

At the base of a water column, $P_W = \rho_W g_z h_W$:

$$(F_z)_G = \rho_W V_W g_z = \rho_W (h_W w \Delta x) g_z = P_W A_z = P_W w \Delta x \quad (3.2)$$

Combining Equations (3.1) and (3.2) gives a basal buoyancy factor formed by the ratio of P_W to P_I:

$$\frac{P_W}{P_I} = \frac{\rho_W h_W}{\rho_I h_I} \quad (3.3)$$

Sheet flow occurs when the bed is either frozen, so that $P_W / P_I = 0$, or thawed but basal water is insufficient to support the ice overburden, so that $P_W / P_I << 1$. Stream flow occurs over a thawed bed when basal water can support a substantial part of the ice overburden such that $0 < P_W / P_I < 1$. Shelf flow occurs when basal water supports the entire ice overburden, so that $P_W / P_I = 1$.

Longitudinal gravitational pulling force $(F_x)_G$ in a flowband of width w is $(F_x)_G = A_x \Delta P_I$ in Figure 3.8, where $\Delta P_I = \rho_I g_z [\Delta h + 1/2 h_I (1 - \rho_I / \rho_W) (P_W / P_I)^2]$ is the reduced lithostatic pressure on the downslope column area $A_x = w h_I$ normal to x that causes the ice column to move downslope in the longitudinal – x direction of ice flow. Forces $(F_x)_G$ per flowband width w and perpendicular to ice columns in Figure 3.8 are the areas of rectangle 3 and triangles 1, 2, and 4, which show how lithostatic and hydrostatic pressures increase toward the bed in relation to heights h_I and $h_I + \Delta h$ of ice and height h_W of water that would exert basal water pressure P_W on area w Δx. Sheet flow occurs when $h_W = 0$ or $h_W << h_I$, so only triangles 1 and 2 exist. Stream flow occurs when $0 < h_W < (\rho_I / \rho_W) h_I$, so rectangle 3 and triangles 1, 2, and 4 exist. Shelf flow occures when $P_W = P_I$, so that $h_W = (\rho_I / \rho_W) h_I$ in Equation (3.3), and only triangles 1 and 4 exist.

Consider the middle ice column in Figure 3.8. For rectangle 3 and triangles 1, 2, and 4, where the surface slope causes column height to change by Δh in longitudinal distance Δx, the longitudinal gravitational pulling force is the difference between forces pushing on the back and front sides of the column:

$$(F_x)_G = \Delta P_I A_x = P_1 A_1 - P_2 A_2 - P_3 A_3 - P_4 A_4$$

$$= \left[\tfrac{1}{2} \rho_I g_z (h_I + \Delta h)\right] [(w \pm \Delta w)(h_I + \Delta h)]$$

$$- \left[\tfrac{1}{2} \rho_I g_z \{h_I - h_W(\rho_W / \rho_I)\}\right] [w \{h_I - h_W(\rho_W / \rho_I)\}]$$

$$- [\rho_I g_z \{h_I - h_W(\rho_W / \rho_I)\}] [w \{h_W(\rho_W / \rho_I)\}]$$

$$- \left[\tfrac{1}{2} \rho_W g_z h_W\right] [w_I h_W]$$

$$\approx [\rho_I g_z h_I][w \Delta h \pm h_I \Delta w] + \left[\tfrac{1}{2} \rho_I g_z h_I (P_W / P_I)\right] [w h_I (P_W / P_I)]$$

$$- \left[\tfrac{1}{2} \rho_W g_z h_W (P_W / P_I)\right] [w h_W (P_W / P_I)]$$

$$\approx \rho_I g_z w h_I \left[\Delta h \pm (h_I / \Delta w) + \tfrac{1}{2} h_I (1 - \rho_I / \rho_W)(P_W / P_I)^2\right]$$

$$\approx F_D = (\sigma_{xx} - \sigma_{zz}) A_D \quad (3.4)$$

where the term containing $(\Delta h)^2$ is ignored, $h_W (\rho_W / \rho_I)$ is the height of ice that would just float in water of height h_W, Equation (3.3) is used to convert $\rho_W h_W / \rho_I h_I$ to P_W / P_I, and variations in h_W along x occurs in steps of Δx. Note that on the downslope front face of the middle ice column in Figure 3.8, average ice pressure $1/2 \rho_I g_z h_I$, average water pressure $1/2 \rho_W g_z h_W$, and respective areas $w h_I$ and $w h_W$ against which ice and water press, are all reduced by basal buoyancy factor P_W / P_I.

Downslope gravitation pulling force $(F_x)_G = F_D$ reduces lithostatic pressure P in ice on downslope area A_D of this face by amount $\sigma_{xx} - \sigma_{zz}$, where:

$$\sigma_{zz} = - \rho_I g_z (h_I - z) = P \quad (3.5)$$

Therefore, $\sigma_{xx} - \sigma_{zz}$ is a deviator stress pulling on this face of the ice column, Deviator stresses σ'_{ij} are related to strain rates $\dot{\varepsilon}_{ij}$ through the flow law for glacial ice, which is derived in Chapter 4:

$$\dot{\varepsilon}_{ij} = \tfrac{1}{2} (\partial u_i / \partial j + \partial u_j / \partial i)$$

$$= (\sigma_c^{n-1}/A^n)(\sigma_{ij} - \delta_{ij} P) = (\sigma_c^{n-1}/A^n)\sigma'_{ij} \quad (3.6)$$

Following Nye (1953) and Glen (1955), subscripts i, j refer to axes x, y, z in the standard tensor notation, δ_{ij} is the Kronecker delta ($\delta_{ij} = 1$ when i = j and $\delta_{ij} = 0$ when $i \neq j$), σ_c is the effective creep stress, A is an ice hardness parameter that depends mainly on ice temperature and fabric, n is an ice viscoplastic parameter such that n = 1 for viscous creep, $n = \infty$ for plastic creep, and $n \approx 3$ for creep in ice.

Since $\dot{\varepsilon}_{ij} \propto \sigma'_{ij}$ in Equation (3.6), assuming that ice is incompressible, so that $\dot{\varepsilon}_{xx} + \dot{\varepsilon}_{yy} + \dot{\varepsilon}_{zz} = 0$, requires that $\sigma'_{xx} + \sigma'_{yy} + \sigma'_{zz} = 0$. Therefore:

$$\sigma'_{zz} = -(\sigma'_{xx} + \sigma'_{yy}) \quad (3.7)$$

Since normal deviator stresses σ'_{ij}, for which i = j, are deviations of normal stresses σ_{ij} from lithostatic pressure P, Equation (3.7) allows deviator stress $\sigma_{xx} - \sigma_{zz}$ pulling on the front face of the ice column to be written;

$$\sigma_{xx} - \sigma_{zz} = (\sigma'_{xx} - P) - (\sigma'_{zz} - P) = \sigma'_{xx} - \sigma'_{zz} = \sigma'_{xx} + (\sigma'_{xx} + \sigma'_{yy})$$

$$= \left[2 + \left(\sigma'_{yy} / \sigma'_{xx}\right)\right] \sigma'_{xx} = \left[2 + \left(\dot{\varepsilon}_{yy} / \dot{\varepsilon}_{xx}\right)\right] \sigma'_{xx} \qquad (3.8)$$

The force balance requires that gravitational force $(F_x)_G = F_D$ pulling on downslope area A_D of the ice column is balanced by kinematic forces $(F_x)_K$ induced by downslope ice motion, where $(F_x)_K$ consists of longitudinal force F_U acting on upslope area A_U, side shear force F_S acting on side areas A_S, and basal shear force F_B acting on basal area A_B:

$$(F_x)_K = F_U + F_S + F_B$$

$$= [(\sigma_{xx} - \sigma_{zz}) + \Delta(\sigma_{xx} - \sigma_{zz})] A_U + \tau_s A_S + \tau_o A_B$$

$$= \left[2 + \left(\dot{\varepsilon}_{yy} / \dot{\varepsilon}_{xx}\right)\right] \left(\sigma'_{xx} + \Delta\sigma'_{xx}\right) (w \pm \Delta w) (h_I + \Delta h_I)$$

$$+ \tau_s (2 \bar{h}_I \Delta x) + \tau_o (\bar{w} \Delta x)$$

$$= \left[2 + \left(\dot{\varepsilon}_{yy} / \dot{\varepsilon}_{xx}\right)\right] [w h_I + w \Delta h_I \pm h_I \Delta w] \sigma'_{xx}$$

$$+ \left[2 + \left(\dot{\varepsilon}_{yy} / \dot{\varepsilon}_{xx}\right)\right] [w h_I + w \Delta h_I \pm h_I \Delta w] \Delta\sigma'_{xx}$$

$$+ (2 h_I \tau_s + w \tau_o) \Delta x \qquad (3.9)$$

where gradient $\Delta(P_W / P_I)/\Delta x$ causes $(\sigma_{xx} - \sigma_{zz})$ to increase by $\Delta(\sigma_{xx} - \sigma_{zz})$ on upslope area $A_U = (w \pm \Delta w)(h_I + \Delta h_I)$ of the column, side shear stress τ_s acts on side areas $A_S = 2 \bar{h}_I \Delta x$ of the column, basal shear stress τ_o acts on basal area $A_B = \bar{w} \Delta x$ of the column, $\bar{h}_I = h_I + 1/2 \Delta h_I \approx h_I$ and $\bar{w} = w \pm \Delta w \approx w$ are the respective average height h_I and width w of the column, $\Delta h_I \Delta w$ is ignored, and $\Delta h_I = \Delta h$ for a horizontal bed.

Setting $(F_x)_G = (F_x)_K$ in Equations (3.4) and (3.9) and solving for ice surface slope $\Delta h/\Delta x$, but distinguishing it from ice thickness gradient $\Delta h_I / \Delta x$, although they are the same for a horizontal bed:

$$\frac{\Delta h}{\Delta x} = \frac{\left[2 + \left(\dot{\varepsilon}_{yy} / \dot{\varepsilon}_{xx}\right)\right] [w h_I + w \Delta h_I \pm h_I \Delta w] \left[\sigma'_{xx} + \Delta\sigma'_{xx}\right]}{\rho_I g_z w h_I \Delta x}$$

$$+ \frac{2 h_I \tau_s + w \tau_o}{\rho_I g_z w h_I} - \left(1 - \frac{\rho_I}{\rho_W}\right) \left(\frac{P_W}{P_I}\right)^2 \frac{h_I}{2 \Delta x} \pm \left(\frac{h_I}{w}\right) \frac{\Delta w}{\Delta x} \qquad (3.10)$$

Summing terms that become infinite as $\Delta x \to 0$, equating their sum to zero, and solving for σ'_{xx}:

$$\sigma'_{xx} = \frac{\rho_I g_z h_I}{2 \left[2 + \left(\dot{\varepsilon}_{yy} / \dot{\varepsilon}_{xx}\right)\right]} \left(1 - \frac{\rho_I}{\rho_W}\right) \left(\frac{P_W}{P_I}\right)^2 \qquad (3.11)$$

Differentiating gives the longitudinal deviator stress gradient:

$$\frac{\Delta\sigma'_{xx}}{\Delta x} \approx \frac{d\sigma'_{xx}}{dx} = \frac{\rho_I g_z (1 - \rho_I / \rho_W)}{2 \left[2 + \left(\dot{\varepsilon}_{yy} / \dot{\varepsilon}_{xx}\right)\right]} \left[\left(\frac{P_W}{P_I}\right)^2 \frac{\Delta h_I}{\Delta x} + h_I \frac{\Delta}{\Delta x} \left(\frac{P_W}{P_I}\right)^2\right] \qquad (3.12)$$

Substituting Equations (3.11) and (3.12) into Equation (3.10) gives:

$$\frac{\Delta h}{\Delta x} = \left(1 - \frac{\rho_I}{\rho_W}\right) \left(\frac{P_W}{P_I}\right)^2 \left(\frac{\Delta h_I}{\Delta x} \pm \frac{h_I \Delta w}{2 w \Delta x}\right) \pm \frac{h_I \Delta w}{w \Delta x}$$

$$+ \frac{h_I}{2} \left(1 - \frac{\rho_I}{\rho_W}\right) \frac{\Delta}{\Delta x} \left(\frac{P_W}{P_I}\right)^2 + \frac{2 h_I \tau_s + w \tau_o}{\rho_I g_z w h_I} \qquad (3.13)$$

where terms containing products $\Delta w \, \Delta\sigma'_{xx}$ and $\Delta h_I \, \Delta\sigma'_{xx}$ are ignored, and $\Delta h/\Delta x$ is distinguished from $\Delta h_I / \Delta x$ because $\Delta h_I / \Delta x$ is obtained independently from the mass balance, and is then used to relate $\Delta h/\Delta x$ to the mass balance.

The mass balance along an ice-sheet flowband

For a flowband, the mass-balance requirement that changes ice thickness h_I over time t at rate $\partial h_I / \partial t$ is the difference between the rate a of ice accumulation (positive a) or ablation (negative a) and converging or diverging flow of ice having mean velocity u through ice thickness h_I:

$$\frac{\partial h_I}{\partial t} = a - \left(\frac{\partial}{\partial x} + \frac{\partial}{\partial y}\right)(h_I u) = a - h_I \left(\frac{\partial u}{\partial x} + \frac{\partial u}{\partial y}\right) - u \left(\frac{\partial h_I}{\partial x} + \frac{\partial h_I}{\partial y}\right) \qquad (3.14)$$

Assume steady-state equilibrium, so that $\partial h_I / \partial t = 0$. Converging or diverging flow causes flowband width to change from w to $w \pm \Delta w$ along length Δx. If u_x and u_y are the respective longitudinal and transverse components of u, the angular convergence or divergence θ of flow is given by:

$$\tan \theta = \frac{\Delta w}{\Delta x} = \frac{u_y}{u_x} \qquad (3.15)$$

where:

$$u = u_x \cos \theta + u_y \sin \theta \qquad (3.16)$$

The velocity gradients of u along x and y are then:

$$\partial u/\partial x = (\partial u_x / \partial x) \cos \theta + (\partial u_y / \partial x) \sin \theta$$

$$= \dot{\varepsilon}_{xx} \cos \theta + (\partial u_y / \partial x) \sin \theta \qquad (3.17a)$$

$$\partial u/\partial y = (\partial u_x / \partial y) \cos \theta + (\partial u_y / \partial y) \sin \theta$$

$$= (\partial u_x / \partial y) \cos \theta + \dot{\varepsilon}_{yy} \sin \theta \qquad (3.17b)$$

where $\partial u_x/\partial x = \dot{\varepsilon}_{xx}$ and $\partial u_y/\partial y = \dot{\varepsilon}_{yy}$ by Equation (3.6).

For variable accumulation rate a and for x horizontal and positive in the upslope direction along the flowband, let x = 0 at the basal grounding line where marine ice becomes afloat or at the surface equilibrium line where terrestrial ice enters the ablation zone. Let ice width w, thickness h_I, and velocity u in the flowband be w_0, h_0, and u_0 at x = 0. Then the mass continuity condition over length x is:

$$\int_0^{awx} d(a\,w\,x) = \int_{w_0 h_0 u_0}^{w h_I u} d(w\,h_I\,u) = w\,h_I\,u - w_0\,h_0\,u_0 = a\,w\,x \quad (3.18)$$

Solving Equation (3.14) for $\Delta h_I / \Delta x$, using Equations (3.17) and (3.18):

$$\frac{\Delta h_I}{\Delta x} \approx \frac{\partial h_I}{\partial x} = \frac{a}{u} - \frac{h_I}{u}\left(\frac{\partial u}{\partial x} + \frac{\partial u}{\partial y}\right) - \frac{\partial h_I}{\partial y} \approx \frac{a}{u} - \frac{h_I}{u}\left(\frac{\partial u}{\partial x} + \frac{\partial u}{\partial y}\right)$$

$$\approx \frac{w\,h_I\,a}{a\,w\,x + w_0\,h_0\,u_0} - \frac{w\,h_I^2}{a\,w\,x + w_0\,h_0\,u_0}$$

$$\left[\dot\varepsilon_{xx}\cos\theta + \dot\varepsilon_{yy}\sin\theta + \left(\frac{\partial u_x}{\partial y}\right)\cos\theta + \left(\frac{\partial u_y}{\partial x}\right)\sin\theta\right] \quad (3.19)$$

when $\partial h_I/\partial y \ll \partial h_I/\partial x$ and can be is ignored.

Strain rates $\dot\varepsilon_{xx}$ and $\dot\varepsilon_{yy}$ in Equation (3.19) can be related to deviator stresses σ'_{xx} and σ'_{yy} through Equation (3.6), where effective creep stress σ_c as derived in Chapter 4 makes use of Equations (3.6) and (3.7) and is:

$$\sigma_c = \left[\tfrac{1}{2}\left(\sigma'^2_{xx} + \sigma'^2_{yy} + \sigma'^2_{zz} + 2\,\sigma^2_{xy} + 2\,\sigma^2_{yz} + 2\,\sigma^2_{zx}\right)\right]^{1/2}$$

$$= \frac{1}{\sqrt{2}}\left[\sigma'^2_{xx} + \sigma'^2_{yy} + \left(\sigma'_{xx} + \sigma'_{yy}\right)^2 + 2\,\sigma^2_{xy} + 2\,\sigma^2_{yz} + 2\,\sigma^2_{zx}\right]^{1/2}$$

$$= \left[\left(\frac{\sigma'_{xx}}{\sigma'_{xx}}\right)^2 + \left(\frac{\sigma'_{xx}\,\sigma'_{yy}}{\sigma'^2_{xx}}\right) + \left(\frac{\sigma'_{yy}}{\sigma'_{xx}}\right)^2 + \left(\frac{\sigma_{xy}}{\sigma'_{xx}}\right)^2 + \left(\frac{\sigma_{yz}}{\sigma'_{xx}}\right)^2 + \left(\frac{\sigma_{zx}}{\sigma'_{xx}}\right)^2\right]^{1/2} \sigma'_{xx}$$

$$= \left[1 + \left(\frac{\dot\varepsilon_{yy}}{\dot\varepsilon_{xx}}\right) + \left(\frac{\dot\varepsilon_{yy}}{\dot\varepsilon_{xx}}\right)^2 + \left(\frac{\dot\varepsilon_{xy}}{\dot\varepsilon_{xx}}\right)^2 + \left(\frac{\dot\varepsilon_{yz}}{\dot\varepsilon_{xx}}\right)^2 + \left(\frac{\dot\varepsilon_{zx}}{\dot\varepsilon_{xx}}\right)^2\right]^{1/2} \sigma'_{xx}$$

$$= R_{xx}\,\sigma'_{xx} \quad (3.20)$$

Substituting Equation (3.20) into Equation (3.6) for $\dot\varepsilon_{xx}$:

$$\dot\varepsilon_{xx} = \left(\frac{\sigma_c^{\,n-1}}{A^n}\right)\sigma'_{xx} = \left(R_{xx}\,\sigma'_{xx}\right)^{n-1}\frac{\sigma'_{xx}}{A^n}$$

$$= \left[1 + \left(\frac{\dot\varepsilon_{yy}}{\dot\varepsilon_{xx}}\right) + \left(\frac{\dot\varepsilon_{yy}}{\dot\varepsilon_{xx}}\right)^2 + \left(\frac{\dot\varepsilon_{xy}}{\dot\varepsilon_{xx}}\right)^2 + \left(\frac{\dot\varepsilon_{xz}}{\dot\varepsilon_{xx}}\right)^2\right]^{\frac{n-1}{2}} \left(\frac{\sigma'_{xx}}{A}\right)^n$$

$$= R_{xx}^{n-1}\left(\sigma'_{xx}/A\right)^n \quad (3.21)$$

where $\dot\varepsilon_{yz}/\dot\varepsilon_{xx}$ is nil and can be ignored. Substituting Equation (3.11) for σ'_{xx} into Equation (3.21) gives:

$$\dot\varepsilon_{xx} = R_{xx}^{n-1}\left[\frac{\rho_I\,g_z\,h_I}{2\left[2 + \left(\dot\varepsilon_{yy}/\dot\varepsilon_{xx}\right)\right]A}\left(1 - \frac{\rho_I}{\rho_W}\right)\left(\frac{P_W}{P_I}\right)^2\right]^n$$

$$= R\left[\frac{\rho_I\,g_z\,h_I}{2\,A}\left(1 - \frac{\rho_I}{\rho_W}\right)\left(\frac{P_W}{P_I}\right)^2\right]^n \quad (3.22)$$

where:

$$R = \frac{\left[1 + \left(\dot\varepsilon_{yy}/\dot\varepsilon_{xx}\right) + \left(\dot\varepsilon_{yy}/\dot\varepsilon_{xx}\right)^2 + \left(\dot\varepsilon_{xy}/\dot\varepsilon_{xx}\right)^2 + \left(\dot\varepsilon_{xz}/\dot\varepsilon_{xx}\right)^2\right]^{\frac{n-1}{2}}}{\left[2 + \left(\dot\varepsilon_{yy}/\dot\varepsilon_{xx}\right)\right]^n}$$

$$= \frac{R_{xx}^{n-1}}{\left[2 + \left(\dot\varepsilon_{yy}/\dot\varepsilon_{xx}\right)\right]^n} \quad (3.23)$$

Substituting Equation (3.22) for $\dot\varepsilon_{xx}$ in Equation (3.19) gives $\Delta h_I/\Delta x$:

$$\frac{\Delta h_I}{\Delta x} \approx \frac{w\,h_I\,a}{axw + w_0\,h_0\,u_0} - \frac{w\,h_I^2\,R\cos\theta}{axw + w_0\,h_0\,u_0}\left[\frac{\rho_I\,g_z\,h_I}{2\,A}\left(1 - \frac{\rho_I}{\rho_W}\right)\left(\frac{P_W}{P_I}\right)^2\right]^n$$

$$- \frac{w\,h_I^2}{a\,w\,x + w_0\,h_0\,u_0}\left[\dot\varepsilon_{yy}\sin\theta + \left(\frac{\partial u_x}{\partial y}\right)\cos\theta + \left(\frac{\partial u_y}{\partial x}\right)\sin\theta\right] \quad (3.24)$$

Substituting Equation (3.24) for $\Delta h_I/\Delta x$ in Equation (3.13) gives $\Delta h/\Delta x$:

$$\frac{\Delta h}{\Delta x} = \left(1 - \frac{\rho_I}{\rho_W}\right)\left(\frac{P_W}{P_I}\right)^2$$

$$\left\{\frac{w\,h_I\,a}{awx + w_0\,h_0\,u_0} - \frac{w\,h_I^2\,R\cos\theta}{a\,w\,x + w_0\,h_0\,u_0}\left[\frac{\rho_I\,g_z\,h_I}{2\,A}\left(1 - \frac{\rho_I}{\rho_W}\right)\left(\frac{P_W}{P_I}\right)^2\right]^n - \frac{w\,h_I^2}{a\,w\,x + w_0\,h_0\,u_0}\right.$$

$$\left.\left[\dot\varepsilon_{yy}\sin\theta + \left(\frac{\partial u_x}{\partial y}\right)\cos\theta + \left(\frac{\partial u_y}{\partial x}\right)\sin\theta\right]\right\}$$

$$+ \frac{h_I}{w}\left[1 + \frac{1}{2}\left(1 - \frac{\rho_I}{\rho_W}\right)\left(\frac{P_W}{P_I}\right)^2\right]\tan\theta$$

$$+ \frac{h_I}{2}\left(1 - \frac{\rho_I}{\rho_W}\right)\frac{\Delta}{\Delta x}\left(\frac{P_W}{P_I}\right)^2 + \frac{2\,\tau_s}{\rho_I\,g_z\,w} + \frac{\tau_o}{\rho_I\,g_z\,h_I} \quad (3.25)$$

If the flowband has constant width w, then $w = w_0$, $\theta = \Delta w/\Delta x = \partial u_y/\partial x = \dot\varepsilon_{yy} = 0$, $\partial u_x/\partial y = 2\,\dot\varepsilon_{xy}$, and $R = [1 + (\dot\varepsilon_{xy}/\dot\varepsilon_{xx})^2 + (\dot\varepsilon_{xz}/\dot\varepsilon_{xx})^2]^{(n-1)/2}/2^n$. If the flowband curves, bending shear produces strain rate $\dot\varepsilon_{xy}$ (Hughes, 1983). Following Equations (3.6), (3.20), and (3.21), $\dot\varepsilon_{xy}$ is given by:

$$\dot\varepsilon_{xy} = \left[\sigma'^2_{xx} + \sigma'_{xx}\,\sigma'_{yy} + \sigma'^2_{yy} + \sigma^2_{xy} + \sigma^2_{yz} + \sigma^2_{zx}\right]^{\frac{n-1}{2}}\left(\sigma_{xy}/A^n\right)$$

$$= \left[1 + \left(\frac{\dot\varepsilon_{xx}}{\dot\varepsilon_{xy}}\right)^2 + \left(\frac{\dot\varepsilon_{xx}\,\dot\varepsilon_{yy}}{\dot\varepsilon^2_{xy}}\right) + \left(\frac{\dot\varepsilon_{yy}}{\dot\varepsilon_{xy}}\right)^2 + \left(\frac{\dot\varepsilon_{xz}}{\dot\varepsilon_{xy}}\right)^2\right]^{\frac{n-1}{2}} \left(\frac{\sigma_{xy}}{A}\right)^n$$

$$= R_{xy}^{n-1}\left(\sigma_{xy}/A\right)^n \approx R_{xy}^{n-1}\left(\tau_s/A\right)^n \tag{3.26}$$

where σ_{yz} is ignored and τ_s is a yield stress for shear.

A straight flowband having constant width is characteristic of stream flow. Equation (3.25) reduces to:

$$\frac{\Delta h}{\Delta x} = \left(1 - \frac{\rho_I}{\rho_W}\right)\left(\frac{P_W}{P_I}\right)^2$$

$$\left\{\frac{h_I\, a}{a\, x + h_0\, u_0} - \frac{h_I^2\, R}{a\, x + h_0\, u_0}\left[\frac{\rho_I\, g_z\, h_I}{2\, A}\left(1 - \frac{\rho_I}{\rho_W}\right)\left(\frac{P_W}{P_I}\right)^2\right]^n\right\}$$

$$+ \frac{h_I}{2}\left(1 - \frac{\rho_I}{\rho_W}\right)\frac{\Delta}{\Delta x}\left(\frac{P_W}{P_I}\right)^2 + \frac{2\,\tau_s}{\rho_I\, g_z\, w} + \frac{\tau_o}{\rho_I\, g_z\, h_I} \tag{3.27}$$

All terms in Equation (3.27) are important for stream flow, where $\dot{\varepsilon}_{xx} >> \dot{\varepsilon}_{yy}$ for a straight ice stream of constant width, $\dot{\varepsilon}_{xy} = 0$ along its centerline, and $\dot{\varepsilon}_{xx} >> \dot{\varepsilon}_{xz}$ except near its head, so that $R \approx 1/2^n$ along most of its length. If h_W decreases along x as shown in Figure 3.8, then (P_W/P_I) is small and $\Delta(P_W/P_I)/\Delta x$ is large where sheet flow becomes stream flow and (P_W/P_I) is large and $\Delta(P_W/P_I)/\Delta x$ is small where stream flow becomes shelf flow, in which case the second term may be nil. For sheet flow, $(P_W/P_I) = 0$ for a frozen bed, $(P_W/P_I) << 1$ for a thawed bed, and $\tau_s = 0$. Equation (3.27) reduces to (Orowan, 1949):

$$\frac{\Delta h}{\Delta x} = \frac{\tau_o}{\rho_I\, g_z\, h_I} \tag{3.28}$$

For unconfined shelf flow, $(P_W/P_I) = 1$, $\dot{\varepsilon}_{xy} = \dot{\varepsilon}_{yy}$, and $\tau_s = \tau_o = \dot{\varepsilon}_{xy} = \dot{\varepsilon}_{xz} = 0$, so that $R = \sqrt{3}^{n-1}/3^n$. Equation (3.27) reduces to (Weertman, 1957b; Van der Veen, 1983):

$$\frac{\Delta h}{\Delta x} = \left(1 - \frac{\rho_I}{\rho_W}\right)\left\{\frac{a\, h_I}{a\, x + h_0\, u_0} - \frac{\sqrt{3}^{\,n-1}\, h_I^2}{a\, x + h_0\, u_0}\left[\frac{\rho_I\, g_z\, h_I}{6\, A}\left(1 - \frac{\rho_I}{\rho_W}\right)\right]^n\right\} \tag{3.29}$$

For confined shelf flow, pulling stress σ_P is resisted by back stress σ_B and P_W/P_I in Equation (3.25) is replaced by:

$$\frac{P_W}{P_I} = \left(\frac{\sigma_P - \sigma_B}{\sigma_P}\right)^{1/2} \tag{3.30}$$

On the ice shelf R expresses diverging and converging flow through $\dot{\varepsilon}_{yy}$, lateral shear and bending shear through $\dot{\varepsilon}_{xy}$, and shear over basal pinning points through $\dot{\varepsilon}_{xz}$, where $\dot{\varepsilon}_{xx}$, $\dot{\varepsilon}_{yy}$, $\dot{\varepsilon}_{xy}$, $\dot{\varepsilon}_{xz}$, and σ_B are evaluated theoretically and by field studies on confined ice shelves (Thomas, 1973a, 1973b; Thomas and MacAyeal, 1982; Jezek, 1984). In Equation (3.11), $\sigma_P = \sigma'_{xx}$ when $(P_W/P_I) = 1$, so Equation (3.25) is:

$$\frac{\Delta h}{\Delta x} = \left[\left(1 - \frac{\rho_I}{\rho_W}\right) - \left(2 + \frac{\dot{\varepsilon}_{yy}}{\dot{\varepsilon}_{xx}}\right)\frac{2\,\sigma_B}{\rho_I\, g_z\, h_I}\right]\left\{\frac{a\, w\, h_I}{a\, w\, x + w_0\, h_0\, u_0}\right.$$

$$- \frac{w\, h_I^2}{a\, w\, x + w_0\, h_0\, u_0}\left(R\, \cos\theta\left[\frac{\rho_I\, g_z\, h_I}{2\, A}\left(1 - \frac{\rho_I}{\rho_W}\right) - \frac{\sigma_B}{A}\right]^n\right.$$

$$\left. + \dot{\varepsilon}_{yy}\, \sin\theta + \left[\frac{\partial u_x}{\partial y}\right]\cos\theta + \left[\frac{\partial u_y}{\partial x}\right]\sin\theta\right) + \left.\left(\frac{h_I}{2\, w}\right)\tan\theta\right\}$$

$$+ \left(\frac{h_I}{w}\right)\tan\theta + \frac{2\,\tau_s}{\rho_I\, g_z\, w} \tag{3.31}$$

where $\tau_o = 0$ for floating ice and τ_s is caused by shear along the sides of the embayment that confines the ice shelf.

When ice is grounded, basal shear stress τ_o depends on the mass balance and is given by:

$$\tau_o = f_f\, \tau_C + f_t\, \tau_S + f_t\, \tau_D = (1 - f_t)\, \tau_C + f_t\, (\tau_S + \tau_D) \tag{3.32}$$

where for area A_z at the base of an ice column in Figure 3.8, f_f and f_t are the respective frozen and thawed fractions of the bed, τ_C is the basal shear stress for creep in ice overlying the bed, τ_S is the basal shear stress for basal sliding at the interface between ice and the bed, and τ_D is the basal shear stress for a deforming bed of sediments or till.

Variations of τ_o with distance x along a flowline of length L can be obtained from Equation (3.16) for mass-balance continuity and constitutive equations that relate τ_C, τ_S, and τ_D to ice velocity u_x. Equation (3.16) is integrated most easily for steady-state flow and constant a:

$$a\int_x^L dx = \int_{u_x h_I}^{0} d\,(u_x\, h_I) \tag{3.33}$$

where u_x is negative and $u_x = 0$ at $x = L$. Integrating Equation (3.33) for positive u_x and applying it to the accumulation zone, where a is also positive, gives:

$$a\,(L - x) = u_x\, h_I \tag{3.34}$$

Therefore, all quantities in Equation (3.34) are positive.

If the bed is frozen, all longitudinal motion of overlying ice is due to creep in the ice, so that $\tau_o = \tau_C$. At distance z above a horizontal bed, σ_{xz} replaces τ_o, and $h - z$ replaces h_I in Equation (3.28), so that:

$$\sigma_{xz} = \rho_I\, g_z\, (h - z)\, \Delta h/\Delta x \tag{3.35}$$

When σ_{xz} is the dominant deviator stress, following Equations (3.6), (3.20) and (3.21), $\dot{\varepsilon}_{xz}$ is given by:

$$\dot{\varepsilon}_{xz} = \frac{1}{2}\left(\frac{\partial u_x}{\partial z} + \frac{\partial u_z}{\partial x}\right) = \frac{1}{2}\frac{du_x}{dz} = \left(\frac{\sigma_c^{\,n-1}}{A^n}\right)\sigma_{xz}$$

$$= \left[\sigma'^{\,2}_{xx} + \sigma'_{xx}\,\sigma'_{yy} + \sigma'^{\,2}_{yy} + \sigma_{xy}^2 + \sigma_{yz}^2 + \sigma_{zx}^2\right]^{\frac{n-1}{2}}\left(\sigma_{xz}/A^n\right)$$

$$= \left[1 + \left(\frac{\dot{\varepsilon}_{xx}}{\dot{\varepsilon}_{xz}}\right)^2 + \left(\frac{\dot{\varepsilon}_{xx}\,\dot{\varepsilon}_{yy}}{\dot{\varepsilon}_{xz}^2}\right) + \left(\frac{\dot{\varepsilon}_{yy}}{\dot{\varepsilon}_{xz}}\right)^2 + \left(\frac{\dot{\varepsilon}_{xy}}{\dot{\varepsilon}_{xz}}\right)^2\right]^{\frac{n-1}{2}}\left(\frac{\sigma_{xz}}{A}\right)^n$$

$$= R_{xz}^{n-1}\left(\sigma_{xz}/A\right)^n \tag{3.36}$$

where $\partial u_x/\partial z >> \partial u_z/\partial x$ for laminar flow, $\sigma_{xz} = \sigma_{zx}$ by symmetry and σ_{yz} is nil. Equations (3.35) and (3.36) can be combined and integrated to give u_x:

$$u_x = \int_0^{u_x} du_x = \int_0^z 2\,\dot{\varepsilon}_{xz}\,dz = 2\int_0^z R_{xz}^{n-1}\left(\frac{\sigma_{xz}}{A}\right)^n dz$$

$$= 2\,R_{xz}^{n-1}\int_0^z\left[\frac{\rho_I\,g_z\,(h-z)\,\Delta h/\Delta x}{A}\right]^n dz$$

$$= 2\,R_{xz}^{n-1}\left[\frac{\rho_I\,g_z\,\Delta h/\Delta x}{A}\right]^n\left[\frac{h^{\,n+1}-(h-z)^{n+1}}{n+1}\right] \tag{3.37}$$

Taking u_C as the vertically averaged creep velocity:

$$u_C\,h = \int_0^h u_x\,dz \tag{3.38}$$

Substituting Equation (3.37) for u_x and integrating gives:

$$u_C = \frac{2\,R_{xz}^{n-1}\,h}{n+2}\left(\frac{\rho_I\,g_z\,h\,\Delta h/\Delta x}{A}\right)^n = \frac{2\,R_{xz}^{n-1}\,h}{n+2}\left(\frac{\tau_o}{A}\right)^n \tag{3.39}$$

Setting $u_C = u_x$ and $h = h_I$ in Equation (3.34) links u_C to a:

$$u_C = \frac{2\,R_{xz}^{n-1}\,h}{n+2}\left(\frac{\rho_I\,g_z\,h\,\Delta h/\Delta x}{A}\right)^n = \frac{a\,(L-x)}{h} \tag{3.40}$$

Setting $\Delta h/\Delta x = dh/dx$ and separating variables lead to:

$$\int_{h_E}^h h^{\,1+2/n}\,dh = \left(\frac{A}{\rho_I\,g_z}\right)\left[\frac{(n+2)\,a}{2\,R_{xz}^{n-1}}\right]^{1/n}\int_E^x (L-x)^{1/n}\,dx \tag{3.41}$$

Where $x = E$ and $h = h_E$ at the equilibrium line. Integrating Equation (3.41) and solving for h:

$$h = \left\{h_E^{\frac{2n+2}{n}} + \frac{2\,A}{\rho_I\,g_z}\left[\frac{(n+2)\,a}{2\,R_{xz}^{n-1}}\right]^{\frac{1}{n}}\left[(L-E)^{\frac{n+1}{n}} - (L-x)^{\frac{n+1}{n}}\right]\right\}^{\frac{n}{2n+2}} \tag{3.42}$$

Substituting this expression for h in Equation (3.40) and solving Equation (3.40) for surface slope $\Delta h/\Delta x$:

$$\frac{\Delta h}{\Delta x} = \frac{\dfrac{A}{\rho_I\,g_z}\left[\left(\dfrac{n+2}{2\,R_{xz}^{n-1}}\right)a\,(L-x)\right]^{\frac{1}{n}}}{\left\{h_E^{\frac{2n+2}{n}} + \dfrac{2\,A}{\rho_I\,g_z}\left[\dfrac{(n+2)\,a}{2\,R_{xz}^{n-1}}\right]^{\frac{1}{n}}\left[(L-E)^{\frac{n+1}{n}} - (L-x)^{\frac{n+1}{n}}\right]\right\}^{\frac{n+2}{2n+2}}} \tag{3.43}$$

Setting $\tau_o = \tau_C$ and solving Equation (3.28) using these expressions for h and $\Delta h/\Delta x$ gives:

$$\tau_C = \frac{A\left[\left(\dfrac{(n+2)\,a}{2\,R_{xz}^{n-1}}\right)\right]^{\frac{1}{n}}(L-x)^{\frac{1}{n}}}{\left\{h_E^{\frac{2n+2}{n}} + \dfrac{2\,A}{\rho_I\,g_z}\left[\dfrac{(n+2)\,a}{2\,R_{xz}^{n-1}}\right]^{\frac{1}{n}}\left[(L-E)^{\frac{n+1}{n}} - (L-x)^{\frac{n+1}{n}}\right]\right\}^{\frac{1}{n+1}}} \tag{3.44}$$

Equation (3.44) applies for sheet flow over a frozen bed.

If motion of the overlying ice is due mainly to basal sliding, then $\tau_o \approx \tau_S$. Sliding is resisted primarily at "sticky patches" where bedrock outcrops through saturated sediments or till that do not retard sliding. Sliding occurs by a combination of creep and regelation around bedrock bumps having a critical average dimension Λ spaced distance Λ' apart, so that Λ/Λ' is a bed-roughness parameter. A sediment or till layer of thickness λ produces unburied height $\Lambda - \lambda$ of critical bedrock projections that retard sliding. A reduced bed-roughness parameter $(\Lambda - \lambda)/\Lambda'$ is related to P_W/P_I as follows:

$$\frac{\Lambda-\lambda}{\Lambda'} = \frac{\Lambda}{\Lambda'}\left(1-\frac{P_W}{P_I}\right) \tag{3.45}$$

where $\lambda \to 0$ as $P_W \to 0$ and $\lambda \to \Lambda$ as $P_W \to P_I$. Sliding velocity u_S for this situation is derived in Chapter 4 and is:

$$u_S = \left[\frac{\rho_I\,g_z\,h_I\,\Delta h/\Delta x}{B\,(1-P_W/P_I)^2}\right]^m = \left[\frac{\tau_o}{B\,(1-P_W/P_I)^2}\right]^m \tag{3.46}$$

where m is a viscoplastic parameter and B is a basal sliding parameter that includes bed roughness and the hardness of temperate basal ice. In classical sliding theory, $m = \frac{1}{2}(n+1)$ and $B = B_o\,(\Lambda/\Lambda')^2$, where B_o includes ice hardness (Weertman, 1957a). Equation (3.46) reduces to the classical sliding law for ice over bedrock when $\lambda = 0$, because then $P_W << P_I$. Equation (3.46) is indeterminate when $\lambda \geq \Lambda$, because $P_W \to P_I$ and $\tau_o \to 0$ when bedrock is entirely blanketed by water-saturated sediments or till. In general, $P_W << P_I$ for sheet flow, $0 < P_W < P_I$ for stream flow, and $P_W = P_I$ for shelf flow. Setting $u_S = u_x$ in Equation (3.34), and obtaining u_S from Equation (3.46), with $h_I = h$ for a horizontal bed:

$$u_S = \left[\frac{\rho_I\,g_z\,h\,\Delta h/\Delta x}{B\,(1-P_W/P_I)^2}\right]^m = \frac{a\,(L-x)}{h} \tag{3.47}$$

Setting $\Delta h/\Delta x = dh/dx$ and separating variables lead to:

$$\int_{h_E}^h h^{\,1+1/m}\,dh = \frac{a^{1/m}\,B}{\rho_I\,g_z}\left(1-\frac{P_W}{P_I}\right)^2\int_E^x (L-x)^{1/m}\,dx \tag{3.48}$$

Where $x = E$ and $h = h_E$ at the equilibrium line. Integrating Equation (3.48) and solving for h:

$$h = \left\{h_E^{\frac{2m+1}{m}} + \left(\frac{2\,m+1}{m+1}\right)\frac{a^{1/m}\,B}{\rho_I\,g_z}\left(1-\frac{P_W}{P_I}\right)^2\right.$$

$$\left[(L-E)^{\frac{m+1}{m}} - (L-x)^{\frac{m+1}{m}}\right]\Big\}^{\frac{m}{2m+1}} \quad (3.49)$$

Substituting this expression for h in Equation (3.47) and solving Equation (3.47) for surface slope $\Delta h/\Delta x$:

$$\frac{\Delta h}{\Delta x} = \frac{A^{1/m} B}{\rho_I g_z}\left(1 - \frac{P_W}{P_I}\right)^2 (L-x)^{1/m}$$

$$\div \left\{ h_E^{\frac{2m+1}{m}} + \left(\frac{2m+1}{m+1}\right)\frac{a^{1/m} B}{\rho_I g_z}\left(1 - \frac{P_W}{P_I}\right)^2 \right.$$

$$\left.\left[(L-E)^{\frac{m+1}{m}} - (L-x)^{\frac{m+1}{m}}\right]\right\}^{\frac{m+1}{2m+1}} \quad (3.50)$$

Setting $\tau_o = \tau_S$ and solving Equation (3.28) using these expressions for h and $\Delta h/\Delta x$ gives:

$$\tau_S = a^{1/m} B\left(1 - \frac{P_W}{P_I}\right)^2 (L-x)^{1/m}$$

$$\div \left\{ h_E^{\frac{2m+1}{m}} + \left(\frac{2m+1}{m+1}\right)\frac{a^{1/m} B}{\rho_I g_z}\left(1 - \frac{P_W}{P_I}\right)^2 \right.$$

$$\left.\left[(L-E)^{\frac{m+1}{m}} - (L-x)^{\frac{m+1}{m}}\right]\right\}^{\frac{1}{2m+1}} \quad (3.51)$$

Taking $x = E = 0$ and h_E at the grounding line, Equation (3.51) should apply to Ice Stream B in Antarctica, where basal sliding dominates (Kamb, 1996).

If nearly all longitudinal motion of the overlying ice is due to deformation of the bed, then $\tau_o \approx \tau_D$. Deformation in a basal layer of thickness λ_D of sediment or till need not be homogeneous because their properties need not be uniform, especially the water content. Apart from randomly distributed discrete shear bands, which can develop, the most general expression for the variation of longitudinal velocity u_x with vertical distance z above the base of a deforming layer of thickness λ_D is:

$$u_x = u_D\,(z/\lambda_D)^c \quad (3.52)$$

where $u_x = u_D$ at the top of the layer and c is a creep exponent that varies over the range $0 \le c \le \infty$. Shear gradient du_x/dz is confined to the bottom of the layer when $c = 0$, is confined to the top of the layer when $c = \infty$, is constant through the layer when $c = 1$, and varies through the layer for other values of c. Equation (3.52) is plotted in Figure 3.9.

Assume that laminar flow in a layer of sediment or till having variable wetness, and deforming through thickness λ_D according to Equation (3.52), obeys a flow law of the form:

$$\dot\varepsilon_{xz} = \frac{1}{2}\left(\frac{\partial u_x}{\partial z} + \frac{\partial u_z}{\partial x}\right) = \frac{1}{2}\left(\frac{du_x}{dz}\right) = \frac{1}{2}\frac{d}{dz}\left[u_D\left(\frac{z}{\lambda_D}\right)^c\right] = \left(\frac{c\,u_D}{2\,\lambda_D^c}\right) z^{c-1}$$

$$= \dot\varepsilon_c\left(\frac{\sigma_{xz}}{\sigma_c}\right)^c = \dot\varepsilon_c\left[\frac{\tau_o}{\sigma_o(1 - P_W/P_I)}\right]^c \quad (3.53)$$

where $\dot\varepsilon_c$ is the effective creep strain rate, σ_c is the effective creep stress, basal shear stress $\sigma_{xz} = \tau_o = \tau_D$ is effectively constant through λ_D because $\lambda_D << h_I$, and σ_o is the yield stress of dry sediment or till in layer λ_D such that $\sigma_c = \sigma_o\,(1 - P_W/P_I)$. The sediment or till is dry when the water table is at the bottom of λ_D, in which case $c \to 0$, $P_W/P_I \to 0$, and $\sigma_c \to \sigma_o$. It is soaked when the water table is at the top of λ_D, in which case case $c \to \infty$, $P_W/P_I \to 1$, and $\sigma_c \to 0$. It is wet but not soaked when $c \to 1$, $0 < P_W/P_I < 1$, and $0 < \sigma_c < \sigma_o$.

Solving Equation (3.53) for u_D when $z = \lambda_D$ gives:

$$u_D = \frac{2\,\lambda_D\,\dot\varepsilon_c}{c}\left(\frac{\tau_o}{\sigma_c}\right)^c = 2\,\lambda_D\left[\frac{\tau_o}{C\,(1 - P_W/P_I)}\right]^c \quad (3.54)$$

where $C = \sigma_o\,(c/\dot\varepsilon_c)^{1/c}$ is a creep parameter. When layer λ_D is soaked, it cannot support a basal shear stress, so $\tau_o \to 0$ and u_D is indeterminate in Equation (3.54). This is also true when the top of layer λ_D is soaked due to basal melting of ice, but $0 < P_W/P_I < 1$ through λ_D because low permeability in the layer and dilation caused by basal sliding keep basal meltwater at the ice-bed interface, just as troweling concrete keeps water at the top surface. In these cases, the velocity of basal ice is controlled by $\dot\varepsilon_{xx}$ when $P_W/P_I \to 1$

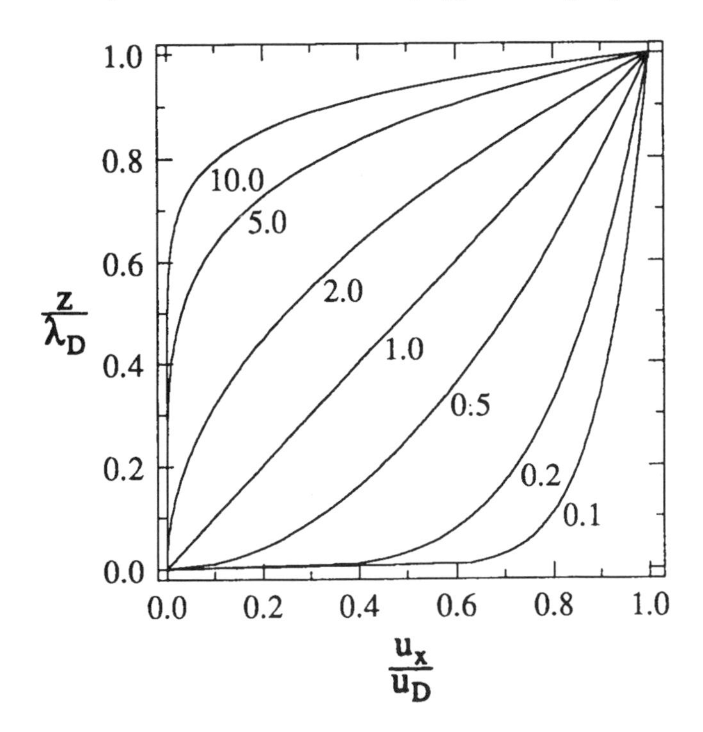

Figure 3.9: Variations of longitudinal velocity u_x with height z through a deforming layer of subglacial sediment or till according to Equation (3.52). Normalized velocity u_x/u_D varies through normalized thickness z/λ_D according to $(u_x/u_D) = (z/\lambda_D)^c$. Curves are shown for selected values of c over the range $0 \le c \le \infty$ that give velocity u_D at the top of thickness λ_D. Plot by Kirk Maasch.

in Equation (3.22) for a floating ice shelf, or by u_S when $P_W/P_I < 1$ in Equation (3.46) because bedrock outcrops through λ_D and creates "sticky spots" that retard basal sliding. Equations (3.53) and (3.54) are entirely intuitive. They are proposed for the purpose of modeling flowline profiles. Physically based models for deforming beds are still tentative and not fully tested by field data. They take their point of departure from soil mechanics (e.g., Iverson, 1985, 1986a, 1986b; Kamb, 1991; Iverson and others, 1995, 1996; Hooke and Iverson, 1995; Hooke and others, 1997; Hooke, 1998).

Setting $u_D = u_x$ in Equation (3.34) and $\tau_o = \rho_I g_z h \Delta h/\Delta x$ in Equation (3.54) gives:

$$u_D = 2\lambda_D \left|\frac{\rho_I g_z h \Delta h/\Delta x}{C(1 - P_W/P_I)}\right|^c = \frac{a(L-x)}{h} \tag{3.55}$$

Setting $\Delta h/\Delta x = dh/dx$ and separating variables gives:

$$\int_{h_E}^{h} h^{1+\frac{1}{c}}\, dh = \frac{C(1-P_W/P_I)}{\rho_I g_z}\left(\frac{a}{2\lambda_D}\right)^{\frac{1}{c}} \int_E^x (L-x)^{\frac{1}{c}}\, dx \tag{3.56}$$

Integrating Equation (3.56) and solving for h:

$$h = \left\{ h_E^{\frac{2c+1}{c}} + \left(\frac{2c+1}{c+1}\right)\frac{C(1-P_W/P_I)}{\rho_I g_z}\left(\frac{a}{2\lambda_D}\right)^{\frac{1}{c}} \left[(L-E)^{\frac{c+1}{c}} - (L-x)^{\frac{c+1}{c}}\right]\right\}^{\frac{c}{2c+1}} \tag{3.57}$$

Substituting Equation (3.57) for h in Equation (3.55) and solving for $\Delta h/\Delta x$:

$$\frac{\Delta h}{\Delta x} = \frac{C}{\rho_I g_z}\left(1-\frac{P_W}{P_I}\right)\left(\frac{a}{2\lambda_D}\right)^{\frac{1}{c}}(L-x)^{\frac{1}{c}} \div \left\{ h_E^{\frac{2c+1}{c}} + \left(\frac{2c+1}{c+1}\right)\frac{C}{\rho_I g_z}\left(1-\frac{P_W}{P_I}\right)\left(\frac{a}{2\lambda_D}\right)^{\frac{1}{c}} \left[(L-E)^{\frac{c+1}{c}} - (L-x)^{\frac{c+1}{c}}\right]\right\}^{\frac{c+1}{2c+1}} \tag{3.58}$$

Setting $\tau_o = \tau_D$ and solving Equation (3.28) using these expressions for $h = h_I$ and $\Delta h/\Delta x$ gives:

$$\tau_D = C\left(1-\frac{P_W}{P_I}\right)\left(\frac{a}{2\lambda_D}\right)^{\frac{1}{c}}(L-x)^{\frac{1}{c}} \div \left\{ h_E^{\frac{2c+1}{c}} + \left(\frac{2c+1}{c+1}\right)\frac{C}{\rho_I g_z}\left(1-\frac{P_W}{P_I}\right)\left(\frac{a}{2\lambda_D}\right)^{\frac{1}{c}} \left[(L-E)^{\frac{c+1}{c}} - (L-x)^{\frac{c+1}{c}}\right]\right\}^{\frac{1}{2c+1}} \tag{3.59}$$

Alley (1992) noted that sliding under ice streams is controlled by rheology of a deformable substrate if the substrate has a high creep strength, but otherwise sliding is controlled by the size and spacing of bedrock bumps that poke through the substrate and penetrate basal ice.

Alternatively, if deforming thickness λ_D of sediment or till is fraction ϕ_D of ice thickness h_I, so that $\lambda_D = \phi_D h_I$, then Equation (3.55) becomes:

$$u_D = \phi_D h_I\left[\frac{\tau_o}{C(1-P_W/P_I)}\right]^c = \phi_D h_I\left[\frac{\rho_I g_z h_I \Delta h/\Delta x}{C(1-P_W/P_I)}\right]^c = \frac{a(L-x)}{h_I} \tag{3.60}$$

Proceeding as in the other cases, the ice thickness is:

$$h = \left\{ h_E^{\frac{2c+2}{c}} + \left(\frac{2c+2}{c+1}\right)\frac{C}{\rho_I g_z}\left(1-\frac{P_W}{P_I}\right)\left(\frac{a}{2\phi_D}\right)^{\frac{1}{c}} \left[(L-E)^{\frac{c+1}{c}} - (L-x)^{\frac{c+1}{c}}\right]\right\}^{\frac{c}{2c+2}} \tag{3.61}$$

The surface slope is:

$$\frac{\Delta h}{\Delta x} = \frac{C}{\rho_I g_z}\left(1-\frac{P_W}{P_I}\right)\left(\frac{a}{2\phi_D}\right)^{\frac{1}{c}}(L-x)^{\frac{1}{c}} \div \left\{ h_E^{\frac{2c+2}{c}} + \left(\frac{2c+2}{c+1}\right)\frac{C}{\rho_I g_z}\left(1-\frac{P_W}{P_I}\right)\left(\frac{a}{2\phi_D}\right)^{\frac{1}{c}} \left[(L-E)^{\frac{c+1}{c}} - (L-x)^{\frac{c+1}{c}}\right]\right\}^{\frac{c+2}{2c+2}} \tag{3.62}$$

Substituting for $h = h_I$ and $\Delta h/\Delta x$ for $\tau_o = \tau_D$ in Equation (3.28):

$$\tau_D = C\left(1-\frac{P_W}{P_I}\right)\left(\frac{a}{2\phi_D}\right)^{\frac{1}{c}}(L-x)^{\frac{1}{c}} \div \left\{ h_E^{\frac{2c+2}{c}} + \left(\frac{2c+2}{c+1}\right)\frac{C}{\rho_I g_z}\left(1-\frac{P_W}{P_I}\right)\left(\frac{a}{2\phi_D}\right)^{\frac{1}{c}} \left[(L-E)^{\frac{c+1}{c}} - (L-x)^{\frac{c+1}{c}}\right]\right\}^{\frac{1}{c+1}} \tag{3.63}$$

Equation (3.63) is useful when λ_D is unknown beneath a marine ice sheet, but glacial erosion presumably causes λ_D to decrease with the decrease of h and the increase of u_D from the ice divide to the grounding line.

Equations (3.44), (3.51), and either (3.59) or (3.63) for τ_C, τ_S, and τ_D in Equation (3.32), thereby give the dependence of τ_o on a in the accumulation zone of the flowline. Equation (3.27) then becomes:

$$\frac{\Delta h}{\Delta x} = \left(1-\frac{\rho_I}{\rho_W}\right)\left(\frac{P_W}{P_I}\right)^2 \left\{\frac{a\, h_I}{a\, x + h_0 u_0} - \frac{h_I^2 R}{a\, x + h_0 u_0}\right.$$

$$\left[\frac{\rho_I\, g_z\, h_I}{2\,A}\left(1-\frac{\rho_I}{\rho_W}\right)\left(\frac{P_W}{P_I}\right)^2\right]^n\Bigg\}$$

$$+\, h_I\left(1-\frac{\rho_I}{\rho_W}\right)\left(\frac{P_W}{P_I}\right)\frac{\Delta}{\Delta x}\left(\frac{P_W}{P_I}\right)+\frac{2\,\tau_s}{\rho_I\, g_z\, w}$$

$$+\,\frac{f_f\,\phi_f\,A}{\rho_I\, g_z\, h_I}\left[\frac{(n+2)\,a}{2\,R_{xz}^{n-1}}\right]^{\frac{1}{n}}(L-x)^{\frac{1}{n}}$$

$$\div\left\{h_E^{\frac{2n+2}{n}}+\frac{2\,A}{\rho_I\, g_z}\left[\frac{(n+2)\,a}{2\,R_{xz}^{n-1}}\right]^{\frac{1}{n}}\left[(L-E)^{\frac{n+1}{n}}-(L-x)^{\frac{n+1}{n}}\right]\right\}^{\frac{1}{n+1}}$$

$$+\,\frac{f_t\,\phi_t\,a^{1/m}\,B}{\rho_I\, g_z\, h_I}\left(1-\frac{P_W}{P_I}\right)^2(L-x)^{1/m}$$

$$\div\Bigg\{h_E^{\frac{2m+2}{m}}+\left(\frac{2m+1}{m+1}\right)\frac{a^{1/m}\,B}{\rho_I\, g_z}\left(1-\frac{P_W}{P_I}\right)^2$$

$$\left[(L-E)^{\frac{m+1}{m}}-(L-x)^{\frac{m+1}{m}}\right]\Bigg\}^{\frac{1}{2m+1}}$$

$$+\,\frac{f_t\,\phi_t\,C}{\rho_I\, g_z\, h_I}\left(1-\frac{P_W}{P_I}\right)\left(\frac{a}{2\,\lambda_D}\right)^{\frac{1}{c}}(L-x)^{\frac{1}{c}}$$

$$\div\Bigg\{h_E^{\frac{2c+2}{c}}+\left(\frac{2c+1}{c+1}\right)\frac{C}{\rho_I\, g_z}\left(1-\frac{P_W}{P_I}\right)\left(\frac{a}{2\,\lambda_D}\right)^{\frac{1}{c}}$$

$$\left[(L-E)^{\frac{c+1}{c}}-(L-x)^{\frac{c+1}{c}}\right]\Bigg\}^{\frac{1}{2c+1}} \qquad (3.64)$$

where Equation (3.59) is used for τ_D since λ_D may not depend on h_I, and factors ϕ_f and ϕ_t for frozen and thawed beds, respectively, include the effect on $\Delta h/\Delta x$ of variable accumulation rate a and flowband width w along L for sheet flow, as shown in Chapter 5, for which $0 \le P_W/P_I << 1$.

Equation (3.64) applies only to the accumulation zone of an ice sheet. In Figure 3.8, the accumulation zone typically extends beyond the grounding line of a floating ice shelf and ends at a calving ice wall, but an ice lobe typically lies in the ablation zone. Nye (1952a) showed how the surface profile of a flowline varied with bed slope β, provided that basal shear stress τ_o is constant. His analysis can be applied to the ablation zone of ice lobes. Let direction x be parallel to a bed of slope β, with $x = 0$ at the lobe terminus, and direction z be normal to the bed, with $z = 0$ at the bed. Balancing longitudinal gravity forces on opposite sides of an ice column normal to the bed gives:

$$(F_x)_G = \left[\tfrac{1}{2}\,\rho_I\, g_z\,(h+\Delta h)\right]\left[w\,(h+\Delta h)\right]$$

$$-\left[\tfrac{1}{2}\,\rho_I\, g_z\, h\right]\left[w\, h\right]+\left[\rho_I\, g_x\, w\left(h+\tfrac{1}{2}\,\Delta h\right)\Delta x\right]$$

$$\approx \rho_I\, g\cos\beta\; w\, h\,\Delta h+\rho_I\, g\sin\beta\; w\, h\,\Delta x \qquad (3.65)$$

where the first two terms are the respective longitudinal gravitational forces on the upslope and downslope sides of the ice column, and the third term is the longitudinal gravitational body force acting on volumn $w\,(h + \tfrac{1}{2}\,\Delta h)\,\Delta x$ of the ice column, w is column width, $h + \tfrac{1}{2}\,\Delta h \approx h$ is mean column height along z, Δx is column length along x, $g_x = g \sin\beta \approx g\,\beta$ and $g_z = g \cos\beta \approx g$ are the respective components of gravity acceleration g resolved along directions x and z, ρ_I is ice density, and $\alpha = \beta + \Delta h/\Delta x$ is ice-surface slope. Longitudinal kinematic force $(F_x)_K$ from the ice motion induced by $(F_x)_G$ is:

$$(F_x)_K = \tau_o\, w\, \Delta x \qquad (3.66)$$

where basal shear stress τ_o is constant and acts on basal area $w\,\Delta x$.

Setting $(F_x)_G = (F_x)_K$ and solving for $\Delta h/\Delta x$ gives:

$$\alpha = \frac{\Delta h}{\Delta x}+\beta = \frac{\tau_o}{\rho_I\, g\, h} \qquad (3.67)$$

Letting $dh/dx = \Delta h/\Delta x$ and and $h_o = \tau_o/\rho_I\, g$, Equation (3.67) can be integrated as follows:

$$x = \int_0^x dx = \int_0^h \frac{h\, dh}{h_o - h\,\beta}$$

$$= \left[-\frac{h}{\beta}-\frac{h_o}{\beta^2}\, ln\,(-h\,\beta+h_o)\right]-\left[-\frac{h_o}{\beta^2}\, ln\,(h_o)\right]$$

$$= \frac{h_o}{\beta^2}\, ln\left(\frac{h_o}{h_o-h\,\beta}\right)-\frac{h}{\beta} = \frac{\tau_o}{\rho_I\, g\,\beta^2}\, ln\left(\frac{\tau_o}{\tau_o-\rho_I\, g\, h\,\beta}\right)-\frac{h}{\beta} \qquad (3.68)$$

If the bed is horizontal, $\beta = 0$ and Equation (3.67) can be integrated to give for the ablation zone:

$$x = \int_0^x dx = \left(\frac{\rho_I\, g_z}{\tau_o}\right)\int_0^h h\, dh = \left(\frac{\rho_I\, g_z}{2\,\tau_o}\right)h^2 \qquad (3.69)$$

Note that bed slope β has an important effect on flowline profiles.

In the accumulation zone, Reeh (1982) showed that when τ_o, a, and w are constant along a flowline, τ_o can be related to accumulation rate a. At the ice divide, $x = L$ and $h = h_L$ so Equation (3.40) for the mass balance can be integrated as follows, setting $R_{xz} = 1$ and $\Delta h/\Delta x = dh/dx$:

$$\left(\frac{2}{n+2}\right)^{1/n}\frac{\rho_I\, g_z}{A}\int_0^{h_L} h^{1+2/n}\, dh = a^{1/n}\int_0^L (L-x)^{1/n}\, dx \qquad (3.70)$$

Integrating:

$$\left(\frac{2}{n+2}\right)^{1/n}\frac{\rho_I\, g_z\, h_L^{2+2/n}}{A\,(2+2/n)} = \frac{a^{1/n}\, L^{1+1/n}}{1+1/n} \qquad (3.71)$$

Equation (3.71) for the mass balance can be solved for h_L^2/L:

$$\frac{h_L^2}{L} = \left[\frac{(n+2)^{1/n}\, 2^{1-1/n}\, a^{1/n} A}{\rho_I\, g_z}\right]^{\frac{n}{n+1}} \tag{3.72}$$

Equation (3.69) for the force balance can also be solved for h_L^2/L when $h = h_L$ at $x = L$:

$$\frac{h_L^2}{L} = \frac{2\,\tau_o}{\rho_I\, g_z} \tag{3.73}$$

Equating Equations (3.72) and (3.73), and solving for τ_o:

$$\tau_o = \left[\tfrac{1}{4}(n+2)\,\rho_I\, g_z\, a\, A\right]^{\frac{1}{n+1}} \tag{3.74}$$

Since $n \rightarrow \infty$ when τ_o is constant, the dependence of τ_o on a is nil.

Equation (3.64) was derived for an ice sheet on a horizontal bed so the physical processes could be presented simply, without obscuring the physics with the additional mathematics that must be introduced when the bed is not horizontal. A non-horizontal bed is virtually guaranteed by: (1) glacioisostatic downwarping of the bed by the weight of overlying ice and (2) topography caused by landscape evolution linked to tectonic, erosional, and depositional processes. These complications can now be considered in Equation (3.64), because surface slope $\alpha = \Delta h/\Delta x$ when x is horizontal and z is vertical, and bed slope β can be included because ice thickness h_I is distinguished from ice elevation h.

Glacioisostatic adjustments along an ice-sheet flowline

Glacioisostatic downwarping of an initially horizontal bed beneath an ice sheet can be expressed by isostasy ratio r. Assume that bed height h_M at the ice-sheet margin is not depressed, but at distance x from the ice margin along a flowline, isostatic sinking under the ice load will lower bedrock height above or depth below sea level from h_R to h_R^*, and this will lower the ice surface from h to h^*. These adjustments are described by r as follows:

$$r = \frac{h_R - h_R^*}{h^* - h_R} = \frac{h_M - h_R^*}{h^* - h_M} \tag{3.75}$$

where $h_R = h_M$ for an initially horizontal bed. As shown in Chapter 5, whether an ice margin ends on land or in water does not change how r is incorporated into Equation (3.64), but the mathematical treatment is simplest for sheet flow ending on land. In this case, the longitudinal gravitational force $(F_x)_G$ at x is:

$$(F_x)_G = \left[\tfrac{1}{2}\rho_I\, g_z\,(h - h_M)\right][h - h_M] = \tfrac{1}{2}\rho_I\, g_z\,(h - h_M)^2 \tag{3.76}$$

After isostatic sinking, using Equation (3.75):

$$(F_x)_G = \left[\tfrac{1}{2}\rho_I\, g_z\,(h^* - h_R^*)\right][h^* - h_M] = \tfrac{1}{2}\rho_I\, g_z\,(1 + r)(h^* - h_M)^2 \tag{3.77}$$

Equating Equations (3.76) and (3.77) and solving for h:

$$h = h_M + (1 + r)^{1/2}(h^* - h_M) \tag{3.78}$$

Since r changes more slowly along x than does h or h^*, assume that r can be treated as a constant in distance Δx. If distance x consists of i increments of Δx, where i is an integer such that $i = x/\Delta x$, ice height change Δh in length Δx in Equation (3.64) can be written:

$$\begin{aligned}\Delta h = h_{i+1} - h_i &= \left[h_M + (1 + r_{i+1})^{1/2}(h_{i+1}^* - h_M)\right] \\ &\quad - \left[h_M + (1 + r_i)^{1/2}(h_i^* - h_M)\right] \\ &\approx (1 + r)^{1/2}(h_{i+1}^* - h_i^*) = (1 + r)^{1/2}\,\Delta h^*\end{aligned} \tag{3.79}$$

Incorporating Equation (3.79) into Equation (3.64), but using the more compact form given by Equation (3.27), where $h_I = h$ for an initially horizontal bed:

$$\begin{aligned}\frac{\Delta h^*}{\Delta x} = \frac{(1 - \rho_I/\rho_W)}{(1+r)^{1/2}}\left(\frac{P_W}{P_I}\right)^2 &\left\{\frac{a\left[h_M + (1+r)^{1/2}(h^* - h_M)\right]}{a\,x + h_0\, u_0}\right. \\ &\left. - \frac{R\left[h_M + (1+r)^{1/2}(h^* - h_M)\right]^{n+2}}{a\,x + h_0\, u_0}\left[\frac{\rho_I\, g_z}{2A}\left(1 - \frac{\rho_I}{\rho_W}\right)\left(\frac{P_W}{P_I}\right)^2\right]^n\right\} \\ &+ \frac{\left[h_M + (1+r)^{1/2}(h^* - h_M)\right]}{(1+r)^{1/2}}\left(1 - \frac{\rho_I}{\rho_W}\right)\left(\frac{P_W}{P_I}\right)\frac{\Delta}{\Delta x}\left(\frac{P_W}{P_I}\right) \\ &+ \frac{2\,\tau_s}{(1+r)^{1/2}\rho_I\, g_z\, w} \\ &+ \frac{\tau_o}{(1+r)^{1/2}\rho_I\, g_z\left[h_M + (1+r)^{1/2}(h^* - h_M)\right]}\end{aligned} \tag{3.80}$$

where $\Delta h^*/\Delta x = (h^*_{i+1} - h_i^*)/\Delta x$ and h_I in Equation (3.27) is given by Equation (3.78) for $h = h_I$.

The bed stops sinking when $r = r_o$ for glacioisostatic equilibrium. In this case, vertical gravitational force $(F_z)_G$ at depth $h_R - h_R^*$ below the unglaciated horizontal bed is related to r_o through Equation (3.75) and density ρ_R of rocks:

$$(F_z)_G = \rho_R\, g_z\,(h_R - h_R^*) = \rho_R\, g_z\, r_o\,(h^* - h_R) \tag{3.81}$$

When isostatic equilibrium is attained for the downwarped bed beneath the ice sheet, $(F_z)_G$ is given by setting $r = r_o$ in Equation (3.75), where $h_R = h_M$ at the undepressed ice margin:

$$(F_z)_G = \rho_I\, g_z\,(h^* - h_R^*) = \rho_I\, g_z\,(1 - r_o)(h^* - h_R) \tag{3.82}$$

Equating Equations (3.81) and (3.82) for glacioisostatic equilibrium and solving for r_o:

$$r_o = \frac{\rho_I}{\rho_R - \rho_I} \tag{3.83}$$

Mantle density is used for ρ_R because the ice sheet only downwarps crustal rock, whereas downwarping displaces mantle rock horizontally from beneath the ice sheet. Taking $\rho_R = 3600$ kg m^{-1} for mantle rock and $\rho_I = 917$ kg m^{-1} for glacial ice gives $r_o \approx 1/3$ in Equation (3.83). The range of glacioisostatic corrections in Equation (3.78) is therefore $1 \leq (1 + r)^{1/2} \leq 2/\sqrt{3}$ along an ice-sheet flowline, assuming that $r = 0$ at the ice margin and $r = r_o$ at the ice divide. This supports the assumption that r varies slowly along the flowline, compared to the variation of h^*.

Variations of bed topography along an ice-sheet flowline

Bed topographic variations along ice-sheet flowlines are incorporated into Equations (3.27) and (3.64) by making the substitution:

$$h_I = h - h_R \tag{3.84}$$

where h_R is the bed height above (positive h_R) or depth below (negative h_R) sea level, and h_R is obtained from digitized present-day topographic or bathymetric maps, assuming isostatic equilibrium exists. Equation (3.80) then becomes:

$$\begin{aligned}\frac{\Delta h^*}{\Delta x} = \frac{(1+\rho_I/\rho_W)}{(1+r)^{1/2}}\left(\frac{P_W}{P_I}\right)^2 \Bigg\{ & \frac{a\left[h_M + (1+r)^{1/2}(h^* - h_M) - h_R\right]}{a\,x + h_0 u_0} \\ & - \frac{R^{n-1}\left[h_M + (1+r)^{1/2}(h^* - h_M) - h_R\right]^{n+2}}{a\,x + h_0 u_0} \\ & \left[\frac{\rho_I g_z}{2A}\left(1 - \frac{\rho_I}{\rho_W}\right)\left(\frac{P_W}{P_I}\right)^2\right]^n \Bigg\} \\ & + \frac{\left[h_M + (1+r)^{1/2}(h^* - h_M) - h_R\right]}{(1+r)^{1/2}} \\ & \left(1 - \frac{\rho_I}{\rho_W}\right)\left(\frac{P_W}{P_I}\right)\frac{\Delta}{\Delta x}\left(\frac{P_W}{P_I}\right) \\ & + \frac{2\,\tau_s}{(1+r)^{1/2}\rho_I g_z w} \\ & + \frac{\tau_o}{(1+r)^{1/2}\rho_I g_z\left[h_M + (1+r)^{1/2}(h^* - h_M) - h_R\right]}\end{aligned} \tag{3.85}$$

Equation (3.85) can be integrated using various numerical methods. The simplest is the Euler method, in which $\Delta h^* = h_{i+1}^* - h_i^*$ and variables along x that must be specified for each step $i = x/\Delta x$ include P_W/P_I, a, h_R, R, A, w, τ_s, and τ_o. In deriving Equation (3.85), a, A, and w were taken as constants, and no variation of τ_s was considered. Euler solutions convert Equation (3.85) into an initial-value, finite-difference, recursive formula, in which variables are specified at $i = 0$ for the initial value, Δx is the finite-difference between steps i and i + 1, and h^* is calculated at step i + 1 and all subsequent steps in a recursive manner until step $i = L/\Delta x$ at the ice divide of a flowline of length L. Cumulative errors inherent in calculating h^* using the Euler method are minimized by keeping Δx small.

Evaluating conditions at grounded ice-sheet margins

Ice sheets have terrestrial, marine, and lacustrine margins. For terrestrial margins, $h_I^* = h^* - h_R^* = 0$, so that $\Delta h^*/\Delta x$ is infinite in Equation (3.85). Therefore, Equations (3.68) or (3.69) are used to compute h along x in the ablation zone, where ice is thin so that $h \approx h^*$ because $r \approx 0$. Then $i = 0$ is assigned to the equilibrium line, where $x = E$ and $h = h_E$ so that Equation (3.69) gives as the initial value in Equation (3.85) for a horizontal bed:

$$h_E = \left(\frac{2\,\tau_o E}{\rho_I g_z}\right)^{1/2} \tag{3.86}$$

where E is a specified distance from terminal moraines (Clark, 1992). For marine margins, the most convenient location for $i = 0$ is the grounding line of ice shelves, since P_W/P_I satisfies the buoyancy requirement in Equation (3.3). Equation (3.3) is then used to obtain ice elevation h_G at the grounding line, for which $i = 0$ in Equation (3.85):

$$h_G = h_I - h_W = h_W\left(\frac{\rho_W}{\rho_I} - 1\right) \tag{3.87}$$

where h_W is known from bathymetric data at sites of characteristic grounding-line glacial deposits. For marine or lacustrine margins, when an ice wall is grounded in water of variable depth, h^* at $i = 0$ in Equation (3.85) is typically from 30 m to 70 m above the water surface (Brown and others, 1982).

Summarizing one-dimensional ice-sheet flowline models

One-dimensional ice-sheet flowline models present the essential physics of ice sheets without excessive mathematics, so that changes over time of ice-sheets along specified flowlines can be understood. Although the one-dimensional model presented here is for steady-state flow, it can be made time-dependent by specifying that $\partial h_I/\partial t \neq 0$ in Equation (3.14). Even the steady-state solution provides an empirical way for studying variations over time.

The most important variable introduced by the force balance is the basal buoyancy factor P_W/P_I. As seen in Equation (3.64), the flowline surface profile converts from concave to convex to nearly zero as $P_W/P_I \approx 0$ for sheet flow increases toward unity to produce ice streams during stream flow, and becomes $P_W/P_I \approx 1$ for shelf flow. Therefore, the three major flow regimes of an ice sheet can be specified as variations in this single parameter which reflects the single physical process of melting and freezing at the bed, which constitutes the basal mass balance along the flowline. The surge cycle of an ice stream can be mimicked by allowing P_W/P_I to vary with time t over the range from zero to unity. The surface profile of a flowline reveals the surge as the length of the flowline along which the surface is concave, which is determined by the variation of P_W/P_I along its length over time. Although P_W/P_I can be calculated by solving the equations of heat and mass transport simultaneously in a time-dependent solution, this is impossible in one-dimensional models because heat and mass transport vary in all directions over time. This transport affects the vertically averaged ice hardness parameter A, the basal sliding parameter B, and the bed deformation parameter C, as well as P_W/P_I, in Equation (3.64).

In addition, τ_s should change from the head to the foot of an ice stream, due to thermal and strain softening in the lateral shear zones alongside ice streams. Changes in τ_s over distance and time depend on heat and mass transport in the shear zones, including dissipation of frictional heat generated by shear deformation (Raymond, 1996). However, varying τ_s over its expected range in sensitivity experiments of Equation (3.64) show that these variations have little effect on $\Delta h/\Delta x$.

Sensitivity experiments for sheet flow in Equation (3.64) show that f_t is an important variable in controlling $\Delta h/\Delta x$. Glacial geology shows clearly that thawed fraction f_t of the bed beneath former ice sheets varied over the full range $0 \leq f_t \leq 1$ along its flowlines (Kleman and Borström, 1994, 1996; Hughes, 1995; Menzies, 1995a). Although these variations for sheet flow occur for $P_W/P_I \approx 0$ in Equation (3.64), calculating f_t variations in space and time also depends on simultaneous solutions of the equations for heat and mass transport, for both the ice sheet and for a deformable bed (Zotikov, 1986; Menzies, 1995b, 1995c).

The most important variable that appears in Equation (3.64) from the mass balance is accumulation rate a. In Equation (3.14) for the mass balance, a is not allowed to vary with distance along the flowline or with time. Ignoring variations of a along flowlines seems reasonable, however, because $h \propto a^{1/8}$ for n = 3 in Equation (3.42) when $f_t = 0$, and $h \propto a^{1/5}$ for m = 2 in Equation (3.49) when $f_t = 1$. These are standard values for n (Nye, 1953) and m (Weertman, 1957a). This is an encouraging result, because Equation (3.64) is for one-dimensional flow along a flowband of constant width w. In ice sheets, however, the mass balance should be calculated for flowbands (a flowband is bounded by two flowlines) of variable width, with flowlines being merely the centerlines of flowbands. Flowband widths can vary greatly along their length, with w widening downslope from ice domes on the ice divide and from ice domes or ridges between ice streams, due to diverging flow, and w narrowing downslope from ice saddles on the ice divide and at the heads of ice streams, due to converging flow. Therefore a more realistic mass-balance calculation would include variations in both a and w along flowbands. This is accomplished by factors ϕ_f and ϕ_t in Equation (3.64), as shown in Chapter 5, and by allowing R and R_{xz} to vary accordingly.

Converting a flowline model into a flowband model, so that variations in a and w along x can be included, is straightforward and is done in Chapter 5. Sensitivity tests for constant a along the flowband, but with w being (1) constant, (2) linearly increasing downslope from an ice dome, (3) linearly decreasing downslope from an ice saddle, and (4) linearly increasing from an ice dome and decreasing toward an ice stream, show that the maximum variation in ice elevation along the flowband is only about five percent of the elevation for contant w (Hughes, 1995). Likewise, sensitivity tests for constant w along the flowband were conducted with a (1) being a constant accumulation rate to the ice margin, (2) being a constant accumulation rate in the accumulation zone and changing abruptly at the equilibrium line to a greater constant ablation rate in the ablation zone, (3) decreasing linearly from the ice divide to zero at the equilibrium line and decreasing linearly more sharply to the ice margin, and (4) increasing linearly from zero at the ice divide to a peak in the accumulation zone and decreasing linearly more sharply from the peak to the ice margin, with a = 0 at the equilibrium line (Hughes, 1995). Except for constant a along the entire flowband, these tests show that maximum variations in ice elevation along the flowband are only about five percent for these extreme variations in a along the flowband.

More serious than ignoring variations of a and w along flowlines in Equation (3.64), after its modification to include glacioisostatic and topographic variations in Equation (3.86), are variations of h* and h_I* with time. Ice sheets never attain steady-state equilibrium, so the mass-balance continuity condition expressed by Equation (3.14) when $\partial h_I/\partial t = 0$ should be replaced by using increments δ for time and Δ for space:

$$\frac{\delta h_I^*}{\delta t} = a - \frac{\Delta(u_x h_I^*)}{\Delta x} = a - u_x \frac{\Delta h_I^*}{\Delta x} - h_I^* \frac{\Delta u_x}{\Delta x}$$

$$= \frac{\delta(h^* - h_R^*)}{\delta t} = a - u_x \frac{\Delta(h^* - h_R^*)}{\Delta x} - (h^* - h_R^*) \frac{\Delta u_x}{\Delta x} \qquad (3.88)$$

where $\partial(h_I u)/\partial y$ is ignored, $u = u_x$, and h_R* is obtained by noting that $h_I = h - h_R$ in Equation (3.64) becomes $h_I^* = h^* - h_R^* = (1 + r)^{1/2} [h_M + (1 + r)^{1/2} (h^* - h_M) - h_R]$ in Equation (3.85). Solving this expression for h_R*gives:

$$h_R^* = h^* - (1 + r)(h^* - h_M) + (1 + r)^{1/2} (h_R - h_M) \qquad (3.89)$$

Therefore calculating $\delta h_R^*/\delta t$ in Equation (3.88) requires knowing $\delta h^*/\delta t$ and $\delta r/\delta t$ in Equation (3.89). In Equations (3.64) and (3.85), a should be replaced by effective accumulation rate a* defined as:

$$a^* = a - (\partial h^*/\partial t) + (\partial a/\partial t)\, \delta t \qquad (3.90)$$

where $a - (\partial h^*/\partial t) = \partial(u h^*)/\partial x$ from Equation (3.88) and $(\partial a/\partial t)$ is especially important during abrupt changes in a glaciation cycle (Alley and others, 1993).

Flowline models make maximum use of glacial geology. For terrestrial ice-sheet margins where ice streams become terminal ice lobes, E in Equation (3.64) is the distance from terminal moraines at the end of lobes to the beginning of lateral moraines on the sides of lobes, because moraines form in ablation zones (Clark, 1992). In Equation (3.86), E gives h_E, which is h at i = 0 when Equation (3.64) is written as an initial-value, finite-difference, recursive formula using Equation (3.79). In addition to specifying E, calculating h_E from Equation (3.86) also requires knowing τ_o in the ablation zone. Glacial geology, especially terminal and lateral moraines looping around ice lobes at the ends of ice streams, provides a means for calculating τ_o in the ablation zone of former ice sheets (Matthews, 1974; Clark, 1992). This possibility exists because Equation (3.64), using Equation (3.68) or Equation (3.69) for the ablation zone along terrestrial ice margins, calculates flowline profiles from grounded ice margins to ice divides of ice sheets. This approach is used to take maximum advantage of glacial geology in specifying variations in P_W/P_I and f_t when these are not calculated directly by simultaneously solving the equations of heat and mass transport. However, if glacial geology is not interpreted correctly, this approach does not produce reliable flowline profiles. The price for mathematical simplicity obtained by not including simultaneous solutions of the equations for heat and mass transport is paid by the effort expended in understanding glacial geology.

When using Equation (3.85) to reconstruct ice sheets through a glaciation cycle having abrupt transitions between stadials and interstadials, keeping track of glacioisostasy ratio r becomes virtually impossible. For a given advance of an ice sheet with relaxation time t_o:

$$r = r_o\,[1 - exp\,(-\,t/t_o)] \tag{3.91}$$

where ice covered rock for time t at a given site, and for a given retreat of the ice sheet:

$$r = r_a\,exp\,(-\,t/t_o) \tag{3.92}$$

where t is the time since the site was deglaciated and $r = r_a$ at t = 0 in Equation (3.92), with $r = r_a$ obtained from Equation (3.91) at time t when the advance ended. However, after multiple advances and retreats, it is unrealistic to keep track of r using Equations (3.91) and (3.92). These are empirical expressions that apply to a single advance or retreat. They do not contain the physics needed to specify r for multiple advances and retreats of ice-sheet margins that thin and thicken interior ice without removing it. The necessary physics requires a comprehensive rheological model for Earth's crust and mantle that responds to changing ice and water loads on Earth's surface, and that allows three-dimensional creep (e.g., Cathless, 1975; Peltier, 1989).

Introducing a two-dimensional ice-sheet gridpoint model

Two alternative research strategies to flowline or flowband models have been developed for reconstructing ice sheets during a glaciation cycle. One strategy is to develop physically-based three-dimensional time-dependent ice-sheet models that include climate dynamics to specify boundary conditions, the surface mass balance in particular (e.g., Budd and Smith, 1987; Lindstrom, 1990; Huybrechts, 1990; Budd and others, 1994; Fastook and Prentice, 1994; Huybrechts and T'siobel, 1995). The other strategy is to abandon ice-sheet modeling altogether, and use only the known areal extent of ice sheets, and known rates of changing sea level worldwide, as input to a model of crustal and mantle rheology that produces ice thicknesses and elevations as output over the areas occupied by ice sheets during the last deglaciation (Peltier, 1994).

Ice-sheet modeling has progressed from one-dimensional flowline models to partially two-dimensional flowband models to fully two-dimensional gridpoint models, to partially three-dimensional layered flowband models to fully three-dimensional layered gridpoint models. An important intermediate model is therefore the time-dependent two-dimensional gridpoint model. As developed by Fastook (1990), this model has great versatility that is ideally suited for incorporating any constitutive relationship, treating all boundary conditions, and conducting sensitivity experiments in minimal computer time. The mass balance of an ice sheet is determined by the mass continuity equation for map-plane coordinates x, y that is the two-dimensional time-dependent counterpart of Equation (3.16):

$$\partial h_I\,(x, y)/\partial t = a\,(x, y) - \nabla \bullet \sigma\,(x, y) \tag{3.93}$$

where a is the (positive) accumulation rate or (negative) ablation rate and $\nabla \bullet \sigma\,(x, y)$ is the divergence of ice flux density $\sigma\,(x, y)$. Equation (3.93) equates changes in ice thickness over time, $h_I\,(x, y)/\partial t$, to the net flux into a region of the ice sheet.

In two-dimensional ice flow, $\sigma\,(x, y) = u\,(x,y)\,h_I\,(x, y)$, and axes x, y are Cartesian coordinates or lines of longitude and latitude, but ice velocity u is best represented by variable coordinates L, T, where axis L is in the longitudinal direction of ice flow and axis T is transverse to the flow direction. Only the vertical z axis is common to the two coordinate systems. If θ is the variable angle between the constant x direction and the variable L direction, the relationships between ice velocity u in the two coordinate systems is (Fastook and others, 1995):

$$tan\,\theta = u_L/u_x \tag{3.94a}$$

$$u_L = u_x\,cos\,\theta + u_y\,sin\,\theta \tag{3.94b}$$

$$u_T = -\,u_x\,sin\,\theta + u_y\,cos\,\theta \tag{3.94c}$$

Strain rates obtained from the gradients of u_L and u_T are:

$$\dot{\varepsilon}_{LL} = \partial u_L/\partial L = \left(\partial u_L/\partial x\right) cos\,\theta + \left(\partial u_L/\partial y\right) sin\,\theta \tag{3.95a}$$

$$\dot{\varepsilon}_{TT} = \partial u_T/\partial T = \left(\partial u_T/\partial x\right) sin\,\theta + \left(\partial u_T/\partial y\right) cos\,\theta \tag{3.95b}$$

$$\dot{\varepsilon}_{LT} = \tfrac{1}{2}\left(\partial u_L/\partial T + \partial u_T/\partial L\right)$$

$$= \tfrac{1}{2}\left[-\left(\partial u_L/\partial x\right) \sin\theta + \left(\partial u_L/\partial y\right)\cos\theta\right]$$

$$+ \tfrac{1}{2}\left[\left(\partial u_T/\partial x\right)\cos\theta + \left(\partial u_T/\partial y\right)\sin\theta\right] \quad (3.95c)$$

$$\dot{\varepsilon}_{ZZ} = -\left(\dot{\varepsilon}_{LL} + \dot{\varepsilon}_{TT}\right) \quad (3.95d)$$

The dominant stresses in the ice sheet are σ'_{LZ} (x, y) for sheet flow, σ_{LZ} (x, y), and σ'_{LL} (x, y) for stream flow, and σ'_{LL} (x, y), σ'_{TT} (x, y), and σ_{LT} (x, y) for shelf flow. The two-dimensional counterpart of Equation (3.11) for stream flow and shelf flow is:

$$\sigma'_{LL} = \frac{\tfrac{1}{2}\rho_I g_z h_I (1 - \rho_I/\rho_W)}{2 + \left(\dot{\varepsilon}_{TT}/\dot{\varepsilon}_{LL}\right)}\left(\frac{P_W}{P_I}\right)^2 \quad (3.96)$$

where $\sigma'_{LL} = \sigma'_{LL}$ (x, y), $h_I = h_I$ (x, y), $P_W = P_W$ (x, y), and $P_I = P_I$ (x, y). The two-dimensional counterpart of Equation (3.28) for sheet flow and stream flow is:

$$\sigma_{LZ} = \tau_o = \rho_I g_z h_I |\nabla h| \quad (3.97)$$

where $\tau_o = \tau_o$ (x, y), $h_I = h_I$ (x, y), and $\nabla h = \nabla h$ (x, y). Extracting $P_I = \rho_I g_z h_I$ from Equations (3.96) and (3.97), the generalized constitutive equation that relates ice flux density $\sigma = \sigma$ (x, y) = u (x, y) h_I (x, y) to basal buoyancy factor P_W/P_I and ice-surface gradient ∇h is:

$$\sigma = u h_I = -k\left[\tfrac{1}{2}(1 - \rho_I/\rho_W)(P_W/P_I)^2/(2 + \dot{\varepsilon}_{TT}/\dot{\varepsilon}_{LL}) + |\nabla h|\right] \quad (3.98)$$

where k = k (x, y) is a physical coefficient that depends on specific creep laws for the ice sheet, specific sliding laws at the ice-bed interface, and specific flow laws for a deforming bed, and u = u (x, y) is the vertically integrated longitudinal ice velocity. Solving Equation (3.98) for u:

$$u(x, y) = -\frac{k(x, y)}{h_I(x, y)}\left[\frac{(1 - \rho_I/\rho_W)}{2(2 + \dot{\varepsilon}_{TT}/\dot{\varepsilon}_{LL})}\left(\frac{P_W(x, y)}{P_I(x, y)}\right)^2 + |\nabla h|(x, y)\right] \quad (3.99)$$

where u consists of velocities u_L for linear creep in ice, u_C for laminar creep in ice, u_S for sliding at the ice-bed interface, and u_D for bed deformation. If f_L, f_C, f_S, and f_D are the respective fractional contributions of u_L, u_C, u_S, and u_D to u:

$$u = f_L u_L + f_C u_C + f_S u_S + f_D u_D \quad (3.100)$$

where $f_L + f_C + f_S + f_D = 1$ and u is in the L direction.

Creep velocity u_L arises from longitudinal strain rate $\dot{\varepsilon}_{LL}$ over longitudinal incremental length ΔL of creep. If u_o is ice velocity entering increment ΔL, by analog with Equation (3.22):

$$u_L = u_o + \dot{\varepsilon}_{LL}\,\Delta L = u_o + R\left[\frac{\rho_I g_z h_I}{2A}\left(1 - \frac{\rho_I}{\rho_W}\right)\left(\frac{P_W}{P_I}\right)^2\right]^n \Delta L \quad (3.101)$$

In Equation (3.101), u_o is positive and increases along L, whereas u_0 in Equation (3.27) is negative and located at the grounded ice margin where x = 0. By analogy to R in Equation (3.23), R in Equation (3.101) is:

$$R = \frac{\left[1 + (\dot{\varepsilon}_{TT}/\dot{\varepsilon}_{LL}) + (\dot{\varepsilon}_{TT}/\dot{\varepsilon}_{LL})^2 + (\dot{\varepsilon}_{LT}/\dot{\varepsilon}_{LL})^2 + (\dot{\varepsilon}_{LZ}/\dot{\varepsilon}_{LL})^2\right]^{(n-1)/2}}{\left[2 + (\dot{\varepsilon}_{TT}/\dot{\varepsilon}_{LL})\right]^n} \quad (3.102)$$

For the particular creep law for glacial ice given by the two-dimensional counterpart to Equation (3.39):

$$u_C = \frac{2 R_{LZ} h_I}{n + 2}\left(\frac{\tau_o}{A}\right)^n = \frac{2 R_{LZ} h_I}{n + 2}\left[\frac{\rho_I g_z h_I |\nabla h|}{A}\right]^n \quad (3.103)$$

For the particular sliding law between ice and the bed given by the two-dimensional counterpart to Equation (3.46):

$$u_S = \left[\frac{\tau_o}{B(1 - P_W/P_I)^2}\right]^m = \left[\frac{\rho_I g_z h_I |\nabla h|}{B(1 - P_W/P_I)^2}\right]^m \quad (3.104)$$

For the particular creep law for a deforming bed given by the two-dimensional counterpart to Equation (3.54):

$$u_D = 2\lambda_D\left[\frac{\tau_o}{C(1 - P_W/P_I)}\right]^c = 2\lambda_D\left[\frac{\rho_I g_z h_I |\nabla h|}{C(1 - P_W/P_I)}\right]^c \quad (3.105)$$

Physical coefficient k (x, y) in Equation (3.99) can now be obtained by substituting Equations (3.101), (3.103), (3.104), and (3.105):

$$k(x, y) = -\left[\frac{u(x, y)\, h_I(x, y)}{\dfrac{(1 - \rho_I/\rho_W)}{2(2 + \dot{\varepsilon}_{TT}/\dot{\varepsilon}_{LL})}\left(\dfrac{P_W(x, y)}{P_I(x, y)}\right)^2 + |\nabla h|(x, y)}\right]$$

$$= -\left[\frac{(f_L u_L + f_C u_C + f_S u_S + f_D u_D)\, h_I}{\dfrac{(1 - \rho_I/\rho_W)}{2(2 + \dot{\varepsilon}_{TT}/\dot{\varepsilon}_{LL})}\left(\dfrac{P_W}{P_I}\right)^2 + |\nabla h|}\right]$$

$$= -f_L\left[\frac{u_o h_I + R\, h_I\left[\left(\dfrac{\rho_I g_z h_I}{2A}\right)\left(1 - \dfrac{\rho_I}{\rho_W}\right)\left(\dfrac{P_W}{P_I}\right)^2\right]^n \Delta L}{\dfrac{(1 - \rho_I/\rho_W)}{2(2 + \dot{\varepsilon}_{TT}/\dot{\varepsilon}_{LL})}\left(\dfrac{P_W}{P_I}\right)^2 + |\nabla h|}\right]$$

$$- f_C\left[\frac{\left(\dfrac{2}{n + 2}\right) R_{LZ} h_I^2 \left(\dfrac{\rho_I g_z h_I |\nabla h|}{A}\right)^n}{\dfrac{(1 - \rho_I/\rho_W)}{2(2 + \dot{\varepsilon}_{TT}/\dot{\varepsilon}_{LL})}\left(\dfrac{P_W}{P_I}\right)^2 + |\nabla h|}\right]$$

$$- f_S\left[\frac{h_I\left(\dfrac{\rho_I g_z h_I |\nabla h|}{B(1 - P_W/P_I)^2}\right)^m}{\dfrac{(1 - \rho_I/\rho_W)}{2(2 + \dot{\varepsilon}_{TT}/\dot{\varepsilon}_{LL})}\left(\dfrac{P_W}{P_I}\right)^2 + |\nabla h|}\right]$$

$$- f_D \left[\frac{2 \lambda_D h_I \left(\frac{\rho_I g_z h_I |\nabla h|}{C (1 - P_W / P_I)} \right)^c}{\frac{(1 - \rho_I / \rho_W)}{2 (2 + \dot{\varepsilon}_{TT} / \dot{\varepsilon}_{LL})} \left(\frac{P_W}{P_I} \right)^2 + |\nabla h|} \right] \quad (3.106)$$

The two-dimensional counterpart of Equation (3.14) for mass-balance continuity is:

$$\partial h_I (x, y) / \partial t = a (x, y) - \nabla \bullet \sigma (x, y)$$

$$= a - \nabla \bullet \left\{ -k \left[\frac{(1 - \rho_I / \rho_W)}{2 (2 + \dot{\varepsilon}_{TT} / \dot{\varepsilon}_{LL})} \left(\frac{P_W}{P_I} \right)^2 + h_I |\nabla h| \right] \right\} \quad (3.107)$$

where $\sigma (x, y)$ is given by Equations (3.98) and (3.106). All terms in Equation (3.106) for k (x, y) are retained for stream flow in Equation (3.107). For sheet flow, $P_W << P_I$ and $f_L \approx 0$, so k (x, y) reduces to:

$$k (x, y) = - f_C \left[\frac{2 R_{LZ}}{n + 2} \left(\frac{\rho_I g_z}{A} \right)^n h_I^{n+2} |\nabla h|^{n-1} \right]$$

$$- f_S \left[\left(\frac{\rho_I g_z}{B (1 - P_W / P_I)^2} \right)^m h_I^{m+1} |\nabla h|^{m-1} \right]$$

$$- f_D \left[2 \lambda_D \left(\frac{\rho_I g_z}{C (1 - P_W / P_I)} \right)^c h_I^{c+1} |\nabla h|^{c-1} \right] \quad (3.108)$$

For shelf flow, $P_W \approx P_I$, $f_L \approx 1$, $f_C + f_S + f_D \approx 0$, and $\nabla h \approx 0$, so k (x, y) reduces to:

$$k (x, y) = -2 \left(2 + \frac{\dot{\varepsilon}_{TT}}{\dot{\varepsilon}_{LL}} \right) \left\{ \frac{u_o h_I}{(1 - \rho_I / \rho_W)} \right.$$

$$\left. + R \left(\frac{\rho_I g_z}{2 A} \right)^n \left(1 - \frac{\rho_I}{\rho_W} \right)^{n-1} h_I^{n+1} \Delta L \right\} \quad (3.109)$$

Fastook (1990) has obtained solutions of Equations (3.107) using the finite-element method for $f_L = 0$ and $P_W << P_I$.

The increase in sliding velocity as the bed thaws

The relationship between fraction f_t of a thawed bed and f_S of basal sliding over bedrock is obtained by combining Equations (3.32), (3.39), (3.46), and (3.100) when $P_W << P_I$ and $\tau_D = f_L = f_D = 0$ for sheet flow along axis L = x, so that:

$$u = f_C u_C + f_S u_S = f_C [2 h_I / (n + 2)] [\tau_o / A]^n + f_S [\tau_o / B]^m$$

$$= f_C \frac{2 h_I}{n + 2} \left[\frac{f_t \tau_S + f_f \tau_C}{A} \right]^n + f_S \left[\frac{f_t \tau_S + f_f \tau_C}{B} \right]^m \quad (3.110)$$

Setting $f_C = 1 - f_S$ and $f_f = 1 - f_t$, and solving for f_S:

$$f_S = \frac{\frac{2 h_I}{n + 2} \left[\frac{f_t \tau_S + (1 - f_t) \tau_C}{A} \right]^n - u_x}{\frac{2 h_I}{n + 2} \left[\frac{f_t \tau_S + (1 - f_t) \tau_C}{A} \right]^n - \left[\frac{f_t \tau_S + (1 - f_t) \tau_C}{B} \right]^m} \quad (3.111)$$

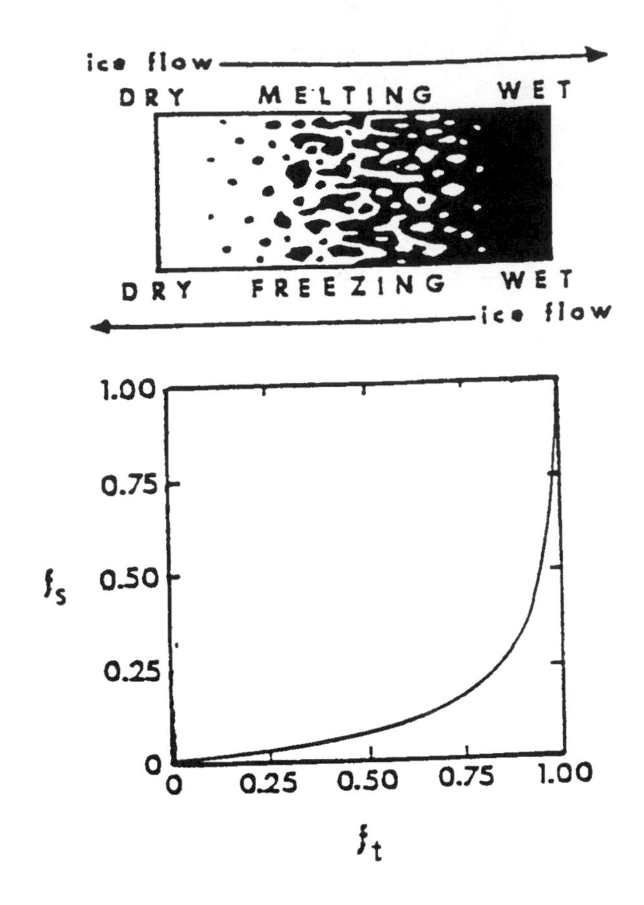

Figure 3.10: The variation of fraction f_S of sliding velocity to total velocity with fraction f_t of the bed that is thawed across zones of basal melting and freezing beneath an ice sheet. *Top*: white areas are frozen (dry) and black areas are thawed (wet) across basal melting and freezing zones. *Bottom*: The highly nonlinear variation of f_S with f_t shows that a few frozen patches on the bed can prevent sliding from dominating ice motion.

Figure 3.10 is a plot of f_S versus f_t obtained from Equation (3.111) for n = 2, m = 3, A = 4.70 bar $a^{1/3}$, and B = 0.02 bar $a^{1/2}$ $m^{-1/2}$ for a flowband of constant width w on a flat bed and specified values of h_I and u across the melting or freezing zone. The f_S versus f_t curve is strongly nonlinear. A bed 50 percent thawed (f_t = 0.5) permits only 5 percent sliding (f_S = 0.05), and creep predominates over sliding until almost 90 percent of the bed is thawed. This theoretical conclusion is consistent with stream flow occurring along only about 10 percent of the Antarctic perimeter, even though ice streams drain some 90 percent of the Antarctic interior. This result shows that a few frozen patches on the bed are highly effective in shutting down glacial sliding, so that an ice sheet becomes highly unstable as the last frozen patches thaw. Gravitational collapse will lower the ice elevation 20 to 25 percent in the interior area of sheet flow (Hughes, 1981a), and stream flow will advance the ice-sheet margin rapidly (Hughes, 1996a). It also emphasizes the importance of frozen patches shown in Figure 3.10 or other "sticky spots" that retard sliding, as analyzed by MacAyeal and others (1995) for Ice Stream E.

Equation (3.111) shows that the relationship between f_S and f_t depends strongly on creep constants A and n and on sliding constants B and m; indeed, on the very form of creep and sliding laws for glacial ice. An important glaciological problem is to examine this dependence by combining the equations of heat transport and mass transport to produce a time-dependent three-dimensional model in which n = 3 can be assumed and a vertically averaged A can be calculated at each gridpoint from the computed profiles of temperature T and the observed development of ice fabric, which combine to decrease A near the bed (Budd and Jacka, 1989). Values of B at each gridpoint depend on what sliding law is used, which in turn determines the relationship between f_S and f_t. Provided that sheet flow is primarily over rigid bedrock and stream flow is primarily over a deformable substrate, published B and m values based on theoretical and experimental studies of ice sliding over bedrock are acceptable (Paterson, 1994). For stream flow, these values can be applied to bedrock "sticky spots" that outcrop through the deformable substrate.

Inverse and forward applications of Equations (3.64) and (3.107) are possible. In inverse applications, the surface elevations of present ice sheets are duplicated in order to calculate the distribution of f_t or f_S over the bed for sheet flow, where $f_L = f_D = 0$ is assumed (Fastook and others, 1995). In forward applications, the distribution of f_t or f_S beneath a former ice sheet is specified, perhaps by interpreting glacial geology, in order to calculate the surface elevations of the ice sheet (Denton and Hughes, 1981).

Modeling Antarctic Ice-Sheet Stability

Stability of the Antarctic Ice Sheet can be modeled successfully only by employing both inverse and forward modeling strategies and only by using a three-dimensional time-dependent computer model that couples the equations of mass transport and heat transport in order to correctly simulate sheet flow, stream flow, and shelf flow components of the ice sheet. Work on this task has extended over four decades.

Pioneering work by Robin (1955), Zotikov (1963), and Budd and others (1971) calculated temperature profiles through the Antarctic Ice Sheet from ice thicknesses, ice accumulation rates, and ice-surface temperatures for basal geothermal heating rates at a bed assumed to be frozen. If the calculated basal temperature was above the melting point of ice, the excess sensible heat was converted into latent heat to give a basal melting rate. This approach permitted only frozen and thawed beds, with basal ice temperature calculated for the former and basal melting rates calculated for the latter. No provision was made for the refreezing of basal meltwater until Hughes (1973) and Sugden (1977) placed basal freezing zones between inner melting zones and outer frozen zones in the numerical flowline model developed by Budd and others (1971). The bed was wholly thawed in basal melting and freezing zones, which were separated by a basal equilibrium line that was analogous to the surface equilibrium line that separates accumulation and ablation zones.

A different conception of basal freezing and melting zones was proposed by Hughes (1981a) and applied to a flowline model for the Antarctic Ice Sheet at the last glacial maximum (Hughes and others, 1981) and during subsequent partial deglaciation (Stuiver and others, 1981; Fastook, 1984). In this view, which was based on an interpretation of the glacial geology produced by former Northern Hemisphere ice sheets, a wholly thawed inner zone is separated from a wholly frozen outer zone by a freezing zone that consists of a mosaic of frozen and thawed patches which begins with isolated frozen patches on a thawed bed and ends with isolated thawed patches on a frozen bed; conversely, a wholly frozen inner zone is separated from a wholly thawed outer zone by a melting zone that consists of a mosaic of thawed and frozen patches which begins with isolated thawed patches on a frozen bed and ends with isolated frozen patches on a thawed bed. Figure 3.10 relates these frozen (dry), thawed (wet), melting, and freezing basal thermal zones to the direction of ice flow. Basal sliding occurs where the bed is thawed, so that glacial erosion is possible. In the classification of glacial erosion by Sugden and John (1976), areal scouring predominates on a wholly thawed bed, quarrying by freeze-thaw processes occurs in thawed patches of melting or freezing zones and produces a pitted lake-studded deglaciated landscape, and selective linear erosion predominates beneath ice streams or outlet glaciers, producing troughs or fjords in the deglaciated landscape. In a thawed patch, basal ice melts and the basal meltwater refreezes as ice slides over the patch. For equilibrium condi-tions, basal melting and freezing rates balance, so the thawed patch deepens by erosion but does not change in area. For transient conditions, thawed patches grow when melting rates exceed freezing rates and shrink when freezing rates exceed melting rates. This causes basal melting and freezing zones to migrate. In Antarctica, the East Antarctic Ice Sheet is close to equilibrium, so basal thawed, melting, freezing, and frozen zones should be relatively stable, but the West Antarctic Ice Sheet seems to be in a transient condition, so these basal zones are either expanding or contracting. In order to assess the relative stabilities of the East and West Antarctic Ice Sheets, it is necessary to (1) map these Ant-arctic basal thermal zones and (2) relate the zones to relative amounts of internal creep and basal sliding for both sheet flow and stream flow. This information is input for a finite-difference model (FDM) that is based on the assumption that basal traction controls ice-sheet stability, or for a finite-ele-ment model (FEM) that is based on the assumption that basal sliding controls ice-sheet stability.

Mapping basal thermal zones

Mapping basal thermal zones is an inverse application of Equation (3.64) that exploits f_t as the dominant variable for sheet flow. In general, P_W/P_I is a dominant variable because $0 \leq P_W/P_I \leq 1$, but $P_W << P_I$ and $0 \leq f_t \leq 1$ for sheet flow, so f_t becomes the dominant variable. Bed topography and surface elevations, accumulation rates, and mean annual temperatures have been mapped in enough detail to allow basal thermal zones for sheet flow to be mapped in the Wilkes Land sector of East Antarctica and in all of West Antarctica except the Antarctic Peninsula. These data were distributed at gridpoints spaced 20 km apart and have been provided through the generosity of William F. Budd, at the University of Tasmania in Australia. Maps presenting these data appear in the Antarctic Map Folio published by Scott Polar Research Institute, Cambridge, England (Drewry, 1983), with more recent mass balance data by Giovinetto and others (1989) and ice thickness data by Shabtaie and Bentley (1988). These maps allow basal thawed, melting, freezing, and frozen thermal zones to be mapped using the FDM and allow the contributions of internal creep and basal sliding to be mapped using the FEM.

One of the maps in the Antarctic Map Folio (Drewry, 1983) shows the distribution of basal shear stress τ_o. A vertical ice column having density ρ_I, height h_I, and basal area A_z experiences gravitational acceleration g_z in the vertical z direction and exerts a vertical gravitational force $F_z = \rho_I h_I A_z g_z$ on basal area A_z. The ice column is moved in the horizontal x direction of maximum surface slope α by a horizontal gravitational force F_x given by ice mass $\rho_I h_I A_z$ times downslope component $g_z \alpha$ of gravitational acceleration. The basal shear stress resisting the downslope motion is:

$$\tau_o = \frac{F_x}{A_z} = \frac{\rho_I h_I A_z g_z \alpha}{A_z} = \rho_I h_I g_z \alpha \qquad (3.112)$$

Equation (3.27), and therefore Equation (3.64), reduces to Equation (3.112) when $P_W << P_I$ and $\tau_s = 0$ for sheet flow. Equation (3.112) holds when creep is dominated by laminar flow in ice. The finite-difference solution of Equation (3.27) writes Equation (3.112) as an initial-value, finite-difference, recursive formula for ice elevation change Δh in horizontal distance Δx along an ice flowline (Nye, 1952b):

$$h_{i+1} = h_i + [\tau_o /(h - h_R)]_i \Delta x / \rho_I g_z \qquad (3.113)$$

where i is the number of steps of length Δx from an initial position $i = 0$ along an ice flowline, $\Delta h = h_{i+1} - h_i$, $\alpha = \Delta h / \Delta x$, and $h_R = h - h_I$ is bedrock elevation above or depth below sea level. Note that τ_o, h, and h_R must be known at step i in order to compute h at step i + 1 for specified values of Δx, ρ_I, and g_z. The h_R data cited earlier allow Δx = 20 km as the minimum step length along a flowline. A spline-fit program giving smooth variations of ice height h and bedrock height h_R can be employed to obtain $h - h_R$ and Δh at smaller steps Δx. Equation (3.113) can be solved numerically using a modified Euler method or a Runge Kutta method to determine what maximum Δx gives the same results for both methods.

Since our data sources allow ice flowlines to be drawn normal to ice elevation contour lines, Equation (3.113) can be used to compute τ_o in the finite-difference model at each step i from the known variations of h and h_R along flowlines. In applying the model, a decision must be made on whether basal sliding or bed deformation dominates when the bed is thawed. In Equation (3.32), $\tau_S >> \tau_D$ when sliding dominates because there is little coupling between ice and the bed, but $\tau_D >> \tau_S$ when stronger coupling suppresses sliding and allows the bed to deform. What is important in both cases is that basal ice acquires a velocity, whether due to basal sliding or to bed deformation. Whether the bed under the ice sheet is bedrock or sediment or till is generally unknown, and the thickness, wetness, and deformability of sediment or till are also usually unknown. Therefore, only ice creep over a frozen bed and ice sliding over a thawed bed will be considered, so that Equation (3.32) reduces to:

$$\tau_o = f_f \tau_C + f_t \tau_S = (1 - f_t) \tau_C + f_t \tau_S \qquad (3.114)$$

where f_f and f_t are the respective frozen and thawed fractions of the bed at each Δx step, τ_C is τ_o for ice creeping over a frozen bed, τ_S is τ_o for ice sliding over a thawed bed, and $f_f + f_t = 1$ by definition. Equation (3.114) is the basis for mapping contours of f_t for West Antarctica and the Wilkes Land sector of East Antarctica.

Expressions for τ_C and τ_S have been derived from creep and sliding laws of ice . The creep law, also called the flow law, gives the creep velocity u_C averaged through ice of thickness h_I:

$$u_C = \frac{2 h_I}{n + 2} \left(\frac{\tau_o}{A}\right)^n \qquad (3.115)$$

where A is the vertically averaged ice hardness parameter that depends primarily on ice temperature and ice fabric, and n is a viscoplastic parameter that lies between n = 1 for purely viscous flow and n = ∞ for purely plastic flow (Nye, 1953; Glenn, 1955, 1958; however, see Alley, 1992). A number of sliding laws have been proposed (e.g., Weertman, 1957a; Lliboutry, 1968; Budd and others, 1979; Alley, 1989a, 1989b; Kamb, 1991;however, see Alley 1990). Most of them give a sliding velocity u_S that can be expressed by a generalized sliding law for sheet flow when $P_W << P_I$ in Equation (3.46) so that:

$$u_C = (\tau_o / B)^m \qquad (3.116)$$

where u_S is basal sliding velocity, τ_o is basal shear stress, B is a basal sliding parameter that depends primarily on high-stress creep in temperate basal ice, the hardness and roughness of bedrock in contact with basal ice, and the permeability and deformability of wet till or sediments between basal ice and bedrock, and m is a viscoplastic parameter for temperate basal ice. The variation of A through the ice thickness and the formulation of B for different beds are difficult theoretical problems. In practical applications, column-averaged values of A for a given n, and bed-averaged values of B for a given m for beds of known properties are available from empirical studies, some of which are cited above. Equations (3.115) and (3.116) appear in Equation (3.110) that relates f_S to f_t.

A reasonable assumption for sheet flow is

$|a(x)| >> |\partial h(x)/\partial t|$ in Equation (3.14), especially in East Antarctica. This assumption will not affect the calculation of f_t because ice surface slopes are not strongly affected by accumulation or thinning rates, with $\Delta h/\Delta x \propto a^{1/4}$ for n = 3 in the creep law and $\Delta h/\Delta x \propto a^{2/5}$ for m = 2 in the sliding law when $P_W << P_I$, a is constant and $\partial h/\partial t = 0$ in Equation (3.64). The FDM uses the finite-difference solution of Equation (3.64), ice and bed heights, and the constants A, B, n, and m as the sensitive variables for computing f_t.

Figure 3.11 shows how Equation (3.64) in the FDM can be applied to the Antarctic flowbands shown in Figure 3.7, with $P_W << P_I$, C = 0 for yield stress $\sigma_o << \tau_o$ in basal sed-iment or till (Kamb, 1991), and f_t calculated along the central flowlines of these flowbands. All the flowbands shown in Figure 3.7 dovetail into ice streams identified by numbers 1 through 30. Wide flowbands, such as the ice drainage basin for Byrd Glacier (number 18), can be divided into several narrower flowbands in order to map f_t contours more accurately. An example of f_t variations along flow-bands computed from the FDM is shown in Figure 3.11 for Mullock Glacier (number 19). In this figure, τ_o is the solid curve computed from Equation (3.112), and it is bracketed by the upper τ_C and the lower τ_S dashed curves computed from Equations (3.44) and (3.51), respectively. Variations of f_t along the central flowline of each flowband are then calculated from Equation (3.64). In these calculations, n = 3, m = 2, A = 4.70 bar $a^{1/3}$, and B = 0.02 bar $a^{1/2}$ $m^{-1/2}$ were used initially, assuming column-averaged ice temperatures along Antarctic flowlines computed by Budd and others (1971), and sliding over bedrock according to the Weertman (1957a) model, as generalized by Kamb (1970) and Nye (1970). Then A and B are adjusted until the respective theo-retical τ_C and τ_S curves make point contact with the mea-sured τ_o curve, as shown in Figure 3.11. Other values of τ_o then lie between the τ_C and τ_S curves, and depend only on f_t. In moving from the ice divide to the ice mar-gin along flowlines, a frozen bed exists where $\tau_o = \tau_C$, a thawed bed exists where $\tau_o = \tau_S$, a melting bed exists where τ_o moves from τ_C toward τ_S, and a freezing bed exists where τ_o moves from τ_S toward τ_C. A rapid drop of τ_o near the ice margin occurs when $f_t = 1$ and P_W/P_I becomes dominant, so that sheet flow becomes stream flow.

Following this procedure, a map of the thawed fraction f_t of the bed was obtained by solving Equation (3.64) along the central flowlines of selected flowbands in Figure 3.7, us-ing known variations of ice accumulation rates and flowband widths to determine ϕ_f and ϕ_t as described in Chapter 5. Figures 3.12 through 3.17 present the inverse modeling study by Ellen Wilch, who applied Equation (3.64), reduced to Equations (3.113) through (3.116), to selected Antarctic flowlines shown in Figures 3.12 and 3.13 to compute the percent thawed bed in Figures 3.14 and 3.15 for comparison with bed topography in Figures 3.16 and 3.17. In general, East Antarctic ice has a mostly frozen bed on subglacial highlands beneath Dome Argus and other subglacial high-lands, a mostly thawed bed in the subglacial basin beneath Dome Circe and other subglacial basins, and a wholly thaw-

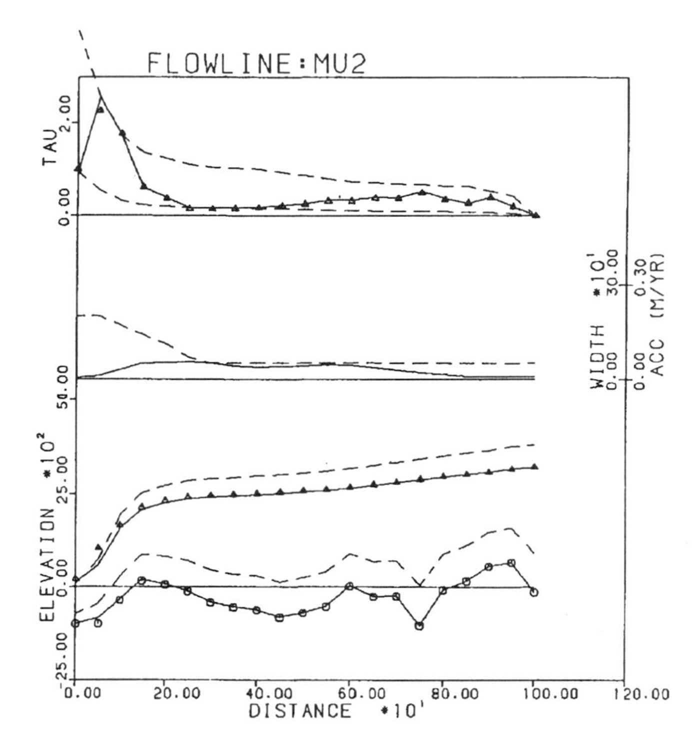

Figure 3.11: The basal thermal regime along the central flowline of the Mulock Glacier ice drainage basin. *Top*: The actual τ_o variation (solid curve with triangles denoting 50 km steps) lies between the theoretical variations τ_C for creep over a dry frozen bed (upper dashed curve) and τ_S for sliding over a wet thawed bed (lower dashed curve), with f_t being the fraction that τ_o lies between the τ_C and τ_S curves at each step in order to satisfy Equation (3.114). *Center*: Variations in flowband width (solid curve) and accumulation rate (dashed curve) along the flowline. *Bottom*: The spline fitted bedrock topography (curve through circled dots) that was used with f_t obtained from the τ_o, τ_C, and τ_S curves to compute a theoretical ice surface profile (solid curve) that provides a best fit with the actual ice surface profile (triangles). The dashed curves are for theoretical isostat-ic rebound of the surface and the bed after deglaciation

ed bed beneath ice streams. The same generalities are found in West Antarctica, but the mostly thawed bed along the ice divide between flowlines R7 and R13 may be due to a high geothermal flux from subglacial volcanoes, as reported by Blankenship and others (1993). Domes (D) along ice divides and ice margins are identified.

Mapping creep and sliding velocities

The FEM developed by Fastook (1990) applies the data base used in the FDM to the gridpoints 20 km apart where data were digitized by William F. Budd, instead of applying the data along specified flowlines. When $P_W << P_I$ and $f_L = f_D = 0$ in Equation (3.106), Equation (3.107) in the FEM calculates ice thicknesses from the relative contributions of

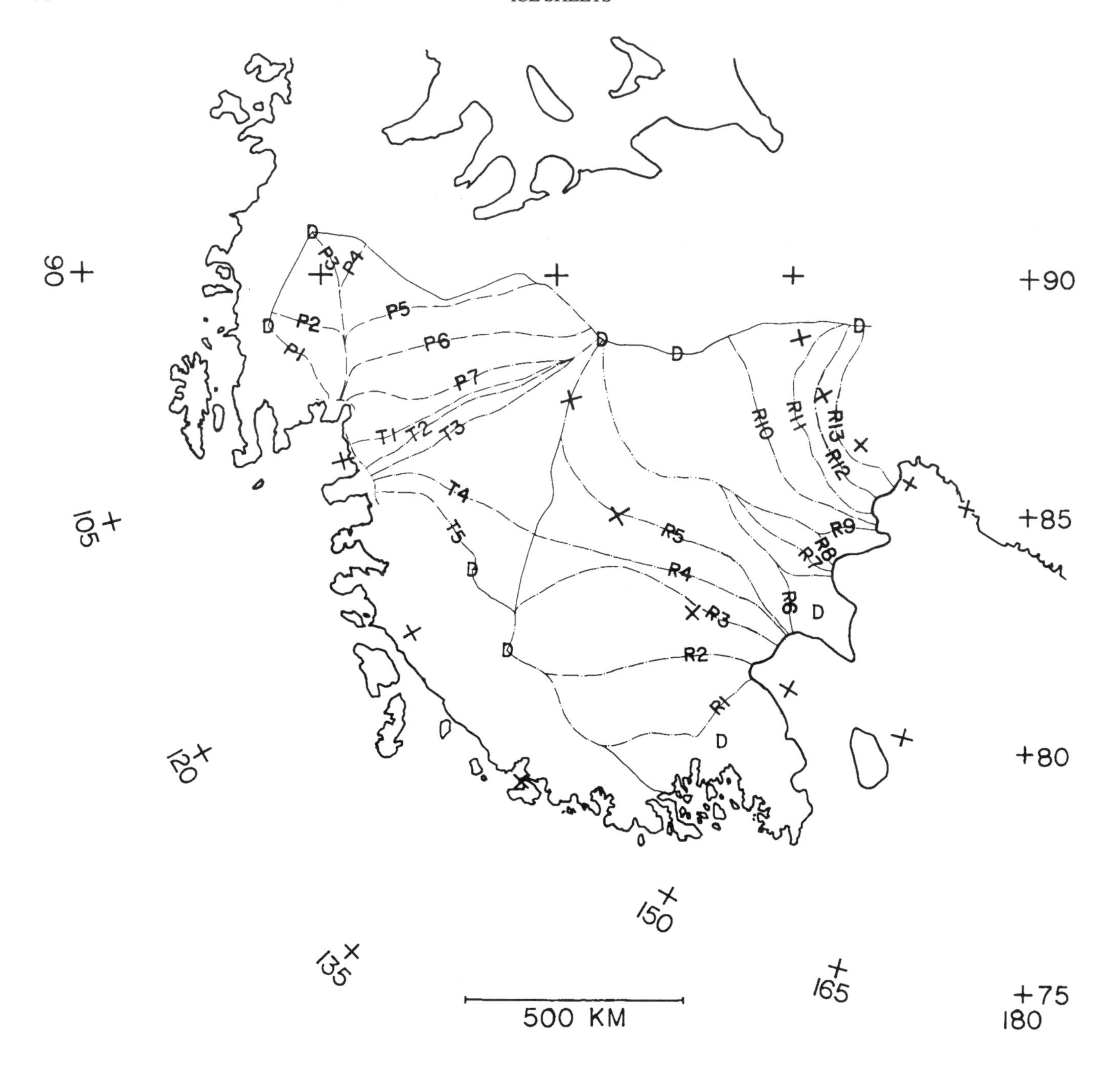

Figure 3.12: Flowlines of selected flowbands in West Antarctica along which the percent of thawed bed was calculated using Equations (3.113) through (3.116). West Antarctic flowlines are for Pine Island Glacier (PI through P7), Thwaites Glacier (T1 through T5), and ice streams that supply the Ross Ice Shelf (R1 through R13).

internal creep and basal sliding to ice velocity, whereas Equation (3.64) in the FDM calculates these thicknesses from the relative contributions of basal frozen and thawed patches to bed traction, as measured by the basal shear stress. These are two different ways for understanding variations in surface topography of the Antarctic Ice Sheet. The contribution of bed topography to surface topography is included in both models, but it alone is inadequate to account for surface topography. The FEM uses the finite-element method to obtain a two-dimensional time-dependent numerical solution to the continuity equation. Mass conservation requires that the net ice flux into a region be balanced by changes in the ice thickness of that region. The mass balance of an ice sheet is determined by Equation (3.93) for mass continuity in the horizontal map-plane using Cartesian coordinates x, y and digitized bed topography.

Solutions of Equation (3.93) are obtained from Equation (3.107) and are independent of how k (x, y) is formulated, so

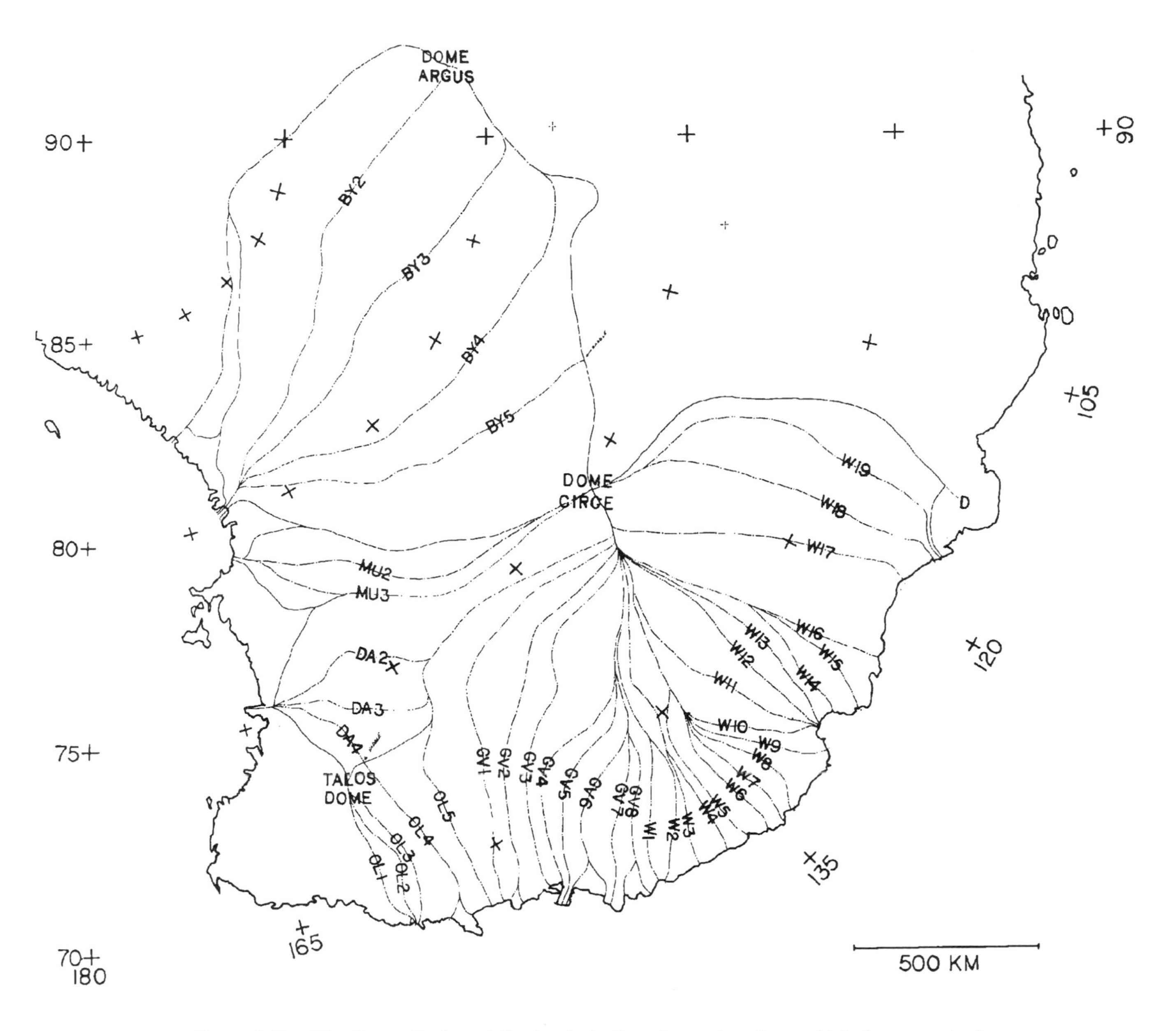

Figure 3.13: Flowlines of selected flowbands in East Antarctica along which the percent of thawed bed was calculated using Equations (3.113) through (3.116). East Antarctic flow-lines are for Byrd Glacier (BY2 through BY5), Mullock Glacier (MU2 and MU3), David Glacier (DA2 through DA4), Oakes Land (OL1 through OL5), George V Land (GV1 through GV8), and Wilkes Land (W1 through W19).

a variety of creep and sliding "laws" can be employed. This is important because it universalizes the FEM so that it can be applied to any new breakthroughs in understanding creep and sliding of glacial ice and the dynamics of a deformable bed. Figure 3.18, provided by James Fastook, is a map of the sliding fraction f_S of ice velocity obtained by solving Equation (3.107) at gridpoints for sheet flow when $f_D = 0$, for comparison with f_t in Figure 3.11 with these conditions.

Figure 3.19, provided by James Fastook, demonstrates how vectors for mass-balance ice velocity u are affected by sliding fraction f_S at each node of a 20 x 20 grid of nodes 1000 km on a side. The bed is flat, the left side has zero ice thickness, an ice flux of 80,000 m^2/yr crosses the right side, and the surface has a uniform accumulation rate of 0.05 m/yr. Equilibrium conditions are displayed, and are attained when 10,000 years of 100-yr time steps gives $\partial h_I (x, y)/\partial t = 0$ for sheet flow when $f_D = 0$ in Equation (3.107). Sliding occurs at 0, 10, 20, 30, 40, 50, 75, and 100 percent of the nodes in successive panels of Figure 3.19, with sliding occurring in the square boxes around each node, as displayed in

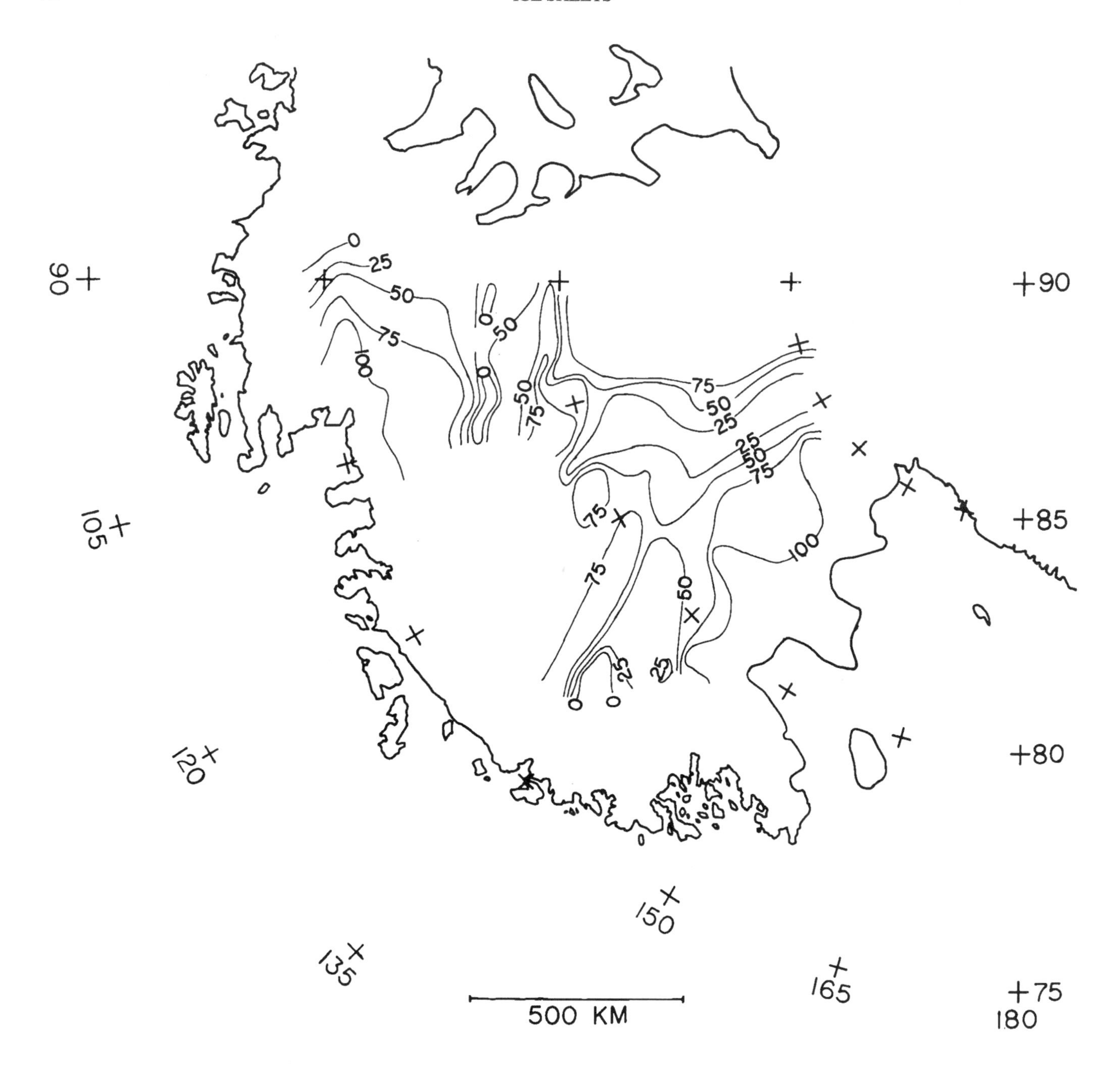

Figure 3.14: Isopleths of percent thawed bed in West Antarctica calculated along the flowlines identified in Figure 3.12. Calculations and map by Ellen Wilch.

the left panels. Between 20 and 30 percent, the velocity field in the right panels assumes a definite braided pattern evocative of the pattern of streamlines converging at the heads of Antarctic ice streams, notably outlet glaciers through the Transantarctic Mountains. As f_S increases, this pattern disappears, as would be the case in the main trunk of an ice stream. Therefore, the FEM can be used to model the sheet–flow–to–stream–flow transition, even when P_W/P_I is ignored, especially for high-resolution bed topography.

Mapping basal water pressure for stream flow

Mapping basal water pressure P_W along ice streams is an inverse application of Equation (3.64) that exploits P_W/P_I as the dominant variable for stream flow, in which $0 < P_W/P_I < 1$, $f_f = 0$, $f_t = 1$, and $\tau_s = A_s\, \dot{\varepsilon}_s^{1/n}$, where A_s and $\dot{\varepsilon}_s$ are the respective values of A and $\dot{\varepsilon}_{xy}$ in the lateral shear zones of thermally softened and strain softened ice (Raymond, 1996). Since P_I is determined by h_I in Equation (3.1),

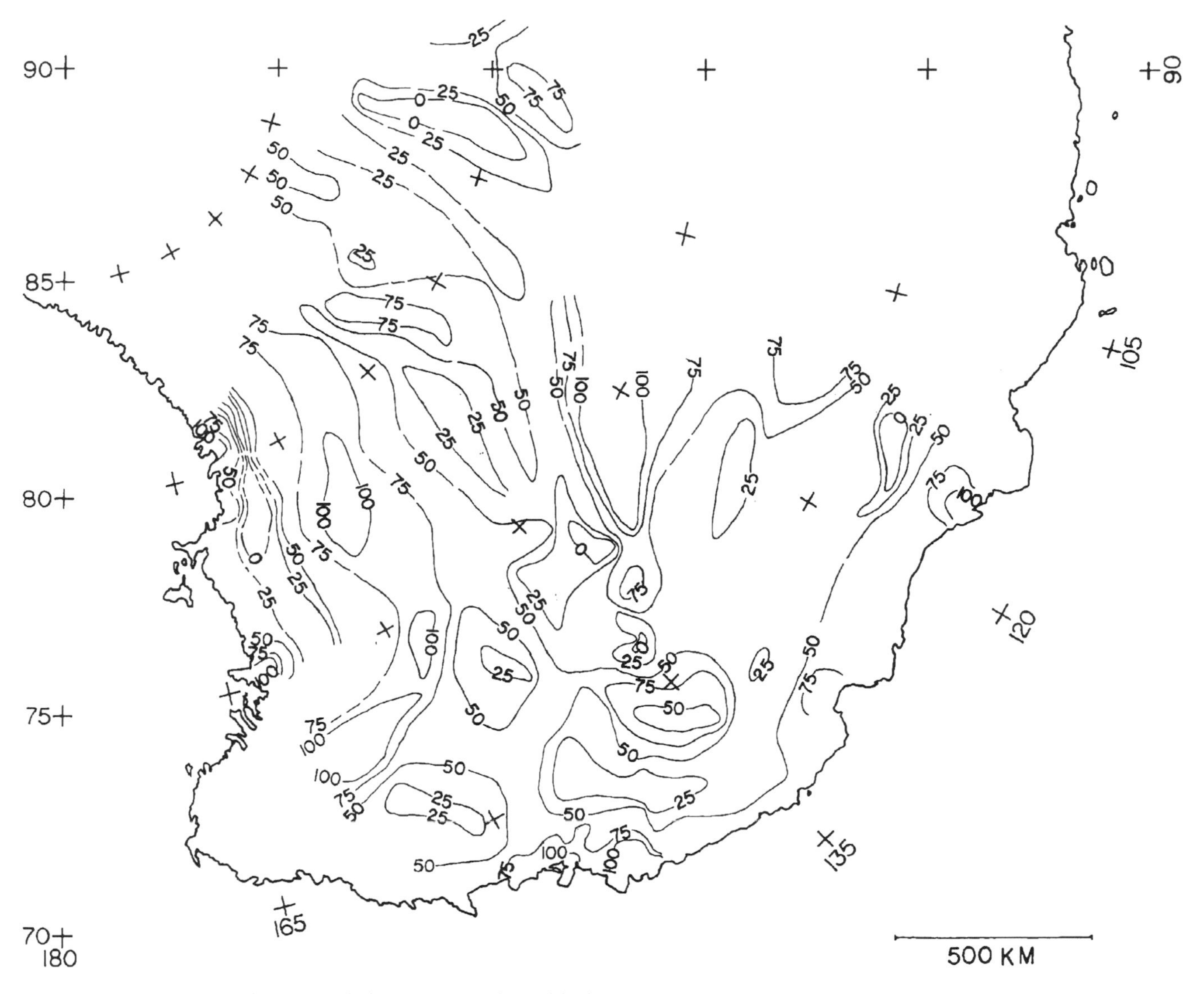

Figure 3.15: Isopleths of percent thawed bed in East Antarctica calculated along the flowlines identified in Figure 3.13. Calculations and map by Ellen Wilch.

P_W can be calculated when h_I, h_0, u_0, A_s, $\dot{\varepsilon}_s$, A, B, C, n, m, c, a, R, ϕ_t and $\Delta h/\Delta x$ in Equation (3.64) are determined from field studies on ice streams in Antarctica and Greenland. In Antarctica, field data are almost adequate to compute P_W along Byrd Glacier and Ice Stream B, which are ice streams 14 and 18 in Figure 3.7. A value of P_W/P_I can be determined for each Δx step such that the surface profile given by Equation (3.64) will match the actual surface profile of the ice stream. If these P_W/P_I values fall within the range $0 < P_W/P_I < 1$, the ice-stream model will be vindicated at least in this fundamental test. Actual P_W/P_I ratios can be measured by drilling through ice streams, as has already been done for Ice Stream B (Kamb, 1991; 1996).

Byrd Glacier is the largest and fastest ice stream supplying the Ross Ice Shelf. It seems to be in equilibrium (Brecher, 1986). We placed photogrammetric ground control on Byrd Glacier and along the sidewalls of Byrd Glacier fjord during the 1978–1979 Antarctic summer in order to map surface ice elevations and velocities photogrammetrically (Hughes and Fastook, 1981). Results were published by Brecher (1982), and were later used by Whillans and others (1989) to compute the force balance. Three radio-echo flights were flown along Byrd Glacier, but only one gave a fairly continuous bottom reflection, see Figure 3.20. The rugged bed topography favors bedrock outcrops through subglacial till. All of these data were used by Scofield and others (1991) to calculate the fraction of measured surface ice velocity due to basal sliding at various distances from the grounding line of the Ross Ice Shelf in Byrd Glacier fjord,see Figure 3.21. Scofield (1988) also calculated surface strain rates $\dot{\varepsilon}_{xx}$, $\dot{\varepsilon}_{yy}$, and $\dot{\varepsilon}_{xy}$ from the surface velocity field measured by Brecher (1986). These results are presented as principal strain rates $\dot{\varepsilon}_1$ and $\dot{\varepsilon}_2$ in Figure 3.22. Zhao (1990) densified the surface velocity and strain rate measurements in

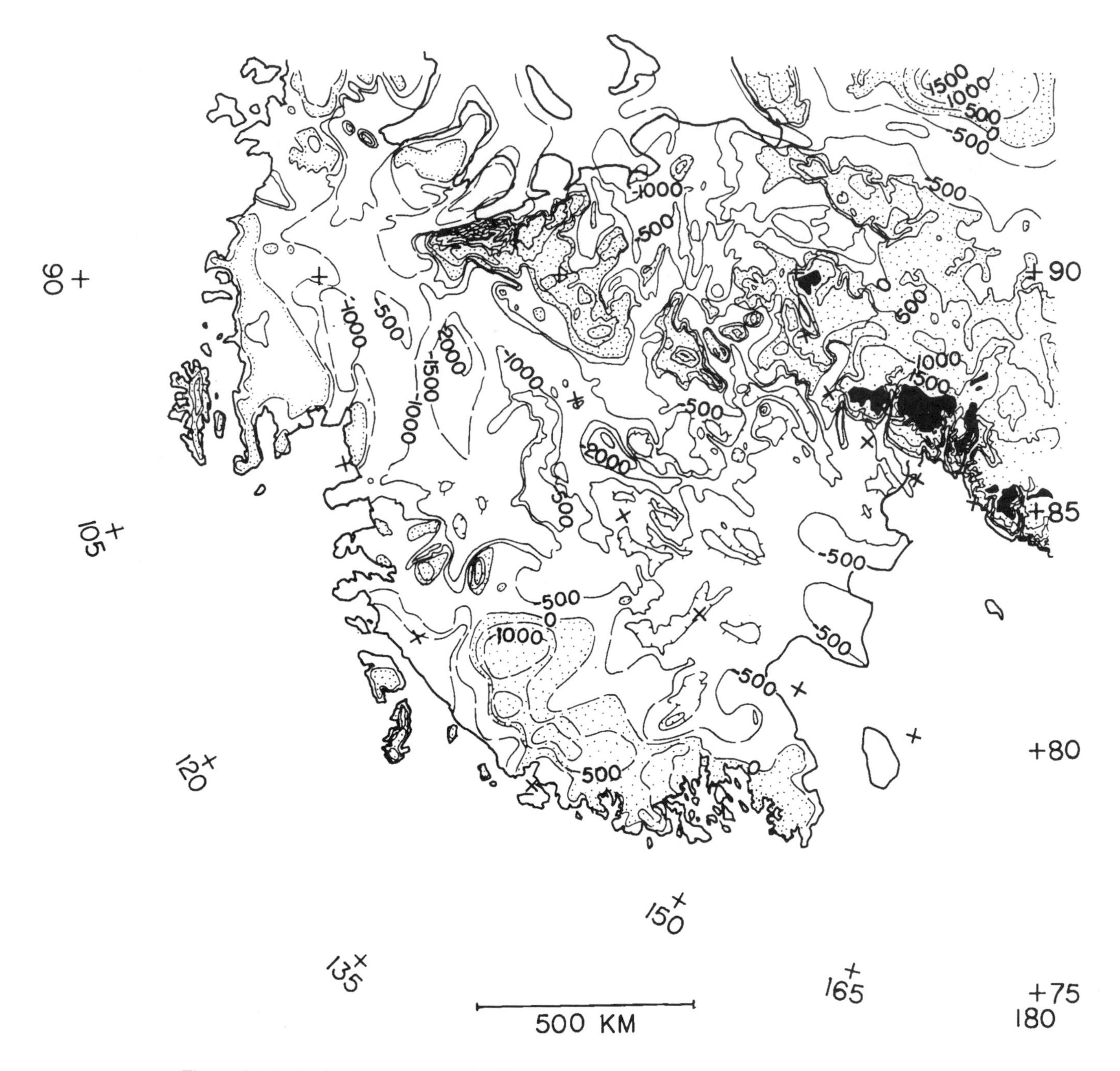

Figure 3.16: Bedrock topography in West Antarctica above the ice surface (black areas) and contoured every 500 m above sea level (shaded areas) or below sea level (white areas) for comparison with the percent of thawed bed in Figure 3.14.

the north shear zone of Byrd Glacier fjord. These data are adequate to compute P_W/P_I variations along Byrd Glacier using Equation (3.64), as previously discussed, if ice slides on bedrock so that C and c can be ignored. Velocity profiles shown in Figure 3.21 will be increasingly dominated by creep as $P_W/P_I \rightarrow 0$ and by sliding as $P_W/P_I \rightarrow 1$. This will provide an independent test of the theoretical ice-stream model, since velocity profiles in Figure 3.21 were obtained from conservation of mass flux, not from a balance of forces. An independent test for the force balance is provided by the model of Whillans and others (1989). Of particular interest is the role of side shear, as measured in detail by Zhao (1990), in constraining ice flow, since Whillans and Van der Veen (1993a, 1993b) maintain that side shear is the major constraint on the velocity of Ice Stream B.

Ice Stream B is nearly as large and as fast as Byrd Glacier, but it is longer, has a much lower slope, is unconstrained by bedrock sidewalls, and slides on a soft till substrate (Blankenship and others, 1987) that deforms near the plastic end of the viscoplastic creep spectrum (Kamb, 1991).

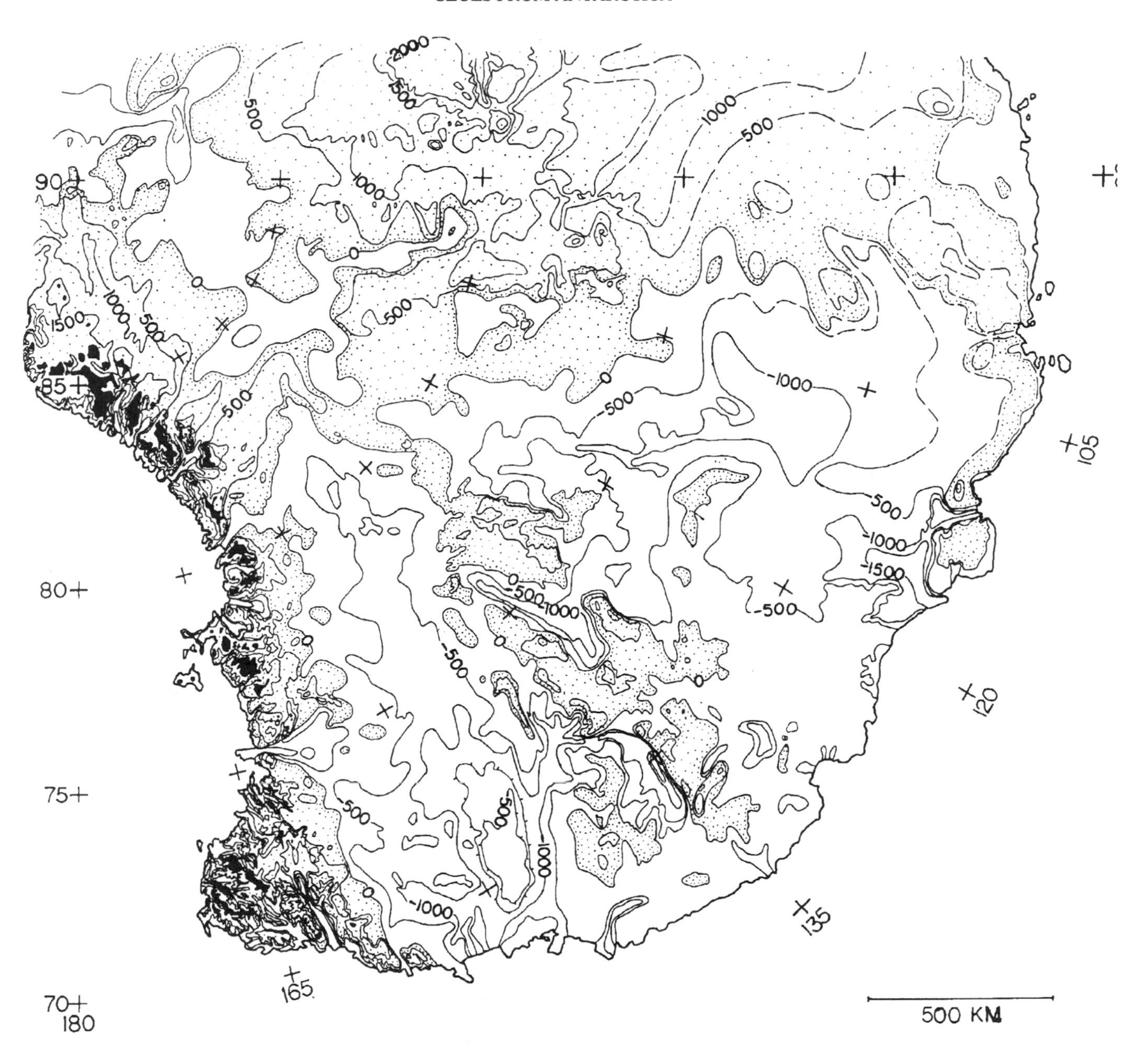

Figure 3.17: Bedrock topography in East Antarctica above the ice surface (black areas) and contoured every 500 m above sea level (shaded areas) or below sea level (white areas) for comparison with the percent of thawed bed in Figure 3.15.

However, surface crevasse fields imply "sticky spots" on the bed where bedrock may poke through the substrate and penetrate basal ice (Whillans and others, 1987). It is possible that the unusual length of Ice Stream B is a result of subglacial volcanism in a rift zone near the West Antarctic ice divide that provides water-soaked ash as a basal lubricant (Blankenship and others, 1993), but ash was not recovered from a corehole through Ice Stream B (Engelhardt and others, 1990). The substrate was a wet, clay-rich till with a plastic yield stress of 1.6 x 10^{-2} bar (Kamb, 1991), so it can be assumed that resistance to sliding is due to bedrock " sticky spots" or side shear rather than the creep strength of the substrate. This result allows C and c to be ignored in applying Equation (3.64) to Ice Stream B. Unlike Byrd Glacier, Ice Stream B has not eroded a deep fjord, its bed is not rugged, and it is not in equilibrium. As on Byrd Glacier, intensive aerial photogrammetric measurements have been conducted on Ice Stream B for the purpose of determining surface strain rates (Brecher, 1986). The mass balance has been studied over the catchment area (Whillans and Bindschadler, 1988) and length (Shabtaie and others, 1988) of Ice Stream B, its surface and subglacial topography have been

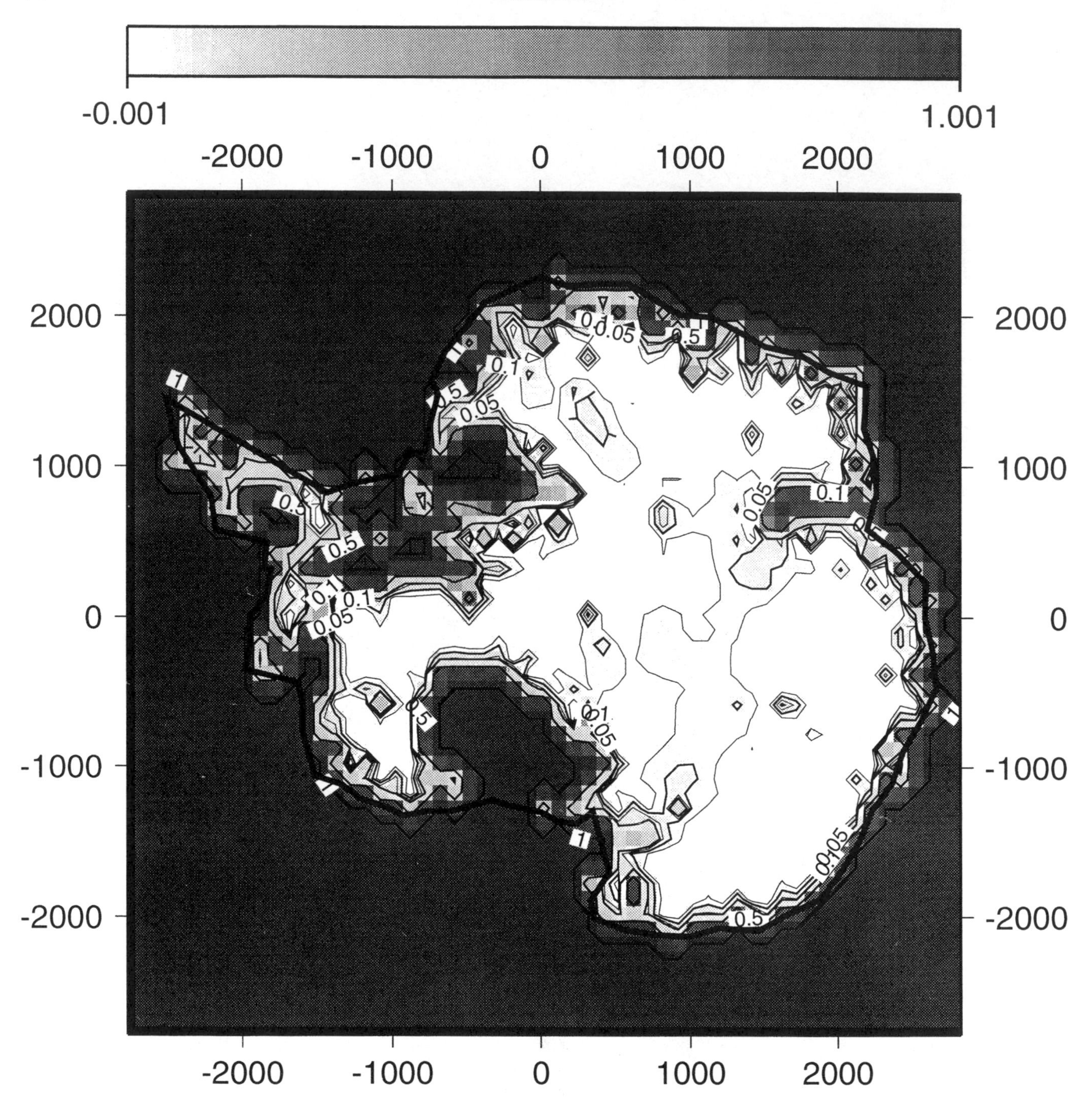

Figure 3.18: Fraction of ice velocity due to basal sliding computed at gridpoints 20 km apart using Equations (3.107) and (3.108) for $f_D = 0$, so that $f_C = 1 - f_S$. Shading grades from no sliding (white end) to only sliding (black end). Note that sliding dominates over much more restricted areas than thawing in Figures 3.14 and 3.15, as quantified in Figure 3.10. Computer plot by James L. Fastook

mapped (Shabtaie and Bentley, 1988), its interaction with the Ross Ice Shelf has been investigated (Bindschadler and others, 1987), deformation of its bed has been studied both seismically (Alley and others, 1987a, 1987b) and by deep drilling (Engelhardt and others, 1990; Kamb, 1991), and erosion of its deforming substrate has been linked to stability of its grounding line(Alley and others, 1989). In contrast to the FDM in which stream flow is controlled by bedrock "sticky spots" projecting through a slippery substrate, MacAyeal (1989) modeled Ice Stream B using a FEM in which the substrate was stiff enough to control stream flow. Other approaches include those by Bindschadler and Gore (1982),

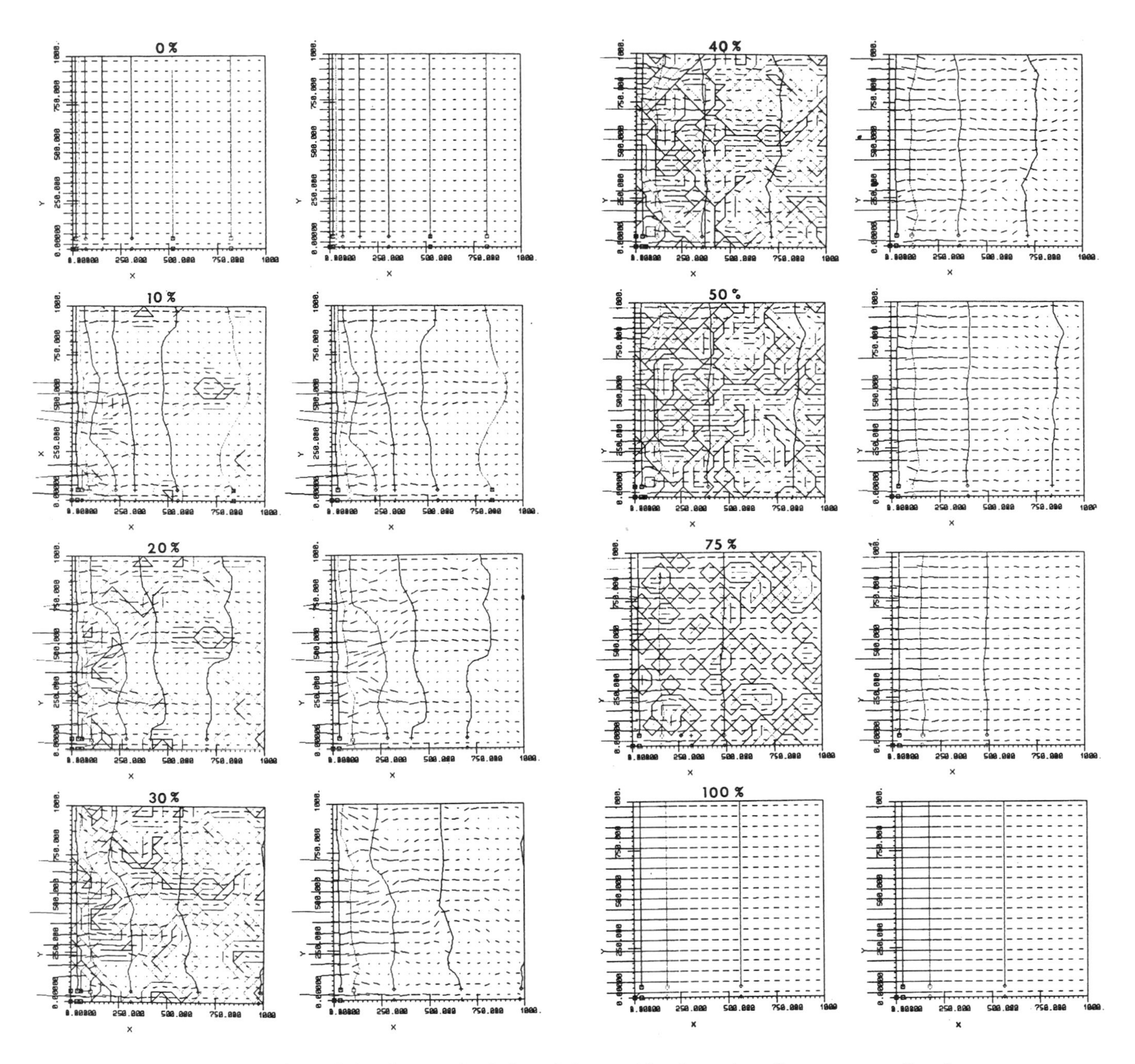

Figure 3.19: A finite-element simulation of the transition from sheet flow to stream flow by randomly specifying sliding at nodes of a 20 × 20 grid on a flat bed. Sliding occurs at 0, 10, 20, 30, 40, 50, 75, and 100 percent of the nodes, each inside a square box as shown. Surface ice elevation contours increase from left to right at 500 m intervals, and are crossed at right angles by surface ice velocity vectors. All lines were plotted using color-coded graphics and therefore appear as shades of gray. Map-plane axes x, y are shown in km. Surface accumulation is 5 cm/yr, and an ice flux of 80,000 m^2/yr enters the grid from the right. Both columns show ice surface elevations and ice surface velocities. Left column also shows basal areas where glacial sliding dominated. Computer plot by James L. Fastook.

Gore (1982), Alley (1984), Alley and Whillans (1984), McInnes and Budd (1984), Echelmeyer and Kamb (1986), Muszynski and Birchfield (1987), van der Veen and Whillans (1989), and Whillans and others (1989).

The Antarctic Ice Sheet at the Last Glacial Maximum

Calculating former flowline profiles of the Antarctic Ice Sheet for specified basal conditions is a forward application

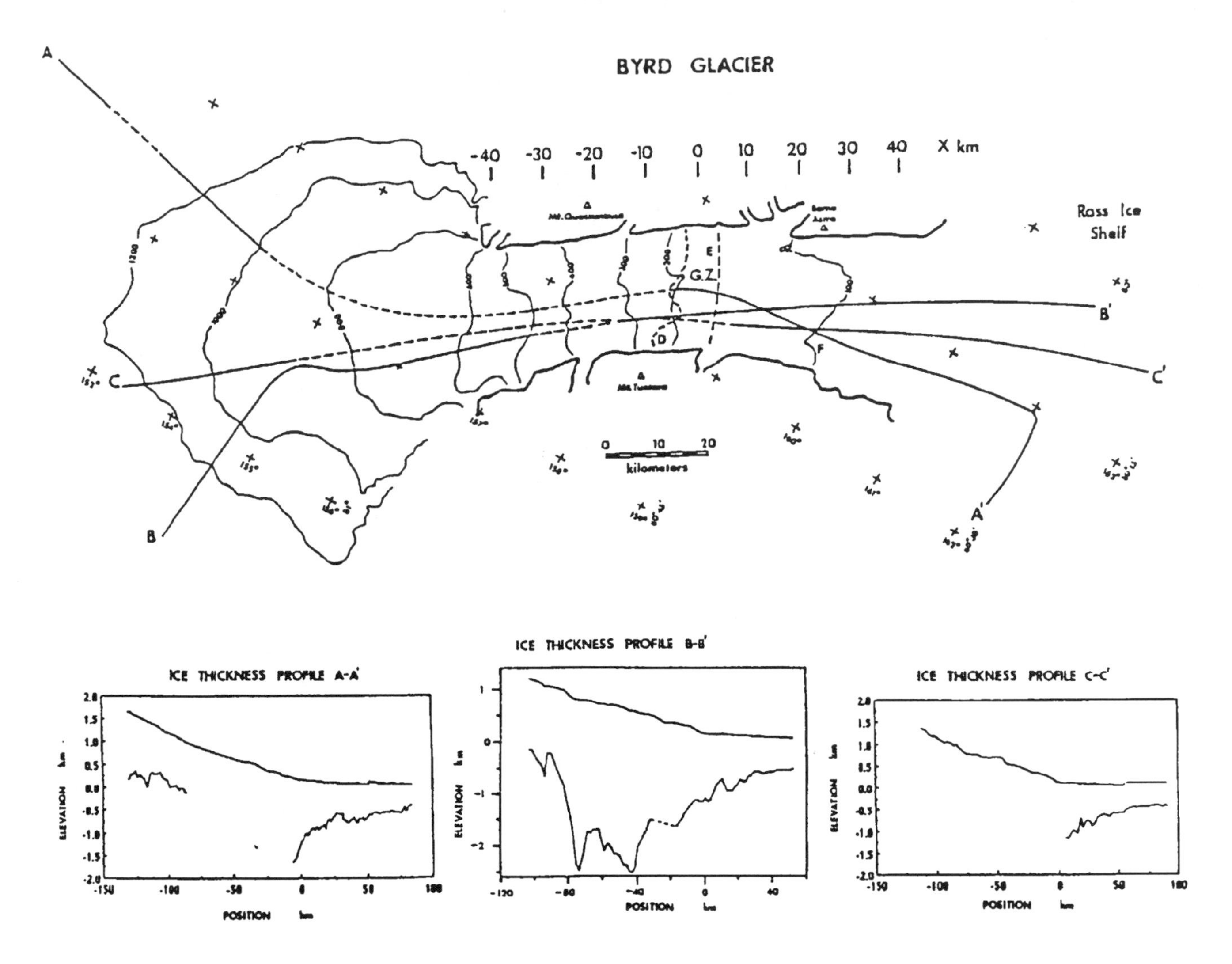

Figure 3.20: Radar sounding of the Byrd Glacier area. *Top*: Radio-echo flightlines on Byrd Glacier. *Bottom*: Ice thickness of Byrd Glacier along the radio-echo flightlines. Flow is from left to right, the fjord entrance is located near x = 40 km. Mass-balance calculations were made along a flowband that includes radio-echo flightline B-B' in the half of Byrd Glacier fjord where ice is grounded. Note the coordinates for the x axis, and the location of the grounding zone (G.Z.), between transverse broken lines. Data on bed topography are absent for the dashed portions of the flightlines (Scofield and others, 1991). Reproduced with permission.

of Equation (3.64), in which the percent of thawed bed mapped in Figures 3.14 and 3.15 from the inverse application is used to compute ice-surface elevations at the last glacial maximum. The Antarctic Ice Sheet has experienced substantial gravitational collapse since the last glacial maximum, especially the marine portions in West Antarctica that were most vulnerable to rising sea level caused by the collapse of Northern Hemisphere ice sheets. Active and passive responses of the ice sheet to changing sea level can be studied by fitting known former ice elevations along Antarctic flowlines during collapse to flowline profiles for active stream flow and passive sheet flow using a generalized equation that allows sheet flow to become stream flow in a flowband from the ice divide to the grounding line (Kellogg and others, 1996). To reconstruct former flowline profiles in the FDM, Equation (3.64) must be modified to include glacio-isostatic depression of the bed from h_R to $h_R{}^*$ beneath the former ice load, which lowered the surface from h to h*. Equation (3.85) is the result, where τ_o is given by Equations (3.32), (3.44), (3.51), and (3.57), and r is given by Equations (3.83) and (3.92) for $r_a = r_o$:

$$r = \frac{h_R - h_R{}^*}{h^* - h_R} = \left(\frac{\rho_I}{\rho_R - \rho_I}\right) exp\left(-\frac{t}{t_o}\right) \tag{3.117}$$

where t is the time since gravitational collapse began, t_o is the relaxation time, and ρ_R is the average rock density in Earth's mantle. At the last glacial maximum, t = 0 and

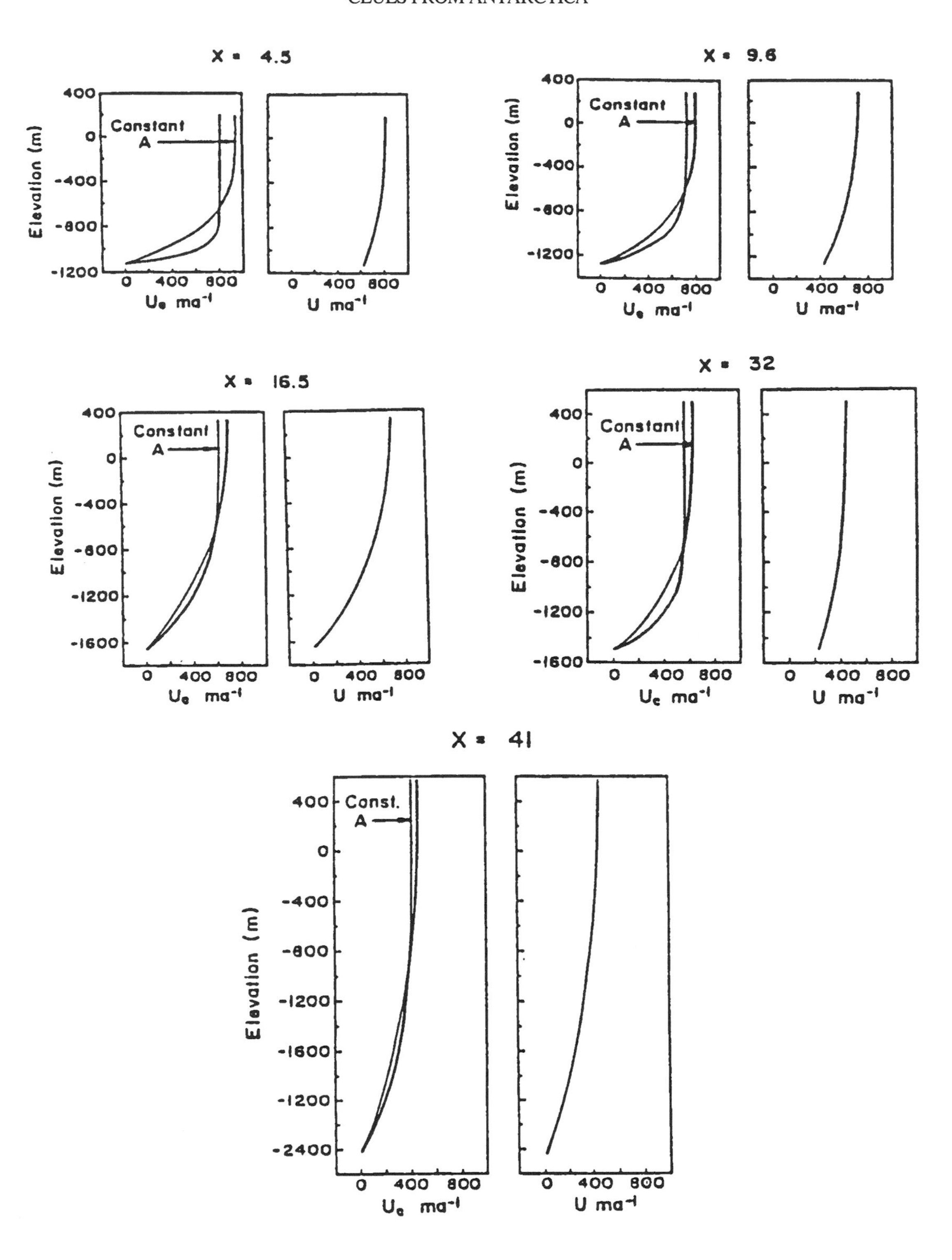

Figure 3.21: Vertical velocity profiles through Byrd Glacier computed for frozen (left) and thawed (right) beds at x = 4.5, 9.6, 16.5, 32, and 41 km upstream from the grounding line for flightline B-B' in Figure 3.20. Velocity profiles for a frozen bed are shown for laminar flow in which the measured surface velocity is reproduced for variable A, but not for constant A, when n = 3 in the flow law of ice, $\dot{\varepsilon}_c = (\sigma_c / A)^n$. Velocity profiles for a thawed bed reproduce the measured surface velocity for variable A, constant effective strain rate, and a linear temperature profile, when n = 3 (Scofield and others, 1991, Figure 4). Reproduced with permission.

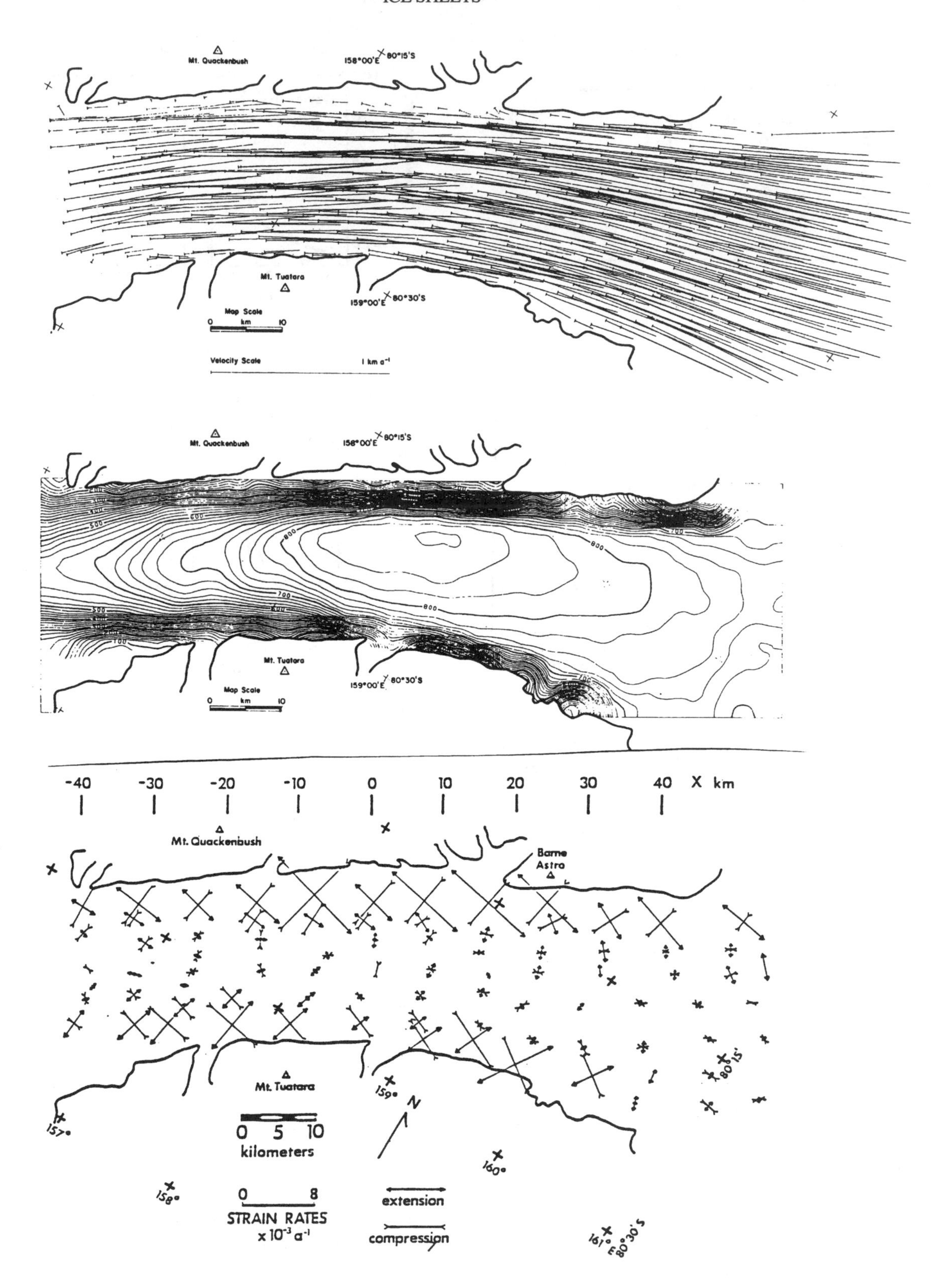

Figure 3.22: Surface velocities and strain rates on Byrd Glacier. *Top*: Surface velocity vectors (from Brecher, 1986). *Center*: Isopleths of surface velocity at contour intervals of 250 m/a (from Brecher, 1986). *Bottom*: Principal strain rates obtained from 82 clusters of from 4 to 10 measured surface velocity vectors (from Scofield, 1988). Reproduced with permission.

$r = \rho_I/(\rho_R - \rho_I)$ for glacioisostatic equilibrium. In Equation (3.85), $\Delta h^* = h_{i+1}^* - h_i^*$, Equations (3.32), (3.44), (3.51), and (3.59) are used to express τ_o, a^* replaces a to account for ice thinning rate $\delta h_I^*/\delta t$ over time t as given by Equation (3.90) written as:

$$a^* = a - \delta h_I^*/\delta t = a - \delta\,(h^* - h_R^*)/\delta t \qquad (3.118)$$

Therefore, a^* is an effective ice accumulation rate to account for reduced flow when ice thickens (positive $\delta h_I^*/\delta t$) and increased flow when ice thins (negative $\delta h_I^*/\delta t$).

The resulting initial-value finite-difference recursive formula, taking i = 0 at the grounding line as the initial value, is:

$$h_{i+1}^* = h_i^* + \frac{(1-\rho_I/\rho_W)}{(1+r)^{1/2}}\left(\frac{P_W}{P_I}\right)^2 \left\{\frac{a^*\left[h_B + (1+r)^{1/2}(h^* - h_B) - h_R\right]_i}{a^*\,x + h_0\,u_0} - \frac{\left[h_B + (1+r)^{1/2}(h^* - h_B) - h_R\right]_i^{n+2} R}{a^*\,x + h_0\,u_0}\left[\frac{\rho_I\,g_z}{2\,A}\left(1 - \frac{\rho_I}{\rho_W}\right)\left(\frac{P_W}{P_I}\right)^2\right]^n\right\}\Delta x + \left\{\frac{\left[h_B + (1+r)^{1/2}(h^* - h_B) - h_R\right]_i}{2}\left(1 - \frac{\rho_I}{\rho_W}\right)\left(\frac{P_W}{P_I}\right)\frac{\Delta}{\Delta x}\left(\frac{P_W}{P_I}\right)\right\}\Delta x + \left\{\frac{(1-f_t)\,\tau_C + f_t\,(\tau_S + \tau_D)}{\rho_I\,g_z\left[h_B + (1+r)^{1/2}(h^* - h_B) - h_R\right]_i} + \frac{2\,\tau_s}{\rho_I\,g_z\,w_I}\right\}\Delta x \qquad (3.119)$$

At the grounding line of Antarctic ice shelves, glacio-isostatic depression is assumed to be nil, so that $h = h^* = h_S$ is the ice surface elevation and $h_B = h_B^* = h_G$ is the bed depth. From the buoyancy requirement for floating ice:

$$h_S = h_G + h_0 = h_G - (\rho_W/\rho_I)\,h_G = (1 - \rho_W/\rho_I)\,h_G \qquad (3.120)$$

where h_G is negative because it is depth below sea level and $h_0 = h_S - h_G$ is ice thickness at i = 0.

Equation (3.119) gives flowline profiles for sheet flow when $P_W << P_I$ and $\tau_s = 0$, for stream flow when $f_t = 1$, and for shelf flow when $P_W = P_I$ and $f_t = 1$, with r varying from $r = \rho_I/(\rho_R - \rho_I)$ when glacioisostatic depression attains equilibrium at the last glacial maximum to r = 0 when glacioisostatic rebound is complete following deglaciation. Variations of r have little effect in Equation (3.119), since $(1 + r)^{1/2} = 2\sqrt{3} = 1.15$ when $\rho_I = 4\,\rho_I$ gives r = 1/3 and $(1 + r)^{1/2} = 1$ when r = 0. As shown by Equations (3.44), (3.51), and (3.59), τ_C, τ_S, and τ_D in Equation (3.119) depend weakly on surface accumulation rates, for constant flowband widths. Each flowline is the centerline of a flowband, and adjacent flowlines reveal convergence and divergence of flow that changes the width of the flowband. Chapter 5 shows how this can be included in Equation (3.119) so that it becomes a fully generalized expression for calculating ice elevations along ice-sheet flowlines.

Variations of P_W/P_I along a flowline can be calculated by simultaneously solving the equations of heat transport and of mass transport, using numerical methods. In particular, models for sheet flow (e.g., Huybrechts, 1990), and for stream flow (e.g., Funk and others, 1994) must be combined. Alternatively, variations of P_W/P_I along flow-lines can be obtained by fitting Equation (3.119) to ice elevations that are known from glacial geology. Former ice elevations are known only where the West Antarctic Ice Sheet has partially or totally collapsed, notably along the Ellsworth Mountains and the Transantarctic Mountains in the Weddell Sea and Ross Sea embayments. These former ice elevations are compatible with nearly stagnant ice that thickened slowly over a frozen bed after lowering sea level grounded the Ronne and Ross Ice Shelves or with active ice streams that occupied marine troughs containing water-soaked sediments. Both possibilities can be modeled. Former ice elevations are known because nunataks are widespread even in the Antarctic interior, and many outlet glaciers pass through the Transantarctic Mountains, which bisect Antarctica. Depositional moraines and erosional "trimlines" locate former inland ice elevations at these sites (e.g., Mercer, 1968a, 1972; Carrara, 1979, 1981; Stuiver and others, 1981; Denton and others, 1986a, 1986b, 1989, 1991, 1992; Orombelli and others, 1990). These former ice elevations cannot be used to uniquely distinguish between sheet flow over a frozen bed and stream flow over a thawed bed because the glacial geology is often discontinuous and undated, so the accuracy needed for inverse modeling is lacking. Under these circumstances, forward modeling is possible using a simple empirical expression for the variation of P_W/P_I along a flowline:

$$\frac{P_W}{P_I} = \left(1 - \frac{x}{L_S}\right)^c \qquad (3.121)$$

where x = 0 at the grounding line, L_S is the length of an ice stream, $c = \infty$ for sheet flow, $0 < c < \infty$ for stream flow, and c = 0 for shelf flow. Differentiating Equation (3.121) gives the gradient of P_W/P_I along a flowline:

$$\frac{\Delta}{\Delta x}\left(\frac{P_W}{P_I}\right) = -\frac{c}{L_S}\left(1 - \frac{x}{L_S}\right)^{c-1} \qquad (3.122)$$

Note that L_S must be specified in Equation (3.122). Substituting Equations (3.121) and (3.122) into Equation (3.119) locates the inflection point where the flowline surface profile changes from concave for stream flow to convex for sheet flow. This should coincide with L_S.

An empirical expression for which both P_W/P_I and $\partial(P_W/P_I)/\partial x$ are continuous at the head ($x = L_S$) and the foot

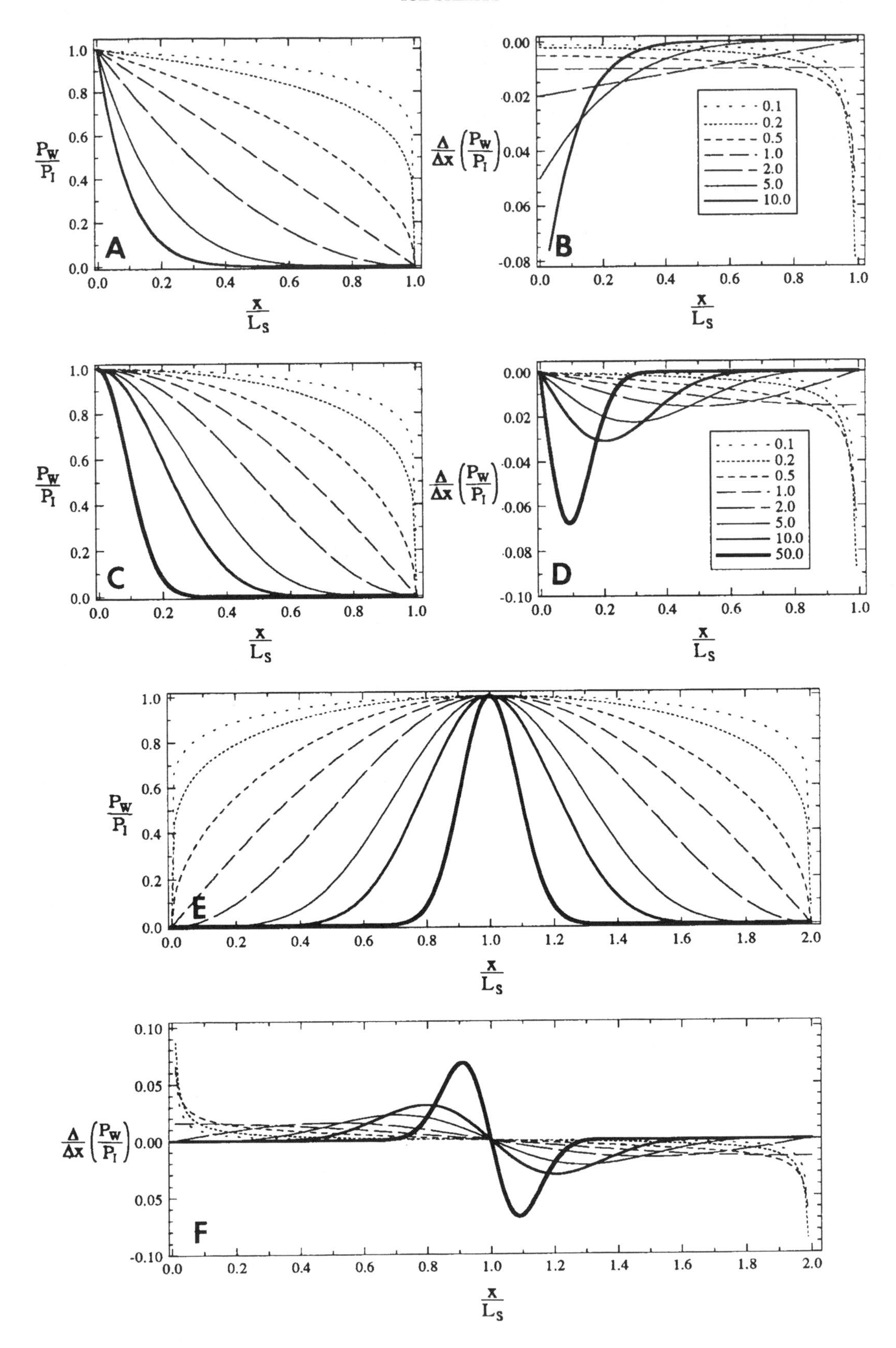

Figure 3.23: Plots of Equations (3.121) through (3.126), provided by Kirk Maasch.
A. Equation (3.121) B. Equation (3.122) C. Equation (3.123)
D. Equation (3.124) E. Equation (3.125) F. Equation (3.126)

(x = 0) of a marine ice stream is:

$$\frac{P_W}{P_I} = cos^c\left(\frac{\pi x}{2 L_S}\right) \qquad (3.123)$$

where x = 0 at the grounding line for an ice stream of length L_S, and $0 < c < \infty$ for stream flow, with $c = \infty$ for sheet flow and c = 0 for shelf flow. The gradient of P_W / P_I obtained by differentiating Equation (3.123) is:

$$\frac{\Delta}{\Delta x}\left(\frac{P_W}{P_I}\right) = -\left(\frac{c\,\pi}{2 L_S}\right) cos^{c-1}\left(\frac{\pi x}{2 L_S}\right) sin\left(\frac{\pi x}{2 L_S}\right) \qquad (3.124)$$

Equations (3.121) through (3.124) are for marine ice streams, which drain the Antarctic Ice Sheet. Terrestrial ice streams that drained the landward flank of former Quaternary ice sheets in the Northern Hemisphere ended as terminal ice lobes. For these terrestrial ice streams, the counterpart of Equation (3.123) is:

$$\frac{P_W}{P_I} = sin^c\left(\frac{\pi x}{2 L_S}\right) \qquad (3.125)$$

where x = 0 at the terminus of an ice lobe of length L_S that is the continuation of an ice stream of length L_S, $x = 2\,L_S$ at the head of the ice stream, and $0 < c < \infty$ for stream flow, including the ice lobe, with c = 0 for sheet flow. The gradient of P_W / P_I obtained from Equation (3.125) is:

$$\frac{\Delta}{\Delta x}\left(\frac{P_W}{P_I}\right) = \left(\frac{c\,\pi}{2 L_S}\right) sin^{c-1}\left(\frac{\pi x}{2 L_S}\right) cos\left(\frac{\pi x}{2 L_S}\right) \qquad (3.126)$$

Equations (3.121) through (3.126) are plotted in Figure 3.23.

A disadvantage of Equation (3.125) is that the concave ice stream is the same length as its convex ice lobe, because of symmetry. The physical basis for this symmetry is that it reflects the collapse of ice that produces the concave ice stream profile and the forward redistribution of collapsed ice to produce the convex ice lobe surface in the absence of ice ablation. With ablation, a more probable profile for the ice lobe is obtained by integrating Equation (3.112) for $\alpha = dh/dx$ and $h_I = h - h_R$ when τ_o and h_R are constant:

$$h = h_R + (2\,\tau_o\,x/\rho_I\,g_z)^{1/2} \qquad (3.127)$$

where τ_o varies from 0.1 bar for a low-traction bed such as wet sediments or till to 1.0 bar for a high-traction bed such as permafrost or bedrock.

Equation (3.120) provides the initial ice thickness needed to reconstruct the Antarctic Ice Sheet at the last glacial maximum (21 ka BP) in a forward application of Equation (3.119). Figure 3.24 shows the distribution of ice accumulation over Antarctica, based on the present-day pattern of accumulation mapped by Giovinetto and others (1989) but modified for conditions at the last glacial maximum using the method developed by Fastook and Prentice (1994). Figure 3.25 shows the distribution of thawed basal conditions, based on Figures 3.14 and 3.15 for present-day conditions. Figure 3.26 shows the distribution of surface flowlines obtained by extending present-day flowlines along submarine troughs to the edge of the continental shelf (Drewry, 1983). These data and former ice elevations recorded by glacial geology should be used to solve Equation (3.119) for f_t when $P_W / P_I << 1$ for sheet flow and for P_W / P_I when $f_t = 1$ for stream flow, because these variables range between zero and unity, and therefore have the greatest influence on determining former ice elevations. In this forward application, τ_D and τ_s can be ignored, and τ_C and τ_S are linked to ice accumulation rate a through Equation (3.44) and (3.51). Even using the highest glacial geological features and assuming that they all were created during the last glacial maximum, they are compatible only with very gently sloping flowline profiles in the Ross Sea and Weddell Sea embayments. Two scenarios can produce such low profiles.

In the slow-flow scenario, Antarctic ice shelves became grounded on the continental shelf as sea level fell during the last glaciation cycle. Heat was rapidly conducted upward through this initially thin ice, causing the bed to freeze. Newly grounded ice would creep very slowly because the surface slope was very low and the bed was frozen. Therefore, $P_W / P_I = 0$ for a frozen bed and a* is minimized because $a \approx \delta h_I{}^*/\delta t$ in Equation (3.118) for nearly stagnant ice. Setting $P_W / P_I = f_t = \tau_s = 0$ in Equation (3.119) for the newly grounded ice, and keeping $\Delta h/\Delta x$ low enough to allow calculated flowline profiles to conform with the glacial geology, the appropriate values of a* can then be calculated by substituting a* for a in Equation (3.44). Calculated values of a* can be compared with a mapped in Figure 3.24, so that $\delta h_I{}^*/\delta t$ can be calculated. In this scenario, the ice sheet has a passive response to changing sea level because sheet flow over frozen submarine sediments or till is a passive response to surface conditions.

In the fast-flow scenario, Antarctic ice streams advanced along submarine troughs to the edge of the continental shelf as sea level fell during the last glaciation cycle. Frictional heat from basal sliding beneath the ice streams kept the bed thawed, so the ice streams advanced rapidly over water-soaked marine sediments or till. Therefore, $f_t = 1$, $\tau_s \approx 1.0$ bar, τ_S is computed from Equation (3.51) using a in Figure 3.19 because $\delta\, h_I{}^*/\delta t \approx 0$ in Equation (3.118), and P_W / P_I can be varied along these ice streams to give the best fit between flowline profiles calculated from Equation (3.119) and the glacial geology. In this scenario, the ice sheet has an active response to rising sea level because stream flow over thawed sediments or till in the troughs is an active response to bed conditions.

Figure 3.27 shows ice surface elevation contours that are obtained for both the slow-flow and fast-flow scenarios by solving Equation (3.119) along the flowlines in Figure 3.26. Both solutions are controlled by the known glacial geology, so they are forward modeling applications of Equation (3.119). Since the known terrestrial glacial geology allows both of these end members in the spectrum of glacial flow, intermediate flow conditions would also be compatible with that glacial geology. However, marine glacial geolo gy studied by coring and seismic profiling revealed extensive

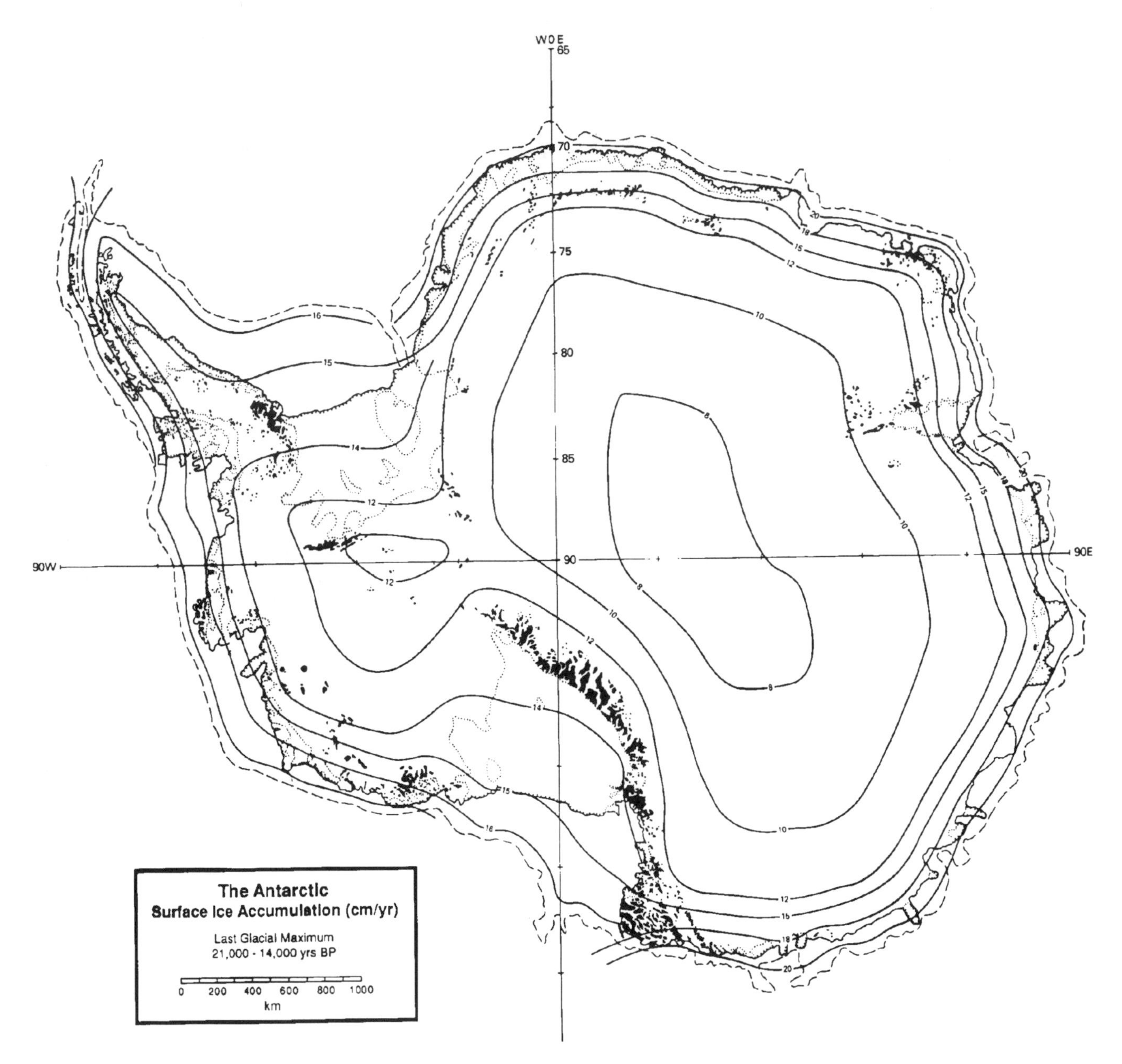

Figure 3.24: Isopleths of surface accumulation rates a (cm/a) over the Antarctic Ice Sheet at the last glacial maximum based on present-day accumulation rates (Giovinetto and others, 1989) adjusted for full glacial conditions (Fastook and Prentice, 1994). From work performed for Battelle, Pacific Northwest Laboratories. Reproduced with permission.

glacial erosion in the submarine troughs and revealed the provenance of glacial till on the continental shelf and slope (Anderson and others, 1980, 1992; Anderson and Thomas, 1991; Elvehøi, 1981; Licht and others, 1996). This indicates that the fast-flow scenario prevailed.

Gravitational Collapse of the Antarctic Ice Sheet

Gravitational collapse of the Antarctic Ice Sheet since the last glacial maximum left either sea ice or ice shelves on much of the Antarctic continental shelf that had been glaciated with grounded ice. The dotted areas in Figure 3.28 show the extent of deglaciation. Deglaciation has reduced the area of grounded ice by more than 70 percent in West Antarctica, compared to less than 10 percent in East Antarctica. In West Antarctica, gravitational collapse has produced two large embayments, one in the Ross Sea (e.g., Anderson and others, 1992; Denton and others, 1989; Licht and others, 1996; Kellogg and others, 1996) and one in the Weddell Sea (e.g., Carrara, 1979, 1981; Anderson and others, 1980; Elverøi, 1981), and a small embayment in the Amundsen Sea (e.g.,

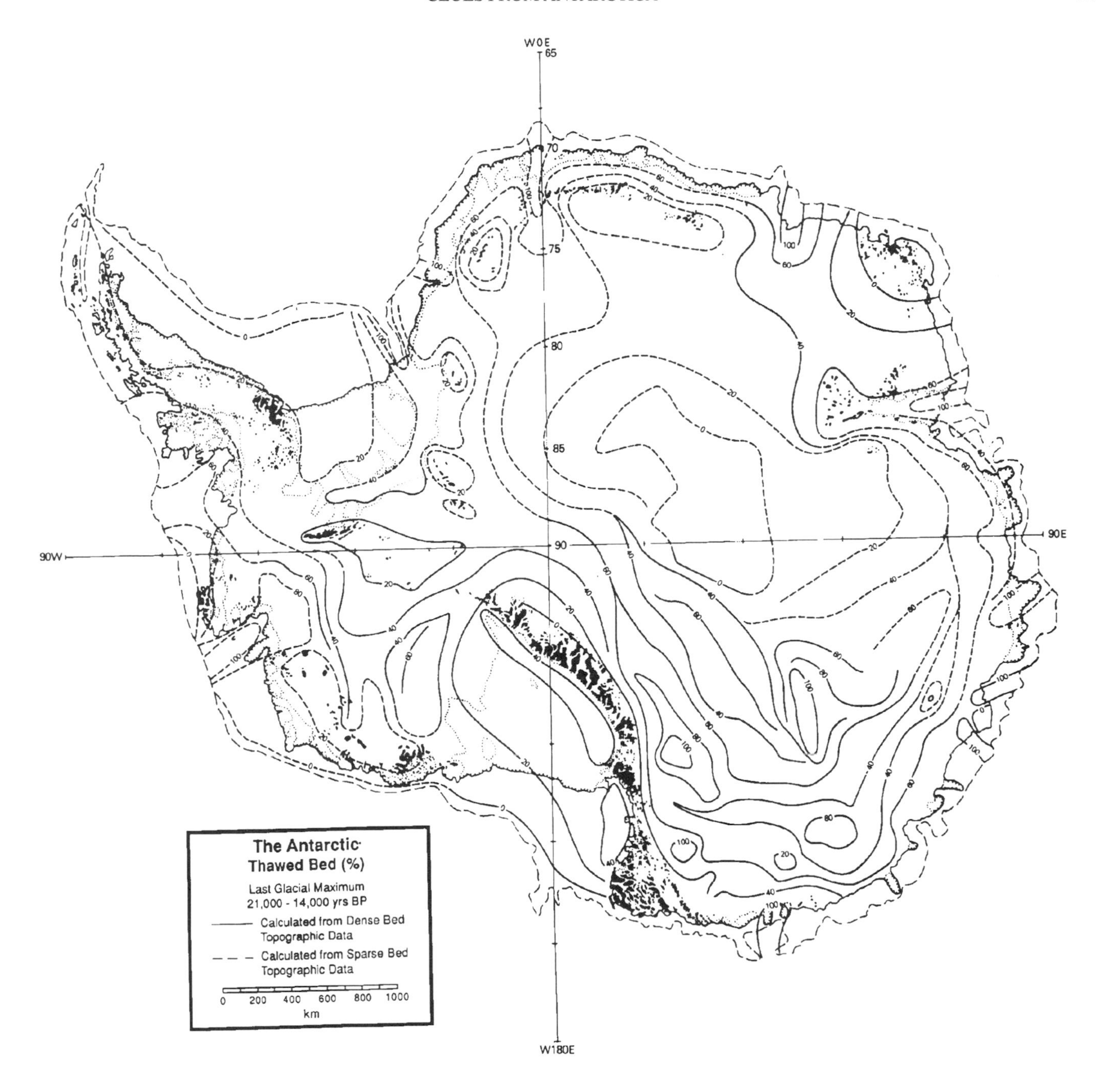

Figure 3.25: Isopleths of the percentage of thawed bed beneath the Antarctic Ice Sheet at the last glacial maximum based on present-day percentages calculated by Ellen Wilch (unpublished) from Equations (3.113) through (3.116) and mapped in Figures 3.12 and 3.13, but modified for lower accumulation rates during full glacial conditions. From work performed for Battelle, Pacific Northwest Laboratories. Reproduced with permission.

Kellogg and others, 1985; Kellogg and Kellogg, 1987a, 1987b, 1987c). Floating ice shelves now occupy the southern half of the Ross Sea and Weddell Sea embayments. However, the southern part of the Amundsen Sea embayment is a polynya, in Pine Island Bay, which is kept ice-free by strong katabatic winds that flow down Pine Island Glacier and Thwaites Glacier. These three embayments are shown in Figure 3.29. In all three, submarine troughs extend from present-day West Antarctic ice streams to the edge of the West Antarctic continental shelf. Marine glacial geology indicates that these ice streams occupied the submarine troughs and reached the edge of the continental shelf at the last glacial maximum (Anderson and others, 1980, 1992; Elverhøi, 1981; Kellogg and Kellogg, 1987a, 1987b, 1987c;

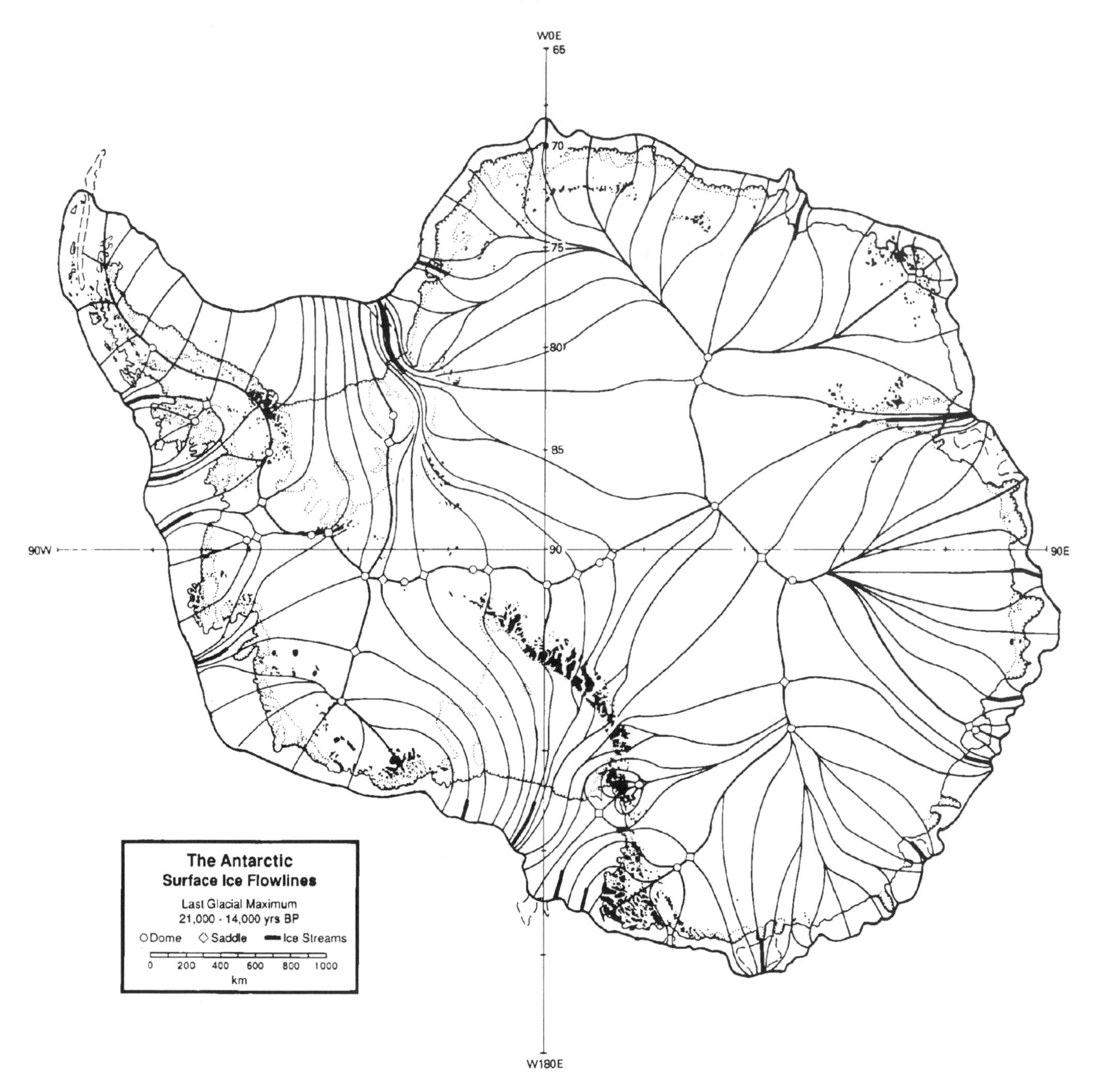

Figure 3.26: Selected surface flowlines of the Antarctic Ice Sheet at the last glacial maximum obtained by extending present-day flowlines in Figure 3.7 along submarine troughs to the edge of the Antarctic continental shelf. From work performed for Battelle, Pacific Northwest Laboratories. Reproduced by permission.

Kellogg and others, 1996; Licht and others, 1996). In that case, the fast-flow scenario existed at the last glacial maximum and the Antarctic Ice Sheet had an active response to the subsequent rise in sea level.

Figure 2.13 shows plan and profile views of flowlines down opposite flanks of the high saddle of the West Antarctic Ice Sheet. The flowline to the end of Thwaites Glacier has a high, mostly convex sheet-flow surface over a deep subglacial basin, and Thwaites Glacier calves into open water. Its grounding line is on a bed that slopes steeply down into this basin. As shown in Figure 2.2, this is a situation in which much gravitational potential energy is poised precariously for imminent release if the grounding line retreats. The flowline to the end of Ice Stream D has a low, mostly concave stream-flow surface over a shallow subglacial basin, and Ice Stream D becomes afloat as part of the

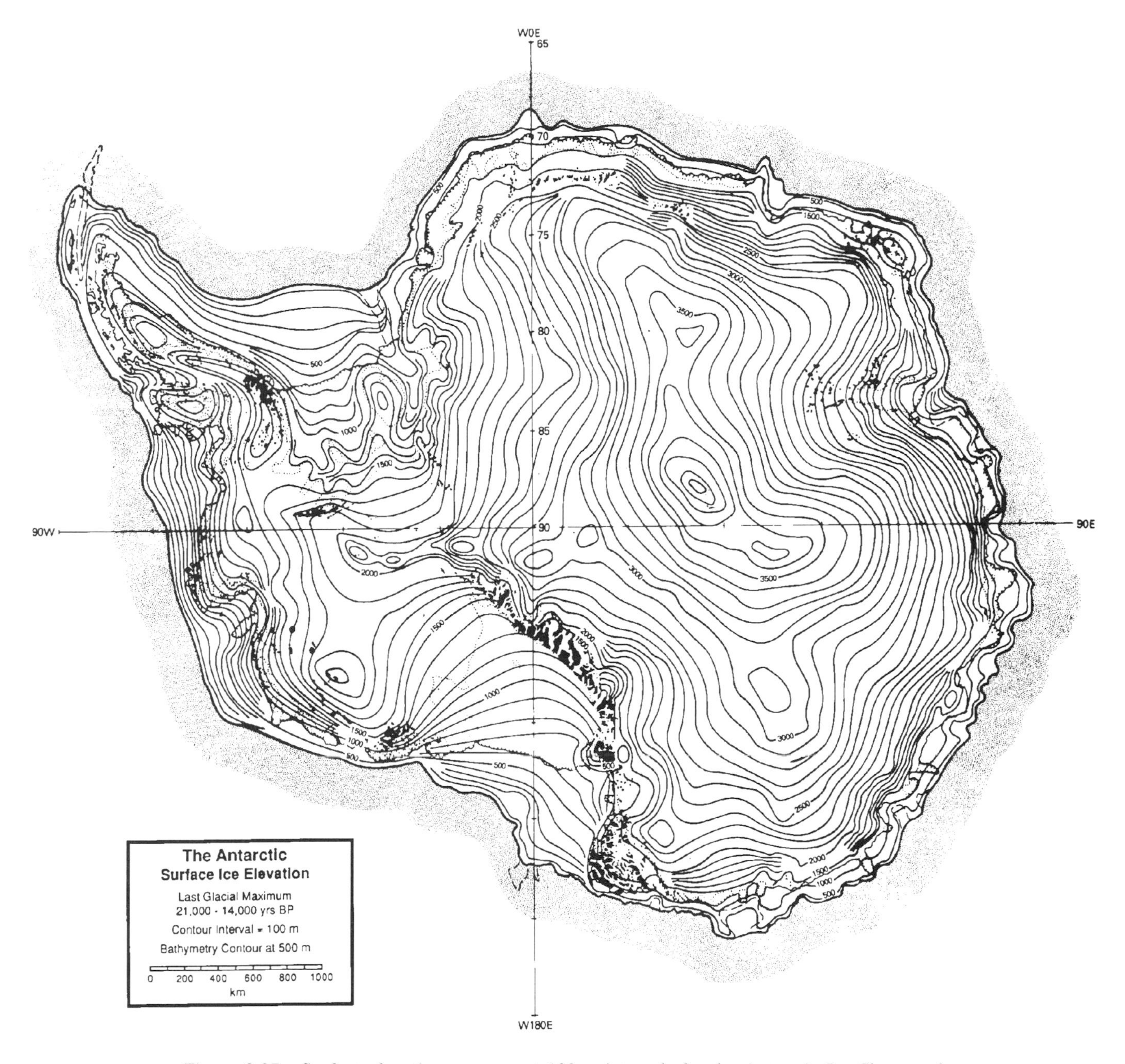

Figure 3.27: Surface elevation contours at 100 m intervals for the Antarctic Ice Sheet at the last glacial maximum obtained by fitting elevations calculated from Equation (3.25) to known former elevations recorded by glacial geology. From work performed for Battelle, Pacific Northwest Laboratories. Reproduced with permission.

Ross Ice Shelf. Its grounding line is on a nearly horizontal bed. This is the situation after gravitational collapse has lowered the ice surface and glacioisostatic rebound has raised the bed, thereby producing the ice shelf, as shown in Figure 2.2. Both situations also account for the fact that the flowline to the Ross Ice Shelf is much longer than the flowline to Pine Island Bay, as if the ice divide had migrated toward Pine Island Bay during gravitational collapse that produced the Ross Sea embayment. Impending gravitational collapse in Pine Island Bay, therefore, may produce an embayment in the Amundsen Sea as large as the Ross Sea embayment. Moreover, formation of the Ross Sea embay-ment may be a blueprint for formation of a similar embay-ment in the Amundsen Sea that further lowers the ice divide and reverses its migration, as shown in Figure 3.28.

Two models for past Holocene collapse of the West Antarctic Ice Sheet in the Ross Sea embayment have been proposed. These models are presented in Figure 3.30. Both models placed ice streams in the linear troughs that extend north to the edge of the continental shelf, as shown in Fig-

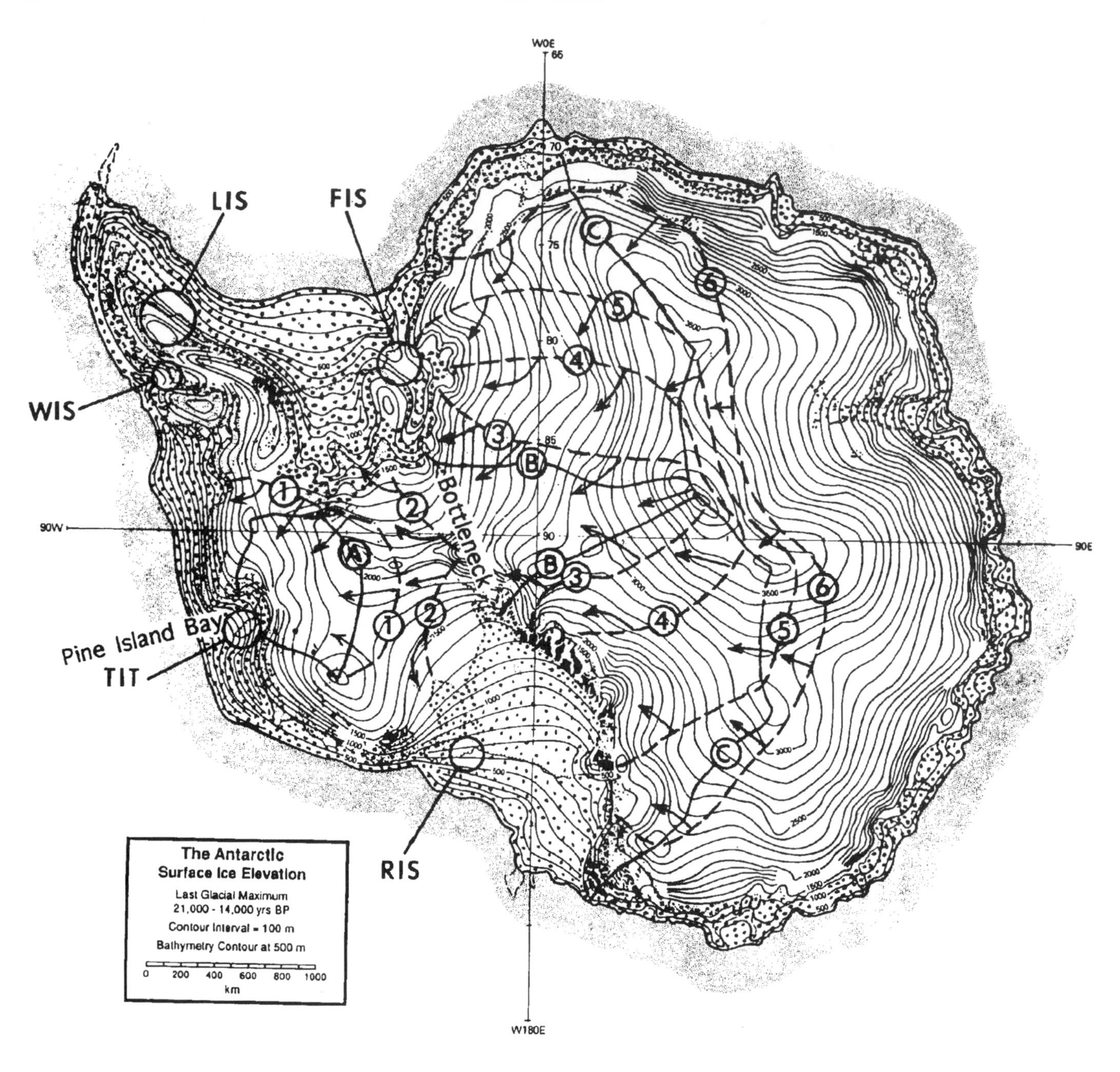

Figure 3.28: Past, present, and future of the Antarctic Ice Sheet in the worst-case deglaciation scenario. Ice elevation contour lines at 100 m intervals are from a computer reconstruction of the last glacial maximum, controlled by glacial geology, see Figure 3.20. Dotted areas are portions that have subsequently undergone complete gravitational collapse. Black areas are present-day nunataks. Circled areas are sites of recent iceberg outbursts from the Ross Ice Shelf (RIS), Thwaites Iceberg Tongue (TIT), Wordie Ice Shelf (WIS), Larsen Ice Shelf (LIS), and Filchner Ice Shelf (FIS). Heavy solid lines are present-day ice divides for ice entering Pine Island Bay (A), the Bottleneck (B), and West Antarctica from East Antarctica (C). Heavy broken lines (1 through 6) show sequential retreats of present-day ice divides (A through C) that might accompany gravitational collapse of West Antarctic ice entering Pine Island Bay and of East Antarctic ice entering the Bottleneck. Arrows show how ice may be rerouted during collapse.

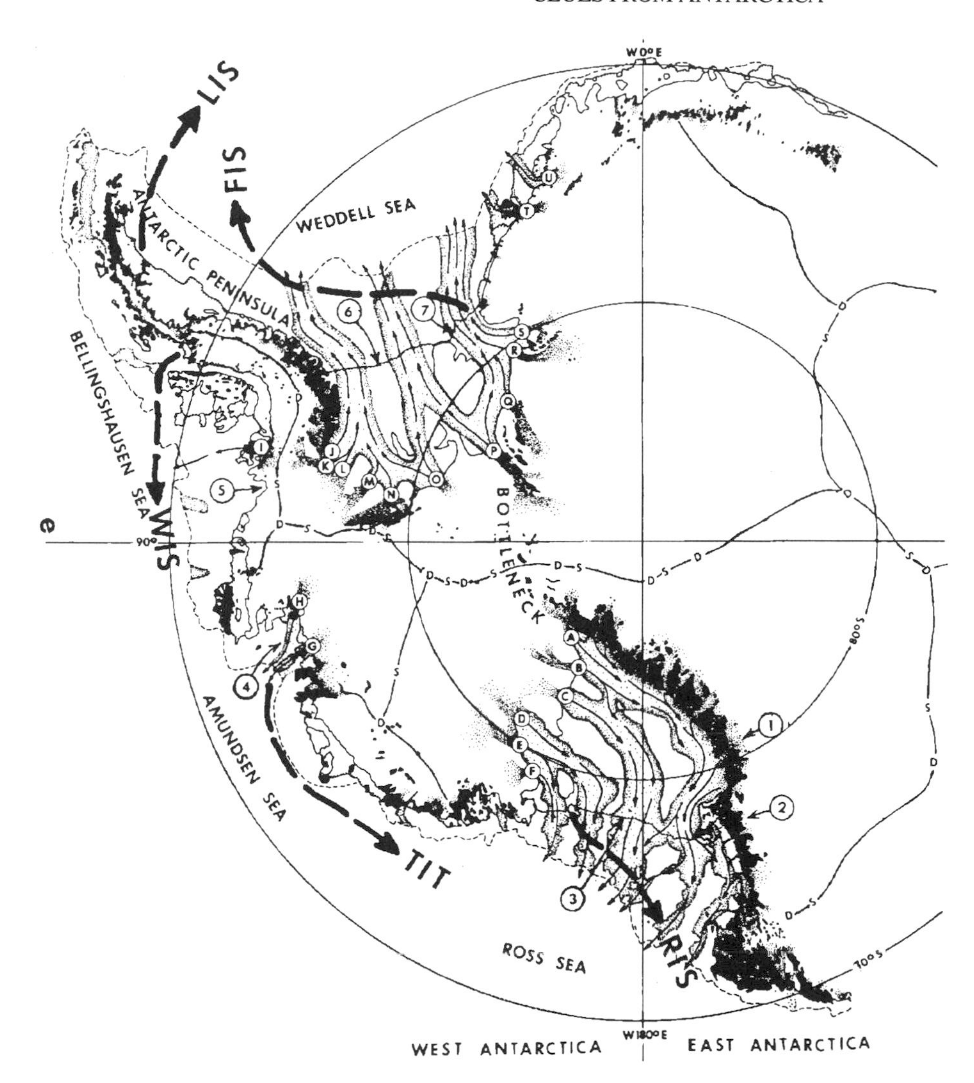

A. Ice Stream A
B. Ice Stream B
C. Ice Stream C
D. Ice Stream D
E. Ice Stream E
F. Ice Stream F
G. Thwaites Glacier
H. Pine Island Glacier
I. Berg Ice Stream
J. unnmamed ice stream
K. Evans Ice Stream
L. unnmamed ice stream
M. Carlson Ice Stream
N. Rutford Ice Stream
O. Institute Ice Stream
P. Foundation Ice Stream
Q. Support Force Glacier
R. Recovery Glacier
S. Slessor Glacier
T. Stancomb-Wills Glacier
U. Endurance Glacier

1. Byrd Glacier
2. Dry Valleys
3. Ross Ice Shelf
4. Pine Island Glacier
5. Eltanin Bay
6. Ronne Ice Shelf
7. Filclhner Ice Shelf

Figure 3.29: The role of ice streams in disintegration of the West Antarctic Ice Sheet during the Holocene. Letters A through U identify the major West Antarctic ice streams presently active, the letters being located at the ice-shelf grounding lines. Dotted areas on the ice stream side of the grounding lines identify the concave surge basins of the ice streams. Dotted areas on the ice-shelf side of the grounding lines identify the bedrock troughs eroded by the surging ice streams when they retreated from the continental-shelf margin. The dashed line identifies the continental-shelf margin of West Antarctica and the approx-imate grounded limit of the West Antarctic Ice Sheet at the last glacial maximum. The solid line identifies the present grounded limit of the West Antarctic Ice Sheet. The hatchured lines identify the present calving front of West Antarctic ice shelves. Black areas identify unglaciated regions and regions where glaciated mountains project above the ice sheet. Arrows show the continuity between past and present ice streams during disintegration of the West Antarctic ice sheet. These arrows emphasize the fact that present East Antarctic outlet glaciers are remnants of huge ice streams that extended to the continental-shelf margin during the late Pleistocene and retreated during the Holocene to the Transantarctic Mountains from marine portions of the West Antarctic Ice Sheet. Thin solid lines identify ice divides, with D denoting domes and S denoting saddles on ice divides. Thick dashed lines show tracks of recent iceberg outbursts from Ross Ice Shelf (RIS) in the Bay of Whales, from Thwaites Iceberg Tongue (TIT) in Pine Island Bay, from Wordie Ice Shelf (WIS) on the western Antarctic Peninsula, from Larsen Ice Shelf (LIS) on the eastern Antarctic Peninsula, and from Filchner Ice Shelf (FIS) north of the Grand Chasms.

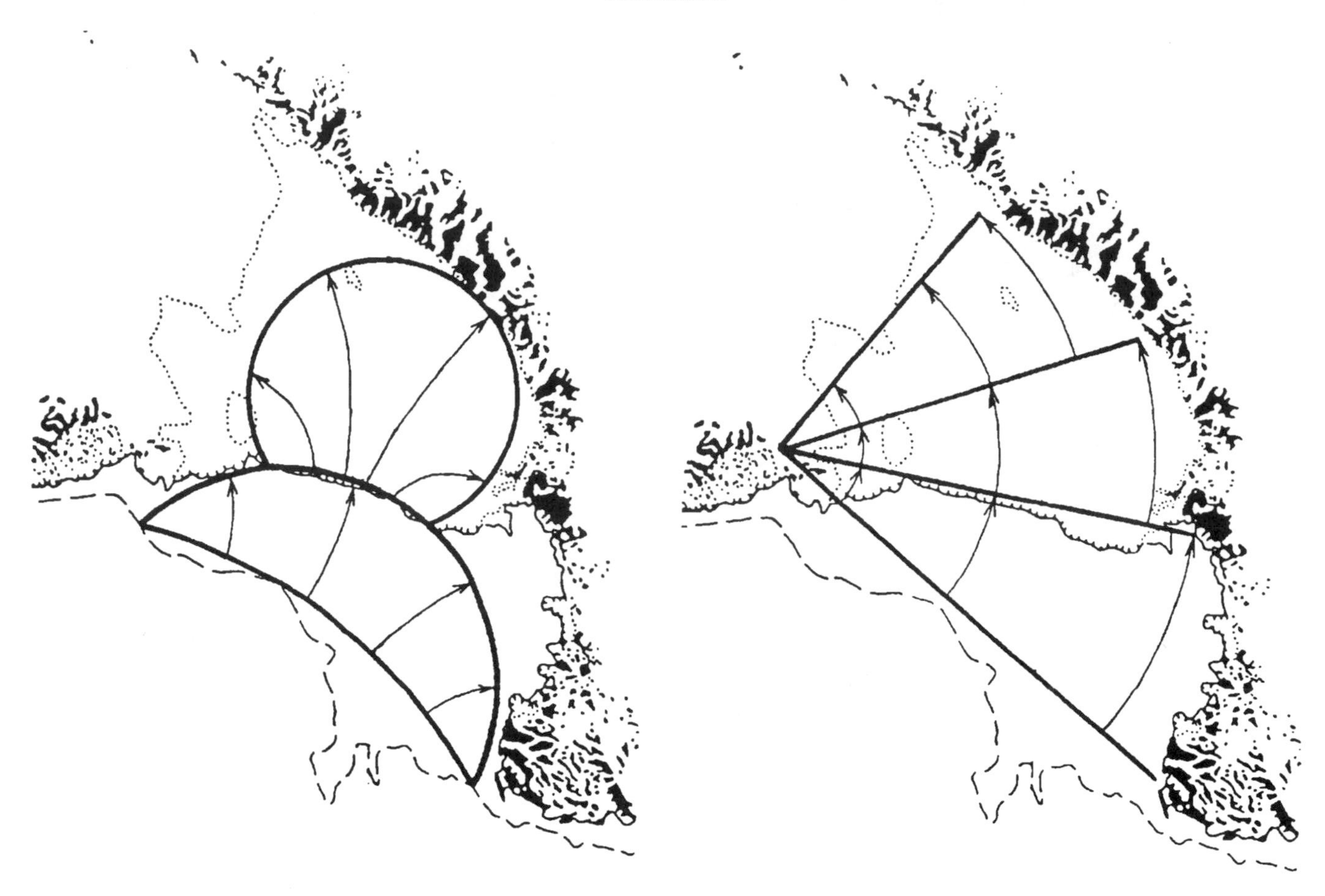

Figure 3.30: Cartoons showing two simplified versions of how the Ross Ice Shelf grounding line retreated from 18,000 ka BP to the present. Shown are the present-day grounding lines (dotted lines) and calving front (hatchured line) of the Ross Ice Shelf, the edge of the continental shelf (dashed line), mountain ranges (black areas), the directions (arrows), and stages (heavy lines) of grounding-line retreat for the two versions. *Left*: The central spreading version of Thomas and Bentley (1978). *Right*: The swinging gate version of Stuiver and others (1981).

ure 3.29. Both models produced the Ross Ice Shelf as rising late glacial sea level caused ice-shelf grounding lines to retreat south, initially along these troughs. In the model by Thomas and Bentley (1978), the central Ross Sea embayment became afloat first, after which grounding lines retreated equally to the east, west, and south. In the model by Stuiver and others (1981), the grounding line followed the troughs to the West Antarctic ice streams which occupy these troughs today, so that the grounding line retreated like a swinging gate that was hinged in the Rockefeller Mountains (78°E, 155°E), with retreat being fastest in the deep trough along the Transantarctic Mountains.

According to the model by Stuiver and others (1981), the grounding line is still retreating along this trough,which continues into the Weddell Sea embayment as the deep Crary Trough. Today, the trough is occupied by Foundation Ice Stream in the Weddell Sea embayment and by Ice Stream A in the Ross Sea embayment. Maps of bed topography by Drewry (1983) and by Shabtaie and Bentley (1988) show that this trough has numerous branches that extend from these ice streams through wide gaps in the Transantarctic Mountains. Trough branches underlying Ice Stream A, Ice Stream B, and Ridge AB are shown in Figure 3.31. The ice surface of Ice Stream A, Ice Stream B, and Ice Ridge AB between these ice streams has the rough surface associated with the high stress gradients in stream flow, as illustrated by the longitudinal gradients of P_W / P_I in Figure 3.23 (Shabtaie and others, 1987). Figure 3.31 suggests that Ice Stream A recently included or is now incorporating Ridge AB, giving it a rough surface. Then a super ice stream consisting of Ice Stream A, Ridge AB, and Ice Stream B could draw down the West Antarctic ice divide from the west, while Foundation Ice Stream is drawing it down from the east. Eventually, the Ross and Weddell grounding lines from these ice streams will meet, and the West Antarctic Ice Sheet will be isolated from the much larger East Antarctic Ice Sheet. As shown in Figure 3.29, the West Antarctic Ice Sheet acts like a plug that prevents East Antarctic ice from

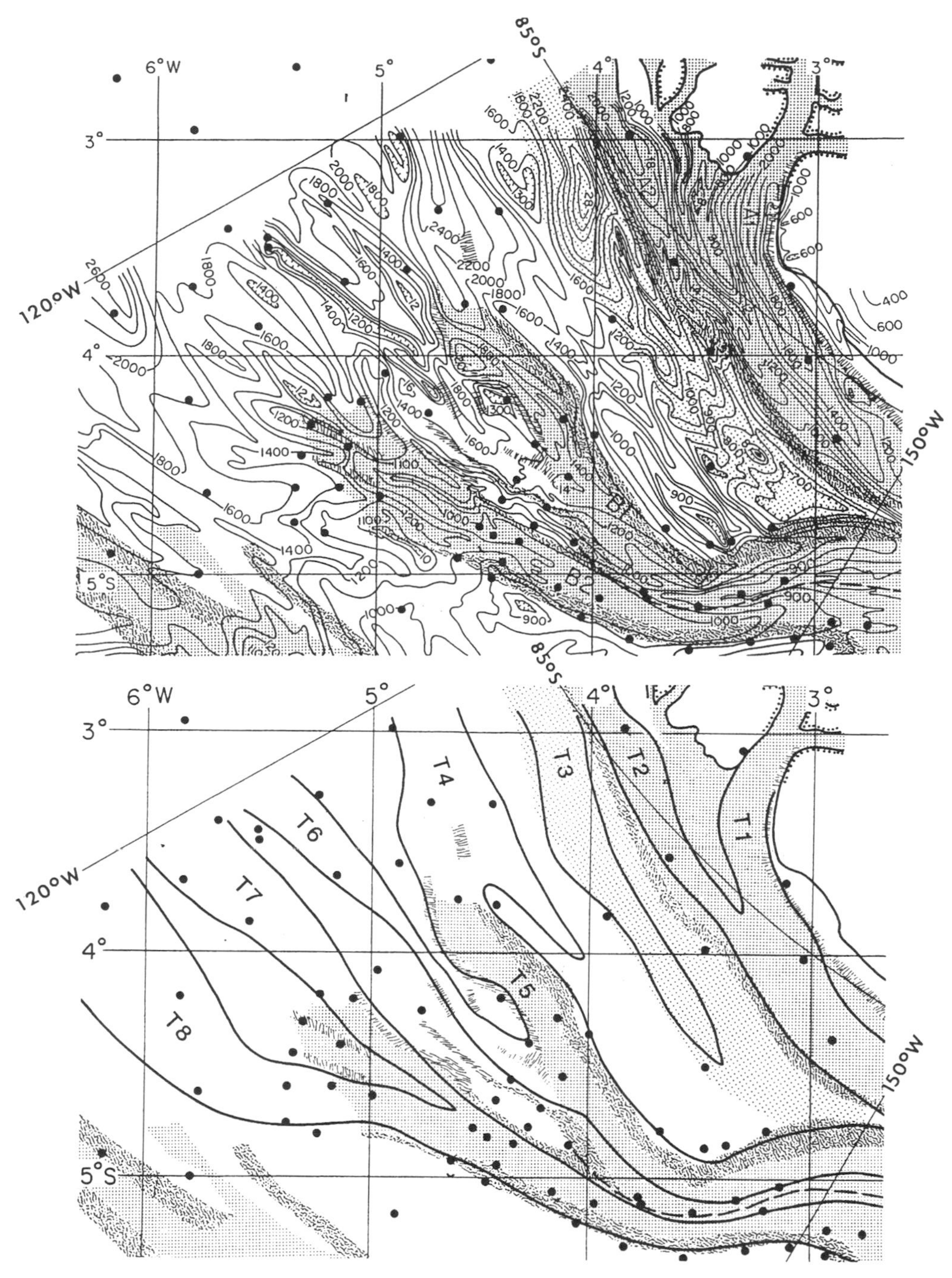

Figure 3.31: The bottleneck of the West Antarctic Ice Sheet for ice flowing into the southeastern Ross Ice Shelf. *Top*: Ice thickness contour lines (m). *Bottom:* Subglacial troughs (T1 through T8). Solid circles are surface control stations for radar flights lines. Darker shading denotes Ice Streams A, B, and C, and their branches, notably A1, A2, and B1, B2. Lighter shading denotes possible widening of Ice Stream A into Ridge AB. Mottled bands denote shear crevasses alongside Ice Streams A, B, and C, and other crevasse fields. Heavy lines with dotted borders denote East Antarctic outlet glaciers through the Transantarctic Mountains that merge with Ice Stream A. Reedy Glacier (A1) occupies trough T1 and Shirase Ice Stream (A2) occupies trough T2. They merge to become Ice Stream A. Heavy dashed line denotes the B1 and B2 portions of Ice Stream B. From Shabtaie and Bentley (1988). Reproduced with permission.

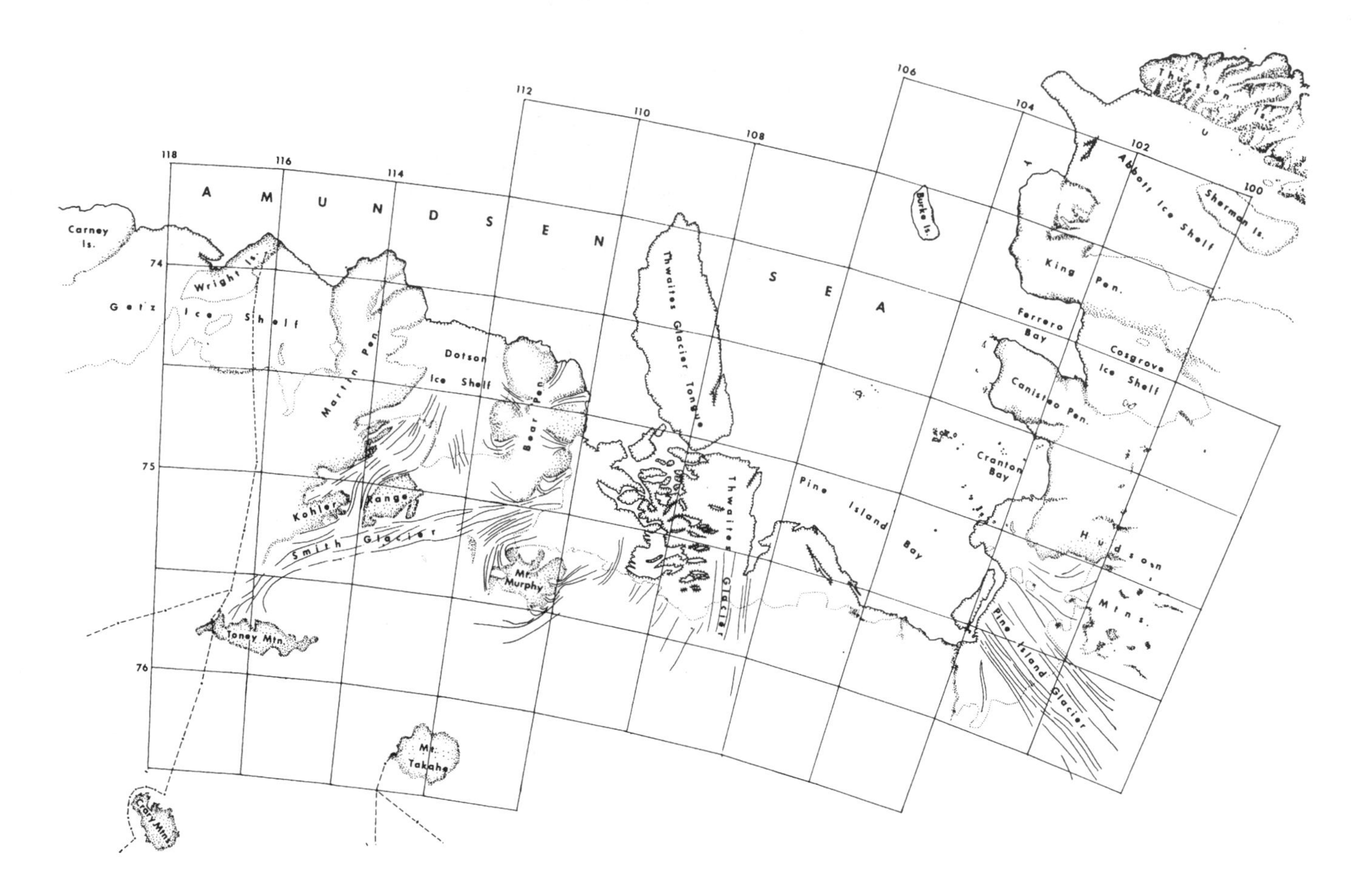

Figure 3.32: The Pine Island Bay sector of the Amundsen Sea. Tidewater ice margins are denoted by solid lines, ice-shelf calving fronts are denoted by hachured lines, ice-shelf grounding lines are denoted by dotted lines, and nunataks are denoted by dotted areas. Thwaites Glacier (75.0°S, 107.0°W) has punched through a narrow ice shelf so its floating terminus has no side shear. Pine Island Glacier (75.0°S, 101.0°W) remains imbedded in a narrow ice shelf, so its floating terminus is constrained by side shear.

surging through the wide gaps in the southern Transantarctic Mountains. When that plug is pulled, part of the 60 m of higher sea level now locked in the East Antarctic Ice Sheet will be released, in addition to the 6 m locked in the West Antarctic Ice Sheet. Already, the narrowest neck of the West Antarctic Ice Sheet lies between Ice Stream A and Foundation Ice Stream. In addition, Ice Stream A and Ice Stream B have a negative mass balance, so their grounding lines retreat as the ice divide lowers (Shabtaie and others, 1988).

This situation is our most promising opportunity to study the collapse of Holocene ice sheets by the same internally driven mechanisms that collapsed Pleistocene ice sheets during the last deglaciation in the Northern Hemisphere. Whether gravitational collapse of the Antarctic Ice Sheet is ongoing, especially in West Antarctica, is the outstanding glaciological question of our time (Hughes, 1973; Weertman, 1976; Mercer, 1978). In seeking an answer to this question, interest has focused on Pine Island Bay shown in Figure 3.32, and its ice drainage basin shown in Figure 3.33 (Thomas and others, 1979; Hughes, 1981b; Stuiver and others, 1981; Fastook, 1984). This region will be studied as part of the West Antarctic Ice Sheet Initiative (WAIS).

Focusing WAIS research on the worst-case scenario

The West Antarctic Ice Sheet Initiative (WAIS) is designed "to answer two critical and interrelated climate questions: How will the potentially unstable West Antarctic ice sheet affect future sea level? How do rapid global climatic changes occur?" (Bindschadler, 1995). With the National Science Foundation (NSF) and all federal agencies facing uncertain budgets, NSF programs will be subjected to increased congressional scrutiny. Therefore, it is prudent for WAIS to deliver at least tentative answers to these questions without undue delay. In particular, the first question should be addressed immediately, since it is the more tractable, it has the highest public interest, and rising sea level will have the most drastic social, political, and economic impact on a global scale. The following discussion presents a research strat

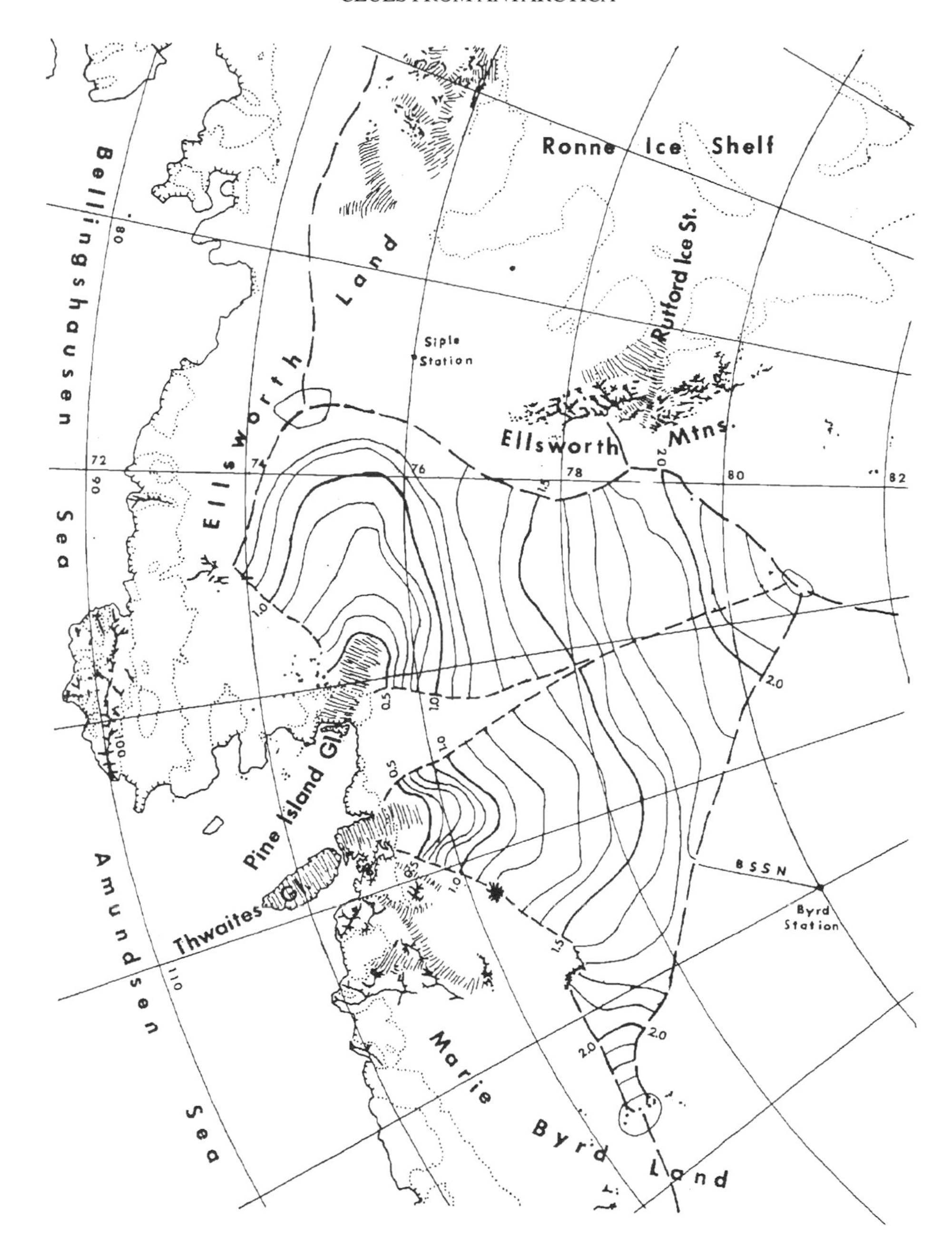

Figure 3.33: The ice catchment area of Thwaites Glacier and Pine Island Glacier in West Antarctica. The ice elevation contour interval is 0.1 km. BSSN is the Byrd Station Strain Network.

egy designed to answer this question and field research designed to implement the strategy.

Figure 3.28 illustrates the sea-level problem in the context of how "greenhouse" climatic warming might impact the West Antarctic Ice Sheet and, by extension, the much larger East Antarctic Ice Sheet. This is important because, whereas the West Antarctic Ice Sheet can contribute up to 6 m of rising sea level, the East Antarctic Ice Sheet can contribute up to 60 m. However, climate warming, by removing much of the winter sea ice surrounding Antarctica and thereby bringing moisture sources closer to the continent, may actually enhance precipitation over the ice sheet and lower sea level. Alternatively, Antarctica may be so isolated climatically that climatic warming will have no significant effect on the ice sheet. These two possibilities will not have the drastic global consequences that rapidly rising sea level would have. Therefore, the research strategy should address how instabilities in the West Antarctic Ice Sheet could cause a rapid and large rise in sea level as the worst-case scenario. What we learn from examining this scenario will automatically provide data that address the other two relatively benign scenarios. The worst-case scenario shown in Figure 3.28 is based on three observations: (1) iceberg outbursts as an early-warming system of ice-sheet collapse in West Antarctica, (2) Pine Island Bay as the site of the most rapid future collapse and rising sea level, and (3) the Bottleneck as the site sustained collapse and rising sea level.

Iceberg outbursts as an early warning system for collapse

Iceberg outbursts may be an early-warning system for impending further gravitational collapse of the West Antarctic Ice Sheet, as it was for Pleistocene ice sheets in the Northern Hemisphere at the end of the Younger Dryas 10,000 years ago (e.g., Andrews and Tedesco, 1992; Broecker, 1994). If so, those alarm bells have begun ringing all around West Antarctica during the last decade. In the Ross Sea, Iceberg B9 was released from Ross Ice Shelf at the Bay of Whales in 1987. In the Amundsen Sea, Thwaites Iceberg Tongue moved out of Pine Island Bay in 1986 after decades of immobility. Along the Antarctic Peninsula, Wordie Ice Shelf began disintegrating in 1979 and is now largely gone, while disintegration of the much larger Larsen Ice Shelf began in earnest in 1995. In the Weddell Sea, Filchner Ice Shelf north of the Grand Chasms disintegrated in 1986.

As seen in Figure 3.29, these recent iceberg outbursts are occurring all around West Antarctica, and in different climatic regimes. The Ross Ice Shelf (RIS) at the Bay of Whales lies in a mixed polar continental/maritime climatic regime, depending on seasonal prevailing winds. Thwaites Iceberg Tongue (TIT) was in Pine Island Bay, a polar maritime climatic regime that is frequently an ice-free polynya. Wordie Ice Shelf (WIS), being on the west side of the Antarctic Peninsula and exposed to moisture-bearing westerly winds, was in a subpolar maritime climatic regime. Larson Ice Shelf (LIS), lying in a precipitation shadow on the east side of the Antarctic Peninsula, is in a subpolar continental climatic regime. Filchner Ice Shelf (FIS) lies between Berkner Island and the East Antarctic Ice Sheet in an essentially polar continental climatic regime.

What we know about the dynamics of these climatic regimes does not reveal a mechanism for atmospheric or oceanic circulation that can account for massive iceberg outbursts that are virtually simultaneous all around West Antarctica. Thwaites Iceberg Tongue was 100 km long and immobile for at least 50 years. The largest iceberg released north of the Grand Chasms was as big as Connecticut. Climatic warming as a cause can be argued plausibly only for the Antarctic Peninsula, where the disintegration of ice shelves seems to be progressing south with the –5°C mean annual air temperature isotherm (Vaughan and Doake, 1996). The only other explanations are that these widespread iceberg outbursts have been nearly simultaneous only coincidentally, or that the whole West Antarctic dynamic system, atmosphere, ocean, and ice sheet, is experiencing destabilizations that we simply do not understand. What these iceberg outbursts have in common is that they surround an ice sheet that Weertman (1974) showed may be inherently unstable. Weertman (1976) called this instability "glaciology's grand unsolved problem." Basal melting is up to 12 m/a under the floating part of Pine Island Glacier (Jacobs and

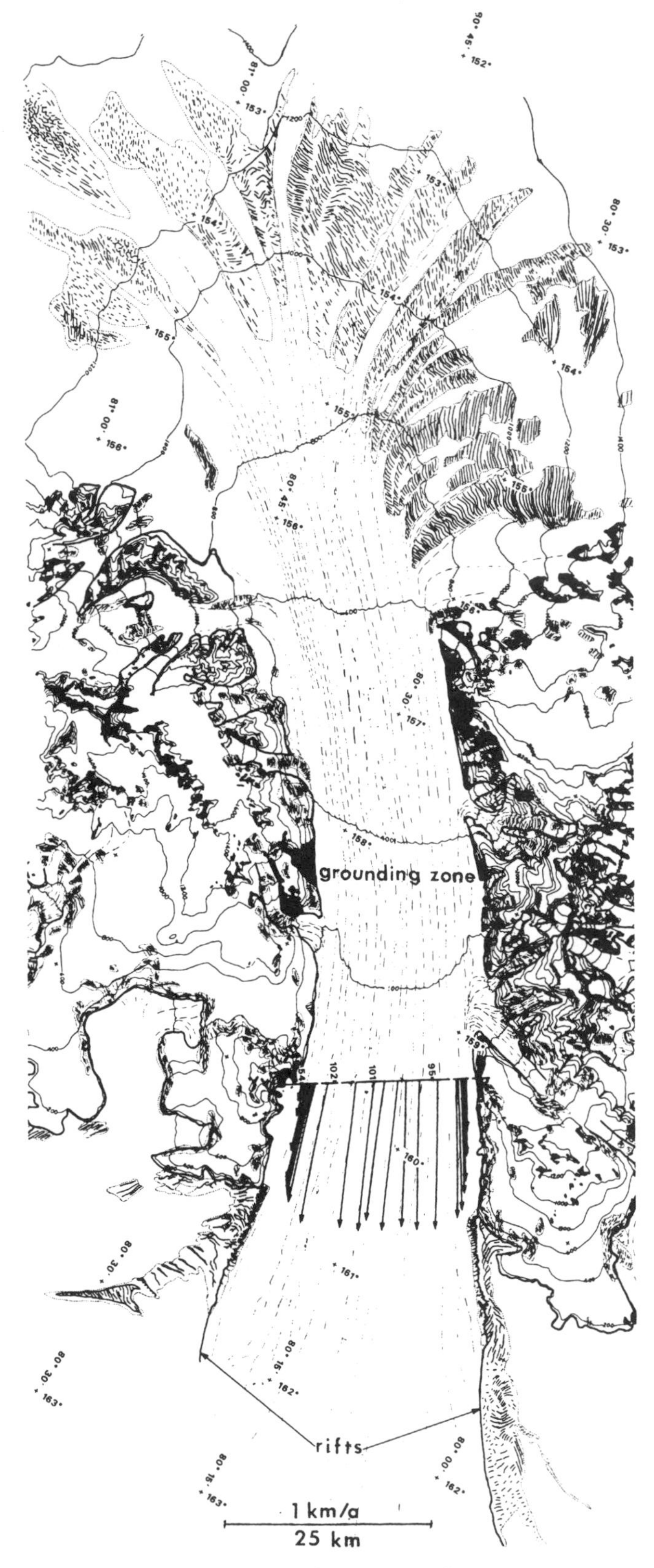

Figure 3.34: Byrd Glacier (80.5°S, 158.0°E) From the Mount Olympus and Cape Selbourne quadrangle maps, U.S. Geological Survey. Ice velocity vectors are from Swithinbank (1963).

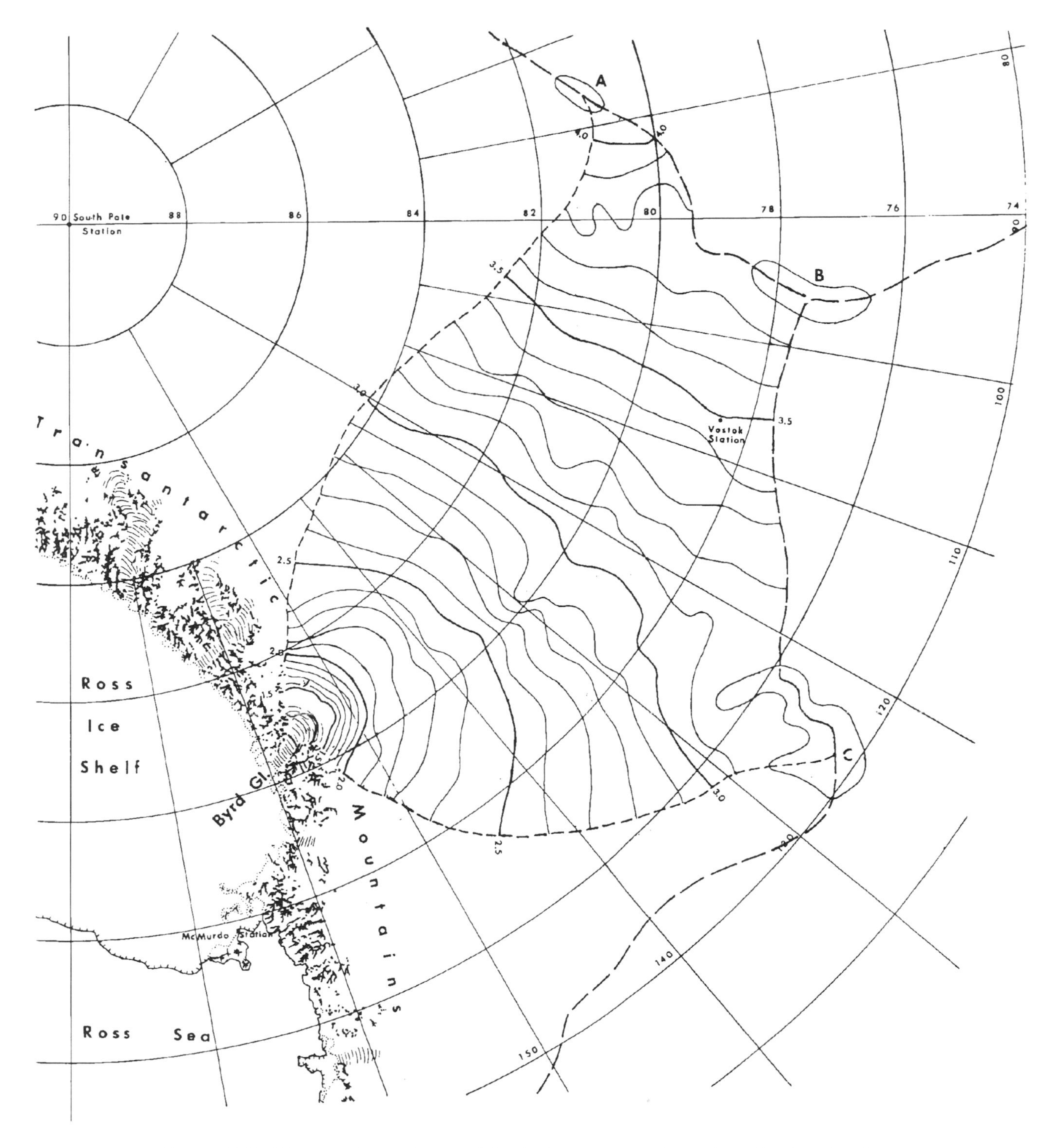

Figure 3.35 The ice catchment area of Byrd Glacier in East Antarctica. The ice elevation contour interval is 0.1 km.

others, 1996; Jenkins and others, 1997). If rapid basal melting of ice in Pine Island Bay is recent, perhaps it caused the recent ungrounding of Thwaites Iceberg Tongue. If so, basal melting may promote iceberg outbursts as an early-warning system for impending gravitational collapse.

The role of Pine Island Bay in rapidly rising sea level

If climatic warming is indeed proceeding south, as recorded by the disintegration of ice shelves along the Antarctic Peninsula, the first major ice streams draining the West Antarctic Ice Sheet that will be affected are Pine Island Glacier and Thwaites Glacier, as shown in Figure 3.32. These ice streams enter Pine Island Bay and drain one-third of the West Antarctic Ice Sheet, as shown in Figure 3.33. They are also the fastest Antarctic ice streams. Thwaites Glacier moves up to 3.4 km/a and Pine Island Glacier moves up to 2.5 km/a (Ferrigno and others, 1993; Luchitta and others, 1995, 1996). In contrast, Byrd Glacier and most other Antarctic ice streams move less than 1 km/a and the fastest-known ice stream, Jakobshavns Isbræ in Greenland, moves 7 km/a. Velocities increase as constraints to stream flow are removed. As seen in Figures 3.32 and 3.33, Thwaites Glacier becomes a freely floating ice tongue with no lateral or forward constraints and Pine Island Glacier becomes a floating ice tongue with side constraints but no forward constraint. As seen in Figures 3.34 and 3.35, Byrd Glacier also becomes afloat but it has both lateral and forward constraints, the walls of Byrd Glacier fjord on the sides and the Ross Ice Shelf in front. Ice-shelf disintegration removes these constraints on stream flow and hastens collapse of the West Antarctic Ice Sheet. Pine Island Glacier and Thwaites Glacier drain ice over the deepest subglacial basins in Antarctica, Byrd Subglacial Basin and Bentley Subglacial Trench, which is up to 3 km below present-day sea level. As shown in Figure 2.2, retreat of their grounding lines will also contribute to collapse of the West Antarctic Ice Sheet.

Available radio-echo data show that the floating termini of Pine Island Glacier and Thwaites Glacier, especially Thwaites Glacier, have grounding lines that lie on beds that slope steeply down toward Bentley Subglacial Trench (Drewry, 1983). As shown by Thomas and others (1979), an ice-stream grounding line on a downsloping bed is highly unstable. Slight increases in surface or basal melting rates, increases in sea level, increases in longitudinal strain rate, or increases in bed depth due to glacioisostatic depression or glacial erosion can cause irreversible retreat of the grounding line that can be checked only by increased ice velocity or formation of a buttressing ice shelf. The polynya in Pine Island Bay seems to preclude formation of an ice shelf. Indeed, sediment analyses in Pine Island Bay point to recent disintegration of an ice shelf (Kellogg and Kellogg, 1987a, 1987b, 1987c).

The fast velocities of Pine Island Glacier and Thwaites Glacier would impart a strongly negative mass balance to the third of the West Antarctic Ice Sheet that drains into Pine Island Bay if convective storm systems did not give this ice drainage basin the highest accumulation rates in Antarctica (Giovinetto and others, 1989). High ice discharge rates precariously balanced by high accumulation rates creates an unstable system, as perturbations of either discharge or accumulation can cause major shifts of ice-stream grounding lines on inwardly downsloping beds, and such shifts can be irreversible. A negative shift in mass balance would cause retreat of the ice divide for ice draining into Pine Island Bay (solid line A in Figure 3.28). Several consequences follow.

Initially, the saddle on the ice divide between Pine Island Glacier and Rutford Ice Stream, already the lowest point on the West Antarctic ice divide at 1250 m above sea level (see Figure 2.13), could lower to sea level, isolating the Antarctic Peninsula from the West Antarctic Ice Sheet, and the ice divide between Thwaites Glacier and Ice Stream D, as shown in Figure 2.13, would migrate toward Ross Ice Shelf (broken line 1 in Figure 3.28). This would impart a negative mass balance to West Antarctic ice draining into the Weddell Sea and Ross Sea embayments, allowing grounding-line retreat for ice shelves floating in these embayments. Grounding line retreat in Pine Island Bay would create an ice shelf floating over Byrd Subglacial Basin and Bentley Subglacial Trench, located in Figure 2.15.

Finally, the West Antarctic ice divide would retreat to the Bottleneck (broken line 2 in Figure 3.28), with only East Antarctic ice flooding through the Bottleneck supplying West Antarctic ice shelves floating in the Weddell Sea and Ross Sea embayments, and the new embayment over Byrd Subglacial Basin and Bentley Subglacial Trench. Modeling studies show that collapse of the West Antarctic Ice Sheet into Pine Island Bay as it expands south could occur rapidly, perhaps in only a few centuries (e. g., Stuiver and others, 1981; Fastook, 1984). However, the resulting 6 m rise in global sea level may be only a fraction of the long-term rise in sea level from East Antarctic ice flooding through the Bottleneck.

The role of the Bottleneck in sustaining rising sea level

The West Antarctic Ice Sheet buttresses the East Antarctic Ice Sheet at the Bottleneck. As seen in Figures 3.28 and 3.29, the Bottleneck is also the narrowest part of the West Antarctic Ice Sheet. Therefore, the West Antarctic Ice Sheet is the "cork" in the "bottle" of the East Antarctic Ice Sheet, and pulling this cork through Pine Island Bay will allow East Antarctic ice to flood into West Antarctica. Even now, East Antarctic ice reaching to the center of the East Antarctic Ice Sheet, Dome Argus over the subglacial Gamburtsev Mountains, enters West Antarctica through the Bottleneck (between solid lines B in Figure 3.28). Available radio-echo data, although meager, indicate that East Antarctic ice passes through the Bottleneck by way of several wide and deep subglacial troughs, some extending well below present-day sea level (Drewry, 1983). East Antarctic ice is nearly 3 km above sea level as it approaches the Bottleneck, and it rises to more than 4 km above sea level at Dome Argus. This represents a tremendous reservoir of gravitational potential energy that could be converted into kinetic energy of motion. It would cause a sustained rise in global sea level if the West Antarctic Ice Sheet were to collapse, a rise substantially greater than the 6 m that accompanied the gravitational collapse of West Antarctic ice.

As shown in Figure 3.28, a sustained rise in global sea level, as East Antarctic ice floods through the Bottleneck following collapse of West Antarctic ice, is accomplished by expanding the drainage basin of ice entering the Bottleneck (broken lines 3 through 6 and arrows showing ice flow rerouted toward the Bottleneck in Figure 3.28). Ultimately, the present-day ice divide for East Antarctic ice flowing toward West Antarctica (solid line C in Figure 3.28) could retreat (broken lines 5 and 6 in Figure 3.28), thereby allowing West Antarctic ice shelves to be the major recipients of East Antarctic ice and, therefore, of rising sea level.

This ice-divide retreat would impart a negative mass balance to East Antarctic ice flowing north into the Southern Ocean through ice streams that occupy subglacial channels. The channels extend from the outer continental shelf into large subglacial basins beneath the East Antarctic Ice Sheet, as shown in Figure 2.15. Rising sea level could destabilize these East Antarctic ice streams, causing their grounding lines to retreat along the channels and into the subglacial basins. Retreat would reinforce the negative mass balance by increasing ice discharge even as the area of ice accumulation basins decreases. This would also sustain the rise in global sea level.

At present, most East Antarctic ice entering West Antarctica through the Bottleneck is transported by Foundation Ice Stream and Ice Stream A. Foundation Ice Stream is heavily crevassed, which implies a fast velocity and possibly a negative mass balance, and enters Filchner Ice Shelf, which has disintegrated north of the Grand Chasms. As seen in Figure 3.31, Reedy Glacier and Shirase Ice Stream merge to become Ice Stream A, which enters Ross Ice Shelf. Reedy Glacier drains East Antarctic ice. Shirase Ice Stream drains both East Antarctic and West Antarctic ice. At present, Ice Stream A seems to have a negative mass balance (Shabtaie and others, 1988). However, Shabtaie and Bentley (1988) show that Ice Stream A, Ice Stream B, and intervening Ice Ridge AB are all underlain by subglacial troughs, and the half of Ice Ridge AB adjacent to Ice Stream A has a rough surface of snow-buried crevasses. This indicates that Ice Stream A can widen to include at least half of Ice Ridge AB, or even merge with Ice Stream B, to create a super ice stream that, together with Foundation Ice Stream, would collapse the Bottleneck.

Recommended field studies

Initial field studies needed to examine the worst-case scenario for the West Antarctic Ice Sheet fall into three categories.

1. This category studies ice-shelf disintegration and iceberg outbursts as early-warning systems for impending further gravitational collapse of the West Antarctic Ice Sheet. This study is underway (Williams and others, 1995). It is best conducted using satellite imagery to monitor the proliferation of surface melting and crevasse propagation on West Antarctic ice shelves and to monitor the volumes and paths of icebergs released from these ice shelves and from West Antarctic ice streams that supply ice to the ice shelves. Of particular interest are icebergs that are carried toward the equator by ocean currents that curve north from the Antarctic Circumpolar Current, as shown in Figure 3.5. Melting of these icebergs may destabilize climate beyond the Antarctic. Changes in the extent of sea ice and cloud cover signal changing accumulation patterns over West Antarctica.

2. This category studies the Pine Island Bay ice drainage basin. It includes detailed radio-echo mapping of subglacial topography and seismic mapping of subglacial sediments and till beneath Pine Island Glacier and Thwaites Glacier, and velocity measurements of these ice streams. It includes a study of the history of these ice streams and their drainage basin, as recorded by sediments and till in Pine Island Bay and by glacial geology in nunataks within and along the ice divide of the drainage basin. It includes coring and drilling through these two ice streams and through subglacial sediments and till to investigate their past and present ice dynamics. It includes mapping the surface accumulation and ablation rates in their ice drainage basin, locating bottom crevasses after the ice streams become afloat, and studies of grounding line dynamics and ice calving dynamics for the floating tongues of these ice streams. Meteorological studies that include katabatic winds flowing down the ice streams are an important part of the mass-balance studies.

3. This category studies the Bottleneck. It includes detailed radio-echo mapping of subglacial topography in the bottleneck, especially of the wide and deep troughs that direct East Antarctic ice into West Antarctica, and of the velocity of crevasses in ice moving along these troughs. Surface accumulation rates from the Bottleneck to Dome Argus should also be determined. The dynamics of Foundation Ice Stream and Ice Stream A should be studied by mapping surface ice velocities from moving crevasses, mapping bed topography and till distribution by radar and seismic studies, and drilling to recover ice and till cores. These studies of Ice Stream A should include Reedy Glacier, Shirase Ice Stream, and Ice Ridge AB. Glacial history in the Bottleneck can be studied by mapping glacial geology in nunataks and along fjords occupied by ice streams. For example, much higher ice elevations are recorded by perched moraines on the fjord walls above Reedy Glacier (Mercer, 1968b). The role of katabatic winds in scouring the surface of ice streams draining the Bottleneck, thereby reducing ice accumulation, is also important (Bromwich and others, 1994).

Antarctic Glaciation and Climatic Change

The Antarctic Ice Sheet contains a record of paleoclimates and samples of both paleo-atmospheres and paleo-oceans. Paleoclimatic information is contained in stable isotope ratios which reflect precipitation temperatures, radioactive isotope horizons, which give total precipitation between atmospheric nuclear detonations, microparticle variations which give seasonal variations in precipitation, ash horizons which reflect atmospheric turbidity following major volcanic eruptions, and bulk ion concentrations which vary with distance from open seas. Paleo-atmospheres are preserved inside the air bubbles created when snow becomes ice during firnification. Paleo-oceans are preserved beneath parts of the ice sheet created when floating ice shelves become grounded nonuniformly, so that marine sediments become ice-cemented permafrost.

Oxygen isotope ratios provide a good example of the

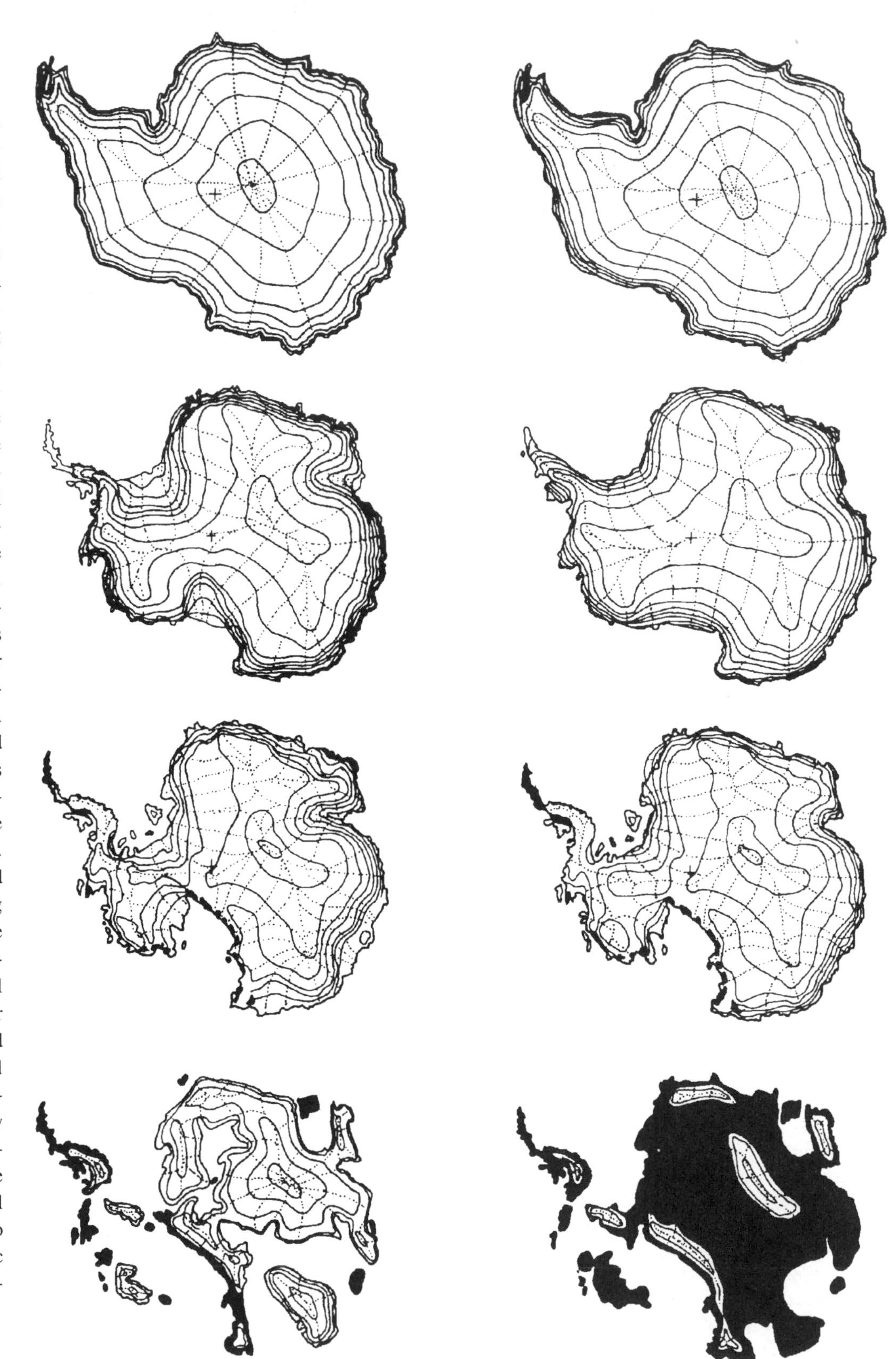

Figure 3.36 Disintegration of the Antarctic ice sheet resulting from a surge (*left*) and from a negative mass balance (*right*). Surges occur preferentially in portions of the ice sheet grounded below sea level. If isostatic rebound cannot keep pace with decreasing ice thickness, all such portions might eventually become ungrounded and ice thickness may be halved in the remaining portions. Surge ice-sheet disintegration will probably be nonuniform because only some ice drainage basins are inherently unstable. A negative mass balance caused by less precipitation occurs preferentially over the interior of the ice sheet where precipitation is always lowest. A negative mass balance caused by higher temperature occurs preferentially over the periphery of the ice sheet where temperature is always highest. In both cases isostatic rebound can keep pace with decreasing ice thickness and only the ice sheet covering land permanently below sea level will become ungrounded. However the ice sheet covering land permanently above sea level may also completely disintegrate, perhaps leaving only the mountainous regions glaciated. Negative mass balance ice sheet disintegration will probably be uniform owing to the symmetry of the Antarctic cryosphere-hydrosphere-atmosphere system.

difficulties which arise. These ratios vary with precipitation temperature, and the variations are preserved in varying degrees over an ice sheet. Seasonal variations reflect seasonal temperatures, and the distance between maxima and minima in isotope ratios measured down an ice-sheet corehole gives the annual ice accumulation via precipitation, until compression and diffusion in ice obscures seasonal horizons. However, seasonal variations can also become obliterated by windblown drift if annual snowfall is too light and by percolation melting if summer temperatures are too high. In such cases, a Fourier analysis of isotope ratios might still show variations which reflect longer-term climatic fluctuations. Even if isotope ratios are preserved, do they reflect (1) changes in atmospheric circulation, (2) global climatic changes, (3) regional climatic fluctuations, (4) local climatic variations due to changes in the ice-sheet elevation, or (5) some combination of these? The answer to this question involves the time-scale of the change and the problem of gradual versus catastrophic change (such as ice-sheet surges).

An answer is possible in principle by combining measurements of oxygen isotope ratios down coreholes with the corehole temperature profile and the variation with depth of total gas content observed mainly in the size, shape, and pressure of trapped air bubbles. Oxygen isotope ratios relate directly to atmospheric conditions during precipitation. Deep-ice temperatures are determined mainly by surface temperature and precipitation after the ice was deposited, especially at ice divides. Total gas content is a measure of the atmospheric pressure at the elevation above sea level where the gas became trapped during firnification. Theoretically, these data will enable researchers to determine changes in past ice-sheet surface temperatures, accumulation rates, elevations, and thicknesses with time. These changes in turn will help researchers to distinguish among real cli- matic changes independent of the ice sheet, spatial ice-sheet elevation changes caused by downstream equilibrium flow, and temporal ice-sheet elevation changes caused by nonequilibrium flow due to climatic or ice-sheet instability (e.g., Robin and Weertman, 1973). Coreholes at domes and saddles along the West Antarctic ice divide (see Figure 3.24) should distinguish between climatic changes and changes in ice dynamics. Domes are usually associated with nunataks and subglacial highlands, so ice cores would record climatic changes primarily. Saddles are usually over subglacial lowlands or basins and are downdrawn by flanking major ice streams, so ice cores would record changes in ice-stream dynamics primarily. The International Trans-Antarctic Scientific Expedition (ITASE) envisions a traverse along the West Antarctic ice divide that includes coring at domes and saddles for this purpose (Mayewski, 1996). Past climatic conditions prevailing over the interior Antarctic Ice Sheet, as deduced from ice-core analyses along ice divides, can be correlated with past conditions along the ice-sheet margin, as deduced from the glacial geology of de-glaciated landscapes, the marine geology on the floors of de-glaciated embayments and the surrounding seas, and climate records obtained from coreholes in local ice domes between major ice streams. Present climatic conditions for the Antarctic cryosphere-hydrosphere-atmosphere system can be understood by studying its present dynamics. A combined study can use the dynamics of the present system to help understand the causes of past climatic changes recorded in ice cores, in sediment cores, and by glacial geology so that future changes can be predicted. Important problems are (1) stability of the West Antarctic Ice Sheet, because its unique marine character and past history strongly indicate that it is inherently unstable and may be presently collaps- ing, (2) how collapse of the West Antarctic Ice Sheet influences stability of the much larger East Antarctic Ice Sheet, and (3) how collapse of the West Antarctic Ice Sheet affects the stability of pack ice and circulation in the Southern Ocean.

These problems address two scenarios for Antarctic deglaciation. When the internally triggered surge instability and the externally triggered mass-balance instability are applied to the Antarctic Ice Sheet, the result might be the disintegration sequences depicted in Figure 3.36. Assuming the entire Antarctic Ice Sheet is affected, the main difference between the two instability mechanism is that the surge would cause rapid shrinkage but about half of the ice sheet would survive (Weertman, 1966). The surge instability could cause a rapid 30-m rise in worldwide sea level and the mass-balance instability could cause a slower 60-m rise. One would be as calamitous as the other. If instability were confined to West Antarctica, halving the ice thickness would unground the remaining ice so both mechanisms would cause the same sea-level rise (about 6 m), but the increase would be much more rapid with a surge. However, it is likely that a West Antarctic instability would propagate into East Antarctica through the Bottleneck. The ultimate rise in sea level from a surge-like instability is indeterminate. The largest and fastest rise would result from an ice-sheet surge, which would require a sustained and largely simultaneous surge of many major ice streams. Therefore, the crucial role of ice streams in ice-sheet stability is the ultimate clue from Antarctica for understanding how ice sheets impact global climate change.

4

THE DYNAMICS OF ICE SHEETS

The Crystal Structure of Ice

The covalent bond in molecular water requires an angle of 104° between hydrogen and oxygen atoms. In crystalline ice, his angular requirement is most closely approximated by bonding hydrogen and oxygen atoms as tetrahedra linked to form a very open crystal lattice. This lattice is represented by the unit cell of ice, shown in Figure 4.1, which is the smallest unit of ice that can be repeated in three-dimensions to produce the crystal structure of ice within each ice crystal.

The H_2O molecule is formed as water by bonding two hydrogen atoms to one oxygen atom, but it is formed as ice by bonding 24 hydrogen atoms to 12 oxygen atoms in the unit cell. An ice sheet spreads under its own weight when gravitational and thermal forces break the tetrahedral bonds. Spreading is by creep when gravitational forces break and

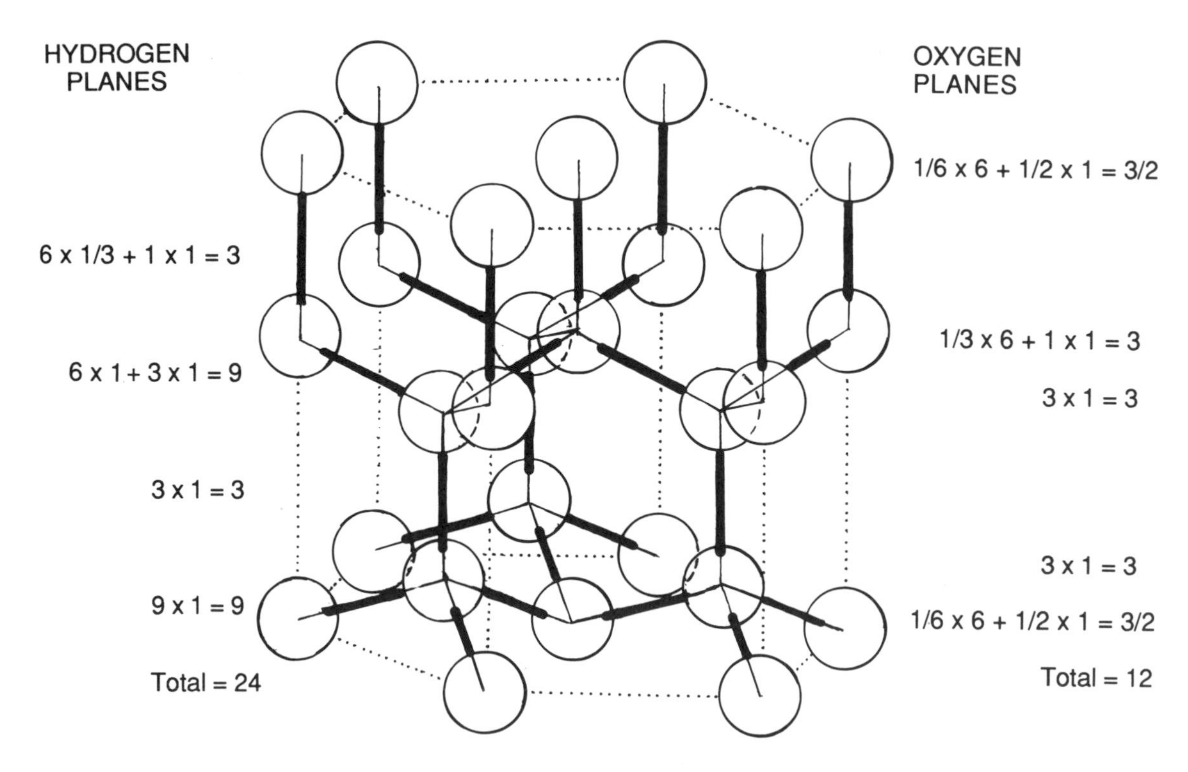

Figure 4.1: The hexagonal unit cell for ice. Circles represent oxygen atoms, which are close-packed in basal planes. Thick lines represent hydrogen atoms as bonds between oxygen atoms at tetrahedral angles. Dotted lines define the unit cell, such that all or parts of hydrogen and oxygen atoms are within the unit cell, as show for planes of hydrogen atoms and oxygen atoms. These atoms in whole or in part are added on all planes to give the chemical formula H_2O.

reform bonds sequentially as linear lattice distortions called dislocations sweep through the crystal lattice. Dislocation motion is enhanced by thermal vibrational energy that breaks bonds at individual lattice sites, creating lattice vacancies and interstitial atoms that can migrate to and from dislocations by diffusion. When basal ice reaches its melting temperature, thermal distortion of the lattice is great enough for the gravitational force of overlying ice to crush the roomy crystal lattice of ice into the more dense water phase. The 36 atoms in a unit cell of ice occupy about 10 percent less volume as 12 water molecules. The ice sheet can then spread by sliding over its own basal meltwater layer.

Dislocations form and move most easily in the hexagonal basal planes of the unit cell, rather than in prismatic or pyramidal planes, because fewer bonds need to be broken. As seen in Figure 4.1, basel layers of oxygen atoms are arranged in pairs held closely together by 9 hydrogen bonds, and these paired layers are kept at maximum separation from other paired layers by 3 hydrogen bonds. Dislocations cause shearing between basal planes by break-ing the 3 bonds between widely separated basal planes in the unit cell. Breaking these bonds in shear is easy, like snapping a match stick in two. Breaking the 9 bonds is much more difficult because there are more of them and breaking them requires tension and compression forces along the bonds in addition to a shear force across them.

Flow of ice sheets can be understood as Nature's attempt to segregate ice in its proper niche between water and air in the concentric strati-fication of Earth's constituents according to their density. In ice sheets, transitions from sheet flow to stream flow to shelf flow are responses to Nature's call. If ice did not melt, gravity would try to convert present-day ice sheets into a layer some 82 m thick over Earth's oceans.

Tensors

All glaciological properties and processes can be described by tensors (Nye, 1960). Tensors are grouped into ranks, depending on the combinations of coordinate axes needed to describe them. The first three ranks suffice for most glaciological applications. Tensors above zero rank have suffixes i, j, k, l, and so on, each one of which can refer to any one of rectilinear axes x, y, z. Zero-rank tensors have magnitude only, so they are scalars. First-rank tensors have both magnitude and direction, so they are vectors. Second-rank tensors have magnitudes that vary in both longitudinal and transverse directions. Common glaciological tensors are density ρ and temperature T for zero rank, force F_i and velocity u_i for first rank, and stress σ_{ij} and strain rate $\dot{\varepsilon}_{ij}$ for second rank. The number of components of a tensor referred to the three axes x, y, z is the cube of the number of suffixes needed to describe the tensor. Hence, tensors of the first five ranks, t, t_i, t_{ij}, t_{ijk}, and t_{ijkl} have $3^0 = 1$, $3^1 = 3$, $3^2 = 9$, $3^3 = 27$, and $3^4 = 81$ components, respectively, obtained by letting i, j, k, l be x, y, z successively. The number of components can be reduced substantially for tensors that are symmetrical. Symmetry reduces the components of stress and strain rate from nine to six because $\sigma_{ij} = \sigma_{ji}$ and $\dot{\varepsilon}_{ij} = \dot{\varepsilon}_{ji}$.

Stresses in glaciology are of two kinds: lithostatic pressure, which changes glacier size, and deviator stresses, which change glacier shape. The size change affects glacier volume, not mass, and the shape change causes the glacier to flow. Lithostatic pressure P is a scalar defined as the average of the three axial stresses:

$$P = \frac{1}{3}\left(\sigma_{xx} + \sigma_{yy} + \sigma_{zz}\right) = \frac{1}{3}\sigma_{kk} \tag{4.1}$$

In glaciology, paired suffixes kk denote the sum of tensor components along all three rectilinear axes, whereas suffixes ij denote a single tensor component specified by these axes. Lithostatic pressure causes two kinds of size change: the volume decrease when gravitational compression converts snow into firn and firn into ice beneath the surface of a glacier and the volume decrease when the crystal structure of ice is crushed into the molecular structure of water at the base of a glacier. Thermal expansion toward the base of an ice sheet is a volume increase. Deviator stress σ_{ij}' deviates from lithostatic pressure P, and the deviation causes ice to flow. Hence, σ_{ij}' is defined by subtracting P from applied stress σ_{ij}:

$$\sigma_{ij}' = \sigma_{ij} - \delta_{ij} P = \sigma_{ij} - \frac{1}{3}\delta_{ij}\sigma_{kk} \tag{4.2}$$

where δ_{ij} is the Kronecker delta, for which $\delta_{ij} = 1$ when $i = j$ and $\delta_{ij} = 0$ when $i \neq j$. The Kronecker delta is necessary because axial stresses ($i = j$) contribute to P but shear stresses ($i \neq j$) do not. It functions as an on-off switch in the binary arithmetic of computers.

Stresses result from applied forces. The applied force in glaciers is a gravitational pulling force that compresses ice and causes glaciers to flow. Flow produces kinematic forces related to the hardness of ice and traction at the bed. Gravitational force F_i is the product of ice mass m_I experiencing gravity acceleration g_i resolved in the i direction:

$$F_i = m_I \, g_i \tag{4.3}$$

where m_I is a scalar and g_i is a vector, so F_i is a vector that is strongest when i is the vertical direction. Since mass is the product of density and volume, and volume is the product of area and length normal to the area, a gravitational pulling force exists in direction i normal to area A_i experiencing lithostatic pressure P such that:

$$F_i = P \, A_i \tag{4.4}$$

where P is a scalar and A_i is a vector whose length is the magnitude of the area and whose direction is normal to the area. Kinematic forces F_i result from deviator stresses σ_{ij}' acting on areas A_j such that:

$$F_i = \sigma_{ij}' \, A_j \tag{4.5}$$

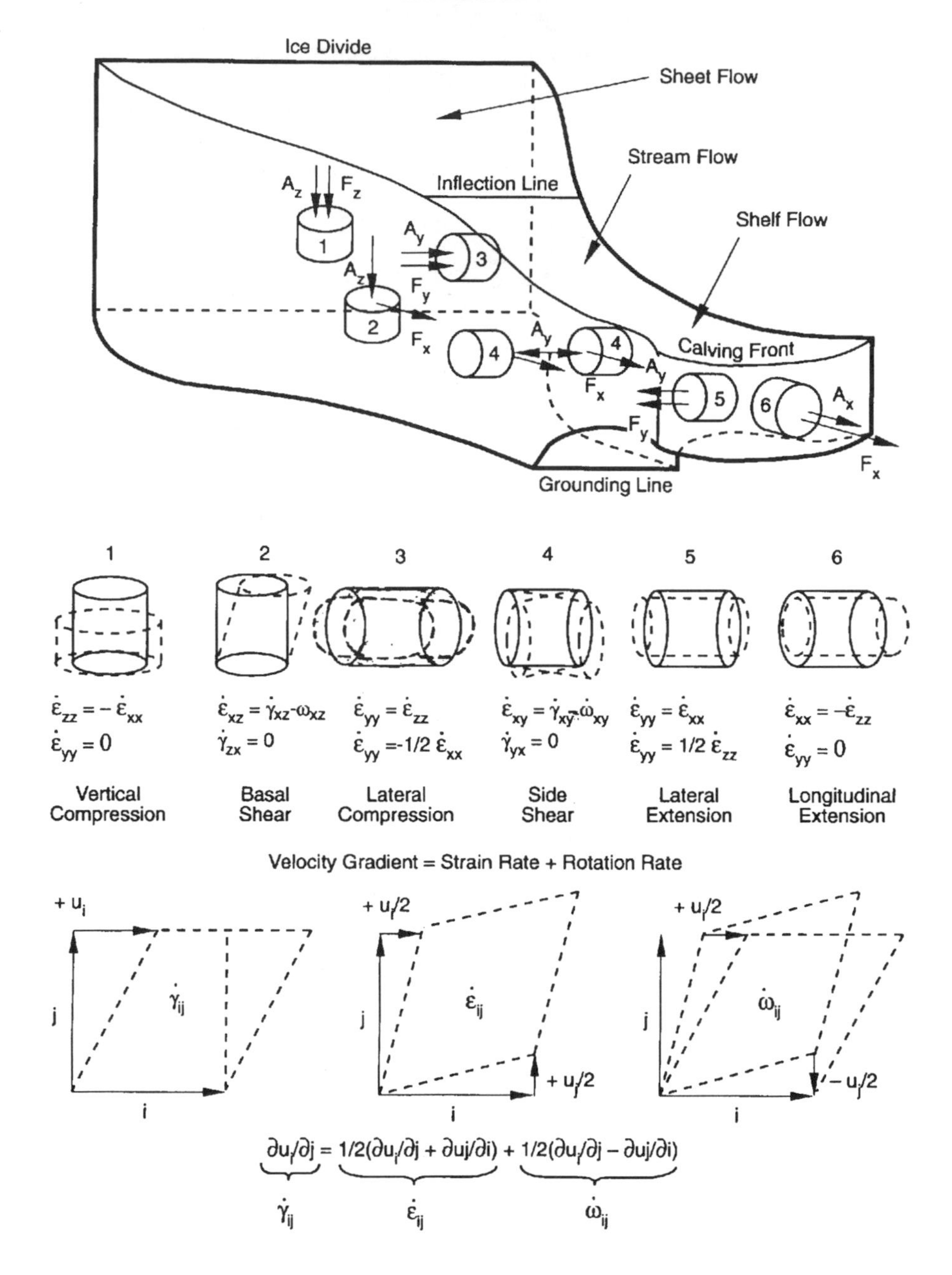

Figure 4.2: The major deviator stresses from the ice divide to the calving front for a flowband along which sheet flow becomes stream flow and stream flow becomes shelf flow. The cylinders of ice are greatly enlarged and are deformed by deviator stresses σ_{ij}'. In plan view, axis x is along the flowband, axis y is transverse to the flowband, and axis z is vertical. *Cylinder 1*: For vertical compression beneath a linear ice divide, $\sigma_{zz}' = F_z / A_z$, $\dot{\varepsilon}_{zz} = -\dot{\varepsilon}_{xx}$, and $\dot{\varepsilon}_{yy} = 0$. *Cylinder 2*: For simple shear at the bed, $\sigma_{xz} = F_x / A_z$, $\dot{\varepsilon}_{xz} = \dot{\gamma}_{xz} - \dot{\omega}_{xz}$, and $\dot{\gamma}_{zx} = 0$. *Cylinder 3*: For transverse compression from converging flow, $\sigma_{yy}' = F_y / A_y$, $\dot{\varepsilon}_{yy} = \dot{\varepsilon}_{zz}$, and $\dot{\varepsilon}_{yy} = -1/2\ \dot{\varepsilon}_{xx}$. *Cylinder 4*: For simple shear at the sides, $\sigma_{xy}' = F_x / A_y$, $\dot{\varepsilon}_{xy} = \dot{\gamma}_{xy} - \dot{\omega}_{xy}$, and $\dot{\gamma}_{yx} = 0$. *Cylinder 5*: For transverse extension from diverging flow, $\sigma_{yy}' = F_y / A_y$, $\dot{\varepsilon}_{yy} = \dot{\varepsilon}_{xx}$, and $\dot{\varepsilon}_{yy} = -\ 1/2\ \dot{\varepsilon}_{zz}$. *Cylinder 6*: For longitudinal extension along a linear calving front, $\sigma_{xx}' = F_x / A_x$, $\dot{\varepsilon}_{xx} = -\dot{\varepsilon}_{zz}$, and $\dot{\varepsilon}_{yy} = 0$. By definition, for velocities u_i and u_j in directions i and j, $\dot{\gamma}_{ij} = \partial u_i / \partial j$ is the velocity gradient, $\dot{\varepsilon}_{ij} = 1/2\ (\partial u_i / \partial j + \partial u_j / \partial i)$ is the strain rate, and $\dot{\omega}_{ij} = 1/2\ (\partial u_i / \partial j + \partial u_j / \partial i)$ is the rotation rate, so $\dot{\gamma}_{ij} = \dot{\varepsilon}_{ij} + \dot{\omega}_{ij}$.

where F_i is a normal force when i = j and a shear force when i ≠ j. Normal stresses σ_{ii}' act perpendicular to area A_i for i = j and shear stresses σ_{ij}' act parallel to area A_j i ≠ j.

Figure 4.2 shows the major deviator stresses and strain rates in a flowband of an ice sheet for sheet flow from the ice divide to the surface inflection line, stream flow from the surface inflection line to the basal grounding line, and shelf flow from the basal grounding line to the calving front of the ice shelf. Rectilinear axes x, y, z have their origin on the bed beneath the ice divide, with x horizontal and positive toward the calving front, y horizontal and directed along the ice divide, and z vertical and positive upward. Horizontal velocity increases from the bed to the surface, as long as basal traction exists. Horizontal velocity also increases from the ice divide to the calving front, provided that ablation is primarily by calving.

The dominant stresses are σ_{zz} induced by vertical gravitational compression everywhere, with σ_{xz} induced by basal traction for sheet flow, σ_{xx}', σ_{yy}', and σ_{xy} induced by sheet flow converging to become stream flow, σ_{xx}' and σ_{xy} induced by reduced basal traction for stream flow, and σ_{xx}' and σ_{yy}' induced by longitudinal and transverse spreading for unconfined shelf flow. Kinematic stresses $\sigma_{xx}' = F_x / A_x$, $\sigma_{yy}' = F_y / A_y$, $\sigma_{xy} = F_x / A_y$, and $\sigma_{xz} = F_x / A_z$ result from velocity gradients induced by deviations from gravitational stress $\sigma_{zz} = F_z / A_z$, as illustrated in Figure 4.2. The velocity gradient tensor $\dot{\gamma}_{ij} = \partial u_i / \partial j$ is the sum of a symmetrical strain rate tensor $\dot{\varepsilon}_{ij} = 1/2\ (\partial u_i / \partial j + \partial u_j / \partial i)$ and an anti-symmetrical rotation rate tensor $\dot{\omega}_{ij} = 1/2\ (\partial u_i / \partial j - \partial u_j / \partial i)$:

$$\dot{\gamma}_{ij} = \dot{\varepsilon}_{ij} + \dot{\omega}_{ij} = \frac{1}{2}\left(\partial u_i / \partial j + \partial u_j / \partial i\right) + \frac{1}{2}\left(\partial u_i / \partial j - \partial u_j / \partial i\right) = \partial u_i / \partial j \tag{4.6}$$

where u_i and u_j are ice velocities in the i and j directions, i = j for axial velocity gradients, and i ≠ j for shear velocity gradients. Pure shear exists when $\partial u_i / \partial j = \partial u_j / \partial i$. Simple shear exists when $\partial u_i / \partial j > 0$ and $\partial u_j / \partial i = 0$. Pure axial extension or compression produces pure shear resolved on planes at 45° to that axis. Basal or side traction produces simple shear consisting of pure shear plus rigid rotation.

Glacial flow is described by linking kinematic deviator stresses σ_{ij}' induced by strain rates $\dot{\varepsilon}_{ij}$ and velocities u_i when gravitational potential energy is converted into kinetic energy of motion. The flow law of ice provides the linkage between $\dot{\varepsilon}_{ij}$ and σ_{ij}' within an ice sheet where glacial creep occurs. The sliding law of ice provides the linkage between u_i and σ_{ij}' at the ice-bed interface where glacial sliding occurs.

Creeping Ice

The flow law for creep of ice is derived by first balancing forces to obtain force equilibrium. Figure 4.3 is a lump of ice in which internal stresses σ_{ij} resist external stress σ. The applied force is σA_s, where A_s is surface area JKL which has direction cosines l, m, and n with rectilinear axes, x, y, z. Hence, planes OKL, OJK, and OJL have areas $A_s l = A_x$, $A_s m = A_y$, and $A_s n = A_z$. The sum of forces in the x, y and z directions is zero for equilibrium:

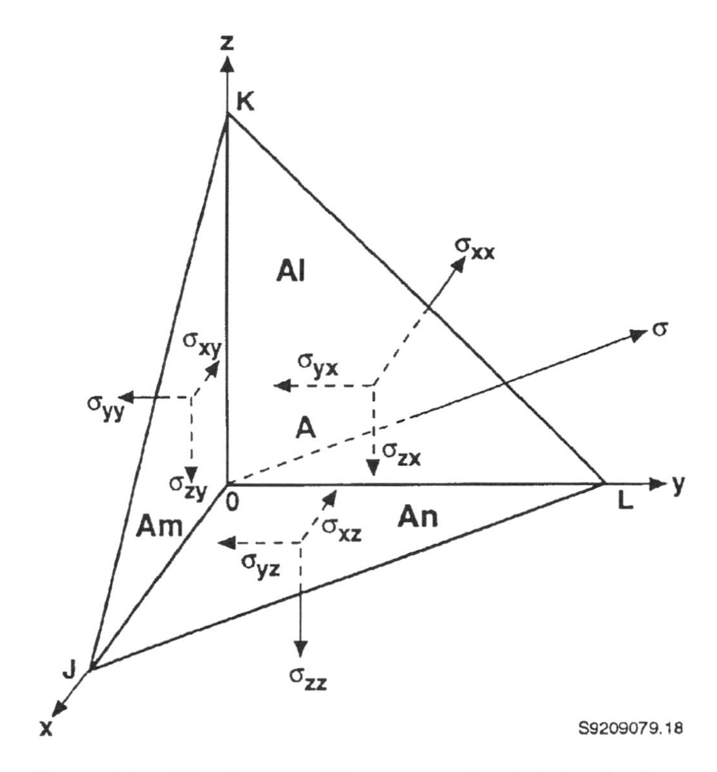

Figure 4.3: The force equilibrium used for deriving the flow law of ice. Stress σ normal to area A is resisted by stresses σ_{ij}, normal (i = j), and parallel (i ≠ j) to areas Al, Am, and An normal to axes x, y, and z that have direction cosines l, m, and n with respect to the direction of σ. In deriving the flow law, stresses σ and σ_{ij} are replaced by deviator stresses σ' and σ_{ij}', where i, j = x, y, z in the usual tensor notation.

$$(\sigma l)\, A_s - \sigma_{xx} (A_s l) - \sigma_{xy} (A_s m) - \sigma_{xz} (A_s n) = 0 \tag{4.7a}$$

$$(\sigma m)\, A_s - \sigma_{yx} (A_s l) - \sigma_{yy} (A_s m) - \sigma_{yz} (A_s n) = 0 \tag{4.7b}$$

$$(\sigma n)\, A_s - \sigma_{zx} (A_s l) - \sigma_{zy} (A_s m) - \sigma_{zz} (A_s n) = 0 \tag{4.7c}$$

where the first subscript in σ_{ij} is the direction of the stress and the second subscript is the normal to the plane upon which the stress acts. Axial stresses (i = j) are normal to the plane and shear stresses (i ≠ j) are parallel to the plane. A solution of Equations (4.7) is obtained by setting the determinant of the coefficients of l, m, and n equal to zero:

$$\begin{vmatrix} (\sigma - \sigma_{xx}) & (-\sigma_{xy}) & (-\sigma_{xz}) \\ (-\sigma_{yx}) & (\sigma - \sigma_{yy}) & (-\sigma_{yz}) \\ (-\sigma_{zx}) & (-\sigma_{zy}) & (\sigma - \sigma_{zz}) \end{vmatrix} = 0 \tag{4.8}$$

The solution of Equation (4.8) has terms that can be grouped as coefficients of σ:

$$\sigma^3 - (\sigma_{xx} + \sigma_{yy} + \sigma_{zz})\,\sigma^2 + (\sigma_{xx}\,\sigma_{yy} + \sigma_{yy}\,\sigma_{zz} + \sigma_{zz}\,\sigma_{xx} - \sigma_{xy}{}^2 - \sigma_{yz}{}^2 - \sigma_{zx}{}^2)\,\sigma - (\sigma_{xx}\,\sigma_{yy}\,\sigma_{zz} + 2\sigma_{xy}\,\sigma_{yz}\,\sigma_{zx} - \sigma_{xx}\,\sigma_{yz}{}^2 - \sigma_{yy}\,\sigma_{zx}{}^2 - \sigma_{zz}\,\sigma_{xy}{}^2) = 0 \quad (4.9)$$

where $\sigma_{ij} = \sigma_{ji}$ by symmetry. The coefficients of σ are:

$$\Sigma_1 = (\sigma_{xx} + \sigma_{yy} + \sigma_{zz}) \quad (4.10a)$$

$$\Sigma_2 = -\,(\sigma_{xx}\,\sigma_{yy} + \sigma_{yy}\,\sigma_{zz} + \sigma_{zz}\,\sigma_{xx} - \sigma_{xy}{}^2 - \sigma_{yz}{}^2 - \sigma_{zx}{}^2) \quad (4.10b)$$

$$\Sigma_3 = (\sigma_{xx}\,\sigma_{yy}\,\sigma_{zz} + 2\sigma_{xy}\,\sigma_{yz}\,\sigma_{zx} - \sigma_{xx}\,\sigma_{yz}{}^2 - \sigma_{yy}\,\sigma_{zx}{}^2 - \sigma_{zz}\,\sigma_{xy}{}^2) \quad (4.10c)$$

where Σ_1, Σ_2, and Σ_3 are the first, second, and third invariants of the stress tensor and are scalars. Therefore, Equation (4.9) is a cubic algebraic equation having three solutions for σ:

$$\sigma^3 - \Sigma_1\,\sigma^2 - \Sigma_2\,\sigma - \Sigma_3 = 0 \quad (4.11)$$

The three solutions of σ are the three principal stresses σ_1, σ_2, σ_3 acting along principal axes, 1, 2, 3, which have direction cosines l, m, n with respect to general axes x, y, z (Dieter, 1961).

A force balance can also be applied to components σ_{ij}' of the deviator stress tensor. Using Equation (4.2), the three invariants of the deviator stress tensor are shown to be:

$$\Sigma_1' = \sigma_{xx}' + \sigma_{yy}' + \sigma_{zz}' = \sigma_{kk}' = 0 \quad (4.12a)$$

$$\Sigma_2' = -\,(\sigma_{xx}'\,\sigma_{yy}' + \sigma_{yy}'\,\sigma_{zz}' + \sigma_{zz}'\,\sigma_{xx}' - \sigma_{xy}{}^2 - \sigma_{yz}{}^2 - \sigma_{zx}{}^2) = \frac{1}{2}(\sigma_{xx}'^2 + \sigma_{yy}'^2 + \sigma_{zz}'^2 + 2\sigma_{xy}{}^2 + 2\sigma_{yz}{}^2 + 2\sigma_{zx}{}^2) = \frac{1}{2}\,\sigma_{ij}'\,\sigma_{ij}' \quad (4.12b)$$

$$\Sigma_3' = \begin{vmatrix} \sigma_{xx}' & \sigma_{xy} & \sigma_{xz} \\ \sigma_{yx} & \sigma_{yy}' & \sigma_{yz} \\ \sigma_{zx} & \sigma_{zy} & \sigma_{zz}' \end{vmatrix} = \frac{1}{3}\,\sigma_{ij}'\,\sigma_{jk}'\,\sigma_{ki}' \quad (4.12c)$$

Symmetry reduces the nine terms in Σ_2' to six terms, and the twenty-seven terms in Σ_3' to ten terms.

Invariants can be extracted from any second-rank tensor. The invariants for strain rate are:

$$E_1 = \dot{\varepsilon}_{kk} = 0 \quad (4.13a)$$

$$E_2 = \frac{1}{2}\,\dot{\varepsilon}_{ij}\,\dot{\varepsilon}_{ij} \quad (4.13b)$$

$$E_3 = \frac{1}{3}\,\dot{\varepsilon}_{ij}\,\dot{\varepsilon}_{jk}\,\dot{\varepsilon}_{ki} \quad (4.13c)$$

where $E_1 = 0$ requires that ice is incompressible. This is a good approximation for analyzing glacial flow because $\Sigma_1' = 0$. Since glacial strain rates are small, $E_2 >> E_3$ is also a good approximation because strain rates can be positive or negative, so summing products of the same two strain rates having the same sign is then much larger than summing products of three strain rates, which can differ in sign. The six terms in E_2 are all positive, whereas half of the ten terms in E_3 are likely to be negative, making $E_2 >> E_3$. This implies that $\Sigma_2' >> \Sigma_3'$ because strain rates are caused by deviator stresses. The algebraic equations corresponding to Equation (4.11) then reduce to:

$$\sigma'^3 - \Sigma_2'\,\sigma' = 0 \quad (4.14a)$$

$$\dot{\varepsilon}^3 - \Sigma_2\,\dot{\varepsilon} = 0 \quad (4.14b)$$

Therefore, since second invariants must be positive and second deviator invariants express creep deformation:

$$\sigma_c = \sigma' \equiv \left(\Sigma_2'\right)^{1/2} = \left(\frac{1}{2}\,\sigma_{ij}'\,\sigma_{ij}'\right)^{1/2} \quad (4.15a)$$

$$\dot{\varepsilon}_c = \dot{\varepsilon} \equiv \left(E_2\right)^{1/2} = \left(\frac{1}{2}\,\dot{\varepsilon}_{ij}\,\dot{\varepsilon}_{ij}\right)^{1/2} \quad (4.15b)$$

where σ_c is the effective creep stress, $\dot{\varepsilon}_c$ is the effective creep rate, and both are scalars.

A logarithmic plot of σ_c versus $\dot{\varepsilon}_c$ from creep experiments on ice is linear for steady-state creep, so the flow law of ice has the form (Glen, 1958):

$$\dot{\varepsilon}_c = (\sigma_c/A)^n \quad (4.16)$$

where A is a measure of ice hardness that depends on temperature and the orientation of ice crystals, and n locates ice in the viscoplastic creep spectrum for polycrystalline materials. Creep in ice is usually described when $n = 3$, which lies between $n = 1$ for viscous creep and $n = \infty$ for plastic creep.

The flow law for ice can also be written for individual components of strain rate and deviator stress. For single components, $\dot{\varepsilon}_{ij}/\sigma_{ij}' = \dot{\varepsilon}_c/\sigma_c$, so Equation (4.16) becomes:

$$\dot{\varepsilon}_{ij} = \dot{\varepsilon}_c\left(\sigma_{ij}'/\sigma_c\right) = \left(\sigma_c/A\right)^n\left(\sigma_{ij}'/\sigma_c\right) = \left(\sigma_c^{n-1}/A^n\right)\sigma_{ij}' \quad (4.17)$$

Equation (4.17) shows that the flow law is based upon two

assumptions. First, strain rate component $\dot{\varepsilon}_{ij}$ is proportional to the corresponding deviator stress component σ_{ij}'. Second, the proportionality constant (σ_c^{n-1}/A^n) is a function of the second invariant of deviator stresses. In laboratory creep experiments, it is easier to relate strain rates to applied stresses using Equation (4.17). On glaciers, it is easier to relate stresses to measured strain rates using Equation (4.16) written for a single deviator stress component:

$$\sigma_{ij}' = \sigma_c\left(\dot{\varepsilon}_{ij}/\dot{\varepsilon}_c\right) = \left(A\ \dot{\varepsilon}_c^{\ 1/n}\right)\left(\dot{\varepsilon}_{ij}/\dot{\varepsilon}_c\right) = \left(A/\dot{\varepsilon}_c^{\ 1\text{-}1/n}\right)\dot{\varepsilon}_{ij} \qquad (4.18)$$

Note that σ_c and $\dot{\varepsilon}_c$ contain all components of deviator stress and strain rate. All these components must appear in σ_c and $\dot{\varepsilon}_c$ when σ_c^{n-1} and $\dot{\varepsilon}_c^{\ 1-1/n}$ are evaluated in Equations (4.17) and (4.18).

Basal Sliding

When P is great enough to crush basal ice into water, the glacier can slide over its bed. In the Weertman (1957a, 1964) analysis of glacier sliding, the basal water layer is a lubricating film that eliminates basal traction except at bedrock bumps of average dimension Λ separated by average distance Λ'. His analysis can allow bedrock projections through soft, deformable, water-soaked sediments or till. A bed like this provides little traction to resist sliding ice. The only significant resistance to sliding is provided by longitudinal deviator stress σ_{xx}', which is compressive on the stoss side and tensile on the lee side of bumps. Taking Λ^2 as the cross-sectional area of bumps and Λ'^2 as the area of reduced-traction bed surrounding each bump, balancing the compressive and traction forces gives $\Lambda^2\ \sigma_{xx}' = \Lambda'^2\ \tau_o$, so that:

$$\sigma_{xx}' = (\Lambda'/\Lambda)^2\ \tau_o \qquad (4.19)$$

where interstitial water in the sediment or till supports, by buoyancy, most of the overlying ice burden, and bedrock projections resist, by traction, most of the ice motion (see Figure 4.3). Sliding theory utilizes the thermal constants in Table 4.1.

Longitudinal compressive deviator stress σ_{xx}' on the stoss side of bumps causes divergence of ice around the bumps and convergence of ice in the lee side of bumps. Compressive ice motion in the x direction increases basal lithostatic pressure P_o by ΔP_o causing the pressure melting point T_M to change by:

Table 4.1: Thermal Constants for Pure Ice

Symbol	Meaning	Value
c_H	Specific heat capcity	2009 J kg^{-1} K^{-1}
H_M	Latent heat of fusion	335 kJ kg^{-1}
K	Thermal conductivity	2.10 W m^{-1} K^{-1}
κ	Thermal diffusivity	1.15 x 10^{-6} m^2 a^{-1}

$$\Delta T_M = (\partial T_M/\partial P)\ \Delta P_o = C\ \sigma_{xx}' = C(\Lambda'/\Lambda)^2\ \tau_o \qquad (4.20)$$

where $(\partial T_M/\partial P) = C$ is constant for small ΔT_M changes and, since most bumps are squat, more ice flows over them than around them so that $\sigma_{yy} \approx P_o$ and $\sigma_{xx}' = \sigma_{xx} - 1/3\ (\sigma_{xx} + \sigma_{yy} + \sigma_{zz}) \approx (P_o + \Delta P_o) - 1/3\ [(P_o + \Delta P_o) + P_o + (P_o - \Delta P_o)] = \Delta P_o$. If ice also flows around bumps, $\sigma_{xx} = P_o + \Delta P_o$ and $\sigma_{yy} \approx \sigma_{zz} \approx P_o - 1/2\ \Delta P_o$.

If u_r is the freezing rate over area Λ^2 on the lee side of bumps, then volume flux $u_r\ \Lambda^2$ of ice having density ρ_I and latent heat of melting H_M releases latent heat at rate $u_r\ \Lambda^2\ \rho_I\ H_M$. This heat is then conducted longitudinally through the bump having thermal conductivity K, along temperature gradient $\Delta T_M/\Lambda$, to supply sensible heat at rate $(K\ \Delta T_M/\Lambda)\ \Lambda^2$ to transverse area Λ^2 on the stoss side of the bump, where sensible heat is converted into latent heat of melting for meltwater flowing over and around the bump to refreeze on the lee side. Equating the latent heat rate with the sensible heat rate and solving for u_r gives the regelation velocity of ice moving past the bump by this melting/freezing regelation mechanism:

$$u_r = K\ \Delta T_M/\Lambda\ \rho_I\ H_M = K\ C(\Lambda'/\Lambda)^2\ \tau_o/\Lambda\ \rho_I\ H_M \qquad (4.21)$$

where ΔT_M is given by Equation (4.20) and u_r is the regelation velocity.

Longitudinal stress σ_{xx}' is a deviator stress that causes creep deformation at strain rate $\dot{\varepsilon}_{xx}$ in ice near the bump. From the flow law of ice and Equation (4.19):

$$\dot{\varepsilon}_{xx} = (\sigma_c^{n-1}/A^n)\ \sigma_{xx}' \approx (\sigma_{xx}'/A_M)^n = [(\Lambda'/\Lambda)^2\ \tau_o]^n/A_M^{\ n} \qquad (4.22)$$

where $A = A_M$ at the pressure melting point of ice and $\sigma_{xx}' >> \tau_o$ because $\Lambda' >> \Lambda$. Since most bumps are squat, ice should flow over them more than around them. Therefore, $\sigma_{yy}' << \sigma_{zz}'$ so that $\sigma_{xx}' \approx -\sigma_{zz}' \approx \sigma_c$ in Equation (4.22), see Equations (4.12a) and (4.12b). If $\dot{\varepsilon}_{xx}$ occurs within distance Λ of the bump, Equation (4.6) gives:

$$\dot{\varepsilon}_{xx} = \partial u_x/\partial x \approx u_c/\Lambda \qquad (4.23)$$

Equating Equations (4.22) and (4.23) and solving for u_c gives the creep velocity of ice around the bump:

$$u_c = \Lambda\ \dot{\varepsilon}_{xx} = \Lambda(\Lambda'/\Lambda)^{2n}\ (\tau_o/A_M)^n \qquad (4.24)$$

where Λ'/Λ is a bed smoothness factor and u_c is the creep velocity.

When only a few bedrock bumps project above the soft sediment or till blanket, $\Lambda'/\Lambda >> 1$, but Λ'/Λ decreases when many bumps project. As glacial sliding strips away the soft blanket, Λ'/Λ decreases because bedrock projections proliferate. Note that u_r is inversely proportional to Λ, and u_c is directly proportional to Λ. For pyramid-shaped bumps,

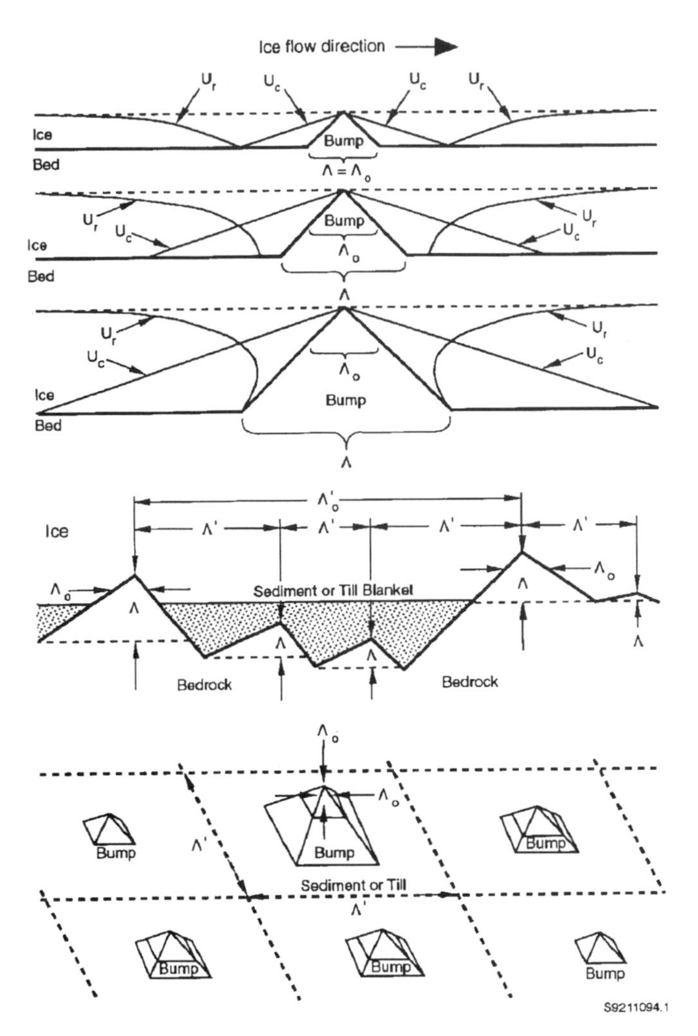

Figure 4.4: The bed conditions used for deriving the sliding law of ice. Bed roughness is represented by bedrock pyramids having average height Λ above their bases and average separation Λ' between their apexes. Regelation velocity u_r and creep velocity u_c are equal where Λ_o is the distance through the pyramid in the direction of ice flow and Λ_o^2 is the transverse cross-sectional area normal to the direction of ice flow. A blanket of sediment or till can decrease Λ and increase Λ', thereby allowing stream flow to replace sheet flow if the blanket is saturated with water and cannot support a basal shear stress.

therefore, the upper part offers little resistance to sliding because u_r is rapid, and the lower part offers little resistance to sliding because u_C is rapid, as shown in Figure 4.4. Maximum resistance to sliding occurs at a distance above the base of bumps where both u_r and u_c are equally affected by the bump cross-section. Setting $u_r = u_c$ and solving Equations (4.21) and (4.24) for the rate-controlling bump dimension Λ_o for bedrock bumps having average spacing Λ_o' between each Λ_o projecting above the sediment or till blanket:

$$\Lambda_o = \left(\frac{\Lambda_o}{\Lambda_o'}\right)^{n-1} \left[\frac{K C A_M^n}{\rho_I H_M \tau_o^{\ n-1}}\right]^{1/2} \tag{4.25}$$

where Λ_o/Λ_o' is the effective bed roughness factor and is the inverse of effective bed smoothness factor Λ_o'/Λ_o.

Let a wet, porous sediment or till blanket of thickness λ bury bedrock projections of height Λ such that only height $\Lambda - \lambda$ penetrates ice above the blanket. Projections having unburied heights of $\Lambda - \lambda$ or more are then separated by average distance Λ'. If water pressure in the porous blanket is P_W and the basal pressure of overlying ice is P_I, the effective bed roughness factor becomes $(\Lambda - \lambda)/\Lambda' = \Lambda/\Lambda'\ (1 - P_W/P_I)$. Then $(\Lambda - \lambda)/\Lambda' = \Lambda/\Lambda'$ when $\lambda << \Lambda$ or the blanket is unsaturated, so that $P_W << P_I$; and $(\Lambda - \lambda)/\Lambda' = 0$ when $\lambda = \Lambda$ and the blanket is saturated or supersaturated, so that $P_W = P_I$. Setting $\Lambda/\Lambda' = \Lambda_o/\Lambda_o'$ for fractal geometry, substituting Λ_o from Equation (4.25) into Equations (4.21) and (4.24), and replacing Λ_o/Λ_o' with $(\Lambda_o/\Lambda_o')\ (1 - P_W/P_I)$ gives $u_S = f\,(u_r + u_c)$ as the sliding velocity, where $1/2 \leq f \leq 1$ according to Nye (1969), so that:

$$u_S = 2\,f\,u_r = 2\,f\,u_c = 2\,f \left[\frac{\left[(K\,C/\rho_I\,H_M)\left(\tau_o^{\ n+1}/A_M^{\ n}\right)\right]^{1/2}}{\left[(\Lambda_o/\Lambda_o')\,(1 - P_W/P_I)\right]^{n+1}}\right]$$

$$= \left[\frac{(\tau_o/B_o)}{\left[(\Lambda_o/\Lambda_o')\,(1 - P_W/P_I)\right]^2}\right]^{\frac{n+1}{2}} = \frac{(\tau_o/B)^m}{(1 - P_W/P_I)^{2m}} \tag{4.26}$$

where $B_o = (\rho_I H_M A_M^n/4 f^2 K\,C)^{1/2m}$, $B = (\Lambda_o/\Lambda_o')^2\,B_o$, and $m = 1/2\,(n + 1)$. Equation (4.26) should apply to stream flow in which bedrock "sticky spots" poke through a slippery blanket of wet sediments or till. Equation (4.26) reduces to $u_S = (\tau_o/B)^m$ for sheet flow over wet bedrock such that $P_W << P_I$ and becomes indeterminate for shelf flow because $\tau_o = 0$ and $P_W = P_I$ for floating ice.

This sliding analysis differs from the Weertman (1957a, 1964) analysis by including wet sediment or till that provides little or no traction to resist sliding and by replacing cubical bedrock bumps with bedrock pyramidal bumps so that all bumps larger than Λ_o have rate-controlling cross-sections that retard sliding. This is illustrated in Figure 4.3 by the wet blanket and by the crossover point of the nonlinear curve for u_r and the straight line for u_c. A sediment or till blanket increases the space between bumps that can control the sliding velocity. Lliboutry (1968, 1975, 1978), Nye (1970), and Kamb (1970) presented the most elaborate theories for sliding velocities controlled by bedrock roughness. Lingle and Brown (1987), Alley and others (1987), MacAyeal (1989), and Alley (1989a, 1989b, 1990) presented theories for sliding velocities controlled by deformable wet sediments or till that, nonetheless, provide traction to resist sliding.

Bed Deformation

Figure 4.4 demonstrates the need to unify theories for ice sliding over rough bedrock and for ice shearing a deformable sediment or till blanket. Deformation of this blanket depends on the range of size and the mineralogy of particles that constitute the sediment or till, the pore water fraction at various depths through this blanket, mobilization of pore water as the blanket is sheared by sliding of the overlying ice, and changes in water volume in the blanket due to its permeability, pressure gradients within it that cause pore water to migrate, and freezing or thawing at the ice-blanket interface. These physical properties and thermo-mechanical processes can change the viscoplastic creep behavior of the blanket. When the water fraction and mobility are high, the blanket tends to deform as a viscous material because water is a viscous fluid. When the water fraction and mobility are low, the blanket may deform as a plastic material that has the consistency of putty.

A vast literature deals with the rheology and hydrology of granular materials, ranging from gouge in faults to landslides. Studies of deforming sediments or till beneath ice sheets are a relatively recent addition to this literature, and draw heavily from it. Three examples of such studies serve to illustrate the range of possibilities.

Iverson (1985, 1986a, 1986b) developed a three-dimensional constitutive equation for modeling and classifying mass-movement processes, with particular attention to landslides moving as laminar flow. He showed how restricting rheological parameters reduced his general equation to the classical equations in general use for viscous, plastic, and viscoplastic rheologies (Iverson, 1985). Equation (3.54), proposed in Chapter 3 for deforming subglacial sediments or till, is a reduced version of this general constitutive equation that includes these standard rheologies for wet or dry conditions. Iverson (1986a) reduced his general constitutive equation to a datum-state equation of motion, which he applied to slow, nearly steady-state landslides. His equation has much in common with the nearly steady-state creep equation for glacial ice (Glen, 1958), but includes the hydraulic pressure of ground water. Iveron (1986b) then investigated perturbations of his datum-state equation to study transient behavior in otherwise steady-state landslides, with special attention to scarp slumping at the headwall, raising the water table at a point between the headwall and the toe, and eroding the toe. In all these cases, he found that local perturbations triggered advective-diffusive behavior in the landslide, with slow advection of kinematic waves from these sites when the rheology is mostly viscous and rapid diffusion of the perturbation when the rheology is mostly plastic. Instabilities propagated downslope from the scarp slump, downslope and upslope from the rising water table, and upslope from the eroded toe.

Transient responses studied by Iverson (1986b) have similarities to kinematic waves induced by mass-balance perturbations on glaciers (Nye, 1960, 1963a, 1963b), but he allows more complex rheologies that he applies to a different problem. Nonetheless, his study brings glaciological applications to mind. Ice streams may be a transient perturbation of nearly steady-state sheet flow. Migration of the surface inflection between the convex sheet flow profile and the concave stream flow profile may be analogous to scarp slumping at the headwall of a landslide. Raising the water-table beneath the landslide induces a kinematic response that may be analogous to converting sheet flow into stream flow. Eroding the toe of a landslide may be analogous to disintegrating an ice shelf that buttresses an ice stream. There is no reason why the approach developed by Iverson (1985, 1986a, 1986b) cannot be applied to ice sheets, with results that are equally rewarding in understanding ice streams.

Kamb (1991) examined several constitutive equations for analyzing laboratory creep data on till obtained from beneath Ice Stream B in West Antarctica. After yielding at stress σ_0, strain rate $\dot{\varepsilon}$ varied with residual stress σ according to $\dot{\varepsilon} = C\,\sigma^n/(P_I - P_W)^m$ and $\dot{\varepsilon} = \dot{\varepsilon}_o\, exp\,(k\,\sigma/\sigma_0)$, where $\sigma_0 = c_o + \mu_o\,(P_I - P_W)$, C is a softness parameter, n is a viscoplastic parameter, P_I is ice overburden pressure at the bed, P_W is basal water pressure, m is a constant, σ_o is the residual strenth at strain rate $\dot{\varepsilon}_o$ below which σ is independent of $\dot{\varepsilon}$, c_o is the cohesion of the till, μ_o is the internal friction of the till, and k is constant. Kamb (1991) showed that $c_o \approx 0$ and $n \approx m \approx k$ for creep of till in general, with $n \approx 100$ and $\sigma_0 \approx 0.02$ bar for the till beneath Ice Stream B, which is creep at the plastic end of the viscoplastic creep spectrum. In sharp contrast, Boulton and Hindmarsh (1987) reported n = 1.3 and m = 1.8 for creep of till beneath an Icelandic glacier. They fitted their creep data to the expression $\dot{\varepsilon} = C\,(\sigma - \sigma_o)^n/(P_I - P_W)^m$, with $\sigma_0 = c_o + \mu_o\,(P_I - P_W)$, to obtain n = 0.6, m = 1.2, c_o = 0.04 bar, and μ_o = 0.6. Hooke and others (1997) discounted these results because the method for obtaining σ was not reported and large variations in σ implied that creep was dominated by large longitudinal stress gradients, which were not included in the analysis.

Kamb (1991) analyzed the subglacial hydrology for the till beneath Ice Stream B, assuming a nonuniform film of water having a spatially varying thickness of order 1 mm between the till and basal ice, as used by Alley (1989a, 1989b). Kamb (1991) computed adjustments of this thickness in response to perturbations in the melting rate of basal ice and flow of basal meltwater through the system in response to gradients in $(P_I - P_W)$, all as "instantaneous" responses to velocity changes in till deformation that were largely independent of basal shear stress because of the strongly nonlinear rheological equation that was deduced from the creep experiments. He found that spatially sinusoidal perturbations grew exponentially when $n > 5$ for perturbations having a 100 km wavelength and $n > 20$ for wavelengths of 30 km. These were unstable responses for which the till provided no effective resistance to ice motion. Kamb (1991) concluded that the fast flow of ice streams must be resisted by basal shear over bedrock "sticky spots" that outcropped through the till and by side shear against slowly moving lateral ice, if n for the till exceeded values between 5 and 20, depending on the wavelength of the basal perturbation. The instability results because basal meltwater is generated faster than it can be dissipated through the subglacial hydrological system, thereby creating a basal water layer too thick to allow mechanical coupling between basal ice and basal till.

Kamb (1991) examined the possibility of long-term damping of the instability by ice dynamics though a feedback between basal sliding and basal melting, following Oerlemans and Van der Veen (1984). The instability was still possible, despite the feedback, but instabilities were bounded by steady-state flow regimes. This led Kamb (1991) to speculate on whether West Antarctic ice streams may themselves be manifestations of till instability, with the lengths of ice streams being determined by the wavelengths of the spatially sinusoidal perturbations. Lliboutry (1968, 1969) held a similar view for surging glaciers. Meltwater that drowned bed roughness bumps on the short-wavelength scale initiated the surge, but the surge was stabilized by undrowned bed roughness bumps on the long-wavelength scale. In ice streams, bedrock outcropping through the till as "sticky spots" would provide the long-wavelength stabilization of stream flow that prevented the major gravitational collapse of ice sheets of the kind modeled by Oerlemans and Van der Veen (1984). West Antarctic ice streams typically consist of a series of terraces having lengths in the 30 km to 100 km range of perturbation wavelengths. Perhaps these terraces are boudins and West Antarctic ice streams are examples of large-scale boundinage controlled by the spacing of frozen and thawed patches in basal till or sediments.

Hooke and others (1997) and Hooke (1998) examined the physical basis for till that deforms near the plastic end of the viscoplastic creep spectrum. They measured till deformation in situ beneath Storglaciären in northern Sweden. Hooke and others (1997) continued studies described by Hooke and others (1992) which included surface mass balance measurements, surface strain networks, and boreholes through the glacier, making this one of the most comprehensive glaciological field studies. Instrumentation for the boreholes through the glacier, making this one of the most comprehensive glaciological field studies. Instrumentation for the borehole and early results were described by Iverson and others (1994, 1995). Till was sampled from the boreholes to determine its physical properties. Contributions to surface ice velocity from creep through some 100 m of ice, sliding at the ice-till interface, and till deformation were determined. In situ till deformation was measured using tiltmeters, dragometers, and ploughmeters. Pressure transducers placed in the boreholes and the till recorded basal water pressure and pore water pressure. Data from all instruments were correlated over the summer months from several boreholes. Diurnal variations of the drag force, tilt rate, and effective pressure ($P_I - P_W$) in the till varied out-of-phase with or were independent of ice-surface velocity. Therefore, till deformation did not control ice velocity. Instead, ice velocity increased as $P_I - P_W$ decreased, as expected when basal sliding controls ice velocity. In addition, till strength varied directly with $P_I - P_W$, but did not correlate with ice surface velocity, which implies till deforming near the plastic end of the viscoplastic creep spectrum. Hooke and others (1997) favored a constitutive equation in which $\dot{\varepsilon} = \dot{\varepsilon}_o \, exp \, (k \, \sigma/\sigma_O)$. Setting $\sigma_O = c_o + \mu_o \, (P_I - P_W)$, as did Kamb (1991), was not appropriate because internal friction μ_o was itself found to increase as $P_I - P_W$ increased.

Hooke (1998) presented a theoretical model to explain till deformation beneath Storglaciären and, perhaps, ice streams. Laminar flow induced by simple shear of the till is subjected to a principal compressive stress at 45° to the shear couple. The compressive stress forces grains of rock to remain in contact in this direction, even as pore water pressure tends to relieve contact stresses between grains in other directions. This tendency to maintain contact between grains in lines angled upstream at about 45° to the bed produces a network of grain "bridges" that provide resistance to till deformation in simple shear. Failure of the bridges allows till deformation and occurs when the principal compressive stress fractures rock grains or the shear couple dislodges rock grains in the bridges. Failure is retarded most when a size range of rock grains provides multiple points of contact between grains and satisfies the fractal relationship $N \, (d) = N_o \, (d_o / d)^m$, where $N \, (d)$ is the number of particles having mean diameter d, $N_o \, (d_o)$ is the number having reference diameter d_o, and m is the fractal dimension. Hence, the fractal size distribution is independent of the scale of rock grains. This justifies applying principles of statistical mechanics in which strain rate $\dot{\varepsilon}$ in till is proportional to the probability p that a grain bridge will fail when axial force F is applied to the bridge, where $p \propto exp \, (F)$ expresses the statistical probability of failure. Since $p \propto \dot{\varepsilon}$ and $F \propto \sigma$ because of the fractal geometry, the constitutive expression for till deformation is $\dot{\varepsilon} = \dot{\varepsilon}_o \, exp \, (k \, \sigma/\sigma_O)$, where $\dot{\varepsilon}_o$ is a reference strain rate, σ_O is the yield stress that causes initial failure of the grain bridges, and σ is the post-yielding residual stress that maintains a steady-state rate of bridge failure when laminar flow continuously creates new bridges.

This brief review of deforming beds is not comprehensive, but it does aspire to convey an appreciation of the possibilities that understanding the rheology and hydrology of deforming sediment or till may bring to understanding the dynamics of ice sheets, and especially of transient behavior that may produce the ice streams that discharge most of the ice. Although the best of these studies are largely inconclusive, owing to the complexity of the problem, some conclusions can be drawn.

A major control on the stability of deforming beds is effective basal pressure $P_e = P_I - P_W$, where P_I and P_W are the respective pressures of ice and water at the bed. Kamb (1991) proposed that P_e in a discontinuous water layer between the bottom of the ice and the top of the deformable blanket allows storage of this water in inverse proportion to P_e^c, with $c = 1$ when the water layer forms interconnected conduits that link water cavities (Kamb, 1987) and $c \approx 12$ when the water layer is a continuous film of variable thickness (Alley, 1989a, 1989b). The deforming sediment or till blanket becomes unstable somewhere between these extremes, with the instability occurring in a transition from relatively viscous stable creep ($n < 5$) to relatively plastic unstable creep (n > 20). However, for viscous creep at a low viscosity, or for plastic creep at a low yield stress, the sediment or till blanket provides little or no traction for the overlying ice. In thse cases, the only resistance to glacial sliding is provided by bedrock bumps that penetrate through the blanket and into basal ice. For these bumps, only dimension Λ_0 through a bump provides effective resistance to sliding ice. Therefore, this portion of a bump acts like a "sticky spot" retarding ice motion, and glacial sliding is controlled by the number and distribution of these sticky spots per unit area of the sediment or till blanket. If the sediment or till has a high viscosity or a high yield stress, it deforms slowly and its creep rate restricts the sliding velocity of overlying ice more than the number and distribution of bedrock sticky spots that penetrate through the blanket and into the ice.

A unified theory of glacial sliding and bed deformation requires a synthesis of (1) the regelation and creep mechanism that controls sliding at dimension Λ_0 through bedrock bumps and makes bumps behave as sticky spots, (2) the creep rheology of any sediment or till blanket that covers bedrock to a variable depth having variable properties between projecting bedrock sticky spots, (3) the hydrology and thickness of a water layer between the ice and the blanket that regulates frictional coupling across the water interface and the resulting creep behavior of the blanket, and (4) glacial erosion and deposition processes that change the number and distribution of bedrock sticky spots, the thickness of the sediment or till blanket, and the hydrology of the intervening water layer. Developing this theory is a major glaciological priority because it will quantify instabilities in the glacial sliding velocity. This is especially important in ice streams, because the instabilities can turn stream flow on or off, thereby regulating life cycles of ice streams. Since ice streams drain more than 90 percent of ice from the present-day Antarctic and Greenland ice sheets, and probably drained a similar percentage of ice from former ice sheets, quantifying this unstable aspect of ice stream dynamics will quantify the most important internal instability of ice sheets.

Since a unified theory of glacial sliding and bed deformation does not now exist, the assumption will be made that bedrock bumps projecting through a sediment or till blanket and acting as sticky spots exert the major control on basal ice velocity, including ice streams. The theory for this process is well established, and leads to Equation (4.26) for the basal ice velocity. Kamb (1991) argues that this assumption is valid for Ice Stream B, located in Figure 2.14. Ice stream B is a typical Antarctic marine ice stream.

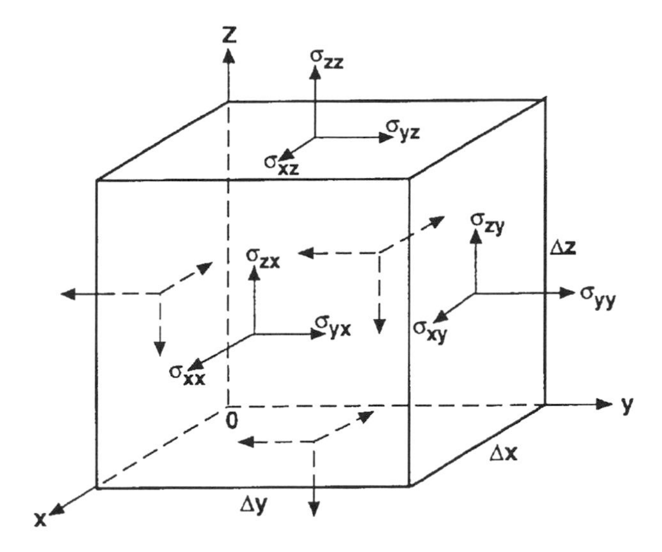

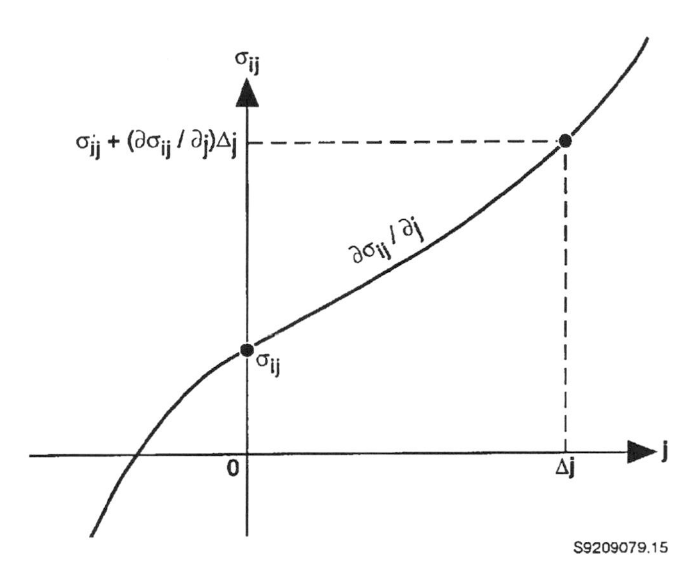

Figure 4.5: The force of balance used for deriving the equilibrium equations. Forces from kinematic stresses σ_{ij} acting normal and parallel to faces of a body having lengths Δx, Δy, Δz along rectilinear axes x, y, z are balanced by gravitational body forces and forces from kinematic stresses $\sigma_{ij} + (\partial\sigma_{ij}/\partial j)\,\Delta j$ acting normal and parallel to the opposite faces, where stress gradients $\partial\sigma_{ij}/\partial j$ are constant within the body and i, j = x, y, z in the usual tensor notation. Stresses σ_{ij} on a given face (top) are stress differences $[\sigma_{ij} + (\partial\sigma_{ij}/\partial j)\,\Delta j] - \sigma_{ij}$ across opposite faces (bottom).

The Equilibrium Equations

The equilibrium equations are derived by balancing forces on a volume element of ice having sides of length Δx, Δy, and Δz along axes x, y, and z. The forces acting on this element consist of normal forces, shear forces, and body forces. As shown in Figure 4.5, normal and shear forces result from stress gradients along x, y, and z, and forces in the x direction are:

$$(F_x)_N = \Delta\sigma_{xx}\,\Delta y\,\Delta z = [(\partial\sigma_{xx}/\partial x)\,\Delta x]\,\Delta y\,\Delta z \qquad (4.27a)$$

$$(F_x)_S = \Delta\sigma_{xy}\,\Delta x\,\Delta z + \Delta\sigma_{xz}\,\Delta x\,\Delta y$$

$$= [(\partial\sigma_{xy}/\partial y)\ \Delta y]\ \Delta x\ \Delta z + [(\partial\sigma_{xz}/\partial z)\ \Delta z]\ \Delta x\ \Delta y \tag{4.27b}$$

$$(F_x)_B = m_I\, g_x = (\rho_I\, \Delta x\ \Delta y\ \Delta z)\, g_x \tag{4.27c}$$

Subscripts N, S, and B identify normal, shear, and body forces, and $m_I = \rho_I\, \Delta x\ \Delta y\ \Delta z$ is the mass of the ice volume element. Equilibrium in the x direction is attained by setting the sum of these forces equal to zero, and similar force summations can be made in the y and z directions, giving:

$$\partial\sigma_{xx}/\partial x + \partial\sigma_{xy}/\partial y + \partial\sigma_{xz}/\partial z + \rho_I\, g_x = 0 \tag{4.28a}$$

$$\partial\sigma_{yx}/\partial x + \partial\sigma_{yy}/\partial y + \partial\sigma_{yz}/\partial z + \rho_I\, g_y = 0 \tag{4.28b}$$

$$\partial\sigma_{zx}/\partial x + \partial\sigma_{zy}/\partial y + \partial\sigma_{zz}/\partial z + \rho_I\, g_z = 0 \tag{4.28c}$$

Using the suffix notation of tensors, the compact form of Equations (4.28) is:

$$\partial\sigma_{ij}/\partial j + \rho_I\, g_i = 0 \tag{4.29}$$

where i and j are x, y, and z in succession.

Exact analytical solutions of the equilibrium equations are possible for lithostatic pressure, pure sheet flow, pure stream flow, and pure shelf flow in an ice sheet.

Lithostatic Pressure

The largest stresses in an ice sheet are σ_{xx}, σ_{yy}, and σ_{zz}, with $\sigma_{xx} \approx \sigma_{yy} \approx \sigma_{zz}$ because deviator stresses are much smaller, except near the ice surface (usually within 10 meters). Therefore, Equation (4.1) becomes:

$$P = \tfrac{1}{3}\left(\sigma_{xx} + \sigma_{yy} + \sigma_{zz}\right) \approx \tfrac{1}{3}\left(3\sigma_{zz}\right) = \sigma_{zz} \tag{4.30}$$

Since σ_{zx} and σ_{zy} and their gradients are small compared to σ_{zz} and its gradient, Equation (4.28c) reduces to:

$$\partial\sigma_{zz}/\partial z + \rho_I\, g_z \approx 0 \tag{4.31}$$

Integrating Equation (4.31) gives the lithostatic pressure in ice of height h_I:

$$P_I = \int_{\sigma_{zz}}^{0} d\sigma_{zz} = -\sigma_{zz} \approx \rho_I\, g_z \int_{z}^{h_I} dz = \rho_I\, g_z \left(h_I - z\right) \tag{4.32}$$

where z = 0 at the base and $z = h_I$ at the surface of the ice sheet. Setting z = 0 in Equation (4.32) gives:

$$P_I = \rho_I\, g_z\, h_I \tag{4.33}$$

where P_I is the basal lithostatic pressure.

Pure Sheet Flow

In pure sheet flow, σ_{xz} is the dominant deviator stress and $\partial\sigma_{xz}/\partial z$ is the dominant deviator stress gradient resisting flow in the downslope x direction. Equation (4.28a) then reduces to:

$$\partial\sigma_{xz}/\partial z + \rho_I\, g_x \approx 0 \tag{4.34}$$

which integrates to:

$$\sigma_{xz} = \int_0^{\sigma_{xz}} d\sigma_{xz} = -\rho_I\, g_x \int_{h_I}^{z} dz \approx \rho_I\, g_z\, \alpha \left(h_I - z\right) = P_I\, \alpha \tag{4.35}$$

where $g_x \approx g_z\, \alpha$ when z is normal to the ice surface and α is the surface slope, which is steepest in the x direction. Setting z = 0 in Equation (4.35) gives:

$$\tau_o \approx \rho_I\, g_z\, h_I\, \alpha = P_I\, \alpha \tag{4.36}$$

where τ_o is the basal shear stress. Note that $\sigma_{xz} = \tau_o = 0$ when $\alpha = 0$.

Laminar flow prevails in pure sheet flow, so $\sigma_c \approx \sigma_{xz}$ in the flow law and Equation (4.17) becomes:

$$\dot{\varepsilon}_{xz} = \left(\sigma_c^{\,n-1}/A^{\,n}\right)\sigma_{xz} \approx \left(\sigma_{xz}/A\right)^{n} = \left[\rho_I\, g_z\, \alpha \left(h_I - z\right)\right]^{n}/A^{\,n}$$
$$\approx \left[P_o\, \alpha \left(1 - z/h_I\right)/A\right]^{n} \approx \left[\tau_o\left(1 - z/h_I\right)/A\right]^{n} \tag{4.37}$$

where A is constant. Note that $\dot{\varepsilon}_{xz} = 0$ when $\alpha = 0$.

Pure Stream Flow

In pure stream flow, σ_{xx}' is the dominant deviator stress, and it resists ice flowing in the x direction. With no significant basal or side traction, $\sigma_{xz} \approx \sigma_{xy} \approx 0$, so that $\alpha \approx 0$. Therefore, only Equation (4.28c) contains a significant gravitational body force per unit volume of ice, because $g_x \approx g_y = 0$. Integrating Equation (4.28c) gives the gravitational lithostatic pressure P in Equation (4.32). Since the width of an ice stream is constant for pure stream flow, $\dot{\varepsilon}_{yy} = 0$ and, therefore, from Equation (4.18), $\sigma_{yy}' = 0$. From Equation (4.2):

$$\sigma_{yy}' = \sigma_{yy} - \tfrac{1}{3}\left(\sigma_{xx} + \sigma_{yy} + \sigma_{zz}\right) = \tfrac{2}{3}\,\sigma_{yy} - \tfrac{1}{3}\left(\sigma_{xx} + \sigma_{zz}\right) = 0 \tag{4.38}$$

Solving Equation (4.38) for σ_{yy}:

$$\sigma_{yy} = \tfrac{1}{2}\left(\sigma_{xx} + \sigma_{zz}\right) \tag{4.39}$$

The effective stress σ_c defined by Equation (4.15a) can be expressed in terms of stresses σ_{xx}, σ_{yy}, and σ_{zz} by applying Equation (4.2) to Equation (4.12b):

$$\sigma_c = \left[\Sigma_2'\right]^{1/2} = \left[\tfrac{1}{6}\left(\sigma_{xx} - \sigma_{yy}\right)^2 + \tfrac{1}{6}\left(\sigma_{yy} - \sigma_{zz}\right)^2 + \tfrac{1}{6}\left(\sigma_{zz} - \sigma_{xx}\right)^2 + \sigma_{xy}^2 + \sigma_{yz}^2 + \sigma_{zx}^2\right]^{1/2} \tag{4.40}$$

Since $\sigma_{xy} \approx \sigma_{yz} \approx \sigma_{zx} \approx 0$ for pure stream flow, applying Equation (4.39) to Equation (4.40) gives:

$$\sigma_c = \tfrac{1}{2}(\sigma_{xx} - \sigma_{zz}) \tag{4.41}$$

Pure stream flow causes extensional strain rate $\dot{\varepsilon}_{xx}$, which is proportional to σ_{xx}' in the flow law of ice expressed by Equation (4.17). Equation (4.2) and (4.39) combine to give :

$$\sigma_{xx}' = \tfrac{1}{2}(\sigma_{xx} - \sigma_{zz}) \tag{4.42}$$

Substituting Equations (4.41) and (4.42) into Equation (4.17) gives:

$$\dot{\varepsilon}_{xx} = \left[\left(\sigma_{xx} - \sigma_{zz}\right)/2A\right]^n = \left(\sigma_{xx}'/A\right)^n \tag{4.43}$$

Solving Equation (4.43) for σ_{xx}, with σ_{zz} given by Equation (4.32):

$$\sigma_{xx} = 2A\,\dot{\varepsilon}_{xx}^{\,1/n} - \rho_I g_z (h_I - z) \tag{4.44}$$

If basal traction is nil, so that $\tau_o \approx 0$, the lithostatic pressure P_o exerted by the overlying ice must equal the hydrostatic pressure P_W in the basal water. Otherwise some of the ice overburden is supported by bedrock, with or without an intervening layer of sediment or till. Basal water pressure equivalent to water of height h_W and density ρ_W is:

$$P_W = \int_{\sigma_{zz}}^{0} d\sigma_{zz} = -\sigma_{zz} = \int_0^{h_W} \rho_W g_z\, dz = \rho_W g_z h_W \tag{4.45}$$

Since $(F_z)_I = P_I A_z$ and $(F_z)_W = P_W A_z$ are vertical lithostatic forces acting on area A_z beneath the ice stream, setting $(F_z)_I - (F_z)_W = 0$ and solving for h_W gives:

$$h_W = (\rho_I/\rho_W)\, h_I \tag{4.46}$$

as the water depth that would produce P_W.

Ice hardness parameter A averaged through ice of height h_I is $\bar{A}$, defined by the expression:

$$\bar{A}\, h_I = \int_0^{h_I} A\, dz \tag{4.47}$$

where $\bar{A}$ appears in the balance of longitudinal gravitational forces, from which $\dot{\varepsilon}_{xx}$ is computed.

Longitudinal gravitational forces produced by P_I and P_W are in equilibrium when basal traction is nil, so that $\dot{\varepsilon}_{xx}$ is constant with z for pure stream flow. The longitudinal gravitational force per unit width of an ice stream is:

$$\left(F_x\right)_I = \int_0^{h_I} \sigma_{xx}\, dz = \int_0^{h_I} \left[2A\,\dot{\varepsilon}_{xx}^{\,1/n} - \rho_I g_z \left(h_I - z\right)\right] dz$$

$$= 2\,\dot{\varepsilon}_{xx}^{\,1/n} \int_0^{h_I} A\, dz - \tfrac{1}{2}\rho_I g_z h_I^2 = 2\,\dot{\varepsilon}_{xx}^{\,1/n}\bar{A}\, h_I - \tfrac{1}{2}\rho_I g_z h_I^2$$

$$= \left(2\sigma_{xx}' - \bar{P}_I\right) h_I = \left(2\sigma_{xx}' - \tfrac{1}{2}P_I\right) h_I$$

$$= \left(2\sigma_{xx}' - \tfrac{1}{2}P_I\right)\left(A_x\right)_I \tag{4.48}$$

where $\bar{P}_I = \frac{1}{2}P_I$ is the average lithostatic pressure of ice and $(A_x)_I$ is the transverse area of the ice stream. Equation (4.48) combines Equation (4.5) for a longitudinal kinematic force with Equation (4.4) for a longitudinal gravity force. The longitudinal gravitational force per unit width for water below the ice stream, using Equations (4.33) and (4.46):

$$\left(F_x\right)_W = \int_0^{h_W} \sigma_{xx}\, dz = \int_0^{h_W} \sigma_{zz}\, dz = -\int_0^{h_W} \rho_W g_z \left(h_W - z\right) dz$$

$$= -\tfrac{1}{2}\rho_W g_z h_W^2 = -\bar{P}_W h_W = -\tfrac{1}{2}P_W\left(A_x\right)_W \tag{4.49}$$

where $\sigma_{xx} = \sigma_{zz}$ for water, $\bar{P}_W = \frac{1}{2}P_W$ is the average hydrostatic pressure of water, and $(A_x)_W$ is the transverse area of the ice stream if melting reduced its height from h_I to h_W. Equation (4.49) is an application of Equation (4.4) for the longitudinal gravity force exerted by the water column. Setting $(F_x)_I - (F_x)_W = 0$ for equilibrium, substituting $h_W = (\rho_I/\rho_W)\, h_I$ because basal traction is nil, and solving for $\dot{\varepsilon}_{xx}$ gives:

$$\dot{\varepsilon}_{xx} = \left[\frac{\rho_I g_z h_I^2 - \rho_W g_z h_W^2}{4\bar{A}\, h_I}\right]^n = \left[\frac{\rho_I g_z h_I}{4\bar{A}}\left(1 - \frac{\rho_I}{\rho_W}\right)\right]^n$$

$$= \left[P_I\left(1 - \rho_I/\rho_W\right)/4\bar{A}\right]^n \tag{4.50}$$

This longitudinal extending strain rate for pure stream flow is also produced by linear shelf flow, as analyzed by Weertman (1957b), who also analyzed pure shelf flow.

Pure Shelf Flow

In pure shelf flow, the ice shelf is unconfined, and it is free to flow in both the x and the y directions, or in any other horizontal direction. Therefore, $\dot{\varepsilon}_{xx} = \dot{\varepsilon}_{yy}$ and the incompressibility condition expressed by Equation (4.13a) gives:

$$\dot{\varepsilon}_{xx} + \dot{\varepsilon}_{yy} + \dot{\varepsilon}_{zz} = 2\,\dot{\varepsilon}_{xx} + \dot{\varepsilon}_{zz} = 0 \tag{4.51}$$

From the flow law expressed by Equation (4.18), the relationship between the corresponding deviator stresses is:

$$\sigma_{xx}' = \sigma_{yy}' = -\tfrac{1}{2}\sigma_{zz}' \tag{4.52}$$

so that $\sigma_{xx} = \sigma_{yy}$ from Equation (4.2). The effective stress given by Equation (4.40) is therefore:

$$\sigma_c = \left[\tfrac{1}{3}\left(\sigma_{xx} - \sigma_{zz}\right)^2\right]^{1/2} = \left(\sigma_{xx} - \sigma_{zz}\right)/\sqrt{3} \tag{4.53}$$

Deviator stress σ_{xx}' given by Equation (4.2) is:

$$\sigma_{xx}' = \sigma_{xx} - \tfrac{1}{3}\left(\sigma_{xx} + \sigma_{yy} + \sigma_{zz}\right) = \left(\sigma_{xx} - \sigma_{zz}\right)/3 \tag{4.54}$$

The flow law for $\dot{\varepsilon}_{xx}$ given by Equation (4.17) is:

$$\dot{\varepsilon}_{xx} = \left(\sigma_c^{\,n-1}/A^{\,n}\right)\sigma_{xx}' = \left[\left(\sigma_{xx} - \sigma_{zz}\right)/\sqrt{3}\right]^{n-1}\left(\sigma_{xx} - \sigma_{zz}\right)/3\,A^{\,n}$$

$$= \sqrt{3}^{\;n-1}\left[\left(\sigma_{xx} - \sigma_{zz}\right)/3\,A\right]^{n} = \sqrt{3}^{\;n-1}\left(\sigma_{xx}'/A\right)^{n} \tag{4.55}$$

Solving for σ_{xx}, with σ_{zz} given by Equation (4.32):

$$\sigma_{xx} = \sqrt{3}^{\;(n+1)/n} A\,\dot{\varepsilon}_{xx}^{\,1/n} - \rho_I\, g_z\left(h_I - z\right) \tag{4.56}$$

The longitudinal gravitational force for ice in a flowband of unit width is:

$$\left(F_x\right)_I = \int_0^{h_I} \sigma_{xx}\,dz = \int_0^{h_I}\left[\sqrt{3}^{\;(n+1)/n} A\,\dot{\varepsilon}_{xx}^{\,1/n} - \rho_I\, g_z\left(h_I - z\right)\right]dz$$

$$= \sqrt{3}^{\;(n+1)/n}\,\dot{\varepsilon}_{xx}^{\,1/n}\,\bar{A}\,h_I - \tfrac{1}{2}\rho_I\, g_z\, h_I^2 = \left(3\,\sigma_{xx}' - \bar{P}_I\right)h_I$$

$$= \left(3\,\sigma_{xx}' - \tfrac{1}{2}P_o\right)h_I = \left(3\,\sigma_{xx}' - \tfrac{1}{2}P_o\right)\left(a_x\right)_I \tag{4.57}$$

As does Equation (4.48), Equation (4.57) combines Equations (4.5) and (4.4) for longitudinal kinematic and gravity forces. Setting $(F_z)_I - (F_z)_W = 0$ in Equation (4.57) and (4.49), substituting $h_W = (\rho_I/\rho_W)\,h_I$ for floating ice, and solving for $\dot{\varepsilon}_{xx}$ with $m = (n + 1)/2$ gives:

$$\dot{\varepsilon}_{xx} = \left[\frac{\rho_I\, g_z\, h_I^2 - \rho_W\, g_z\, h_W^2}{2\sqrt{3}^{\;(n+1)/n}\,\bar{A}\,h_I}\right]^{n} = \left[\frac{\rho_I\, g_z\, h_I}{2\,(3^{\,m/n})\,\bar{A}}\left(1 - \frac{\rho_I}{\rho_W}\right)\right]^{n}$$

$$\approx \left[\bar{P}_I\left(1 - \rho_I/\rho_W\right)/2\left(3^{\,m/n}\right)\bar{A}\right]^{n} \tag{4.58}$$

The Sheet-Flow to Stream-Flow Transition

Pure sheet flow converges to become pure stream flow in a transition zone through which a progressive reduction in basal traction causes σ_{xx}' to progressively replace σ_{xz} as the dominant longitudinal deviator stress along the flowband. Converging flow caused by a reduction in basal traction causes longitudinal extension and vertical thinning of ice, in addition to narrowing of the flowband and development of side traction along the flowband. Transverse strain rate $\dot{\varepsilon}_{yy}$ will be minimal and side shear strain rate $\dot{\varepsilon}_{xy}$ will be zero along the centerline of the flowband. Assume that the transition from sheet flow to stream flow occurs in a region where longitudinal stress gradients can be ignored. Therefore, the transition from sheet flow to stream flow along the centerline can be approximated by a plane strain solution in which:

$$\dot{\varepsilon}_{xy} = \dot{\varepsilon}_{yy} = \dot{\varepsilon}_{zy} = 0 \tag{4.59a}$$

$$\sigma_{xy} = \sigma_{yy}' = \sigma_{zy} = 0 \tag{4.59b}$$

$$\partial\sigma_{xx}/\partial x = \partial\sigma_{xy}/\partial x = \partial\sigma_{xz}/\partial x = 0 \tag{4.59c}$$

where Equation (4.59a) is the plane strain condition, Equation (4.59b) follows because $\dot{\varepsilon}_{ij} \propto \sigma_{ij}'$, and Equation (4.59c) precludes any longitudinal stress gradients. The equilibrium equations for axis x parallel to the ice surface and positive downslope, and axis z perpendicular to the ice surface and positive downward, then reduce to:

$$(\partial\sigma_{xz}/\partial z) + \rho_I\, g_x = 0 \tag{4.60a}$$

$$(\partial\sigma_{zz}/\partial z) + \rho_I\, g_z = 0 \tag{4.60b}$$

Integrating the equilibrium equations:

$$\sigma_{xz} = \int_0^{\sigma_{xz}} d\sigma_{xz} = \int_0^{z} \rho_I\, g_x\, dz = \rho_I\, g_x\, z \tag{4.61a}$$

$$\sigma_{zz} = \int_0^{\sigma_{zz}} d\sigma_{zz} = \int_0^{z} \rho_I\, g_z\, dz = \rho_I\, g_z\, z \tag{4.61b}$$

For these coordinates:

$$g_x = g\,\mathit{sin}\,\alpha \tag{4.62a}$$

$$g_z = g\cos\alpha \tag{4.62b}$$

Surface slope α along the centerline is a maximum in the stream-flow to shelf-flow transition zone.

Since the transition zone is defined as a region in which lengthening along x is almost entirely due to thinning along z, then $\dot{\varepsilon}_{xx} = -\dot{\varepsilon}_{zz}$ and $\dot{\varepsilon}_{yy} = 0$ are good approximations. Since $\dot{\varepsilon}_{ij} \propto \sigma_{ij}'$:

$$\sigma_{yy}' = \sigma_{yy} - 1/3\,(\sigma_{xx} + \sigma_{yy} + \sigma_{zz}) = 0 \tag{4.63}$$

which gives:

$$\sigma_{yy} = 1/2\,(\sigma_{xx} + \sigma_{zz}) \tag{4.64a}$$

$$\sigma_{xx}' = \sigma_{xx} - 1/3\,(\sigma_{xx} + \sigma_{yy} + \sigma_{zz})$$

$$= \sigma_{xx} - 1/3\,[\sigma_{xx} + 1/2\,(\sigma_{xx} + \sigma_{zz}) + \sigma_{zz}]$$

$$= 1/2\,(\sigma_{xx} - \sigma_{zz}) \tag{4.64b}$$

$$\sigma_{zz}' = \sigma_{zz} - 1/3\,(\sigma_{xx} + \sigma_{yy} + \sigma_{zz})$$

$$= \sigma_{zz} - 1/3\,[\sigma_{xx} + 1/2\,(\sigma_{xx} + \sigma_{zz}) + \sigma_{zz}]$$

$$= 1/2\ (\sigma_{zz} - \sigma_{xx}) = -\sigma_{xx}' \qquad (4.64c)$$

The effective creep rate and effective creep stress are readily evaluated for $\dot{\varepsilon}_{xx} = -\dot{\varepsilon}_{zz}$, $\sigma_{xx}' = -\sigma_{zz}'$, and $\dot{\varepsilon}_{xy} = \dot{\varepsilon}_{yy} = \dot{\varepsilon}_{zy} = \sigma_{xy}' = \sigma_{yy}' = \sigma_{zy}' = 0$:

$$\dot{\varepsilon}_c = [1/2\ (\dot{\varepsilon}_{xx}^2 + \dot{\varepsilon}_{yy}^2 + \dot{\varepsilon}_{zz}^2 + 2\ \dot{\varepsilon}_{xy}^2 + 2\ \dot{\varepsilon}_{yz}^2 + 2\ \dot{\varepsilon}_{zx}^2)]^{1/2}$$

$$= [1/2\ \dot{\varepsilon}_{xx}^2 + 1/2\ (-\dot{\varepsilon}_{xx})^2 + \dot{\varepsilon}_{xz}^2]^{1/2}$$

$$= (\dot{\varepsilon}_{xx}^2 + \dot{\varepsilon}_{xz}^2)^{1/2} \qquad (4.65a)$$

$$\sigma_c = [1/2\ (\sigma'^2_{xx} + \sigma'^2_{yy} + \sigma'^2_{zz} + 2\ \sigma_{xy}^2 + 2\ \sigma_{yz}^2 + 2\ \sigma_{zx}^2)]^{1/2}$$

$$= [1/2\ \sigma'^2_{xx} + 1/2\ (-\sigma'_{xx})^2 + \sigma_{xz}^2]^{1/2} = (\sigma'^2_{xx} + \sigma_{xz}^2)^{1/2}$$

$$= [1/4\ (\sigma_{xx} - \sigma_{zz})^2 + \sigma_{xz}^2]^{1/2} \qquad (4.65b)$$

Equations (4.16) and (4.17) can now be combined as follows:

$$\dot{\varepsilon}_c^2 = \dot{\varepsilon}_{xx}^2 + \dot{\varepsilon}_{xz}^2 = \dot{\varepsilon}_{xx}^2 + (\sigma_c^{n-1}\ \sigma_{xz}/A^n)^2 = (\sigma_c/A)^{2n} \qquad (4.66)$$

Solving Equation (4.66) for $\dot{\varepsilon}_{xx} = -\dot{\varepsilon}_{zz}$:

$$\dot{\varepsilon}_{xx} = \pm\left[\left(\frac{\sigma_c}{A}\right)^{2n} - \left(\frac{\sigma_c^{n-1}\ \sigma_{xz}}{A^n}\right)^2\right]^{1/2} = \pm\left(\frac{\sigma_c}{A}\right)^n\left[1 - \left(\frac{\sigma_{xz}}{\sigma_c}\right)^2\right]^{1/2} = -\dot{\varepsilon}_{zz} \qquad (4.67)$$

Solving Equations (4.66) and (4.67) for $\dot{\varepsilon}_{xz}$:

$$\dot{\varepsilon}_{xz} = \frac{\sigma_c^{n-1}\ \sigma_{xz}}{A^n} = \left(\frac{\sigma_c}{A}\right)^n \frac{\sigma_{xz}}{\sigma_c} = \pm\frac{\dot{\varepsilon}_{xx}\left(\sigma_{xz}/\sigma_c\right)}{\left[1 - \left(\sigma_{xz}/\sigma_c\right)^2\right]^{1/2}}$$

$$= \frac{\pm\ \dot{\varepsilon}_{xx}}{\left[\left(\sigma_c/\sigma_{xz}\right)^2 - 1\right]^{1/2}} \qquad (4.68)$$

Downslope velocity in the transition zone increases with the reduction in basal traction. Basal traction causes strain rate $\dot{\varepsilon}_{xz}$ and the downslope reduction of basal traction causes strain rate $\dot{\varepsilon}_{xx}$, where volume conservation in plane strain requires that $\dot{\varepsilon}_{xx} = -\dot{\varepsilon}_{zz}$. These strain rates are defined by velocity gradients according to Equation (4.6) as follows:

$$\dot{\varepsilon}_{xx} = \frac{1}{2}\left(\frac{\partial u_x}{\partial x} + \frac{\partial u_x}{\partial x}\right) = \frac{\partial u_x}{\partial x} = -\dot{\varepsilon}_{zz} \qquad (4.69a)$$

$$\dot{\varepsilon}_{xz} = \frac{1}{2}\left(\frac{\partial u_x}{\partial z} + \frac{\partial u_z}{\partial x}\right) = \frac{1}{2}\frac{\partial u_x}{\partial z} = \dot{\varepsilon}_{zx} \qquad (4.69b)$$

$$\dot{\varepsilon}_{zz} = \frac{1}{2}\left(\frac{\partial u_z}{\partial z} + \frac{\partial u_z}{\partial z}\right) = \frac{\partial u_z}{\partial z} = -\dot{\varepsilon}_{xx} \qquad (4.69c)$$

The downslope velocity is a function of both surface distance from the ice divide and ice depth below the surface, as determined by strain rates $\dot{\varepsilon}_{xx}$ and $\dot{\varepsilon}_{xz}$. If ice surface velocity is u_o at the origin of axes x, z, then Equations (4.67) through (4.69) give:

$$u_x = u_o + \int_0^x \left(\partial u_x/\partial x\right) dx + \int_0^z \left(\partial u_x/\partial z\right) dz$$

$$= u_o + \int_0^x \dot{\varepsilon}_{xx}\, dx + \int_0^z 2\ \dot{\varepsilon}_{xz}\, dz$$

$$= u_o + \dot{\varepsilon}_{xx}\, x - 2\int_0^z \left[\frac{\dot{\varepsilon}_{xx}\ \sigma_{xz}}{\left(\sigma_c^2 - \sigma_{xz}^2\right)^{1/2}}\right] dz \qquad (4.70)$$

Strain rate $\dot{\varepsilon}_{xx}$ in the integral of Equation (4.70) is an absolute strain rate because deformation of the slab produces a positive rate of work. Equation (4.59c) precludes variations of $\dot{\varepsilon}_{xx}$ along x, but variations of $\dot{\varepsilon}_{xx}$ along z occur because $\dot{\varepsilon}_{xx} \propto \sigma_{xx}' = 1/2\ (\sigma_{xx} - \rho_I\, g_z\, z)$. Velocity u_o is zero at the ice divide and increases with increasing distance downslope from the ice divide.

Equation (4.70) can be integrated for unit strain rate $\dot{\varepsilon}_u$, unit effective stress σ_u, unit lengths $x_u = z_u$, unit velocity u_u, a dimensionless effective stress $\hat{\sigma}_c$, dimensionless lengths $\hat{x}$ and $\hat{z}$, dimensionless velocities $\hat{u}_x$ and $\hat{u}_z$, and a dimensionless constant c. These quantities are related to the actual effective stress σ_c, lengths x and z, and velocities u_x and u_z as follows:

$$\hat{\sigma}_c = \sigma_c/\sigma_u \qquad (4.71a)$$

$$\hat{x} = x/x_u \qquad (4.71b)$$

$$\hat{z} = z/z_u \qquad (4.71c)$$

$$\hat{u}_x = u_x/u_u \qquad (4.71d)$$

$$\hat{u}_z = u_z/u_u \qquad (4.71e)$$

Additional relationships are:

$$\dot{\varepsilon}_{xx} = \pm\, c\ \dot{\varepsilon}_u \qquad (4.72a)$$

$$\dot{\varepsilon}_u = (\sigma_u/A)^n \qquad (4.72b)$$

$$\sigma_u = -\rho_I\ g_x\ z_u = \sigma_{xz}\ (z_u/z) \qquad (4.72c)$$

$$\sigma_u = \dot{\varepsilon}_u\ x_u = \dot{\varepsilon}_u\ z_u \qquad (4.72d)$$

Terms in Equation (4.70) become:

$$\dot{\varepsilon}_{xx}\, x = (c\ \dot{\varepsilon}_u)\ (x_u\ \hat{x}) = c\ (u_c/x_u)\ x_u\ \hat{x} = c\ u_u\ \hat{x} \qquad (4.73a)$$

$$\dot{\varepsilon}_{xx}\ \sigma_{xz}\, dz = (c\ \dot{\varepsilon}_u)\ (\sigma_u\ z/z_u)\ d\,(z_u\ \hat{z}) = (c\ u_c/x_u)\ (\sigma_u\ \hat{z})\ d\,(z_u\ \hat{z})$$

$$= c\ u_u\ \sigma_u\ (z_u/x_u)\ \hat{z}\, d\hat{z} = c\ u_u\ \sigma_u\ \hat{z}\, d\hat{z} \qquad (4.73b)$$

Equation (4.70) can now be written in normalized form:

$$u_x = u_o + \dot{\varepsilon}_{xx}\, x - 2\int_0^z \frac{\dot{\varepsilon}_{xx}\,\sigma_{xx}\,dz}{\left(\sigma_c^2 - \sigma_{xz}^2\right)^{1/2}} \tag{4.74a}$$

$$u_u\,\hat{u}_x = u_u\,\hat{u}_o \pm \left(c\,u_u/x_u\right)\left(x_u\,\hat{x}\right) - 2\int_0^{\hat{z}} \frac{\left(c\,u_u/z_u\right)\left(\sigma_u\,\hat{z}\right)\left(z_u\,d\hat{z}\right)}{\left[\left(\sigma_u\,\hat{\sigma}_c\right)^2 - \left(\sigma_u\,\hat{z}\right)^2\right]^{1/2}} \tag{4.74b}$$

$$\hat{u}_x = \hat{u}_o \pm c\,\hat{x} - 2\,c\int_0^{\hat{z}} \frac{\hat{z}\,d\hat{z}}{\left(\hat{\sigma}_c^{\,2} - \hat{z}^2\right)^{1/2}} \tag{4.74c}$$

A relationship between $\hat{\sigma}_c$ and $\hat{z}$ can be obtained from Equations (4.67) and (4.72). Since c cannot be negative:

$$c = \frac{\dot{\varepsilon}_{xx}}{\dot{\varepsilon}_u} = \frac{\left(\sigma_c/A\right)^n}{\dot{\varepsilon}_u}\left[1 - \left(\frac{\sigma_{xz}}{\sigma_c}\right)^2\right]^{1/2}$$

$$= \frac{\left(\sigma_c/A\right)^n}{\left(\sigma_u/A\right)^n}\left[1 - \left(\frac{\sigma_u\,z/z_u}{\sigma_c\,\hat{\sigma}_c}\right)^2\right]^{1/2}$$

$$= \frac{\left(\sigma_c/\sigma_u\right)^n}{\hat{\sigma}_c}\left[\hat{\sigma}_c^{\,2} - \left(z/z_u\right)^2\right]^{1/2}$$

$$= \frac{\hat{\sigma}_c^{\,n}}{\hat{\sigma}_c}\left[\hat{\sigma}_c^{\,2} - \left(\frac{z_u\,\hat{z}}{z_u}\right)^2\right]^{1/2}$$

$$= \hat{\sigma}_c^{\,n-1}\left(\hat{\sigma}_c^{\,2} - \hat{z}^2\right)^{1/2} \tag{4.75}$$

Like the Kronecker delta, c is an on-off switch that has values c = 0 or c = 1. Solving Equation (4.75) for $\hat{z}$ and then differentiating:

$$\hat{z} = \hat{\sigma}_c^{\,2} - c^2\,\hat{\sigma}_c^{\,2-2n} \tag{4.76a}$$

$$2\,\hat{z}\,d\hat{z} = 2\,\hat{\sigma}_c\,d\hat{\sigma}_c - (2-2n)\,c^2\,\hat{\sigma}_c^{\,1-2n}\,d\hat{\sigma}_c \tag{4.76b}$$

Equations (4.76) can be substituted into Equation (4.74) and an integration performed by changing variables. Solving Equation (4.76a) for c = 0 gives the change in variable for the upper limit of integration:

$$\hat{z} = \hat{\sigma}_c \tag{4.77}$$

Solving Equation (4.76a) for $\hat{z} = 0$ and c = 1 gives the change in variable for the lower limit of integration:

$$\hat{\sigma}_c = c^{1/n} \tag{4.78}$$

Integrating Equation (4.74) using Equations (4.76) through (4.78) gives the normalized velocity:

$$\hat{u}_x = \hat{u}_o \pm c\,\hat{x} - \int_0^{\hat{z}} \frac{2\,c\,\hat{z}\,d\hat{z}}{\left(\hat{\sigma}_c^{\,2} - \hat{z}^2\right)^{1/2}}$$

$$= \hat{u}_o \pm c\,\hat{x} - \int_{c^{1/n}}^{\hat{\sigma}_c} \frac{2\,c\,\hat{\sigma}_c\,d\hat{\sigma}_c - (2-2n)\,c^3\,\hat{\sigma}_c^{\,1-2n}\,d\hat{\sigma}_c}{\left[\hat{\sigma}_c^{\,2} - \left(\hat{\sigma}_c^{\,2} - c^2\,\hat{\sigma}_c^{\,2-2n}\right)\right]^{1/2}}$$

$$= \hat{u}_o \pm c\,\hat{x} - 2\int_{c^{1/n}}^{\hat{\sigma}_c} \hat{\sigma}_c^{\,n}\,d\hat{\sigma}_c + 2\,(1-n)\,c^2\int_{c^{1/n}}^{\hat{\sigma}_c} \hat{\sigma}_c^{\,-n}\,d\hat{\sigma}_c$$

$$= \hat{u}_o \pm c\,\hat{x} - 2\left[\frac{\hat{\sigma}_c^{\,1+n} - c^{\frac{1+n}{n}}}{1+n}\right] + 2\,(1-n)\,c^2\left[\frac{\hat{\sigma}_c^{\,1-n} - c^{\frac{1-n}{n}}}{(1-n)}\right]$$

$$= \hat{u}_o \pm c\,\hat{x} - \frac{2}{1+n}\left[\hat{\sigma}_c^{\,1+n} - (1+n)\,c^2\,\hat{\sigma}_c^{\,1-n} + n\,c^{\frac{1+n}{n}}\right] \tag{4.79}$$

Equation (4.79) was first derived by Nye (1957) for an application to glaciers. It clearly illustrates the effect that n has on the variation of downslope velocity for various values of n in the viscoplastic creep spectrum $1 \le n \le \infty$. Note the substantial difference that a viscous rheology with n = 1 and a plastic rheology with $n \to \infty$ has on the velocity. The value n = 3 is commonly observed for ice.

Denormalizing Equation (4.79) gives the actual velocity:

$$u_x = u_u\,\hat{u}_x = u_u\left\{\hat{u}_o \pm c\,\hat{x} - \frac{2}{1+n}\left[\hat{\sigma}_c^{\,1+n} - (1+n)\,c^2\,\hat{\sigma}_c^{\,1-n} + n\,c^{\frac{1+n}{n}}\right]\right\}$$

$$= u_u\left\{\frac{u_o}{u_u} + \frac{\dot{\varepsilon}_{xx}\,x}{\dot{\varepsilon}_u\,x_u} - \frac{2}{1+n}\right.$$

$$\left.\left[\left(\frac{\sigma_c}{\sigma_u}\right)^{1+n} - (1+n)\left(\frac{\dot{\varepsilon}_{xx}}{\dot{\varepsilon}_u}\right)^2\left(\frac{\sigma_c}{\sigma_u}\right)^{1-n} + n\left(\frac{\dot{\varepsilon}_{xx}}{\dot{\varepsilon}_u}\right)^{\frac{1+n}{n}}\right]\right\}$$

$$= u_o + \dot{\varepsilon}_{xx}\,x - \frac{2\,u_u}{n+1}\left(\frac{\sigma_c}{\sigma_u}\right)^{n+1}$$

$$\left[1 - (n+1)\left(\frac{\dot{\varepsilon}_{xx}}{\dot{\varepsilon}_u}\right)^2\left(\frac{\sigma_u}{\sigma_c}\right)^{2n} + n\left(\frac{\dot{\varepsilon}_{xx}}{\dot{\varepsilon}_u}\right)^{\frac{n}{n+1}}\left(\frac{\sigma_u}{\sigma_c}\right)^{n+1}\right]$$

$$= u_o + \dot{\varepsilon}_{xx}\,x - \frac{2\dot{\varepsilon}_u\,x_u}{n+1}\left(\frac{\sigma_c}{\sigma_u}\right)^{n+1}$$

$$\left[1 - (n+1)\left(\frac{\dot{\varepsilon}_{xx}}{\dot{\varepsilon}_u}\right)^2\left(\frac{\sigma_c^n}{\dot{\varepsilon}_u}\right)^2 + n\left(\frac{\dot{\varepsilon}_{xx}}{\sigma_c^n}\right)^{\frac{n+1}{n}}\left(\frac{\sigma_u^n}{\dot{\varepsilon}_u}\right)^{\frac{n+1}{n}}\right]$$

$$= u_o + \dot{\varepsilon}_{xx}\, x - \frac{2\, x_u}{n+1}\left(\frac{\sigma_u}{A}\right)\left(\frac{\sigma_c}{\sigma_u}\right)^{n+1}$$

$$\left[1-(n+1)\left(\frac{\dot{\varepsilon}_{xx}}{\sigma_c^n}\right)^2 A^{2n} + n\left(\frac{\dot{\varepsilon}_{xx}}{\sigma_c^n}\right)^{\frac{n+1}{n}} A^{n+1}\right]$$

$$= u_o + \dot{\varepsilon}_{xx}\, x - \frac{2\, z_u}{n+1}\left(\frac{\sigma_c^{n+1}\, z}{A^n \sigma_{xz}\, z_u}\right)$$

$$\left[1-(n+1)\left(\frac{\dot{\varepsilon}_{xx} A^n}{\sigma_c^n}\right)^2 + n\left(\frac{\dot{\varepsilon}_{xx} A^n}{\sigma_c^n}\right)^{\frac{n+1}{n}}\right]$$

$$= u_o + \dot{\varepsilon}_{xx}\, x - \frac{2\, z}{n+1}\left(\frac{\sigma_{xz}}{A}\right)^n\left(\frac{\sigma_c}{\sigma_{xz}}\right)^{n+1}$$

$$\left[1-(n+1)\left(\frac{\sigma'_{xx}}{\sigma_c}\right)^2 + n\left(\frac{\sigma'_{xx}}{\sigma_c}\right)^{\frac{n+1}{n}}\right] \tag{4.80}$$

The part of Equation (4.80) that shows the effect of longitudinal strain rate $\dot{\varepsilon}_{xx}$ on the vertical velocity profile of laminar flow is the factor ψ, defined as (Raymond, 1980):

$$\psi = \left(\sigma_c/\sigma_{xz}\right)^{n+1}\left[1-(n+1)\left(\sigma'_{xx}/\sigma_c\right)^2 + n\left(\sigma'_{xx}/\sigma_c\right)^{\frac{n+1}{n}}\right]$$

$$= \left[1+\left(\frac{\sigma'_{xx}}{\sigma_{xz}}\right)^2\right]^{\frac{n+1}{n}}\left[1-\frac{(n+1)}{\left[1+\left(\sigma'_{xx}/\sigma_{xz}\right)^2\right]}+\frac{n}{\left[1+\left(\sigma'_{xx}/\sigma_{xz}\right)\right]^{\frac{n+1}{2n}}}\right] \tag{4.81}$$

where $\sigma_c = (\sigma'^{\,2}_{xx} + \sigma_{xz}^2)^{1/2}$ is obtained from Equation

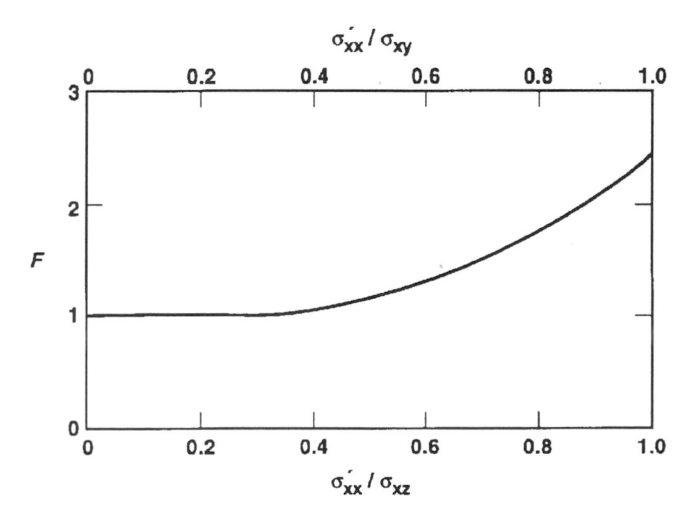

Figure 4.6: A plot of Equation (4.81) for n = 3. Figure 6 in Raymond (1980). Reproduced with permission .

(4.65b). Figure 4.6 is a plot of ψ versus σ'_{xx}/σ_{xz} for n = 3 in Equation (4.81). The effect of $\dot{\varepsilon}_{xx}$ is unimportant so long as $\sigma_{xz} > 2\,\sigma'_{xx}$ in absolute value.

If $\dot{\varepsilon}_{xx}$ is constant in z, deviator stress σ'_{xx} is evaluated at the ice surface where $\dot{\varepsilon}_{xx}$ can be measured directly. The constitutive relationship between $\dot{\varepsilon}_{xx}$ and σ'_{xx} is obtained from Equation (4.18):

$$\sigma'_{xx} = \left(A/\dot{\varepsilon}_c^{\frac{n-1}{n}}\right)\dot{\varepsilon}_{xx} = \left[A/\left(2\,\dot{\varepsilon}_{xx}^2 + \dot{\varepsilon}_{xz}^2\right)^{\frac{n-1}{n}}\right]\dot{\varepsilon}_{xx} \tag{4.82}$$

where $\dot{\varepsilon}_{xz}$ could be measured along z from the tilt rate of a corehole. Ice velocity in terms of ψ and α is:

$$u_x = u_o + \dot{\varepsilon}_{xx}\, x - \left(\frac{2\psi}{n+1}\right)\left(\frac{\sigma_{xz}}{A}\right)^n z$$

$$= u_o + \dot{\varepsilon}_{xx}\, x - \left(\frac{2\psi}{n+1}\right)\left(\frac{\rho_I\, g \sin\alpha}{A}\right)^n z^{n+1} \tag{4.83}$$

where Equations (4.61a) and (4.62a) relate σ_{xz} to α. Equation (4.83) is valid only if hardness coefficient A is constant with depth through the ice, which is not the case if thermal and strain softening vary with depth. Figure 4.6 shows that u_x in Equation (4.83) increases as σ'_{xx} replaces σ_{xz} in the transition zone from sheet flow to stream flow. Along the ice stream, σ_{xy} continues to decrease and becomes zero when the ice stream enters an ice shelf.

The Stream-Flow to Shelf-Flow Transition

As with the Ross Ice Shelf, many ice streams often enter an ice shelf in a confined embayment. This can approximate plane strain in which converging flow in the xy plane causes σ_{xx} to progressively dominate over σ_{xy} toward the calving front. Assume that transverse convergence increases longitudinal extension with minimal vertical thinning or thickening, and that longitudinal stress gradients can be ignored. With $\dot{\varepsilon}_{xy}$ replacing $\dot{\varepsilon}_{xz}$ when ice becomes afloat in an embayment such that $\dot{\varepsilon}_{xx} \approx -\dot{\varepsilon}_{yy}$, the transition from stream flow to shelf flow can be approximated by a plane strain solution in which:

$$\dot{\varepsilon}_{xz} = \dot{\varepsilon}_{yz} = \dot{\varepsilon}_{zz} = 0 \tag{4.84a}$$

$$\sigma_{xz} = \sigma_{yz} = \sigma'_{zz} = 0 \tag{4.84b}$$

$$\partial\sigma_{xx}/\partial x = \partial\sigma_{xy}/\partial x = \partial\sigma_{xz}/\partial x = 0 \tag{4.84c}$$

where Equation (4.84a) is the plane strain condition, Equation (4.84b) follows because $\dot{\varepsilon}_{ij} \propto \sigma'_{ij}$, and Equation (4.84c) precludes any longitudinal stress gradients. The equilibrium equations then reduce to:

$$(\partial\sigma_{xy}/\partial y) + \rho_I\, g_x = 0 \tag{4.85a}$$

$$(\partial \sigma_{zz} / \partial z) + \rho_I g_z = 0 \tag{4.85b}$$

Integrating the equilibrium equations for the origin of coordinates on the centerline of stream flow above the grounding line, with the x axis positive downslope, the y axis positive along the grounding line, and the z axis positive downward:

$$\sigma_{xy} = \int_0^{\sigma_{xy}} d\sigma_{xy} = \int_0^{y} \rho_I g_x \, dy = \rho_I g \sin \alpha \, y \tag{4.86a}$$

$$\sigma_{zz} = \int_0^{\sigma_{zz}} d\sigma_{zz} = \int_0^{z} \rho_I g_z \, dz = \rho_I g \cos \alpha \, y \tag{4.86b}$$

Surface slope α along the centerline is a minimum in the stream-flow to shelf-flow transition zone because $\dot{\varepsilon}_{zz}$ is a minimum.

Since the transition zone is defined as a region in which lengthening along y is almost entirely due to shortening along x, then $\dot{\varepsilon}_{xx} = -\dot{\varepsilon}_{yy}$ and $\dot{\varepsilon}_{zz} = 0$ are good approximations. Since $\dot{\varepsilon}_{ij} \propto \sigma'_{ij}$:

$$\sigma'_{zz} = \sigma_{zz} - 1/3\,(\sigma_{xx} + \sigma_{yy} + \sigma_{zz}) = 0 \tag{4.87}$$

for which:

$$\sigma_{zz} = 1/2\,(\sigma_{xx} + \sigma_{yy}) \tag{4.88a}$$

$$\sigma'_{yy} = \sigma_{yy} - 1/3\,(\sigma_{xx} + \sigma_{yy} + \sigma_{zz}) = 1/2\,(\sigma_{yy} - \sigma_{xx}) \tag{4.88b}$$

$$\sigma'_{xx} = -\sigma'_{yy} = 1/2\,(\sigma_{xx} - \sigma_{yy}) \tag{4.88c}$$

The effective creep rate and the effective creep stress are evaluated for $\dot{\varepsilon}_{xx} = -\dot{\varepsilon}_{yy}$, $\sigma'_{xx} = -\sigma'_{yy}$, and $\dot{\varepsilon}_{zz} = \dot{\varepsilon}_{zy} = \dot{\varepsilon}_{zx} = \sigma'_{zz} = \sigma_{zy} = \sigma_{zx} = 0$:

$$\dot{\varepsilon}_c = \left[\tfrac{1}{2}\left(\dot{\varepsilon}_{xx}^2 + \dot{\varepsilon}_{yy}^2 + \dot{\varepsilon}_{zz}^2 + 2\,\dot{\varepsilon}_{xy}^2 + 2\,\dot{\varepsilon}_{yz}^2 + 2\,\dot{\varepsilon}_{zx}^2\right)\right]^{1/2}$$

$$= \left(\dot{\varepsilon}_{xx}^2 + \dot{\varepsilon}_{xy}^2\right)^{1/2} \tag{4.89a}$$

$$\sigma_c = \left[\tfrac{1}{2}\left(\sigma'^2_{xx} + \sigma'^2_{yy} + \sigma'^2_{zz} + 2\,\sigma_{xy}^2 + 2\,\sigma_{yz}^2 + 2\,\sigma_{zx}^2\right)\right]^{1/2}$$

$$= \left(\sigma'^2_{xx} + \sigma_{xy}^2\right)^{1/2} \tag{4.89b}$$

The flow law is:

$$\dot{\varepsilon}_c^2 = \dot{\varepsilon}_{xx}^2 + \dot{\varepsilon}_{xy}^2 = \dot{\varepsilon}_{xx}^2 + \left(\sigma_c^{n-1}\,\sigma_{xy} / A^n\right)^2 = \left(\sigma_c / A\right)^{2n} \tag{4.90}$$

Solving for $\dot{\varepsilon}_{xx}$:

$$\dot{\varepsilon}_{xx} = \pm\left[\left(\frac{\sigma_c}{A}\right)^{2n} - \left(\frac{\sigma_c^{n-1}\,\sigma_{xy}}{A^n}\right)^2\right]^{1/2} = \pm\left(\frac{\sigma_c}{A}\right)^n\left[1 - \left(\frac{\sigma_{xy}}{\sigma_c}\right)^2\right]^{1/2}$$

$$= \frac{1}{2}\left(\frac{\partial u_x}{\partial x} + \frac{\partial u_x}{\partial x}\right) = \frac{\partial u_x}{\partial x} \tag{4.91}$$

Solving for $\dot{\varepsilon}_{xy}$:

$$\dot{\varepsilon}_{xy} = \frac{\sigma_c^{n-1}\,\sigma_{xy}}{A^n} = \left(\frac{\sigma_c}{A}\right)^n \frac{\sigma_{xy}}{\sigma_c} = \pm\frac{\dot{\varepsilon}_{xx}\left(\sigma_{xy}/\sigma_c\right)}{\left[1 - \left(\sigma_{xy}/\sigma_c\right)^2\right]^{1/2}}$$

$$= \frac{\pm\,\dot{\varepsilon}_{xx}}{\left[\left(\sigma_c/\sigma_{xy}\right)^2 - 1\right]^{1/2}} = \frac{1}{2}\left(\frac{\partial u_x}{\partial y} + \frac{\partial u_y}{\partial x}\right) = \frac{1}{2}\frac{\partial u_x}{\partial y} \tag{4.92}$$

Downslope surface velocity u_x is a function of distance x from the grounding line and of distance y from the centerline, and it is determined by strain rates $\dot{\varepsilon}_{xx}$ and $\dot{\varepsilon}_{xy}$.

Let u_o be the centerline ice velocity across the grounding line. Integrating Equations (4.91) and (4.92):

$$u_x = u_o + \int_0^x \left(\partial u_x / \partial x\right) dx + \int_0^y \left(\partial u_x / \partial y\right) dy$$

$$= u_o + \int_0^x \dot{\varepsilon}_{xx}\, dx + \int_0^y 2\,\dot{\varepsilon}_{xy}\, dy$$

$$= u_o + \dot{\varepsilon}_{xx}\, x - 2\int_0^y \left[\frac{\dot{\varepsilon}_{xx}\,\sigma_{xy}}{\left(\sigma^2 - \sigma_{xy}^2\right)^{1/2}}\right] dy \tag{4.93}$$

where $\dot{\varepsilon}_{xx}$ and $\dot{\varepsilon}_{xy}$ are negative.

The dimensionless version of Equation (4.93) is integrated by defining dimensionless values, $\hat{\sigma}_c$, $\hat{x}$, $\hat{y}$, $\hat{u}_x$, and $\hat{u}_y$ as ratios of actual values σ, x, y, u_x, and u_y to unit values σ_u, x_u, y_u, and u_u:

$$\hat{\sigma}_c = \sigma_c / \sigma_u \tag{4.94a}$$

$$\hat{x} = x / x_u \tag{4.94b}$$

$$\hat{y} = y / y_u \tag{4.94c}$$

$$\hat{u}_x = u_x / u_u \tag{4.94d}$$

$$\hat{u}_y = u_y / u_u \tag{4.94e}$$

where unit values obey the relationships:

$$\dot{\varepsilon}_{xx} = \pm\, c\, \dot{\varepsilon}_u \tag{4.95a}$$

$$\dot{\varepsilon}_u = (\sigma_u / A)^n \tag{4.95b}$$

$$\sigma_u = -\rho_I g_x y_u = \sigma_{xy}\,(y_u / y) \tag{4.95c}$$

$$u_u = \dot{\varepsilon}_u x_u = \dot{\varepsilon}_u y_u \tag{4.95d}$$

In Equations (4.94) and (4.95), $x_u = y_u$ and c is either unity or zero. These equations are used to obtain the dimensionless form of Equation (4.93):

$$u_x = u_o + \dot{\varepsilon}_{xx}\, x - 2\int_0^y \frac{\dot{\varepsilon}_{xx}\,\sigma_{xy}\, dy}{\left(\sigma_c^2 - \sigma_{xy}^2\right)^{1/2}}$$

$$u_u\, \hat{u}_x = u_u\, \hat{u}_o \pm \left(c\, u_u/x_u\right)\left(x_u\, \hat{x}\right) - 2\int_0^{\hat{y}} \frac{\left(c\, u_u / y_u\right)\left(\sigma_u\, \hat{y}\right)\left(y_u\, d\hat{y}\right)}{\left[\left(\sigma_u\, \hat{\sigma}_c\right)^2 - \left(\sigma_u\, \hat{y}\right)^2\right]^{1/2}}$$

$$\hat{u}_x = \hat{u}_o \pm c\, \hat{x} - 2\, c\int_0^{\hat{y}} \frac{\hat{y}\, d\hat{y}}{\left(\hat{\sigma}_c^{\,2} - \hat{y}^2\right)^{1/2}} \tag{4.96}$$

Equations (4.94) and (4.95) also give a dimensionless version of Equation (4.91):

$$c = \frac{\dot{\varepsilon}_{xx}}{\dot{\varepsilon}_u} = \frac{(\sigma_c / A)^n}{\dot{\varepsilon}_u}\left[1 - \frac{\sigma_{xy}}{\sigma_c}\right]^{1/2} = \hat{\sigma}_c^{\,n-1}\left(\hat{\sigma}_c^{\,2} - \hat{y}^2\right)^{1/2} \tag{4.97}$$

Equation (4.97) is solved for $\hat{y}$ and then differentiated:

$$\hat{y}^2 = \hat{\sigma}_c^{\,2} - c^2\, \hat{\sigma}_c^{\,2-2n} \tag{4.98a}$$

$$2\, \hat{y}\, d\hat{y} = 2\, \hat{\sigma}_c\, d\hat{\sigma}_c - (2 - 2\, n)\, c^2\, \hat{\sigma}_c^{\,1-2n}\, d\hat{\sigma}_c \tag{4.98b}$$

Equation (4.96) is integrated by changing the variable from $\hat{y}$ to $\hat{\sigma}_c$ using Equations (4.98). The changes in the limits of integration are obtained from Equation (4.97), which gives $\hat{y} = \hat{\sigma}_c$ for $c = 0$ and finite $\hat{y}$, and $\hat{\sigma}_c = c^{1/n}$ for $c = 1$ and $\hat{y} = 0$. The result is:

$$\hat{u}_x = \hat{u}_o \pm c\, \hat{x} - \int_0^{\hat{y}} \frac{2\, c\, \hat{y}\, d\hat{y}}{\left(\hat{\sigma}_c^{\,2} - \hat{y}^2\right)^{1/2}}$$

$$= \hat{u}_o \pm c\, \hat{x} - \int_{c^{1/n}}^{\hat{\sigma}} \frac{2\, c\, \hat{\sigma}_c\, d\hat{\sigma}_c - (2 - 2\, n)\, c^3\, \hat{\sigma}_c^{\,1-2n}\, d\hat{\sigma}_c}{\left[\hat{\sigma}_c^{\,2} - \left(\hat{\sigma}_c^{\,2} - c^2\, \hat{\sigma}_c^{\,2-2n}\right)\right]^{1/2}}$$

$$= \hat{u}_o \pm c\, \hat{x} - \frac{2}{1+n}\left[\hat{\sigma}_c^{\,1+n} - (1 + n)\, c^2\, \hat{\sigma}_c^{\,1-n} + n\, c^{\frac{1+n}{n}}\right] \tag{4.99}$$

Denormalizing Equation (4.99), using Equations (4.94) and (4.95) gives the actual velocity:

$$u_x = u_o + \dot{\varepsilon}_{xx}\, x - \left(\frac{2\,\psi}{n+1}\right)\left(\frac{\sigma_{xy}}{A}\right)^n y$$

$$= u_o + \dot{\varepsilon}_{xx}\, x - \left(\frac{2\,\psi}{n+1}\right)\left(\frac{\rho_I\, g \sin\alpha}{A}\right)^n y^{\,n+1} \tag{4.100}$$

where:

$$\psi = \left(\sigma / \sigma_{xy}\right)^{n+1}\left[1 - (n+1)\left(\sigma'_{xx} / \sigma_c\right)^2 + n\left(\sigma'_{xx} / \sigma_c\right)^{\frac{n+1}{n}}\right]$$

$$= \left[1 + \left(\frac{\sigma'_{xx}}{\sigma_{xy}}\right)^2\right]^{\frac{n+1}{2}}\left[1 - \frac{(n+1)}{\left[1 + \left(\sigma'_{xx} / \sigma_{xy}\right)^2\right]} + \frac{n}{\left[1 + \left(\sigma'_{xx} / \sigma_{xy}\right)^2\right]^{\frac{n+1}{2n}}}\right] \tag{4.101}$$

Equation (4.89b) is used to obtain the second form of Equation (4.101). The effect of σ'_{xx} on ψ is the same for σ_{xy} in Equation (4.101) as it was for σ_{xz} in Equation (4.81), which is plotted in Figure 4.6.

Dislocations and Ice Hardness

Analytical solutions for sheet flow, stream flow, and shelf flow and for transitions from sheet flow to stream flow and from stream flow to shelf flow require that ice hardness parameter A in the flow law of ice is constant, or nearly so, compared to the stress variations considered in these solutions. Ice gets harder as it gets colder and as strain energy increases in individual ice crystals. Both thermal hardening and strain hardening cause A to increase, so strain rate decreases at a given stress in the flow law. Strain rate is decreased because the mobility of dislocations through the crystal lattice of ice is retarded by decreasing temperature, which reduces the rate of atomic or molecular diffusion, and by increasing strain energy , which distorts the crystal lattice. Creep deformation depends on keeping dislocations mobile. Crystals deform by the migration of dislocations along high-density planes of atoms, which can slide easily past one another. Such planes are called slip planes, or planes of easy glide. Lacking an organized crystal lattice structure, fluids cannot have dislocations and no crystallographic slip planes exist in fluid flow. The consequent distinction between fluid flow and crystalline deformation is profound. In a flow law like Equations (4.16), fluid flow exists when $n = 1$ and crystalline deformation exists when $n > 1$, with $n = 3$ for ice.

Dislocations are Nature's way of easing the difficult task of slipping one plane of atoms over another. The hard way of doing this requires simultaneously breaking all of the atomic bonds that link the two planes together. Nature greatly simplifies this task by passing a dislocation across the interface between the two planes. A dislocation is a line defect in which the regular spacing of atoms in a crystal is disrupted in one dimension. It contrasts with an essentially zero-dimensional point defect in which an atomic site in the crystal lattice structure is left vacant or is occupied by a foreign atom, with an interfacial defect in two dimensions where the crystal lattice becomes warped at the boundaries between crystal grains having differently oriented lattices, and with a volumetric defect in three dimensions that consists of atomic or molecular clusters imbedded in the crystal

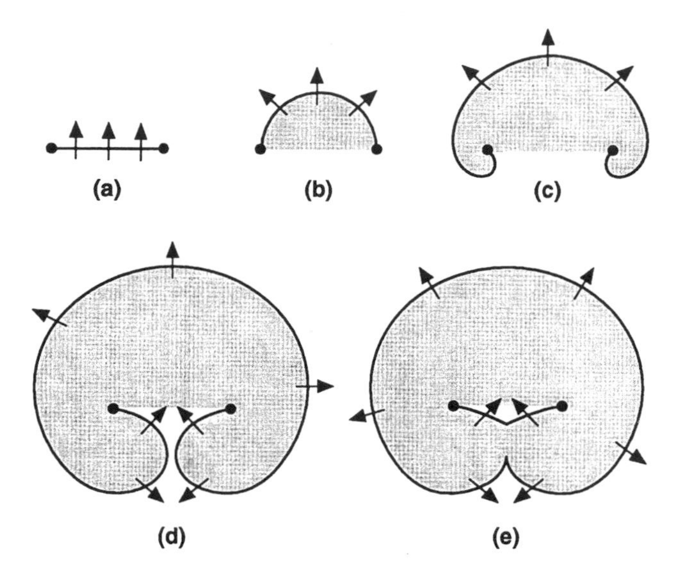

Figure 4.7: The Frank-Read mechanism for generating dislocations in a crystalline material (Read, 1950). In (a) through (e), a shear force acting on a line defect between two point defects bows the line defect in the direction of the force, causing the line defect to wind around the point defects anchoring it, closing the dislocation loop and thereby initiating a new dislocation loop. Atoms are offset by one lattice site across the plane of the loop for each loop that forms. The loops expand to the free surface of the crystal, causing shear deformation of the crystal in the plane of the dislocation loops.

lattice. The task of shearing the crystal lattice along a slip plane can be likened to the task of moving a carpet across a floor. The hard way is to grab one end and pull the entire carpet all at once. This is analogous to simultaneously breaking all the atomic bonds across the slip plane. The easy way is to put a wrinkle in one end of the carpet and to sweep the wrinkle across the carpet to the other end. This is analogous to breaking atomic bonds across the slip plane in only one line and then sweeping that line across the slip plane. The wrinkle in the carpet is a line defect, a dislocation.

Figure 4.7 illustrates an important mechanism for generating dislocations and sweeping them across a slip plane. Being a line defect, the dislocation can be either of finite length or a closed loop. A dislocation of finite length often connects two point defects. When the crystal lattice is stressed, the dislocation line is bowed forward by the stress field, winds around each point defect, and the two coils join each other behind the point defects to form a closed loop that expands outward to the crystal grain boundaries and a new line segment that bows forward between the point defects to repeat the entire process. Each cycle of the process allows a shear displacement of one atomic lattice spacing across the slip plane and within each dislocation loop. The crystal grain is sheared in successive steps as each dislocation loop reaches the grain boundary.

Figure 4.8 shows a quadrant of a dislocation loop in a crystal lattice. The portion of the loop that lies across the direction of the shear stress is an extra half-plane of atoms that rises vertically above the slip plane. The portion of the loop that lies parallel to the direction of the shear stress is the axis of planes that spiral in the slip direction. The end of the extra half plane is an edge dislocation, the spiral axis is a screw dislocation, and they are connected by a mixed edge-screw dislocation. This configuration is necessary if the crystal lattice inside the dislocation loop is to be displaced one atomic spacing in the shear direction compared to the crystal lattice outside the dislocation loop. Moving a dislocation is like tearing a sheet of paper. The edge is the offset between the torn portions, and the screw is the ramp axis where the torn portions meet. The edge offset moves parallel to the applied shear, whereas the screw ramp moves transverse to the shear. Offset is complete when the tear reaches the opposite edge of the paper.

Grain boundaries thicken as successive dislocation loops pile up against them, particularly where the misalignment of the crystal lattice in adjacent grains is great. If the misalignment is small, dislocations generated in one grain are able to cross the low-angle grain boundaries and continue on into the neighboring grains. If the misalignment is large, dislocations that pile up where several grains meet adjust to the stress field so that shear stresses are maximized across their easy-glide slip planes. These dislocation pileups then become the nuclei of new grains in which strain energy is minimized, thereby satisfying Neumann's Principle, which requires that strain energy in a crystal is minimized when its crystallographic axes of symmetry become aligned with the principal axes of the representation quadric of the strain field. Having the stable alignment, these new grains grow at the expense of the old grains until a new grain fabric is created in which the number of small-angle grain boundaries is maximized for a given stress-strain field. The process where-

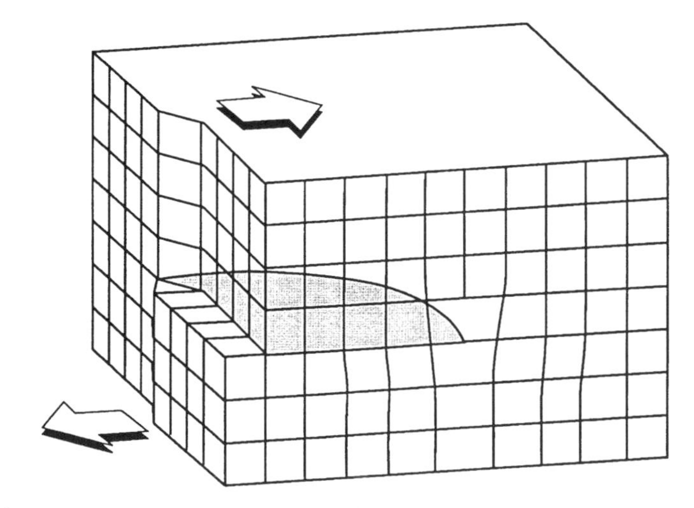

Figure 4.8: A perspective view of one quadrant of a dislocation loop. The applied shear force (shown by arrows) offsets the lattice across the plane of the dislocation loop, with a screw dislocation (the axis of a spiraling plane of atoms) being parallel to the force and an edge dislocation (the end of a half-plane of atoms) being normal to the force.

by new grains absorb old grains to minimize the strain energy is recrystallization.

Thermal Activation of Creep

Thermal activation is expressed by A in Equation (4.16), and it is a major feature of steady-state creep, with A decreasing rapidly as ice temperature increases. This is thermal softening. Weertman (1970) proposed a way to estimate the temperature and pressure dependence of the creep rate, based on theoretical and empirical studies showing that the coefficient D of atomic or molecular diffusion is related to internal energy E, activation volume V, lithostatic pressure P, absolute temperature T, and the absolute melting temperature T_M in crystals consisting of a single atomic or molecular species. The relationship is:

$$D = D_o \, exp[-(E + PV)/R\,T] = D_o \, exp(-Q/RT) \approx D_o \, exp(-C\,T_M/T) \qquad (4.102)$$

where R = 1.987 cal/mol is the ideal gas constant, Q = E + PV is the activation energy, D_o and C are constants whose values are determined by the crystal structure and the atomic or molecular species, and T/T_M is the homologous temperature. Typically, D_o varies between 10^{-1} cm^2/sec and 10^{-2} cm^2/sec, and C varies between 15 and 25.

Activation volume V is generally understood as the volume of atoms or molecules, individually or in clusters, that retards the motion of dislocations through a crystal lattice; and which must therefore be removed by atomic or molecular diffusion before deformation by moving dislocations can continue. Stress and temperature provide the energy for diffusion. Diffusion proceeds by moving atoms or molecules into vacant lattice sites, the number of which increases rapidly with increasing stress and temperature. Diffusion removes atoms individually when atomic bonds are nondirectional, such as in metals, but molecules may be the diffusing species when atomic bonds are highly directional, as in covalent minerals such as ice. The activation volume for dislocation creep in Equation (4.102) probably lies within the unit cell in Figure 4.1.

Determinations of V have been made up to pressures of about 10 kbars, over which T_M varies with P in an almost linear manner. Analytically,

$$T_M \approx (T_M)_0 + (dT_M/dP)\,P \qquad (4.103)$$

where $(T_M)_0$ is T_M at P = 0 and P is taken as positive for hydrostatic compression. For most crystals dT_M/dP is positive. It is negative for those structures, such as ice (Rigsby, 1958), whose solid phase is less dense than the liquid phase.

Equations (4.102) and (4.103) imply that:

$$C = \frac{E + P\,V}{R\,T_M} = \frac{(E + P\,V)/R}{(T_M)_0 + (dT_M/dP)\,P} \qquad (4.104)$$

At P = 0, Equation (4.104) reduces to:

$$C = C_0 = \frac{E}{R\,(T_M)_0} \qquad (4.105)$$

At P >> 0, Equation (4.104) reduces to:

$$C = C_P = \frac{V}{R}\frac{dP}{dT_M} \qquad (4.106)$$

Using experimental data, Weertman (1970) showed that $C_0 \approx C_P$. Equations (4.105) and (4.106) then give an expression for the activation volume:

$$V = \frac{E}{(T_M)_0}\frac{dT_M}{dP} \qquad (4.107)$$

In practice, E is measured in diffusion experiments at atmospheric pressure, and V can then be estimated from Equation (4.107) if Equation (4.103) is known. Provided that the activation energy for thermal diffusion is also the activation energy for steady-state creep, Equations (4.16) and (4.17), with stress $\sigma = \sigma_c = \sigma_{ij}$ and strain rate $\dot{\varepsilon} = \dot{\varepsilon}_c = \dot{\varepsilon}_{ij}$, assumes the form:

$$\dot{\varepsilon} = (\sigma/A)^n = (\sigma/A_o)^n \, exp\,[-(E + P\,V)/R\,T] = (\sigma/A_o)^n \, exp\,(-C\,T_M/T) \qquad (4.108)$$

where $A = A_o \, exp\,(C\,T_M/nT)$ and A_o is a constant that includes D_o and the effects of grain fabric and grain size on strain hardening or softening. Thermal hardening or softening depends only on homologous temperature T/T_M.

Strain Hardening

Strain hardening begins at the elastic limit, when dislocations are few, so they generally can move through the crystal lattice with no mutual interference, and continues as moving dislocations proliferate and increasingly produce traffic jams where dislocations pile up on intersecting slip planes and at grain boundaries. Since ice has one dominant set of slip planes, the basal planes having hexagonal symmetry, dislocation pileups are at grain boundaries much more than on intersecting slip planes. Strain hardening causes A_o in Equation (4.108) to increase, so strain rate $\dot{\varepsilon}$ decreases.

In laboratory studies of strain hardening in which strain rate $\dot{\varepsilon}$ is kept constant, such as a uniaxial tension test, measured strain ε is observed to vary with applied stress σ according to the empirical expression:

$$\varepsilon = (\sigma/\sigma_s)^s \qquad (4.109)$$

where σ_s is a strength parameter and s is a strength exponent. If the specimen has length L_z and cross-sectional area A_z, and applied stress σ_{zz} produces strain ε_{zz}, the constant strain rate $\dot{\varepsilon}_{zz}$ obtained from Equation (4.109) for conservation of volume $L_z\,A_z$ is:

$$\dot{\varepsilon}_{zz} = \frac{d L_z}{L_z d t} = - \frac{d A_z}{A_z d t} = \frac{d \varepsilon_{zz}}{d t} = \frac{s \sigma_{zz}^{s-1}}{\sigma_s^s} \frac{d \sigma_{zz}}{d t} \qquad (4.110)$$

where t is time. Near the end of strain hardening, relatively strain-free grains begin to nucleate and grow in grain boundaries, causing localized necking of a cylindrical specimen being pulled in uniaxial tension, so that area A_z decreases as stress σ_{zz} increases such that tensile force $F_z = \sigma_{zz} A_z$ is constant when σ_{zz} reaches the viscoplastic yield stress σ_v. Setting $dF_z /dt = 0$ when $\sigma_{zz} = \sigma_v$ gives:

$$\frac{dF_z}{dt} = \frac{d(\sigma_v A_z)}{dt} = \frac{A_z d\sigma_v}{dt} + \frac{\sigma_v d A_z}{dt} = 0 \qquad (4.111)$$

From Equations (4.110) and (4.111):

$$\dot{\varepsilon}_{zz} = - \frac{dA_z /A_z}{dt} = \frac{d\sigma_v /\sigma_v}{dt} = \dot{\varepsilon}_v \qquad (4.112)$$

Combining Equations (4.110) and (4.112) for $\sigma_{zz} = \sigma_v$ gives:

$$\sigma_s^s = s \, \sigma_v^s \qquad (4.113)$$

Comparing Equation (4.109) with Equation (4.113) for $\varepsilon = \varepsilon_v$ and $\sigma = \sigma_v$ shows that:

$$\varepsilon_v = 1/s \qquad (4.114)$$

Equation (4.114) predicts that strain hardening ends at a viscoplastic strain of 1/s.

Since s = 2 is commonly observed for strain hardening in metals, Equation (4.109) is often called the parabolic strain hardening law, and it implies that strain hardening ends at a critical viscoplastic strain of $\varepsilon_v \approx 0.5$ in these metals. Tensile tests on ice specimens at atmospheric pressure quickly end in brittle fracture, so tensile tests at high lithostatic pressures are needed to determine if Equation (4.109) applies for glacial ice, and what value of ε_v would be obtained.

States of Equilibrium

Strain hardening is a state of metastable equilibrium that becomes a state of unstable equilibrium at the beginning of recrystallization when strain hardening in unrecrystallized grains equals strain softening in recrystallized grains, with strain softening producing a state of stable equilibrium when recrystallization is complete. These three states of equilibrium are illustrated in Figure 4.9, which shows how lattice strain ε varies with lattice energy E, varies with stress σ at constant strain rate $\dot{\varepsilon}$, and varies with time t at constant stress. The lattice strain energy E has three equilibrium values for which $dE /d\varepsilon = 0$: an intermediate metastable energy E_M at strains $\varepsilon \leq \varepsilon_M$, a maximum unstable energy E_U at strain ε_U, and a minimum stable energy E_S at strains $\varepsilon \geq \varepsilon_S$. By increasing lattice distortion, strain hardening increases E so that $dE /d\varepsilon > 0$. By decreasing lattice distortion, strain softening decreases E so that $dE /d\varepsilon < 0$. The energy increase from E_M to E_U is greatest for maximum strain hardening. The energy decrease from E_U to E_S is least for minimum strain hardening and minimum strain softening . The energy decrease from E_M to E_S is least for minimum strain softening. Strain hardening followed by strain softening takes place when the strain field of an ice sheet changes. Plots of stress versus strain are called flow curves and plots of strain versus time are called creep curves. Figure 4.10 shows how increasing stress and temperature changes creep curves. When either stress or temperature increase, the distinction between ε_S and ε_U becomes increasingly blurred until it vanishes. Eventually ε_M, ε_S, and ε_U are all indistinguishable, and distinct states of equilibrium no longer exist.

The dominant stress fields in an ice sheet are uniaxial tension or compression near the surface and simple shear near the bed for sheet flow, uniaxial tension near the centerline and simple shear near the sides for stream flow, uniaxial tension or compression for a confined ice shelf, and biaxial tension for an unconfined ice shelf. Uniaxial tension and simple shear are applied to cylindrical specimens of polycrystalline ice in Figure 4.11 to illustrate how recrystallization converts a random orientation of ice grains into an ordered orientation that has the crystal symmetry axes aligned with the principal axes of stress and strain, as required by Neumann's Principle (Nye, 1957, pp. 20–24, 104). The grain fabric is the distribution of poles normal to easy-glide slip planes. In ice, easy-glide slip planes are basal planes having hexagonal symmetry, and their pole is the crystallographic c-axis, also called the optic axis. Prismatic and pyramidal planes in ice are not active slip planes because few if any dislocations move in them. The grain fabric is revealed by centering the specimen in a sphere, extending the poles from each grain until they intersect the spherical surface, and then projecting these points of intersection onto a plane that is tangent to the south pole of the sphere. The pattern of points reveals the grain fabric. The points will be random if the orientation of crystal grains is random, resulting in a random grain fabric.

In uniaxial tension (or compression), as shown in Figure 4.12, shear stresses are maximized on planes at 45° to the tension (or compression) axis. For the polycrystalline ice specimen in Figure 4.11, shear stresses are maximized on the surface of two right circular cones placed end to end, where the tension (or compression) axis coincides with the conical axes and subtends a 45° angle with the conical surfaces, so pure shear exists on these surfaces. If the tension (or compression) axis coincides with the north pole of the sphere, poles of recrystallized grains are normal to the conical surfaces and intersect the sphere along small circles midway between the equator and the north and south poles of the sphere. These intersections create a small circle fabric on the planar projection of the sphere. Grain rotation also produces this fabric (Van der Veen and Whillans, 1994).

In simple shear, shear stresses are maximized on sur-

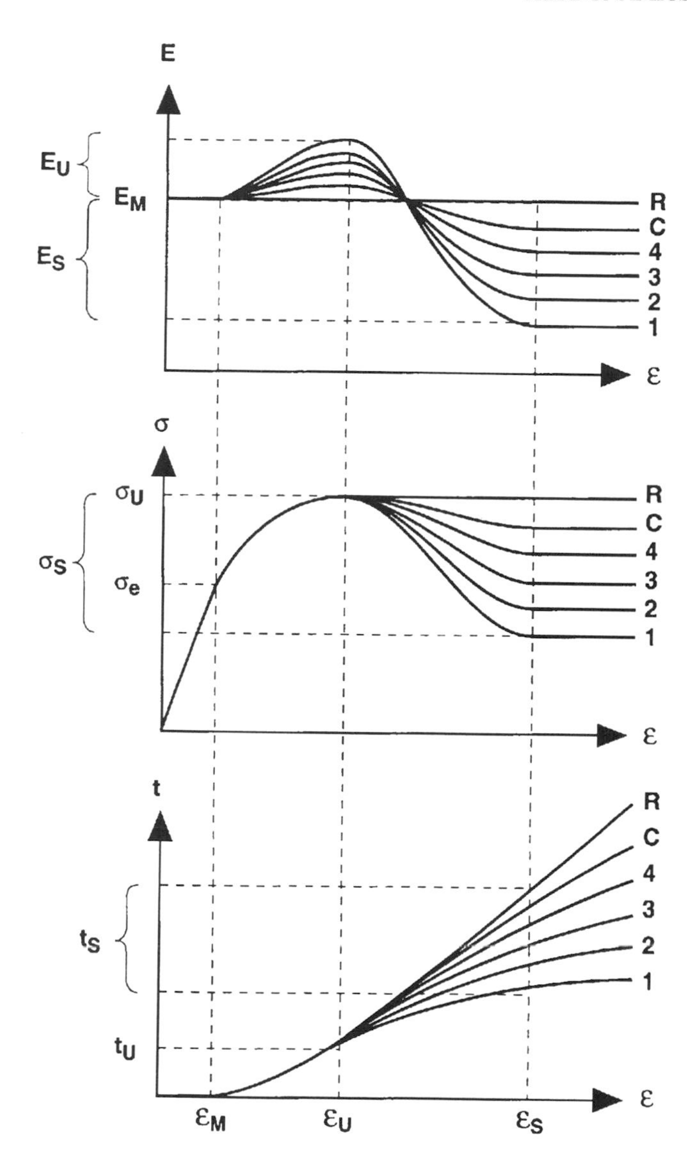

Figure 4.9: Equilibrium states in relation to various stable grain fabrics produced by recrystallization. Variations with stream ε are shown for strain energy E (*top*), stress σ at constant strain rate $\dot{\varepsilon}$ (*center*), and time t at constant σ (*bottom*). Subscripts denote metastable (M), unstable (U), and stable (S) conditions as ε increases. Depending on the strain field, recrystallization produces grain fabrics that are random R, form small circles C and form quadruple maxima 4, triple maxima 3, double maxima 2, and a single maximum 1.

faces parallel to the traction surface. If the traction surface coincides with the equator of the sphere, poles of recrystallized grains intersect the north and south poles of the sphere, and create a point or single-maximum fabric at the center of the planar projection of the sphere. Grain rotation produces this fabric as well (Van der Veen and Whillans, 1994).

The presence of small angles between slip planes across grain boundaries in the recrystallized polycrystalline aggre-

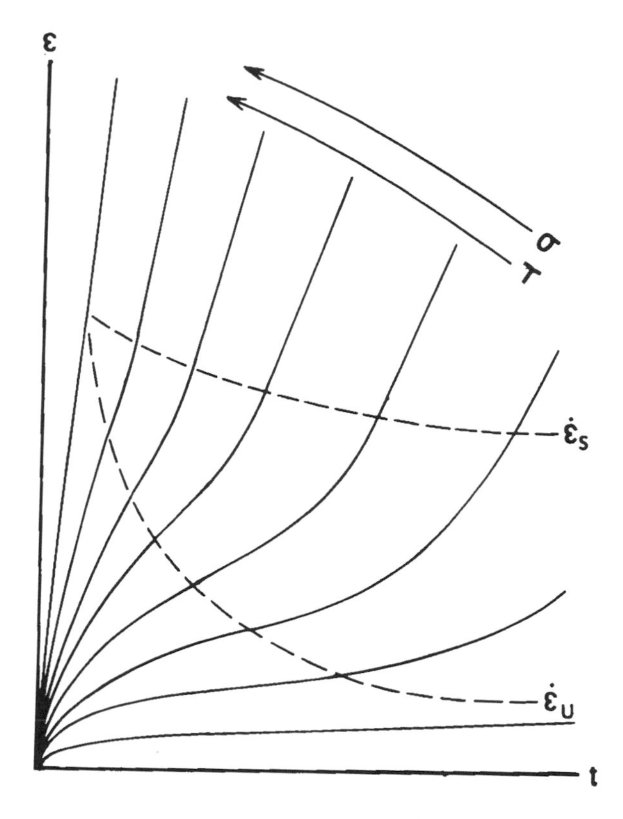

Figure 4.10: The effect of stress and temperature on creep curves. Creep curves plot strain ε versus time t at constant stress σ and temperature T. Each value of σ and T gives a separate creep curve. As σ and T increase, strain rate $\dot{\varepsilon}$ increases all along the creep curve, with a minimum strean rate $\dot{\varepsilon}_U$ for unstable equilibrium just before recrystallization (lowered dashed line) to a maximum strain rate $\dot{\varepsilon}_S$ for stable equilibrium just after recrystallization (upper dashed line), with $\dot{\varepsilon}_M = \dot{\varepsilon}_U = \dot{\varepsilon}_S$ for metastable equilibrium for continuous recrystallization at high σ and T.

gate gives an annular area to the circle fabric and a circular area to the point fabric. As seen in Figure 4.11, nearly all of the recrystallized grains have small-angle grain boundaries after deformation by shear traction, whereas only about half have small-angle grain boundaries after deforma-tion by uniaxial tension (or compression), even though nearly all grains are oriented at 45° to the tensile (or compression) axis after recrystallization. Consequently, hard glide, which is caused by both high-angle grain boundaries and by low values of τ_R with respect to σ_{zz} on planes of easy glide in Figure 4.12, is still important after a uniaxial stress has resulted in recrystallization. The gravitational pulling force that causes an ice sheet to spread induces an internal kinematic force that resists spreading primarily by hard glide in ice crystals and an external kinematic force that resists spread-ing primarily by basal traction. After recrystallization, the small circle fabric created by pure shear is intermediate between the single maximum fabric created by simple shear and the random fabric that existed before recrystallization. The change from pure shear to simple shear toward the bed of an ice sheet softens the ice (Paterson, 1991).

Recrystallization produces a unique grain fabric for each state of stress. In Figure 4.9, starting with a random grain fabric R, recrystallization produces grain fabrics consisting of a small circle C, quadruple maxima 4, triple maxima 3, double maxima 2, and a single maximum 1, depending on the state of stress, with strain energy decreasing as the number of maxima in the fabric decreases. The fewer the maxima, the more the fabric promotes easy glide of dislocations through the crystal lattice. The decrease of maxima in the grain fabric causes a decreases of viscoplastic yield stress σ_v in the flow curves and an increase of strain rate $\dot{\varepsilon}$ in the creep curves during recrystallization, as shown in Figure 4.9. The flow curve has upper and lower viscoplastic yield stresses σ_v where $d\sigma / d\varepsilon = 0$. These correspond to minimum and maximum steady-state viscoplastic strain rates $\dot{\varepsilon}_v$ where $d\dot{\varepsilon}/dt = d^2\varepsilon / dt^2 = 0$. The upper σ_v and minimum $\dot{\varepsilon}_v$ reflect unstable equilibrium, and the lower σ_v and maximum $\dot{\varepsilon}_v$ reflect stable equilibrium, before and after recrystallization, respectively.

Dynamic recrystallization is a condition of metastable equilibrium produced when rates of strain hardening and strain softening are in a precarious balance, with dislocations piling up against grain boundaries as fast as new grains nucleate and grow in these boundaries, so that dislocations are released from the pileups at a controlled rate that allows no net change in strain energy E during strain hardening and strain softening, as shown for curve R in Figure 4.9. Consequently, viscoplastic yield stress σ_v remains at its upper limit in flow curve R, and viscoplastic strain rate $\dot{\varepsilon}_v$ remains at its minimum value in creep curve R, in Figure 4.9. As long as dynamic equilibrium is sustained, the ice crystal fabric will retain random fabric R and A_0 will remain constant in Equation (4.108). If a fabric other than R exists at $\varepsilon = 0$ in Figure 4.9, that other fabric will be retained during dynamic recrystallization.

Steady-state creep exists where strain rate $\dot{\varepsilon} = d\varepsilon/dt$ is constant in Figure 4.9. Unstable steady-state creep occurs at an unstable upper viscoplastic yield stress $\sigma_v = \sigma_U$ where $\dot{\varepsilon}_U$ is the unstable minimum strain rate before recrystallization, and stable steady-state creep occurs at a stable lower viscoplastic yield stress $\sigma_v = \sigma_S$ where $\dot{\varepsilon}_S$ is the stable maximum strain rate after recrystallization, as shown in Figure 4.9. If creep experiments are conducted over a range of stresses and temperatures, creep curves for each constant stress or temperature display the spectrum shown in Figure 4.10. As stress or temperature increases, $\dot{\varepsilon}_U$ and $\dot{\varepsilon}_S$ occur at shorter times and move closer together until $\dot{\varepsilon}_U = \dot{\varepsilon}_S$ at high stresses and tem-peratures, which is the condition for dynamic recrystallization when $\sigma_U = \sigma_S$. Dislocations are driven by stress and temperature, with stress creating dislocations by deforming the crystal lattice, and temperature keeping dislocations mobile by thermal diffusion at grain boundaries and other sites where dislocation pileups occur. Dynamic recrystallization occurs at high stresses and temperatures because pileups are obliterated as fast as they form so that strain hardening and strain softening occur simultaneously. Therefore, $\sigma_U = \sigma_S$ and $\dot{\varepsilon}_U = \dot{\varepsilon}_S$.

Superplasticity

A given effective creep rate $\dot{\varepsilon}_c$ at a given effective creep stress σ_c in Equation (4.16) can be produced by various combination of hardness parameter A and viscoplastic parameter n. Therefore, changes in $\dot{\varepsilon}_c$ due to thermal and strain hardening or softening that changes A for a given n can also be produced by varying n and keeping A constant. In particular, strain rates caused by thermal and strain softening that localize substantial reduction of A to particular regions in ice are sustained if A is held constant and n is increased substantially. In laboratory creep experiments on polycrystalline ice specimens and in field studies on glaciers, $n = 3$ is commonly observed, but A is greatest in grain boundaries where dislocations pile up and is least in the basal shear zone of sheet flow and in the lateral shear zones of stream flow where thermal and strain softening are concentrated. However, the observed creep behavior is reproduced by Equation (4.16) if A is held constant and $n >> 3$. This condition is superplasticity. In the limiting extreme, $n = \infty$ and perfectly plastic flow exists.

Strain hardening during laboratory tensile experiments on cylinders of polycrystalline ice can be analyzed as superplastic behavior. Figure 4.13 shows polycrystalline ice cylinders with grains joined in series and in parallel. For i grains linked in series along the tensile z axis, tensile stress σ_{zz} is identical in each grain but tensile strain ε_{zz} is distributed over all grains. In analytical terms:

$$\sigma_{zz} = \sigma_1 = \sigma_2 = \sigma_3 = \sigma_i \tag{4.115a}$$

$$\varepsilon_{zz} = i^{-1} (\varepsilon_1 + \varepsilon_2 + \varepsilon_3 + \ldots + \varepsilon_i) = i^{-1} \sum_i \varepsilon_i \tag{4.115b}$$

For i grains linked in parallel along the tensile axis, σ_{zz} is distributed over all grains but ε_{zz} is the same in each grain. Analytically:

$$\sigma_{zz} = i^{-1} (\sigma_1 + \sigma_2 + \sigma_3 + \ldots + \sigma_i) = i^{-1} \sum_i \sigma_i \tag{4.116a}$$

$$\varepsilon_{zz} = \varepsilon_1 = \varepsilon_2 = \varepsilon_3 = \varepsilon_i \tag{4.116b}$$

Polycrystalline ice, of course, generally has grains that are linked both in series and in parallel with respect to the tensile axis. Which linkage is the most important in determining the shape of the flow curve for uniaxial tension?

If the crystal structure is mechanically anisotropic, as ice is, each grain will have a preferred slip system that will be activated first by the tensile stress. As seen in Figure 4.12, each grain i will have an angle factor Ω_i that resolves axial stress σ_i into a shear stress τ_i in the preferred slip sys

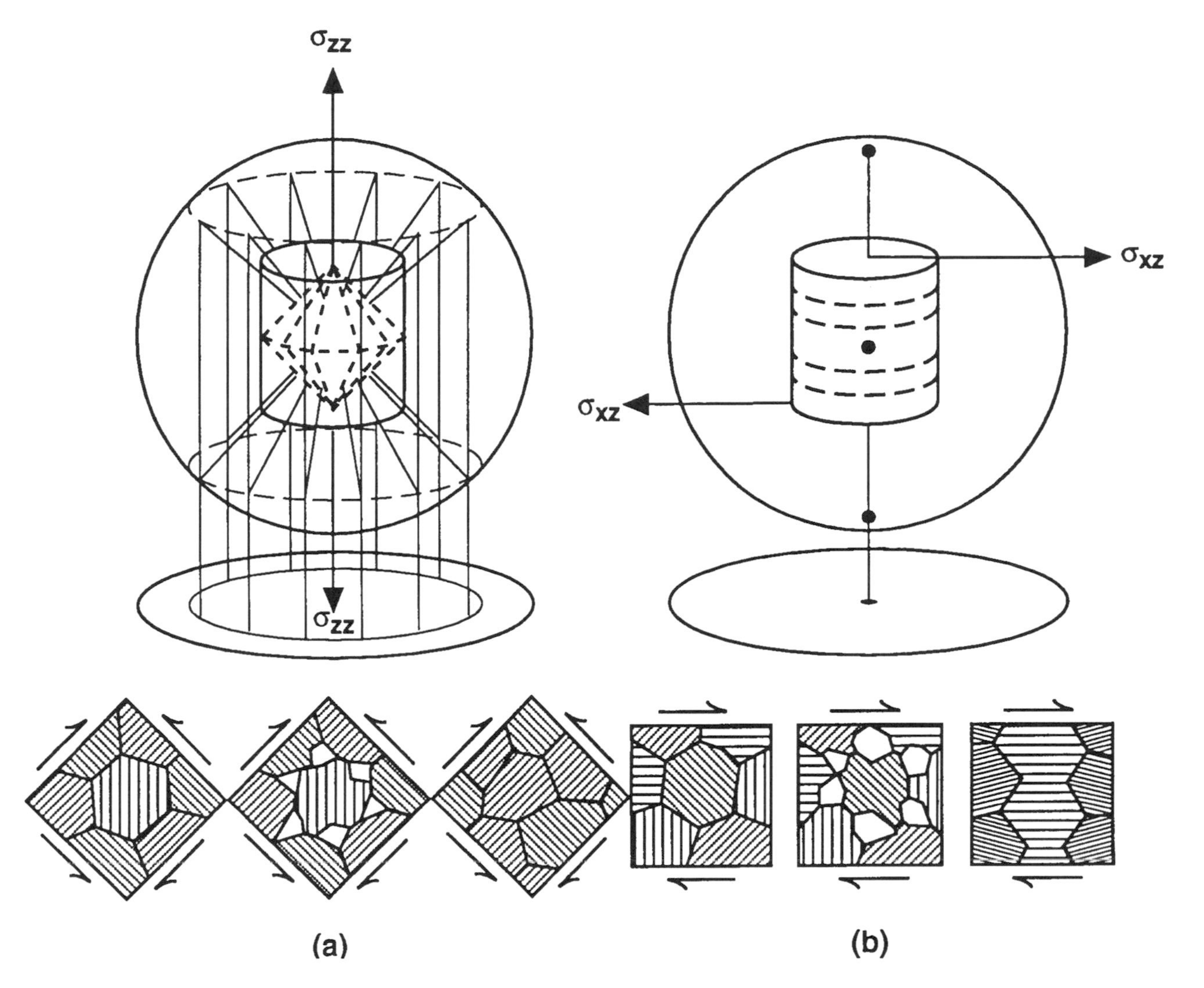

Figure 4.11: Viscoplastic instability and recrystallization from random to ordered fabrics for a cylindrical specimen deformed by uniaxial tension and by shear traction. Uniaxial stress σ_{zz} in (*a*) produces pure shear strain on the faces of right circular cones stacked parallel to the z axis. Traction stress σ_{xz} in (*b*) produces simple shear strain on the faces of discs stacked parallel to the z axis. For thin sections in the xz plane, both cause dislocations moving on slip planes (parallel lines) in each grain to pile up at grain boundaries, causing strain hardening. Recrystallization reduces strain energy by nucleating new grains that allow dislocations to cross grain boundaries, allowing strain softening. These new grains (shown without slip lines when they are nuclei) grow and absorb misaligned old grains. Pure shear produces a small circle fabric because slip planes in neighboring recrystallized grains are subparallel in only about half the grains. Simple shear produces a single maximum fabric because slip planes in neighboring recrystallized grains are subparallel in nearly all of the grains.

tem, where $\tau = \sigma \cos\alpha \cos\beta = \sigma/\Omega$, so $\Omega = (\cos\alpha \cos\beta)^{-1}$ by definition. As shown in Figure 4.13 for grains linked in series, yielding of the cylindrical specimen will occur first in the grain with the lowest value of Ω. This grain then necks, and necking causes the preferred slip system to rotate toward the tensile axis, so Ω must increase. Figure 4.13 also shows that necking and the consequent increase in Ω are avoided when grains are linked in parallel. Even though one grain will have the lowest Ω, all or most of its vertical faces are attached to neighboring grains instead of being free surfaces, as was the case for grains linked in series. Consequently, strain cannot be localized in a single grain or group of grains with a low Ω; strain must be distributed uniformly among all the grains, and necking is suppressed. This conclusion holds even in a polycrystalline specimen in which grains are linked in both series and parallel with respect to the tensile axis. The mere presence of parallel linkages to neighboring grains promotes a uniform strain in all grains, although

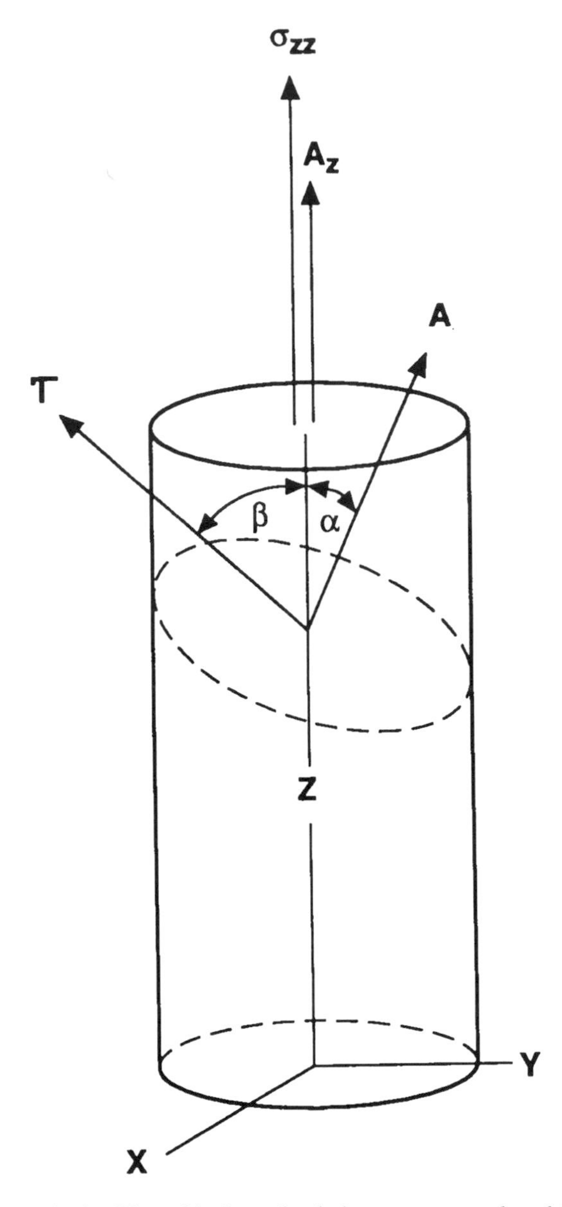

Figure 4.12: The critical resolved shear stress produced by an axial stress. Axial stress σ_{zz} acting normal to surface plane A_z produces shear stress τ resolved tangent to interior plane A. Angle α subtends area vectors A_z and A. Angle β subtends stress vectors σ_{zz} and τ. Axial force $F_z = \sigma_{zz} A_z$ produces shear force $F = \tau A$ resolved in the direction of τ and acting tangent to plane of area A. The critical resolved shear stress for $F = F_z \cos\beta$ tangent to area $A = A_z / \cos\alpha$ is therefore $\tau = \frac{F}{A} = \frac{F_z \cos\beta}{A_z/\cos\alpha} = \sigma_{zz} \cos\alpha \cos\beta$. The maximum resolved shear stress occurs when $\alpha = \beta = 45^\circ$.

grains also possess series linkages with neighboring grains. Equations (4.116a) and (4.116b), therefore, are the pertinent stress-strain relationships for uniaxial tension in polycrystalline ice. Distributing strain uniformly among all grains suppresses shear stresses between grains, since shear stresses of this type produce differential strain between grains.

Successive yielding of individual grains in polycrystalline ice can be examined most easily, while still preserving the physical basis of the process, by assuming for simplicity that the elastic modulus σ_e is the same in all grains and that slip in each grain begins at a plastic yield stress $\sigma_o = \tau$. Plastic yielding occurs for $n = \infty$ in Equation (4.16). This has the advantage of screening out strain hardening in individual grains so that strain hardening caused by successive yielding of grains in polycrystalline ice can be examined. Screening out strain hardening in individual grains keeps A constant, while letting the strain hardening that actually exists be taken into account by increasing n. Therefore, polycrystalline ice under uniaxial tension exhibits superplasticity. To illustrate the process, consider a tricrystal in which the three grains are aligned in parallel with tensile axis z. The elastic modulus σ_e and the plastic yield stress σ_o are the same in each grain, and yielding occurs when $\tau = \sigma_o$ in Figure 4.12. Tensile stresses σ_1, σ_2, and σ_3 in grains 1, 2, and 3 are $\sigma_1 = \Omega_1 \tau$, $\sigma_2 = \Omega_2 \tau$, and $\sigma_3 = \Omega_3 \tau$, where $\Omega_1 < \Omega_2 < \Omega_3$. Consequently grain 1 yields first, grain 2 yields second, and grain 3 yields last. Total strain ε is the sum of elastic strain ε' and plastic strain ε''. Before any grains yield, however, strain is purely elastic in all three grains. The stresses and strains in the grains for elastic deformation are:

$$\begin{aligned}\sigma_{zz} &= 1/3\,(\sigma_1 + \sigma_2 + \sigma_3)\\ &= 1/3\,(\sigma_e\,\varepsilon_1' + \sigma_e\,\varepsilon_2' + \sigma_e\,\varepsilon_3')\\ &= 1/3\,\sigma_e\,(\varepsilon_{zz} + \varepsilon_{zz} + \varepsilon_{zz})\\ &= \sigma_e\,\varepsilon_{zz} \end{aligned} \tag{4.117a}$$

$$\begin{aligned}\varepsilon_{zz} &= \varepsilon_1 = \varepsilon_1' = \sigma_1/\sigma_e\\ &= \varepsilon_2 = \varepsilon_2' = \sigma_2/\sigma_e\\ &= \varepsilon_3 = \varepsilon_3' = \sigma_3/\sigma_e \end{aligned} \tag{4.117b}$$

After the first grain yields, it deforms plastically at constant yield stress $\sigma_1 = \Omega_1 \sigma_o$ while grains 2 and 3 continue to deform elastically as σ_{zz} increases above σ_1. The stresses and strains in the grains are:

$$\begin{aligned}\sigma_{zz} &= 1/3\,(\sigma_1 + \sigma_2 + \sigma_3)\\ &= 1/3\,(\sigma_e\,\varepsilon_1' + \sigma_e\,\varepsilon_2' + \sigma_e\,\varepsilon_3')\\ &= 1/3\,\sigma_e\,[(\varepsilon_1 - \varepsilon_1'') + \varepsilon_2 + \varepsilon_3]\\ &= 1/3\,\sigma_e\,(3\,\varepsilon_{zz} - \varepsilon_1'')\\ &= \sigma_e\,\varepsilon_{zz} - 1/3\,\sigma_e\,\varepsilon_1''\end{aligned}$$

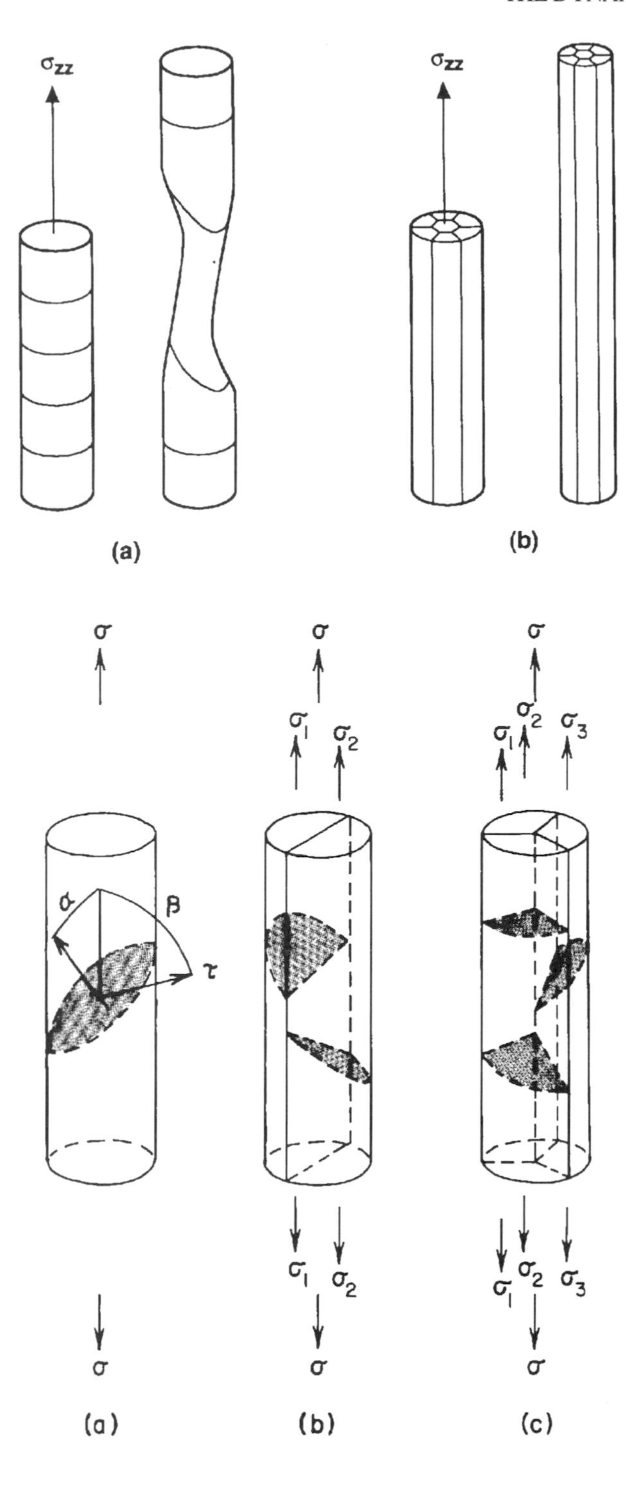

Figure 4.13: Deformation of polycrystalline cylindrical specimens pulled in uniaxial tension. *Top*: In (a) crystals are pulled in series by tensile stress σ_{zz} and deformation is concentrated in the crystal having the lowest critical resolved shear stress. In (b) crystals are pulled in parallel by tensile stress σ_{zz} and deformation is equally distributed among all crystals. *Bottom*: Easy glide slip planes (shaded) in a single crystal (a), a bicrystal (b), and a tricrystal (c) pulled in uniaxial tension.

$$= \sigma_e \, \varepsilon_{zz} - 1/3 \, \sigma_e \, (\varepsilon_{zz} - \varepsilon_1')$$

$$= \sigma_e \, \varepsilon_{zz} - 1/3 \, \sigma_e \, (\varepsilon_{zz} - \sigma_1 / \sigma_e)$$

$$= 2/3 \, \sigma_e \, \varepsilon_{zz} + 1/3 \, \Omega_1 \, \sigma_o \tag{4.118a}$$

$$\varepsilon_{zz} = \varepsilon_1' + \varepsilon_1'' = \sigma_1 / \sigma_e + \varepsilon_1'' = \Omega_1 \, \sigma_o / \sigma_e + \varepsilon_1''$$

$$= \varepsilon_2 = \varepsilon_2' = \sigma_2 / \sigma_e$$

$$= \varepsilon_3 = \varepsilon_3' = \sigma_3 / \sigma_e \tag{4.118b}$$

After the second grain yields, grains 1 and 2 deform plastically at constant stresses $\sigma_1 = \Omega_1 \sigma_o$ and $\sigma_2 = \Omega_2 \sigma_o$ while grain 3 continues to deform elastically as σ_{zz} increases above σ_2. The stresses and strains in the grains are:

$$\sigma_{zz} = 1/3 \, (\sigma_1 + \sigma_2 + \sigma_3)$$

$$= 1/3 \, (\sigma_e \, \varepsilon_1' + \sigma_e \, \varepsilon_2' + \sigma_e \, \varepsilon_3')$$

$$= 1/3 \, \sigma_e \, [(\varepsilon_1 - \varepsilon_1'') + (\varepsilon_2 - \varepsilon_2'') + \varepsilon_3]$$

$$= 1/3 \, \sigma_e \, (3 \, \varepsilon_{zz} - \varepsilon_1'' - \varepsilon_2'')$$

$$= \sigma_e \, \varepsilon_{zz} - 1/3 \, \sigma_e \, (\varepsilon_1'' + \varepsilon_2'')$$

$$= \sigma_e \, \varepsilon_{zz} - 1/3 \, \sigma_e \, [(\varepsilon_{zz} - \varepsilon_1') + (\varepsilon_{zz} - \varepsilon_2')]$$

$$= \sigma_e \, \varepsilon_{zz} - 1/3 \, \sigma_e \, [(\varepsilon_{zz} - \sigma_1 / \sigma_e) + (\varepsilon_{zz} - \sigma_2 / \sigma_e)]$$

$$= 1/3 \, \sigma_e \, \varepsilon_{zz} + 1/3 \, (\Omega_1 + \Omega_2) \, \sigma_o \tag{4.119a}$$

$$\varepsilon_{zz} = \varepsilon_1 = \varepsilon_1' + \varepsilon_1'' = \sigma_1 / \sigma_e + \varepsilon_1'' = \Omega_1 \, \sigma_o / \sigma_e + \varepsilon_1''$$

$$= \varepsilon_2 = \varepsilon_2' + \varepsilon_2'' = \sigma_2 / \sigma_e + \varepsilon_2'' = \Omega_2 \, \sigma_o / \sigma_e + \varepsilon_2''$$

$$= \varepsilon_3 = \varepsilon_3' = \sigma_3 / \sigma_e \tag{4.119b}$$

After the third grain yields, all three grains deform plastically at constant stresses $\sigma_1 = \Omega_1 \sigma_o$, $\sigma_2 = \Omega_2 \sigma_o$, and $\sigma_3 = \Omega_3 \sigma_o$, and no elastic deformation occurs in any of the grains because σ_{zz} does not increase above σ_3 after all grains have yielded plastically. The stresses and strains in the grains are:

$$\sigma_{zz} = 1/3 \, (\sigma_1 + \sigma_2 + \sigma_3)$$

$$= 1/3 \, (\sigma_e \, \varepsilon_1' + \sigma_e \, \varepsilon_2' + \sigma_e \, \varepsilon_3')$$

$$= 1/3 \, \sigma_e \, [(\varepsilon_1 - \varepsilon_1'') + (\varepsilon_2 - \varepsilon_2'') + (\varepsilon_3 - \varepsilon_3'')]$$

$$= 1/3 \, \sigma_e \, (3 \, \varepsilon_{zz} - \varepsilon_1'' - \varepsilon_2'' - \varepsilon_3'')$$

$$= \sigma_e \varepsilon_{zz} - 1/3\, \sigma_e (\varepsilon_1'' + \varepsilon_2'' + \varepsilon_3'')$$

$$= \sigma_e \varepsilon_{zz} - 1/3\, \sigma_e [(\varepsilon_{zz} - \varepsilon_1') + (\varepsilon_{zz} - \varepsilon_2') + (\varepsilon_{zz} - \varepsilon_3')]$$

$$= \sigma_e \varepsilon_{zz} - 1/3\, \sigma_e [(\varepsilon_{zz} - \sigma_1/\sigma_e) + (\varepsilon_{zz} - \sigma_2/\sigma_e) + (\varepsilon_{zz} - \sigma_3/\sigma_e)]$$

$$= 1/3\,(\Omega_1 + \Omega_2 + \Omega_3)\,\sigma_o \quad (4.120a)$$

$$\varepsilon_{zz} = \varepsilon_1 = \varepsilon_1' + \varepsilon_1'' = \sigma_1/\sigma_e + \varepsilon_1'' = \Omega_1 \sigma_o/\sigma_e + \varepsilon_1$$

$$= \varepsilon_2 = \varepsilon_2' + \varepsilon_2'' = \sigma_2/\sigma_e + \varepsilon_2'' = \Omega_2 \sigma_o/\sigma_e + \varepsilon_2''$$

$$= \varepsilon_3 = \varepsilon_3' + \varepsilon_3'' = \sigma_3/\sigma_e + \varepsilon_3'' = \Omega_3 \sigma_o/\sigma_e + \varepsilon_3'' \quad (4.120b)$$

Equations (4.117) through (4.120) are plotted in Figure 4.14, which shows how successive yielding causes successive decreases in the slope of the flow curve. For plastic yielding, with $n = \infty$ in Equation (4.16), the slope becomes zero after the last grain has yielded. For viscoplastic yielding, in which strain hardening occurs in each grain and $n \approx 3$, the slope is finite but continues to decrease until the last grain had yielded, according to Equation (4.109). It is easy to imagine that if a specimen having a very large number of grains were pulled in uniaxial tension, a flow curve having a smoothly decreasing slope would be produced. After the last grain had yielded, the slope would be zero for two situations: plastic yielding with no recrystallization and viscoplastic yielding with dynamic recrystallization. The first situation is hypothetical superplastic behavior, and it mimics the second situation, which is actual viscoplastic behavior. Figure 4.14 corrects an error that appeared in Figures 8 and 9 of the original analysis of successive yielding (Hughes, 1977).

Strain softening in lateral shear zones alongside ice streams can be analyzed as superplastic behavior. Byrd Glacier, a major Antarctic ice stream shown in Figure 3.34 and located in Figure 3.35, exhibits this kind of superplasticity. The flow law of ice given by Equation (4.17) for lateral shear stress σ_{xy} is:

$$\dot{\varepsilon}_{xy} = (\sigma_c^{\,n-1}/A^n)\,\sigma_{xy} \quad (4.121)$$

where x is positive along the centerline of Byrd Glacier and y is positive toward its lateral shear zones. Since $\dot{\varepsilon}_{ij} \propto \sigma_{ij}'$ and $\dot{\varepsilon}_{xx} + \dot{\varepsilon}_{yy} + \dot{\varepsilon}_{zz} = 0$, effective creep stress σ_c can be written as follows, taking $\dot{\varepsilon}_{zz} = -\dot{\varepsilon}_{xx} - \dot{\varepsilon}_{yy}$ for constant volume:

$$\sigma_c = [1/2\,(\sigma'^2_{xx} + \sigma'^2_{yy} + \sigma'^2_{zz} + 2\sigma^2_{xy} + 2\sigma^2_{yz} + 2\sigma^2_{zx})]^{1/2}$$

$$= [1/2\,(\dot{\varepsilon}_{xx}/\dot{\varepsilon}_{xy})^2 \sigma^2_{xy} + 1/2\,(\dot{\varepsilon}_{yy}/\dot{\varepsilon}_{xy})^2 \sigma^2_{xy} + 1/2\,(\dot{\varepsilon}_{zz}/\dot{\varepsilon}_{xy})^2 \sigma^2_{xy} + \sigma^2_{xy} + (\dot{\varepsilon}_{yz}/\dot{\varepsilon}_{xy})^2 \sigma^2_{xy} + (\dot{\varepsilon}_{zx}/\dot{\varepsilon}_{xy})^2 \sigma^2_{xy}]^{1/2}$$

$$= [1 + (\dot{\varepsilon}_{xx}/\dot{\varepsilon}_{xy})^2 + (\dot{\varepsilon}_{yy}/\dot{\varepsilon}_{xy})^2 + (\dot{\varepsilon}_{xx}\dot{\varepsilon}_{yy}/\dot{\varepsilon}^2_{xy})]^{1/2}\,\sigma_{xy}$$

$$= R_{xy}\,\sigma_{xy} \quad (4.122)$$

where $R_{xy} = \left[1 + (\dot{\varepsilon}_{xx}/\dot{\varepsilon}_{xy})^2 + (\dot{\varepsilon}_{yy}/\dot{\varepsilon}_{xy})^2 + (\dot{\varepsilon}_{xx}\dot{\varepsilon}_{yy}/\dot{\varepsilon}_{xy})^2\right]^{1/2}$

and $\dot{\varepsilon}_{yz} = \dot{\varepsilon}_{zx} = 0$ for the floating part of Byrd Glacier where Swithinbank (1963) measured the transverse profile of longitudinal velocity, shown in Figure 3.34. Strain rates $\dot{\varepsilon}_{xx}$, $\dot{\varepsilon}_{xy}$, and $\dot{\varepsilon}_{yy}$ were calculated over the entire surface of Byrd Glacier by Whillans and others (1989) and are shown in Figure 4.15. Combining Equations (4.121) and (4.122) gives:

$$\dot{\varepsilon}_{xy} = R_{xy}^{n-1}(\sigma_{xy}/A)^n \quad (4.123)$$

Let σ_{xy} increase linearly along y from $\sigma_{xy} = 0$ at $y = 0$ on the centerline to upper viscoplastic yield stress $\sigma_v = \sigma_U$ at $y = \Lambda$ on each side of the ice stream:

$$\sigma_{xy} = \sigma_U\,(y/\Lambda) \quad (4.124)$$

Equation (4.123) can then be integrated to give the following velocity profile:

$$u_x = \int_y^{\Lambda} 2\,\dot{\varepsilon}_{xy}\,dy = \int_y^{\Lambda} 2 R_{xy}^{n-1}(\sigma_U\, y/A\,\Lambda)^n\,dy$$

$$= 2 R_{xy}^{n-1}(\sigma_U/A\,\Lambda)^n(\Lambda^{n+1} - y^{n+1})/(n+1)$$

$$= u_o\,[1 - (y/\Lambda)^{n+1}] \quad (4.125)$$

where $\sigma_U = 0.53$ bars is the unstable upper viscoplastic yield stress, $A = 4.7$ bar $a^{1/3}$, and centerline ice velocity u_o is:

$$u_o = 2 R_{xy}^{n-1}(\sigma_U/A)^n\,\Lambda/(n+1) \quad (4.126)$$

The velocity profiles computed from Equation (4.125) for $n = 3$ and $n = 9$ are compared with the velocity profile measured by Swithinbank (1963) in Figure 4.16. Agreement is poor for $n = 3$ and good for $n = 9$. Hence, Byrd Glacier has superplastic flow, even though $n \approx 3$ in the flow law of ice given by Equation (4.17).

A good fit between computed and measured velocities in Figure 4.16 can be obtained for $n = 3$ in Equation (4.125) if A is given a high value for strain-hardened ice over distance $0 \le y \le \lambda$ in the central part of Byrd Glacier, A is given a low value for strain-softened ice over lateral shear zones of width W, and A decreases smoothly from its high value to its low value in transition zones of width $\lambda \le y \le (\Lambda - W)$ where recrystallization produces strain softening, as shown in Figure 4.16. As a consequence of strain softening, the unstable upper viscoplastic yield stress $\sigma_v = \sigma_U = 0.53$ bars at $y = \Lambda$ is replaced by the stable lower yield stress $\sigma_v = \sigma_S = 0.42$ bars at $y = \Lambda - W$, and A reduces from 4.7 bar $a^{1/3}$ for σ_U to 0.75 bar $a^{1/3}$ for σ_S. Over distance $0 \le y \le \lambda$ where strain hardened ice exists, with $\sigma_{xy} = \sigma_S$ at $y = \lambda$ for a transition zone of width $\Lambda - W - \lambda$:

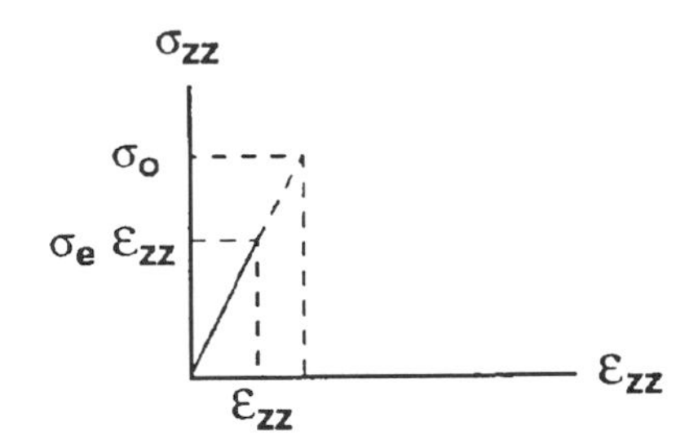

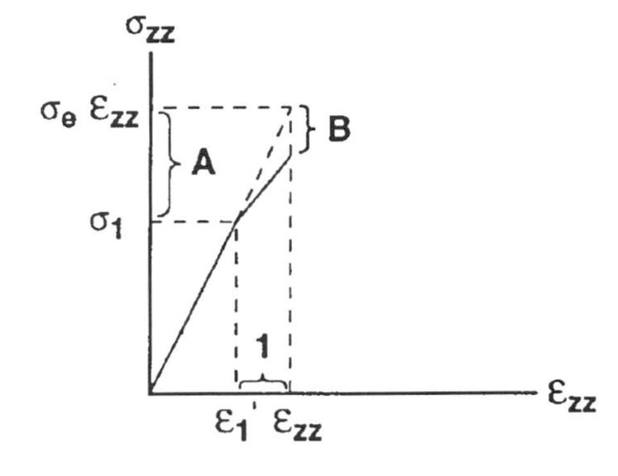

A $\quad \sigma_e \varepsilon_{zz} - \sigma_1 = \sigma_e (\varepsilon_{zz} - \sigma_1/\sigma_e)$

B $\quad 1/3\ \sigma_e \varepsilon_1'' = 1/3\ \sigma_e (\varepsilon_{zz} - \sigma_1/\sigma_e)$

1 $\quad \varepsilon_1'' = \varepsilon_{zz} - \varepsilon_1' = \varepsilon_{zz} - \sigma_1/\sigma_e$

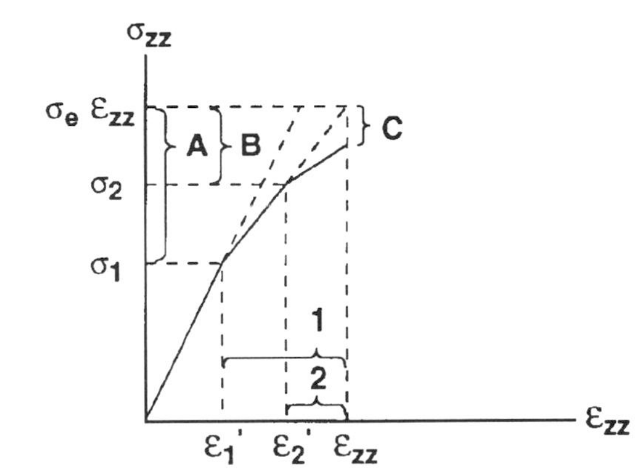

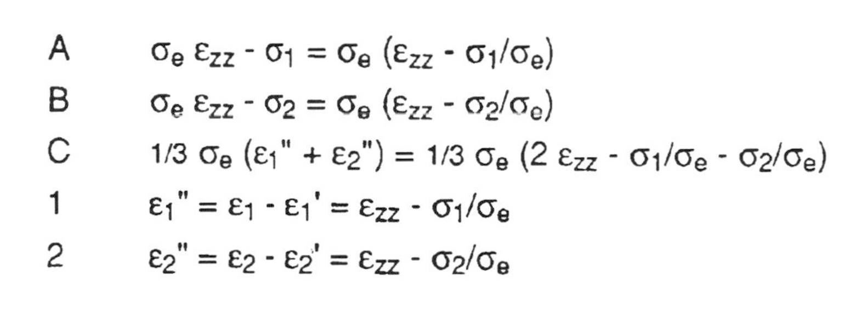

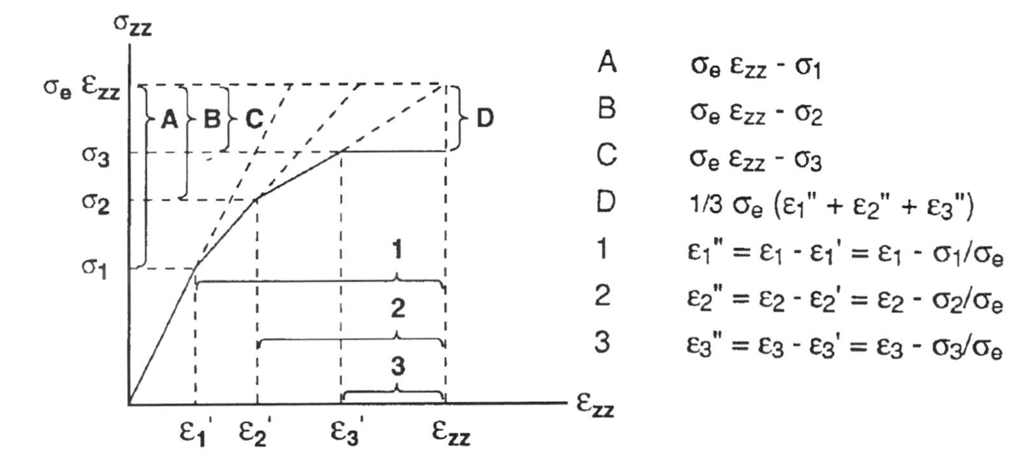

Figure 4.14: Strain hardening in randomly oriented polycrystalline ice simulated by successive yielding of grains linked in parallel.

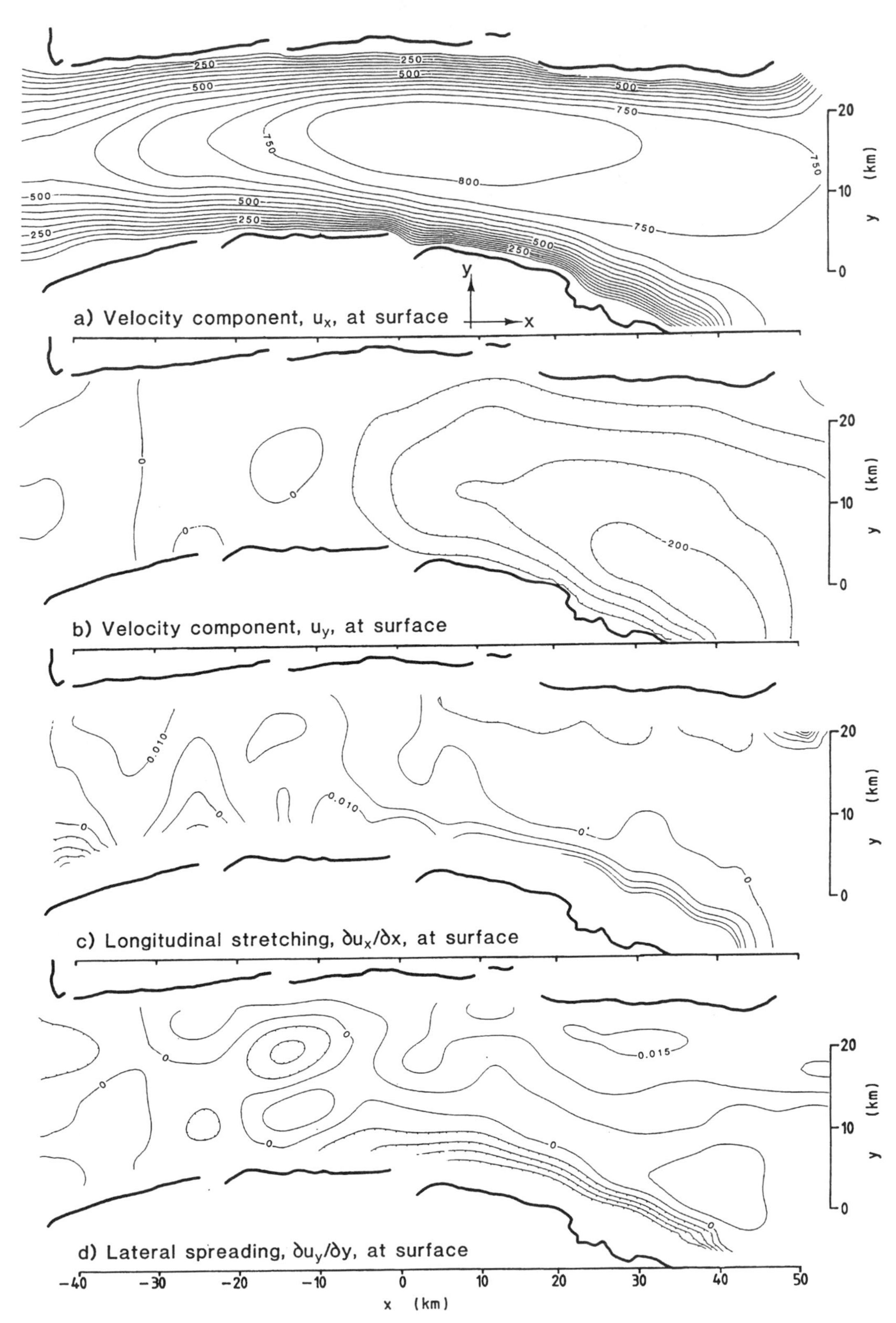

Figure 4.15a: Surface velocities, velocity gradients, turning rates, and effective strain rates on Byrd Glacier. From Whillans and others (1989). Reproduced with permission.

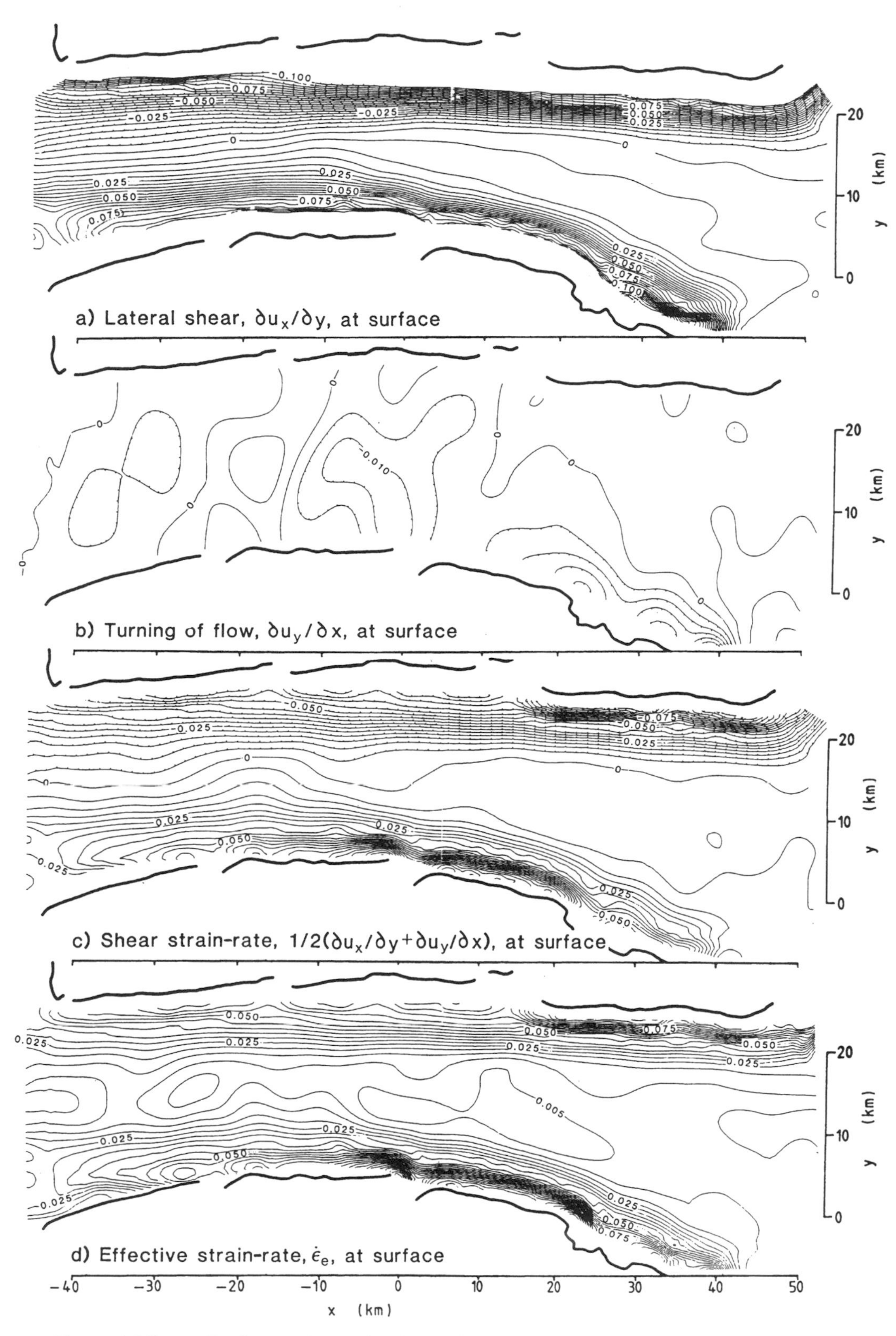

Figure 4.15b: Further surface velocities, velocity gradients, turning rates, and effective strain rates on Byrd Glacier. From Whillans and others (1989). Reproduced with permission.

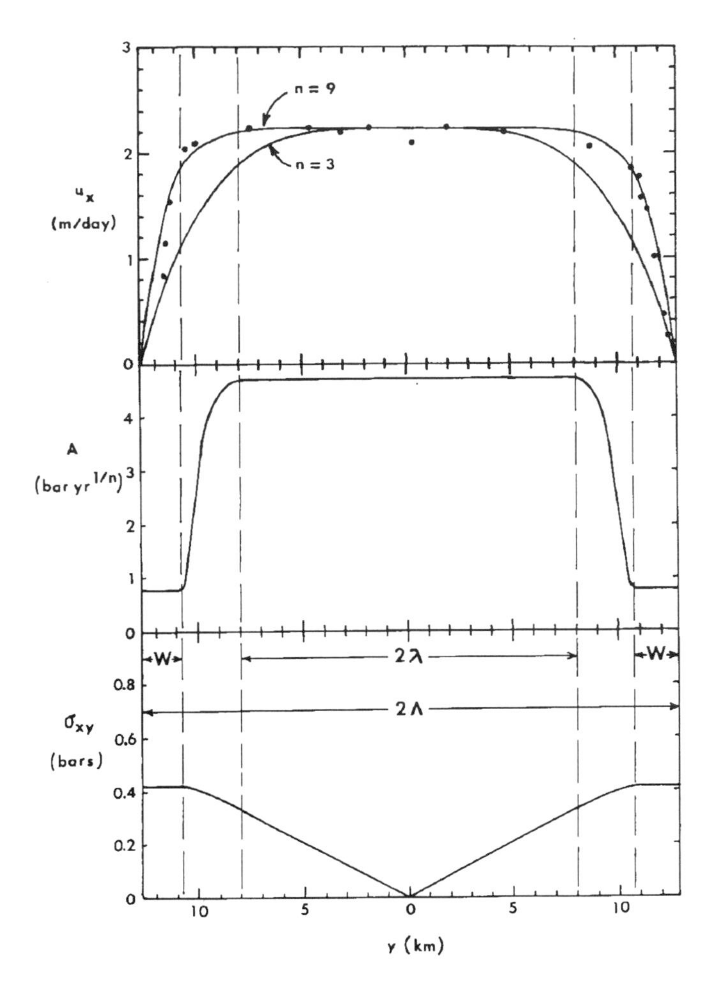

Figure 4.16: The Byrd Glacier velocity data related to viscoplastic instability and strain softening alongside an ice stream. Longitudinal surface velocities u_X across section Y –Y' in Figure 2.19, reported by Swithinbank (1963) and shown as solid circles, are compared with theoretical curves for homogeneous flow, in which the hardness parameter A is constant and the shear stress σ_{xy} varies linearly across the ice stream. Agreement is poor for n = 3, the measured viscoplastic parameter of ice, and fair for n = 9, a nearly plastic viscoplastic parameter. Agreement is best for n = 3 if A and σ_{xy} vary as shown, which is expected if viscoplastic instability creates a small-circle ice fabric in zone 2λ, a single-pole ice fabric in zones W, and transitional ice fabrics in between. As shown in Figure 4.11, strain softening is much more pronounced in a single-pole fabric than in a small-circle fabric.

$$\sigma_{xy} = \sigma_S \, (y/\lambda) \tag{4.127}$$

Ice velocity over distance λ where A = 4.7 bar $a^{1/3}$ is therefore:

$$u_x = \int_y^\lambda 2\,\dot{\varepsilon}_{xy}\,dy = \int_y^\lambda 2\,R_{xy}^{n-1}\,(\sigma_S\, y/A\,\lambda)^n\,dy$$

$$= 2\,R_{xy}^{n-1}\,(\sigma_S/A)^n\,\lambda\left[1-(y/\lambda)^{n+1}\right]/(n+1)$$

$$= u_o\,[1-(y/\lambda)^{n+1}] \tag{4.128}$$

Across width W:

$$\sigma_{xy} = \sigma_S \tag{4.129}$$

Ice velocity across width W where A = 0.75 bar $a^{1/3}$ is therefore:

$$u_x = \int_y^\Lambda 2\,\dot{\varepsilon}_{xy}\,dy = \int_y^\Lambda 2\,R_{xy}^{n-1}\,(\sigma_S/A)^n\,dy$$

$$= 2\,R_{xy}^{n-1}\,(\sigma_S/A)^n\,\Lambda\,[1-(y/\Lambda)]$$

$$= u_o\,(\sigma_S/\sigma_U)^n\,(n+1)\,[1-(y/\Lambda)]$$

$$= u_\lambda\,(\Lambda/W)\,[1-(y/\Lambda)] \tag{4.130}$$

where $u_\lambda = 2\,R_{xy}^{n-1}\,(\sigma_S/A)^n\,W$ is obtained at y = Λ – W. As Figure 4.16 shows, the measured profile of u_x across Byrd Glacier is approximated equally well by Equation (4.126) with n = 9 for superplasticity and by Equations (4.128) and (4.130) with n = 3 for strain softening such that transition width Λ – W – λ is chosen to give the best fit to the velocity data.

The success of superplasticity in simulating strain hardening in a polycrystalline ice cylinder pulled in uniaxial tension and in simulating strain softening in lateral shear zones alongside an ice stream is of great significance in ice dynamics. It implies that analytical solutions of sheet flow, stream flow, shelf flow, and transitional flow from sheet flow to stream flow and from stream flow to shelf flow, in which A is held constant, might give reasonable results if n >> 3 is used, n = 9 for example. In such cases, superplasticity mimics the effects of thermal hardening or softening and strain hardening or softening that cause A to change in various directions through the ice, notably variations with depth for sheet flow and shelf flow and variations with width and depth for stream flow. In sheet flow, both thermal and strain softening occur near the bed. In shelf flow, thermal softening occurs near the underside. In stream flow, thermal softening occurs near the bed and strain softening occurs in lateral shear zones, where frictional heat generates thermal softening in very fast ice streams.

Superplasticity in these situations encourages applications of plasticity theory to ice dynamics. Plasticity theory, for which n = ∞ in the flow law of ice, allows the slipline field to be computed for perfectly plastic flow. Sliplines are lines of maximum shear stress in plane strain solutions. In sheet flow, sliplines are those for a plastic slab spreading under its own weight over a high-traction rigid bed. In the transition from sheet flow to stream flow, sliplines are those for a plastic slab being pulled through a rigid die that reduces its thickness. In stream flow, sliplines are those for a plastic slab between high-traction rigid plates that are pulled together by extending flow in the slab. In the transition from stream flow to shelf flow, sliplines are those for a rigid punch indenting a plastic slab. In shelf flow, sliplines are those for a plastic slab spreading under its own weight on a frictionless bed. These slipline solutions are shown individually in Figure 4.17 and then combined to approximate sliplines for Byrd Glacier in Figure 3.34. The sliplines were computed by Nye (1951, 1967) for sheet flow, by Weertman (1957b) for shelf flow, and by Hill (1964) for extrusion through a die, extension between plates, and inden-tation by a punch. Figure 4.18 shows the slipline field of Byrd Glacier for a high yield stress across width 2λ and a low yield stress across widths W so boundary $y = \Lambda$ provides traction but boundary $y = \lambda$ does not, where $\Lambda = \lambda + W$.

Plasticity theory has also been used successfully to construct surface flowlines and to compute ice thicknesses along flowlines for the Greenland Ice Sheet (Reeh, 1982) and to assess the stability of ice-age ice sheets (Weertman, 1961). These successes, despite the fact that plastic flow is an end-member of the viscoplastic creep spectrum, are possible because superplasticity mimics the effects of thermal and strain hardening or softening on the dynamics of ice sheets.

The Viscoplastic Creep Spectrum

The viscoplastic creep spectrum shows the effect of viscoplastic parameter n on the flow law for steady-state creep. Steady-state creep exists when creep rate $\dot{\varepsilon} = d\varepsilon/dt$ is constant for creep curves shown in Figures 4.9 and 4.10. Two steady states generally exist: an unstable steady state giving an unstable minimum creep rate $\dot{\varepsilon}_U$ before recrystallization, and a stable steady state giving a stable maximum creep rate $\dot{\varepsilon}_S$ after recrystallization, with $\dot{\varepsilon}_U = \dot{\varepsilon}_S$ during dynamic recrystallization at high stresses or temperatures.

If values of $\dot{\varepsilon}_U$ and $\dot{\varepsilon}_S$ are plotted at each stress that produces a creep curve in Figure 4.10 for constant temperature, points for each $\dot{\varepsilon}$ and σ pair lie along a straight line that satisfies the flow law of ice given by Equation (4.108) in logarithmic form:

$$ln\ \dot{\varepsilon} = ln\ (\sigma/A)^n = n\ ln\ \sigma - n\ ln\ A \qquad (4.131)$$

If Equation (4.131) is written separately for points $\dot{\varepsilon}_1$, σ_1, and $\dot{\varepsilon}_2$, σ_2 on the straight line, the second equation can be subtracted from the first equation, and the resulting expression can be solved for n to give:

$$n = \frac{ln\ \dot{\varepsilon}_1 - ln\ \dot{\varepsilon}_2}{ln\ \sigma_1 - ln\ \sigma_1} = \frac{ln\ (\dot{\varepsilon}_1/\dot{\varepsilon}_2)}{ln\ (\sigma_1/\sigma_2)} \qquad (4.132)$$

The value of n obtained from Equation (4.132) can then be used in Equation (4.131) at either point $\dot{\varepsilon}_1$, σ_1, or points $\dot{\varepsilon}_2$, σ_2 to obtain A.

If values of $\dot{\varepsilon}_U$ and $\dot{\varepsilon}_S$ are plotted at each temperature that produces a creep curve in Figure 4.10 for constant stress, points for each $\dot{\varepsilon}$ and T pair lie along a straight line that satisfies the flow law of ice given by Equation (4.108) in semilogarithmic form:

$$ln\ \dot{\varepsilon} = ln\ [(\sigma/A_o)^n\ e^{-CT_M/T}] = ln\ (\sigma/A_o)^n - C\ T_M/T \qquad (4.133)$$

If Equation (4.133) is written separately for points $\dot{\varepsilon}_1$, T_1, and points $\dot{\varepsilon}_2$, T_2 on the straight line, the second equation can be subtracted from the first equation, and the resulting expression can be solved for C:

$$C = \frac{ln\ \dot{\varepsilon}_1 - ln\ \dot{\varepsilon}_2}{(T_M/T_2) - (T_M/T_1)} = \frac{ln\ (\dot{\varepsilon}_1/\dot{\varepsilon}_2)}{(T_M/T_2) - (T_M/T_1)} \qquad (4.134)$$

With Equations (4.131) and (4.133) giving n, A, and C, Equation (4.108) can be used to compute A_o:

$$A_o = A\ exp\ (-C\ T_M/n\ T) \qquad (4.135)$$

Constants n, C, A, and A_o were obtained at steady-state creep rates $\dot{\varepsilon}_U$ and $\dot{\varepsilon}_S$ for unstable and stable equilibrium in order to establish Equation (4.108) as the flow law of ice. Recrystallization does not affect n and C but does affect A_o and therefore A. Before recrystallization, a maximum A_o minimizes $\dot{\varepsilon}_U$. After recrystallization, a minimum A_o maximizes $\dot{\varepsilon}_S$.

The viscoplastic creep spectrum is displayed best if Equation (4.108) is written in a form that allows for superplasticity by including plastic yield stress σ_o:

$$\dot{\varepsilon} = \dot{\varepsilon}_o\ (\sigma/\sigma_o)^n = \dot{\varepsilon}_o^*\ exp\ (-C\ T_M/T)\ (\sigma/\sigma_o)^n \qquad (4.136)$$

where $\dot{\varepsilon} = \dot{\varepsilon}_o$ when $\sigma = \sigma_o$ and $\dot{\varepsilon} = \dot{\varepsilon}_o^*\ exp\ (-C)$ when $\sigma = \sigma_o$ at $T = T_M$. Comparing Equations (4.108) and (4.136):

$$\dot{\varepsilon}_o = (\sigma_o/A)^n = (\sigma_o/A_o)^n\ exp\ (-C\ T_M/T) = \dot{\varepsilon}_o^*\ exp\ (-C\ T_M/T) \qquad (4.137)$$

From Equation (4.137):

$$\dot{\varepsilon}_o^* = (\sigma_o/A_o)^n \qquad (4.138)$$

Figure 4.19 displays the viscoplastic creep spectrum by plotting $\dot{\varepsilon}/\dot{\varepsilon}_o$ versus σ/σ_o for $1 \le n \le \infty$ in Equation (4.136). All curves in the spectrum intersect at the point $\dot{\varepsilon}_o$, σ_o. The curves become straight lines for purely viscous flow at $n = 1$ and for purely plastic flow at $n = \infty$. The other curves bend for viscoplastic flow at $1 < n < \infty$, with the bends getting sharper as n increases. Viscoplastic creep in ice is usually observed at $n = 3$.

Bends in the curves of Figure 4.19 have a localized "knee" when $n \ge 3$. This knee occurs at viscoplastic yield stress σ_v. The knee appears at lower strain rates and higher stresses, while becoming more distinct, as n increases, until it becomes a right-angle at $\dot{\varepsilon} = 0$, $\sigma = \sigma_o$, and $n = \infty$. For this condition strain is purely elastic when $\sigma < \sigma_o$ and purely plastic when $\sigma = \sigma_o$.

Plastic yield stress σ_o can be computed from the elastic strain energy of distortion Q_s that changes the shape of an

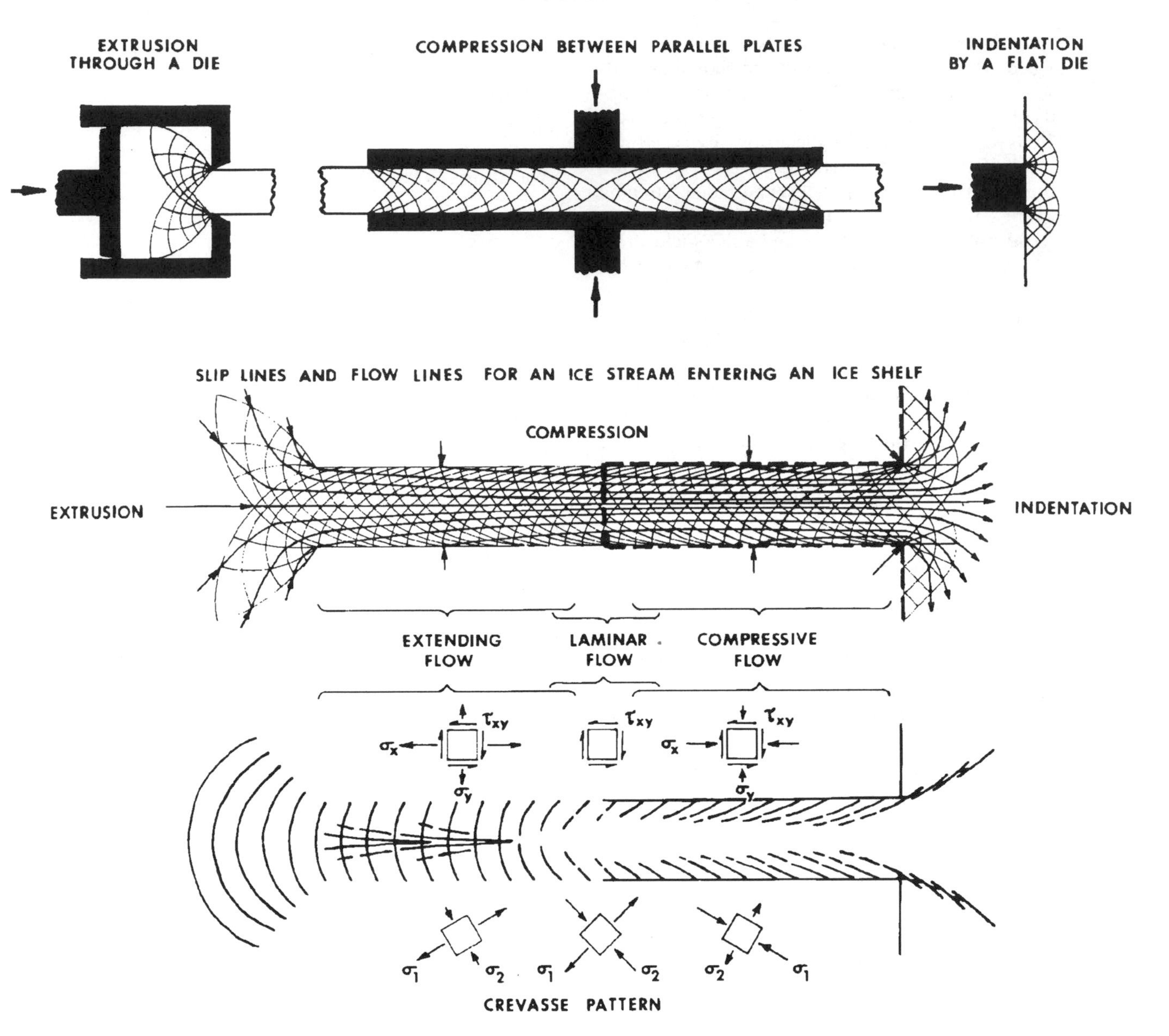

Figure 4.17: The slipline field for an ice stream idealized to represent Byrd Glacier in Figure 3.34. Sliplines are lines of maximum shear stress or of the yield stress in plasticity theory (Hill, 1964). *Top left*: Converging flow at the head of an ice stream is analogous to extrusion of plastic material through a die, except that converging ice is pulled into the ice stream. *Top center*: Extending flow within an ice stream is analogous to compression of a plastic slab between high-traction parallel plates, except that extending flow draws ice along the sides into the ice stream. *Top right*: Compressive flow at the foot of an ice stream entering an ice shelf is analogous to indentation of a plastic plate by a flat rigid dye, except that the ice stream is not rigid. *Middle*: These sliplines can be combined to produce the idealized slipline field for an ice stream that drains an ice sheet and enters an ice shelf, grounded along the thick dashed lines. Flowlines are drawn at 45° angles to the sliplines. *Bottom*: Arcurate transverse crevasses at the head of the ice stream demonstrate that ice is pulled into the ice stream, shear crevasses alongside the ice stream and longitudinal crevasses along the centerline demonstrate that ice is pulled into the ice stream from the sides; and lateral diverging rifts at the foot of the ice stream demonstrate that ice is pushed into the ice shelf by the ice stream; as shown by the changing stress field represented by arrows acting on square elements in the side shear zones. Crevasses are at 45° angles to sliplines.

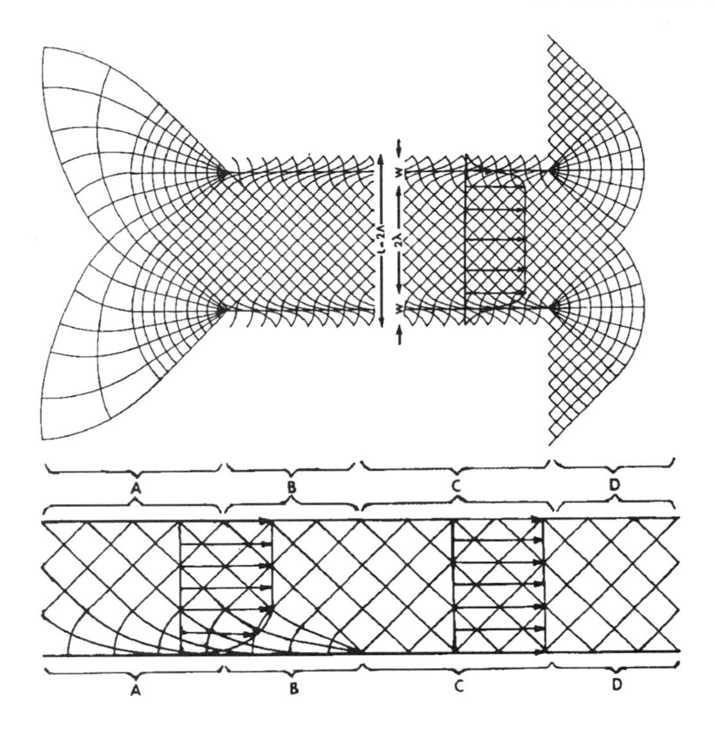

Figure 4.18: A plan view and a longitudinal profile of slipline fields and velocity vectors in Byrd Glacier when strain softened ice in side shear zones and basal shear zones is represented by plasticity theory. *Top*: Viscoplastic yielding in the side shear zones produces an easy-glide ice fabric that greatly reduces side traction. The slipline field in the body of the ice stream (2λ) is for extending flow between low-traction parallel plates and the slipline field in the shear zones (W) is for shear between high-traction parallel plates. The slipline fields for ice entering the ice stream in extending flow and for ice entering the ice shelf in compressive flow are modified from Figure 4.17 as shown. *Bottom*: Viscoplastic yielding in the basal shear zone produces a strain-softened ice fabric that greatly reduces basal traction. The slipline field in ice above the strain-softened ice and in the ice shelf is for flow along the centerline with no basal traction. Bed traction is constant over length A of converging sheet flow, reduces to zero along length B of linear stream flow, and is zero for linear shelf flow over length C and for diverging shelf flow over length D. Lines of maximum shear stress meet rigid boundaries at 0° or 90° angles and meet free bounaries at 45° angles. Hence the slipline field in the ice stream is a continuous change from portions of orthogonal cycloids in the ice sheet to orthogonal straight lines in the ice shelf and reflects progressive ice-bed uncoupling.

ice body. Let Q_e be the total elastic strain energy and Q_v be the volumetric strain energy that changes the size of the ice body, where elastic energy is the area beneath the linear plot of stress σ versus strain ε. In terms of principal stresses σ_k, principal strains ε_k, elastic modulus E_e for uniaxial tension or compression, and Poisson's ratio υ, for k = 1, 2, 3:

$$
\begin{aligned}
Q_e &= \tfrac{1}{2}\sigma_1\varepsilon_1 + \tfrac{1}{2}\sigma_2\varepsilon_2 + \tfrac{1}{2}\sigma_3\varepsilon_3 \\
&= \tfrac{1}{2}\sigma_1\left(\sigma_1/E_e - \upsilon\,\sigma_2/E_e - \upsilon\,\sigma_3/E_e\right) \\
&\quad + \tfrac{1}{2}\sigma_2\left(\sigma_2/E_e - \upsilon\,\sigma_3/E_e - \upsilon\,\sigma_1/E_e\right) \\
&\quad + \tfrac{1}{2}\sigma_3\left(\sigma_3/E_e - \upsilon\,\sigma_1/E_e - \upsilon\,\sigma_2/E_e\right) \\
&= \frac{1}{2E_e}\left(\sigma_1^2 + \sigma_2^2 + \sigma_3^2\right) - \frac{\upsilon}{E_e}\left(\sigma_1\sigma_2 + \sigma_2\sigma_3 + \sigma_3\sigma_1\right)
\end{aligned}
\tag{4.139a}
$$

$$
\begin{aligned}
Q_v &= \tfrac{1}{2}\left[\tfrac{1}{3}\left(\sigma_1 + \sigma_2 + \sigma_3\right)\right]\left[\varepsilon_1 + \varepsilon_2 + \varepsilon_3\right] \\
&= \tfrac{1}{2}\left[\tfrac{1}{3}\left(\sigma_1 + \sigma_2 + \sigma_3\right)\right]\left[\left(\sigma_1/E_e - \upsilon\,\sigma_2/E_e - \upsilon\,\sigma_3/E_e\right)\right] \\
&\quad + \left(\sigma_2/E_e - \upsilon\,\sigma_3/E_e - \upsilon\,\sigma_1/E_e\right) \\
&\quad + \left(\sigma_3/E_e - \upsilon\,\sigma_1/E_e - \upsilon\,\sigma_2/E_e\right) \\
&= \tfrac{1}{2}\left[\frac{3\,(1 - 2\,\upsilon)}{E_e}\right]\left[\tfrac{1}{3}\left(\sigma_1 + \sigma_2 + \sigma_3\right)\right]^2 \\
&= \frac{1}{2\,K_v}\left[\tfrac{1}{3}\left(\sigma_1 + \sigma_2 + \sigma_3\right)\right]^2
\end{aligned}
\tag{4.139b}
$$

$$
\begin{aligned}
Q_s &= Q_e - Q_v \\
&= \frac{1}{2E_e}\left(\sigma_1^2 + \sigma_2^2 + \sigma_3^2\right) - \frac{\upsilon}{E_e}\left(\sigma_1\sigma_2 + \sigma_2\upsilon_3 + \upsilon_3\sigma_1\right) \\
&\quad - \frac{1 - 2\upsilon}{6\,E_e}\left(\sigma_1 + \sigma_2 + \sigma_3\right)^2 \\
&= \left[\frac{1 + \upsilon}{6\,E_e}\right]\left[\left(\sigma_1 - \sigma_2\right)^2 + \left(\sigma_2 - \sigma_3\right)^2 + \left(\sigma_3 - \sigma_1\right)^2\right]
\end{aligned}
\tag{4.139c}
$$

where $K_v = E_e/3\,(1 - 2\,\upsilon)$ is the volumetric bulk modulus.

For a uniaxial state of stress at the elastic limit, where plastic yielding takes place, $\sigma_1 = \sigma_o$, $\sigma_2 = \sigma_3 = 0$, and Equation (4.139c) reduces to:

$$
Q_s = \left[\frac{1 + \upsilon}{3\,E_e}\right]\sigma_o^2 \tag{4.140}
$$

For a triaxial state of stress at the elastic limit, the plastic yield stress is obtained by equating Equations (4.139c) and (4.140) and solving for σ_o:

$$
\sigma_o = \frac{1}{\sqrt{2}}\left[\left(\sigma_1 - \sigma_2\right)^2 + \left(\sigma_2 - \sigma_3\right)^2 + \left(\sigma_3 - \sigma_1\right)^2\right]^{1/2} \tag{4.141}
$$

Plastic yielding occurs for values of σ_1, σ_2, and σ_3 that locate points on the surface of an ellipsoid satisfying Equation (4.141).

Effective creep stress σ_c is defined by Equation (4.15a)

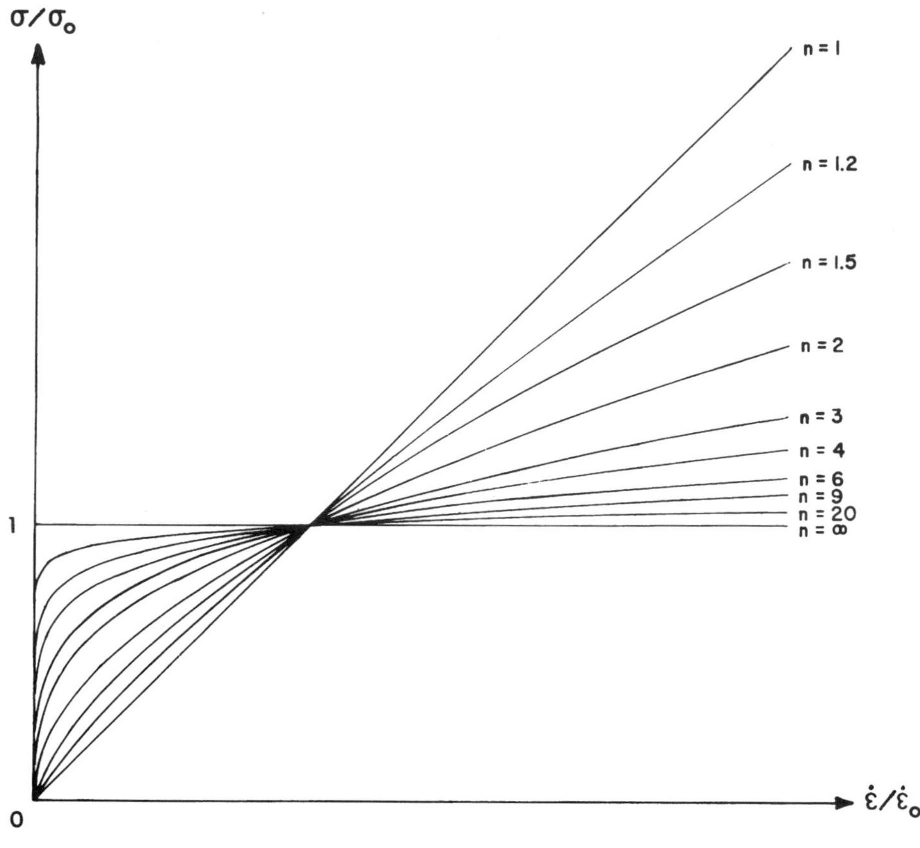

Figure 4.19: The viscoplastic creep spectrum in crystals. The creep law $\dot{\varepsilon} = \dot{\varepsilon}_o (\sigma/\sigma_o)^n$ is plotted for various values of the viscoplastic parameter n, including viscous flow (n = 1), glacial flow (n = 3), and plastic flow (n = ∞). Note that n = 3 is closer to viscous flow, even though 3 is closer to unity than to infinity.

in terms of deviator stress components σ_{ij}' that can be converted to principal stress components σ_k using Equations (4.2) and (4.11):

$$\sigma_c = \left(\tfrac{1}{2}\sigma_{ij}'\sigma_{ij}'\right)^{1/2}$$

$$= \frac{1}{\sqrt{2}}\left(\sigma'^2_{xx} + \sigma'^2_{yy} + \sigma'^2_{zz} + 2\sigma^2_{xy} + 2\sigma^2_{yz} + 2\sigma^2_{zx}\right)^{1/2}$$

$$= \frac{1}{\sqrt{2}}\left[\frac{1}{3}\left(\sigma_{xx} - \sigma_{yy}\right)^2 + \frac{1}{3}\left(\sigma_{yy} - \sigma_{zz}\right)^2 + \frac{1}{3}\left(\sigma_{zz} - \sigma_{xx}\right)^2 + 2\left(\sigma^2_{xy} + \sigma^2_{yz} + \sigma^2_{zx}\right)\right]^{1/2}$$

$$= \frac{1}{\sqrt{6}}\left[\left(\sigma_1 - \sigma_2\right)^2 + \left(\sigma_2 - \sigma_3\right)^2 + \left(\sigma_3 - \sigma_1\right)^2\right]^{1/2} = \sigma_o/\sqrt{3} \quad (4.142)$$

Therefore, plastic yielding occurs when $\sigma_c = \sigma_o/\sqrt{3}$. This is the strain energy of distortion yield criterion (Dieter, 1961, p. 60–62).

The maximum shear stress yield criterion states that plastic yielding occurs when maximum shear stress τ_m is:

$$\tau_m = \tfrac{1}{2}\left(\sigma_1 - \sigma_2\right) = \tfrac{1}{2}\sigma_o \quad (4.143)$$

where σ_1 and σ_2 are the algebraically largest and smallest principal stresses, respectively, and for yielding in uniaxial tension, $\sigma_1 = \sigma_o$ and $\sigma_2 = 0$ (Dieter, 1961, p. 58–59).

Figure 4.20 plots values of σ_o obtained from Equations (4.141) and (4.143) for plane stress, with $\sigma_3 = 0$, so that the strain energy of distortion and maximum shear stress yield criteria can be compared directly. For both criteria, maximum yielding occurs with equal biaxial stress for which $\sigma_1 = \sigma_2$, minimum yielding occurs with pure shear for which $\sigma_1 = -\sigma_2$, and uniaxial yielding occurs with uniaxial stress for which $\sigma_1 = \sigma_o$ and $\sigma_2 = 0$. The case $\sigma_2 = \sigma_o$ and $\sigma_1 = 0$ is irrelevant when σ_1 is defined as the maximum principal stress.

Flow is essentially plastic for n > 9, as is also evident in Figure 4.19. As seen in Figure 4.19, the large increases in $\dot{\varepsilon}$ at high values of n occur for small increases in σ. Superplasticity, therefore, can be invoked to overcome the necessity to keep A constant in Equation (4.16) in order to obtain analytical solutions for pure sheet flow, pure stream flow, pure shelf flow, transitional sheet flow to stream flow, and transitional stream flow to shelf flow. Using n = 9, for example, and keeping A constant may have the same effect on $\dot{\varepsilon}$ as using n = 3 and letting A decrease near the bottom for sheet flow and shelf flow and near the bottom and near lateral shear zones for stream flow. The superplastic value of n that allows A to be kept con-stant is determined by how A actually varies through the ice, a variation that can be computed by determining the temperature and the fabric of ice in coreholes and that is demonstrated for Byrd Glacier in Figure 4.16. For n ≥ 3, the rapid decrease in viscoplastic viscosity $\eta_v = \partial\sigma/\partial\dot{\varepsilon}$ at $\sigma = \sigma_v$, the knee of curves in Figure 4.19, suggests that viscoplastic instability leads to creep rupture at these stresses, so σ_v may be a fracture stress that

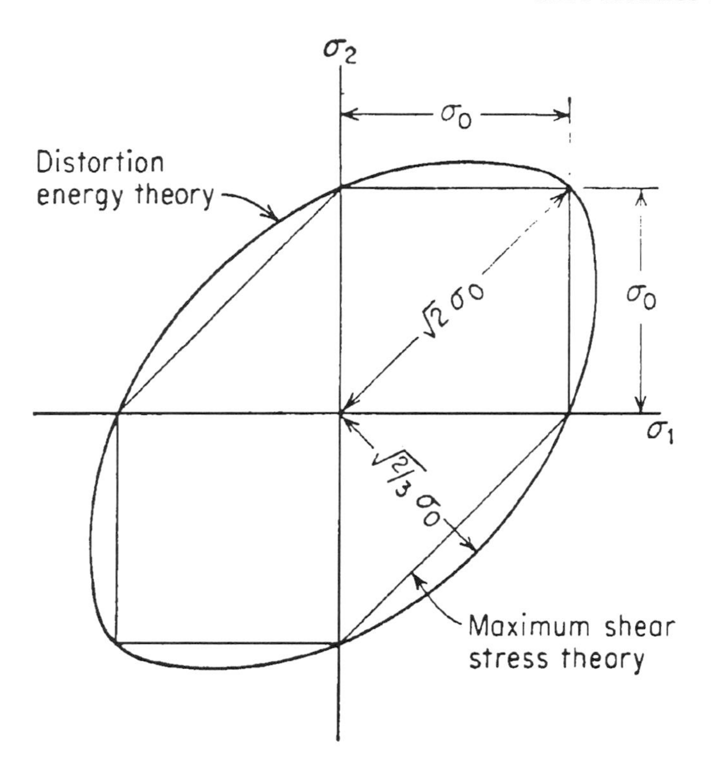

Figure 4.20: A comparison of plastic yielding criteria for plane stress.

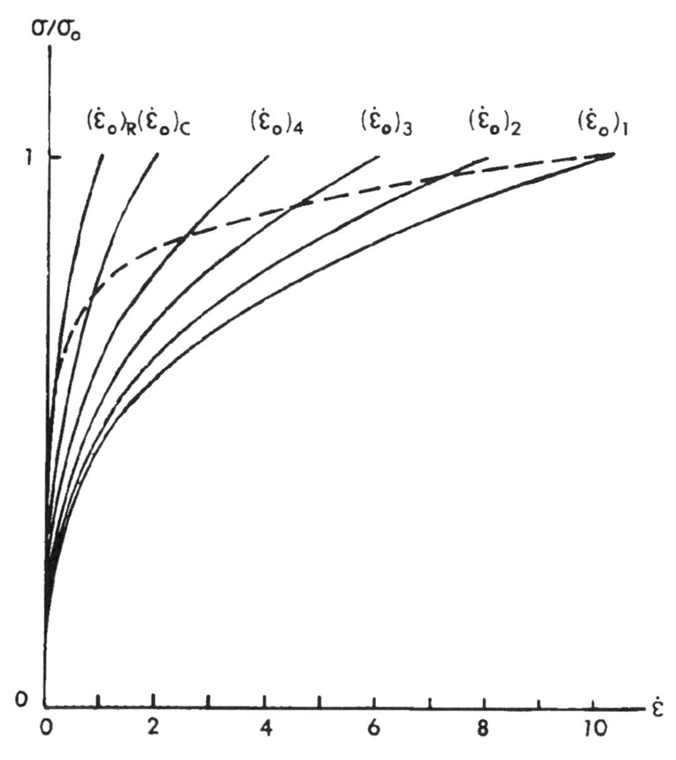

Figure 4.21: Plastic creep mimicked by recrystallization from hard glide to easy glide. Creep near the plastic end of the viscoplastic creep spectrum (n = 9) can be mimicked for glacial creep (n = 3) as recrystallization proceeds from hard glide $(\dot{\varepsilon}_o)_R$ for random ice fabric, to $(\dot{\varepsilon}_o)_C = 2\,(\dot{\varepsilon}_o)_R$ for a circle ice fabric, to $(\dot{\varepsilon}_o)_4 = 4\,(\dot{\varepsilon}_o)_R$ for a four-pole ice fabric, to $(\dot{\varepsilon}_o)_3 = 6\,(\dot{\varepsilon}_o)_R$ for a three-pole ice fabric, to $(\dot{\varepsilon}_o)_2 = 8\,(\dot{\varepsilon}_o)_R$ for a two-pole ice fabric, to easy glide $(\dot{\varepsilon}_o)_1 = 10\,(\dot{\varepsilon}_o)_R$ for a one-pole ice fabric. Creep during recrystallization from $(\dot{\varepsilon}_o)_R$ to $(\dot{\varepsilon}_o)_1$ is shown as a dashed curve that intersects the solid curves for creep when $(\dot{\varepsilon}_o)$ is specified for each ice fabric and n = 3, but the dashed curve mimics $(\dot{\varepsilon}_o) = (\dot{\varepsilon}_o)_R$ and n = 9.

opens crevasses when σ_v is tensile and exceeds lithostatic pressure P in ice. Rifts alongside Byrd Glacier in Figure 3.34 are shear rupture crevasses. The relationship between yielding criteria and crevasse formation has been developed by Vaughan (1993).

Viscoplastic Instability

Since the viscoplastic creep spectrum exists for steady-state creep, $\sigma = \sigma_v$ is the viscoplastic yield stress at the "knee" of curves in Figure 4.19. As seen for the flow curves in Figure 4.8, $\sigma_v = \sigma_U$ is high for unstable equilibrium just before recrystallization, and $\sigma_v = \sigma_S$ is low for stable equilibrium just after recrystallization, with $\sigma_U = \sigma_S$ for dynamic recrystallization at the high temperatures and stresses shown in Figure 4.10. High and low values of $\sigma = \sigma_v$ before and after recrystallization correspond to low and high values of $\dot{\varepsilon}_o$ for random and ordered crystal grain fabrics, thereby giving low and high values of $\dot{\varepsilon}$ in Equation (4.136). This transition from low to high strain rates as recrystallization transforms unstable equilibrium to stable equilibrium is a consequence of viscoplastic instability. Figure 4.21 illustrates how nearly plastic creep with n = 9 can be mimicked by n = 3 in ice when recrystallization transforms hard glide for a random ice fabric into easy glide for a single-maximum fabric.

Stress σ, strain ε, and strain rate $\dot{\varepsilon}$ in Figures 4.9, 4.10, and 4.19 can be effective values σ_c, ε_c, and $\dot{\varepsilon}_c$ when the flow law of ice is expressed by Equation (4.16) or individual values σ_{ij}, ε_{ij}, and $\dot{\varepsilon}_{ij}$ when the flow law is expressed by Equation (4.17). Both the flow and the sliding laws of ice contain the viscoplastic parameter n and the hardness parameter A. In the sliding law, n = 2 m − 1, and $A = A_M$ at the melting point of ice. Therefore, both flow and sliding of glacial ice are subject to viscoplastic instability. When the flow law is written in the form of Equation (4.136), which displays the viscoplastic creep spectrum, with $\sigma = \sigma_c$ and $\varepsilon = \varepsilon_c$, then:

$$\dot{\varepsilon}_c = \dot{\varepsilon}_o\,(\sigma_c/\sigma_o)^n \qquad (4.144)$$

Viscoplastic instability is illustrated by Equation (4.144) when viscoplastic yield stress σ_v becomes more pronounced as n increases, so that a small change in σ_c results in a large change in $\dot{\varepsilon}_c$.

Viscoplastic instability associated with $n \approx 3$ for ice is greatly enhanced by recrystallization of polycrystalline ice from a hard-glide fabric to an easy-glide fabric, as illustrated in Figures 4.9, 4.11, and 4.21. Strain hardening is then replaced by strain softening, and the ice hardness parameter is reduced substantially.

Viscoplastic instability occurs when strain energy E, caused by piling up of dislocations (line defects in the ice crystal structure) at grain boundaries, reaches a critical value. New grains having minimal strain energy then nucleate in the grain boundaries and grow until they completely absorb the old grains. The result is a new grain fabric in which easy-glide crystallographic planes are subparallel from grain to grain, so dislocations can move easily across grain boundaries instead of piling up. This increases $\dot{\varepsilon}_o$ in the flow law, and the rapid increase of $\dot{\varepsilon}_o$ has the same effect on strain rate $\dot{\varepsilon}_c$ as a large n has. Figure 4.9 illustrates this effect by comparing the effect of increasing strain ε with lattice energy E, stress σ_c, and time t. Viscoplastic instability occurs at a critical strain ε_v, beyond which recrystallization reduces E, reduces σ_c for a constant strain rate $\dot{\varepsilon}_c$, and increases $\dot{\varepsilon}_c$ for a constant σ_c, where $\dot{\varepsilon}_c = \partial\varepsilon_c/\partial t$.

A dislocation is Nature's way of minimizing the stress needed for creep strain. As already noted, if we want to move a large carpet across the floor, it is easier to throw a wrinkle in one end and work the wrinkle to the other end than it is to pull the whole carpet from the other end. The wrinkle is a line defect that moves through the carpet, much like a dislocation is a line defect that moves across a crystallographic slip plane. Both produce displacements with minimal applied stress. A hard-glide crystal fabric is be analogous to a carpet in which most wrinkles move in much different directions, whereas an easy-glide fabric is analogous to one in which most wrinkles move in nearly the same direction. Net displacement of the carpet is small in the first case and large in the second case. Viscoplastic instability is analogous to rearranging the wrinkles until, ideally, they all move in the same direction.

Steady-state creep, for which $\partial\dot{\varepsilon}_c/\partial t = 0$ and for which the flow law was derived, is unstable at the strain of viscoplastic instability and becomes stable only after recrystallization has produced an easy-glide ice fabric that is compatible with the stress field. This satisfies Neumann's Principle, which states that strain energy is minimized when the stress and strain ellipsoids have the same symmetry and their principal axes have the same orientation (Nye, 1957, pp. 20–24, 104). Most laboratory creep experiments on polycrystalline ice with a random-grain fabric determine $\dot{\varepsilon}_o$ and n in the flow law at the minimum strain rate. This is near or at the critical strain for viscoplastic instability, and the creep test should be continued until a maximum strain rate after recrystallization is attained. Two values of $\dot{\varepsilon}_o$ can then be determined, a minimum $\dot{\varepsilon}_v$ for strain-hardened ice and a maximum $\dot{\varepsilon}_r$ for strain-softened ice.

Figure 4.22 shows how the creep curve changes during recrystallization, where $\varepsilon = \varepsilon_e$ is elastic strain at t = 0, $\varepsilon = \varepsilon_v$ at t_v when recrystallization begins, and $\varepsilon = \varepsilon_r$ at t_r when recrystallization ends. For transient creep, $\varepsilon \propto t^m$ with $m \approx 1/3$ for polycrystalline ice (Glen, 1955). For steady-state creep, $\varepsilon \propto t$ so that $\partial^2\varepsilon/\partial t^2 = 0$. Taking $\dot{\varepsilon}_o = \dot{\varepsilon}_v$ when $t < t_v$ and $\dot{\varepsilon}_o = \dot{\varepsilon}_r$ when $t > t_r$, during recrystallization $\dot{\varepsilon}_o$ is given by:

$$\dot{\varepsilon}_o = \left(\frac{t_r - t}{t_r - t_v}\right)\dot{\varepsilon}_v + \left(\frac{t - t_v}{t_r - t_v}\right)\dot{\varepsilon}_r \qquad (4.145)$$

For $t < t_v$:

$$\varepsilon = \varepsilon_e + (\dot{\varepsilon}_v\, t)^m\, (\sigma_c/\sigma_o)^n + (\dot{\varepsilon}_v\, t)\, (\sigma_c/\sigma_o)^n \qquad (4.146)$$

For $t_v \le t \le t_r$:

$$\varepsilon = \varepsilon_e + \left(\dot{\varepsilon}_v\, t\right)^m \left(\sigma_c/\sigma_o\right)^n + \left[\left(\frac{t_r - t}{t_r - t_v}\right)\dot{\varepsilon}_v\, t + \left(\frac{t - t_v}{t_r - t_v}\right)\dot{\varepsilon}_r\left(t - t_v\right)\right]\left(\frac{\sigma_c}{\sigma_o}\right)^n \qquad (4.147)$$

For $t > t_r$

$$\varepsilon = \varepsilon_e + (\dot{\varepsilon}_v\, t)^m\, (\sigma_c/\sigma_o)^n + \dot{\varepsilon}_r\, (t - t_v)\, (\sigma_c/\sigma_o)^n \qquad (4.148)$$

Equations (4.145) through (4.148) are consistent with creep experiments by Duval (1976) when m = 1/3 and n = 3. Sunder and Wu (1990) analyze transient creep.

The effect of viscoplastic instability, represented by increasing $\dot{\varepsilon}_o$ during recrystallization in Figure 4.21, is striking for both sheet flow and stream flow. It causes the longitudinal ice velocity u_x to have a strong vertical gradient $\partial u_x/\partial z$ near the bed for sheet flow, and a strong transverse gradient $\partial u_x/\partial y$ across the lateral shear zones of stream flow. To compute these gradients, consider the column of ice in Figure 4.23 of width w, length Δx, and average height $h + 1/2\,\Delta h$. The net longitudinal gravitational force moving the column forward is given by Equation (4.4) and is:

$$(F_x)_G = P_I\, A_x = (P_o + \Delta P_o)\, w\, (h + \Delta h) - P_o\, w\, h$$
$$= 1/2\, \rho_I\, g_z\, w\, (h + \Delta h)^2 - 1/2\, \rho_I\, g_z\, w\, h^2 \approx \rho_I\, g_z\, w\, h\, \Delta h \qquad (4.149)$$

where $P_I = \rho_I\, g_z\, \Delta h$ and $A_x = w\,(h + 1/2\,\Delta h) \approx wh$. Motion of the ice column induces shear traction at its base and along its sides. Ignoring σ_{xx}', the longitudinal kinematic force induced by shear is given by Equation (4.5):

$$(F_x)_K = \sigma_{xz}\, A_z + 2\, \sigma_{xy}\, A_y \approx \tau_o\, w\, \Delta x + 2\, \tau_s\, h\, \Delta x \qquad (4.150)$$

where $\sigma_{xz} = \tau_o$ is the basal shear stress and $A_z = w\,\Delta x$ is the basal area, and $\sigma_{xy} = \tau_s$ is the side shear stress and $A_y = 2\,(h + 1/2\,\Delta h)\,\Delta x \approx 2\,h\,\Delta x$ is the side area. Setting $(F_x)_G - (F_x)_K = 0$ for force equilibrium and solving for surface slope α:

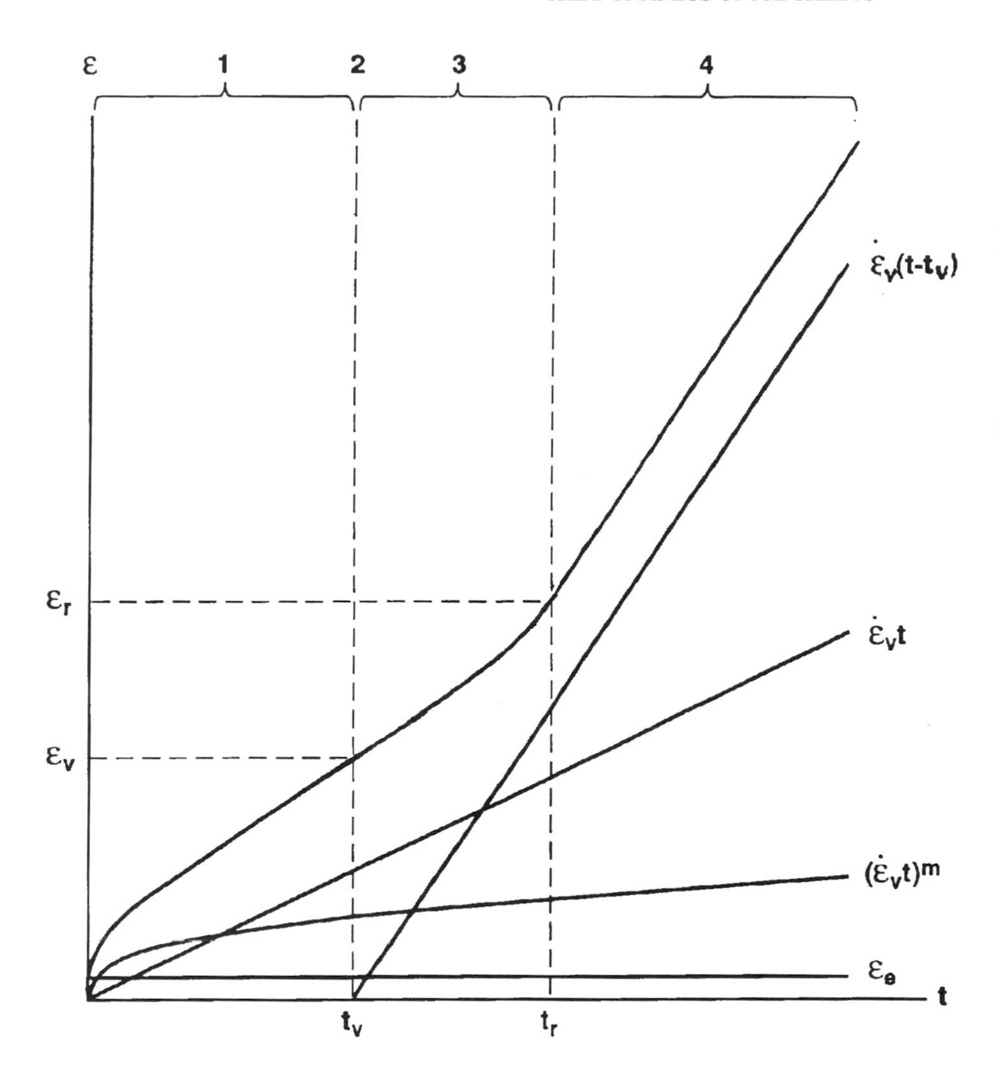

Figure 4.22: Components of the creep curve before and after recrystallization. Strain ε is the sum of elastic strain ε_e, transient strain $(\dot{\varepsilon}_v\ t)^m$, steady-state strain before recrystallization $\dot{\varepsilon}_v$ t, and steady-state strain after recrystallization $\dot{\varepsilon}_v$ (t - t_v), where recrystallization begins at time t_v and ends at time t_r, a timespan over which $\dot{\varepsilon}_v$ (t - t_v) replaces $\dot{\varepsilon}_v$ t. Creep stages are (1) transient, (2) slow unstable steady state, (3) recrystallization, and (4) fast stable steady state.

$$\alpha = \frac{\Delta h}{\Delta x} = \frac{\tau_o}{\rho_I\, g_z\, h} + \frac{2\,\tau_s}{\rho_I\, g_z\, w} \qquad (4.151)$$

For sheet flow, $\sigma_c \approx \sigma_{xz} = \tau_o >> \tau_s$ and $\dot{\varepsilon}_c \approx \dot{\varepsilon}_{xz} = 1/2$ (du_x/dz). Replacing h with h – z in Equation (4.151), and setting $\sigma_c = \sigma_{xz}$ when z < h in Equation (4.144):

$$u_x = \int_0^{u_x} du_x = \int_0^{u_c} 2\,\dot{\varepsilon}_{xz}\,dz \approx \int_0^{z} 2\,(\sigma_{xz}/A)^n\,dz$$

$$\approx \int_0^{z} 2\left[\rho_I\, g_z\,(h - z)\,\alpha\,/A\right]^n dz$$

$$\approx 2\left[\frac{\rho_I\, g_z\,\alpha}{A}\right]^n \frac{h^{\,n+1} - (h - z)^{\,n+1}}{n + 1} \qquad (4.152)$$

where creep velocity u_x in the ice column increases from z = 0 at its base to z = h at its top. The column-averaged creep velocity u_C is:

$$u_C = \int_0^{h} (u_x/h)\,dz = \left(\frac{\rho_I\, g_z\, h\,\alpha}{A}\right)^n \frac{2\,h}{n + 2} \approx \left(\frac{\tau_o}{A}\right)^n \frac{2\,h}{n + 2} \qquad (4.153)$$

Equations (4.152) and (4.153) require A and therefore $\dot{\varepsilon}_o$ to be constant along z, a condition that can be met by using a superplastic value of n instead of n = 3 for ice. The alternative is to abandon analytical solutions for u_C in favor of numerical solutions that allow A to vary through z in response to variations of ice temperature and polycrystalline fabrics.

For stream flow, $\sigma_c \approx \sigma_{xy} = \tau_s >> \tau_o$ and $\dot{\varepsilon} \approx \dot{\varepsilon}_{xy} =$ 1/2 (du_x/dy). The transverse profile of longitudinal velocity has an inflection point halfway across each lateral shear zone, with the velocity gradient decreasing to the center of the ice stream and decreasing to its sides, as shown in Figure 4.18. This symmetry divides the velocity profile into convex and concave portions, with the convex portion lying between inflection points. Since σ_{xy} is the viscoplastic yield stress τ_v at the inflection points where $\sigma_{xy} = \tau_s$, and taking w as the ice stream width between inflection points, computing u_C between these points is accomplished by replac-

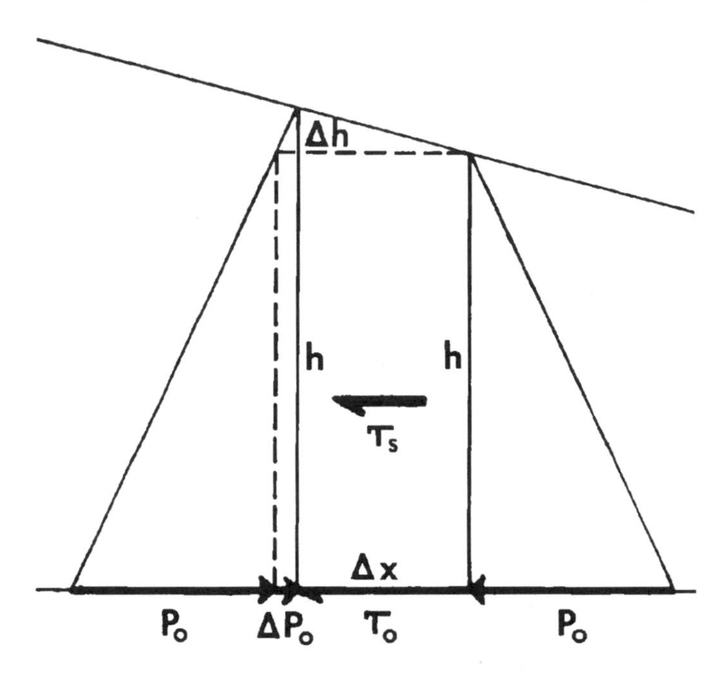

Figure 4.23: Forces acting on an ice column for sheet flow and stream flow. The column has width w, length Δx, and mean height $h + 1/2\ \Delta h$. The horizontal force is obtained from the difference in area between the two triangles, $[1/2\ (P_o + \Delta P_o)(h + \Delta h) - 1/2\ P_o h]\ w$, and is balanced by kinematic forces $\tau_o\ w\ \Delta x$ at the base and $2\ \tau_s\ (h + 1/2\ \Delta h)\ \Delta x$ at the sides of the column, where P_o is basal lithostatic pressure, τ_o is basal shear stress, and τ_s is side shear stress. Area $1/2\ \Delta P_o\ \Delta h$ is ignored.

ing (w/2) with (w/2) – y in Equation (4.151), and setting $\sigma_c = \sigma_{xy}$ when $y < (w/2)$ in Equation (4.144):

$$u_x = \int_0^{u_x} du_x = \int_0^{u_c} 2\,\dot{\varepsilon}_{xy}\,dy \approx \int_{\tau_S}^{\sigma_{xy}} 2\left(\sigma_{xy}/A\right)^n dy$$

$$= \int_0^{y} 2\left\{\rho_I\, g_z\,[(w/2) - y]\,\alpha/A\right\}^n dy$$

$$= 2\left[\frac{\rho_I\, g_z\, \alpha}{A}\right]^n \frac{(w/2)^{n+1} - [(w/2) - y]^{n+1}}{n+1} \tag{4.154}$$

The velocity of stream flow over width 2w increases from zero at the sides where y = 0 to a maximum at the center where y = (w/2). The average creep velocity u_c over width 2w is half the maximum velocity at the centerline, which is obtained by setting y = (w/2) in Equation (4.154):

$$u_C = \int_0^{w/2} (2\,u_x/w)\,dy = \left(\frac{\rho_I\, g_z\, w\,\alpha}{2A}\right)^n \frac{w}{n+2} \approx \left(\frac{\tau_s}{A}\right)^n \frac{w}{n+2} \tag{4.155}$$

Equations (4.154) and (4.155) require A and therefore $\dot{\varepsilon}_o$ to be constant along y.

With viscoplastic instability, A has a separate value for each state of stress. Two states of stress predominate in ice sheets, pure shear produced by longitudinal deviator stress σ_{xx}' and simple shear produced by shear traction stresses σ_{xz} and σ_{xy} for sheet flow and stream flow, respectively. For sheet flow, σ_{xx}' dominates over the upper ice thickness and σ_{xz} dominates over the lower ice thickness. For stream flow, σ_{xx}' dominates over the central ice width and σ_{xy} dominates in the lateral shear zones. These changes in state of stress, accompanied by changes in A, concentrate gradient $\partial u_x/\partial z$ near the bed and gradient $\partial u_x/\partial y$ in the lateral shear zones, so the bulk of sheet flow and stream flow resembles plug flow in which u_x has no strong vertical or transverse gradient. Frictional heating, which is greatest where shear gradients are greatest, reduces A even more in boundary shear zones, thereby further accentuating plug flow in the bulk of an ice sheet or an ice stream.

Creep Instability

In addition to its dependence on grain fabric, A is strongly dependent on temperature. Since creep is thermally activated, Equation (4.16) can be written:

$$\dot{\varepsilon}_c = \left(\frac{\sigma_c}{A}\right)^n = \left(\frac{\sigma_c}{A_o}\right)^n exp\left(\frac{-Q}{RT}\right) = \left(\frac{\sigma_c}{A_o}\right)^n exp\left[\frac{-(E+PV)}{RT}\right] \tag{4.156}$$

where A_o gives the dependence of A on grain fabric, Q is thermal activation energy, E is internal energy, P is lithostatic pressure, V is activation volume, R is the ideal gas constant, and T is absolute temperature (Glen, 1955; Paterson, 1994). Equation (4.156) gives the temperature dependence of A as:

$$A = A_o\, exp\,(Q/n\,RT) \tag{4.157}$$

This strong sensitivity to temperature can give rise to creep instability (Clarke, 1977), especially in ice streams that have surge-like behavior during their life cycles (Paterson and others, 1978).

Creep instability results from the frictional heat of shear deformation, which is greatest at the base of an ice sheet and in the lateral shear zones of ice streams. Robin (1955) first suggested that frictional heat beneath a glacier with a frozen bed would heat the basal ice, thereby increasing the shear strain rate rapidly, according to Equation (4.156), which generates more frictional heat in positive feedback, until the bed thaws and basal sliding causes the glacier to surge. For ice sheets, creep instability might be a mechanism for converting sheet flow to stream flow, especially along basal troughs in the flow direction where frictional heating would be concentrated.

Following Paterson and others (1978), analyzing creep instability is mathematically easiest for two-dimensional flow along a flowline, with x in the downslope direction parallel to the ice surface having slope α and z normal to the surface and positive downward. If heat is transported by vertical conduction much more rapidly than by horizontal advection, and if $\dot{\varepsilon}_{xz}$ is the dominant strain rate component, a good approximation of the heat flow equation is:

$$\frac{\partial T}{\partial t} = \kappa \frac{\partial^2 T}{\partial z^2} - u_z \frac{\partial T}{\partial z} + \frac{2\kappa}{K}\left(\sigma_{xz}\,\dot{\varepsilon}_{xz}\right) \tag{4.158}$$

where K is thermal conductivity, κ is thermal diffusivity, and t is time. Equation (4.158) shows that temperature T at point z in an ice column at time t depends on upward heat conduction, downward ice transport, and frictional heat generation. For an ice column of height h:

$$\sigma_{xz} \approx \rho_I\, g_z\, z\, \alpha = \tau_o\,(z/h) \tag{4.159}$$

If the bed is frozen initially:

$$u_z = a\,(1 - z/h) \tag{4.160}$$

where a is the rate of surface accumulation (positive) or ablation (negative).

Solutions of Equation (4.160) for the boundary conditions $T = T_o$ at $z = 0$ and $\partial T/\partial z = G/K$ at $z = h$, where G is the geothermal heat flux, are obtained from an initial temperature distribution and normalized temperature $\hat{T}$, depth $\hat{h}$, time $\hat{t}$, activation energy $\hat{Q}$, Peclet number *Pe*, stability parameter $\hat{\beta}$, and geothermal parameter $\hat{J}$:

$$\hat{T} = Q\,(T - T_o)/RT_o^2 \tag{4.161a}$$

$$\hat{h} = z/h \tag{4.161b}$$

$$\hat{t} = \kappa\, t/h^2 \tag{4.161c}$$

$$\hat{Q} = Q/RT_o \tag{4.161d}$$

$$Pe = h\,a/\kappa \tag{4.161e}$$

$$\hat{\beta} = \frac{\left(2\,h^2\,Q/A_o^n\right)\hat{t}_o^{\,n+1}\,exp\left(-\,Q/RT_o\right)}{K\,RT_o^2} \tag{4.161f}$$

$$\hat{J} = hEG/KRT_o^2 \tag{4.161g}$$

In analyzing creep instability, $\hat{\beta}$ is the critical parameter because it is proportional to the ratio between frictional heat generation and the rate of conducting this heat up to the surface.

Boundary conditions at time are now $\hat{T} = 0$ at $\hat{h} = 0$ and $\partial\hat{T}/\partial\hat{h} = \hat{J}$ at $\hat{h} = 1$ for any given time $\hat{t}$. Equation (4.158) now becomes:

$$\frac{\partial \hat{T}}{\partial \hat{t}} = \frac{\partial^2 \hat{T}}{\partial \hat{h}^2} - Pe\left(1 - \hat{h}\right)\frac{\partial \hat{T}}{\partial \hat{h}} + \hat{\beta}\hat{h}^{\,n+1}\, exp\left(\frac{\hat{T}}{1 + \hat{T}/\hat{Q}}\right) \tag{4.162}$$

When steady-state heat flow is attained, $\partial\hat{T}/\partial t = 0$, and *Pe* is nil when the surface accumulation or ablation rates are nil. Under these conditions, Equation (4.162) reduces to:

$$\frac{\partial^2 \hat{T}}{\partial \hat{h}^2} + \hat{\beta}\hat{h}^{\,n+1}\, exp\left(\frac{\hat{T}}{1 + \hat{T}/\hat{t}}\right) = 0 \tag{4.163}$$

This equation is nonlinear, so it has multiple solutions, none analytical. Some numerical solutions are displayed in Figure 4.24 as intersections between solid curves showing rates of frictional heating for different values of $\hat{\beta}$ and the dashed curve showing the rate of heat conduction to the surface. Two critical values of $\hat{\beta}$ exist, with a single low temperature solution (intersection 1) below the smaller critical value of $\hat{\beta}'$, a single high-temperature solution (intersection 5) above the larger critical value $\hat{\beta}''$, and three solutions (intersections 2, 3, and 4) between the critical values.

These five solutions are illustrated schematically as points 1, 2, 3, 4, and 5 on the plot of normalized basal ice temperature, $\hat{T}$ at $\hat{h} = 1$, versus $\hat{\beta}$. Point 3 is unstable because a small temperature increase causes a temperature jump to point 4 and a small temperature decrease causes a temperature drop to point 2. Temperatures at or above point 4 are usually above the melting point of ice, so these solutions are unrealistic. Instead, rapid basal melting occurs and may cause a glacier or ice stream to surge if basal meltwater eliminates bed traction. This is creep instability.

Including realistic surface accumulation or ablation rates in the solution of Equation (4.162) merely changes the slope of the dashed heat loss curve in Figure 4.24. Paterson and others (1978) obtained solutions that included downward or upward ice transport rates due to surface accumulation or ablation rates. They concluded that creep instability was possible for parts of the present-day Antarctic Ice Sheet and the former Laurentide Ice Sheet, with time constants for creep instability ranging from 18,000 to 50,000 years. The time constant is the time needed for basal ice temperature to reach the lower critical temperature in Figure 4.25, after which rapid basal melting takes place. Their time constants encompass the cycles of 23,000 and 41,000 years for insolation controlled by Earth's axial precession and tilt. Ice-sheet surges in this timeframe could trigger global climatic changes. Time constants for mountain glaciers were of the order of 100 years in the accumulation zone. Paterson and others (1978) did not compute time constants for ice streams, but time constants between 1000 and 10,000 years would be reasonable, given their size range. Creep instability in ice streams would then be a mechanism for abrupt changes in this time frame that could trigger abrupt climatic changes on a local or regional scale around the perimeter of ice sheets. The time frame encompasses Dansgaard-Oeschger events that record abrupt climate changes shown in Figure 1.1.

A novel version of creep instability has been proposed by Rowden-Rich and Wilson (1996). They noted that rugged bedrock underlies much of the Antarctic Ice Sheet, as seen in radio-echo records. They found that zones of high stress exist in ice above bedrock rises having relief of a few tens of meters, producing sites of strain and thermally softened ice that propagates as lenses in the direction of ice flow. If sites are plentiful, the lenses coalesce to become a layer of high shear strain rates that acts as the dynamic "bed" of the ice sheet.

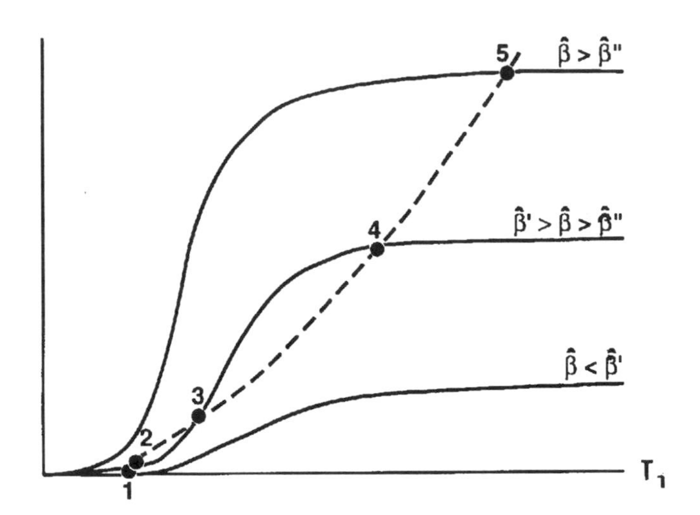

Figure 4.24: Numerical solutions of Equation (4.163) for creep instability. Solutions exist where solid curves giving the rate of basal frictional heating intersects the dashed curve giving the rate of conducting basal frictional heat to the surface, with all curves plotted for different normalized basal ice temperatures $\hat{T} = \hat{T}_1$ at $\hat{h} = 1$. From Clark and others (1977). Reproduced with permission.

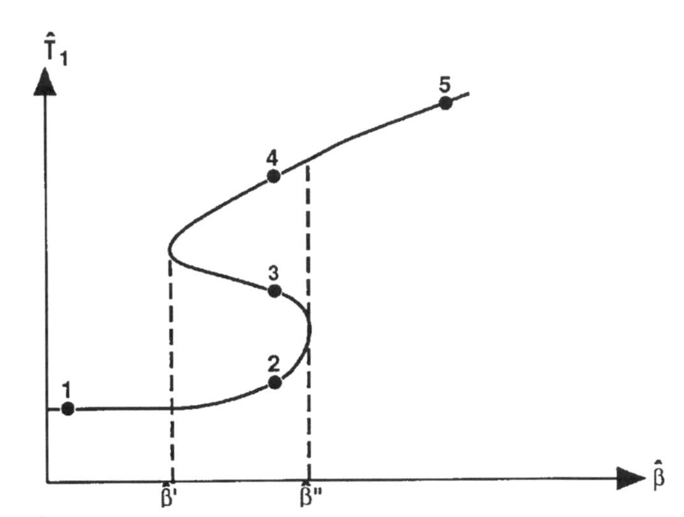

Figure 4.25: The variations of normalized basal ice temperatures $\hat{T} = \hat{T}_1$ at $\hat{h} = 1$ with parameter $\hat{\beta}$ that is proportional to the ratio between the ratio of basal frictional heat generation and the rate at which this heat is conducted upward to the ice surface. Intersection points 1 through 5 in Figure 4.24 appear as points 1 through 5 on this curve. Creep instability is represented by point 3, which jumps to point 4 for a small temperature increase and drops to point 2 for a small temperature decrease, both for the same β. From Clarke and others (1977). Reproduced with permission.

Thermal Instability

Thermal instability in an ice sheet occurs when the vertical buoyancy stress σ_{zz}' caused by thermal expansion of ice near the bed is able to overcome the viscoplastic hardness of basal ice, so that cold ceiling ice sinks *en masse* into warm basement ice upward. In viscous fluids, this thermal basement ice, extruding plumes or curtains of warm instability occurs when the Rayleigh number *Ra* exceeds a critical value *Ra** and thermal convection begins. The vertical thermal buoyancy stress σ_{zz}' in a convecting basal ice layer of thick-ness h_c, through which temperature increase ΔT_c causes thermal expansion Δh_c and density decrease $\Delta\rho_I$, is obtained by equating the vertical buoyancy gravitational force $(F_z)_G = \Delta\rho_I\, h_c\, A_z\, g_z$ due to density decreases $\Delta\rho_I$ in ice volume $h_c\, A_z$ above basal area A_z with the kinematic force $(F_z)_K = \sigma_{zz}'\, A_z$ due to rising convecting flow above A_z:

$$\sigma_{zz}' = \Delta\rho_I\, g_z\, h_c = \rho_I\, g_z\, h_c\, \alpha_c\, \Delta T_c \tag{4.164}$$

where $\alpha_c = \Delta\rho_I / \rho_I\, T_c = \Delta h_c / h_c\, \Delta T_c$ is the thermal expansion coefficient. The thermal buoyancy stress is driven by gravity, and it induces a creep stress σ_c which resists creep deformation in layer h_c having viscosity η_c:

$$\sigma_c = \eta_c\, \dot{\varepsilon}_c = \eta_c\, \kappa / h_c^2 \tag{4.165}$$

where $\dot{\varepsilon}_c = \kappa / h_c^2$ is the creep rate for heat transport by mass transport (thermal convection) that equals heat transport by diffusion of atomic vibrational energy (thermal conduction). The dimensionless Rayleigh number is the ratio of the gravitational and creep stresses (Strutt, 1916):

$$Ra = \sigma_{zz}'/\sigma_c = \rho_I\, g_z\, h_c^3\, \alpha_c\, \Delta T_c / \eta_c\, \kappa \tag{4.166}$$

When conduction cannot transport all the basal heat upward, *Ra* > *Ra** and convection transports the excess heat. Knopoff (1964) obtained *Ra** ≈ 1000 for a convecting fluid layer with semirigid upper and lower boundaries. This would be the case for layer h_c on a thawed bed, because overlying ice could deform by creep and basal ice could deform by sliding.

In computing *Ra* for an ice sheet, η_c is determined for creep in layer h_c. As shown in Figure 4.22, creep has four stages, beginning with $\dot{\varepsilon}_c$ decreasing over time from an infinite elastic strain rate $\dot{\varepsilon}_e$ to a minimum for strain hardened ice just prior to viscoplastic instability and then increasing to a maximum for strain-softened ice after recrystallization is complete. Stage 1 or transient creep exists during strain hardening when $\dot{\varepsilon}_c$ decreases, stage 2 or unstable steady-state creep exists when $\dot{\varepsilon}_c$ is a minimum, stage 3 or tertiary creep exists during strain softening when $\dot{\varepsilon}_c$ increases, and stage 4 or stable steady-state creep exists when $\dot{\varepsilon}_c$ is a maximum and constant over time.

Near the melting temperature T_M of ice, stage 1 tends to go directly to stage 4 because dynamic recrystallization to an easy-glide ice fabric is rapid and begins as soon as creep begins (Glen, 1955). If the bed is thawed, therefore, the creep curve obeys the relationship:

$$\varepsilon = \varepsilon_e + \varepsilon_c = \varepsilon_e + (\dot{\varepsilon}_v t)^m (\sigma_c/\sigma_o)^n + \dot{\varepsilon}_r t (\sigma_c/\sigma_o)^n \quad (4.167)$$

where Glen (1955) found m = 1/3 and n = 3. The creep rate obtained from Equation (4.167) and (4.17) is:

$$\dot{\varepsilon}_{zz} = m \dot{\varepsilon}_v^m t^{m-1} (\sigma_c^{n-1}/\sigma_o^n) \sigma_{zz}' + \dot{\varepsilon}_r (\sigma_c^{n-1}/\sigma_o^n) \sigma_{zz}'$$

$$= [m \dot{\varepsilon}_v^m t^{m-1} + \dot{\varepsilon}_r] (\sigma_c^{n-1}/\sigma_o^n) \sigma_{zz}' \quad (4.168)$$

Effective creep stress σ_c in layer h_c is dominated by stresses σ_{xz} and σ_{zz}' for sheet flow, and σ_{xy} and σ_{zz}' for stream flow. From Equation (4.15a), with $\sigma_{xx}' = \sigma_{yy}' = -\sigma_{zz}'/2$ for ascending plumes with side shear stresses $\sigma_{zx} = \sigma_{zy}$ and $\sigma_{zx} = \sigma_{xz}$ in sheet flow:

$$\sigma_c \approx [3/4\, \sigma_{zz}'^2 + \sigma_{zx}^2 + \sigma_{zy}^2]^{1/2} \approx [3/4\, \sigma_{zz}'^2 + 2\, \sigma_{xz}^2]^{1/2} \quad (4.169)$$

and $\sigma_{xx}' = 0$, $\sigma_{yy}' = -\sigma_{zz}'$ for ascending curtains with side shear stress $\sigma_{zy} << \sigma_{xy}'$ alongside stream flow:

$$\sigma_c \approx [\sigma_{zz}'^2 + \sigma_{xy}^2 + \sigma_{zy}^2]^{1/2}\, [\sigma_{zz}'^2 + \sigma_{xy}^2]^{1/2} \quad (4.170)$$

assuming that easy-glide ice fabrics minimize side shear stresses $\sigma_{zx} = \sigma_{zy}$ in plumes and σ_{zy} in curtains.

By definition, $\eta_c = (\partial\sigma_{zz}'/\partial\dot{\varepsilon}_{zz})$ for a ball sinking in a viscous fluid or a thermal plume rising in an ice sheet. This is obtained by differentiating Equation (4.168). Assuming that $d\sigma_{xy}/d\sigma_{zz}'$ and $d\sigma_{xz}/d\sigma_{zz}'$ can be ignored:

$$\frac{d\dot{\varepsilon}_{zz}}{d\sigma_{zz}'} = \left[\left(\frac{\sigma_c^{n-1}}{\sigma_o^n}\right) + (n-1)\left(\frac{\sigma_c^{n-2}}{\sigma_o^n}\right)\sigma_{zz}'\frac{d\sigma_c}{d\sigma_{zz}'}\right]\left[m\,\dot{\varepsilon}_v^m\, t^{m-1} + \dot{\varepsilon}_r\right]$$

$$= \left[\left(\frac{\sigma_c^{n-1}}{\sigma_o^n}\right) + (n-1)\left(\frac{\sigma_c^{n-2}}{\sigma_o^n}\right)\frac{\sigma_{zz}'^2}{2\sigma_c}\right]\left[m\,\dot{\varepsilon}_v^m\, t^{m-1} + \dot{\varepsilon}_r\right]$$

$$= \left[\sigma_c^{n-1}/\sigma_o^n\right]\left[1 + \tfrac{1}{2}(n-1)\left(\sigma_{zz}'/\sigma_c\right)^2\right]\left[m\,\dot{\varepsilon}_v^m\, t^{m-1} + \dot{\varepsilon}_r\right] \quad (4.171)$$

The effective viscosity is therefore:

$$\eta_c = \frac{d\sigma_{zz}'}{d\dot{\varepsilon}_{zz}} = \frac{2\left(\sigma_o^n/\sigma_c^{n-3}\right)}{\left[2\,\sigma_c^2 + (n-1)\,\sigma_{zz}'^2\right]\left[m\,\dot{\varepsilon}_v^m\, t^{m-1} + \dot{\varepsilon}_r\right]} \quad (4.172)$$

which gives as the Rayleigh number:

$$Ra = \frac{h_c^2\,\sigma_{zz}'}{\kappa\,\eta_c} = \frac{h_c^2\,\sigma_{zz}'\left[2\,\sigma_c^2 + (n-1)\,\sigma_{zz}'^2\right]\left[m\,\dot{\varepsilon}_v^m\, t^{m-1} + \dot{\varepsilon}_r\right]}{2\,\kappa\left(\sigma_o^n/\sigma_c^{n-3}\right)} \quad (4.173)$$

where σ_{zz}' is given by Equation (4.164).

For sheet flow, $\sigma_{zy} = \sigma_{zx} = \sigma_{xz}$ for plumes rising in easy-glide and σ_{xz} is given by Equation (4.151) with $\tau_s = 0$ and σ_{xz} at height $h_o/2$ for the ice overburden at the midpoint of the convecting layer replacing τ_o and $h - h_c/2$ replacing h. Equation (4.169) reduces to:

$$\sigma_c = \left(\left[\tfrac{3}{4}\,\rho_I\, g_z\, h_c\, \alpha_c\, \Delta T_c\right]^2 + \left[2\,\rho_I\, g_z\left(h - h_c/2\right)\alpha\right]^2\right)^{1/2} \quad (4.174)$$

For stream flow, $\sigma_{xy} >> \sigma_{zy}$ for curtains rising in easy-glide and σ_{xy} is given by Equation (4.151) with σ_{xy} replacing τ_s. Equation (4.170) reduces to:

$$\sigma_c = \left(\left[\rho_I\, g_z\, h_c\, \alpha_c\, \Delta T_c\right]^2 + \left[\tfrac{1}{2}\,\rho_I\, g_z\, w\, \alpha\right]^2\right)^{1/2} \quad (4.175)$$

Glen (1955) obtained m = 1/3 and n = 3 from laboratory creep experiments on polycrystalline ice. Equation (4.173) is then:

$$Ra = \frac{h_c^2\,\sigma_{zz}'\left[\sigma_c^2 + \sigma_{zz}'^2\right]\left[\dot{\varepsilon}_v^{1/3} + 3\,\dot{\varepsilon}_r\, t^{2/3}\right]}{3\,\kappa\,\sigma_o^3\, t^{2/3}} \quad (4.176)$$

Thermal instability begins at t = 0, for which $Ra = \infty$. Hence, Ra exceeds any finite critical Rayleigh number such as $Ra^* = 1000$ and thermal convection will continue until time t* when $Ra = Ra^*$. If t* is sufficiently large, steady-state thermal convection takes place for:

$$Ra = h_c^2\,\sigma_{zz}'\left(\sigma_c^2 + \sigma_{zz}'^2\right)\dot{\varepsilon}_r/\kappa\,\sigma_o^3 \quad (4.177)$$

If σ_{zz}' is much greater than σ_{xy} or σ_{xz}, then $\sigma_c \approx \sqrt{3}\,\sigma_{zz}'/2$ for plumes and $\sigma_c \approx \sigma_{zz}'$ for curtains. Equation (4.177) can then be written for vertical creep rate $\dot{\varepsilon}_{zz} = \dot{\varepsilon}_r (\sigma_c^{n-1}/\sigma_o^n)\,\sigma_{zz}' \approx \dot{\varepsilon}_r (\sigma_{zz}'/\sigma_o)^n$:

$$Ra \approx 2\left(h_c^2\,\dot{\varepsilon}_r/\kappa\right)\left(\sigma_{zz}'/\sigma_o\right)^3 = 2\,\dot{\varepsilon}_{zz}\, h_c^2/\kappa \quad (4.178)$$

For purely viscous flow, n = 1 and $Ra = h_c^2\,\dot{\varepsilon}_r\,\sigma_{zz}'/\kappa\,\sigma_o$ in Equation (4.173) when t* is large. This is identical to the Rayleigh number for fluid convection given by Equation (4.166), with $\eta_c = \sigma_{zz}'/\dot{\varepsilon}_{zz} = \sigma_o/\dot{\varepsilon}_r$ and $\sigma_{zz}' = \rho_I\, g_z\, h_c\, \alpha_c\, \Delta T_c$.

Equation (4.173) expresses a fundamental distinction between thermal convection in a polycrystalline solid and in a viscous fluid. In a polycrystalline solid, convection is initiated at t = 0 for any thermal buoyancy stress σ_{zz}' and is sustained until t = t*. During this time, Ra decreases from infinity to Ra^*. Thermal convection will be transient if t* < t_v and will be steady-state if t* ≥ t_v, where t_v is the time when viscoplastic instability sets in, with t_v approaching zero as T approaches the melting point T_M.

The relationship between the creep curve and how Ra varies with time is illustrated in Figure 4.26. Basal ice temperature is T_o, and T_M is the melting temperature. Case 1, for $T_o = T_M$, has relatively free boundaries for layer h_c, so Ra^* is low ($Ra^* = 657$ for free boundaries in a viscous

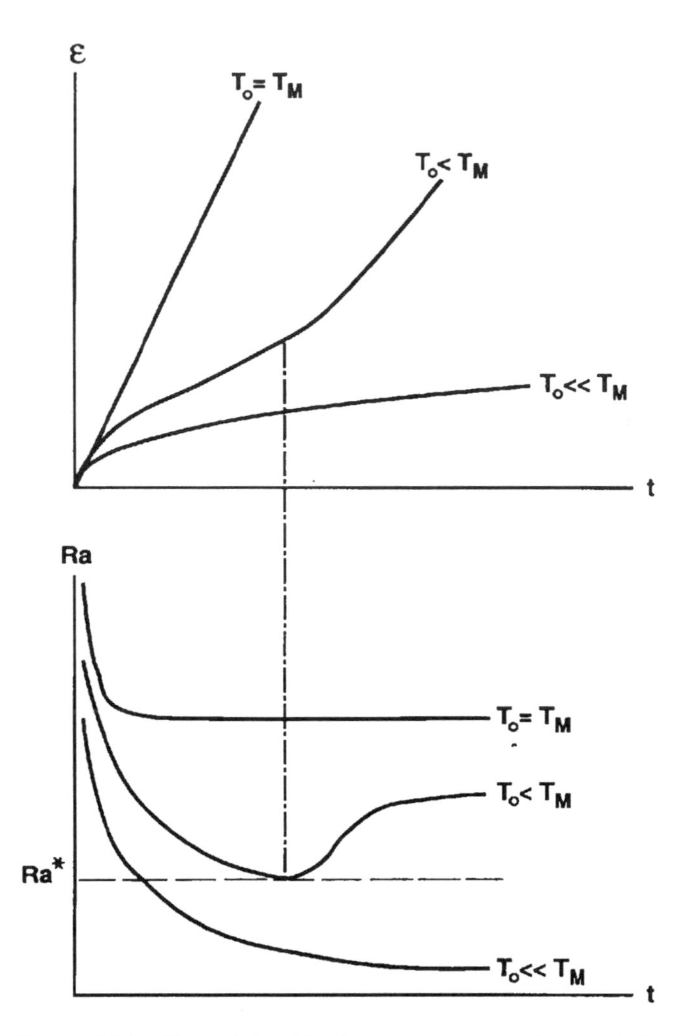

Figure 4.26: The relationship between creep strain ε and the Rayleigh number Ra for thermal convection over time t for various basal ice temperatures T_0. *Top*: Creep curves for dynamic recrystallization in ice above a thawed bed ($T_o = T_M$) with free boundary conditions, for recrystallization at time t_v in ice above a warm frozen bed ($T_o < T_M$) with easy glide boundary conditions, and for no recrystallization in ice above a cold frozen bed ($T_o << T_M$) with hard glide boundary conditions. *Bottom*: When $T = T_M$, transient convection becomes steady-state convection if Ra > Ra* at t > 0; when $T > T_M$, transient convection ends before t_v but begins again and becomes steady-state convection after t_V if Ra < Ra* at t_v but Ra > Ra* after t_v; and when $T_o << T_M$, transient convection ends before t_v if Ra < Ra* before t_v and does not begin again. Note the similarity between Ra versus t and $\dot{\varepsilon}$ versus t, where $\dot{\varepsilon} = \partial\varepsilon/\partial t$.

fluid), t_v is nearly zero, and *Ra* decreases to a constant value greater than *Ra**. Case 2, for $T_o < T_M$, has deformable boundaries for layer h_c, so *Ra** is intermediate (*Ra** = 1101 for one free boundary and one rigid boundary in a viscous fluid), t_v is finite, and *Ra* decreases to a minimum near *Ra** at t_v before rising to a constant value after recrystallization is complete. Case 3, for $T_o << T_M$, has relatively rigid boundaries for layer h_c so *Ra** is high (*Ra** = 1708 for rigid boundaries in a viscous fluid), t_v approaches infinity, and *Ra* decreases to a constant value less than *Ra**. Thermal convection attains steady state in Case 1, might attain steady state in Case 2, and never attains steady state in Case 3. However, all three cases begin with transient thermal convection.

Thermal convection is most likely to be transient in sheet flow, where transient diapiric activity is random in time and space if the bed is smooth and diapiric activity is frequent enough to scramble ice stratigraphy in layer h_c, so that radar waves are not reflected and an echo-free zone results, as shown in Figure 4.27. The top of the echo-free zone should show "near-vertical cusps and fingers" caused by diapiric activity, which Robin and Millar (1982) described but did not explain. Steady-state thermal convection in sheet flow should cause more systematic warping of the radio-echo stratigraphy in ice overlying convecting layer h_c, as also shown in Figure 4.27. Robin and Millar (1982) attributed this warping to an earlier time when the ice moved over a presumably rugged bed.

Thermal convection is most likely to be in a steady-state in the lateral shear zones alongside stream flow. The bed should be thawed beneath the shear zones and ice in the shear zones should have an easy-glide fabric favorable for ascending curtains of ice displaced by colder ice sinking *en masse* between the shear zones, so that $\sigma_{zy} = \sigma_{xy}$. The surface of an ice stream is indeed lower than the surface of flanking ice. Ice advected laterally into the ice stream from adjacent ice ridges carries ascending curtains with it, thereby completing the convection circuit. Convecting ice spirals downstream. However, longitudinal advection of ice may be so rapid in an ice stream that a transverse convection circuit cannot be completed along the length of the ice stream.

In an ice sheet, basal heat is transported up by thermal conduction when hot atoms vibrating about their lattice sites excite larger vibrational amplitudes in neighboring colder atoms that are farther above the heat source. Heat is also transported by thermal convection if the vibrational amplitude is large enough to dislodge lines of hot atoms from their lattice sites, so the thermal buoyancy stress can move them up as line defects or dislocations in the crystal structure, allowing heat transport by mass transport. There is no proof that this actually takes place, and the effect it might have on the dynamics of ice sheets is unknown.

Diapiric thermal convection below the density inversion in ice sheets may explain the breakdown in correlating oxygen isotope stratigraphy between the GRIP corehole to bedrock at the summit of the Greenland Ice Sheet and the GISP 2 corehole to bedrock that is 28 km west of the GRIP corehole. According to the temperature profile down the GRIP corehole, the density inversion occurs about 1500 m down in ice 3000 m thick, with a nearly linear temperature increase in the lowest 1000 m of ice (Gundestrup and others, 1993). Correlations in oxygen isotope stratigraphy break down in the lowest 250 m of ice (Grootes and others, 1993). Steeply inclined ice layers have been observed in basal ice from these coreholes, and Boulton (1993) has interpreted this as folding generated by ice shearing over a rough bed. He

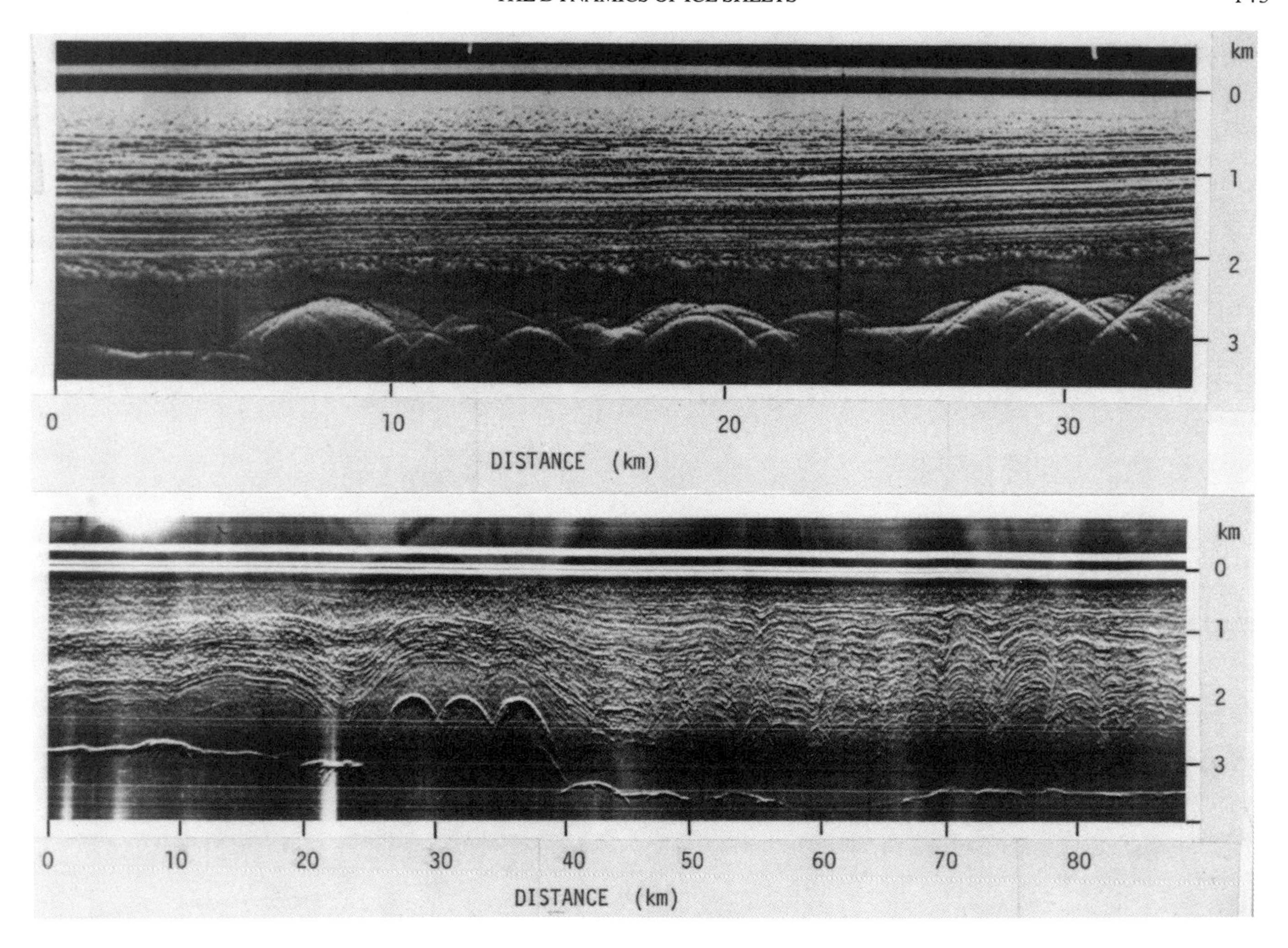

Figure 4.27: Distributions of radio-echo stratigraphy revealing possible thermal convection in the Antarctic Ice Sheet. Radio-echo stratigraphy reported by Robin and Millar (1982) for Antarctica often shows an echo-free zone in ice that is up to 1 km above the bed, which is the approximate height of the density inversion due to thermal expansion of ice near ice divides. *Top profile*: A transition zone of vertical "cusps" and "fingers" that separates the horizontal radio-echo layers from an echo-free zone over a rolling subglacial topography (line T on map). The fingers and cusps may be the tops of transient thermal con-vection diapirs that scramble stratigraphy to produce the echo-free zone. *Bottom profiles*: A transition zone of domed radio-echo layers above an echo-free zone over a smooth subglacial topography (line B on map). The domes may be the tops of steady-state thermal convection cells that scramble stratigraphy to produce the echo-free zone. *Left*: A location map for the top (T) and bottom (B) radio-echo profiles. Robin and Millar (1982) interpreted these features as representing a transition between upper ice flow controlled by the ice-surface slope and lower ice flow controlled by bed topography. From Robin and Millar (1982). Reproduced with permission.

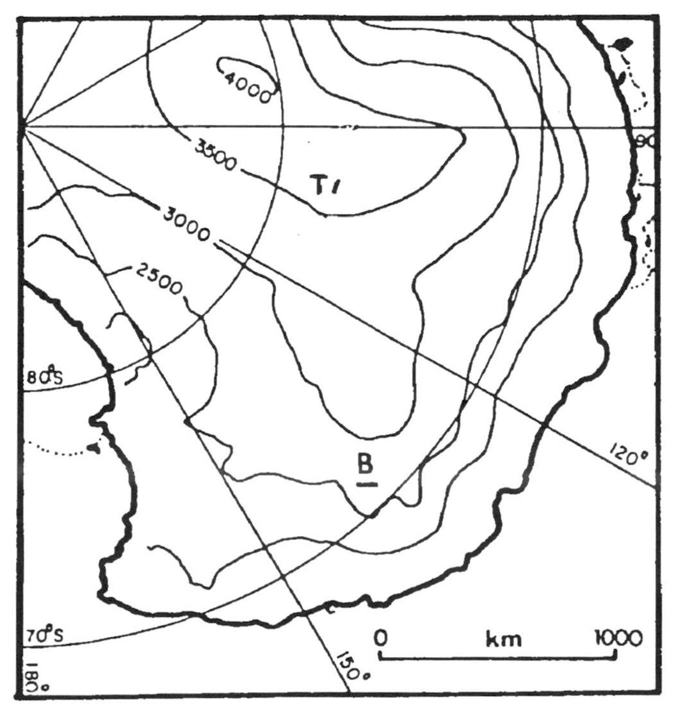

believes that these folds are responsible for the breakdown in correlation of both oxygen isotope ratios and electrical conductivity between the two coreholes. However, the 28 km between the two coreholes is on the ice divide where downslope creep is nil, and the bed is frozen and not particularly rugged, so the folds may not be caused by basal traction variations linked to bed roughness or basal thawing. Thermal convection in this basal ice could produce either diapiric plumes or curtains that would then become tilted in the direction of basal shear. As shown in Figure 3.6, tilting of episodic convection plumes could produce the observed folded ice structures that Boulton (1993) described, but without the need for rough patches or thawed patches on the bed.

From Equation (4.178), the Rayleigh number is:

$$Ra = \frac{2\,\dot{\varepsilon}_{zz}\,h_c^{\,2}}{\kappa} = \frac{2\,\sigma_{zz}'\,h_c^{\,2}}{\eta_c\,\kappa} = \frac{2\,\rho_I\,g_z\,h_c^{\,3}\,\alpha_c\,\Delta T_c}{\eta_c\,\kappa} \tag{4.179}$$

Using measurements from the GRIP corehole (Gundestrup and others, 1993), ρ_I = 917 kg m^{-3}, g_z = 9.8 m s^{-2}, h_c = 1500 m, α_C = 1.53 × 10^{-4} °C^{-1}, ΔT_c = 23 °C, η_c = 8 × 10^{13} kg m^{-1} s^{-1}, and κ = 1.15 × 10^{-6} m^2 s^{-1}, Equation (4.179) gives Ra = 2.3 × 10^3. This should easily sustain transient diapiric thermal convection in ice below the density inversion, and even steady–state thermal convection with rigid upper and lower boundary conditions seems to be possible in ice below the density inversion.

Surface Slope and Bed Slope

Analytical solutions of pure sheet flow, pure stream flow, pure shelf flow, transitional sheet flow to stream flow, transitional stream flow to shelf flow, viscoplastic instability, creep instability, thermal instability, and basal sliding all include the surface slope in order to compute ice velocities, but not the bed slope. The reason for this is illustrated in Figure 4.28 when the bed slopes in the same direction and in the opposite direction as the surface slope. In both cases, a column of ice perpendicular to the bed has basal length Δx, basal width w, height h on its short side, height h + Δh on its tall side, surface slope α, and bed slope β. Axis x is parallel to the bed and positive in the down-slope direction of the surface. Axis z is perpendicular to the bed and positive toward the surface. Forces acting on the ice column are a body force F_B, a lithostatic force F_L, and a traction force F_T. From Equations (4.3), (4.4), and (4.5):

$$F_B = m_I\,g_x = \pm\,[\rho_I\,w\,(h + 1/2\,\Delta h)\,\Delta x]\,g_x \tag{4.180a}$$

$$F_L = \Delta\bar{P}\,A_x = [1/2\,\rho_I\,g_z\,(h + \Delta h)]\,[w\,(h + \Delta h)] - [1/2\,\rho_I\,g_z\,h]\,[w\,h] \tag{4.180b}$$

$$F_T = \sigma_{xz}\,A_z = -\,\tau_o\,w\,\Delta x \tag{4.180c}$$

where m_I is the mass of the ice column, $\Delta\bar{P}$ is the decrease in average lithostatic pressure $\bar{P}$ from the tall side to the short side of the ice column, τ_o is the shear stress at the base of the ice column, and in F_B the plus sign is for a bed that slopes upward along x and the minus sign is for a bed that slopes downward along x. Setting $F_B + F_L + F_T = 0$ and solving for τ_o gives:

$$\begin{aligned}\tau_o &= 1/2\,\rho_I\,g_z\,(2h + \Delta h)\,\Delta h/\Delta x \pm \rho_I\,g_x\,(h + 1/2\,\Delta h)\\ &\approx \rho_I\,(g\cos\beta)\,h\,[\tan(\alpha \mp \beta)] \pm \rho_I\,(g\sin\beta)\,h\\ &\approx \rho_I\,g\,h\,[\cos\beta\,\tan(\alpha \mp \beta) \pm \sin\beta]\end{aligned} \tag{4.181}$$

where g is gravity acceleration, $g_x = g\sin\beta$, $g_z = g\cos\beta$, $\Delta h/\Delta x = \tan(\alpha \mp \beta)$, angle $\alpha + \beta$ is for a bed that slopes upward along x and angle $\alpha - \beta$ is for a bed that slopes downward along x, and h + 1/2 $\Delta h \approx$ h.

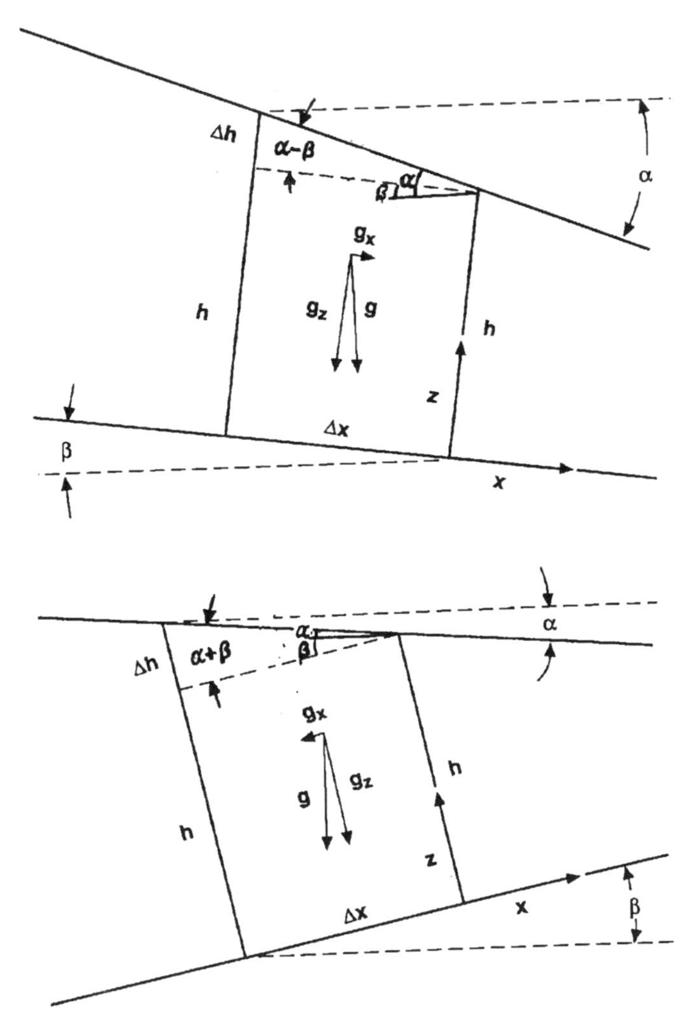

Figure 4.28: Free body diagrams of ice columns for calculating the contributions of surface slope α and bed slope β on the basal shear stress. *Top*: The free body diagram when both the surface and the bed slope downward in the direction of ice flow. *Bottom*: The free body diagram when the surface slopes downward and the bed slopes upward in the direction of ice flow.

Equation (4.181) shows that bed slope has a significant effect on τ_o only when $\beta >> \alpha$, reducing τ_o for an upsloping negative β and increasing τ_o for a downsloping positive β. When $\beta = 0$ in Equation (4.181):

$$\tau_o = \rho_I \, g \, h \, tan \, \alpha \approx \rho_I \, g_z \, h \, \alpha \tag{4.182}$$

which is the usual expression for τ_o because, regardless of β, when $\alpha = 0$:

$$\tau_o = 0 \tag{4.183}$$

Since τ_o is a kinematic stress that exists only when ice moves over the bed and is therefore resisted by bed traction, when $\tau_o = 0$ there is no ice motion. Hence, ice motion is controlled by the surface slope, not the bed slope. Motion is affected by the bed slope only if ice has a surface slope.

Since gravitational driving force $F_x = \rho_I \, g \, h \, A \, tan \, \alpha$ is parallel to and acts on basal area $A = w \, \Delta x$ of ice columns having basal width w, length Δx, and slope β in Figure 4.28, gravitational driving stress $\tau_o = F_x / A$ in Equation (4.182) is also the basal shear stress. In numerical solutions for ice-sheet flowline profiles, vertical ice columns are used for which $A = w \, \Delta x$ is the basal area of width w and length Δx in the horizontal direction of surface slope α. Then when the bed has a non-zero slope β, gravitational driving stress τ_o at the base of a vertical ice column is not parallel to the bed, so τ_o is resolved into components σ_R normal to the bed and τ_R parallel to the bed, where τ_R is the basal shear stress and $\tau_o \approx \tau_R$ in the shallow-ice approximation (Hutter, 1983). Deriving Equation (4.181) for the tilted ice columns in Figure 4.28 demonstrates that the shallow-ice approximation is legitimate. This book uses the shallow-ice approximation and vertical ice columns in calculating the surface slope of ice sheets when the bed slope varies.

If ice columns were perpendicular to the bed, adjacent ice columns would lean across each other where the bed was concave and lean away from each other where the bed was convex. Therefore it would be impossible to compute continuous surface slopes from ice columns that were perpendicular to sloping beds. Instead, vertical ice columns must be parallel to each other and the bed must be approximated by a sequence of upward or downward horizontal steps, with one step at the base of each ice column.

5

SHEET FLOW

Constraints on Flowlines

Sheet flow occurs over nearly all of the grounded part of an ice sheet, even though stream flow discharges nearly all of the grounded ice and feeds much of it to floating ice shelves. In modeling a glaciation cycle, therefore, the first task is to simulate sheet flow. For present-day ice sheets in Antarctica and Greenland, sheet flow is revealed by the pattern of ice flowlines drawn perpendicular to ice-surface contour lines. Ice flows in downslope directions from ice divides to ice margins, with the slowest flow from domes to saddles along the crest of the ice divide and faster flow down the flanks of the ice divide. Most flow down the flanks converges on ice streams. The remainder diverges along ice ridges between ice streams.

In modeling sheet flow, flowlines for former ice sheets can be either output obtained from a specified distribution of surface accumulation and ablation rates or input obtained from a specified distribution of basal erosion and deposition processes. The first choice depends on a reliable general circulation model for the atmosphere when the ice sheet existed. The second choice depends on a reliable interpretation of the glacial geology produced beneath the ice sheet at that time. Both choices present difficulties.

Precipitation rates are the least reliable output from a general circulation model. Model input is insolation at the top of the atmosphere and the distribution of continents and oceans at the bottom of the atmosphere, including albedo and topography. From these, the model sequentially computes surface air temperatures, atmospheric pressure gradients, wind fields, ocean surface currents, cloud production, and, finally, precipitation, with all of the positive and negative feedbacks used to make successive iterations until a steady-state global climate is produced. Being the last thing computed, precipitation rates are subject to all the errors and uncertainties in earlier computations. Yet, the distribution of precipitation over an ice sheet determines its pattern of surface flowlines.

Glacial geology is produced by the pattern of surface flowlines. However, the pattern changes during a glaciation cycle, and the final glacial geology is the result of superimposed mechanisms of glacial erosion and deposition from the whole suite of flow patterns. In general, two patterns are preserved well enough to reconstruct surface flowlines. First-order glacial geology produced during and near the glacial maximum is preserved because its imprint on the landscape is reinforced with each glaciation cycle, especially if successive cycles repeat the basic history of advance and retreat for the ice sheets so that large-scale features are produced. Second-order glacial geology produced during the last deglaciation is preserved because it is the last glacial imprint on the landscape, an imprint of small-scale features that changed continuously, so it had time to overprint the first-order glacial landscape without obliterating it.

In principle, if a general circulation model is able to provide surface temperatures and precipitation rates over an ice sheet, equations of heat and mass flow can be used to compute what parts of the bed are frozen, thawed, freezing, and melting so that specific mechanisms of glacial erosion and deposition can be applied to the bed. Conversely, the glacial geology in principle can be used to compute the surface mass balance needed to produce it. Determining the glacial geology from the mass balance is the forward solution. Determining the mass balance from the glacial geology is the inverse solution. The forward solution has ice-sheet flowlines from ice divides to ice margins as model output. The inverse solution has ice-sheet flowlines from ice margins to ice divides as model input. This book employs the inverse solution. Budd and Smith (1981) pioneered the forward solution.

The inverse solution relates ice-surface elevation upslope along flowlines of a former ice sheet to (1) present-day topography of the deglaciated landscape, (2) glacioisostatic depression of that landscape by the former ice sheet, and (3) ice-bed coupling due to mean annual temperature, ice accumulation rates, ice ablation rates, and converging or diverging flow at the surface that are deduced from a specified pattern of frozen, thawed, freezing, and melting conditions at the bed. These constraints will be examined separately and then combined in a formula that computes increasing ice elevations along a flowline with increasing distance from the ice margin to the ice divide.

Bed Topography

The effect of bed topography on the surface profile of a steady-state ice sheet can be examined by referring to Figure 5.1. The bed is approximated by a series of flat horizontal steps, each of length Δx along a flowband of width w, where w can vary from step to step. The flowband lies between two flowlines. The column of ice above basal area w Δx has height h_I on the downslope side and surface slope $\Delta h/\Delta x$. A

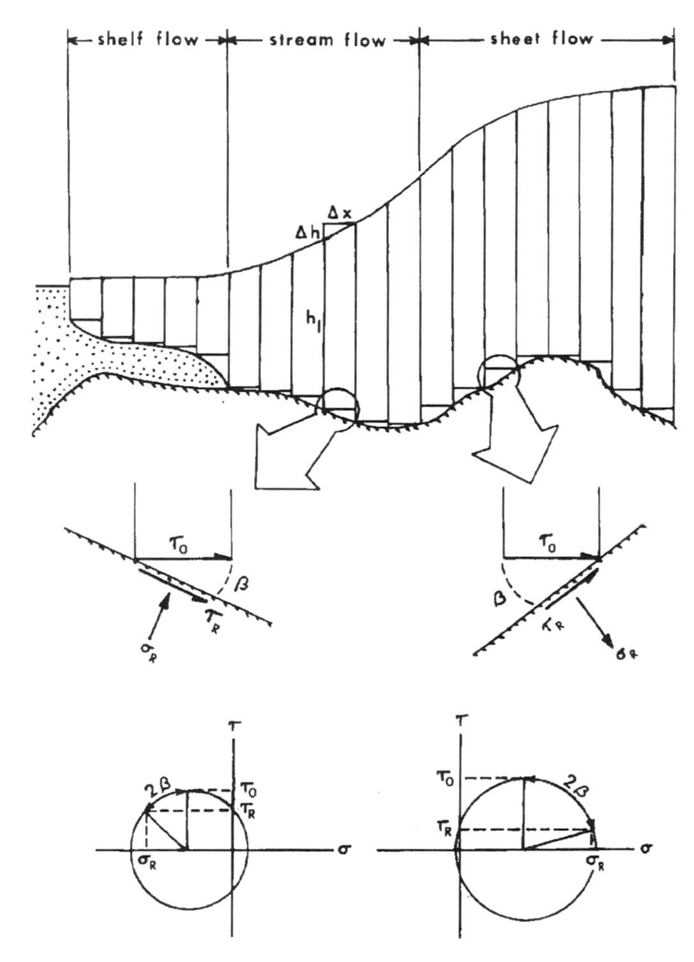

Figure 5.1: Representation of the bed beneath an ice-sheet flowline as a sequence of upward or downward steps. Constant step lengths Δx have variable step heights Δh_R. Ice columns above each step have height h_I on the low side and surface slope $\Delta h/\Delta x$. Basal shear stress τ_o at the base of each ice column can be resolved into two deviator components, a shear stress τ_R parallel to the bed and a normal stress σ_R perpendicular to the bed, using Mohr circles. If the bed slopes upward in the direction of ice flow, σ_R is compressive. If the bed slopes downward in the direction of ice flow, σ_R is tensile. A compressive σ_R deposits lodgement till and a tensile σ_R quarries jointed bedrock.

summation of forces in horizontal distance x along the flowband includes the net horizontal gravitational force $(F_x)_G$ moving the ice column in the downslope direction, a deviator force $(F_x)_D$ caused by extending converging flow or compressive diverging flow along the flowband, traction forces $(F_x)_T$ at the base and the two sides of the ice column induced by ice motion, and a body force $(F_x)_B$ equal to the product of the column's mass and acceleration or deceleration of ice along x as the ice sheet advances or retreats. Let ρ_I be ice density, g_z be gravity acceleration, σ_{xx}' be longitudinal deviator stress, τ_o be basal shear stress, τ_s be side shear stress, u_x be longitudinal velocity, and t be time. Then:

$$(F_x)_G = [1/2\ \rho_I\ g_z\ (h_I + \Delta h)]\ [w\ (h_I + \Delta h)] - [1/2\ \rho_I\ g_z\ (h_I)]\ [w\ h_I] \tag{5.1a}$$

$$(F_x)_D = N\ \sigma_{xx}'\ w\ h_I + N\ [\sigma_{xx}' - (\partial \sigma_{xx}/\partial x)\ \Delta x]\ [w\ (h_I + \Delta h)] \tag{5.1b}$$

$$(F_x)_T = \tau_o\ w\ \Delta x + 2\ \tau_s\ (h_I + 1/2\ \Delta h)\ \Delta x \tag{5.1c}$$

$$(F_x)_B = \rho_I\ [w\ (h_I + 1/2\ \Delta h)\ \Delta x]\ [\partial u_x/\partial t] \tag{5.1d}$$

In Equation (5.1b), Equation (3.22) shows that $N = [2 + (\dot{\varepsilon}_{yy}/\dot{\varepsilon}_{xx})]$, so that $(F_x)_D$ arises from stress difference $(\sigma_{xx} - \sigma_{zz}) = N\ \sigma_{xx}'$, where N = 2 when $\dot{\varepsilon}_{yy} = 0$ with no diverging flow from ice divides to give Equation (4.42), and N = 3 when $\dot{\varepsilon}_{yy} = \dot{\varepsilon}_{xx}$ with radially diverging flow from ice domes to give Equation (4.54). Kamb (1986), Kamb and Echelmeyer (1986a, 1986b) and Echelmeyer and Kamb(1986) give the most comprehensive treatment of these forces. For sheet flow, τ_s and $\partial u_x/\partial t$ are small, and $(\partial \sigma_{xx}/\partial x)\ \Delta x$ is small compared to τ_o if $\Delta x \geq 20\ h_I$ (Paterson, 1981, pp. 85–89, 98–101). Applying these constraints and summing longitudinal forces along x, with $(\partial \sigma_{xx}/\partial x)$ positive for extending converging flow and negative for compressive diverging flow and $(\partial u_x/\partial t)$ negligible even during ice advance and ice retreat:

$$(F_x)_G + (F_x)_D - (F_x)_T + (F_x)_B \approx (F_x)_G - (F_x)_T \approx 0 \tag{5.2}$$

Applying Equations (4.4) and (4.5) to Equations (5.1) and (5.2), when w is unit flowband width:

$$(F_x)_G = \Delta\bar{P} A_x = \rho_I\ g_z\ \Delta h \left(h_I + \tfrac{1}{2}\Delta h\right) \approx \tau_o\ \Delta x = (\sigma_{xz})_o\ A_z = (F_x)_T \tag{5.3}$$

where net average lithostatic pressure $\Delta\bar{P} = \rho_I\ g_z\ \Delta h$ acting on transverse area A_x of height $(h_I + 1/2\ \Delta h)$ and unit width causes ice motion that induces basal shear stress $\tau_o = (\sigma_{xz})_o$ acting on basal area A_z of length Δx and unit width.

Figure 5.2 is a geometric representation of Equation (5.3), in which $(F_x)_G$ is the longitudinal gravitational force causing downslope motion of the ice column and $(F_x)_T$ is the basal traction force induced by the motion. The two components of $(F_x)_G$ are $\rho_I\, g_z\, h_I\, \Delta h$ and $1/2\, \rho_I\, g_z\, \Delta h^2$ and are the areas of the rectangle and the triangle, respectively, containing the black arrowheads in Figure 5.2. The triangle is much smaller than the rectangle, so that component can be ignored in a first approximation. Therefore, Equation (5.3) can be solved for the surface slope α of the moving ice column:

$$\alpha = \frac{\Delta h}{\Delta x} \approx \frac{\tau_o}{\rho_I\, g_z\, h_I} \tag{5.4}$$

where τ_o is zero at the ice divide where α is zero and a maximum near the equilibrium line or at the head of an ice stream where the product $h_I\, \alpha$ is a maximum. A general increase of α and a decrease of h_I from the ice divide to the equilibrium line or the ice stream results in a convex flowline profile. However, τ_o decreases from the equilibrium line to the ice margin for sheet flow because decreasing ice velocity in the ablation zone reduces bed traction. For stream flow, τ_o decreases from a maximum where sheet flow becomes stream flow to zero where stream flow becomes shelf flow, presumably by reducing bed traction as accumulation of basal meltwater increases, culminating in the ice stream becoming afloat when $\tau_o = 0$. The decrease of τ_o in both cases causes h_I to decrease in the flow direction, producing a somewhat linear ramp in the ablation zone for sheet flow and a concave surface for stream flow.

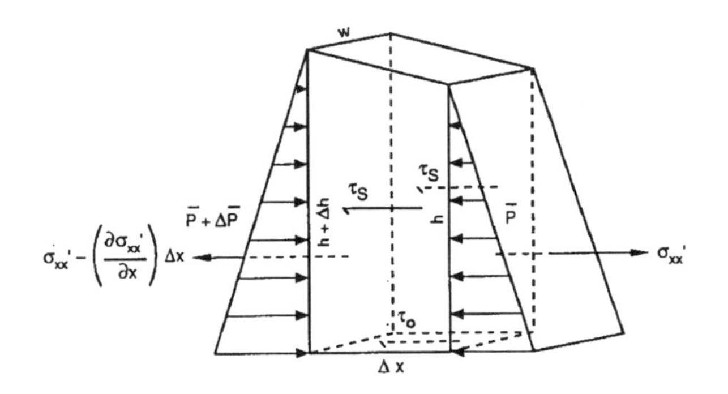

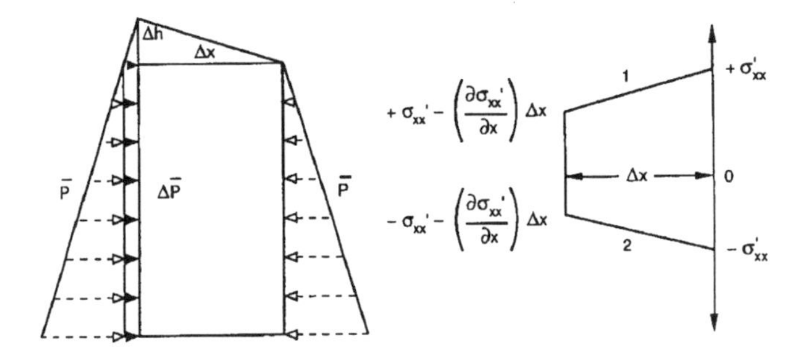

Figure 5.2: The force balance on an ice column for sheet flow. *Top*: Dimensions of the ice column, showing gravitational pressures $\bar{P}$ and $\bar{P} + \Delta\bar{P}$ on downslope and upslope column faces, causing downslope column motion that induces kinematic stresses τ_o for basal shear, τ_S for side shear, and σ_{xx}' for longitudinal extension (or compression). *Lower left*: Net mean gravitational downslope pressure $\Delta\bar{P}$ (black arrows) in excess of mean lithostatic pressure $\bar{P}$ (white arrows). *Lower right*: Longitudinal deviator stress variations along Δx for extending flow in the accumulation zone (line 1) and compressive flow in the ablation zone (line 2).

A first approximation of the convex flowline profile for sheet flow is obtained by integrating Equation (5.4) for constant τ_o and setting $h_I = h - h_R$, with bedrock elevation h_R being constant:

$$h = h_R + \left(2\,\tau_o / \rho_I\, g_z\right)^{1/2} x^{1/2} \tag{5.5}$$

Equation (5.5) gives a parabolic flowline profile.

In order to compute more precise flowline profiles, Equation (5.4) is written as an initial-value finite-difference recursive formula, similar to one proposed by Nye (1952):

$$h_{i+1} = h_i + \left(\frac{\tau_o}{h_I}\right)_i \frac{\Delta x}{\rho_I\, g_z} = \left(\frac{\tau_o}{h - h_R}\right)_i \frac{\Delta x}{\rho_I\, g_z} \tag{5.6}$$

where the flowline is divided into i steps of equal horizontal incremental length Δx, with basal shear stress τ_o, ice thickness h_I, ice elevation h, and bedrock elevation h_R varying along the flowline, for which $h = h_I + h_R$ and $\Delta h = h_{i+1} - h_i$, as shown in Figure 5.3. Orthogonal co-ordinates x, y, and z were chosen such that x extends horizontally along the flowline from $x = 0$ at the grounded ice margin to $x = L$ at the interior ice divide, y extends horizontally and parallels ice elevation contour lines, and z extends vertically up from $z = 0$ at present-day sea level.

The initial value needed to employ Equation (5.6) in reconstructing a flowline of a former ice sheet, step by step for $i = L/\Delta x$ steps, is ice elevation h_o at $i = 0$ for grounded marine margins and ice elevation h_1 at $i = 1$ for terrestrial margins. The value of h_o must be computed from the buoyancy requirement that the former ice sheet becomes afloat at the water depth chosen for the former marine margin. The value of h_1 can be computed at distance x from the ice margin by using Equation (5.5), provided that τ_o and h_R are constant over this distance. Since τ_o results from ice motion, slow ice flow caused by high ablation rates will reduce τ_o, and fast ice flow caused by low ablation rates will increase τ_o. An ice sheet advances for low ablation rates and retreats for high ablation rates, so $h = h_1$ at a given x in Equation (5.5) is higher for an advancing ice sheet than for a retreating ice sheet because τ_o is higher. Figure 5.4 illustrates these conditions.

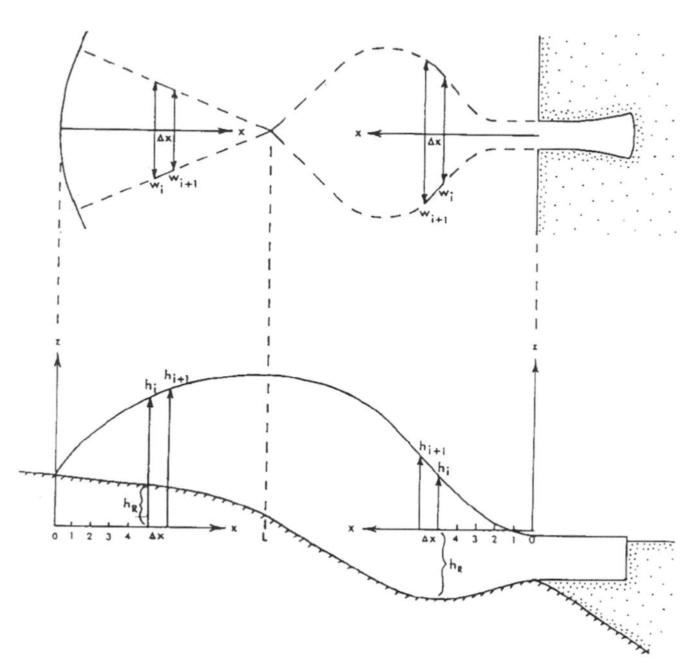

Figure 5.3: A numerical scheme for representing flowbands from the ice divide to terrestrial and marine margins of a former ice sheet. Sheet flow continues to the terrestrial margin (left) but becomes stream flow toward the marine margin (right), beyond which stream flow becomes shelf flow. *Top*: A plan view showing the variation of flowband width w for each step Δx. *Bottom*: A longitudinal cross section showing h as variations of ice-surface elevation and h_R as variations of bedrock elevation above or depth below present-day sea level for each step Δx. Present-day sea level is shown as higher than sea level when the former ice sheet existed. The origin of orthogonal coordinates x and z is at the terrestrial margin for the flowband down the left flank of the ice divide and at the marine grounding line for the flowband down the right flank of the ice divide. Flowbands of length L are divided into L/Δx steps, with integer i representing a given step.

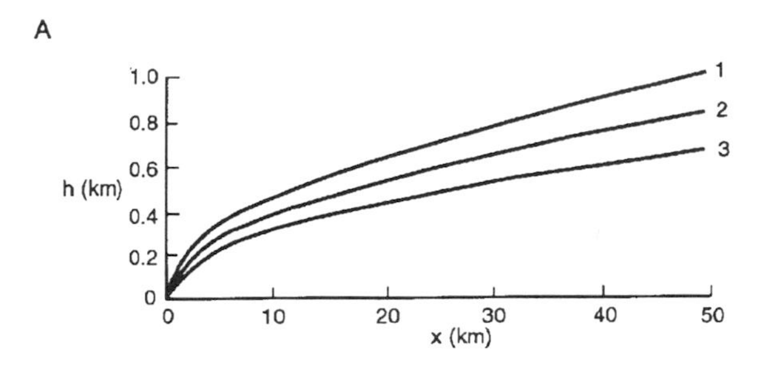

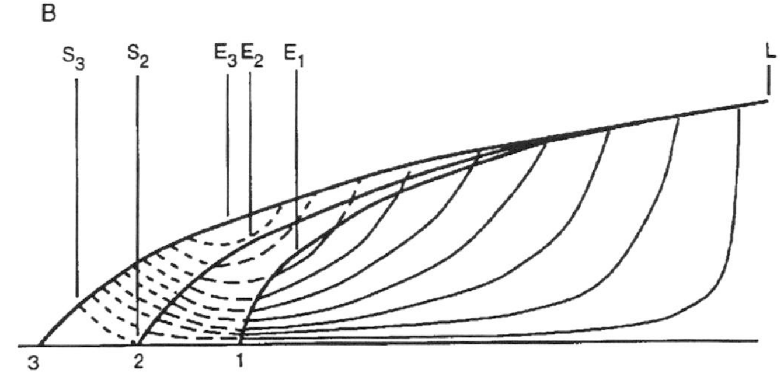

Figure 5.4: Parabolic profiles for a plastic ice sheet on a horizontal bed. (*A*) Profile 1: τ_o = 105 kPa for an advancing ice sheet. Profile 2: τ_o = 66 kPa for an equilibrium ice sheet. Profile 3: τ_o = 42 kPa for a retreating ice sheet. (*B*) Ice trajectories for ice sheets that are advancing (solid curves), in equilibrium (solid and broken curves), and retreating (solid, broken, and dashed curves). Equilibrium points are E_1 for advancing, E_2 for equilibrium, and E_3 for retreating ice sheets. Stagnation points are S_2 for equilibrium and S_3 for retreating ice sheets.

Isostatic Adjustments

Equation (5.6) reconstructs an ice sheet over bedrock without isostatically depressing the bed, with ice elevation h and bedrock elevation h_R referred to present-day sea level. This ice sheet would not be in isostatic equilibrium and, in time, ice elevation h would lower to elevation h* as the ice sheet isostatically depressed bedrock of elevation h_R to elevation h_R*. The counterpart of Equation (5.6) for isostatic depression is therefore:

$$h_{i+1}^* = h_i^* + \left(\frac{\tau_o}{h^* - h_R^*}\right)_i \frac{\Delta x}{\rho_I \, g_z} \tag{5.7}$$

Our task is to express Equation (5.7) in terms of h_R, which is known, instead of h_R*, which is unknown.

Correcting for isostatic adjustments beneath an ice sheet is difficult because equilibrium conditions of ice flow for given boundary conditions are attained much more rapidly than are equilibrium isostatic conditions beneath the ice sheet. This occurs because bedrock is much harder than ice, even for hot rock deep in Earth's mantle. Consequently, steady-state flowline profiles will not be attained by the ice sheet until isostatic equilibrium is attained in the bedrock, and this takes such a long time (many thousands of years) that other boundary conditions (surface temperature and mass balance, in particular) are bound to change owing to insolation variations caused by 23,000- and 41,000-year cycles in tilt and precession, respectively, of Earth's rotation axis. Other less predictable variations, such as atmospheric CO_2 and dust concentrations, can be relatively abrupt because they represent instabilities in Earth's climate machine.

Another problem is how to make isostatic corrections for a rugged subglacial landscape. In particular, thick ice over a valley will also depress adjacent hills covered by thinner ice, so a solution in which a column of ice depresses its basal area w Δx, as if this area were unconnected with adjacent areas, will be flawed. Nevertheless, we have adopted this procedure for the sake of simplicity, but we have used areas w Δx that are great enough to be largely "blind" to topographic variations within distances $w \geq \Delta x = 20$ km.

In our procedure, the effect of isostatic adjustments on ice-sheet elevation h is isolated from bedrock topographic effects h_R by considering isostatic depression of a flat horizontal bed, so that h_R is constant initially. This solution is then adapted to a bed with variable topography. One treatment of isostatic depression is represented schematically by Figure 5.5, in which elevations above present-day sea level have numbered subscripts to distinguish an ice sheet on an initially flat horizontal bed from one on a bed having topographic relief. The origin of axes x, z is at present-day sea level and the grounded margins of the former ice sheet, for both terrestrial and marine components of the ice sheet. Analyses are somewhat different for terrestrial and marine margins, but the results are identical for isostatic equilibrium.

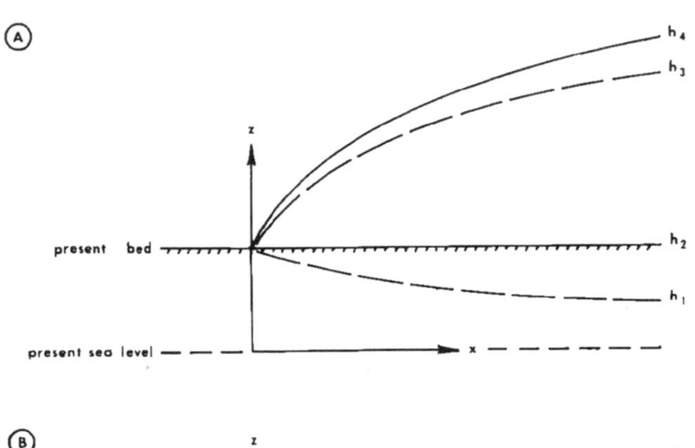

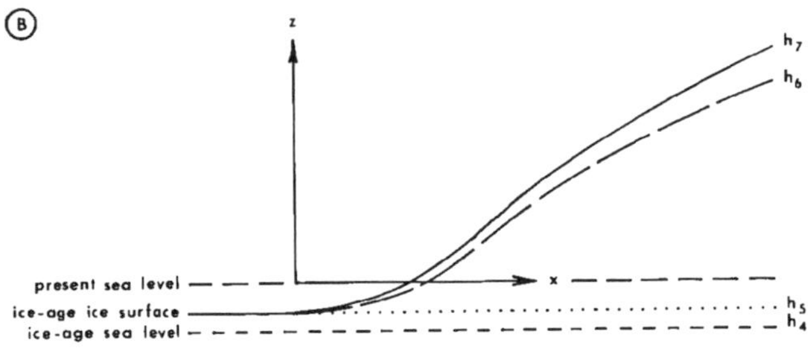

Figure 5.5: Glacioisostatic depression of an initially horizontal bed beneath an ice sheet. *Top*: For a terrestrial ice sheet, depression lowers the bed from h_2 to h_1 and this lowers the surface from h_4 to h_3. *Bottom*: For a marine ice sheet, depression lowers the bed from h_2 to h_1 and this lowers the surface from h_7 to h_6, where sea level has lowered to h_4 so the ice surface is h_5 and the bed is h_3 at the marine grounding line. All values of h are heights above or depths below present-day sea level. Note that $h_5 = h_6 = h_7$ and $h_1 = h_3$ at the marine grounding line.

Terrestrial margins

Consider the ice sheet in Figure 5.5A. An ice load of height $h_4 - h_2$ isostatically depresses the bed from height h_2 to height h_1, which lowers the ice surface from height h_4 to height h_3. The ratio r of lowered bed to lowered surface referred to the original flat horizontal bed is:

$$r = \frac{h_2 - h_1}{h_3 - h_2} \tag{5.8}$$

The longitudinal gravitational force $(F_x)_G$ before isostatic sinking is given by Equation (4.4) for a given point on the flowline:

$$(F_x)_G = \bar{P} A_x = \int_{h_R}^{h} \rho_I g_z (h - h_R)\, dh = \int_{h_2}^{h_4} \rho_I g_z (h_4 - h_2)\, dh_4$$

$$= \frac{1}{2} \rho_I g_z (h_4 - h_2)^2 \tag{5.9}$$

where h_2 is constant along x and transverse cross-sectional area A_x has unit width. After isostatic sinking, $(F_x)_G$ is given by:

$$(F_x)_G = \bar{P} A_x = \int_{h_R}^{h^*} \rho_I g_z (h^* - h_R^*)\, dh^* = \int_{h_2}^{h_3} \rho_I g_z (h_3 - h_1)\, dh_3$$

$$= \frac{1}{2} (1 + r) \rho_I g_z (h_3 - h_2)^2 \tag{5.10}$$

where h_1 varies along x and $h_3 - h_1 = (1 + r)(h_3 - h_2)$ by Equation (5.8). Equating Equations (5.9) and (5.10) relates r to ice-surface elevations above the margin before and after depression of the bed:

$$(h_4 - h_2) = (1 + r)^{1/2} (h_3 - h_2) \tag{5.11}$$

The basic assumption here is that $(F_x)_G$ is not affected by isostatic sinking (Weertman, 1964).

Marine margins

Consider the ice sheet in Figure 5.5B. A marine ice dome forms after sea ice or ice shelves from land ice grounds in response to ice thickening or lowering of sea level. Grounding occurs on a flat horizontal bed at depth h_2 below present-day sea level. Isostatic depression and ice doming begin with further thickening of ice and lowering of sea level. Since thickness $h_5 - h_3$ of grounded ice displace sea water and therefore does not contribute to isostatic depression of the bed, an ice load of height $h_7 - h_5$ above thickness $h_5 - h_3$ isostatically depresses the bed an amount $h_3 - h_1$ below the ice-age bed. Bed depression lowers the ice surface from h_7 to h_6 so the ratio of lowered bed to lowered surface is:

$$r = \frac{h_3 - h_1}{h_6 - h_5} \tag{5.12}$$

Recalling that h_0 is ice elevation at the grounded margins, $(F_x)_G$ for unit flowline width is:

$$(F_x)_G = \bar{P} A_x = \int_{h_0}^{h} \rho_I g_z (h - h_0) \, dh = \int_{h_5}^{h_7} \rho_I g_z (h_7 - h_5) \, dh_7$$

$$= \frac{1}{2} \rho_I g_z (h_7 - h_5)^2 \tag{5.13}$$

where h_5 is constant along x. $(F_x)_G$ after isostatic sinking is:

$$(F_x)_G = \bar{P} A_x = \int_{h_5}^{h_6} \rho_I g_z \left[(h_6 - h_5) + (h_3 - h_1) \right] dh_6$$

$$= \frac{1}{2} \rho_I g_z (1 + r) (h_6 - h_5)^2 \tag{5.14}$$

where h_I varies along x and $h_3 - h_I = r\,(h_6 - h_5)$ by Equation (5.12). Equating Equations (5.13) and (5.14) gives:

$$(h_7 - h_5) = (1 + r)^{1/2} (h_6 - h_5) \tag{5.15}$$

As with terrestrial margins, isostatic sinking is assumed to have no effect on $(F_x)_G$.

Grounding lines

Marine ice sheets become afloat and continue into the sea as floating ice shelves. Grounding lines of ice shelves lower as sea level lowers during growth of ice sheets, and rise as sea level rises during shrinkage of ice sheets, owing to the decrease and increase of the water overburden. Therefore, depth d_o of marine grounding lines below present-day sea-level changes continuously during a glaciation cycle. Referring to Figure 5.5B, the ratio of lowered bed to lowered surface at the grounding line is:

$$r = \frac{h_2 - h_1}{h_5 - h_2} = \frac{h_2 - h_3}{h_5 - h_2} \tag{5.16}$$

where $h_1 = h_3$ at the grounding line. Ice surface and ocean surface positions h_5 and h_4 are related by densities ρ_I and ρ_W of ice and water because the vertical lithostatic force of ice and the vertical hydrostatic force of water on the bed are identical at the grounding line. Therefore, from Equation (4.4):

$$(F_z)_G = P_I A_z = \rho_I (h_5 - h_3) = P_W A_z = \rho_W (h_4 - h_3) \tag{5.17}$$

where A_z is unit basal area at the grounding line. Solving Equations (5.14) and (5.15) to obtain h_o^*:

$$h_o^* = h_5 = h_3 + (\rho_W / \rho_I)(h_4 - h_3)$$

$$= \frac{(1 + r)(1 - \rho_W/\rho_I)\, h_2 + (\rho_W/\rho_I)\, h_4}{1 + r(1 - \rho_W/\rho_I)}$$

$$= \frac{(1 + r)(1 - \rho_W/\rho_I)\, h_R + (\rho_W/\rho_I)\, h_S^*}{1 + r(1 - \rho_W/\rho_I)} \tag{5.18}$$

and combining Equation (5.16) with Equation (5.18) gives h_R^*:

$$h_R^* = h_1 = h_3 = (1 + r)\, h_2 - r\, h_5$$

$$= \frac{(1 + r)\, h_2 - r(\rho_W/\rho_I)\, h_4}{1 + r(1 - \rho_W/\rho_I)}$$

$$= \frac{(1 + r)\, h_R - r(\rho_W/\rho_I)\, h_S^*}{1 + r(1 - \rho_W/\rho_I)} \tag{5.19}$$

where, at the grounding line of an ice-age ice shelf, $h_o^* = h_5$ is the ice-age ice-surface distance below present sea level, $h_R^* = h_1 = h_3$ is the ice-age bedrock distance below present sea level, $h_R = h_2$ is the present bedrock distance below present sea level, and $h_S^* = h_4$ is the ice-age sea-level distance below present sea level.

Isostatic equilibrium

Equations (5.11) and (5.15) for r were obtained by balancing forces in the x-direction. The maximum value of r is r_o, when isostatic equilibrium is complete, and is obtained by balancing vertical gravitational forces $(F_z)_G$ in the z-direction before and after the ice sheet exists. For terrestrial components, using Equation (5.8), hydrostatic pressure P_o acting on unit basal area A_z at depth h_1 is obtained from Equation (4.4) before the ice sheet exists:

$$(F_z)_G = P_o A_z = P_o = \int_{h_1}^{h_2} \rho_R g_z \, dz = \rho_R g_z (h_2 - h_1)$$

$$= \rho_R g_z r_o (h_3 - h_2) \tag{5.20}$$

where ρ_R is bedrock density, and P_o is also obtained from Equation (4.4) after the ice sheet is in isostatic equilibrium:

$$(F_z)_G = P_o A_z = P_o = \int_{h_1}^{h_3} \rho_I g_z \, dz = \int_{h_2}^{h_3} \rho_I g_z \, dz + \int_{h_1}^{h_2} \rho_I g_z \, dz$$

$$= \rho_I g_z \left[(h_3 - h_2) + (h_2 - h_1) \right] = \rho_I g_z (1 + r_o)(h_3 - h_2) \tag{5.21}$$

For marine components, using Equation (5.12), hydrostatic pressure P_o at depth h_1 is obtained from Equation (4.4) before the ice sheet exists:

$$(F_x)_G = P_o A_z = P_o = \int_{h_3}^{h_4} \rho_W g_z dz$$

$$+ \int_{h_1}^{h_3} \rho_R g_z dz = \rho_W g_z (h_4 - h_3) + \rho_R (h_3 - h_1)$$

$$= \rho_W g_z (h_4 - h_3) + \rho_R g_z r_o (h_6 - h_5) \tag{5.22}$$

where ρ_W is sea-water density, and P_o is also obtained from Equation (4.4) after the ice sheet is in isostatic equilibrium:

$$(F_x)_G = P_o A_z = P_o = \int_{h_1}^{h_6} \rho_I g_z dz$$

$$= \int_{h_5}^{h_6} \rho_I g_z dz + \int_{h_3}^{h_5} \rho_I g_z dz + \int_{h_1}^{h_3} \rho_I g_z dz$$

$$= \int_{h_5}^{h_6} \rho_I g_z dz + \int_{h_3}^{h_4} \rho_W g_z dz + \int_{h_1}^{h_3} \rho_I g_z dz$$

$$= \rho_I g_z (h_6 - h_5) + \rho_W g_z (h_4 - h_3) + \rho_I g_z r_o (h_6 - h_5) \tag{5.23}$$

Equating Equation (5.20) with Equation (5.21) and Equation (5.22) with Equation (5.23), both give:

$$r_o = \frac{\rho_I}{\rho_R - \rho_I} \tag{5.24}$$

Note that r_o is a constant.

Isostatic variations

Variations in glacial isostasy along flowlines occur as an ice sheet advances and retreats during a glaciation cycle. If isostatic adjustments begin at time t along a flowline and t_o = 7800 a is the relaxation time for isostatic adjustments (Andrews, 1970, p. 38), for an advancing ice sheet:

$$r = r_o (1 - e^{-t/to}) \tag{5.25}$$

where r = 0 at t = 0 when advance begins and $r = r_o$ at $t = \infty$, and for a retreating ice sheet:

$$r = r_o e^{-t/to} \tag{5.26}$$

where $r = r_o$ at t = 0 when retreat begins and r = 0 at $t = \infty$.

The value of r varies with distance x from the ice margin along a flowline of length L to the ice divide. Since the ice sheet advances and retreats, the ice sheet at its maximum extent is most likely to have $r = r_o$ at its center, where x = L and long-term glaciation favors a fully depressed bed, and is most likely to have r = 0 along its margins, where x = 0 and short-term glaciation favors an undepressed bed. If Earth's crust and mantle have a yield stress σ_o, it is also likely that dr/dx = 0 at x = 0 and x = L. A sine function for the variation of r with x that satisfies these boundary conditions for an advancing ice sheet is:

$$r = r_o \sin^c (\pi x/2 L) (1 - e^{-t/to}) \tag{5.27}$$

and for a retreating ice sheet is:

$$r = r_o \sin^c (\pi x/2 L) e^{-t/to} \tag{5.28}$$

where c = 0 if $\sigma_o = 0$ and $c = \infty$ if $\sigma_o = \infty$. If Earth's crust and mantle have a yield stress below which isostatic depression is minimal, little depression takes place near the margin where the ice load is less than this yield stress. A flow law for creep having this rheological property is given by Equation (4.136) and shown in Figure 4.19. Figure 5.6 shows how r varies with c and t/t_o along a flowline in Equations (5.27) and (5.28). These equations are empirical. Cathless (1975), Peltier (1976), Peltier and Andrews (1976), and Peltier and others (1978) pioneered theoretical glacial isostasy models based on the rheology of Earth's crust and mantle. Constants c and t_o could be computed from such models if they allowed viscoplastic creep in their lithosphere so that it underwent elastic-viscoplastic deformation with a yield stress (Hughes, 1981c).

Combining Topography and Isostasy

We are now ready to apply our isostasy criterion to bedrock having variable topographic relief so that we can reconstruct ice-age ice sheets, both former and future, on a bed having variable amounts of isostatic depression, using present-day bedrock topography as input. The procedure is to use Equations (5.11) and (5.15) to write Equation (5.7), using the present-day h_R values that appear in Equation (5.6). With h_R specified at each Δx step, elevations in Figure 5.5 must be given the following designations: $h = h_4$, $h^* = h_3$, $h_R = h_2$, and $h_R^* = h_1$ in Figure 5.5A and $h = h_7$, $h^* = h_6$, $h_R = h_2$, and $h_R^* = h_1$ in Figure 5.5B. Equations (5.11) and (5.15) then becomes:

$$(h - h_0) = (1 + r)^{1/2} (h^* - h_0) \tag{5.29}$$

where h_0 is the ice-surface elevation at the ice-sheet grounded margin, measured above present-day sea level in Figure 5.5A and below present-day sea level in Figure 5.5B. Solving Equation (5.29) for h and substituting the resulting expression into Equation (5.6) gives:

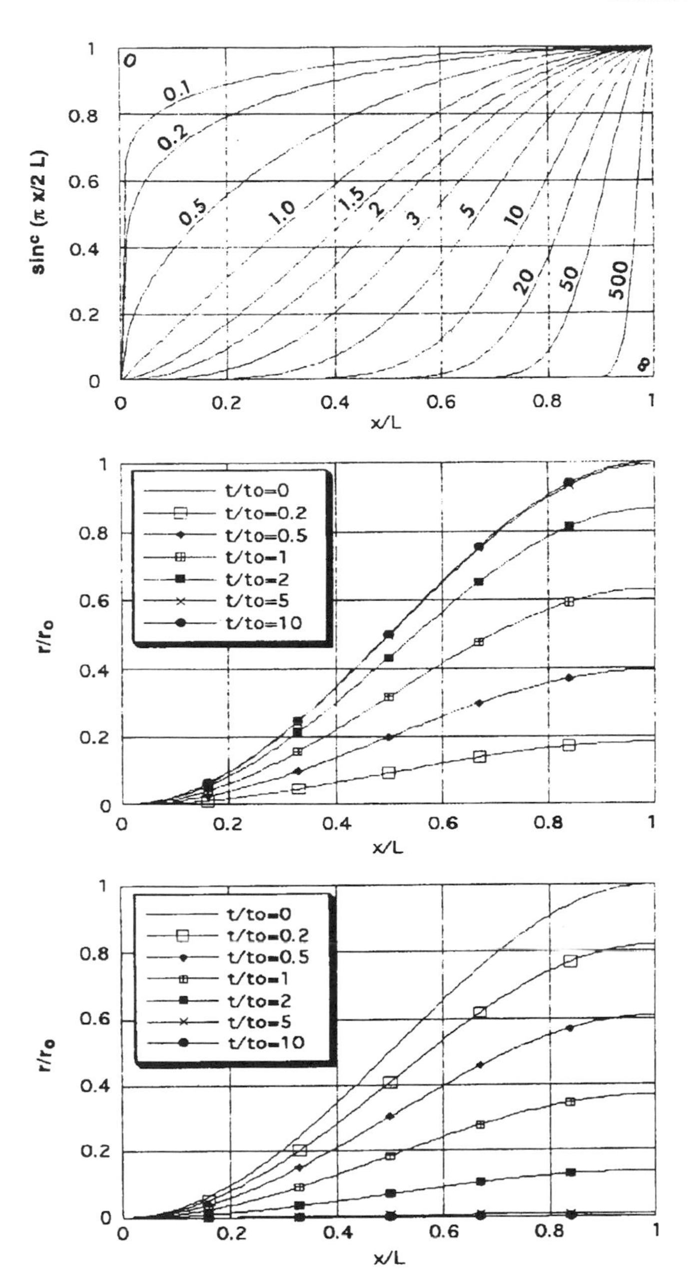

Figure 5.6: Glacioisostatic depression beneath an ice sheet along a flowline for dimensionless length x/L over dimensionless time t/t_o. *Top*: Plot of $\sin^c (\pi\, x/2\, L)$ versus x/L for $0 \leq c \leq \infty$ in Equations (5.27) and (5.28). *Middle*: Plot of Equation (5.27) for an advancing ice sheet when c = 2. *Bottom*: Plot of Equation (5.28) for a retreating ice sheet when c = 2. At isostatic equilibrium, exponents c and t_o both decrease with increasing size of an ice sheet. For small ice caps, $c \to \infty$. For large ice sheets, $c \to 0$. For the Scandinavian Ice Sheet, t_o = 4 ka. For the Laurentide Ice Sheet, t_o = 2 ka. (Reconstructing the last glacial and deglacial ice sheets, EOS, 74(36), 82–84, [22 February 1994].)

$$\left[h_0 + (1+r)^{1/2}\left(h^* - h_0\right)\right]_{i+1} = \left[h_0 + (1+r)^{1/2}\left(h^* - h_0\right)\right]_i$$

$$+ \left[\frac{\tau_o}{\left[h_0 + (1+r)^{1/2}\left(h^* - h_0\right)\right] - h_R}\right]_i \frac{\Delta x}{\rho_I\, g_z} \qquad (5.30a)$$

$$(1+r)^{1/2}\, h_{i+1}^* = (1+r)^{1/2}\, h_i^*$$

$$+ \left[\frac{\tau_o}{(1+r)^{1/2}\left(h^* - h_0\right) - \left(h_R - h_0\right)}\right]_i \frac{\Delta x}{\rho_I\, g_z} \qquad (5.30b)$$

$$h_{i+1}^* \approx h_i^*$$

$$+ \left[\frac{\tau_o}{(1+r)\left(h^* - h_0\right) - (1+r)^{1/2}\left(h_R - h_0\right)}\right]_i \frac{\Delta x}{\rho_I\, g_z} \qquad (5.30c)$$

where $[h_0 - (1+r)^{1/2} h_0]_i \approx [h_0 - (1+r)^{1/2} h_0]_{i+1}$ from the definition of h_0, and for a slow variation of r along x. An expression for h_R^* in terms of h_R is obtained by comparing Equation (5.30c) with Equation (5.7):

$$h_R^* = h^* - [(1+r)(h^* - h_0) - (1+r)^{1/2}(h_R - h_0)]$$

$$= h_0 + (1+r)^{1/2}(h_R - h_0) - r\,(h^* - h_0) \qquad (5.31)$$

Equations (5.30c) and (5.31) give ice-sheet surface elevations over an isostatically depressed bed having variable topography. Values of r entered at each Δx step can be chosen to satisfy any specified isostatic conditions; they need not satisfy Equation (5.27), (5.28), or any other particular formulations for r.

Ice-Bed Coupling

Having removed the influence of bed topography h_R and isostasy r on the basal shear stress τ_o in Equation (5.30c), further variations in τ_o can only reflect variations in ice-bed coupling. These variations are related to variations in ice velocity, which increases for surface accumulation and downslope convergence and decreases for surface ablation and downslope divergence. Convergence or divergence occur as ice thins or thickens and as flowbands narrow or widen. By removing the contributions of topography and isostasy, τ_o reflects ice-bed coupling on a flat horizontal bed, so downslope ice thickening is not allowed.

When basal ice is uniformly or discontinuously at its pressure melting point, τ_o variations reflect the changing amount and distribution of basal meltwater because basal shear stress cannot be transmitted across a basal water layer. Rates of basal melting and freezing at the bed control the amount and distribution of basal meltwater, and these rates are controlled by basal hydrology and the basal heat budget. A positive budget melts basal ice, creating basal meltwater. A negative budget freezes basal meltwater, creating refrozen ice, called regelation ice by glaciologists. Basal melting and freezing rates are determined by the geothermal heat flux, frictional heat from shear deformation in the ice, the ice thickness, the mean annual surface temperature, surface accumulation and ablation rates along the flowband, and advective flow from higher to lower elevations (Budd and others, 1971). By regulating the rate at which basal frictional heat is generated, ice velocity influences basal melting and freezing rates. The amount and distribution of basal meltwater beneath a former ice sheet can be inferred from the glacial erosion and deposition record on the present-day landscape, and appropriate theoretical variations of τ_o can be derived.

The continuity equation

A flowband lies between two adjacent flowlines, and it usually has a variable width. Derivations of τ_o variations along surface flowlines of an ice sheet begin with the continuity equation. For a flowband on a flat horizontal bed at sea level, so that $h = h_I$ in Equation (5.6), the continuity equations is:

$$a\,w\,dx - (\partial h/\partial t)\,w\,dx + d\,(u\,w\,h) = 0 \qquad (5.32)$$

where a is the time-invariant ice accumulation or ablation rate over a flowband of time-invariant width w and height h above the ice-sheet margin, $(\partial h/\partial t)$ is the change of h in time t when the ice sheet is not in equilibrium, and u is the mean horizontal ice velocity in the flowband, all measured at distance x from the ice margin, and a, w, u, and h all varying along x.

In Figure 5.7, with increasing distance from the ice margin at $x = 0$ to the ice divide at $x = L$, a nearly stagnant zone may exist for $0 < x < S$ and an active zone exists for $S < x < L$. The stagnant zone does not exist in a steady-state ice sheet, but it is present as part of the accumulation zone when the ice sheet advances by accretion at the margin and as part of the ablation zone when the ice sheet retreats by melting at the margin. The stagnant zone is absent when the ice sheet advances by flow at the margin. Under these conditions, Equation (5.32) for $x < S$ reduces to:

$$\int_0^x a\,w\,dx = \int_0^x (\partial h/\partial t)\,w\,dx \qquad (5.33)$$

and Equation (5.32) for $x > S$ reduces to:

$$\int_x^L a\,w\,dx = -\int_{uwh}^0 d\,(u\,w\,h) = u\,w\,h \qquad (5.34)$$

Assigning Equation (5.34) to $S < x < L$ treats this region as the steady-state core of an ice sheet. Although the $(\partial h/\partial t)$ w dx term of Equation (5.32) exists in this region, it is not important in determining the profile of the ice sheet so long as $\partial h/\partial t << a$. When accumulation and ablation rates such that $\partial h/\partial t = a$ become general over the ice sheet, then $S \rightarrow L$ and the ice thinning analysis by Paterson (1972) can be used.

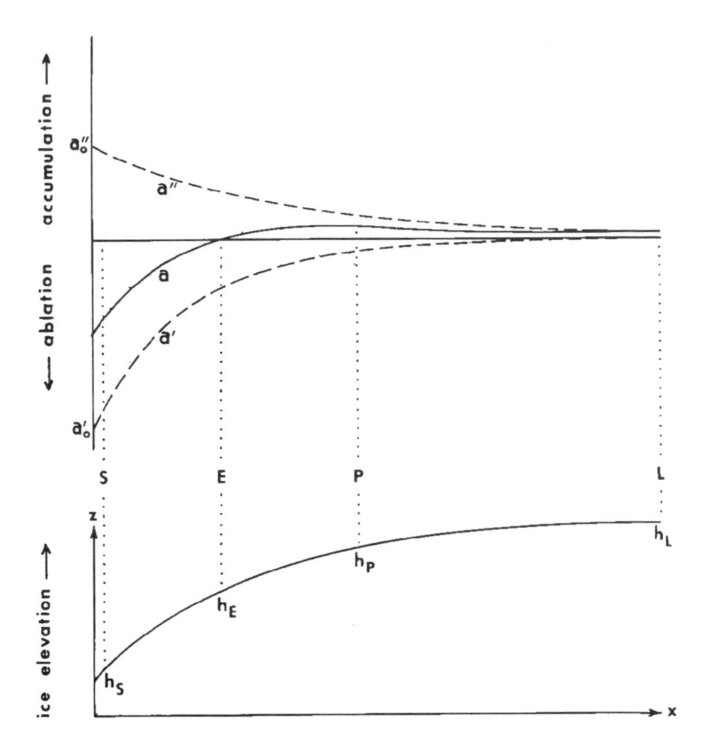

Figure 5.7: The variation of mass balance along a flowline of an ice sheet. *Top*: Net accumulation or ablation rate a is the sum of positive accumulation rate a" and negative ablation rate a', with $a'' = a_o''$ and $a' = a_o'$ at the ice margin. *Bottom*: The net mass balance is $\int_0^S a\,dx = 0$ at stagnation point $x = S$ and stagnation elevation h_S. Accumulation and ablation rates give $a = 0$ at equilibrium point $x = E$ and equilibrium height h_E. Peak accumulation where $da/dx = 0$ occurs at distance $x = P$ and height h_P. The ice divide at $x = L$ has height h_L.

Accumulation and ablation

Ice velocity is increased by accumulation and decreased by ablation. As shown in Figure 5.7, surface accumulation and ablation zones are separated by an equilibrium line at distance $x = E$ from the margin of an ice sheet. The equilibrium line connects the intersections of snowlines with flowlines on the ice-sheet surface, with precipitation falling as snow above the snowline and as rain below the snowline. Precipitation is brought to the ice sheet by convective storm systems which are unable to climb over large ice sheets. They vanish shortly after they release their peak snow precipitation at some distance $x = P$ from the ice-sheet margin. Thereafter, much smaller "no-cloud precipitation" occurs as a continual rain of ice crystals from high-altitude air that is drawn in to replace air flowing down the flanks of the ice divide and funneled into ice streams as katabatic winds.

The changing mass balance along a flowline can be represented as the sum of a negative ablation rate a' and a positive accumulation rate a" which decay exponentially with distance x from the ice-sheet margin (see Figure 5.7) (alternatively, a' and a" could decay exponentially with height h above the ice-sheet margin):

$$a = a' + a'' = a_o' \, exp \, (-\lambda' \, x^k) + a_o'' \, exp \, (-\lambda'' \, x^k) \quad (5.35)$$

where coefficients a_o' and a_o'', damping factors λ' and λ'', and power k are positive constants. Constant k is slightly greater than unity when regelation ice outcrops in the ablation zone and deposits glacially eroded material that suppresses ablation or when katabatic winds transport snow precipitation from the accumulation zone to the foot of the ablation zone. An ice-sheet margin has more windblown snowdrifts during advance and more surface debris during retreat. The position of E for constant w is obtained by setting $a = 0$ at $x = E$ in Equation (5.35) and solving for E:

$$E = \left[\frac{\ln (- a_o'/a_o'')}{\lambda' - \lambda''} \right]^{1/k} \quad (5.36)$$

The position of P for constant w is obtained by setting $da/dx = 0$ at $x = P$ in Equation (5.35) and solving for P:

$$P = \left[\frac{\ln (- \lambda' \, a_o'/\lambda'' \, a_o'')}{\lambda' - \lambda''} \right]^{1/k} \quad (5.37)$$

The position of S for constant w is obtained from mass-balance equilibrium for a flowband of length L when $k = 1$:

$$\int_S^L a \, dx = (a_o'/\lambda') \, (e^{-\lambda' S} - e^{-\lambda' L}) + (a_o''/\lambda'') \, (e^{-\lambda'' S} - e^{-\lambda'' L}) = 0 \quad (5.38)$$

There is no simple expression for S and no analytical solution when $k > 1$. Fortunately, $k = 1$ gives a reasonable fit to most mass-balance data. Numerical integrations are employed to determine E, P, and S when w varies with each Δx step and $k > 1$.

Mass-balance conditions of particular interest are $S < 0$ for an advancing ice sheet, $0 < S < E$ for a retreating ice sheet, $S = 0$ for an equilibrium ice sheet, $S = L$ for a stagnant ice sheet, $P = L$ for an ice sheet low enough to be traversed by convective storm systems, and $E < P < L$ for an ice sheet too high to be traversed by these systems.

Stagnant ice

Stagnant ice exists only for $0 \le x \le S$ in the ablation zone of a retreating ice sheet, a region for which $\partial h/\partial t = a$ and $\tau_o << \sigma_o$. Here τ_o is a constant basal shear stress appearing in Equation (4.136) and σ_o is the plastic yield stress appearing in Equation (4.59). Hence, a stagnating ice margin has a low parabolic profile given by Equation (5.5), with a net ablation rate given by Equation (5.32) that lowers the ice surface and retreats the ice margin over time. Adding Equations (5.5) and (5.32) gives the lowering ice surface over time δt:

$$h + (dh/dt) \, \delta t = h + a \, \delta t = h_R + (2 \, \tau_o \, x/\rho_I \, g_z)^{1/2} + [a_o' \, exp \, (- \lambda' \, x^k) + a_o'' \, exp \, (- \lambda'' \, x^k)] \, \delta t \quad (5.39)$$

where $\partial h/\partial t$ becomes dh/dt, and is negative because a is negative for $0 \le x \le S$.

Active ice

Active ice exists when $S < x < L$, a region where τ_o increases with increasing ice velocity and is comparable to σ_o. Flowband profiles can be reconstructed from the continuity condition in the region $x > S$ where $|a| > |dh/dt|$. Taking $x = 0$ at the ice margin and $x = L$ at the ice divide, a mass–balance function $\Gamma(x)$ can be defined from Equation (5.32):

$$\Gamma(x) = \int_x^L (a - dh/dt) \, w \, dx = \sum_{i=x/\Delta x}^{i=L/\Delta x} \left[(a_i - \delta h_i/\delta t) \, w_i \right] \Delta x \quad (5.40)$$

In general, a, dh/dt, and w all vary along x. Numerical solutions of Equation (5.40) exist for frozen and thawed beds. A numerical solution breaks x into i steps, each of length Δx, with a_i and w_i specified at each step in order for h_i and $\delta h_i/\delta t$ to be determined at the step. A summation over steps replaces integration over x.

A thawed bed permits basal sliding over bedrock and over sediments or till, and permits deformation of the sediment or till. Bed deformation is highly complex and poorly understood, as discussed in Chapter 4. Both basal sliding and bed deformation allow basal ice to move at velocity u_o, which is the important contribution that both make to ice dynamics, whatever the need for distinguishing between the two. For these reasons, only basal sliding will be considered in the analysis for thawed beds. Mass-balance velocity u at distance x measured from the ice margin along a flowline centered in a flowband of variable width w is the sum of basal sliding velocity u_S and average creep velocity u_C in ice of height h above the bed. Using Equation (4.26) for u_S when $P_W << P_I$ during sheet flow and Equation (4.153) for u_C, the ice flux at x obtained from integrating Equation (5.32) is:

$$\Gamma(x) = \int_x^L (a - dh/dt) \, w \, dx = - \int_{uwh}^0 d \, (u \, w \, h) = \int_0^{uwh} d \, (u \, w \, h)$$

$$= u \, w \, h = f_S \, u_S \, w \, h + f_C \, u_C \, w \, h$$

$$= f_S \, w \, h \left(\frac{\tau_o}{B} \right)^m + f_C \frac{2 \, w \, h^2}{n + 2} \left(\frac{\tau_o}{A} \right)^n \quad (5.41)$$

where f_S and f_C are the respective fractions of u due to basal sliding and internal creep, and $f_S + f_C = 1$ by definition.

A frozen bed

For a frozen bed, $f_S = 0$ and $f_C = 1$. Equation (5.41) then reduces to:

$$\Gamma(x) = u_C\, w\, h = \left[\frac{2}{n+2} \left(\frac{\rho_I\, g_z\, \alpha}{A} \right)^n w\, h^{n+2} \right] \qquad (5.42)$$

The surface slope at distance x from the margin is:

$$\alpha = \frac{dh}{dx} = \frac{A}{\rho_I\, g_z} \left[\frac{(n+2)\, \Gamma(x)}{2\, w(x)\, h^{n+2}} \right]^{1/n} \qquad (5.43)$$

where $w = w(x)$ is specified at distance x. Integrating Equation (5.43) for active ice on a flat horizontal bed:

$$\int_{h_S}^{h} h^{\frac{n+2}{n}}\, dh = \int_S^x \frac{A}{\rho_I\, g_z} \left[\frac{(n+2)\, \Gamma(x)}{2\, w(x)} \right]^{1/n} dx = C_f\, \Pi_f(x)$$

$$h = \left[h_S^{\frac{2n+2}{n}} + \left(\frac{2\,n+2}{n} \right) C_f\, \Pi_f(x) \right]^{\frac{n}{2n+2}} \qquad (5.44)$$

where C_f is a frozen bed constant defined by:

$$C_f = \left(\frac{n+2}{2} \right)^{1/n} \frac{A}{\rho_I\, g_z} \qquad (5.45)$$

and $\Pi_D(x)$ is a frozen bed function defined by:

$$\Pi_f(x) = \int_S^x \left[\frac{\Gamma(x)}{w(x)} \right]^{1/n} dx = \sum_{i=S/\Delta x}^{i=x/\Delta x} \left[\frac{\Gamma(i)}{w(i)} \right]^{1/n} \Delta x \qquad (5.46)$$

Basal shear stress $(\tau_o)_C$ for creep over a frozen bed is obtained by substituting Equations (5.43) and (5.44) into Equation (4.36) for $h = h_I$ when the bed is horizontal:

$$(\tau_o)_C = \rho_I\, g_z\, h\, \alpha = \frac{\rho_I\, g_z\, C_f \left[\frac{\Gamma(x)}{w(x)} \right]^{1/n}}{\left[h_S^{\frac{2n+2}{n}} + \frac{2\,n+2}{n} C_f\, \Pi_f(x) \right]^{\frac{1}{n+1}}} \qquad (5.47)$$

where $\tau_o = (\tau_o)_C$ because mass-balance velocity u results from creep in ice.

A thawed bed

For a thawed bed, $f_S > 0$ and $f_C < 1$. Equation (5.41) as $f_S \rightarrow 1$ and $f_C \rightarrow 0$ becomes:

$$\Gamma(x) \approx u_S\, w\, h = \left(\frac{\rho_I\, g_z\, \alpha}{B} \right)^m w\, h^{m+1} \qquad (5.48)$$

the surface slope at distance x from the margin is:

$$\alpha = \frac{dh}{dx} = \frac{B}{\rho_I\, g_z} \left[\frac{\Gamma(x)}{w(x)\, h^{m+1}} \right]^{1/m} \qquad (5.49)$$

where $w = w(x)$ is specified at distance x. Integrating Equation (5.49) for a flat horizontal bed:

$$\int_{h_S}^{h} h^{\frac{m+1}{m}}\, dh = \int_S^x \frac{B}{\rho_I\, g_z} \left[\frac{\Gamma(x)}{w(x)} \right]^{1/m} dx = C_t\, \Pi_t(x)$$

$$h = \left[h_S^{\frac{2m+1}{m}} + \left(\frac{2\,m+1}{m} \right) C_t\, \Pi_t(x) \right]^{\frac{m}{2m+1}} \qquad (5.50)$$

where C_t is a thawed bed constant defined by:

$$C_t = \frac{B}{\rho_I\, g_z} \qquad (5.51)$$

and $\Pi_t(x)$ is a thawed bed function defined by:

$$\Pi_t(x) = \int_S^x \left[\frac{\Gamma(x)}{w(x)} \right]^{1/m} dx = \sum_{i=S/\Delta x}^{i=x/\Delta x} \left[\frac{\Gamma(i)}{w(i)} \right]^{1/m} \Delta x \qquad (5.52)$$

Basal shear stress $(\tau_o)_S$ for sliding over a thawed bed is obtained by substituting Equations (5.49) and (5.50) into Equation (4.36) for $h = h_I$ when the bed is horizontal:

$$(\tau_o)_S = \rho_I\, g_z\, h\, \alpha = \frac{\rho_I\, g_z\, C_t \left[\frac{\Gamma(x)}{w(x)} \right]^{1/m}}{\left[h_S^{\frac{2m+1}{m}} + \left(\frac{2\,m+1}{m} \right) C_t\, \Pi_t(x) \right]^{\frac{1}{2m+1}}} \qquad (5.53)$$

where $(\tau_o)_S$ is the part of τ_o that results from basal traction when the bed is thawed. The remainder of τ_o is $(\tau_o)_C$ and causes creep in ice above the bed.

A requirement of Equations (5.47) and (5.53) is that $\tau_o \geq 0$, which requires that $\Gamma(x) \geq 0$. This is achieved by setting $\Gamma(x) = 0$ and solving Equation (5.40) for $x = S$. Since $\Gamma(x) > 0$ for $S < x \leq L$, negative values of $\Gamma(x)$ are eliminated by moving S toward the ice divide until $\Gamma(x) = 0$. For an ice sheet in equilibrium, $S = 0$ and $dh/dt = 0$ so that from Equations (5.35) and (5.40) when $k = 1$ and w is constant:

$$\begin{aligned} \Gamma(x) &= \int_x^L a\, dx = -\int_S^x a\, dx \\ &= (a_o'/\lambda')\,(e^{-\lambda' x} - e^{-\lambda' S}) + (a_o''/\lambda'')\,(e^{-\lambda'' x} - e^{-\lambda'' S}) \\ &= (a_o'/\lambda')\,(e^{-\lambda' x} - 1) + (a_o''/\lambda'')\,(e^{-\lambda'' x} - 1) \end{aligned} \qquad (5.54)$$

Constants a_o', a_o'', λ', and λ'' locate the equilibrium line and the peak accumulation line through Equations (5.36) and (5.37) and locate stagnation line S through Equation (5.38).

Melting and freezing beds

Ice moving from a frozen bed to a thawed bed passes over a melting bed that begins as scattered thawed patches and ends as scattered frozen patches. Ice moving from a thawed bed to a frozen bed passes over a freezing bed that begins as scattered frozen patches and ends as scattered thawed patches. Basal melting or freezing occur as heat, primarily the frictional heat that varies with ice velocity, is added or removed from basal ice. Consequently, a melting bed is linked to surface accumulation and converging flow, whereas a freezing bed is linked to surface ablation and diverging flow. Sheet flow from the ice divide diverges from domes and converges from saddles and converges on ice streams but diverges along ice ridges between ice streams.

In sheet flow, ice moving down the flanks of the ice divide can cross a melting bed that is already partly thawed at the ice divide, and basal freezing can begin before basal melting is complete. Under these general conditions, the fractional part of the bed that is thawed or frozen must be specified at each Δx step, just as flowband width and surface accumulation or ablation rates must be so specified. The appropriate expression for τ_o at each Δx step in Equation (5.30c) is then:

$$\tau_o = f_t\,(\tau_o)_S + f_f\,(\tau_o)_C \tag{5.55}$$

where f_t is the thawed fraction, f_f is the frozen fraction, and $f_t + f_f = 1$ by definition. These fractions must be determined from the glacial geological record when a former ice sheet is being reconstructed. Assuming that basal sliding is possible only when the bed is thawed, $f_t = 1$ over regions where exposed bedrock is striated and $f_f = 0$ over regions having unstriated bedrock. Equation (5.7) now computes ice-flowline profiles for sheet flow that take into account subglacial topography h_R, glacioisostatic adjustments r, and basal shear stress τ_o caused by variable ice-bed coupling, where ice-bed coupling is linked to variable surface accumulation or ablation rates and flowband widths that determine whether the bed is frozen, thawed, or melting. With these variables included, Equation (5.7) becomes:

$$h_{i+1}^* = h_i^* + \left[\frac{f_t\,(\tau_o)_S + (1-f_t)\,(\tau_o)_C}{(1+r)\,(h^* - h_o) - (1+r)^{1/2}\,(h_R - h_o)}\right]_i \frac{\Delta x}{\rho_I\, g_z} \tag{5.56}$$

where $(\tau_o)_C$ and $(\tau_o)_S$ are given by Equations (5.47) and (5.53), respectively.

Sliding on a thawed bed

Even though basal shear stress τ_o cannot be transmitted across a basal water layer on a smooth bed, bed roughness provides basal traction because bedrock bumps retard sliding by projecting up into basal ice. The sliding law given by Equation (4.26) quantifies this bed traction as a basal shear stress $(\tau_o)_S$ that resists gravitational sliding. Therefore, although a frozen bed supports only a τ_o that resists gravitational creep in the overlying ice, a thawed bed supports a τ_o that consists of a sliding component $(\tau_o)_S$ and a creep component $(\tau_o)_C$. The mass-balance function $\Gamma(x)$ defined by Equation (5.41) equals mass flux u w h, with u having components u_S and u_C due to basal sliding and internal creep and h being less for a low-traction thawed bed than for a high-traction frozen bed. Mass-balance velocity u is therefore:

$$u = f_S\, u_S + f_C\, u_C = \Gamma(x)/w\,h$$
$$= f_S \left(\frac{\tau_o}{B}\right)^m + f_C \frac{2\,h}{n+2}\left(\frac{\tau_o}{A}\right)^n \tag{5.57}$$

where f_S and f_C are fractions of u due to sliding and creep and u_C is u_x averaged through h. If f_t and f_f are thawed and frozen fractions of a horizontal bed, Equation (5.55) and (4.36) give:

$$\tau_o = f_t\,(\tau_o)_S + f_f\,(\tau_o)_C = \rho_I\, g_z\, h\, \alpha \tag{5.58}$$

By definition, $f_S + f_C = 1$ and $f_t + f_f = 1$. Since a thawed bed still provides traction, and therefore u_C occurs in addition to u_S, $f_S < 1$ even when $f_t = 1$ for sheet flow. However, $f_S \rightarrow 1$ for stream flow and $f_S = 1$ for shelf flow.

A relationship between f_S and f_t is desirable for sheet flow because f_t determines how much of the bed can be subjected to glacial erosion and deposition, whereas f_S determines the rates of glacial erosion and deposition. Glacial erosion and deposition strong enough to produce glacial geology occur only when ice slides on a thawed bed. Substituting Equation (5.58) into Equation (5.57):

$$u = f_S\left[\frac{f_t\,(\tau_o)_S + f_f\,(\tau_o)_C}{B}\right]^m + f_C\,\frac{2\,h}{n+2}\left[\frac{f_t\,(\tau_o)_S + f_f\,(\tau_o)_C}{A}\right]^n = \frac{\Gamma(x)}{w\,h} \tag{5.59}$$

Setting $f_C = 1 - f_S$ and $f_f = 1 - f_t$ and solving for f_S:

$$f_S = \frac{\dfrac{2\,h}{n+2}\left[\dfrac{f_t\,(\tau_o)_S + (1-f_t)\,(\tau_o)_C}{A}\right]^n - \dfrac{\Gamma(x)}{w\,h}}{\dfrac{2\,h}{n+2}\left[\dfrac{f_t\,(\tau_o)_S + (1-f_t)\,(\tau_o)_C}{A}\right]^n - \left[\dfrac{f_t\,(\tau_o)_S + (1-f_t)\,(\tau_o)_C}{B}\right]^m} \tag{5.60}$$

When $f_S = 0$, Equation (5.60) reduces to:

$$\frac{\Gamma(x)}{w\,h} = \frac{2\,h}{n+2}\left[\frac{(\tau_o)_C}{A}\right]^n \tag{5.61}$$

When $f_S = 1$, Equation (5.60) reduces to:

$$\frac{\Gamma(x)}{w\,h} = \left[\frac{(\tau_o)_S}{B}\right]^m \quad (5.62)$$

Since $(\tau_o)_C > 0$ when $f_S = 1$, Equation (5.62) is a limiting case of sheet flow for which $(\tau_o)_C \to 0$ as $f_S \to 1$. This limiting case exists near ice divides, where $\Gamma(x) \to 0$.

To illustrate the relationship between f_S and f_t, consider an ice sheet in which frictional heat caused by ice shearing and sliding over the bed controls the distribution of frozen and thawed patches on the bed and melting and freezing rates in the thawed patches. A melting zone, across which isolated thawed patches grade into isolated frozen patches, exists from the ice divide to the equilibrium line because ice velocity increases in the accumulation zone and generates more frictional heat at the bed. A freezing zone, across which isolated frozen patches grade into isolated thawed patches, exists from the equilibrium line to the ice margin because ice velocity decreases in the ablation zone and generates less frictional heat at the bed. Melting and freezing zones consist of a mosaic of frozen and thawed patches, with melting taking place in the back part and freezing taking place in the front part of a thawed patch. This is necessary to keep the patches from growing until the whole bed is thawed or shrinking until the whole bed is frozen. Growth or shrinkage of patches continues until they attain a stable size and distribution compatible with an ice sheet in steady-state equilibrium. Since an ice sheet is never completely in equilibrium, the size and distribution of frozen and thawed patches changes constantly, with faster departure from equilibrium allowing faster changes.

Figure 5.8 is a plot of Equation (5.60) for an ice sheet in steady-state equilibrium for which the equilibrium line at x = E is one-third of the distance to the ice divide at x = L, using Equation (5.41) for dh/dt = 0 and Equation (5.35) for w = 1, k = 1, $a_O' = -450$ cm/a, $a_O'' = +150$ cm/a, $\lambda' = 0.003$, $\lambda'' = 0.0007$, and L = 3 E = 1500 km. Equations (5.47) and (5.53) give $(\tau_o)_C$ and $(\tau_o)_S$ variations along x for $h_S = 0$ when the ice sheet is in equilibrium. Note the strong non-linearity between f_S and f_t, with $f_S = f_t$ at x = 0 and x = L, but $f_S < f_t$ when $0 < x < L$. Basal sliding becomes more important than internal creep ($f_S > 0.5$) when more than 90 percent of the bed is thawed ($f_t > 0.9$). This condition exists in the vicinity of the equilibrium line, where the mass-balance velocity is greatest.

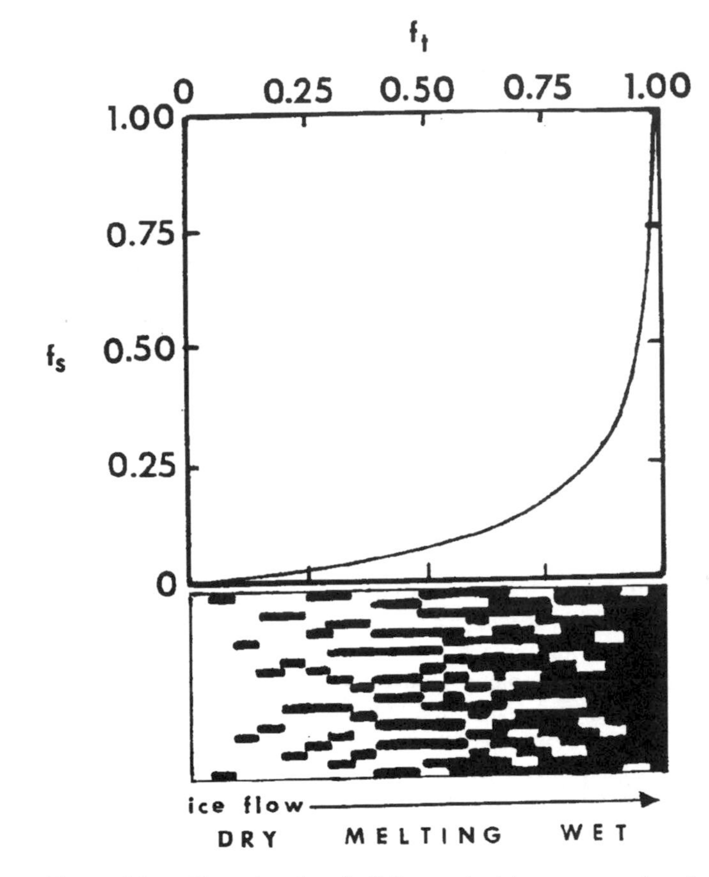

Figure 5.8: Changing basal sliding velocities across a basal melting zone. *Top*: The variation of sliding fraction f_S of ice velocity with thawed fraction f_t of the bed beneath an ice sheet for $f_S = f_t = 0$ at the ice divide and $f_S = f_t = 1$ at the equilibrium line in Equation (5.59). In Equation (5.59), n = 2, m = 3, A = 470 kPa $a^{1/3}$, B = 2 kPa $a^{1/2}$ $m^{-1/2}$, w = 1 m, h is given by Equation (5.56) for r = h_o = 0, and $\Gamma(x)$ is given by Equation (5.54). *Bottom*: Thawed patches (black) grow across a melting zone and merge to produce frozen patches (white). Ice flow would be from right to left across a freezing zone (compare with Figure 3.10, where h and u are constant).

Analytical Solutions

A model is most valuable if it adequately represents reality and can be used easily. This balance between complexity and simplicity is difficult to attain but is a worthwhile goal in any modeling program. For Quaternary glacial geologists who work in the field and who have little or no knowledge of computer programing, the best model for reconstructing flowline profiles of former ice sheets is one with simple equations that can be solved using a hand calculator. A step toward this ideal is the CLIMAP ice-sheet reconstruction model. The major complexities which are sacrificed for simplicity in the CLIMAP model are the effects that converging or diverging flow and accumulation or ablation can have on flowline profiles. These effects are treated quite accurately by introducing the functions $\Gamma(x)$ and $\Pi(x)$ to obtain values of basal shear stress τ_o in Equations (5.47) and (5.53). However, $\Gamma(x)$ and $\Pi(x)$ must be computed numerically at each Δx step along the flowline (Fastook, 1984; Hughes, 1985).

Simple analytical solutions of $\Gamma(x)$ and $\Pi(x)$ can be obtained for radially converging or diverging flow and for constant accumulation or ablation rates. Field geologists and atmospheric scientists have only an approximate understanding of these conditions for former ice sheets, based on the glacial geological record and on general circulation models, so approximate treatments of this input to the model are acceptable.

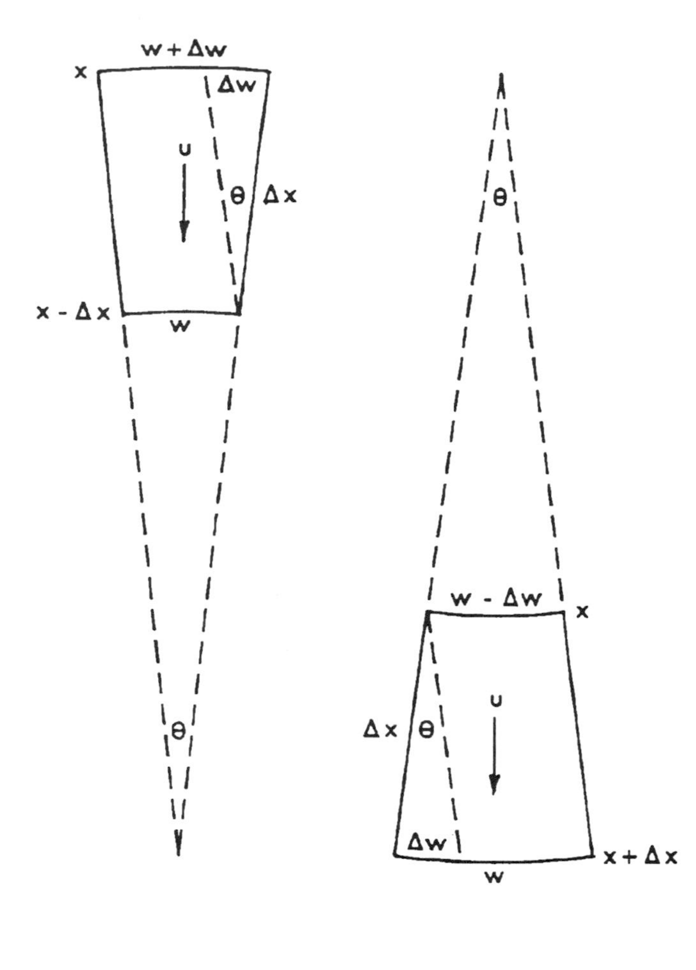

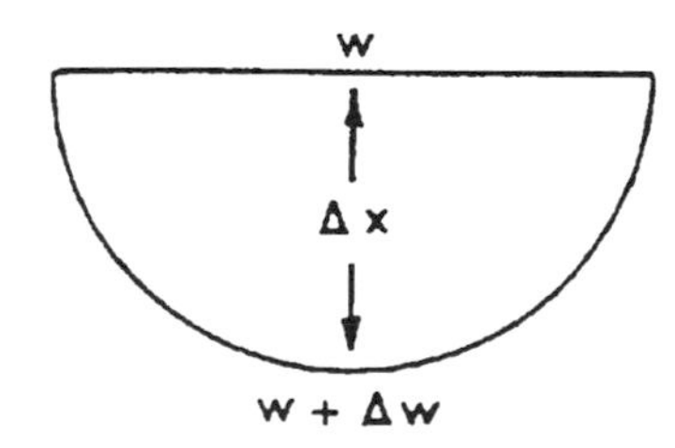

Figure 5.9: Variable flowband width for radial divergence from a dome on the ice divide and for radial convergence from a saddle on the ice divide. *Upper left*: Radial convergence down the flank of a saddle. *Upper right*: Radial divergence down the flank of a dome. *Bottom*: Radial divergence from the crest of an ice divide at a saddle.

The analytical solutions are exact treatments of approximations to reality, whereas the numerical solution is an approximate treatment of reality as it is. Functions $\Gamma(x)$ and $\Pi(x)$ in the numerical solution treat the effects of convergence or divergence by specifying the flowband width w at each Δx step and treat the effects of accumulation or ablation by specifying the accumulation or ablation rate a at each Δx step. Analytical solutions for which radial convergence or divergence approximates actual convergence or divergence in three basic kinds of ice-sheet flowbands will be presented. A flowband for sheet flow from a saddle converges steadily from the ice divide to the ice margin. A flowband for sheet flow from a dome diverges steadily from the ice divide to the ice margin. A flowband for sheet flow from a dome to an ice stream first diverges and then converges from the ice divide to the ice margin. Analytical solutions showing the degree to which constant accumulation and ablation rates approximate the actual distribution of accumulation and ablation rates over an ice sheet for peak accumulation at the ice divide, and between the equilibrium line and the ice divide, will also be presented.

Radial convergence and divergence

Equation (5.47) is for two-dimensional flow along a flowline if w is constant. The three-dimensional effect of lateral convergence and divergence can be incorporated into two-dimensional flow by including a convergence-divergence factor which causes a flowband of width w to narrow or widen an amount Δw in a Δx step of Equation (5.47). As shown in Figure 5.9, if w and $w \pm \Delta w$ curve to be parallel with surface-elevation contours, the angle θ between the sides of the flowband is:

$$\theta = \frac{w \pm \Delta w}{x} \approx \frac{\Delta w}{\Delta x} \tag{5.63}$$

where x is measured from the projected apex of a flowband segment Δx that widens or narrows an amount Δw. The plus sign in Equation (5.63) is for convergence and the minus sign is for divergence.

An ice accumulation or ablation rate a over the flowband segment causes a mass flux change $\Delta(u\,w\,h)$ in length Δx. The area of the flowband segment is the difference between areas of the large and small sectors of the circles having radii x and $x - \Delta x$ for converging flow and radii $x + \Delta x$ and x for diverging flow in Figure 5.9. Mass balance exists if ice flux for both converging and diverging flow in distance Δx changes by:

$$\begin{aligned}
\Delta(u\,w\,h) &= w\,\Delta(u\,h) + u\,h\,\Delta w \\
&= a\,[1/2\,(w + \Delta w)\,x - 1/2\,w\,(x - \Delta x)] \\
&= 1/2\,a\,(w\,\Delta x + x\,\Delta w) \\
&= 1/2\,a\,w\,(\Delta x + x\,\Delta w/w) \\
&= 1/2\,a\,w\,[\Delta x + (w \pm \Delta w)\,\Delta x/w] \\
&= a\,w\,\Delta x\,(1 \pm \Delta w/2\,w)
\end{aligned} \tag{5.64}$$

Equation (5.63) is used to obtain Equation (5.64). The plus sign is for converging flow, and the minus sign is for diverging flow.

Equation (5.64) can be used to incorporate the effect of convergence or divergence into Equation (5.32) for two-dimensional flow. By definition, $u = 0$ at $x = L$, so Equation (5.64) has the integral form:

$$u\,w\,h = -\int_{uwh}^{0} d(u\,w\,h) = -\int_{x}^{L} (1 \pm \Delta w/2\,w)\,a\,w\,dx \qquad (5.65)$$

where u is negative because x increases toward the ice divide. Factoring out a reference width and approximating the integral by a summation:

$$u\,h \approx -\sum_{i=x/\Delta x}^{i=L/\Delta x} (1 \pm \Delta w/2\,w)_i\,a_i\,\Delta x = -\sum_{i=x/\Delta x}^{i=L/\Delta x} \Phi_i\,a_i\,\Delta x \qquad (5.66)$$

where the convergence-divergence factor Φ, see Equations (4.81) and (4.101), is defined as:

$$\Phi = (1 \pm \Delta w/2\,w) \qquad (5.67)$$

The value of Δw for each w can be specified for each Δx step. If no convergence or divergence occurs, $\Delta w = 0$ and $\Phi = 1$. If radial convergence or divergence occurs in which $\Delta w = w$ for each Δx step, $\Phi = 3/2$ for convergence and $\Phi = 1/2$ for divergence. Maximum divergence and convergence of flow, 180°, exists along the ice divide at saddles, where flow from the adjacent domes meets and then parts down opposite flanks of the ice divide. As shown in Figure 5.9, $w = 2\,\Delta x$ and $w + \Delta w = \pi\,w$ at the saddle. This gives $\Phi \approx 2$ for ice meeting along the crest of the ice divide and $\Phi \approx 0$ for ice parting down opposite flanks of the ice divide. For radial flow, Φ is constant at all Δx steps, but it decreases abruptly when diverging flow becomes converging. Most ice radiating from interior ice domes converges on ice streams near the margin.

Constant accumulation and ablation

Basal shear stress τ_o cannot be negative in Equations (5.47) and (5.53), so functions $\Gamma(x)$ and $\Pi(x)$ cannot be negative. This requirement limits analytical evaluations of $\Gamma(x)$ and $\Pi(x)$ to three situations: (1) a steady-state ice sheet in mass-balance equilibrium, (2) an ice sheet that is actively advancing, and (3) the active portion between the stagnation line and the ice divide of an ice sheet that is retreating. In the model, x is measured along an ice-sheet flowline, such that x = 0 at the ice margin and x = L at the ice divide. At a given position x on the flowline, $\Gamma(x)$ is integrated from x to L and $\Pi(x)$ is integrated from S to x, where S is the stagnation line and S = 0 for a stationary or an advancing ice sheet.

A flowband is associated with each flowline of an ice sheet. In the three-dimensional simulation, these flowbands have variable widths, whereas they have constant widths in the two-dimensional simulation. Consider a flowband of length L and constant width w. An equilibrium line crosses the flowband at distance E from the ice margin. The ablation rate a' is constant over $0 \le x \le E$ and the accumulation rate a" is constant over $E \le x \le L$. The mass balance of the flowband is:

$$\bar{a}\,L = a'\,E + a''\,(L - E) \qquad (5.68)$$

where $\bar{a}$ is the net accumulation or ablation rate for the entire flowband; so that $\bar{a} > 0$ for an advancing ice sheet, $\bar{a} = 0$ for a stationary ice sheet, and $\bar{a} < 0$ for a retreating ice sheet. A stagnation line crosses the flowband at distance S from the ice margin when the ice sheet is retreating.

In the ablation zone ($x \le E$):

$$\Gamma(x) = \int_{x}^{E} a'\,w\,dx + \int_{E}^{L} a''\,w\,dx = a'\,w\,(E - x) + a''\,w\,(L - E) \qquad (5.69)$$

$$\Pi_f(x) = \int_{S}^{x} [\Gamma(x)/w]^{1/n}\,dx = \int_{S}^{x} [a'\,(E - x) + a''\,(L - E)]^{1/n}\,dx$$

$$= \left(\frac{n}{n+1}\right) a'^{\,1/n} \left[\left(\frac{\bar{a}}{a'}L - S\right)^{\frac{n+1}{n}} - \left(\frac{\bar{a}}{a'}L - x\right)^{\frac{n+1}{n}}\right] \qquad (5.70)$$

$$\Pi_t(x) = \int_{S}^{x} [\Gamma(x)/w]^{1/m}\,dx = \int_{S}^{x} [a'\,(E - x) + a''\,(L - E)]^{1/m}\,dx$$

$$= \left(\frac{m}{m+1}\right) a'^{\,1/m} \left[\left(\frac{\bar{a}}{a'}L - S\right)^{\frac{m+1}{m}} - \left(\frac{\bar{a}}{a'}L - x\right)^{\frac{m+1}{m}}\right] \qquad (5.71)$$

Equation (5.68) is used so that $\bar{a}$ can be included in the expression for $\Pi(x)$. However, negative values of $\bar{a}$ are not allowed because $\Pi_f(x)$ and $\Pi_t(x)$ cannot be negative. Negative values of $\Pi(x)$ are prevented by integrating over the range $x \ge S$, which automatically eliminates the range $x < S$ where $\Gamma(x)$ would be negative. The three possible combinations of S and $\bar{a}$ in Equations (5.70) and (5.71) are S = 0 and $\bar{a} = 0$ for a steady-state ice sheet in mass-balance equilibrium, S = 0 and $\bar{a} > 0$ for an advancing ice sheet, and S > 0 and $\bar{a} = 0$ for a retreating ice sheet.

In the accumulation zone ($x \ge E$):

$$\Gamma(x) = \int_{x}^{L} a''\,w\,dx = a''\,w\,(L - x) \qquad (5.72)$$

$$\Pi_f(x) = \int_{S}^{E} [a''\,(L - E) + a'\,(E - x)]^{1/n}\,dx + \int_{E}^{x} [a''\,(L - x)]^{1/n}\,dx$$

$$= \left(\frac{n}{n+1}\right) \left\{ a''^{\,1/n} \left[\left(1 - \frac{a''}{a'}\right)(L - E)^{\frac{n+1}{n}} - (L - x)^{\frac{n+1}{n}}\right] + a'^{\,1/n} \left[\frac{\bar{a}}{a'}L - S\right]^{\frac{n+1}{n}} \right\} \qquad (5.73)$$

$$\Pi_t(x) = \int_{S}^{E} [a''\,(L - E) + a'\,(E - x)]^{1/m}\,dx + \int_{E}^{x} [a''\,(L - x)]^{1/m}\,dx$$

$$= \left(\frac{m}{m+1}\right)\left\{ a''^{\,1/m}\left[\left(1-\frac{a''}{a'}\right)(L-E)^{\frac{m+1}{m}} - (L-x)^{\frac{m+1}{m}}\right] + a'^{\,1/m}\left[\frac{\bar{a}}{a'}L - S\right]^{\frac{m+1}{m}}\right\} \tag{5.74}$$

Again, Equation (5.68) is used so that $\bar{a}$ can be included in the expressions for $\Pi(x)$, and the three possible combinations of S and $\bar{a}$ in Equations (5.73) and (5.74) are $S = 0$ and $\bar{a} = 0$ for a steady-state ice sheet in mass-balance equilibrium, $S = 0$ and $\bar{a} > 0$ for an advancing ice sheet, and $S > 0$ and $\bar{a} = 0$ for a retreating ice sheet.

At the equilibrium line $(x - E)$ the expressions for $\Gamma(x)$ and $\Pi(x)$ reduce to:

$$\Gamma(x) = a''\, w\,(L-E) \tag{5.75}$$

$$\Pi_f(x) = \left(\frac{n}{n+1}\right)\left\{ a'^{\,1/n}\left[\frac{\bar{a}}{a'}L - S\right]^{\frac{n+1}{n}} - a''^{\,1/n}\left[\frac{a''}{a'}(L-E)\right]^{\frac{n+1}{n}}\right\} \tag{5.76}$$

$$\Pi_t(x) = \left(\frac{m}{m+1}\right)\left\{ a'^{\,1/m}\left[\frac{\bar{a}}{a'}L - S\right]^{\frac{m+1}{m}} - a''^{\,1/m}\left[\frac{a''}{a'}(L-E)\right]^{\frac{m+1}{m}}\right\} \tag{5.77}$$

For a steady-state ice sheet, the first term in Equations (5.76) and (5.77) vanishes because $S = \bar{a} = 0$.

Solutions for frozen and thawed beds

Equations (5.47) and (5.53) can now be solved analytically for radial convergence and divergence by incorporating Equation (5.66) and for constant accumulation and ablation by incorporating Equations (5.69) through (5.74). For a frozen (dry) bed, $(\tau_o)_f$ equals $(\tau_o)_C$ because none of the ice velocity is due to basal sliding:

$$(\tau_o)_f = (\tau_o)_C = \frac{\rho_I\, g_z\, C_f\left[\frac{\Phi}{w}\int_x^L a\,dx\right]^{1/n}}{\left[h_S^{\frac{2n+2}{n}} + \left(\frac{2n+2}{n}\right)C_f\int_S^x\left\{\frac{\Phi}{w}\int_x^L a\,dx\right\}^{1/n}dx\right]^{\frac{1}{n+1}}}$$

$$= \frac{\rho_I\, g_z\, C_f\,\Phi^{1/n}\left[\frac{\Gamma(x)}{w}\right]^{1/n}}{\left[h_S^{\frac{2n+2}{n}} + \left(\frac{2n+2}{n}\right)C_f\,\Phi^{1/n}\,\Pi_f(x)\right]^{\frac{1}{n+1}}} \tag{5.78}$$

For a thawed (wet) bed, $(\tau_o)_t$ is only approximately $(\tau_o)_S$ because some of the ice velocity is due to internal shear through the total ice thickness:

$$(\tau_o)_t \approx (\tau_o)_S = \frac{\rho_I\, g_z\, C_t\left[\frac{\Phi}{w}\int_x^L a\,dx\right]^{1/m}}{\left[h_S^{\frac{2m+2}{m}} + \left(\frac{2m+1}{m}\right)C_t\int_S^x\left\{\frac{\Phi}{w}\int_x^L a\,dx\right\}^{1/m}dx\right]^{\frac{1}{2m+1}}}$$

$$= \frac{\rho_I\, g_z\, C_t\,\Phi^{1/m}\left[\frac{\Gamma(x)}{w}\right]^{1/m}}{\left[h_S^{\frac{2m+2}{m}} + \left(\frac{2m+1}{m}\right)C_t\,\Phi^{1/m}\,\Pi_t(x)\right]^{\frac{1}{2m+1}}} \tag{5.79}$$

Using Equations (5.69) through (5.71) for the ablation zone $(x \le E)$:

$$(\tau_o)_f = (\tau_o)_C = \frac{\rho_I\, g_z\, C_f\,\Phi^{1/n}\,[a'(E-x) + a''(L-E)]^{1/n}}{\left\{h_S^{\frac{2n+2}{n}} + 2\,C_f\,\Phi^{1/n}\left[\left(\frac{\bar{a}}{a'}L - S\right)^{\frac{n+1}{n}} - \left(\frac{\bar{a}}{a'}L - x\right)^{\frac{n+1}{n}}\right]\right\}^{\frac{1}{n+1}}} \tag{5.80}$$

$$(\tau_o)_t \approx (\tau_o)_S = \frac{\rho_I\, g_z\, C_t\,\Phi^{1/m}\,[a'(E-x) + a''(L-E)]^{1/m}}{\left\{h_S^{\frac{2m+2}{m}} + \left(\frac{2m+1}{m+1}\right)C_t\,\Phi^{1/m}\left[\left(\frac{\bar{a}}{a'}L - S\right)^{\frac{m+1}{m}} - \left(\frac{\bar{a}}{a'}L - x\right)^{\frac{m+1}{m}}\right]\right\}^{\frac{1}{2m+1}}} \tag{5.81}$$

Using Equations (5.72) through (5.74) for the accumulation zone $(x \ge E)$:

$$(\tau_o)_f = (\tau_o)_C = \rho_I\, g_z\, C_f\,\Phi^{1/n}\,[a''(L-x)]^{1/n} \div \left\{h_S^{\frac{2n+2}{n}} + 2C_f\Phi^{1/n}\left[a''^{\,1/n}\left(1-\frac{a''}{a'}\right)(L-E)^{\frac{n+1}{n}} - a''^{\,1/n}(L-x)^{\frac{n+1}{n}} + a'^{\,1/n}\left(\frac{\bar{a}}{a'}L-S\right)^{\frac{n+1}{n}}\right]\right\}^{\frac{1}{n+1}} \tag{5.82}$$

$$(\tau_o)_t \approx (\tau_o)_S = \rho_I\, g_z\, C_t\,\Phi^{1/m}[a''(L-x)]^{1/m} \div \left\{h_S^{\frac{2m+1}{m}} + \left(\frac{2m+1}{m+1}\right)C_t\Phi^{1/m}\left[a''^{\,1/m}\left(1-\frac{a''}{a'}\right)(L-E)^{\frac{m+1}{m}} - a''^{\,1/m}(L-x)^{\frac{m+1}{m}} + a'^{\,1/m}\left(\frac{\bar{a}}{a'}L-S\right)^{\frac{m+1}{m}}\right]\right\}^{\frac{1}{2m+1}} \tag{5.83}$$

Solutions for a steady-state ice sheet

For a steady-state ice sheet in mass-balance equilibrium, $S = h_S = \bar{a} = 0$. In the ablation zone ($x \le E$):

$$(\tau_o)_f = (\tau_o)_C = \frac{\rho_I g_z C_f^{\frac{n+1}{n}} \Phi^{n+1} [a'(E-x) + a''(L-E)]^{1/n}}{\left(2x^{\frac{n+1}{n}}\right)^{\frac{1}{n+1}}} \quad (5.84)$$

$$(\tau_o)_t \approx (\tau_o)_S = \frac{\rho_I g_z C_t^{\frac{2m}{2m+1}} \Phi^{\frac{2}{2m+1}} [a'(E-x) + a''(L-E)]^{1/m}}{\left[\left(\frac{2m+1}{m+1}\right) x^{\frac{m+1}{m}}\right]^{\frac{1}{2m+1}}} \quad (5.85)$$

In the accumulation zone ($x \ge E$):

$$(\tau_o)_f = (\tau_o)_C = \frac{\rho_I g_z \left[\frac{1}{2} C_f^n \Phi a''\right]^{\frac{1}{n+1}} (L-x)^{1/n}}{\left[\left(1 - \frac{a''}{a'}\right)(L-E)^{\frac{n+1}{n}} - (L-x)^{\frac{n+1}{n}}\right]^{\frac{1}{n+1}}} \quad (5.86)$$

$$(\tau_o)_t \approx (\tau_o)_S = \frac{\rho_I g_z \left[\left(\frac{m+1}{2m+1}\right) C_t^{2m} \Phi^2 a''^2\right]^{\frac{1}{2m+1}} (L-x)^{1/m}}{\left[\left(1 - \frac{a''}{a'}\right)(L-E)^{\frac{m+1}{m}} - (L-x)^{\frac{m+1}{m}}\right]^{\frac{1}{2m+1}}} \quad (5.87)$$

Solutions for an advancing ice sheet

For an advancing ice sheet, $S = h_S = 0$ and $\bar{a} > 0$. In the ablation zone ($x \le E$):

$$(\tau_o)_f = (\tau_o)_C = \frac{\rho_I g_z \left(\frac{1}{2} C_f^n \Phi\right)^{\frac{1}{n+1}} [a'(E-x) + a''(L-E)]^{1/n}}{\left[\left(\frac{\bar{a}}{a'} L\right)^{\frac{n+1}{n}} - \left(\frac{\bar{a}}{a'} L - x\right)^{\frac{n+1}{n}}\right]^{\frac{1}{n+1}}} \quad (5.88)$$

$(\tau_o)_t \approx (\tau_o)_S$

$$= \frac{\rho_I g_z \left[\left(\frac{m+1}{2m+1}\right) C_t^{2m} \Phi^2\right]^{\frac{1}{2m+1}} [a'(E-x) + a''(L-E)]^{1/m}}{\left[\left(\frac{\bar{a}}{a'} L\right)^{\frac{m+1}{m}} - \left(\frac{\bar{a}}{a'} L - x\right)^{\frac{m+1}{m}}\right]^{\frac{1}{2m+1}}} \quad (5.89)$$

In the accumulation zone ($x \ge E$):

$(\tau_o)_f = (\tau_o)_C$

$$= \frac{\rho_I g_z \left(\frac{1}{2} C_f^n \Phi a''\right)^{\frac{1}{n+1}} (L-x)^{1/n}}{\left[\left(1 - \frac{a''}{a'}\right)(L-E)^{\frac{n+1}{n}} - (L-x)^{\frac{n+1}{n}} + \left(\frac{a'}{a''}\right)^{\frac{1}{n}} \left(\frac{\bar{a}}{a'} L\right)^{\frac{n+1}{n}}\right]^{\frac{1}{n+1}}} \quad (5.90)$$

$(\tau_o)_t \approx (\tau_o)_S$

$$= \frac{\rho_I g_z \left[\left(\frac{m+1}{2m+1}\right) C_t^{2m} \Phi^2 a''^2\right]^{\frac{1}{2m+1}} (L-x)^{1/m}}{\left[\left(1 - \frac{a''}{a'}\right)(L-E)^{\frac{m+1}{m}} - (L-x)^{\frac{m+1}{m}} + \left(\frac{a'}{a''}\right)^{\frac{1}{m}} \left(\frac{\bar{a}}{a'} L\right)^{\frac{m+1}{m}}\right]^{\frac{1}{2m+1}}} \quad (5.91)$$

Solutions for a retreating ice sheet

For a retreating ice sheet, $\bar{a} < 0$. In the ablation zone ($x \le E$):

$(\tau_o)_f = (\tau_o)_C$

$$= \frac{\rho_I g_z C_f \Phi^{1/n} [a'(E-x) + a''(L-E)]^{1/n}}{\left\{h_S^{\frac{2n+2}{n}} + 2 C_f \Phi^{1/n} \left[\left(\frac{\bar{a}}{a'} L - S\right)^{\frac{n+1}{n}} - \left(\frac{\bar{a}}{a'} L - x\right)^{\frac{n+1}{n}}\right]\right\}^{\frac{1}{n+1}}} \quad (5.92)$$

$$(\tau_o)_t \approx (\tau_o)_S = \rho_I g_z C_t \Phi^{1/m} [a'(E-x) + a''(L-E)]^{1/m} \div \left\{h_S^{\frac{2m+2}{m}} + \left(\frac{2m+1}{m+1}\right) C_t \Phi^{1/m} \left[\left(\frac{\bar{a}}{a'} L - S\right)^{\frac{m+1}{m}} - \left(\frac{\bar{a}}{a'} L - x\right)^{\frac{m+1}{m}}\right]\right\}^{\frac{1}{2m+1}} \quad (5.93)$$

In accumulation zone ($x \ge E$):

$$(\tau_o)_f = (\tau_o)_C = \rho_I g_z C_f \Phi^{1/n} a''^{1/n} (L-x)^{1/n} \div \left\{h_S^{\frac{2n+2}{n}} + 2 C_f \Phi^{1/n} a''^{1/n} \left[\left(1 - \frac{a''}{a'}\right)(L-E)^{\frac{n+1}{n}} - (L-x)^{\frac{n+1}{n}} + \left(\frac{a'}{a''}\right)^{1/n} \frac{\bar{a}}{a'} (L-S)^{\frac{n+1}{n}}\right]\right\}^{\frac{1}{n+1}} \quad (5.94)$$

$$(\tau_o)_t \approx (\tau_o)_S = \rho_I\, g_z\, C_t\, \Phi^{1/m}\, a''^{1/m}\, (L-x)^{1/m}$$

$$\div \left\{ h_S^{\frac{2m+1}{m}} + \left(\frac{2m+1}{m+1}\right) C_t\, \Phi^{1/m}\, a''^{1/m} \left[\left(1 - \frac{a''}{a'}\right)(L-E)^{\frac{m+1}{m}} \right.\right.$$

$$\left.\left. - (L-x)^{\frac{m+1}{m}} - \left(\frac{a'}{a''}\right)^{1/m} \frac{\bar{a}}{a'} (L-S)^{\frac{m+1}{m}} \right] \right\}^{\frac{1}{2m+1}} \quad (5.95)$$

Solutions for the CLIMAP ice sheets

The CLIMAP ice-sheet reconstruction model is a special case of a steady-state ice sheet in which $\Phi = 1$ for two-dimensional flow, a' is unspecified for $x < \Delta x$, and a" is constant for $x > \Delta x$. These conditions apply for ablation by iceberg calving along marine margins where $E = 0$ and $a'\, h_0 = a''\, L$ so that $a' >> a''$ because $h_0 << L$ and by surface melting along terrestrial margins where $E << L$ and $a'\, E = a''\, (L - E)$ so that $a' >> a''$ because $E << L$. Our new ice-sheet reconstruction model reproduces these conditions for $a' >> a'' \approx a$ and either $x = 0$ at $E = 0$ or $x = 0$ at $E << L$ in Equations (5.86) and (5.87), where $h_i = h_0$ or $h_i = h_E$ is entered at $i = 0$ in Equation (5.56). In this case:

$$(\tau_o)_f = (\tau_o)_C = \frac{\rho_I\, g_z \left(\frac{1}{2} C_f^{\,n}\, \Phi\, a\right)^{\frac{1}{n+1}} (L-x)^{1/n}}{\left[L^{\frac{n+1}{n}} - (L-x)^{\frac{n+1}{n}}\right]^{\frac{1}{n+1}}} = \frac{K_f\, (L-x)^{1/n}}{\left[L^{\frac{n+1}{n}} - (L-x)^{\frac{n+1}{n}}\right]^{\frac{1}{n+1}}} \quad (5.96)$$

$$(\tau_o)_t \approx (\tau_o)_S = \frac{\rho_I\, g_z \left[\left(\frac{m+1}{2m+1}\right) C_t^{\,2m}\, \Phi^2\, a^2\right]^{\frac{1}{2m+1}} (L-x)^{1/m}}{\left[L^{\frac{m+1}{m}} - (L-x)^{\frac{m+1}{m}}\right]^{\frac{1}{2m+1}}}$$

$$= \frac{K_t\, (L-x)^{1/m}}{\left[L^{\frac{m+1}{m}} - (L-x)^{\frac{m+1}{m}}\right]^{\frac{1}{2m+1}}} \quad (5.97)$$

The frozen-bed constant is evaluated using Equation (5.45):

$$K_f = [1/4\, (n+2)\, \rho_I\, g_z\, A^n\, \Phi\, a]^{\frac{1}{n+1}} \quad (5.98)$$

The thawed-bed constant is evaluated using Equation (5.51):

$$K_t = \left[\left(\frac{m+1}{2m+1}\right) \rho_I\, g_z B^{2m}\, \Phi^2\, a^2\right]^{\frac{1}{2m+1}} \quad (5.99)$$

Two-dimensional flow, as specified in the CLIMAP model, occurs when $\Phi = 1$.

These equations can be used to illustrate the effort that variable flowband widths w and variable accumulation and ablation rates a along flowbands have on ice elevation h and basal shear stress τ_o.

Figure 5.10 shows the effect on h and τ_o when the bed is horizontal and w varies for constant $a = a''$ and a' is due to calving at the ice-sheet margin, where h_E is the initial value needed to solve Equation (5.6) as an initial-value finite-difference recursive formula with $h = h_I$ for a horizontal bed. Flowbands are shown having a constant width down the flank of a linear ice divide of constant elevation, a linearly decreasing width from a saddle where the ice divide is downdrawn by an ice stream at the ice margin, a linearly increasing width from an ice dome to an ice margin with no ice stream, and a linearly increasing width from an ice dome that becomes a linearly decreasing width to an ice stream at the ice margin. Variations in h along these flowbands are generally within 5 percent of h and τ_o for constant w.

Figure 5.11 shows the effect on h and τ_o when the bed is horizontal and a varies for constant w. Variations are shown for constant $a = a''$ as in Figure 5.10, for a decreasing linearly from the ice divide at $x = L$ to $a = 0$ at $x = E = L/3$ and then decreasing linearly to the ice margin for low ice divides that can be crossed by convective storm systems, for a increasing linearly from zero at the ice divide to a peak at $E < x < L$ and then decreasing linearly to the ice margin with $a = 0$ at $x = E = L/3$ for high ice divides that cannot be crossed by convective storm systems, and constant $a = a'$ for $0 \le x < E$, and $a = a''$ for $E < x \le L$. The h and τ_o curves in Figure 5.11 are for steady-state equilibrium. When ablation is restricted to calving so that h_E occurs at $x = E = 0$, the initial increase of h along x is substantially greater than when ablation occurs along $0 < x < E = L/3$. Variations in h along flowbands are generally within 5 percent for these latter three cases.

A convenient way to combine the effects of variables w and a along flowbands is using factors ϕ_f for creep over a frozen bed and ϕ_t for sliding over a thawed bed, such that τ_o for creep becomes:

$$(\tau_o)_C \cdot \phi_f\, \tau_C \quad (5.100)$$

and τ_o for sliding becomes:

$$(\tau_o)_S = \phi_t\, \tau_S \quad (5.101)$$

Let w and a vary along x. When the bed is frozen, $\tau_o = (\tau_o)_C$ and is given by Equations (5.45) and (5.47) for $h_S = 0$:

$$(\tau_o)_C = \left[\frac{n\,(n+2)\,\rho_I\, g_z\, A^n\, [\Gamma(x)/w(x)]^{\frac{n+1}{n}}}{4\,(n+1)\,\Pi_f(x)}\right]^{\frac{1}{n+1}} \quad (5.102)$$

When the bed is thawed, $\tau_o = (\tau_o)_S$ and is given by Equations (5.51) and (5.53) for $h_S = 0$:

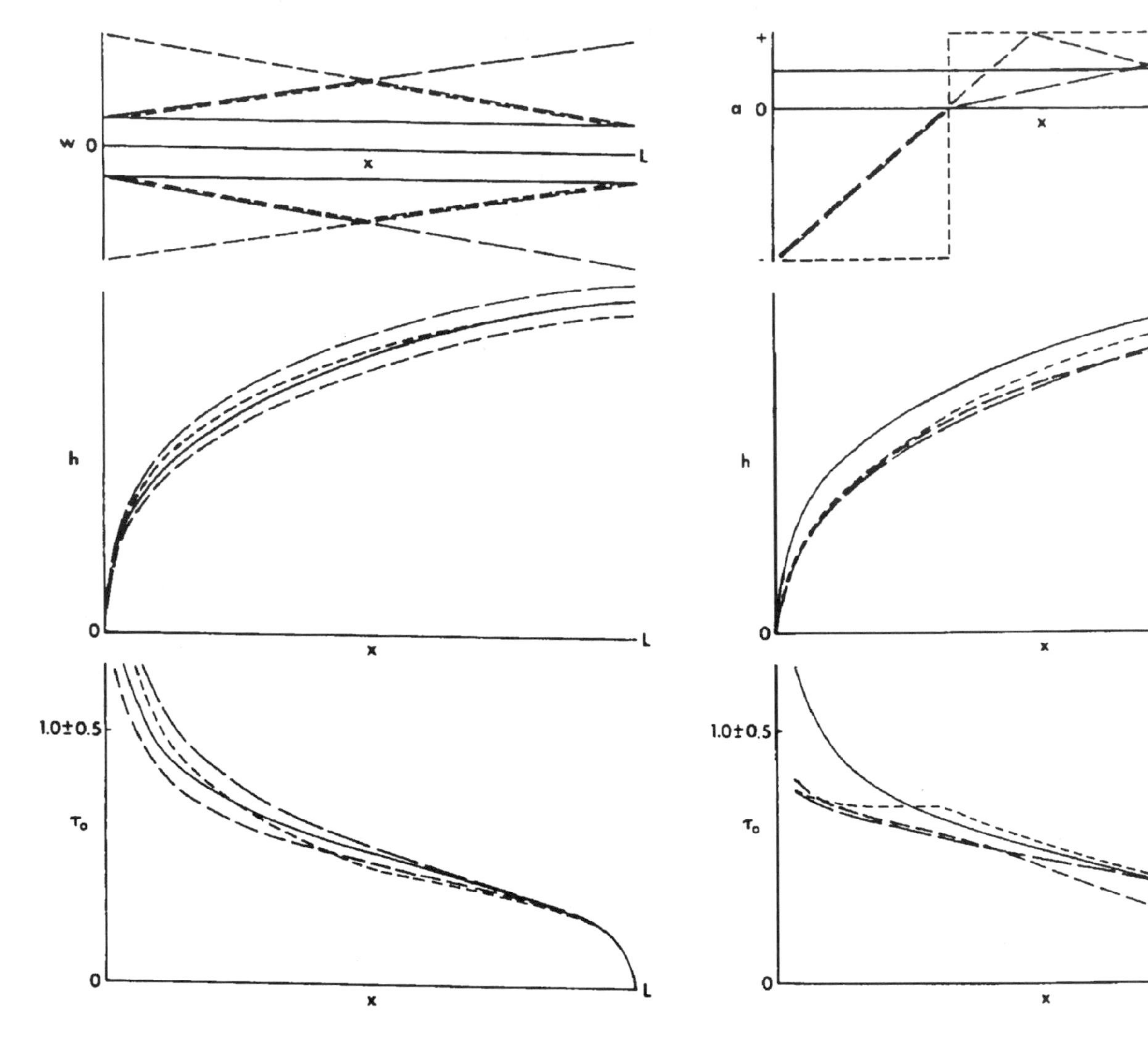

Figure 5.10: The effect on ice sheets of variable flowband widths for constant accumulation rates. Lines with dashes of different lengths are used to show the effect of parallel (solid), converging (long dashes), diverging (intermediate dashes), and initially diverging followed by subsequent converging (short dashes) conditions of flowbands (*top*) on ice elevation profiles above a horizontal bed (*center*) and the basal shear stress (*bottom*). Units are kilometers for x, bars for τ_o and τ_o curves are for a frozen bed. For a thawed bed, 1.0 ± 0.5 bar should be replaced by 0.5 ± 0.3 bar for these τ_o curves.

Figure 5.11: Teffect on ice sheets of variable accumulation and ablation rates for constant flowband widths. Lines with dashes of different lengths are used to show the effect of constant and linear accumulation rates and the equilibrium-line position (*top*) on ice elevation profiles above a horizontal bed (*center*) and the basal shear stress (*bottom*). Mass-balance equilibrium exists in all cases, with the equilibrium line at either the ice margin or one-third the flowline length in from the ice magin. Units are kilometers for x, bars for τ_o and τ_o curves are for a frozen bed. For a thawed bed, 1.0 ± 0.5 bar should be replaced by 0.5 ± 0.3 bar for these τ_o curves.

$$(\tau_o)_S = \left[\frac{m\,\rho_I\,g_z\,B^{2m}\,[\Gamma(x)/w(x)]^{\frac{2m+1}{m}}}{(2\,m+1)\,\Pi_t(x)}\right]^{\frac{1}{2m+1}} \qquad (5.103)$$

Let w and a be constant along x. When the bed is frozen, $\tau_o = \tau_C$ and is given by Equation (5.96) and (5.98) for $\Phi = 1$:

$$\tau_C = \left[\frac{(n+2)\,\rho_I\,g_z\,A^{\,n}\,a\,(L-x)^{\frac{n+1}{n}}}{4\left[L^{\frac{n+1}{n}} - (L-x)^{\frac{n+1}{n}}\right]}\right]^{\frac{1}{n+1}} \qquad (5.104)$$

When the bed is thawed, $\tau_o = \tau_S$ and is given by Equations (5.97) and (5.99) for $\Phi = 1$:

$$\tau_S = \left[\frac{(m+1)\,\rho_I\,g_z\,B^{\,2m}\,a^{\,2}\,(L-x)^{\frac{2m+1}{m}}}{(2\,m+1)\left[L^{\frac{m+1}{m}} - (L-x)^{\frac{m+1}{m}}\right]}\right]^{\frac{1}{2m+1}} \qquad (5.105)$$

Entering Equations (5.102) and (5.104) in Equation (5.100):

$$\phi_f = \frac{(\tau_o)_C}{\tau_C} = \left[\frac{n\,[\Gamma(x)/w(x)]^{\frac{n+1}{n}} \left[L^{\frac{n+1}{n}} - (L-x)^{\frac{n+1}{n}} \right]}{(n+1)\, a\, \Pi_f(x)\, (L-x)^{\frac{n+1}{n}}} \right]^{\frac{1}{n+1}} \qquad (5.106)$$

Entering Equations (5.103) and (5.105) in Equation (5.101):

$$\phi_t = \frac{(\tau_o)_S}{\tau_S} = \left[\frac{m\,[\Gamma(x)/w(x)]^{\frac{2m+1}{m}} \left[L^{\frac{m+1}{m}} - (L-x)^{\frac{m+1}{m}} \right]}{(m+1)\, a^2\, \Pi_t(x)\, (L-x)^{\frac{2m+1}{m}}} \right]^{\frac{1}{2m+1}} \qquad (5.107)$$

Note that $\phi_f = \phi_t = 1$ if w and a are constant. However, since w and a vary, ϕ_f and ϕ_t should be calculated at each Δx step along a flowband. For a flowline or a flowband of unit width, this requires knowing the average value $\bar{a}$ of a for use in Equations (5.104) and (5.105). In the accumulation zone:

$$\bar{a}\,(L-E) = \int_E^L a\, dx = \sum_{i=E/\Delta x}^{i=L/\Delta x} a_i\, \Delta x \qquad (5.108)$$

If a is given by Equation (5.35):

$$\bar{a} = \frac{(a_o'/\lambda')\,(e^{-\lambda' E} - e^{-\lambda' L}) + (a_o''/\lambda'')\,(e^{-\lambda'' E} - e^{-\lambda'' L})}{L-E} \qquad (5.109)$$

Computing $\bar{a}$ is comparable to computing $\Gamma(x)$, $\Pi_f(x)$, and $\Pi_t(x)$. Equations (5.106) and (5.107) then become cumbersome and it is easiest to compute $(\tau_o)_S$ and $(\tau_o)_C$ directly from Equations (5.47) and (5.53).

The reason for introducing ϕ_f and ϕ_t is instructional. Downdraw by major ice streams tends to produce saddles on the ice divide, so w decreases toward these ice streams because of converging flow. Also, the maximum in surface slope at the heads of ice streams tends to trigger a peak in accumulation, so a tends to decrease toward the ice divide. The increase in w and the decrease in a from ice streams to ice divides is such that the product w a is often fairly constant and $\phi_f = \phi_t = 1$ becomes a reasonable approximation, especially since large variations in w and a cause relatively small variations in h, as seen in Figures 5.10 and 5.11. Note that ϕ_f and ϕ_t appear in Equation (3.64). This exercise shows that setting ϕ_f and $\phi_t = 1$ introduces no serious errors in calculating $\Delta h/\Delta x$ or in calculating h by numerically integrating Equation (3.64), especially along flowlines that converge on ice streams, which is the usual case.

6

STREAM FLOW

From Sheet Flow to Stream Flow

The "Big Bang" Theory of the universe can be imagined as matter rushing out into space because Nature abhors a vacuum. The densities of ice and air are 917 kg/m^3 and 1.3 kg/m^3, respectively, so air beyond the edge of an ice sheet behaves like a vacuum that exerts a horizontal gravitational pulling force on the ice sheet, that increases with the ice elevation above land or sea and causes the ice sheet to spread and thin. Spreading over land is resisted by basal traction, which gives the ice sheet a surface slope, unlike tractionless spreading over water. Ice flow induced by the gravitational force occurs in directions where the ice-surface slope is greatest. Ice-surface flowlines are therefore perpendicular to ice-surface elevation contour lines. Taking x horizontal along flowlines, y horizontal transverse to flowlines, and z vertical, only lithostatic pressure P_I would exist if there were no ice motion due to longitudinal gravitational pulling along flowlines. Then $\sigma_{xx} = \sigma_{yy} = \sigma_{zz}$, where stresses σ_{ii} are axial compressive stresses. However, because an ice sheet "sees" the relative vacuum of air in the downslope direction, σ_{xx} is slightly less compressive than σ_{yy} and σ_{zz} because ice moves forward to occupy the vacuum. This forward motion is most evident in ice streams, where ice flowlines are nearly parallel, so that transverse deviator stress $\sigma_{yy}' = \sigma_{yy} - P_I = 0$. Therefore, $\sigma_{yy} = P_I = 1/3\ (\sigma_{xx} + \sigma_{yy} + \sigma_{zz})$, from which $\sigma_{yy} = 1/2\ (\sigma_{xx} + \sigma_{zz})$. For ice streams, then, the longitudinal deviator stress is $\sigma_{xx}' = \sigma_{xx} - P_I = \sigma_{xx} - 1/3\ (\sigma_{xx} + \sigma_{yy} + \sigma_{zz}) = \sigma_{xx} - 1/3\ [\sigma_{xx} + 1/2\ (\sigma_{xx} + \sigma_{zz}) + \sigma_{zz}] = 1/2\ (\sigma_{xx} - \sigma_{zz})$. Although σ_{xx} and σ_{zz} are both negative below the depth of transverse crevasses, across which $\sigma_{xx} = 0$, σ_{xx} is less negative than σ_{zz} because lithostatic pressure is relieved in the x direction by ice moving into the surrounding relative vacuum of air. Therefore, σ_{xx}' is a positive tensile pulling stress. Beneath ice streams, basal shear stress τ_o is reduced by basal meltwater that effectively decouples ice from the bed, thereby converting sheet flow into stream flow, allowing ice to move into the air surrounding an ice sheet more rapidly through ice streams. The fact that huge transverse crevasses arc around the zone of converging flow at the heads of ice streams is proof that ice streams pull ice out of ice sheets.

Ice streams are fast currents of ice that develop near the margins of an ice sheet and which discharge ninety percent or more of the ice. Just as precipitation over a continent collects within watersheds of rivers ranging in size from large rivers that drain huge interior regions to small rivers that drain only coastal plains, so precipitation over an ice sheet collects within ice divides of ice streams ranging in size from large ice streams that drain huge interior regions to small ice streams that drain only marginal regions. Just as rivers discharge water at discrete sites along the shorelines of continents, so ice streams discharge ice at discrete sites along the margins of ice sheets.

As defined by Bader (1961), "An 'ice stream' is something akin to a mountain glacier consisting of a broad accumulation basin and a narrower valley glacier, but a mountain glacier is laterally hemmed in by rock slopes, while the ice stream is contained by slower moving surrounding ice. The edges of the ice stream are often crevassed, and the surface tends to be concave as the ice is 'funnelled' down. Many of the large outlet glaciers in Greenland, particularly in the south, are the narrow outlets of large ice streams which reach back many scores of miles into the ice sheet."

Ice is funneled down by gravitational pulling force F_X that acts along longitudinal axis x of the ice stream. An example of a horizontal gravitational pulling force is observed when the spillway is lowered on a dam. Water rushes horizontally over the spillway because air beyond the spillway has a much lower density than water, and water surges forward to displace the air. As water rushes forward, it acquires a surface slope that changes from concave to convex as ice crosses the spillway. The glaciological counterparts of water rushing over a spillway are an icefall when ice moves over a bedrock sill or step and an ice stream when ice moves from a high traction bed to a low traction bed. The ice surface is convex over the high-traction bed and concave over the low-traction bed. Gravitational pulling opens transverse crevasses at the transition from the high-traction bed to the low-traction bed. If the ice stream becomes afloat, gravity pulls ice outward over the non-traction water so as to stratify water, ice, and air according to their densities. That ice is pulled over water is revealed by the plucking of icebergs from the advancing ice front.

Figure 6.1 illustrates the distinction between pushing and pulling. Figure 6.1a shows horizontal spreading of an ice sheet that accompanies vertical lowering caused by a downward gravitational pulling force F_z equal to mass m_I of the ice sheet multiplied by gravity acceleration g_z, as ex pressed by Equation (4.3). When the ice sheet is in steady-state equilibrium, the surface profile is unchanged because accumulation rates match the lowering rate and ablation rates

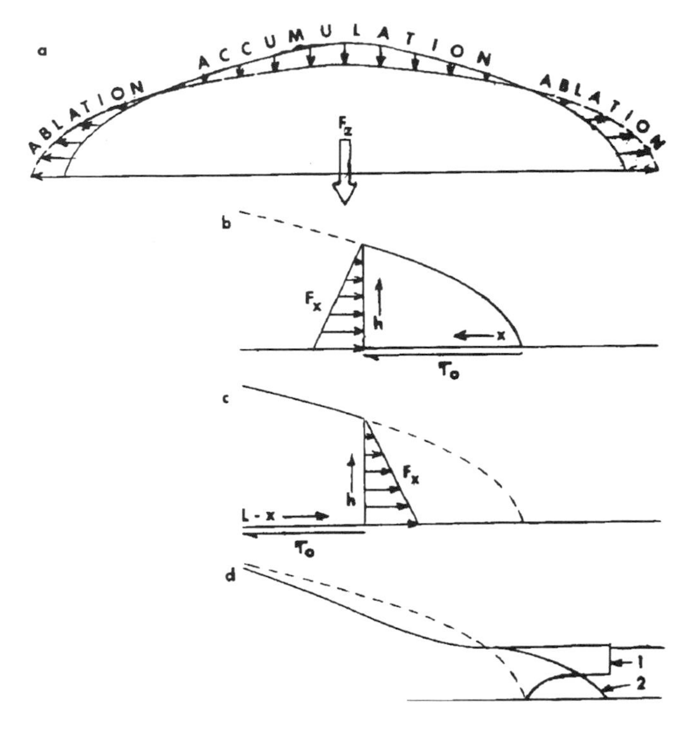

Figure 6.1: Vertical and horizontal gravitational forces on an ice sheet. In *a*, vertical gravitational force F_z pulls the ice surface downward, causing ice to be pushed out laterally. Ice accumulation in the interior counters the decrease in ice-surface elevation, and ice ablation around the perimeter counters the increase in ice areal extent. In *b*, horizontal gravitational force F_x pushing the ice margin forward is resisted by basal traction force τ_o x per unit flowband width. In *c*, net horizontal gravitational force ΔF_x pulling the ice interior forward is resisted by basal traction force τ_o (L − x) per unit flowband width. In *d*, when basal shear stress τ_o is low toward the ice margin and high toward the ice interior, the ice surface will be downdrawn by F_z and pulled forward by ΔF_x to produce an ice shelf if the ice margin becomes afloat (terminus 1) or an ice lobe if the ice margin spreads over land that restores τ_o (terminus 2).

match the spreading rate. Vertical pulling by gravity causes horizontal pushing, with extending flow in the accumulation zone and compressive flow in the ablation zone (Nye, 1951). As seen in Figure 6.1b, if the ice sheet beyond distance x from the ice margin is removed, it can be replaced by longitudinal gravitational force $F_x = \bar{P}_I A_x = (1/2\ \rho_I\ g_z\ h)\ w\ h$, which is resisted by basal traction force $F_x = (\sigma_{xz})_o A_z = \tau_o$ w x over distance x in a flowband of constant width w, where ρ_I is ice density, h is ice height at position x, and τ_o is basal shear stress over distance x. Note that gravitational force F_x is the product of mean lithostatic pressure $\bar{P}_I$, a scalar, and area A_x, a vector pointing toward the ice margin, as expressed by Equation (4.4). Therefore, F_x is directed toward the remaining ice mass, so by definition it is a pushing force. If τ_o is constant, equating the pushing force with the traction force gives $x = (\rho_I\ g_z / 2\ \tau_o)\ h^2$. This is the parabolic convex surface profile of an ice sheet predicted by plasticity theory, in which τ_O is the yield stress, given by Equation (5.5) for $h_R = 0$.

Horizontal pulling can be induced by removing the ice sheet over distance x so that net longitudinal gravitational force $F_x = \bar{P}_I A_x = 1/2\ \rho_I\ g_z\ w\ h^2$ becomes a pulling force resisted by basal traction force $F_x = (\sigma_{xz})_o A_z = \tau_o$ w (L − x) over distance L − x, where h_L is ice height at the ice divide and L is the length of the flowband from the ice divide to the ice margin, as shown in Figure 6.1c. Note that gravitational force F_x is the product of mean lithostatic pressure $\bar{P}_I$, a scalar, and area A_x, a vector pointing toward the ice margin. Therefore, F_x is a pulling force because it is directed away from the remaining ice mass. Pulling extends only to the ice divide because τ_O reverses sign across the ice divide. Equating the forces gives $L - x = (\rho_I\ g_z / 2\ \tau_o)\ h^2$. This is not an equilibrium profile. An equilibrium profile that exists when $h = h_L$ at x = 0 undergoes gravitational collapse as x increases, with collapse beginning slowly and ending rapidly until h = 0 at x = L. Gravitational collapse causes the ice divide to lower and retreat to the opposite ice-sheet margin. With F_x a pulling force instead of a pushing force, therefore, the ice surface over distance L − x is quickly pulled down by F_z as ice is pulled forward by F_x, thereby converting the convex profile into a concave profile. A similar result is obtained by not removing the ice sheet over distance x but, instead, having τ_o be large over L − x and small over x, which produces the concave surface of stream flow in Figure 6.1d. An ice stream can end as a floating ice shelf, beneath which τ_o is zero (shown as terminal region 1), or as a grounded ice lobe, beneath which τ_O increases to the ice margin (shown as terminal region 2). A marine ice stream supplies an ice shelf. A terrestrial ice stream supplies an ice lobe. In Figure 6.1d, the pulling force at distance x from the ice-stream grounding line or the ice lobe terminus increases as basal traction decreases. Basal traction decreases as more of the overlying ice is supported by basal water that either drowns small bumps in the bedrock or mobilizes permeable till or sediment that blankets bedrock, thereby decoupling ice from the bed.

As seen in Figure 6.1, the mass balance of an ice sheet increases ice elevation in the accumulation zone and decreases ice areal extent in the ablation zone. Therefore, accumulation reduces or reverses vertical gravitational thinning and ablation reduces or reverses horizontal gravitational spreading. For steady-state equilibrium, the mass balance exactly offsets the lowering and spreading caused by the force balance, so the ice sheet changes neither its size nor its shape. The size and shape of an ice sheet can be determined by calculating the lengths and profiles of its surface flowlines. Ice motion along flowlines induce kinematic stresses that resist the motion. These are all deviator stresses that change the shape of an ice sheet and therefore change the profiles of its surface flowlines. For sheet flow, the dominant kinematic stress is σ_{xz}, which attains its maximum value τ_o at the base of the ice sheet. For stream flow, σ_{xz} is less important than side shear stress σ_{xy}, which attains its maximum value τ_s along the sides of the ice stream, and longitudinal tensile stress σ_{xx}' which attains its maximum value along the central flowline of the ice stream.

The mass balance and the force balance will be calculated for an ice sheet on a flat horizontal bed. This simplifies the calculations greatly without compromising the results. Surface slope depends on the mass balance, the force balance, and the bed slope. The effect of each will be examined separately, and their combined effect will then be determined. Ice thickness gradient $\Delta h_I / \Delta x$ is related to surface slope $\Delta h / \Delta x$ and bed slope $\Delta h_R / \Delta x$ as follows:

$$\frac{\Delta h_I}{\Delta x} = \frac{\Delta(h - h_R)}{\Delta x} = \frac{\Delta h}{\Delta x} - \frac{\Delta h_R}{\Delta x} \tag{6.1}$$

where h and h_R are the respective heights of the ice surface and the ice bed above present-day sea level. Negative heights are depths below sea level. For an ice stream on a horizontal bed, $\Delta h_R / \Delta x = 0$. In the absence of bed topography, the only effect the bed has on the surface slope is basal traction. Surface slope decreases with decreasing basal traction. Basal traction decreases as more of the overlying ice is supported by basal water, either in a layer or in the pores of a deformable substrate, so basal ice is effectively decoupled from bedrock. This decoupling can be quantified as a basal buoyancy factor.

The Basal Buoyancy Factor

Figure 6.1 shows that horizontal gravitational force F_x is a pushing force for sheet flow and a pulling force for stream flow, both inducing traction forces that resist gravitational motion. As presented in Figure 6.1, stream flow is generated from sheet flow by reducing ice-bed coupling toward the terminus of a flowband. If that is the case, marine ice streams are the major dynamic links connecting a marine ice sheet with its floating ice shelves, and stream flow is transitional between sheet flow and shelf flow. However, stream flow becomes sluggish as reduced ice-bed coupling is increasingly offset by increased ice-shelf buttressing. To illustrate this point, consider an ice cube in a pan. If water is added to the pan, an ice cube of height h_I will begin to float in water of height h_W when the lithostatic pressure of ice P_I and the hydrostatic pressure of water P_W give the same overburden pressure P_o on the bottom of the pan:

$$P_o = P_I = \rho_I g_z h_I = \rho_W g_z h_W = P_W \tag{6.2}$$

where ρ_I is ice density, ρ_W is water density, and g_z is gravity acceleration. Ice-bed coupling is nil when Equation (6.2) is satisfied. Now imagine that the ice cube is a large tabular iceberg and the pan is a larger lake. Elimination of ice-bed coupling allows pulling force F_x to spread the iceberg radially, in the manner analyzed by Weertman (1957b) and quantified by Equation (4.58), until the iceberg grounds along the shoreline of the lake, thereby producing a braking force that cancels the pulling force. Hence, ice-bed coupling is removed in the center of the lake by a gain of basal buoyancy, and ice-shelf buttressing is attained around the perimeter of the lake by a loss of basal buoyancy. This condition exists for subglacial lakes near the dome of the Antarctic Ice Sheet in Wilkes Land (Oswald and Robin, 1973) and subglacial Lake Vostok (Kapista and others, 1996).

These considerations suggest that gravitational pulling in an ice stream is controlled by basal buoyancy. Pulling power is increased when ice-bed coupling beneath an ice stream is decreased by a gain of basal buoyancy. Pulling power is decreased when ice-shelf or ice-lobe buttressing beyond an ice stream is increased by a loss of basal buoyancy due to grounding of the ice shelf or the lobe. Both ice-bed coupling and ice-shelf or ice-lobe buttressing can therefore be quantified by a basal buoyancy factor defined as the ratio of basal water pressure P_W and basal ice pressure P_I:

$$\frac{P_W}{P_I} = \frac{\rho_W g_z h_W}{\rho_I g_z h_I} = \frac{\rho_W h_W}{\rho_I h_I} \tag{6.3}$$

where basal water would rise to height h_W in imaginary temperate boreholes through ice of height h_I along length L_S of an ice stream so that $0 \leq P_W \leq P_I$, as shown in Figure 6.2. Ratio P_W/P_I varies along L_S, and the ice stream is buttressed when P_W/P_I ratios are lower downstream (more grounding) than upstream (less grounding). Buttressing is caused by grounding beneath, alongside, or in front of the ice stream. These grounding conditions cause basal traction and lateral shear in the ice stream, in an ice shelf beyond a marine ice stream, and in an ice lobe beyond a terrestrial ice stream. For an ice shelf, basal traction exists at basal pinning points and lateral shear exists along the sides of a confining embayment. For an ice lobe, basal traction exists when basal meltwater freezes beneath the lobe or discharges along the lobe perimeter.

Since basal water can exist as a thin film at the ice-rock interface, can flow in a network of subglacial channels, can be ponded in subglacial lakes, and can occupy interstitial pores in subglacial till or sediments, height h_W of basal water standing in imaginary temperate boreholes, as employed in Equation (6.3), is a statistical averaging of all these forms of basal water across width w_I of the ice stream at distance x from its terminus. In particular, it will be assumed that $h_W = 0$ where basal water is a thin film on rough bedrock or does not saturate subglacial till or sediments. In these cases the bed still provides traction that resists sliding and creep of overlying ice. These conditions are incorporated into Equation (6.3) by setting $h_W = 0$ in imaginary boreholes drilled down to the bed because h_W has meaning for pulling power only insofar as ice-bed coupling resists stream flow. Hence, $h_W = 0$ for all undrowned bedrock projections and all frozen or unsaturated patches of till or sediments because these resist stream flow by retaining ice-bed coupling that retards ice motion. This is compatible

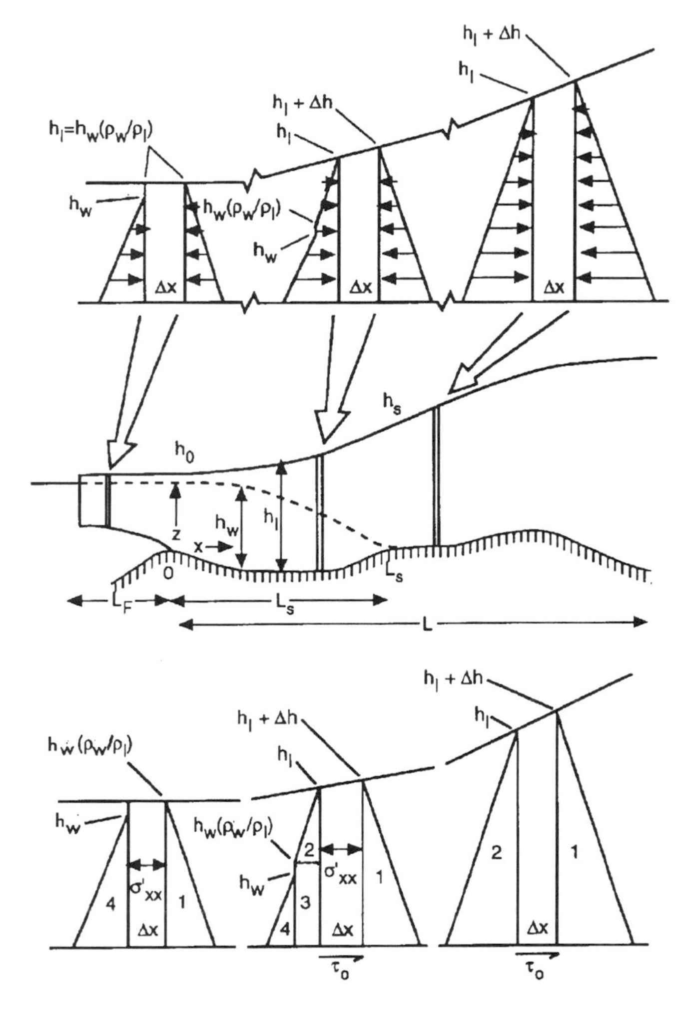

Figure 6.2: Horizontal gravitational forces used to compute the basal buoyancy factor. *Top*: Horizontal gravitational forces acting on ice columns for sheet flow (right), stream flow (center), and shelf flow (left). *Middle*: A possible variation of the height (dashed line) to which basal meltwater would rise in imaginary boreholes, with boreholes (vertical parallel lines) shown for sheet flow, stream flow, and shelf flow. *Bottom*: The areas of triangles 1, 2, and 4 and of rectangle 2 represent the horizontal gravitational forces acting on ice columns for sheet flow, stream flow, and shelf flow.

with the Weertman (1957a) sliding theory for sheet flow in which bed roughness is unaffected by basal water pressure and provides traction that retards sliding. This under standing of h_W can be generalized in Equation (6.3) by specifying that $h_W = 0$ gives $P_W/P_I = 0$ for sheet flow with no basal buoyancy, $h_W = (\rho_I / \rho_W)\, h_I$ gives $P_W/P_I = 1$ for shelf flow with full basal buoyancy, and intermediate values of h_W give $0 < P_W / P_I < 1$ for stream flow with partial basal buoyancy.

Downstream buttressing of an ice stream is determined by downstream values of P_W/P_I in Equation (6.3), such that $P_W/P_I = 0$ brings stream flow to a halt and $P_W/P_I = 1$ allows stream flow to become unconfined linear shelf flow. Beyond ice streams, buttressing is caused by P_W/P_I values for an ice shelf or an ice lobe. For confined nonlinear ice shelves, $P_W/P_I = 0$ at grounding lines between ice streams and at interior pinning points where the ice shelf is grounded locally, but $P_W/P_I = 1$ at ice-stream grounding lines and along the calving front. In Figure 2.14, examples of $P_W/P_I = 0$ are the "pseudo ice shelves" of ice stream C (Robin and others, 1970), which is now stagnant; examples of $P_W/P_I = 1$ are the floating ice tongues of Thwaites, David, Ninnis, and Mertz Glaciers; and examples of $0 < P_W/P_I < 1$ are active ice streams and the Ross and Ronne Ice Shelves. When $0 < P_W/P_I < 1$, ice flow velocity increases in the direction where P_W/P_I increases.

Refer to Figure 6.2, for which P_W/P_I changes along an ice stream of streaming length L_S and $P_W/P_I = 1$ for an ice shelf of floating length L_F. Over length $L - L_S$, sheet flow with $P_W/P_I = 0$ exists when the bed is frozen, or thawed but with only a thin water film that wets bedrock or wets but does not mobilize permeable basal till or sediments. Over length L_S, stream flow with $P_W/P_I = 1$ exists when basal water is thick enough to drown all bedrock projections comparable to the controlling obstacle size in the Weertman (1957a) theory of basal sliding and saturates basal till or sediments enough so they are unable to support a basal shear stress. Combinations of these conditions determine how h_W varies along L_S, where h_W is statistically averaged for basal water conditions across a given width w_I of the ice stream. Stream flow is slow when $P_W/P_I \rightarrow 0$ and fast when $P_W/P_I \rightarrow 1$. Basal buoyancy factor P_W/P_I can be linked to the mass balance and the force balance of an ice stream, from which the surface profile is determined.

The Mass Balance

Following the Van der Veen (1983) analysis for the thickness profile of linear ice shelves, the effect of the mass balance on the surface profile of an ice stream on a frictionless horizontal bed can be computed from the condition for mass continuity expressed by Equation (5.32). When width w_I is constant along length L_S, which is reasonable for ice streams, Equation (5.32) for steady-state stream flow is written:

$$\partial(u\, h_I) / \partial x - a = 0 \qquad (6.4)$$

where u is the average velocity of the ice column for mass-balance equilibrium. Integrating Equation (6.4) for constant accumulation rate a over length L_S and noting that u is negative when x is positive upslope from x = 0 at the foot of the ice stream:

$$\int_0^x a\, dx = a\, x = \int_{u_0 h_0}^{u h_I} d\,(u\, h_I) = h_I\, u - h_0\, u_0 \qquad (6.5)$$

where h_0 and u_0 are values of h_I and u at x = 0. Equation (6.4) also yields:

$$a = h_I\, \partial u/\partial x + u\, \partial h_I/\partial x = h_I\, \dot{\varepsilon}_{xx} + u\, \Delta h_I/\Delta x \qquad (6.6)$$

where $\Delta h_I/\Delta x$ is the ice-thickness gradient and longitudinal strain rate $\dot{\varepsilon}_{xx} = \partial u/\partial x$, defined by Equation (4.6), is related to effective creep stress σ_c and longitudinal deviator stress σ_{xx}' by Equation (4.17), the flow law of ice. Solving Equations (6.5) and (6.6) simultaneously for the thickness gradient gives:

$$\frac{\Delta h_I}{\Delta x} = \frac{a - h_I\, \dot{\varepsilon}_{xx}}{u} = \frac{h_I\, a - h_I^2\, \dot{\varepsilon}_{xx}}{a\, x + h_0\, u_0} \qquad (6.7)$$

Note that $\Delta h_I/\Delta x = \Delta h/\Delta x$ for a horizontal bed, given by Equation (6.1) when $\Delta h_R/\Delta x = 0$. The flow law of ice provides an expression for $\dot{\varepsilon}_{xx}$ in Equation (6.7).

Computing $\dot{\varepsilon}_{xx}$ from the flow law of ice, Equation (4.17), requires knowing σ_c. Since $\sigma_c = (1/2\, \sigma_{ij}'\, \sigma_{ij}')^{1/2}$, $\sigma_{ij}' \propto \dot{\varepsilon}_{ij}$ and $\sigma_{kk}' = 0$, effective creep stress σ_c can be written in terms of σ_{xx}' as follows:

$$\sigma_c = \left[\frac{1}{2}\left(\sigma_{xx}'^2 + \sigma_{yy}'^2 + \sigma_{zz}'^2 + 2\,\sigma_{xy}^2 + 2\,\sigma_{yz}^2 + 2\,\sigma_{zx}^2\right)\right]^{1/2}$$

$$= \left[\frac{1}{2}\sigma_{xx}'^2 + \frac{1}{2}\sigma_{yy}'^2 + \frac{1}{2}(-\sigma_{xx}' - \sigma_{yy}')^2 + \sigma_{xy}^2 + \sigma_{yz}^2 + \sigma_{zx}^2\right]^{1/2}$$

$$= \left[\sigma_{xx}'^2 + \sigma_{xx}'\,\sigma_{yy}' + \sigma_{yy}'^2 + \sigma_{xy}^2 + \sigma_{yz}^2 + \sigma_{zx}^2\right]^{1/2}$$

$$= \left[\sigma_{xx}'^2 + \left(\dot{\varepsilon}_{yy}/\dot{\varepsilon}_{xx}\right)\sigma_{xx}'^2 + \left(\dot{\varepsilon}_{yy}/\dot{\varepsilon}_{xx}\right)^2\sigma_{xx}'^2\right.$$

$$\left. + \left(\dot{\varepsilon}_{xy}/\dot{\varepsilon}_{xx}\right)^2\sigma_{xx}'^2 + \left(\dot{\varepsilon}_{yz}/\dot{\varepsilon}_{xx}\right)\sigma_{xx}'^2 + \left(\dot{\varepsilon}_{zx}/\dot{\varepsilon}_{xx}\right)\sigma_{xx}'^2\right]^{1/2}$$

$$\left[1 + \left(\dot{\varepsilon}_{yy}/\dot{\varepsilon}_{xx}\right) + \left(\dot{\varepsilon}_{yy}/\dot{\varepsilon}_{xx}\right)^2 + \left(\dot{\varepsilon}_{xy}/\dot{\varepsilon}_{xx}\right)^2 + \left(\dot{\varepsilon}_{xz}/\dot{\varepsilon}_{xx}\right)^2\right]^{1/2}\sigma_{xx}'$$

$$= R_{xx}\,\sigma_{xx}' \qquad (6.8)$$

where $\dot{\varepsilon}_{yz}$ is nil and by definition:

$$R_{xx} = \left[1 + \left(\dot{\varepsilon}_{yy}/\dot{\varepsilon}_{xx}\right) + \left(\dot{\varepsilon}_{yy}/\dot{\varepsilon}_{xx}\right)^2 + \left(\dot{\varepsilon}_{xy}/\dot{\varepsilon}_{xx}\right)^2 + \left(\dot{\varepsilon}_{xz}/\dot{\varepsilon}_{xx}\right)^2\right]^{1/2} \qquad (6.9)$$

For ice streams of constant width and negligible laterally inflowing ice, $\dot{\varepsilon}_{yy} \approx 0$. In practice, R_{xx} is evaluated from values of $\dot{\varepsilon}_{ij}$ measured from surface strain rates (Thomas, 1973a, 1973b; Whillans and others, 1989).

The flow law of ice can now be written for $\dot{\varepsilon}_{xx}$ using Equations (4.17), (6.8), and (6.9):

$$\dot{\varepsilon}_{xx} = \left(\sigma_c^{\,n-1}/A^{\,n}\right)\sigma_{xx}' = \left(R_{xx}^{\,n-1}/A^{\,n}\right)\sigma_{xx}'^{\,n} \qquad (6.10)$$

Substituting Equation (6.10) for $\dot{\varepsilon}_{xx}$ in Equation (6.7) and solving for $\Delta h_I/\Delta x$:

$$\frac{\Delta h_I}{\Delta x} = \frac{h_I\, a - h_I^2\left(R_{xx}^{\,n-1}/A^{\,n}\right)\sigma_{xx}'^{\,n}}{a\, x + h_0\, u_0} \qquad (6.11)$$

where σ_{xx}' is obtained from the force balance.

The Force Balance

The force balance for stream flow is computed for dynamic equilibrium. According to Newton's second law, gravitational force F_z results from gravity acceleration g_z acting on mass m_I of an ice column, as expressed by Equation (4.3). Gravitational force F_x results from average differential pressure $\Delta\bar{P}_I$ exerted on opposite faces of the ice column that are normal to the x direction, as expressed by Equation (4.4). The kinematic force resisting F_x arises from motion or deformation of the ice column and consists of the sum of stresses resisting the motion or deformation multiplied by the area of the ice column upon which these stresses act, as expressed by Equation (4.5).

Longitudinal gravitational force $(F_x)_G$ causes motion in an ice stream and is obtained by balancing the horizontal lithostatic and hydrostatic forces acting on the ice column in Figure 6.2 (bottom). The vertical ice column has downslope height h_I and upslope height $h_I + \Delta h$ for ice-surface elevation change Δh along incremental length Δx, over which basal water rises in temperate boreholes to an average height h_W above the bed. The average h_W is based on the fractions of the bed in basal area $w_I\,\Delta x$ having no buoyancy ($P_W/P_I = 0$) and full buoyancy ($P_W/P_I = 1$). The longitudinal gravitational force per unit width is the area of force triangle 1 minus the combined areas of force triangle 2, force rectangle 3, and force triangle 4 in Figure 6.2. Therefore, for an ice stream of width w_I, and introducing P_W/P_I from Equation (6.3), the expression for $(F_x)_G$ is:

$$(F_x)_G = -\,1/2\,\rho_I\, g_z\,(h_I + \Delta h)\, w_I\,(h_I + \Delta h)$$

$$+\,1/2\,\rho_I\, g_z\,[h_I - h_W\,(\rho_W/\rho_I)]\, w_I\,[h_I - h_W\,(\rho_W/\rho_I)]$$

$$+\,\rho_I\, g_z\,[h_I - h_W\,(\rho_W/\rho_I)]\, w_I\, h_W\,(\rho_W/\rho_I)$$

$$+\,(1/2\,\rho_W\, g_z\, h_W)\, w_I\, h_W$$

$$= -\,[\rho_I\, g_z\,\Delta h + 1/2\,\rho_I\, g_z\, h_I\,(1 - \rho_I/\rho_W)\,(P_W/P_I)^2]\, w_I\, h_I \qquad (6.12)$$

where $[\rho_I g_z \Delta h + 1/2 \rho_I g_z h_I (1 - \rho_I/\rho_W)(P_W/P_I)^2]$ is the reduction in lithostatic pressure experienced by front area $w_I h_I$ of the ice column. It leads to the Weertman (1957b) linear ice-shelf solution given by Equation (4.58) when $\Delta h = 0$ and $P_W/P_I = 1$.

In Equation (6.12), downslope pulling by the first right-hand term is partly offset by upslope pushing by the second, third, and fourth right-hand terms, to give a net forward pulling. The first term is the mean lithostatic ice pressure exerted on rear column area $w_I (h_I + \Delta h)$, to produce the horizontal pulling force per unit width given by the area of triangle 1 in Figure 6.2. The second term is the mean lithostatic ice pressure above height $h_W (\rho_W/\rho_I)$ exerted on front column area $w_I [h_I - h_W (\rho_W/\rho_I)]$ above this height, to produce the horizontal pushing force per unit width given by the area of triangle 2 in Figure 6.2. The third term is the lithostatic ice pressure at height $h_W (\rho_W/\rho_I)$ exerted on front column area $w_I h_W (\rho_W/\rho_I)$ below this height to produce the horizontal pushing force per unit width given by the area of rectangle 3 in Figure 6.2. The fourth term is the mean hydrostatic water pressure below height h_W exerted on front column area $w_I h_W$ below this height to produce the horizontal pushing force per unit width given by the area of triangle 4 in Figure 6.2. Notice that the horizontal hydrostatic force per unit width exerted by water, the area of triangle 4 in Figure 6.2, is shown only on the downslope side of ice columns for which $h_W > 0$. This happens because a marine ice stream and its floating ice-shelf extension interface with water beyond the seaward side of the ice column, so subglacial water can escape downstream but not upstream. Similarly, for a terrestrial ice stream ending as an ice lobe, subglacial water can escape around the lobe perimeter downstream but not upstream. Gradient $\partial h_W/\partial x$ along length L_S of stream flow in Figure 6.2 affects bed traction, which produces surface slope $\Delta h/\Delta x$ in the force balance.

Equation (6.12) for stream flow reduces to laminar sheet flow when $P_W/P_I = 0$ and to linear shelf flow when $P_W/P_I = 1$ and $\Delta h = 0$ along length Δx. As demonstrated by Equation (4.180), grounded ice motion depends on surface slope, not bed slope, so h_I is ice thickness, but Δh is a change in ice elevation along Δx in Equation (6.4). That stream flow is transitional between sheet flow and shelf flow is seen in Figure 6.2. Moving $h_W (\rho_W/\rho_I)$ downward increases Δh by causing triangle 2 to encroach upon rectangle 3 and triangle 4 until sheet flow occurs at $h_W (\rho_W/\rho_I) = 0$, reducing Equation (6.4) to sheet flow for which $P_W/P_I = 0$. Moving $h_W (\rho_W/\rho_I)$ upward decreases Δh by causing triangle 4 to encroach upon rectangle 3 and triangle 2 until shelf flow occurs at $h_W (\rho_W/\rho_I) = h_I$, reducing Equation (6.4) to shelf flow for which $P_W/P_I = 1$ and $\Delta h = 0$.

Kinematic force $(F_x)_K$ resists gravitational force $(F_x)_G$. In the ice column, pulling gravitational motion induces creep controlled by deviator tensile stress $(\sigma_{xx} - \sigma_{zz})$ and deviator shear stresses σ_{xy} and σ_{xz}, with maximum values $\sigma_{xy} = \tau_S$ for side traction and $\sigma_{xz} = \tau_o$ for basal traction in the ice stream. As seen in the derivation of Equation (4.42), when $\sigma_{yy}' = 0$ and w_I is constant in an ice stream, $(\sigma_{xx} - \sigma_{zz}) = 2\sigma_{xx}'$ is a longitudinal tensile stress that exists because longitudinal compressive stress σ_{xx} is slightly less than vertical compressive stress σ_{zz} when both are averaged through ice thickness h_I. Reduced lithostatic pressure along x is caused by ice rushing forward to fill the relative vacuum in air beyond the ice stream. In Equation (4.32), $\sigma_{zz} = -\rho_I g_z (h - z)$ is compressive lithostatic pressure P_I, so $(\sigma_{xx} - \sigma_{zz})$ is a longitudinal tensile deviation from P_I for linear sheet flow (Robin, 1967) and for linear shelf flow (Weertman, 1957b). For example, $(\sigma_{xx} - \sigma_{zz}) = (\sigma_{xx} - 0) = 2\sigma_{xx}'$ for $z = h$ at the surface and $(\sigma_{xx} - \sigma_{zz}) = [(2\sigma_{xx}' - \rho_I g_z h) - (-\rho_I g_z h)] = 2\sigma_{xx}'$ for $z = 0$ at the bed. Multiplying deviator stresses by the areas against which they act gives $(F_x)_K$. Taking $\bar{h}_I = h_I + 1/2\,\Delta h_I$ as the mean ice thickness along Δx and $\Delta\sigma_{xx}'$ as the change in σ_{xx}' along Δx, $(F_x)_G$ pulling on front area $w_I h_I$ in Equation (6.12) is resisted by τ_s acting on side areas $2\bar{h}_I \Delta x$, by τ_o acting on basal area $w_I \Delta x$, and by $2(\sigma_{xx}' + \Delta\sigma_{xx}')$ acting on rear area $w_I (h_I + \Delta h_I)$ of the ice column to give $(F_x)_K$:

$$\begin{aligned}(F_x)_K &= 2\tau_s \bar{h}_I \Delta x + \tau_o w_I \Delta x + 2(\sigma_{xx}' + \Delta\sigma_{xx}')\, w_I (h_I + \Delta h_I) \\ &= (2\tau_s h_I + \tau_s \Delta h_I + \tau_o w_I)\Delta x \\ &\quad + 2 w_I (\sigma_{xx}' h_I + \Delta\sigma_{xx}' \Delta h_I + \sigma_{xx}' \Delta h_I + h_I \Delta\sigma_{xx}') \\ &\approx (2\tau_s h_I + \tau_o w_I)\Delta x + 2 w_I [\sigma_{xx}' h_I + \Delta(\sigma_{xx}' h_I)] \end{aligned} \tag{6.13}$$

where the term containing $\Delta\sigma_{xx}'\Delta h_I$ can be ignored and $2\tau_s h_I >> \tau_s \Delta h_I$.

When the ice column is in dynamic equilibrium $(F_x)_G = (F_x)_K$, and equating Equations (6.12) and (6.13) gives the following expression for $\Delta h/\Delta x$:

$$\begin{aligned}\frac{\Delta h}{\Delta x} &= \frac{2\sigma_{xx}'}{\rho_I g_z \Delta x} - \frac{h_I}{2\Delta x}\left(1 - \frac{\rho_I}{\rho_W}\right)\left(\frac{P_W}{P_I}\right)^2 \\ &\quad + \frac{2\Delta(h_I \sigma_{xx}')}{\rho_I g_z h_I \Delta x} + \frac{2\tau_s}{\rho_I g_z w_I} + \frac{\tau_o}{\rho_I g_z h_I} \\ &= \frac{2\sigma_{xx}' \Delta h_I}{\rho_I g_z h_I \Delta x} + \frac{2\Delta\sigma_{xx}'}{\rho_I g_z \Delta x} + \frac{2\tau_s}{\rho_I g_z w_I} + \frac{\tau_o}{\rho_I g_z h_I}\end{aligned} \tag{6.14}$$

where the first two terms, containing σ_{xx}' and P_W/P_I with Δx in their denominators, do not become infinite as Δx approaches zero if they are equated to each other so that:

$$\sigma_{xx}' = \frac{\rho_I g_z h_I}{4}\left(1 - \frac{\rho_I}{\rho_W}\right)\left(\frac{P_W}{P_I}\right)^2 \tag{6.15}$$

Combining the Mass Balance and the Force Balance

Ice-surface slope $\Delta h / \Delta x$ is obtained from Equation (6.14) by substituting Equation (6.11), obtained from the mass balance, for ice-thickness gradient $\Delta h_I / \Delta x$ in Equations (6.14) and (6.15), obtained from the force balance. Equation (6.15) is differentiated along x to extract $\Delta h_I / \Delta x$:

$$\frac{\Delta \sigma_{xx}'}{\Delta x} = \frac{\rho_I g_z}{4}\left(1 - \frac{\rho_I}{\rho_W}\right)\left[2 h_I \left(\frac{P_W}{P_I}\right)\frac{\Delta}{\Delta x}\left(\frac{P_W}{P_I}\right) + \left(\frac{P_W}{P_I}\right)^2 \frac{\Delta h_I}{\Delta x}\right] \quad (6.16)$$

Substituting Equation (6.16) into Equation (6.14), and collecting the terms containing $\Delta h_I / \Delta x$:

$$\frac{\Delta h}{\Delta x} = \left[\frac{2\sigma_{xx}'}{\rho_I g_z h_I} + \frac{1}{2}\left(1 - \frac{\rho_I}{\rho_W}\right)\left(\frac{P_W}{P_I}\right)^2\right]\frac{\Delta h_I}{\Delta x}$$

$$+ h_I\left(1 - \frac{\rho_I}{\rho_W}\right)\left(\frac{P_W}{P_I}\right)\frac{\Delta}{\Delta x}\left(\frac{P_W}{P_I}\right) + \frac{2\tau_s}{\rho_I g_z w_I} + \frac{\tau_o}{\rho_I g_z h_I} \quad (6.17)$$

Substituting Equation (6.11) for $\Delta h_I / \Delta x$:

$$\frac{\Delta h}{\Delta x} = \left[\frac{2\sigma_{xx}'}{\rho_I g_z h_I} + \frac{1}{2}\left(1 - \frac{\rho_I}{\rho_W}\right)\left(\frac{P_W}{P_I}\right)^2\right]\left[\frac{h_I a - h_I^2 \left(R_{xx}^{n-1}/A^n\right)\sigma_{xx}'^{\,n}}{a x + h_0 u_0}\right]$$

$$+ h_I\left(1 - \frac{\rho_I}{\rho_W}\right)\left(\frac{P_W}{P_I}\right)\frac{\Delta}{\Delta x}\left(\frac{P_W}{P_I}\right) + \frac{2\tau_s}{\rho_I g_z w_I} + \frac{\tau_o}{\rho_I g_z h_I} \quad (6.18)$$

Substituting Equation (6.15) for σ_{xx}':

$$\frac{\Delta h}{\Delta x} = \left(1 - \frac{\rho_I}{\rho_W}\right)\left(\frac{P_W}{P_I}\right)^2$$

$$\left\{\frac{h_I a}{a x + h_0 u_0} - \frac{h_I^2 \left(R_{xx}^{n-1}/A^n\right)}{a x + h_0 u_0}\left[\frac{\rho_I g_z h_I}{4}\left(1 - \frac{\rho_I}{\rho_W}\right)\left(\frac{P_W}{P_I}\right)^2\right]^n\right\}$$

$$+ h_I\left(1 - \frac{\rho_I}{\rho_W}\right)\left(\frac{P_W}{P_I}\right)\frac{\Delta}{\Delta x}\left(\frac{P_W}{P_I}\right) + \frac{2\tau_s}{\rho_I g_z w_I} + \frac{\tau_o}{\rho_I g_z h_I} \quad (6.19)$$

For linear shelf flow, $P_W / P_I = 1$, $\tau_s = \tau_o = 0$, and Equation (6.19) reduces to the Van der Veen (1983) solution, given by Equation (6.11):

$$\frac{\Delta h}{\Delta x} = \left(1 - \frac{\rho_I}{\rho_W}\right)\left\{\frac{h_I a}{ax + h_0 u_0} - \frac{h_I^2 R_{xx}^{n-1}}{ax + h_0 u_0}\left[\frac{\rho_I g_z h_I}{4A}\left(1 - \frac{\rho_I}{\rho_W}\right)\right]^n\right\}$$

$$= \left(1 - \frac{\rho_I}{\rho_W}\right)\frac{\Delta h_I}{\Delta x} \quad (6.20)$$

For linear sheet flow, $P_W / P_I = 0$, $\tau_s = 0$, and Equation (6.19) reduces to the Nye (1952) solution, given by Equation (5.4):

$$\frac{\Delta h}{\Delta x} = \frac{\tau_o}{\rho_I g_z h_I} \quad (6.21)$$

Basal Sliding in Ice Streams

Reduced basal coupling along an ice stream reduces τ_o in Equation (6.16) in a way that can be quantified by basal buoyancy factor P_W / P_I. Weertman (1957a) assumed that bedrock beneath an ice sheet is rigid, impermeable, and rough in the manner specified in his basal sliding theory. Weertman (1986) also proposed that an ice-sheet surge could occur if basal meltwater drowned bedrock projections with height Λ_o that controlled the rate of basal sliding. Fast-sliding velocities should also be possible if obstacles of height Λ_o were buried by a layer of water-saturated till or sediments of thickness λ that provided little or no traction for the sliding ice. These are the conditions beneath an ice stream. Let bedrock bumps of height Λ_o project a distance $\Lambda_o - \lambda$ above a layer of basal water or water-saturated till or sediments such that, using Equation (6.3):

$$\lambda = \Lambda_o (P_W / P_I) \quad (6.22)$$

where λ varies from $\lambda \approx 0$ at the head of an ice stream, where $P_W << P_I$, to $\lambda \approx \Lambda_o$ at the foot of an ice stream, where $P_W \approx P_I$. If bedrock bumps projecting distance $\Lambda_o - \lambda$ into the ice are the "sticky spots" that resist sliding and these sticky spots have an average separation Λ_o', then bedrock projecting above layer λ has an effective roughness less than bedrock below the layer, and the bed roughness factor for an ice stream is:

$$\frac{\Lambda_o - \lambda}{\Lambda_o'} = \frac{\Lambda_o}{\Lambda_o'}\left(1 - \frac{P_W}{P_I}\right) \quad (6.23)$$

Sheet flow is converted to stream flow by substituting bed roughness factor $(\Lambda_o - \lambda) / \Lambda_o'$ given by Equation (6.23) for bed roughness factor Λ_o / Λ_o' given by Equation (4.26) in the Weertman (1957a) theory for basal sliding. Basal sliding velocity u_s for sheet flow given by Equation (4.26) then becomes u_o for stream flow by substituting $(\Lambda_o / \Lambda_o')(1 - P_W/P_I)$:

$$u_o = \left[\frac{\tau_o}{B_o (\Lambda_o / \Lambda_o')^2 (1 - P_W/P_I)^2}\right]^m$$

$$= \frac{(\tau_o / B)^m}{(1 - P_W / P_I)^{2m}} = \frac{u_s}{(1 - P_W / P_I)^{2m}} \quad (6.24)$$

In Equation (6.24), u_o becomes infinite as P_W/P_I approaches unity, provided that u_s is finite. However, u_s depends on τ_o because sheet flow requires bed traction and $P_W / P_I = 1$ only for floating tabular icebergs with ice-surface slope $\alpha = 0$. Therefore, $\tau_o = \rho_I g_z h_I \alpha = 0$ when $P_W / P_I = 1$ in Equation (6.24), and u_o is indeterminate. In this limiting case, linear ice spreading is determined by longitudinal deviator stress σ_{xx}' defined by Equation (6.15) when $P_W / P_I = 1$, not by basal shear stress τ_o, and the longitudinal ice velocity is obtained by integrating Equation (4.50) for a linear slab of ice (Weertman, 1957b). If the linear slab is kept grounded at one end, as in the unconfined floating tongue of a marine ice stream, Van der Veen (1983) showed that the surface slope is $\alpha = (1 - \rho_I/\rho_W)\ \Delta h_I/\Delta x$, where $\Delta h_I/\Delta x$ is the ice thickness gradient given by Equation (6.11). The significance of Equation (6.19) is that it must replace the dominance of τ_o with the dominance of σ_{xx}' and τ_s as stream flow replaces sheet flow.

Basal Traction in Ice Streams

Replacing τ_o for sheet flow with σ_{xx}' for stream flow is accomplished by Equations (6.23) and (6.24). When $P_W/P_I < 1$, the variation of τ_o along an ice stream is then determined by the continuity condition expressed by Equation (5.32) for mass continuity in a flowband along which sheet flow over a thawed bed becomes stream flow. The solution is Equation (5.53) modified for stream flow by setting ice stagnation height $h_S = 0$ because ice streams have no mass-balance stagnation height, and by setting $C_t = \rho_I g_z B = \rho_I g_z B_o (\Lambda_o/\Lambda_o')^2 (1 - P_W/P_I)^2$ to conform with Equation (6.23) for drowning bedrock projections. The resulting expression is:

$$\tau_o = \frac{\rho_I g_z C_t [\Gamma(x)/w(x)]^{\frac{1}{m}}}{\left[\left(\frac{2m+1}{m}\right) C_t \Pi_t(x)\right]^{\frac{1}{2m+1}}}$$

$$= \left[\frac{m(\rho_I g_z)^{2m+1} C_t^{2m} [\Gamma(x)/w(x)]^{\frac{2m+1}{m}}}{(2m+1)\Pi_t(x)}\right]^{\frac{1}{2m+1}}$$

$$= \left[\frac{m \rho_I g_z B_o^2 (\Lambda_o/\Lambda_o')^{4m} [\Gamma(x)/w(x)]^{\frac{2m+1}{m}}}{(2m+1)\Pi_t(x)}\right]^{\frac{1}{2m+1}}$$

$$\left(1 - \frac{P_W}{P_I}\right)^{\frac{4m}{2m+1}} = (\tau_o)_S \left(1 - P_W/P_I\right)^{\frac{4m}{2m+1}} \tag{6.25}$$

where $(\tau_o)_S$ is for sheet flow sliding over a thawed bed, as given by Equation (5.103). Equation (6.25) reduces to τ_o for sliding sheet flow when $P_W \ll P_I$. Therefore, $P_W/P_I = 0$ is used for flowband length $L - L_S$ of sheet flow, and $0 < P_W/P_I < 1$ is used for flowband length L_S of stream flow. Equation (6.25) is substituted for τ_o in Equation (6.19). The τ_o term in Equation (6.19) will then give the concave surface for stream flow as $P_W/P_I = 0$ at $x = L_S$ increases to $P_W/P_I = 1$ at $x = 0$.

Ice Ridges between Ice Streams

Fast flow in ice streams is resisted by slow flow in ice ridges between ice streams. This resistance produces side shear stress τ_S in Equation (6.19). Flow in ice ridges is slow because $P_W \ll P_I$ if the bed is thawed or $P_W = 0$ if the bed is frozen, so high bed traction produces a high basal shear stress τ_o beneath ice ridges. Following Paterson (1972) and Reeh (1982), assume that τ_o beneath ridges is constant along a horizontal bed. Ice flowlines then have the parabolic profile obtained when horizontal gravitational force $F_x = \bar{P}_I A_x = (1/2\ \rho_I g_z h)(w\ h)$ is balanced by horizontal kinematic force $F_x = (\sigma_{xz})_o A_z = \tau_o\ w\ x$, where x is measured from the ice margin for an ice ridge of constant width:

$$h^2 = [2\tau_o/\rho_I g_z]\ x \tag{6.26}$$

Let the ice ridge and its flanking ice streams all have length L_R. Setting $x = L_R$ at $h = h_R$, Equation (6.26) can be written:

$$\frac{h_R^2}{L_R} = \frac{2\tau_o}{\rho_I g_z} \tag{6.27}$$

The mass-balance continuity condition for creep in an ice ridge is given by replacing L with L_R in Equation (5.42), assuming that accumulation a supplying the ice ridge extends only over $0 \le x \le L_R$ because accumulation over $L_R \le x \le L$ enters the flanking ice streams. Then:

$$\Gamma(x) = \int_x^{L_R} (a - dh/dt)\ w\ dx = \bar{a}\ w_R (L_R - x)$$

$$= \bar{u}_x w_R h = \frac{2 w_R h^2}{n+2}\left[\frac{\tau_o}{A}\right]^n \tag{6.28}$$

where $dh/dt = 0$ for steady-state flow and $\bar{a}$ is the mean accumulation rate along length L_R of an ice ridge having width w_R. Assume that Equation (6.28) is satisfied even though ice flow along the crest of the ice ridge is strongly divergent into the flanking ice streams, as is usually the case, and $\bar{u}_x$ is the mean horizontal creep velocity of ice given by Equation (4.153).

Separating the variables in Equation (6.28) and integrating over ice ridge length L_R:

$$\bar{a}^{1/n} \int_0^{L_R} (L_R - x)^{1/n} dx = \left(\frac{2}{n+2}\right)^{1/n} \frac{\rho_I g_z}{A} \int_0^{h_R} h^{1+2/n}\ dh$$

$$= \frac{\bar{a}^{1/n} L_R^{1+1/n}}{1 + 1/n} = \left(\frac{2}{n+2}\right)^{1/n} \frac{\rho_I g_z h_R^{2+2/n}}{A(2 + 2/n)} \tag{6.29}$$

Solving Equation (6.29) for h_R^2/R:

$$\frac{h_R^2}{L_R} = \left[\frac{(n+2)^{1/n} 2^{1-1/n} \bar{a}^{1/n} A}{\rho_I g_z}\right]^{\frac{n}{n+1}} \tag{6.30}$$

Equating the mass balance in Equation (6.30) with the force balance in Equation (6.27) and solving when τ_o is the viscoplastic yield stress $(\tau_o)_v$:

$$\tau_o = (\tau_o)_v = \left[\frac{(n+2)\,\rho_I\,g_z\,\bar{a}\,A^n}{4}\right]^{\frac{1}{n+1}} \tag{6.31}$$

Note that τ_o becomes independent of $\bar{a}$ as $n \to \infty$ for plastic flow, in which case ignoring strongly diverging flow and variations in a along an ice ridge is reasonable.

Side Shear between Ice Streams and Ice Ridges

Equation (6.26) for constant τ_o gives the profile of an ice ridge on a horizontal bed. The corresponding profile for the flanking ice streams on a horizontal bed obtained by intergrating Equations (6.11) and (6.17) for $h = h_I$. Replacing σ'_{xx} in these equations with Equation (6.15), separate expressions for $\Delta h/\Delta x$ are obtained from the mass balance and the force balance when $h = h_I$ and $\Delta h/\Delta x = \Delta h_I/\Delta x$:

$$\frac{\Delta h}{\Delta x} = \frac{h_I\,a}{a\,x + h_0\,u_0} - \frac{h_I^2\left(R_{xx}^{\,n-1}/A^n\right)}{a\,x + h_0\,u_0}\left[\frac{\rho_I\,g_z\,h_I}{4}\left(1 - \frac{\rho_I}{\rho_W}\right)\left(\frac{P_W}{P_I}\right)^2\right]^n$$

$$= \frac{h_I\left(1 - \frac{\rho_I}{\rho_W}\right)\left(\frac{P_W}{P_I}\right)\frac{\Delta}{\Delta x}\left(\frac{P_W}{P_I}\right) + \frac{2\,\tau_s}{\rho_I\,g_z\,w_I} + \frac{\tau_o}{\rho_I\,g_z\,h_I}}{1 - \left(1 - \frac{\rho_I}{\rho_W}\right)\left(\frac{P_W}{P_I}\right)^2} \tag{6.32}$$

Equation (6.32) can be solved to give the dependence of τ_s on a and (P_W/P_I), with the a and (P_W/P_I) dependence of τ_o obtained from Equation (6.25) when w and a are constant, as in Equation 5.105):

$$\tau_s = \rho_I\,g_z\,w_I\left\{1 - \left(1 - \frac{\rho_I}{\rho_W}\right)\left(\frac{P_W}{P_I}\right)^2\right\}\left\{\frac{h_I\,a}{a\,x + h_0\,u_0}\right.$$

$$\left. - \frac{h_I^2\left(R_{xx}^{\,n-1}/A^n\right)}{a\,x + h_0\,u_0}\left[\frac{\rho_I\,g_z\,h_I}{4}\left(1 - \frac{\rho_I}{\rho_W}\right)\left(\frac{P_W}{P_I}\right)^2\right]^n\right\}$$

$$- \rho_I\,g_z\,w_I\,h_I\left(1 - \frac{\rho_I}{\rho_W}\right)\frac{\Delta}{\Delta x}\left(\frac{P_W}{P_I}\right)^2 - \frac{w_I}{h_I}$$

$$\left[\frac{(m+1)\,\rho_I\,g_z\,B^{2m}\,a^2\,(L-x)^{\frac{2m+1}{m}}}{(2m+1)\left[L^{\frac{m+1}{m}} - (L-x)^{\frac{m+1}{m}}\right]}\right]^{\frac{1}{2m+1}}\left(1 - \frac{P_W}{P_I}\right)^{\frac{4m}{2m+1}} \tag{6.33}$$

A simpler expression for τ_s is possible if plastic flow occurs in side shear zones. As τ_o is reduced by increasing P_W/P_I along an ice stream, τ_s and σ_{xx}' replace τ_o in resisting the gravitational pulling force. Applying the creep behavior depicted in Figure 4.8 to a generalized ice stream in Figure 6.3, the variation of side viscoplastic yield stress $(\tau_s)_v$ with x along length L_S of an ice stream can be represented as a flow curve during recrystallization, as shown in Figure 6.4. During recrystallization, a smooth decrease from an unstable upper viscoplastic yield stress τ_U at $x = L_S$, where converging flow has produced strain-hardened ice at the head of the ice stream, leads to a stable lower viscoplastic yield stress τ_S at $x = 0$, where laminar flow has produced strain-softened ice at the foot of the ice stream. An expression that gives this smooth variation of $(\tau_s)_v$ along the ice stream and allows τ_s to increase as τ_o decreases is:

$$\tau_s = [\tau_U\,sin^2\,(\pi\,x/2\,L_S) + \tau_S\,cos^2\,(\pi\,x/2\,L_S)]\,[P_W/P_I]^{\nu}$$

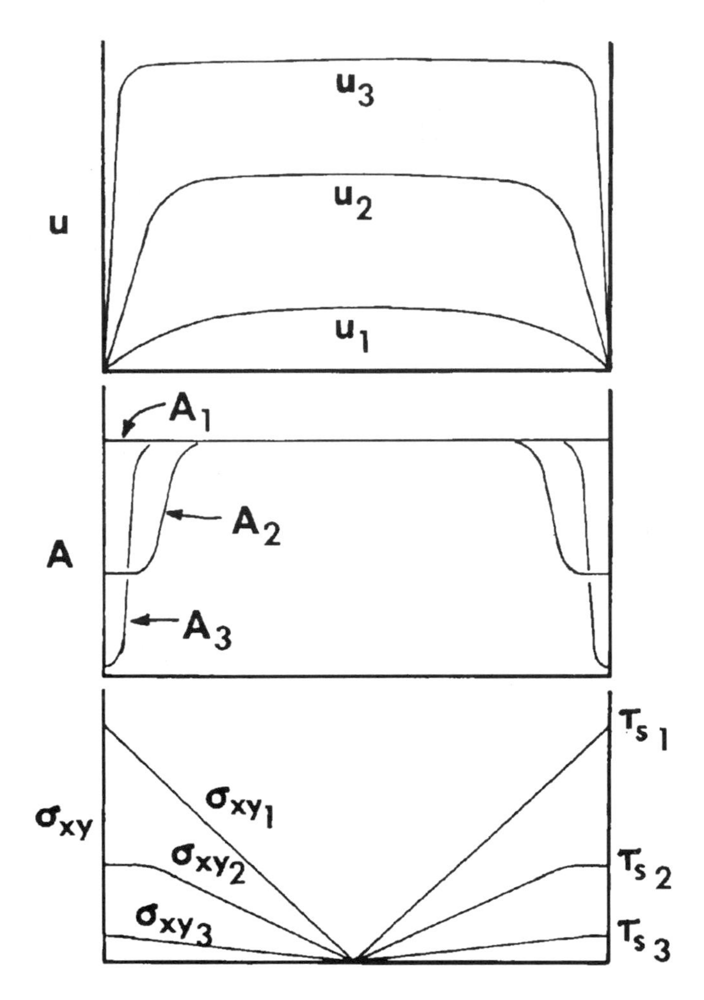

Figure 6.3: Surface velocities on an ice stream showing strain softening in the lateral shear zones. Subscripts 1, 2, and 3 refer to the ice before, during, and after recrystallization in the lateral shear zones. *Top*: Ice-surface velocity. *Middle*: Ice-hardness parameter. *Bottom*: Ice side-shear stress. The transverse profile of longitudinal velocity resembles the n = 3 curve in Figure 4.15 for no strain softening when ice enters Byrd Glacier fjord and resembles the n = 9 curve in Figure 4.15 for strain softened ice in the lateral shear zones when ice leaves Byrd Glacier fjord, with a smooth transition through the fjord.

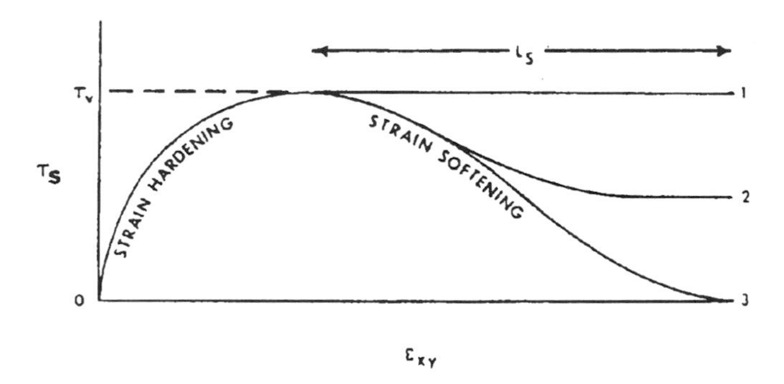

Figure 6.4: Possible flow curves for lateral shear in an ice stream. Strain hardening in the zone of converging flow at the head of the ice stream causes an increase in maximum side-shear stress τ_s and transverse shear strain ε_{xy} until viscoplastic yield stress τ_v is reached. Over length L_S of stream flow, τ_s is constant if strain hardening and softening rates are in balance (curve 1), τ_s decreases if strain softening dominates (curves 2 and 3), and τ_s drops to zero if strain softening leads to shear rupture (curve 3).

$$= (\tau_s)_v \, (P_W/P_I)^{\nu} \qquad (6.34)$$

where $(\tau_s)_v = \tau_U \, sin^2 \, (\pi \, x \, /2 \, L_S) + \tau_S \, cos^2 \, (\pi \, x \, /2 \, L_S)$ and ν is a constant in the range $1 \le \nu \le 4 \, m/(2 \, m + 1)$, with $\nu \approx 1$ if the reduction in τ_o is offset primarily by an increase in σ_{xx}' and $\nu \approx 4 \, m/(2 \, m + 1)$ if the reduction in τ_o is offset primarily by an increase in τ_s. Equation (6.34) is substituted for τ_s in Equation (6.19). If viscoplastic instability occurs in incremental length $\Delta x << L_S$ at the beginning of stream flow, then $(\tau_s)_v = \tau_S$ for strain softened ice along all of length L_S. If stream flow culminates with lateral shear rupture that produces rifts alongside the ice stream, then $\tau_S = 0$. Figure 3.34 shows lateral rifts where Byrd Glacier enters the Ross Ice Shelf in Antarctica. Top and bottom shear crevasses extend down and up until net lithostatic pressure $P_I - P_W$ equals the shear rupture stress (see Chapter 8) and rifts form when these crevasses meet in lateral shear zones. If $\tau_U = \tau_S$, then the decrease in τ_o is accommodated primarily by increasing τ_s. Setting $\upsilon = 4 \, m/(2 \, m + 1)$ and $2 \, (\tau_s)_v \, /w_I = (\tau_o)_s, \, h_I$, Equation (6.34) now becomes:

$$\tau_s = (\tau_s)_v \, (P_W/P_I)^{\frac{4m}{2m+1}} = (w_I/2\,h_I)\,(\tau_o)_s\,(P_W/P_I)^{\frac{4m}{2m+1}} = \frac{w_I}{2\,h_I} \left[\frac{m\,\rho_I\,g_z\,B_o^2\,(\Lambda_o/\Lambda'_o)^{4m}\,[\Gamma(x)/w(x)]^{\frac{2m+1}{m}}}{(2\,m+1)\,\Pi_I(x)} \right]^{\frac{1}{2m+1}} \left(\frac{P_W}{P_I}\right)^{\frac{4m}{2m+1}} \qquad (6.35)$$

This assumes that, because τ_o and τ_s produce laminar flow but σ_{xx}' does not, increased side shear will be more able to generate the laminar flow lost by decreased basal shear. Therefore, the increase in τ_s should more closely mirror the decrease in τ_o (Whillans and Van der Veen, 1997). Note that $\tau_s = 0$ for sheet flow over $L_S \le x \le L$ when $P_W = 0$ for a frozen bed and $P_W << P_I$ for a thawed bed.

Variable Basal Buoyancy Along Ice Streams

Substituting Equations (6.25) and (6.35) into Equation (6.19) incorporates basal buoyancy factor P_W/P_I into every term affecting surface slope $\Delta h/\Delta x$ for an ice stream:

$$\frac{\Delta h}{\Delta x} \approx \left(1 - \frac{\rho_I}{\rho_W}\right)\left(\frac{P_W}{P_I}\right)^2 \left\{ \frac{h_I\,a}{a\,x + h_o\,u_o} - \frac{h_I^2\,R_{xx}^{n-1}}{a\,x + h_o\,u_o} \left[\frac{\rho_I\,g_z\,h_I}{4\,A} \left(1 - \frac{\rho_I}{\rho_W}\right)\left(\frac{P_W}{P_I}\right)^2 \right]^n \right\}$$
$$+ \frac{h_I}{2}\left(1 - \frac{\rho_I}{\rho_W}\right)\frac{\Delta}{\Delta x}\left(\frac{P_W}{P_I}\right)^2 + \left[\left(\frac{m+1}{2\,m+1}\right)\left(\frac{B}{\rho_I\,g_z\,h_I}\right)^{2m}\left(\frac{a^2}{h_I}\right) \frac{(L-x)^{\frac{2m+1}{m}}}{\left[L^{\frac{m+1}{m}}\,(L-x)^{\frac{m+1}{m}}\right]} \right]^{\frac{1}{2m+1}} \left[\left(\frac{P_W}{P_I}\right)^{\frac{4m}{2m+1}} + \left(1 - \frac{P_W}{P_I}\right)^{\frac{4m}{2m+1}} \right] \qquad (6.36)$$

where w and a are taken to be constant for stream flow and $B = B_o \, (\Lambda_o \, /\Lambda'_o)^2$.

A solution of Equation (6.36) for an ice stream requires knowing how P_W/P_I varies along length L_S of an ice stream. These variations are unknown at present, but they should all depend on basal coupling. An expression that displays the spectrum of how basal coupling, defined by P_W/P_I, might vary with x along length L_S is:

$$\frac{P_W}{P_I} = \left(1 - \frac{x}{L_S}\right)^c \qquad (6.37)$$

where $c = \infty$ for sheet flow, $0 < c < \infty$ for stream flow, and $c = 0$ for shelf flow. Figure 6.5 is a plot of Equation (6.37) over this range of c values. Equation (6.37) gives the following gradient of P_W/P_I along length L_S of an ice stream:

$$\frac{\Delta}{\Delta x}\left(\frac{P_W}{P_I}\right) = \frac{\Delta}{\Delta x}\left(1 - \frac{x}{L_S}\right)^c$$
$$= -\frac{c}{L_S}\left(1 - \frac{x}{L_S}\right)^{c-1} = -\frac{c}{L_S - x}\left(1 - \frac{x}{L_S}\right)^c \qquad (6.38)$$

Substituting Equations (6.37) and (6.38) into Equation (6.36):

$$\frac{\Delta h}{\Delta x} \approx \left(1 - \frac{\rho_I}{\rho_W}\right)\left(1 - \frac{x}{L_S}\right)^{2c} \left\{ \frac{h_I\,a}{a\,x + h_o\,u_o} - \frac{h_I^2\,R_{xx}^{n-1}}{a\,x + h_o\,u_o} \left[\frac{\rho_I\,g_z\,h_I}{4\,A}\left(1 - \frac{\rho_I}{\rho_W}\right)\left(1 - \frac{x}{L_S}\right)^{2c} \right]^n \right\}$$

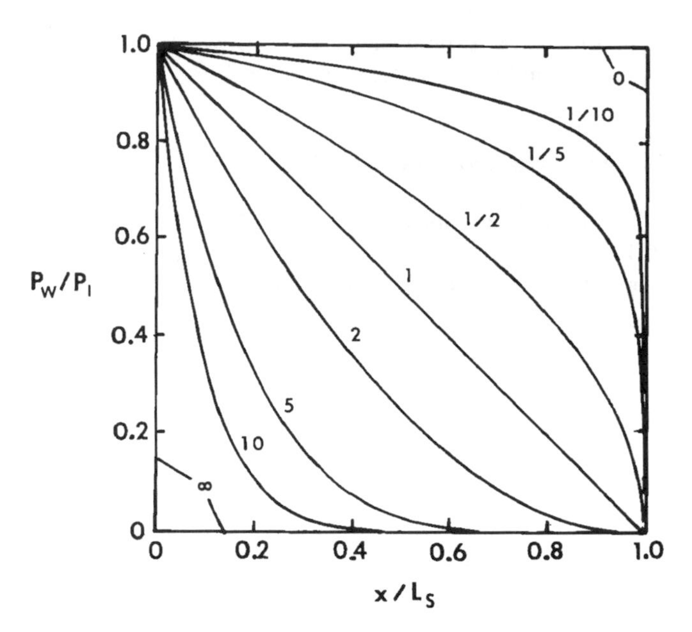

Figure 6.5: A plot of Equation (6.37) for the variation of basal buoyancy factor P_W/P_I with distance x along length L_S of stream flow between sheet flow ($c = \infty$) and shelf flow ($c = 0$).

$$- \frac{c\, h_I}{L_S - x}\left(1 - \frac{\rho_I}{\rho_W}\right)\left(1 - \frac{x}{L_S}\right)^{2c} + \frac{2\,(\tau_s)_v}{\rho_I\, g_z\, w_I}\left(1 - \frac{x}{L_S}\right)^{\frac{4mc}{2m+1}}$$

$$+ \frac{(\tau_o)_S}{\rho_I\, g_z\, h_I}\left[1 - \left(1 - \frac{x}{L_S}\right)^{c}\right]^{\frac{4m}{2m+1}} \qquad (6.39)$$

Equation (6.39) gives $\Delta h/\Delta x$ along L_S if bed topography is known beneath the ice streams.

Components of the Surface Slope

The contribution to the surface slope of an ice stream made by each term in Equation (6.39) can be determined by computing $\Delta h/\Delta x$ when $h_I = h$ for a horizontal bed. These terms are then:

$$\left(\frac{\Delta h}{\Delta x}\right)_1 \approx \left(1 - \frac{\rho_I}{\rho_W}\right)\left(1 - \frac{x}{L_S}\right)^{2c}\left\{\frac{a\, h}{a\, x + h_0\, u_0} - \frac{h^2 R_{xx}^{\,n-1}}{a\, x + h_0\, u_0}\left[\frac{\rho_I\, g_z\, h}{4A}\left(1 - \frac{\rho_I}{\rho_W}\right)\left(1 - \frac{x}{L_S}\right)^{2c}\right]^{n}\right\} \qquad (6.40a)$$

$$\left(\frac{\Delta h}{\Delta x}\right)_2 = - \frac{c\, h}{L_S - x}\left(1 - \frac{\rho_I}{\rho_W}\right)\left(1 - \frac{x}{L_S}\right)^{2c} \qquad (6.40b)$$

$$\left(\frac{\Delta h}{\Delta x}\right)_3 = + \frac{2\,(\tau_s)_v}{\rho_I\, g_z\, w_I}\left(1 - \frac{x}{L_S}\right)^{\frac{4mc}{2m+1}} \qquad (6.40c)$$

$$\left(\frac{\Delta h}{\Delta x}\right)_4 = + \frac{(\tau_o)_S}{\rho_I\, g_z\, h}\left[1 - \left(1 - \frac{x}{L_S}\right)^{c}\right]^{\frac{4m}{2m+1}} \qquad (6.40d)$$

Terms $(\Delta h/\Delta x)_1$ through $(\Delta h/\Delta x)_4$ are the respective contribution to $\Delta h/\Delta x$ from σ_{xx}', $\Delta\sigma_{xx}'/\Delta x$, τ_s, and τ_o.

Variable Bed Topography along Ice Streams

The effect of variable bed topography on surface slope $\Delta h/\Delta x$ for an ice stream is included by writing Equation (6.39) as an initial-value finite-difference recursive formula in which $\Delta h = h_{i+1} - h_i$ is the surface elevation change in constant length increment Δx for each of i steps measured from $i = 0$ at $x = 0$ to $i = L_S/\Delta x$ at $x = L_S$, where $h_I = h - h_R$ is ice thickness for surface height h and bedrock height or depth h_R above or below present-day sea level, and i is an integer:

$$h_{i+1} = h_i + \left(1 - \frac{\rho_I}{\rho_W}\right)\left(1 - \frac{x}{L_S}\right)^{2c}\left\{\frac{a\,(h - h_R)_i}{a\, x + h_0\, u_0} - \frac{R_{xx}^{\,n-1}(h - h_R)_i^{\,n+2}}{a\, x + h_0\, u_0}\left[\frac{\rho_I\, g_z}{4A}\left(1 - \frac{\rho_I}{\rho_W}\right)\left(1 - \frac{x}{L_S}\right)^{2c}\right]^{n}\right\}\Delta x$$

$$- \left[\frac{c\,(h - h_R)_i}{L_S - x}\left(1 - \frac{\rho_I}{\rho_W}\right)\left(1 - \frac{x}{L_S}\right)^{2c}\right]\Delta x$$

$$+ \left[\frac{2\,(\tau_s)_v}{\rho_I\, g_z\, w_I}\left(1 - \frac{x}{L_S}\right)^{\frac{4mc}{2m+1}}\right]\Delta x$$

$$+ \frac{(\tau_o)_v}{\rho_I\, g_z\,(h - h_R)_i}\left[1 - \left(1 - \frac{x}{L_S}\right)^{c}\right]^{\frac{4m}{2m+1}}\Delta x \qquad (6.41)$$

Equation (6.40) shows that $(\Delta h/\Delta x)_2$ is unimportant except when $x \to 0$, and $(\Delta h/\Delta x)_3 \ll (\Delta h/\Delta x)_4$ when $(\tau_s)_v \approx (\tau_o)_S$ and $w_I \gg h$. In general, therefore, the most important terms in Equation (6.40) are $(\Delta h/\Delta x)_1$ and $(\Delta h/\Delta x)_4$. A reasonable approximation of Equation (6.41) is obtained from just these two terms, with $(\tau_o)_S = \tau_S$ given by Equation (5.105) for constant w and a:

$$h_{i+1} = h_i + \left(1 - \frac{\rho_I}{\rho_W}\right)\left(1 - \frac{x}{L_S}\right)^{2c}\left\{\frac{a\,(h - h_R)_i}{a\, x + h_0\, u_0} - \frac{R_{xx}^{\,n-1}(h - h_R)_i^{\,n+2}}{a\, x + h_0\, u_0}\left[\frac{\rho_I\, g_z}{4A}\left(1 - \frac{\rho_I}{\rho_W}\right)\left(1 - \frac{x}{L_S}\right)^{2c}\right]^{n}\right\}\Delta x$$

$$+ \left\{\left[\left(\frac{m+1}{2m+1}\right)\left(\frac{B}{\rho_I\, g_z}\right)^{2m}\frac{a^2}{(h - h_R)^{2m+1}}\frac{(L - x)^{\frac{2m+1}{m}}}{\left[L^{\frac{m+1}{m}} - (L - x)^{\frac{m+1}{m}}\right]}\right]^{\frac{1}{2m+1}}\left[1 - \left(1 - \frac{x}{L_S}\right)^{c}\right]^{\frac{4m}{2m+1}}\right\}\Delta x \qquad (6.42)$$

Equation (6.42) has the advantage of specifying L_S as the surface inflection point along x where a concave profile becomes convex, thereby converting L_S from input in Equation (6.41) to output in Equation (6.42).

In Equation (6.42), if a is the average accumulation rate $\bar{a}$, it is obtained by numerically integrating Equation (6.28) over the entire length L of the flowband:

$$\bar{a} = \frac{\Gamma(x)}{w_I L} = \int_0^L \frac{(a - dh/dx)\, w\, dx}{w_I L} = \frac{\Delta x}{w_I L} \sum_{i=0}^{i=L/\Delta x} (a_i - \delta h_i / \delta t)\, w_i \tag{6.43}$$

Ice velocity u_0 at $x = 0$ in Equation (6.42) is given by numerically integrating Equation (5.41) for a flowband of variable width w for sheet flow that supplies an ice stream of constant width $w_I = w_0$:

$$u_0 = \frac{\Gamma(x)}{w_0 h_0} = \int_0^L \frac{(a - dh/dx)\, w\, dx}{w_0 h_0} = \frac{\Delta x}{w_0 h_0} \sum_{i=0}^{i=L/\Delta x} \left(a_i - \delta h_i / \delta t\right) w_i \tag{6.44}$$

where u_0 is negative because x is positive upstream. The actual accumulation or ablation rate minus the actual thickening or thinning rate, (a – dh/dt), can be replaced by an effective accumulation or ablation rate.

The Equilibrium Length of Ice Streams

Downslope buttressing determines length L_S of an ice stream. In computing downslope buttressing, let L_S replace Δx and $h_S - h_0$ replace Δh in the force diagram in Figure 6.2, where $x = 0$ at $h_I = h_0$ and $x = L_S$ at $h_I = h_S$. If stream flow is caused by longitudinal gravitational deviator stress σ_{xx}', then L_S is defined as the point along x where $\sigma_{xx}' = 0$. The resulting longitudinal force balance is shown in Figure 6.6. By analogy with Equation (6.12), and using Equation (6.3), the horizontal gravitational force is:

$$\begin{aligned}(F_x)_G = &-\tfrac{1}{2}\rho_I g_z h_S (w_I h_S) \\ &+ \tfrac{1}{2}\rho_I g_z [h_0 - h_W(\rho_W/\rho_I)]\, w_I [h_0 - h_W(\rho_W/\rho_I)] \\ &+ \rho_I g_z [h_0 - h_W(\rho_W/\rho_I)]\, w\, h_W(\rho_W/\rho_I) \\ &+ \tfrac{1}{2}\rho_W g_z h_W (w_I h_W) \\ = &-\tfrac{1}{2}\rho_I g_z w_I \left(h_S^2 - h_0^2\right) \\ &- \tfrac{1}{2}\rho_I g_z w_I h_0^2 (1 - \rho_I/\rho_W)(P_W/P_I)_0^2 \end{aligned} \tag{6.45}$$

and the horizontal kinematic force, by analogy with Equation (6.13), is:

$$\begin{aligned}(F_x)_K &= 2\sigma_S w_I h_S + 2\bar{\tau}_s \bar{h}_I L_S + \bar{\tau}_o w_I L_S \\ &\approx \bar{\tau}_s (h_S + h_0) L_S + \bar{\tau}_o w_I L_S \end{aligned} \tag{6.46}$$

where $\bar{\tau}_o$ is the average basal shear stress along L_S, $\bar{\tau}_s$ is the average side shear stress along L_S, $\bar{h}_I \approx 1/2\,(h_S + h_0)$ is the average ice thickness along L_S, $(P_W/P_I)_0$ is P_W/P_I at $x = 0$, and, by definition, $\sigma_S = (\sigma_{xx}')_S = 1/2\,(\sigma_{xx} - \sigma_{zz})_S = 0$ at $x = L_S$. Setting $(F_x)_G + (F_x)_K = 0$ and solving for L_S:

$$L_S \approx \frac{\rho_I g_z w_I \left(h_S^2 - h_0^2\right) + \rho_I g_z w_I h_0^2 (1 - \rho_I/\rho_W)(P_W/P_I)_0^2}{2\bar{\tau}_s (h_S + h_0) + 2\bar{\tau}_c w_I} \tag{6.47}$$

If the ice stream ends as a freely floating ice tongue, $(P_W/P_I)_0 = 1$ and longitudinal stress σ_0 at $x = 0$ is pulling stress σ_P given by Equation (6.15) at $h_I = h_0$:

$$\sigma_P = 1/4\, \rho_I g_z h_0 (1 - \rho_I/\rho_W) \tag{6.48}$$

If the ice stream ends as an ice tongue imbedded in an ice shelf that is confined in an embayment and is pinned to islands and shoals, or ends as an ice lobe grounded on land, then σ_P is resisted by kinematic buttressing stress σ_B needed to push the ice shelf or ice lobe forward. In these cases σ_0 is:

$$\begin{aligned}\sigma_0 &= 1/4\, \rho_I g_z h_0 (1 - \rho_I/\rho_W)(P_W/P_I)_0^2 \\ &= 1/4\, \rho_I g_z h_0 (1 - \rho_I/\rho_W) - \sigma_B = \sigma_P - \sigma_B \end{aligned} \tag{6.49}$$

where σ_B produces an effective $(P_W/P_I)_0$ caused by a constraining ice shelf or a grounded ice lobe. Ice-shelf constraints result from grounding at pinning points and along the sides of a confining embayment, places where $P_W/P_I = 0$. Solving Equation (6.49) for $(P_W/P_I)_0$:

$$\left(\frac{P_W}{P_I}\right)_0 = \left[\frac{\sigma_0}{1/4\, \rho_I g_z h_0 (1 - \rho_I/\rho_W)}\right]^{1/2}$$

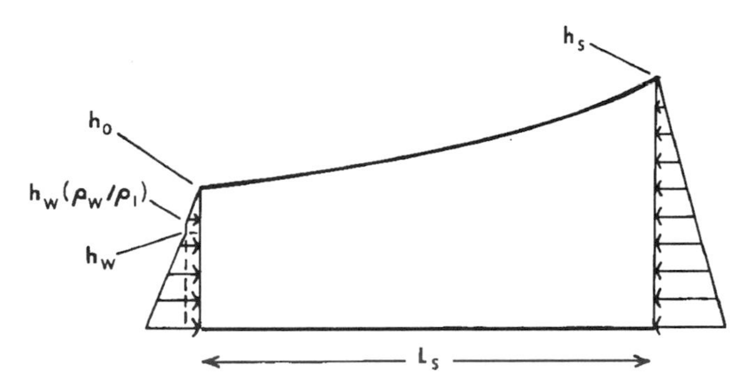

Figure 6.6: The longitudinal force balance on an ice stream. Horizontal gravitational force $(F_x)_S$ at the head of the ice stream is resisted by horizontal gravitational force $(F_x)_0$ at the foot of the ice stream and by kinematic traction forces $2\,\bar{\tau}_s \bar{h}_I L_S$ for side shear and $\bar{\tau}_o w_I L_S$ for basal shear along an ice stream of length L_S, width w_I, and mean thickness $\bar{h}_I$, mean side shear stress $\bar{\tau}_s$ and mean basal shear stress $\bar{\tau}_o$.

$$= \left[\frac{1/4\,\rho_I\,g_z\,h_0\,(1-\rho_I/\rho_W)-\sigma_B}{1/4\,\rho_I\,g_z\,h_0\,(1-\rho_I/\rho_W)}\right]^{1/2} = \left(\frac{\sigma_P-\sigma_B}{\sigma_P}\right)^{1/2} \quad (6.50)$$

Equation (6.47) can now be written to include σ_B:

$$L_S \approx \frac{\rho_I\,g_z\,w_I\left(h_S^2-h_0^2\right)+\rho_I\,g_z\,w_I\,h_0^2\,(1-\rho_I/\rho_W)-4\,w_I\,h_0\,\sigma_B}{2\,\bar{\tau}_s\,(h_S+h_0)+2\,\bar{\tau}_o\,w_I}$$

$$\approx \frac{\rho_I\,g_z\,w_I\left[h_S^2-(\rho_I/\rho_W)\,h_0^2\right]-4\,w_I\,h_0\,\sigma_B}{2\,\bar{\tau}_s\,(h_S+h_0)+2\,\bar{\tau}_o\,w_I} \quad (6.51)$$

It is difficult to calculate $\bar{\tau}_o$ when τ_o has the complex variation along x given by Equation (6.25). However, a simplified expression that has the correct qualitative variation is obtained if τ_o decreases in proportion to the ice overburden unsupported by the bed, so that $\tau_o = (\tau_o)_v\ (1 - P_W/P_I)$, where proportionality constant $(\tau_O)_V$ is the viscoplastic yield stress of basal ice given by Equation (6.31). Letting P_W/P_I vary along x according to Equation (6.37):

$$\bar{\tau}_o\,L_S = \int_0^{L_S} \tau_o\,dx = \int_0^{L_S} (\tau_o)_v\,(1-P_W/P_I)\,dx$$

$$= (\tau_o)_v \int_0^{L_S} \left[1-(1-x/L_S)^c\right] dx$$

$$= \frac{c}{c+1}\,(\tau_o)_v\,L_S = \frac{c}{c+1}\,\sigma_o\,L_S \quad (6.52)$$

This is equivalent to setting 4 m/(2 m + 1) = 1 when $(\tau_o)_S = (\tau_o)_v$ in Equation (6.25) so that m = 1/2. Since m = (n + 1)/2 in the Weertman (1957a) theory of glacial sliding, m = 1/2 requires that n = 0 in the flow law written $\dot{\varepsilon} = \dot{\varepsilon}_o\,(\sigma/\sigma_o)^n$ in Equation (4.136) so that $\dot{\varepsilon} = \dot{\varepsilon}_o$. For all values of n, $\dot{\varepsilon} = \dot{\varepsilon}_o$ when $\sigma = \sigma_o$ is the plastic yield stress. Therefore, Equation (6.52) requires that $(\tau_o)_v = \sigma_o$.

It is easier to calculate $\bar{\tau}_s$ when τ_s varies along x according to Equation (6.35). However, a simplification comparable to that used for τ_o in Equation (6.52) is to allow side traction to increase in proportion to the decrease in basal traction so that $\tau_s = (\tau_s)_v\ (P_W/P_I)$, where proportionality constant $(\tau_s)_v$ is the viscoplastic yield stress in the side shear zones. Letting P_W/P_I vary along x according to Equation (6.37):

$$\bar{\tau}_s\,L_S = \int_0^{L_S} \tau_s\,dx$$

$$= \int_0^{L_S} (\tau_s)_v\,(P_W/P_I)\,dx = (\tau_s)_v \int_0^{L_S} (1-x/L_S)^c\,dx$$

$$= \frac{1}{c+1}\,(\tau_s)_v\,L_S = \frac{1}{c+1}\,\sigma_o\,L_S \quad (6.53)$$

This is equivalent to setting 4 m/(2 m + 1) = 1 in Equation (6.35) so that $(\tau_s)_v = \sigma_o$.

Substituting Equations (6.52) and (6.53) into Equation (6.51):

$$L_S \approx \frac{\rho_I\,g_z\,w_I\left[h_S^2-(\rho_I/\rho_W)\,h_0^2\right]-4\,w_I\,h_0\,\sigma_B}{2\left(\frac{1}{c+1}\right)\sigma_o\,(h_S+h_0)+2\left(\frac{c}{c+1}\right)\sigma_o\,w_I} \quad (6.54)$$

Setting $L_S = 0$ and $h_S = h_0$ for maximum buttressing, solving Equation (6.54) for the maximum σ_B gives:

$$\sigma_B \approx 1/4\,\rho_I\,g_z\,h_0\,(1-\rho_I/\rho_W) \quad (6.55)$$

Note that σ_P in Equation (6.48) is the same as σ_B in Equation (6.55), as required by Equation (6.50), in which $\sigma_P = \sigma_B$ when $(P_W/P_I)_0 = 0$. Setting $\rho_I = 917$ kg m^{-3}, $\rho_W = 1020$ kg m^{-3}, $g_z = 9.8$ m s^{-2}, and $h_0 = 0.7$ km in Equation (6.55) gives $\sigma_B = 16$ bars = 1600 kPa. Maximum buttressing stresses for the Ross Ice Shelf in Antarctica are $\sigma_B \approx 300$ kPa (Thomas and MacAyeal, 1982; Jezek, 1984), for which $h_0 \approx 133$ m in Equation (6.55). Therefore σ_B is of minor importance in determining the length of an ice stream, so it does not need a robust theoretical formulation. Figure 6.7 is a plot of L_S versus h_S in Equation (6.54) for $\rho_I = 917$ kg m^{-3}, $\rho_W = 1020$ kg m^{-3}, $g_z = 9.8$ m s^{-2}, $w_I = 30$ km, $h_0 = 0.7$ km, $\sigma_o = 100$ k Pa, $\sigma_B = 0$, and $0 \le c \le \infty$.

An expression for σ_B for a marine ice stream depends on the form drag and the dynamic drag of the ice shelf (MacAyeal, 1987). As an approximation, σ_B will reduce maximum pulling stress σ_P at the grounding line in proportion to the ratio of grounded perimeter p_G to total perimeter p of the ice shelf beyond the grounding line of the ice stream:

$$\sigma_B = \sigma_P\,(p_G/p) = 1/4\,\rho_I\,g_z\,h_0\,(1-\rho_I/\rho_W)\,p_G/p \quad (6.56)$$

Ice-shelf buttressing is absent when $p_G = 0$ and total when $p_G = p$, where p_G includes grounding lines between ice streams supplying the ice shelf and grounding lines around islands, ice rises, and ice rumples within the ice shelf, and p includes the calving front of the ice shelf.

An expression for σ_B for a terrestrial ice stream depends on the variation of basal coupling along its terminal ice lobe. Mean basal shear stress $\bar{\tau}_o$ exerted beneath an ice lobe of length L_L and mean width $\bar{w}_L$ produces a kinematic traction force $(F_x)_K = \bar{\tau}_o\,\bar{w}_L\,L_L$ that creates buttressing force $\sigma_B\,w_0\,h_0$ at x = 0. If $\bar{\tau}_o$ beneath the ice lobe has the same spectrum of variations as has $\bar{\tau}_o$ beneath the ice stream, then σ_B is given by:

$$\sigma_B = \left(\frac{\bar{w}_L\,L_L}{w_0\,h_0}\right)\bar{\tau}_o = \frac{C}{C+1}\left(\frac{\bar{w}_L\,L_L}{w_0\,h_0}\right)(\tau_o)_v \quad (6.57)$$

where $0 \le C \le \infty$, by analogy with Equation (6.52), but C for the ice lobe can be different from c for the ice stream. Ice-lobe buttressing is absent when C = 0 and a maximum when C = ∞.

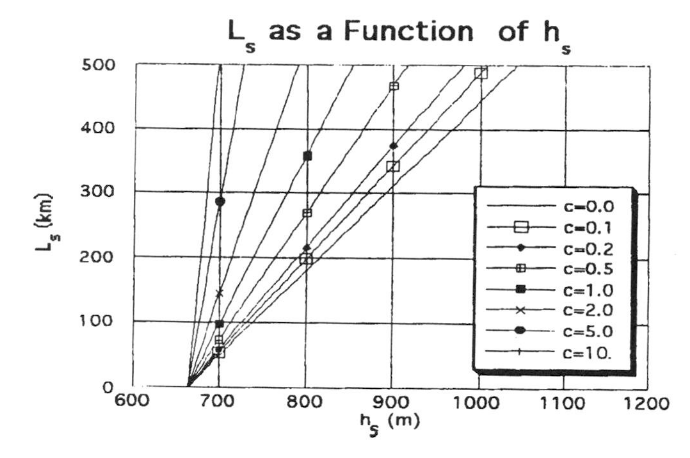

Figure 6.7: A plot of Equation (6.54) showing the variation of ice stream length L_S with h_S when basal buoyancy factor P_W / P_I is specified by Equation (6.37) for $0 \leq c \leq 10$. Specified quantities are $\rho_I = 917$ kg m^{-3}, $\rho_W = 1020$ kg m^{-3}, $g_z = 9.8$ m s^{-2}, $h_0 = 700$ m, $w_I = 30$ km, $\sigma_o = 100$ kPa, and $\sigma_B = 0$.

If an ice stream develops when a layer of ice-cemented sediments that blankets bedrock becomes thawed, the layer will provide little or no traction for the overlying ice, so that $c \approx 0$. Frictional basal heat, being the product $\tau_o u_S$, will be greatest at the head of the ice stream where τ_o is a maximum, and this will force the basal melting isotherm to migrate toward the ice divide, so L_S increases. The thawed sediment layer will be eroded by the ice stream, and eventually the ice will meet the higher points of bedrock, so basal traction increases and c therefore increases. As glacial erosion strips away the soft sediment layer, c continues to increase and causes L_S to decrease as shown in Figure 6.7.

The Changing Surface of Ice Streams

The spectrum of c values in Equation (6.37) produces a spectrum of ice stream surface profiles. The four contributions to surface slope are given by Equations (6.40). Take $\rho_I = 917$ kg m^{-3}, $\rho_W = 1020$ kg m^{-3}, $g_z = 9.8$ m s^{-2}, $a = 0.2$ m a^{-1}, $u_0 = -1000$ m a^{-1}, $h_0 = 0.7$ km, $w_I = 30$ km, $R_{xx} = 1$, $n = 3$, $m = 2$, $(\tau_s)_v = (\tau_o)_v = \sigma_o = 100$ k Pa, and $A = 470$ k Pa a$^{1/n}$ for $A^{-n} = 3.1 \times 10^{-16}$ k Pa^{-3} s^{-1} when -15°C is the mean ice temperature in Table 3.3 of Paterson (1981). By definition, 1 k Pa = 10^3 kg m^{-1} s^{-2}. Figure 6.8 plots each of these terms when L_S is given by Equation (6.54) for $\sigma_B = 0$ and values of c in the range $0 < c < \infty$. Plots of these terms show that $(\Delta h / \Delta x)_1$ dominates when $c < 0.4$, $(\Delta h / \Delta x)_2$ is unimportant except when $c > 1$, $(\Delta h / \Delta x)_3 \ll (\Delta h / \Delta x)_4$ except when $c \approx 0.5$ and $x < 100$ km, and $(\Delta h / \Delta x)_4$ dominates when $c > 0.4$.

Figure 6.9 is a plot of Equation (6.41) for $h_R = 0$ and various values of c and σ_B. The depression near the grounding line for $5 < c < 10$ is due to $\Delta\sigma_{xx}'/\Delta x$, as given by Equation (6.40b). A depression of this kind was observed by Swithinbank (1955) and Robin (1958) along the grounding line of Antarctic ice shelves where sheet flow extended to the ice shelf ($c > 1$). Note the rapid change from a concave to a convex profile in the range $0.4 \leq c \leq 0.5$. This implies that the transition from stream flow to sheet flow is highly sensitive to P_W / P_I. Also, note that as c increases from $c = 0$ to $c = \infty$, the ice stream first lengthens and then shortens, with stream flow producing a concave surface profile.

Figure 6.10 presents plots of Equation (6.42) for $m = 2$ and $m = 0.5$. In both cases, the slight depression at the foot of the ice stream is gone because the $\Delta\sigma_{xx}'/\Delta x$ contribution is excluded. Profiles for $m = 2$ are based on viscoplastic creep ($n = 3$) around bedrock bumps, in Equation (4.26). Profiles for $m = 1/2$ are based on plastic creep ($n = 0$) around these bumps, as shown by Equations (4.26) and (4.136), where $\dot{\varepsilon} = \dot{\varepsilon}_o$ for both $n = 0$ and $n = \infty$ when $\sigma = \sigma_o$ is the plastic yield stress. Plastic creep around bumps lowers the range of c where stream flow becomes sheet flow from $0.4 \leq c \leq 0.5$ to $0.3 \leq c \leq 0.4$.

Pulling Power during the Life Cycle of Ice Streams

Pulling power is the ability of an ice stream to reach deep into an ice sheet and pull out the ice. Formally, pulling power P_x at some position on an ice stream is the product of the mean velocity $\bar{u}_x$ and pulling force F_x at the point, both measured horizontally along direction x of flow:

$$P_x = \bar{u}_x F_x \tag{6.58}$$

where $\bar{u}_x$ and F_x are averaged through ice of height h in a flowband of width w. For steady-state flow, $\bar{u}_x$ is the equilibrium mass-balance velocity u. The pulling power of ice streams enables ice streams to discharge some 90 percent of the ice from ice sheets. Sheet flow pushes ice forward, but stream flow pulls ice forward. A life cycle for ice streams can be attributed to temporal variations in their pulling power P_x, defined by Equation (6.58).

Average ice velocity $\bar{u}_x$ at distance x upstream from the foot of an ice stream is equal to mass-balance velocity u given by Equation (5.41) integrated from 0 to x for constant accumulation rate a and width w_I:

$$\bar{u}_x = u = \frac{u_0 w_I h_0 - \int_0^x \left(a - dh_I/dt\right) w_I \, dx}{w_I h_I}$$

$$= \left(u_0 h_0 / h_I\right) - \left(a x / h_I\right) + \left(\Delta x / h_I\right) \sum_{i=0}^{i=x/\Delta x} \left(\delta h_I / \delta t\right)_i \tag{6.59}$$

where u_0 is negative for x positive in the upslope direction, $dh_I/dt > 0$ for a thickening ice sheet, $dh_I/dt = 0$ for a steady-state ice sheet, and $dh_I/dt < 0$ for a thinning ice sheet.

The gravitational pulling force at distance x from the foot of an ice stream, where it either becomes a marine ice

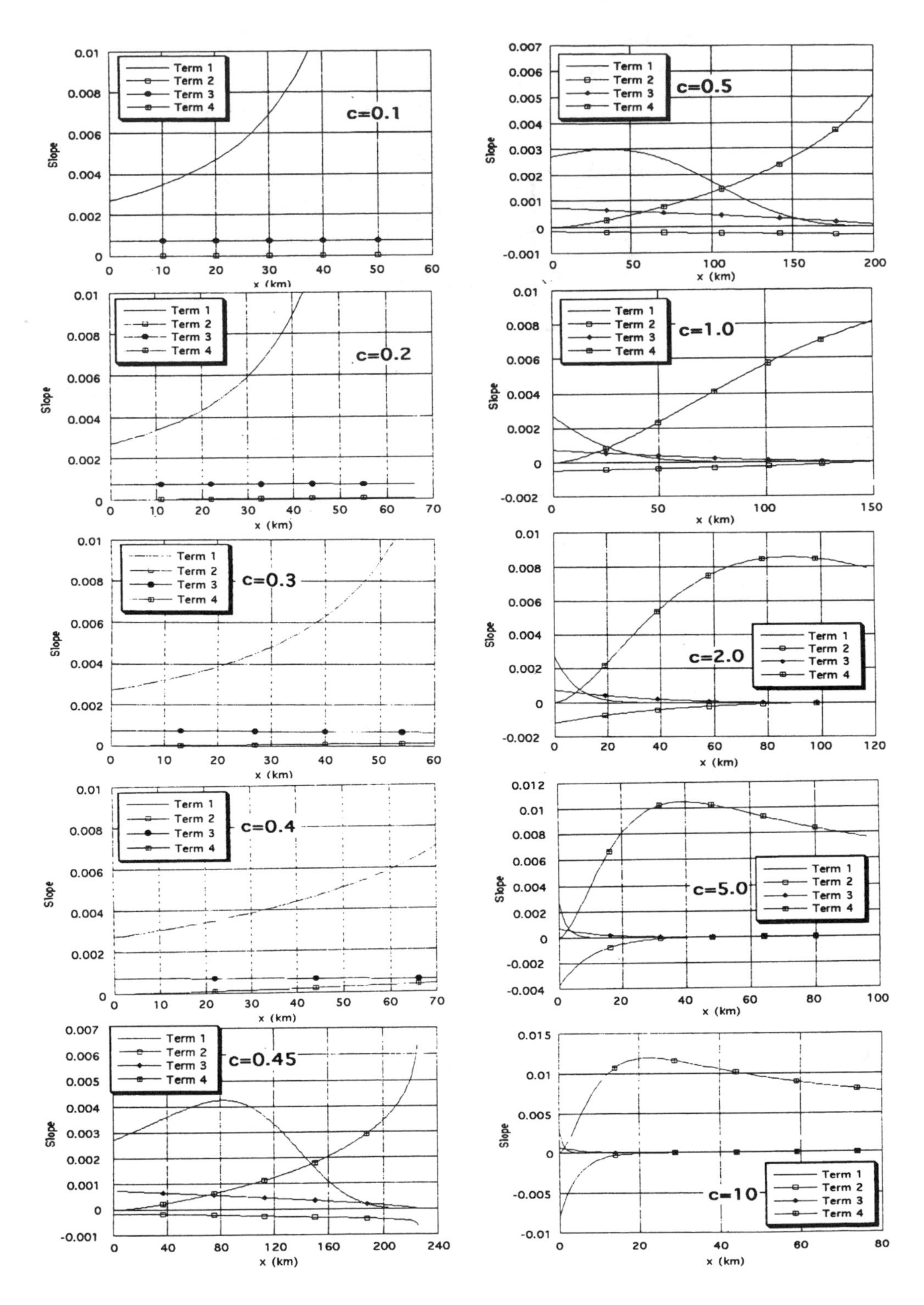

Figure 6.8: A plot of Equation (6.40) for an ice stream on a horizontal bed. Plotted separately are the contributions of (a) Term 1: longitudinal stress σ'_{xx}, (b) Term 2: longitudinal stress gradient $\partial\sigma'_{xx}/\partial x$, (c) Term 3: side shear stress τ_s, and (d) Term 4: basal shear stress τ_o to surface slope $\Delta h/\Delta x$ of an ice stream along length L_S for c in the range $0 \leq c \leq \infty$ in Equation (6.37). Specified quantities are $\rho_I = 917$ kg m^{-3}, $\rho_W = 1020$ kg m^{-3}, $g_z = 9.8$ m s^{-2}, $w_I = 30$ km, $\Delta x = 1$ km, $h_R = 0$, $h_O = 700$ m, $u_O = -1000$ m a^{-1}, $a = 0.2$ m a^{-1}, $n = 3$, $m = 2$, $A = 470$ kPa a$^{1/3}$, $(\tau_s)_v = (\tau_o)_S = 100$ kPa, $R_{xx}^* = 1$, and L_S given by Equation (6.54) for $\sigma_B = 0$. Plots by Paul Prescott.

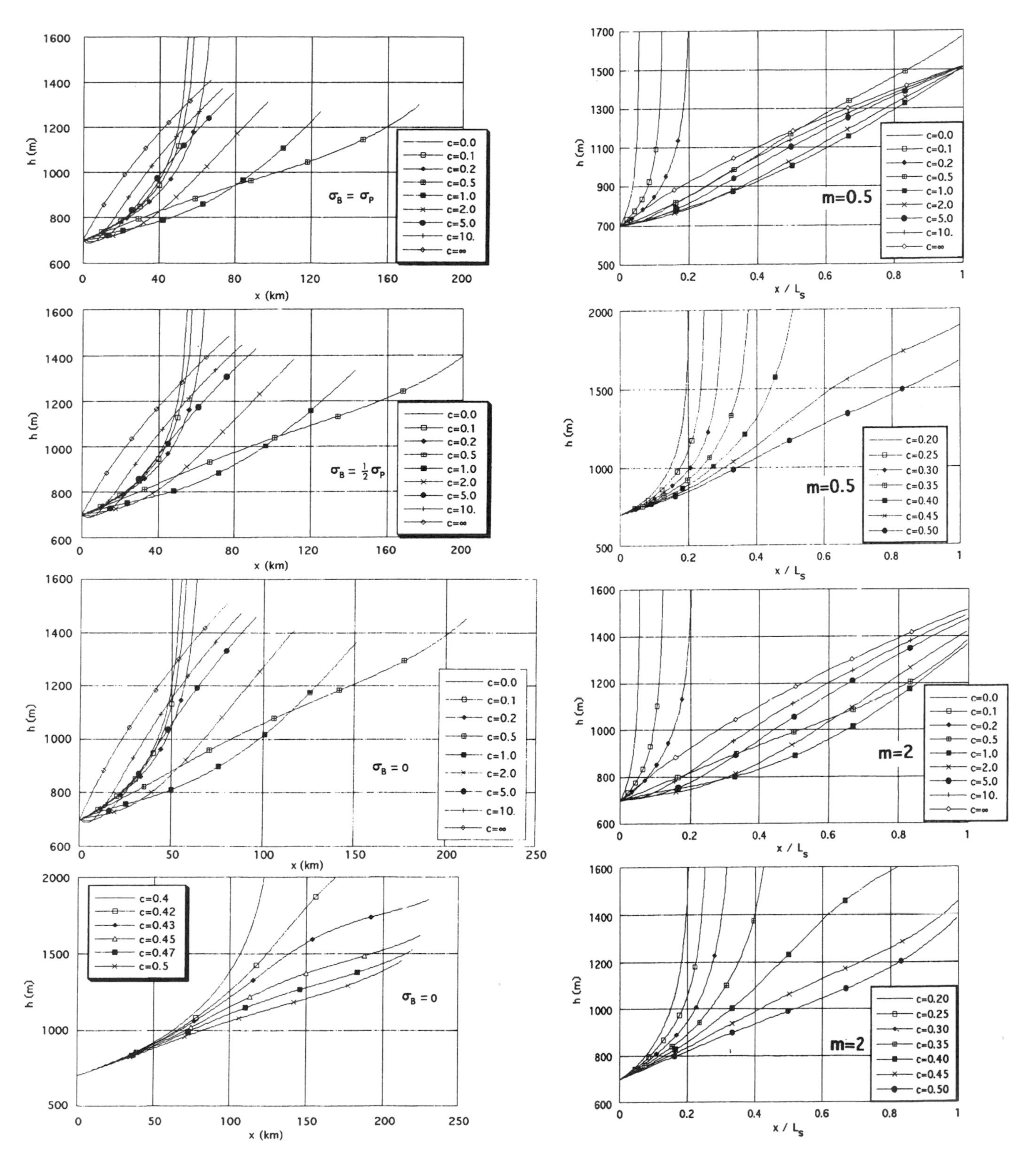

Figure 6.9: Plots of Equation (6.41). Values are those in Figure 6.8, with L_S given by Equation (6.54) for $\sigma_B = \sigma_P$ (top), $\sigma_B = 1/2\ \sigma_P$(middle), $\sigma_B = 0$ (bottom), and $0 \leq c \leq 10$. Plots by Paul Prescott.

Figure 6.10: Plots of Equation (6.42). Values are those in Figure 6.8, with m = 1/2 (top) and m = 2 (bottom) Surface profiles change most rapidly for $0.4 \leq c \leq 0.5$ in Figure 6.9 and for $0.3 \leq c \leq 0.5$ in Figure 6.10. Plots by Paul Prescott.

shelf or a terrestrial ice lobe, is obtained from the force balance in Figure 6.7 but with x replacing L_S and h_I replacing h_S so that Equation (6.45) becomes:

$$F_x = -(F_x)_G = \tfrac{1}{2}\,\rho_I\,g_z\,w_I\left(h_I^{\,2} - h_0^{\,2}\right) + \tfrac{1}{2}\,\rho_I\,g_z\,w_I\,h_0^{\,2}\,(1 - \rho_I/\rho_W)\,(P_W/P_I)_0^2 \qquad (6.60)$$

where $h_I = h$ for a horizontal bed and h is given by Equation (6.42) with $h_R = 0$ and $P_W/P_I = (1 - x/L_S)^c$.

In Equation (6.60), $(F_x)_G$ depends on (P_W/P_I) at $x > 0$ through h_I and on $(P_W/P_I)_o$ at $x = 0$. Therefore, (P_W/P_I) depends on basal buoyancy under the ice stream and $(P_W/P_I)_0$ depends on ice-shelf or ice-lobe buttressing beyond the ice stream, each of which may have different time constants over its range from zero to one. Following Equation (6.37), let (P_W/P_I) at time t during cycling time t_c for the life cycle of an ice stream be:

$$\frac{P_W}{P_I} = \left(1 - \frac{x}{L_S}\right)^c = \left(1 - \frac{x}{L_S}\right)^{\frac{t}{t_c - t}} = \left(1 - \frac{x}{L_S}\right)^{\frac{t/t_c}{1 - t/t_c}} \qquad (6.61)$$

where $0 \le c \le \infty$ as $0 \le t \le t_c$. Following Equation (6.50), let $(P_W/P_I)_0$ at time t during survival time t_s for a buttressing ice shelf or ice lobe be:

$$\left(\frac{P_W}{P_I}\right)_0 = \left(\frac{\sigma_P - \sigma_B}{\sigma_P}\right)^{1/2} = \left(1 - \frac{\sigma_B}{\sigma_P}\right)^{1/2} = \left(1 - \frac{t}{t_s}\right)^{1/2} \qquad (6.62)$$

where $0 \le \sigma_B/\sigma_P \le 1$ as $0 \le t \le t_s$.

In computing P_x from Equations (6.58), (6.59), and (6.60), ice thinning rate $(\delta h_I/\delta_t)_i$ at Δx step i along the ice stream is obtained from the changing ice thickness profile from $t = 0$ to $t = t_c$, with h_I determined from the change in surface slope at each Δx step over time t according to:

$$c = \frac{t}{t_c - t} = \frac{t/t_c}{1 - t/t_c} \qquad (6.63)$$

for an ice stream during its life cycle of duration t_c. No theories or observations are available to test these predictions. A simpler but less accurate formulation of F_x is:

$$F_x = -(F_x)_G \approx (\sigma_{xx} - \sigma_{zz})\,w_I\,h_I = 2\,\sigma_{xx}'\,w_I\,h_I$$

$$\approx 2\,[1/4\,\rho_I\,g_z\,h_I\,(1 - \rho_I/\rho_W)\,(P_W/P_I)^2]\,w_I\,h_I$$

$$\approx 1/2\,\rho_I\,g_z\,w_I\,h_I^2\,(1 - \rho_I/\rho_W)\,(P_W/P_I)_0\,(P_W/P_I) \qquad (6.64)$$

The pulling power obtained from Equations (6.59) and (6.64) is:

$$P_x = \bar{u}_x\,F_x \approx 1/2\,u\,\rho_I\,g_z\,w_I\,h_I^2\,(1 - \rho_I/\rho_W)\,(P_W/P_I)_0\,(P_W/P_I) \qquad (6.65)$$

Normalized pulling power $\hat{P}_x$ is therefore:

$$\hat{P}_x = \left[\frac{2\,P_x}{u\,\rho_I\,g_z\,w_I\,h_I^{\,2}\,(1 - \rho_I/\rho_W)}\right] = \left(\frac{P_W}{P_I}\right)_0\left(\frac{P_W}{P_I}\right) = \left(1 - \frac{t}{t_s}\right)^{1/2}\left(1 - \frac{x}{L_S}\right)^{\frac{t/t_c}{1 - t/t_c}} \qquad (6.66)$$

If many ice streams supply an ice shelf or an ice lobe, t_S and t_C should be quite different. However, for individual ice lobes and for ice shelves supplied by a single ice stream, it is likely that $t_s \approx t_c$ because the ice lobe or ice shelf is produced during the life cycle of the ice stream and cannot be sustained between life cycles. Figure 6.11 is a plot of normalized pulling power $\hat{P}_x$ versus t/t_c given by Equation (6.66) when t_s is t_c, 2 t_c, 3 t_c, 5 t_c, 10 t_c, and 100 t_c. Note that the decrease of $\hat{P}_x$ with time is slow near the foot of an ice stream, especially when $t_s/t_c >> 1$, and rapid near the head of an ice stream, where $\hat{P}_x$ becomes insensitive to t_s/t_c.

Cycling time t_c is sensitive to conditions at both the ice surface and the ice bed. Surface conditions affect basal temperature through the mean annual surface temperatures, surface accumulation or ablation rates, and advection of ice in the direction of maximum surface slope (Robin, 1955; Budd and others, 1971). Surface temperature and mass balance are sensitive to Milankovitch insolation cycles of 23,000 and 41,000 years for Earth's axial precession and axial tilt, respectively. Waves of heat having these periodicities warm the ice surface in the surface accumulation and ablation zones, but basal melting of ice-cemented sediments or till and subsequent refreezing respond to these frequencies in an abrupt manner, with basal freeze-thaw episodes of much shorter duration and no regular frequency (Alley and MacAyeal, 1994).

Basal conditions affecting t_c are rates of glacial erosion and deposition, in addition to rates of basal melting and freezing, which combine to control the thickness and softness of the low-traction layer of basal sediments or till. When this layer is thick and water-saturated, it provides very little traction so that stream flow is likely in the overlying ice (Lingle and Brown, 1987; Clarke, 1987a, b; Boulton and Hindmarsh, 1987; Alley, 1989a; Alley, 1989b; MacAyeal, 1989; Kamb, 1991). Ice-bed uncoupling, and therefore t_c, depends on whether this basal sediment or till layer is frozen or thawed, how water-saturated it is if it is thawed, and whether it is thickening due to net glacial deposition or thinning due to net glacial erosion. Ice lobes along the southern margins of former Northern Hemisphere ice sheets had a lifetime generally between 1000 and 2000 years, with larger lobes lasting longer (Andersen, 1981; Mayewski and others, 1981). This is the usual duration of spikes in the oxygen-isotope record down coreholes through the Greenland and Antarctic ice sheets (Paterson and Hammer, 1987). These spikes may record abrupt changes in ice volume, as well as climate, in which case they may record t_c for major ice streams, given the ability of ice-stream pulling power to downdraw ice sheets.

Survival time t_s of an ice lobe that buttresses a terres-

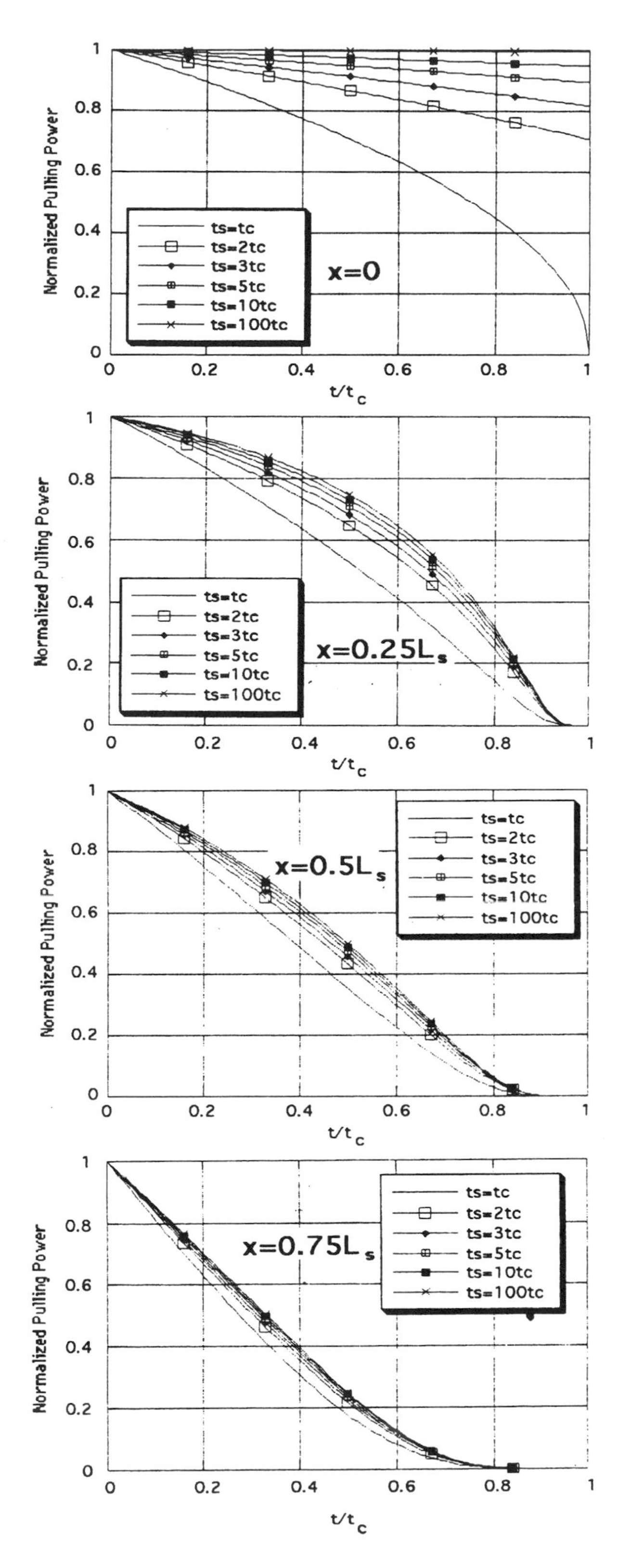

Figure 6.11: Plots of Equation (6.66). Normalized pulling power $\hat{P}$ is greatest near the foot of an ice stream and when survival time t_s for a buttressing ice shelf or ice lobe exceeds cycling time t_c for the life cycle of the ice stream.

trial ice stream, or an ice shelf that buttresses a single marine ice stream, is controlled primarily by ablation rates linked to climate change and to t_c for the ice stream. The survival time for large ice shelves fed by many marine ice streams may correspond to Milankovitch insolation periodicities of 23,000 years for axial precession and 41,000 years for axial tilt because ice shelves are nearly flat, so small snowline elevation changes can reverse their surface mass balance. A large ice shelf in an accumulation zone cannot survive in an ablation zone (Oerlemans and van der Veen, 1984). Flip-flops in ocean circulation (Broecker and Denton, 1989) might accompany the resulting disintegration of a large ice shelf in the subpolar North Atlantic. Ice shelves respond rapidly to mass-balance changes (van der Veen, 1986), so their disintegration and the flip-flops would be abrupt.

A Life-Cycle Classification for Ice Streams

Pulling power provides an analytical foundation for classifying marine ice streams in a hierarchy in which loss of gravitational potential energy over time produces an ice-stream history characterized by inception, growth, mature, declining, and terminal stages of a life cycle but allowing for rejuvenation as well. Pulling power decreases as an ice stream ages and quantifies the vitality of the ice stream at a given stage in its life cycle. The life cycle of an ice stream must be inferred from present-day ice streams and from the glacial geological record of former ice streams. These observations support an hypothesis in which the pulling power of an ice stream is closely related to the degree of ice-bed coupling beneath the ice stream and the degree of ice-shelf or ice-lobe buttressing beyond the ice stream. Letting numbers 1, 2, 3, 4, 5 denote no, weak, moderate, strong, and full coupling, and letting letters A, B, C, D, E denote no, weak, moderate, strong, and full buttressing, Table 6.1 presents a life-cycle classification of ice streams in which various combinations of coupling and buttressing are quantified as values of $P_W/P_I = (1 - x/L_S)^{t/(t_c - t)}$ for coupling, according to Equation (6.61), and $(P_W/P_I)_0 = [(\sigma_P - \sigma_B)/\sigma_P]_0 = (1 - t/t_s)$ for buttressing, according to Equation (6.62). In this classification of an ice-stream life cycle, $P_W/P_I = 1$ for the inception stage, $P_W/P_I = 3/4$ for growth stages, $P_W/P_I = 1/2$ for mature stages, $P_W/P_I = 1/4$ for declining stages, and $P_W/P_I = 0$ for terminal stages. The same is true for $(P_W/P_I)_0$.

A fundamental assumption underlying the ice stream classification in Table 6.1 is that polar continental shelves and embayments are sedimentary basins floored by permafrost at the beginning of a glaciation cycle. Sea ice thickens and grounds on the ice-cemented sediments, and continued thickening produces marine ice domes whose terrestrial margins transgress onto land and whose marine margins encroach onto the outer continental shelf (Hughes, 1987; Lindstrom and MacAyeal, 1987, 1989; Lindstrom, 1990). The life cycle of ice streams follows from this assumption. Marine ice streams develop in straits and interisland channels on the outer continental shelf, and terrestrial ice streams develop in linear lake basins and river valleys beyond crystalline shields. In both cases, the bed beneath ice streams consists of ice-cemented sediments, with stream flow beginning when the sediments thaw and ending when the sediments refreeze or when the ice sheet has collapsed enough to curtail ice-stream pulling power. A glaciation cycle and the life cycles of marine ice streams need not require that marine ice sheets form when sea ice thickens and grounds in shallow water on continental shelves, especially in marine embayments. Terrestrial ice sheets can also advance into this shallow water and end as ice streams in deeper straits or interisland channels.

Inception stage of ice streams

Marine ice accelerates as it enters interisland channels and straits, where the bed is already thawed or becomes thawed by the increased basal frictional heat. Since the thawed bed consists of water saturated sediments, 1A ice streams are nucleated in these channels and straits because ice-bed coupling is nil and only sea ice exists beyond the ice streams. The Wilkes Land margin of the East Antarctic Ice Sheet has the most likely candidates for 1A ice streams, notably Ninnis, Mertz, Totten, and Vanderford Glaciers in Figure 2.14. Terrestrial ice accelerates to become stream flow on the thawed sediments of linear lakes and river valleys that radiate from the edge of crystalline continental shields. These are 1A ice streams until they strip away water-soaked basal sediments and produce buttressing ice lobes.

Growth stage of ice streams

Basal frictional heat is concentrated at the heads of ice streams, where basal traction is a maximum along the basal melting point isotherm that separates frozen from thawed portions of the bed. This is a highly unstable condition because frictional heat causes the isotherm to retreat rapidly. A maximum-slope surface inflection line marks the head of these ice streams and exists above the basal isotherm. Hence the ice streams retreat rapidly, drawing down interior ice domes and thereby raising sea level. Rising sea level floats the grounding line of marine ice streams over the submarine sills that typically exists at the oceanward ends of channels and straits, causing an ice shelf to form as grounding lines retreat. The sills are either bedrock ridges or grounding-line moraines, and they form as the thawed sediments are eroded by the ice streams, causing bedrock to outcrop on the bed in places where sediments were thin. This situation produces 1B, 2A, and 2B ice streams, where grounding-line retreat provides increasing ice-shelf buttressing (1B ice streams) and bedrock outcropping provides increasing ice-bed coupling (2A ice streams), or both (2B ice streams). In Figure 2.14, Pine Island Glacier and Thwaites Glacier may be respective 1B and 2A ice streams in the Pine Island Bay polynya of the Amundsen Sea embayment of the marine West Antarctic Ice Sheet. For terrestrial ice streams draining former Northern Hemisphere ice sheets, bedrock outcrops where linear lakes and river valleys cross the edge of crystalline shields, such as the Great Lakes at the southern end of the Laurentian Shield in North America. Terrestrial ice streams could grow until their retreating basal melting point isotherm passes from sediment to bedrock across this boundary.

Table 6.1: A life cycle classification for ice streams*

Increasing	1A	1B	1C	1D	1E
basal	2A	2B	2C	2D	2E
coupling	3A	3B	3C	3D	3E
↓	4A	4B	4C	4D	4E
	5A	5B	5C	5D	5E
	Increasing forward buttressing →				
	1	3/4	1/2	1/4	0
↑	3/4	3/4	1/2	1/4	0
P_W/P_I	1/2	1/2	1/2	1/4	0
	1/4	1/4	1/4	1/4	0
	0	0	0	0	0
	← $(\sigma_P - \sigma_B)/\sigma_P$				

*Pulling power decreases from stage 1A to stage 5E defined as:
1 Full basal buoyancy along entire length.
2 Basal buoyancy slowly decreasing upstream.
3 Basal buoyancy steadily decreasing upstream.
4 Basal buoyancy rapidly decreasing upstream.
5 No basal buoyancy along entire length.
A No ice shelf or a freely floating ice shelf.
B Weak buttressing by a confined and pinned ice shelf.
C Moderate buttressing by a confined or pinned ice shelf.
D Strong buttressing by a confined and pinned ice shelf.
E Full buttressing by a confined and pinned ice shelf.

Mature stage of ice streams

Pulling power of marine ice streams increases rapidly as grounding lines retreat downslope into the isostatically depressed marine sedimentary basins and decreases rapidly as grounding lines retreat upslope out of these basins, since the pulling force is greatest when the grounding line is furthest below sea level. Hence, the mature stage of marine ice streams exists when a floating ice shelf occupies the seaward half of a marine sedimentary basin and a grounded ice dome occupies the landward half, with ice streams drawing down the marine ice dome and eroding thawed marine sediments beneath the dome. This situation produces 1C, 2C, 3C, 3B, and 3A ice streams, depending on the degree of ice-bed coupling and ice-shelf buttressing for ice streams at various sites in the embayment. The Ross Sea and Weddell Sea embayments of the marine West Antarctic Ice Sheet contain mature ice streams. Mature terrestrial ice streams exist when ice lobes develop beyond linear lakes and in broad river valleys such as the Michigan Lobe beyond Lake Michigan and the Des Moines Lobe in the Des Moines River valley at the southern end of the former Laurentide Ice Sheet.

Declining stage of ice streams

Retreating grounding lines of marine ice streams eventually stabilize when they lie in water too shallow to sustain a significant pulling force or lie against the headwalls of fjords through coastal mountains, while a floating ice shelf fills much of the marine sedimentary basin beyond the grounding line. Retreating inflection lines at the heads of ice streams can continue to retreat, pulling more ice into the marine basins and drawing down terrestrial ice domes. Marine ice streams become terrestrial ice streams at this stage, and they persist so long as accumulation over terrestrial ice domes replaces ice pulled out by stream flow. This loss of marine character is typical of 1D, 2D, 3D, 4D, 4C, 4B, and 4A ice streams and therefore represents the declining stage of marine ice streams. Outlet glaciers through the Transantarctic Mountains are probably 3D, 4D, and 4C ice streams that drain the largely terrestrial East Antarctic Ice Sheet and are buttressed by ice shelves floating in the marine sedimentary basins of West Antarctica. Terrestrial ice streams in linear lakes and in river valleys decline when downdrawn interior ice piles up against the terminal ice lobes, eventually to convert the concave profile of stream flow into the convex profile of sheet flow.

Terminal stage of ice streams

When an ice stream freezes to its bed and is no longer able to deliver ice to the calving front of its ice shelf fast enough to offset the iceberg calving rate or to the melting terminus of its ice lobe fast enough to offset the surface melting rate, the ice stream enters its terminal stage. Ice streams are terminal in stages 1E, 2E, 3E, 4E, 5E, 5D, 5C, 5B, and 5A. For marine ice streams, a calving bay carves away the ice shelf until its calving front coincides with its grounding line to produce a calving ice wall. For terrestrial ice streams, the concave surface of stream flow vanishes when the convex surface of sheet flow at its head coincides with the convex surface of the ice lobe at its foot. A 5E ice stream ideally exists in a fjord floored by rugged bedrock, so ice-bed coupling is optimized; and supplies an ice shelf grounded on all sides, so that ice-shelf buttressing is optimized. Ice stream C in Figure 2.14 is a possible 1E ice stream, because its head seems to be uncoupled from the bed and its foot seems to be buttressed by the Ross Sea Shelf (Whillans and Van der Veen, 1993a). Dibble Glacier and Dalton Glacier in Figure 2.14 are possible 5A ice streams because they are presently inactive, so basal coupling must be complete, and disintegrating iceberg tongues some 100 km long extend beyond them, so they are unbuttressed.

Rejuvenation stage of of ice streams

The life cycle of an ice stream need not proceed irreversibly along paths from 1A to 1E, 2E, 3E, 4E, 5E, 5D, 5C, 5B, or 5A in Table 6.1. Reversals are possible at any combination of number (for basal coupling) and letter (for downstream buttressing) between these extremes. Rejuvenation of a 5E ice stream occurs when a calving bay reduces ice-shelf buttressing or surface melting reduces ice-lobe buttressing, converting it into an ice stream moving from 5E to 5A and thereby causing a great increase in ice-stream velocity. Surface meltwater may reach the bed through crevasses opened when ice velocity increases, and the resulting bed lubrication may move the ice stream from 5A to 1A. This may be happening with Jakobshavns Isbrae (69°N, 50°W) in Greenland, which began a 30-km retreat near the end of the Little Ice Age in 1864, when surface melting probably increased and when a calving bay began migrating up Jakobshavns Isfjord, making it the fastest ice stream known today (Carbonnell and Bauer, 1968). Lambert Glacier, which pulls terrestrial East Antarctic ice into the Amery Ice Shelf in Figure 2.14, may be a rejuvenated ice stream.

Stream Flow on a Deformable Bed

Subglacial sediment or till can deform and have a viscoplastic creep spectrum like that shown in Figure 4.18 for ice and other polycrystalline materials. The role of deformable sediment or till in ice-stream dynamics has been analyzed extensively, notably by Alley (1990) for deformation near the viscous end of the creep spectrum and by Kamb (1991) for deformation near the plastic end of the spectrum. In both analyses, interstitial water plays a major role by allowing the sediment or till to dilate as it undergoes shear deformation imposed by sliding of the overlying ice. Dilation reduces the viscosity or the yield stress, depending on whether deformation occurs at the viscous end or the plastic end of the viscoplastic creep spectrum. In both cases, the sediment or till then provides little or no traction to resist sliding of the overlying ice, so the ice stream flows almost like a linear ice shelf, with a longitudinal strain rate given by Equation (4.50). This is the condition for pure stream flow.

If the subglacial sediment or till deforms near the viscous end of the viscoplastic creep spectrum, the creep velocity of the overlying ice increases with the thickness of the deforming basal layer because:

$$\dot{\varepsilon}_{xz} = u_s / h_D = \tau_o / \eta_N \tag{6.67}$$

where $\dot{\varepsilon}_{xz}$ is the shear strain rate in a basal layer of deforming height h_D which has Newtonian viscosity η_N and through which shearing velocity increases linearly from zero at the bottom to ice sliding velocity u_s at the top, τ_o is the basal shear stress causing $\dot{\varepsilon}_{xz}$, and $h_D << h_I$. Since creep velocity u_c in the ice stream is u_s at its base, a basal deforming sediment or till having constant η_N can regulate u_c by its thickness h_D if there is no slip at the ice-bed interface for a soft bed deforming in simple shear.

If the subglacial sediment or till deforms near the plastic end of the viscoplastic creep spectrum, the only places beneath an ice stream where u_c can be influenced by basal traction is where bedrock bumps penetrate layer thickness h_D

and project into the basal ice or where frozen patches in the layer raise its yield stress σ_o above basal shear stress τ_o so that strain rate $\dot{\varepsilon}_{xz}$ in the layer is zero according to the flow law:

$$\dot{\varepsilon}_{xz} = \dot{\varepsilon}_o (\tau_o / \sigma_o)^n \quad (6.68)$$

where $n = \infty$ for plastic flow. These bedrock bumps and frozen patches are "sticky spots" that behave like the "controlling obstacles" of height $\Lambda_o - \lambda_o$ given by Equation (6.23) in the Weertman (1957a) theory of glacial sliding.

A third kind of deforming basal sediment or till has such a high ice fraction in its permafrost state that water is the matrix component, not the interstitial component, when the permafrost thaws. Whether the thawed basal sediment or till deforms near the viscous or plastic end of the visco-plastic creep spectrum is unimportant, because the thawed sediment or till provides no traction for the overlying ice in any case. Ideally, this is the kind of deforming basal layer invoked in the life-cycle classification of ice streams. When it thaws, it has the soupy consistency described by Kellogg and Kellogg (1987a, b, c) for marine sediments in the Pine Island Bay polynya of the Amundsen Sea, where Pine Island Glacier and Thwaites Glacier are the two fastest ice streams in West Antarctica. The marine West Antarctic Ice Sheet is grounded in Byrd Subglacial Basin, which was probably originally covered by marine sediments much like those now in Pine Island Bay. Gravitational compaction would not have squeezed water out of the upper few meters of sediment, so when this sediment thawed it would have no traction to resist sliding by marine ice streams occupying troughs or interisland channels around the perimeter of Byrd Subglacial Basin, and similar basins beneath former marine ice sheets in the Northern Hemisphere. Likewise, terrestrial ice streams draining former Northern Hemisphere ice sheets occupied elongated lake basins and river bottoms where the uppermost sediments formed a soup in a water matrix. This was especially true where rivers were dammed by the former ice sheets, as was the case along the entire southern perimeter of the Eurasian and North American ice sheets, except during brief periods of maximum advance when ice crossed the watersheds of south-flowing rivers at a few sites.

With these considerations in mind, the importance of deforming sediment or till beneath an ice stream is not so much the limited basal traction it provides but whether it freezes or thaws uniformly or in patches and whether it is able to blanket bedrock bumps, since real traction is provided by frozen patches and outcropping bumps. Both sources of basal traction are controlled by heat flow from the bed to the surface of an ice stream. Rapid heat flow causes patches where basal sediment or till is thinnest to freeze first and, by producing regelation ice laden with frozen-in sediment or till, causes bedrock bumps to outcrop first. Slow heat flow allows these patches to thaw and be blanketed by sediment or till melting out of regelation ice.

Boudinage in Ice Streams

Robin and others (1970) discovered that West Antarctic ice stream C, located in Figure 2.14, consists of several surface terraces that were unrelated to bed topography, but the strong radar reflection from the bed was similar to an ice-water interface instead of an ice-rock interface. Therefore, they called these ice terraces "pseudo ice shelves." In less pronounced form, this condition exists for other long West Antarctic ice streams (Shabtaie and Bentley, 1988), and it may be the reason for their unusual length. Stream flow seems to alternate with sheet flow in these flowbands, causing a succession of ice streams and ice lobes. This condition produces boudinage in the ice stream, with each section of stream flow being a boudin. Whether the boudins remain in place or are migrating down the ice stream is unknown, but mobile variations of basal traction and ice stiffness noted by Whillans and Van der Veen (1993b) are compatible with migrating boudins.

Possible migration and merger of boudins is related to transport of a deformable substrate beneath the ice stream. Boudinage in an ice stream therefore multiplies its life cycle, with one life cycle for each boudin that reaches the terminus. Whether boudins are common or rare in an ice stream is an open question. Boudinage converts an ice stream into a train of ice streams, with one ice stream for each boudin and downdrawn ice between boudins. Boudinage therefore allows the pulling power of ice streams to reach further into an ice sheet. Possible boudins are observed in West Antarctic ice streams supplying the Ross Ice Shelf. Boudinage may therefore account for the partly collapsed state of the West Antarctic Ice Sheet compared to the East Antarctic Ice Sheet, with collapse being greatest in the Ross Sea embayment.

7

SHELF FLOW

Shelf Flow from Thickening Sea Ice

An ice shelf is the floating extension of an ice sheet, so it is thick enough to spread under its own weight. However, this thickness might also be attained from sea ice by snow accumulation on its top surface and water freezing on its bottom surface. Crary (1960) was the first to quantify sea-ice thickening. If h_I is sea-ice thickness, t is time, a is the surface accumulation rate a" (positive) or ablation rate a' (negative) of sea ice, K is the thermal conductivity coefficient for sea ice, ΔT is the temperature increase from the top to the bottom of sea ice, $\dot{Q}_H$ is the heat flux supplied to the bottom of sea ice by ocean currents, H_M is latent heat of melting sea ice, and ρ_I is sea-ice density, the thickening rate of sea ice is:

$$\frac{dh_I}{dt} = a + \frac{(K/h_I)\,\Delta T - \dot{Q}_H}{\rho_I H_M} = \frac{C_1}{h_I} + C_2 \tag{7.1}$$

where:

$$C_1 = K\,\Delta T/\rho_I H_M \tag{7.2a}$$

$$C_2 = a - \dot{Q}_H/\rho_I H_M \tag{7.2b}$$

Setting $dh_I/dt = 0$ in Equation (7.1) and solving for h_I gives the equilibrium sea– ice thickness h_e:

$$h_e = \frac{K\,\Delta T}{\dot{Q}_H - a\,\rho_I H_M} = -\frac{C_1}{C_2} \tag{7.3}$$

Equation (7.3) is plotted in Figure 7.1 for a range of ΔT values, showing how h_e changes with top surface ablation rates a' at $\dot{Q}_H = 0$ and bottom surface heating rates $\dot{Q}_H$ at $a = 0$.

A solution of Equation (7.1) exists for the boundary condition $h_I = 0$ at $t = 0$ when a, Q, and ΔT are constant with time:

$$t = \left[\frac{\rho_I H_M h_I}{a\,\rho_I H_M - \dot{Q}_H}\right] - \left[\frac{\rho_I H_M K\,\Delta T}{\left(a\,\rho_I H_M - \dot{Q}_H\right)^2}\right] ln\left[1 - \frac{\left(a\,\rho_I H_M - \dot{Q}_H\right) h_I}{K\,\Delta T}\right]$$

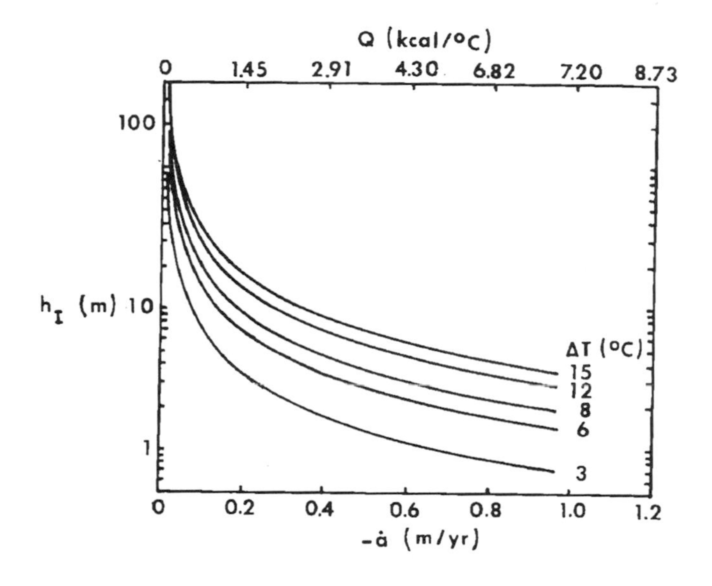

Figure 7.1: Equilibrium thickness of sea ice for average surface temperature of − 17°C and various rates of surface ablation and basal heat input from ocean water. Reproduced from Crary (1960).

$$= \left(\frac{h_I}{C_1}\right) - \left(\frac{C_2}{C_1^{\,2}}\right) ln\left(1 + \frac{C_1 h_I}{C_2}\right) \tag{7.4}$$

Equation (7.4) is plotted in Figure 7.2 for a range of C_1 values, showing how h_I changes with t when C_1 is specified for the combinations of a and $\dot{Q}_H$ at $\Delta T = 15°C$ listed in Table 7.1. In the special case $a = \dot{Q}_H = 0$, Equation (7.1) integrates to give:

$$t = \frac{h_I^{\,2} \rho_I H_M}{2 K\,\Delta T} = \frac{h_I^{\,2}}{2\,C_2} \tag{7.5}$$

Once sea-ice thickness exceeds 10 m, leads that fracture the whole ice thickness become rare, so summer surface meltwater mostly refreezes in the winter rather than draining

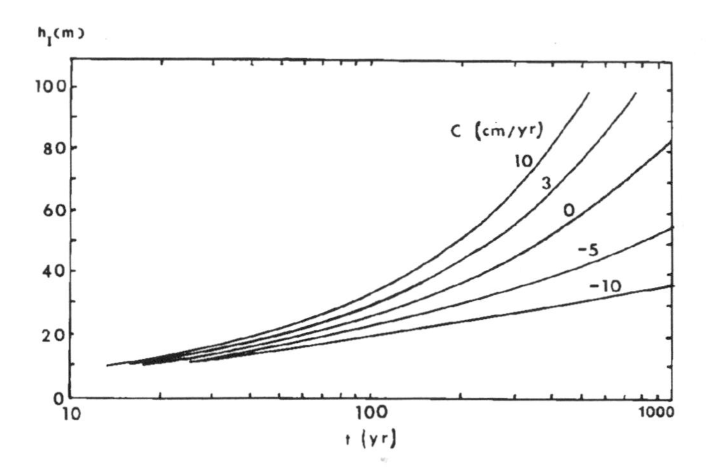

Figure 7.2: Increasing sea-ice thickness with time for various combinations of surface ablation rate and rate of basal heating from ocean water incorporated as the constant C_1 in Equation (7.4). Reproduced from Crary (1960).

away in the summer. Net surface accumulation then replaces net surface ablation, thereby enhancing the ice thickening rate. Once sea-ice thickness exceeds 100 m, it begins to spread significantly under its own weight, with spreading creep rate $\dot{\varepsilon}_{xx}$ given by Equation (3.24), with (P_W / P_I) given by Equation (3.30). For $P_W = P_I$, Equation (3.24) reduces to Equation (4.50) if the sea ice floats in a linear channel, such that $\dot{\varepsilon}_{yy} = 0$, and by Equation (4.58) if the sea ice floats in an open sea, such that $\dot{\varepsilon}_{yy} = \dot{\varepsilon}_{xx}$.

Taking $\dot{\varepsilon}_{zz}$ as the creep thinning rate, the thickening rate when sea ice becomes shelf ice is given by the following modification of Equation (7.1):

$$\frac{dh_I}{dt} = a + \frac{(K/h_I)\,\Delta T - \dot{Q}_H}{\rho_I H_M} - h_I \dot{\varepsilon}_{zz} \tag{7.6}$$

where $\dot{\varepsilon}_{zz} = -\dot{\varepsilon}_{xx}$ and $\dot{\varepsilon}_{yy} = 0$ for an ice shelf in a linear channel, and $\dot{\varepsilon}_{zz} = -2\,\dot{\varepsilon}_{xx}$ and $\dot{\varepsilon}_{yy} = \dot{\varepsilon}_{xx}$ for an ice shelf in the open sea, but grounded at one end.

Table 7.1: Selected values of a (cm yr^{-1}) and Q (kcal cm^{-2} yr^{-1}) used to compute the values of C_1 (cm yr^{-1}) in Figure 7.2 from Equation (7.2a) and (7.4)

C_1	+ 10	+ 3	0	−5	−10
a	+ 10	+ 3	0	−5	−10
Q	0	0	0	0	0
a	0	0	0	0	0
Q	−0.73	−0.22	0	+ 0.37	+ 0.37
a	+ 20	+ 20	+ 20	+ 20	+20
Q	+ 0.73	+ 1.25	+ 1.47	+ 1.85	+ 2.22
a	+ 37.2	+ 30.2	+ 27.2	+ 22.2	+ 17.2
Q	+ 2	+ 2	+ 2	+ 2	+ 2

Shelf Flow from Creep of Ice

Present-day ice shelves of the Antarctic Ice Sheet typically are supplied by ice streams, occupy confined embayments (notably the Ross Sea and Weddell Sea embayments), and calve along a floating front that is less extensive than their grounding lines. Former ice shelves of Arctic ice sheets would have the same constraints if they occupied the Arctic Ocean, Baffin Bay, and the Labrador, Greenland, Norwegian, and Bering Seas. Ice flow would diverge where ice streams entered the ice shelves and converge as floating ice approached calving fronts. Ice streams usually are oblique to calving fronts, so shelf flow normally curves from grounding lines to calving fronts, causing bending shear in flowbands. In addition, ice shelves often shear along grounding lines between ice streams and alongside ice rises, with diverging flow around ice rises and converging flow in the lee of ice rises. Flow is radially divergent in the freely-floating tongues of marine ice streams, which is pure shelf flow.

The complexity of flow in ice shelves can be take into account most easily by measuring principal strain rates $\dot{\varepsilon}_1$ and $\dot{\varepsilon}_2$ in the map lane, which sum to zero with vertical principal strain rate $\dot{\varepsilon}_3$ for conservation of volume:

$$\dot{\varepsilon}_1 + \dot{\varepsilon}_2 + \dot{\varepsilon}_3 = 0 \tag{7.7}$$

Define ration $r = \dot{\varepsilon}_2 / \dot{\varepsilon}_1$ in terms of principal stresses σ_1, σ_2, and σ_3. From Equations (4.2) and (4.17):

$$r = \frac{\dot{\varepsilon}_2}{\dot{\varepsilon}_1} = \frac{\sigma_2'}{\sigma_1'} = \frac{\sigma_2 - \frac{1}{3}(\sigma_1 + \sigma_2 + \sigma_3)}{\sigma_1 - \frac{1}{3}(\sigma_1 + \sigma_2 + \sigma_3)} = \frac{2\,\sigma_2 - \sigma_1 - \sigma_3}{2\,\sigma_1 - \sigma_2 - \sigma_3} \tag{7.8}$$

Solving Equation (7.8) for σ_2:

$$\sigma_2 = \left(\frac{1 + 2\,r}{2 + r}\right)\sigma_1 + \left(\frac{1 - r}{2 + r}\right)\sigma_3 \tag{7.9}$$

Substituting Equation (7.9) for σ_2 in Equation (4.40) for effective creep stress σ_c:

$$\sigma_c = \left\{\tfrac{1}{6}\left[(\sigma_1 - \sigma_2)^2 + (\sigma_2 - \sigma_3)^2 + (\sigma_3 - \sigma_1)^2\right]\right\}^{1/2}$$

$$= \left(1 + r + r^2\right)^{1/2} \frac{\sigma_1 - \sigma_3}{2 + r} \tag{7.10}$$

Substituting Equation (7.9) for σ_2 in Equation (4.2) for principal deviator stress σ_1', so that ij = 1 and k = 1, 2, 3:

$$\sigma_1' = \sigma_1 - \tfrac{1}{3}(\sigma_1 + \sigma_2 + \sigma_3) = \frac{\sigma_1 - \sigma_3}{2 + r} \tag{7.11}$$

Substituting Equations (7.10) and (7.11) into Equation (4.17) for the flow law of ice gives principal strain rates $\dot{\varepsilon}_1$, $\dot{\varepsilon}_2$, and $\dot{\varepsilon}_3$ in terms of principal deviator stress σ_1', and principal stresses σ_1 and σ_3:

$$\dot{\varepsilon}_1 = \frac{\sigma_c^{\,n-1}}{A^n}\,\sigma_1' = \frac{\left(1 + r + r^2\right)^{(n-1)/2}}{(2 + r)^n}\left(\frac{\sigma_1 - \sigma_3}{A}\right)^n \tag{7.12a}$$

$$\dot{\varepsilon}_2 = r\,\dot{\varepsilon}_1 \tag{7.12b}$$

$$\dot{\varepsilon}_3 = -(\dot{\varepsilon}_1 + \dot{\varepsilon}_2) = -\dot{\varepsilon}_1 - r\,\dot{\varepsilon}_2 = -(1+r)\,\dot{\varepsilon}_1 \tag{7.12c}$$

Define R as follows, noting that it is the principal-stress counterpart to Equation (3.25):

$$R = \frac{\left(1 + r + r^2\right)^{(n-1)/2}}{(2+r)^n} = \frac{\left[1 + (\dot{\varepsilon}_2/\dot{\varepsilon}_1) + (\dot{\varepsilon}_2/\dot{\varepsilon}_1)^2\right]^{(n-1)/2}}{\left[2 + (\dot{\varepsilon}_2/\dot{\varepsilon}_1)\right]^n}$$

$$= \frac{\left[1 + \left(\dot{\varepsilon}_{yy}/\dot{\varepsilon}_{xx}\right) + \left(\dot{\varepsilon}_{yy}/\dot{\varepsilon}_{xx}\right)^2 + \left(\dot{\varepsilon}_{xy}/\dot{\varepsilon}_{xx}\right)^2 + \left(\dot{\varepsilon}_{xz}/\dot{\varepsilon}_{xx}\right)^2\right]^{(n-1)/2}}{\left[2 + \left(\dot{\varepsilon}_{yy}/\dot{\varepsilon}_{xx}\right)\right]^n} \tag{7.13}$$

where $\dot{\varepsilon}_1$ and $\dot{\varepsilon}_2$, along with $\dot{\varepsilon}_{xx}$, $\dot{\varepsilon}_{yy}$, and $\dot{\varepsilon}_{xy}$, are constant through the floating ice thickness. Ratio $\dot{\varepsilon}_{xz}/\dot{\varepsilon}_{xx}$ is nil for ice shelves, because $\dot{\varepsilon}_{xz}$ exists only at ice rumples above weak basal pinning points. Surface ice rises are produced above stronger pinning points. Shelf flow passes over ice rumples and around ice rises.

For an ice shelf, combining Equations (7.12a) and (7.13):

$$\dot{\varepsilon}_1 = R\left(\frac{\sigma_1 - \sigma_3}{A}\right)^n \tag{7.14}$$

If $z = 0$ at the base and $z = h_I$ at the surface when z is vertical, then $\sigma_3 = \sigma_{zz}$ and Equation (4.32) gives:

$$\sigma_3 = \rho_I\, g_z\,(h_I - z) \tag{7.15}$$

Substituting Equation (7.15) into Equation (7.14) and solving for σ_1:

$$\sigma_1 = A\,(\dot{\varepsilon}_1/R)^{1/n} + \sigma_3 = A\,(\dot{\varepsilon}_1/R)^{1/n} - \rho_I\, g_z\,(h_I - z) \tag{7.16}$$

Balancing horizontal gravity forces $(F_x)_I$ for ice and $(F_x)_W$ for water, where P_o is the overburden pressure of ice or water at $z = 0$:

$$(F_x)_I - (F_x)_W = (\sigma_1\, A_x)_I - (P_o\, A_x)_W$$

$$= \int_0^{h_I}\left[A\left(\frac{\dot{\varepsilon}_1}{R}\right)^{1/n} - \rho_I\, g_z\,(h - z)\right]dz - \int_0^{h_W} - \rho_W\, g_z\,(h_W - z)\,dz$$

$$= A\, h_I\left(\frac{\dot{\varepsilon}_1}{R}\right)^{1/n} - \rho_I\, g_z\left(\frac{h_I^2}{2}\right) + \rho_W\, g_z\left(\frac{h_W^2}{2}\right) = 0 \tag{7.17}$$

Balancing vertical gravity forces $(F_z)_I$ for ice and $(F_z)_W$ for water at the base of ice and water columns of respective heights h_I and h_W and basal area A_z for floating ice:

$$(F_z)_I - (F_z)_W = (P_o\, A_z)_I - (P_o\, A_z)_W$$

$$= \rho_I\, g_z\, h_I\, A_z - \rho_W\, g_z\, h_W\, A_z = 0 \tag{7.18}$$

Equation (7.18) gives $h_W = h_I\,(\rho_I/\rho_W)$, which can be substituted for h_W in Equation (7.17) so that if x is the flow direction (Hughes, 1983):

$$\dot{\varepsilon}_{xx} = \dot{\varepsilon}_1 = R\left[\frac{\rho_I\, g_z\, h_I}{2A}\left(1 - \frac{\rho_I}{\rho_W}\right)\right]^n \tag{7.19}$$

For linear flow, $\dot{\varepsilon}_2 = 0$, $r = 0$, $R = 1/2^n$, and $\dot{\varepsilon}_3 = -\dot{\varepsilon}_1$. For radial diverging flow, $\dot{\varepsilon}_2 = \dot{\varepsilon}_1$, $r = 1$, $R = 1/3^{(n+1)/2}$, and $\dot{\varepsilon}_3 = -2\,\dot{\varepsilon}_1$. For converging flow in pure shear $\dot{\varepsilon}_2 = -\dot{\varepsilon}_1$, $r = -1$, $R = 1$, and $\dot{\varepsilon}_3 = 0$. Equation (7.19), also given by Equation (3.24) when $P_W/P_I = 1$ for unconstrained floating ice, is the classic expression for creep of ice shelves (Weertman, 1957b). All constraints on the creep of ice shelves are treated by modifying this fundamental expression.

Shelf Flow from Floating Stream Flow

The floating tongue of an ice stream is already thick enough to spread under its own weight. It approximates a freely-floating linear ice shelf when the ice stream becomes afloat in a linear channel, along which shear rupture of ice has eliminated side traction. Taking h_0 and u_0 as ice thickness and ice velocity at the grounding line, with x horizontal and positive seaward, y horizontal and parallel to the grounding line, and z vertical and positive upward, Equation (6.20) with $R_{xx} = 1$ for $\dot{\varepsilon}_{yy} = \dot{\varepsilon}_{xy} = \dot{\varepsilon}_{xz} = 0$ represents this condition and gives the ice thickness gradient as:

$$\frac{dh_I}{dx} = \frac{a\, h_I}{a\,x + h_0\, u_0} - \frac{h_I^{\,n+2}}{a\,x + h_0\, u_0}\left[\frac{\rho_I\, g_z}{4A}\left(1 - \frac{\rho_I}{\rho_W}\right)\right]^n \tag{7.20}$$

Oerlemans and Van der Veen (1984) presented solutions of Equation (7.20) for $a = 0$, $a > 0$, and $a < 0$.

For $a = 0$, Equation (7.20) reduces to:

$$\frac{dh_I}{dx} = -\frac{h_I^{\,n+2}}{h_0\, u_0}\left[\frac{\rho_I\, g_z}{4A}\left(1 - \frac{\rho_I}{\rho_W}\right)\right]^n \tag{7.21}$$

Integrating Equation (7.21) gives:

$$h_I = \left\{h_0^{-(n+1)} + \frac{(n+1)}{h_0\, u_0}\left[\frac{\rho_I\, g_z}{4A}\left(1 - \frac{\rho_I}{\rho_W}\right)\right]^n x\right\}^{\frac{-1}{n+1}} \tag{7.22}$$

As shown in Figure 7.3 for $n = 3$, the ice shelf thins rapidly at short distances from the grounding line and slowly at long distances.

For $a = a''$, net accumulation on the top and bottom surfaces of the ice shelf requires Equation (7.20) to be rewritten for positive a'' as:

$$\frac{dh_I}{dx} = \frac{a''\, h_I}{a''\,x + h_0\, u_0} - \frac{h_I^{\,n+2}}{a''\,x + h_0\, u_0}\left[\frac{\rho_I\, g_z}{4A}\left(1 - \frac{\rho_I}{\rho_W}\right)\right]^n \tag{7.23}$$

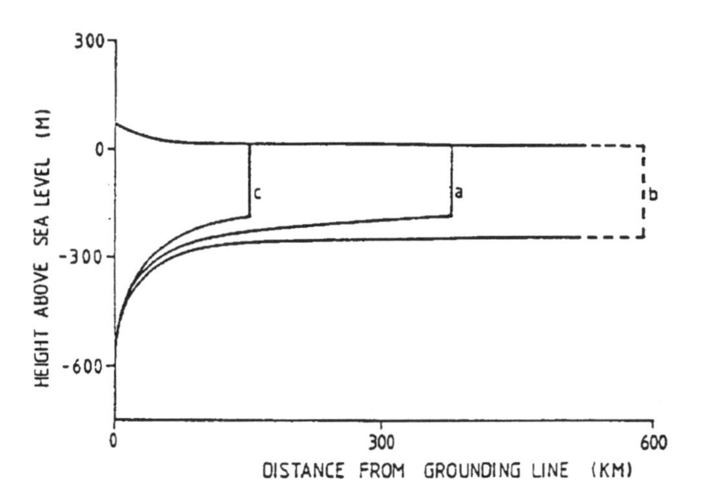

Figure 7.3: Equilibrium profiles for a freely floating ice shelf with a calving front 200 m thick. Profiles are obtained by integrating Equation (7.20) for A = 120 kPa $a^{1/n}$, h_0 = 800 m, and $u_0 = 100$ m a^{-1}. *Profile A*: Equation (7.22) with a = 0. *Profile B*: Equation (7.26) with a = 0.15 m a^{-1}. *Profile C*: Equation (7.31) with a = -0.15 m a^{-1}. Reprinted with permission of Instituut voor Meteorologic en Oceanografgie and C.J. van der Veen, I.M.O.U. publication 83(15), (1983), p. 9, Fig. 2.

Equation (7.23) can be rewritten as:

$$\left[\frac{1}{a'' h_I} + \frac{h_I{}^n/a''\, h_I''{}^{n+1}}{1-\left(h_I/h_I''\right)^{n+1}}\right] dh_I = \frac{dx}{a''\, x + h_0\, u_0} \tag{7.24}$$

where h_I'' is h_I in Equation (7.23) when $dh_I/dx = 0$:

$$h_I'' = \left\{\left[\frac{4\,A}{\rho_I\, g_z\left(1-\rho_I/\rho_W\right)}\right]^n a''\right\}^{\frac{1}{n+1}} \tag{7.25}$$

Critical ice thickness h_I'' is required in the solution in order to prevent positive values of dh_I/dx. Integrating Equation (7.25) gives:

$$h_I = \left\{h_I''^{\,-(n+1)} - \frac{u_0{}^{n+1}\left[\left(h_0/h_I''\right)^{n+1} - 1\right]}{\left(a''\, x + h_0\, u_0\right)^{n+1}}\right\}^{\frac{-1}{n+1}} \tag{7.26}$$

As shown in Figure 7.3 for n = 3, the ice shelf thins less rapidly with increasing distance from the grounding line than when a = 0.

For a = a', net ablation on the top and bottom surfaces of the ice shelf requires Equation (7.20) to be rewritten for negative a' as:

$$\frac{dh_I}{dx} = \frac{-a'\, h_I}{-a'\, x + h_0\, u_0} - \frac{h_I{}^{n+2}}{-a'\, x + h_0\, u_0}\left[\frac{\rho_I\, g_z}{4\,A}\left(1-\frac{\rho_I}{\rho_W}\right)\right]^n \tag{7.27}$$

Equation (7.27) can be rewritten as:

$$\left[\frac{1}{a'\, h_I} + \frac{h_I{}^n/a'\, h_I'{}^{n+1}}{1-\left(h_I/h_I'\right)^{n+1}}\right] dh_I = \frac{dx}{a'\, x + h_0\, u_0} \tag{7.28}$$

where h_I' is h_I in Equation (7.27) when $dh_I/dx = 0$:

$$h_I' = \left\{\left[\frac{4\,A}{\rho_I\, g_z\left(1-\rho_I/\rho_W\right)}\right]^n a'\right\}^{\frac{1}{n+1}} \tag{7.29}$$

In addition to critical ice thickness h_I', critical ice length x' from the grounding line is necessary to prevent dh_I/dx from becoming positive. From Equation (7.27):

$$x' = \frac{h_0\, u_0}{a'} \tag{7.30}$$

Integrating Equation (7.28) gives:

$$h_I = \left\{-h_I'^{\,-(n+1)} + \frac{u_0{}^{n+1}\left[\left(h_0/h_I'\right)^{n+1} + 1\right]}{\left(-a'\, x + h_0\, u_0\right)^{n+1}}\right\}^{\frac{-1}{n+1}} \tag{7.31}$$

As shown in Figure 7.3 for n = 3, the ice shelf thins more rapidly with increasing distance from the grounding line than when a = 0. In addition, the ice shelf is shorter than for a ≥ 0 because it cannot exceed critical length x' given by Equation (7.30).

Shelf Flow from Confluent Stream Flow

The confluence of floating ice-stream tongues can create an ice shelf, since the ice tongues spread laterally and longitudinally at the same rate. Therefore, they eventually merge unless they break up into icebergs first. Ablating ice-stream tongues cannot exceed length x', so they are most likely to disintegrating before they merge. However, accumulating ice-stream tongues are much longer, and merger is more likely. The Ross Ice Shelf in Antarctica is an example of an ice shelf formed by the confluence of ice streams shown in Figure 2.14. Constant width w_I of an ice stream becomes variable width w of flowbands on an ice shelf. In general, these flowbands curve and either widen or narrow between the grounding line and the calving front of an ice shelf, depending on the geometry of the ice shelf.

For an ice shelf confined in an embayment, $\Delta h/\Delta x = (1-\rho_I/\rho_W)\,\Delta h_I/\Delta x$ and $\sigma_P = 1/2\,\rho_I\, g_z\, h_I\,(1-\rho_I/\rho_W)$ are the respective surface slope and gravitational pulling stress in Equation (3.27) and P_W/P_I can be related to buttressing stress σ_B for grounding along the sides of the embayment and at pinning points within the embayment by Equation (3.30). The relationship between P_W/P_I and σ_B/σ_P when ice flow can diverge, converge, and curve is:

$$\frac{P_W}{P_I} = \left[\frac{\frac{1}{2}\rho_I g_z h_I (1-\rho_I/\rho_W) - \sigma_B}{\frac{1}{2}\rho_I g_z h_I (1-\rho_I/\rho_W)}\right]^{1/2} = \left(1 - \frac{\sigma_B}{\sigma_P}\right)^{1/2} \quad (7.32)$$

Equation (3.27), with σ_{xx}' given by Equation (3.23), R given by Equation (3.25), and P_W/P_I given by Equation (3.30), gives for the ice thickness gradient $\Delta h_I/\Delta x$ of an ice shelf, where $\Delta h/\Delta x = (1 + \rho_I \rho_W)\,\Delta h_I/\Delta x$ satisfies the buoyancy condition:

$$\frac{\Delta h_I}{\Delta x} = \left\{\frac{P_W}{P_I}\right\}^2 \left\{\frac{a\,h_I}{a\,x + h_0 u_0} - \frac{R\,h_I^2}{a\,x + h_0 u_0}\right.$$

$$\left.\left[\frac{\rho_I g_z h_I}{2A}\left(1-\frac{\rho_I}{\rho_W}\right)\left(\frac{P_W}{P_I}\right)^2\right]^n\right\} + \frac{2\tau_s}{\rho_I g_z w}$$

$$= \left\{1 - \frac{2\sigma_B}{\rho_I g_z h_I}\left(\frac{\rho_W}{\rho_W - \rho_I}\right)\right\}\left\{\frac{a\,h_I}{a\,x + h_0 u_0} - \frac{R\,h_I^2}{a\,x + h_0 u_0}\right.$$

$$\left.\left[\frac{\rho_I g_z h_I}{2A}\left(1-\frac{\rho_I}{\rho_W}\right) - \frac{\sigma_B}{A}\right]^n\right\} + \frac{2\tau_s}{\rho_I g_z w} \quad (7.33)$$

Gradient $\Delta(P_W/P_I)/\Delta x$ and basal shear stress τ_o are ignored in Equation (7.33), but they are important near and at local basal pinning points that create ice rises and ice rumples on the surface (Thomas, 1973a, 1973b). Equation (7.33) gives the thickness gradient for an ice shelf, as derived by Van der Veen (1983) but including diverging, converging, and curving flow, and back stress σ_B caused by lateral grounding in an embayment and internal grounding at ice rises and ice rumples, as reported by Thomas (1973a, 1973b). In general, pinning points contribute more than the sides of an embayment in determining σ_B (Rommelaere and Ritz, 1996).

The ice thickness profile of the flowband is given by numerically solving Equation (7.33) at i steps of length Δx:

$$(h_I)_{i+1} = (h_I)_i + \left\{1 - \frac{2\sigma_B}{\rho_I g_z h_I}\right\}_i \left\{\frac{a\,h_I}{a\,x + h_0 u_0} - \frac{R\,h_I^2}{a\,x + h_0 u_0}\right.$$

$$\left.-\left[\frac{\rho_I g_z h_I}{2A}\left(1-\frac{\rho_I}{\rho_W}\right) - \frac{\sigma_B}{A}\right]^n + \frac{2\tau_s}{\rho_I g_z w}\right\}_i \Delta x \quad (7.34)$$

where h_I, w, a, R, A, τ_s, and σ_B are specified at each step.

For a flowband of floating length L_F from the grounding line to the calving front, buttressing stress σ_B at point x on the flowband is caused by (1) side shear stress σ_S transferred from the grounded sides of the ice shelf to side area $2\,\bar{h}_I\,(L_F - x)$ of the flowband and by (2) compressive stress σ_C if width w at x becomes width w_F at the calving front. The force balance for kinematic stresses σ_B, σ_S, and σ_C is:

$$F_x = \sigma_B h_I w - 2\sigma_S \bar{h}_I (L_F - x) - \sigma_C \bar{h}_I (w - w_F) = 0 \quad (7.35)$$

where mean ice thickness $\bar{h}_I$ over length $L_F - x$ is approximately the average of thickness h_I at x and h_F at $x = L_F$:

$$\bar{h}_I \approx \tfrac{1}{2}(h_I + h_F) \quad (7.36)$$

Solving Equations (7.35) and (7.36) for σ_B at x:

$$\sigma_B = \frac{\sigma_S}{w}\left(1 + \frac{h_F}{h_I}\right)(L_F - x) + \frac{\sigma_C}{2w}\left(1 + \frac{h_F}{h_I}\right)(w - w_F) \quad (7.37)$$

Except at ice rumples, $\dot{\varepsilon}_{xz} = 0$ for an ice shelf, so R given by Equation (7.13) reduces to:

$$R = \frac{\left[1 + (\dot{\varepsilon}_{yy}/\dot{\varepsilon}_{xx}) + (\dot{\varepsilon}_{yy}/\dot{\varepsilon}_{xx})^2 + (\dot{\varepsilon}_{xy}/\dot{\varepsilon}_{xx})^2\right]^{(n-1)/2}}{\left[2 + (\dot{\varepsilon}_{yy}/\dot{\varepsilon}_{xx})\right]^n} \quad (7.38)$$

Strain rate $\dot{\varepsilon}_{xx}$ for flowband velocity change $\Delta u_x = u_{i+1} - u_i$ in step Δx is:

$$\dot{\varepsilon}_{xx} = \frac{1}{2}\left(\frac{\partial u_x}{\partial x} + \frac{\partial u_x}{\partial x}\right) = \frac{du_x}{dx} = \frac{\Delta u_x}{\Delta x} = \frac{u_{i+1} - u_i}{\Delta x} \quad (7.39)$$

Strain rate $\dot{\varepsilon}_{yy}$ for flowband width change $\Delta w = w_{i+1} - w_i$ in step Δx over time $\delta t = \Delta x/\Delta u_x = 1/\dot{\varepsilon}_{xx}$ is:

$$\dot{\varepsilon}_{yy} = \frac{1}{2}\left(\frac{\partial u_y}{\partial y} + \frac{\partial u_y}{\partial y}\right) = \frac{du_y}{dy} = \frac{1}{dt}\int_{w_i}^{w_{i+1}} \frac{dw}{w} = \dot{\varepsilon}_{xx}\, ln\,(w_{i+1}/w_i) \quad (7.40)$$

Strain rate $\dot{\varepsilon}_{xy}$ for a flowband curving through angle $\delta\theta = \theta_{i+1} - \theta_i$ in step Δx over time $\delta t = 1/\dot{\varepsilon}_{xx}$ is:

$$\dot{\varepsilon}_{xy} = \frac{1}{2}\left(\frac{\partial u_x}{\partial y} + \frac{\partial u_y}{\partial x}\right) = \frac{1}{2}\frac{du_x}{dy} = \frac{1}{2}\frac{\delta\theta}{\delta t} = \frac{1}{2}\dot{\varepsilon}_{xx}(\theta_{i+1} - \theta_i) \quad (7.41)$$

where $\partial u_y/\partial x = 0$ because the flowband curves due to bending in simple shear (Hughes, 1983).

Grounding-Line Migration

The grounding lines of ice shelves are among the primary boundaries that regulate or respond to global climatic change. The ice-sheet surface behind grounding lines is above the flotation height of ice. Therefore, as grounding lines advance or retreat, global sea level falls or rises, thereby decreasing or increasing the heat exchange between the ocean and atmosphere by regulating the ocean surface area. It is therefore important to know if grounding-line migrations are controlled primarily by the dynamics of ice sheets or by the dynamics of other components of Earth's climate machine. Since ice shelves are supplied primarily by ice streams, ice sheets control grounding-line stability through ice streams.

In Figure 7.4, following Thomas (1977), the origin of the coordinates x, y, and z are at present-day sea level at the grounding line at the middle of an ice stream, with x horizontal and positive in the upstream direction, y hori-

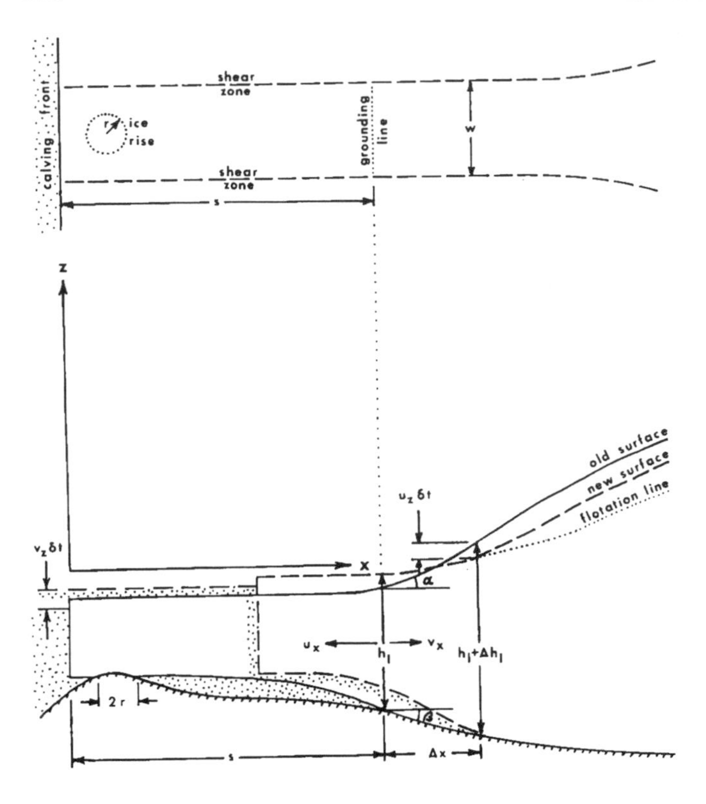

Figure 7.4: Migration of an ice-shelf grounding line along an ice stream in space and time with rising sea level. Shown are the orthogonal coordinates used in our ice-sheet disintegration model (x, z); the rise in sea level that accompanies collapse of the ice sheet; surface (α) and bed (β) slope at the ice-stream grounding line; absolute ice velocity across the grounding line (u_x); absolute grounding-line retreat velocity (v_x); rate of ice surface lowering (u_z) and sea-level increase (v_z) with incremental time (δt); total ice thickness (h_I) at distance (Δx) from the grounding line; and the ice elevation (dotted lines) at which flotation occurs for an incremental rise in sea level ($v_z\,\delta t$), unpinning the ice shelf at the ice rise (r) and increasing the length (s) of the ice shelf as the grounding line retreats an increment Δx. *Top*: Plan view. *Bottom*: Profile view.

zontal and positive toward the lateral shear zones of the ice stream, and z vertical and positive upward, with surface slope α and bed slope β in the x direction. Applying the continuity condition to the grounding line gives the rate of surface elevation change u_Z as the sum of (1) the surface and basal mass balance a, (2) the vertical velocity of creep thinning or thickening $h_I\,\dot{\varepsilon}_{zz}$, and (3) thickening due to advection of upstream ice $(\alpha - \beta)\,u_x$ at velocity u_x:

$$u_z = a + h_I\,\dot{\varepsilon}_{zz} - (\alpha - \beta)\,u_x \qquad (7.42)$$

All quantities in Equation (7.29) are measured at the grounding line. Note that u_x is negative when x is positive upstream.

An ice shelf expands when its grounding line retreats, provided that its calving front does not also retreat at an equal or faster rate. Retreat of ice shelf grounding lines ultimately collapses marine components of an ice sheet. Incremental retreat of the grounding line in steps of equal length Δx generally occurs in unequal time increments δt such that:

$$\Delta x = v_x\,\delta t \qquad (7.43)$$

where v_x is the retreat velocity of the grounding line.

A relationship between u_z and v_x is obtained by balancing vertical gravity forces per unit area of the bed at distance Δx from the grounding line for an ice column when it lowers to the flotation height of ice. As shown in Figure 7.4, this occurs when:

$$\rho_I\,g_z\,(h_I + \Delta h_I + u_z\,\delta t) = \rho_W\,g_z\,(h_W + \Delta h_W + v_z\,\delta t) \qquad (7.44)$$

where sea level rises at velocity v_z in time δt. The ice-thickness change in time δt at distance Δx is:

$$\Delta h_I = (\alpha - \beta)\,\Delta x = (\alpha - \beta)\,v_x\,\delta t \qquad (7.45)$$

and the water-thickness change in time δt at distance Δx is:

$$\Delta h_W = -\,\beta\,\Delta x - \beta\,v_x\,\delta t \qquad (7.46)$$

Substituting Equations (7.45) and (7.46) into Equation (7.44) gives:

$$\rho_I\,[h_I + (\alpha - \beta)\,v_x\,\delta t + u_z\,\delta t] = \rho_W\,[h_W - \beta\,v_x\,\delta t + v_z\,\delta t] \qquad (7.47)$$

Solving Equation (7.47) for h_I when δt = 0 gives the flotation condition at the grounding line:

$$h_W = (\rho_I\,/\rho_W)\,h_I \qquad (7.48)$$

Substituting Equation (7.48) into Equation (7.47) and solving for u_z when δt > 0 gives the ice thinning rate:

$$u_z = (\rho_W\,/\rho_I)\,v_z - [\alpha - \beta\,(1 - \rho_W\,/\rho_I)]\,v_x \qquad (7.49)$$

This is the relationship between u_z and v_x.

Grounding-line retreat rate v_x is given by combining Equations (7.42) and (7.49):

$$v_x = \frac{\left(\rho_W\,/\rho_I\right)v_z + (\alpha - \beta)\,u_x - h_I\,\dot{\varepsilon}_{zz} - a}{\alpha - \beta\left(1 - \rho_W\,/\rho_I\right)} \qquad (7.50)$$

where surface slope α is always positive and ice velocity u_X is always negative. Retreat of the grounding line (positive v_x) is facilitated by rising sea level (positive v_z), a bed that slopes downward upstream (negative β), creep thinning of ice (negative $\dot{\varepsilon}_{zz}$), and net ablation on the top and bottom ice surfaces (negative a). Advance of the grounding line is facilitated by reversing these variables. Ice sheets influence all of the variables: it regulates sea level by its mass balance; it regulates its mass balance by creating its own regional climate; it regulates its ice discharge rate by the life cycles of its ice streams; it regulates its creep thinning rate by adjusting its boundary conditions; and it regulates its bed slope by glacial isostasy and by glacial erosion or deposition processes.

Ice surface and bed slopes are particularly sensitive variables. When the retreat rate of the grounding line is computed in incremental steps of length Δx, the surface slope is computed from the change in surface elevation Δh above present-day sea level from $i = 0$ to $i = x/\Delta x$ over distance Δx, where i is the number of Δx steps in from the grounding line:

$$\alpha = \frac{\left(h_{i=1} - h_{i=0}\right)}{\Delta x} = \frac{\Delta h}{\Delta x} \tag{7.51}$$

and the bed slope is computed from the change in bed depth Δh_R below present-day sea level for this first Δx step:

$$\beta = \frac{\left[\left(h_R\right)_{i=1} - \left(h_R\right)_{i=0}\right]}{\Delta x} = \frac{\Delta h_R}{\Delta x} \tag{7.52}$$

Since α increases steadily along x and is small, whereas β can be highly variable, positive or negative, and large, β is more important than α in regulating grounding line retreat rates.

At the grounding line, $\dot{\varepsilon}_{zz}$ is obtained from Equation (3.24) for $\dot{\varepsilon}_{zz} = \dot{\varepsilon}_3 = -\dot{\varepsilon}_1 = -\dot{\varepsilon}_{xx}$ when $r = 0$ in Equation (7.12c), with $R = 1/2^n$ given by Equation (7.13) and (P_W/P_I) given by Equation (7.32) evaluated at $x = 0$:

$$\dot{\varepsilon}_{zz} = -\dot{\varepsilon}_{xx} \approx -\left[\frac{\rho_I g_z h_I (1 - \rho_I/\rho_W)}{4A}\left(\frac{P_W}{P_I}\right)_0^2\right]^n$$

$$\approx -\left[\frac{\rho_I g_z h_I \left(1 - \rho_I/\rho_W\right)}{4A} - \frac{\sigma_B}{A}\right]^n \tag{7.53}$$

where σ_B is the buttressing stress of the ice shelf given by Equation (7.37) for $h_I = h_0$ and $w = w_0$ at $x = 0$:

$$\sigma_B = \left[\sigma_S\left(L_F/w_0\right) + \sigma_C\left(1 - w_F/w_0\right)\right]\left[1 + h_F/h_0\right] \tag{7.54}$$

In Equation (7.54), σ_S is the mean side shear stress constraining a flowband on the ice shelf over floating length L_F from its grounding line across the ice stream to the calving front, and σ_C is the mean longitudinal compressive stress constraining the flowband over length L_F due to reducing flowband width from w_0 at the grounding line to w_F at the calving front. Stresses σ_S and σ_C are transmitted to the flowband from the side grounding lines of the ice shelf and from grounded ice rises within the ice shelf. These stresses act directly on the flowband itself only when an ice shelf is supplied by a single ice stream, such as the Amery Ice Shelf supplied by Lambert Glacier in Antarctica (see Figure 2.14). Each ice stream supplying the ice shelf provides a flowband on the ice shelf that is defined by velocity vectors on the ice shelf, as determined by Thomas and MacAyeal (1982) for the Ross Ice Shelf in Antarctica. Lingle and others (1989, 1991) have modeled heat and mass transport along one such flowband in response to CO_2–induced climatic warming. They find that the Ross Ice Shelf will not be much affected by anticipated warming. However, their study does not include ice-shelf disintegration induced by fracture mechanisms (Hughes, 1983).

The mass balance of an ice shelf consists of ice accumulation from surface precipitation, basal freezing, and inflowing ice, and ice ablation by calving and by surface and basal melting. An ice shelf disintegrates from fracture mechanisms when either calving rates increase along its calving front, thereby increasing ablation, or when ice streams punch through from its grounding line, thereby decreasing accumulation. Disintegration resulting from increased ablation by calving seems to be related to increased ablation by melting, perhaps because surface meltwater enters crevasses (Weertman, 1973). The recent disintegration of ice shelves along the Antarctic Peninsula has been accompanied by warmer summer air temperatures (e.g., Mercer, 1978; Vaughan and Doake, 1996). Disintegration of an ice shelf after an ice stream punches through occurs because the ice stream thereafter does not contribute to accumulation. Figure 7.5 illustrates this process. As Byrd Glacier crosses the Ross Ice Shelf grounding line, shear rupture between it and the ice shelf produces lateral rifts some 40 km long from side shear stresses σ_S that also open tensile crevasses at about 45° angles to the rifts through principal tensile stresses σ_T. The rifts and crevasses fracture the whole ice thickness, and these gashes through the ice shelf do not heal for nearly 100 km beyond the grounding line. An ice stream larger and faster than Byrd Glacier could rip through an even longer ice shelf. Figure 7.5 indicates that a former ice shelf in Pine Island Bay (Kellogg and Kellogg, 1978a, 1987b, and 1978c) disintegrated after Thwaites Glacier punched through it, depriving it of a major input to its accumulation.

Collapse of a marine ice sheet resulting from retreat of ice-shelf grounding lines up ice streams (see Byrd Glacier in Figure 7.5) is linked to disintegration of its fringing ice shelves, so $\sigma_B = 0$. Disintegration of ice shelves is controlled by iceberg calving rates, so $L_F \rightarrow 0$ and $w_F \rightarrow w_0$ in Equation (7.54). The gravitational pulling force F_x, acting on an unconfined ice shelf of thick-ness h_I, is given by Equation (4.57), in which $F_x \propto h_I^2$ and $P_W/P_I = 1$ because the ice shelf is floating. Figure 2.2 shows geometrically that the pulling force is proportional to the square of floating ice thickness, so the pulling force increases and decreases at an ice-shelf grounding line that retreats down-slope and upslope on the foredeepened, isostatically depressed bed beneath a marine ice stream that drains a marine ice sheet. This pulling force plucks icebergs from the calving front of the ice shelf and is strongest at the ice-shelf grounding line, where floating ice is thickest. Figure 2.3 shows how grounding-line retreat and ice-calving retreat for a floating ice shelf allow the ice shelf to migrate up the ice streams that supply it and cause the grounded ice sheet to collapse. The

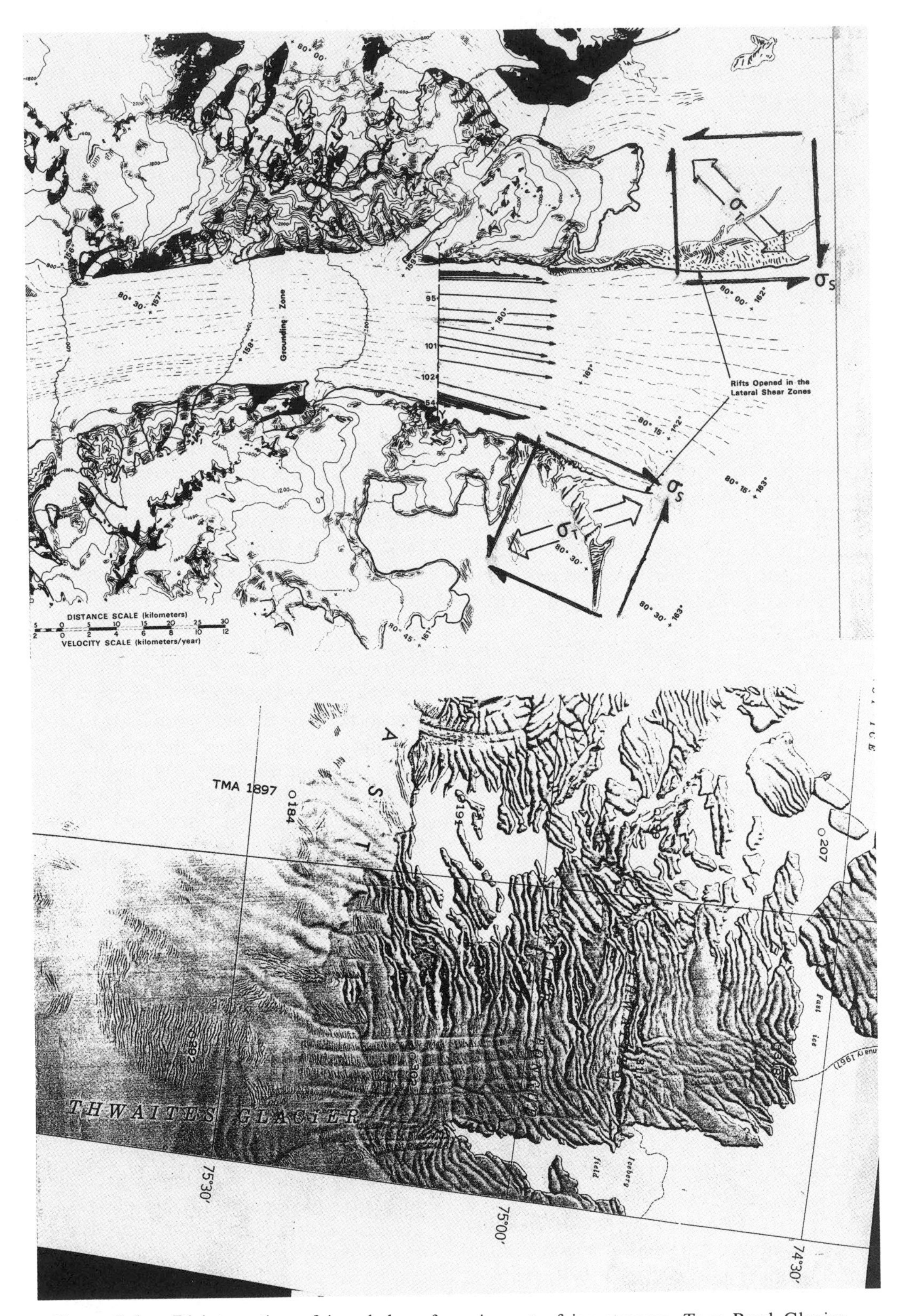

Figure 7.5: Disintegration of ice shelves from impact of ice streams. *Top*: Byrd Glacier shatters the Ross Ice Shelf up to 100 km from its grounding line by opening lateral rifts parallel to side shear stresses σ_S and tensile crevasses perpendicular to principal tensile stress σ_T. Bottom: Thwaites Glacier has punched though an ice shelf in Pine Island Bay that has largely disintegrated after being deprived of ice input from the ice stream.

critical event is floating the ice-shelf grounding line over the basal sill of the foredeepened submarine trough occupied by a marine ice stream. This is accomplished by (a) surge-like life cycles that thin the ice stream crossing the sill, (b) ongoing glacioisostatic depression that lowers the sill, (c) rising sea level that lifts the ice shelf over the sill, (d) surface or basal melting that thins the ice shelf at the sill, and (e) a calving bay that carves away ice over the sill.

After the ice-shelf grounding line crosses ice-stream sills, the grounding line retreats on the downsloping bed of the foredeepened marine trough occupied by the ice stream. Since $F_x \propto h_I^2$, the grounding line retreats at an increasing rate as it moves across the downsloping bed toward the front of the trough and at a decreasing rate as it moves across the upsloping bed toward the rear of the trough. Since F_x plucks icebergs from the calving front, a calving bay follows the retreating grounding line and disintegrates the ice shelf. Estuarine circulation then ferries the discerped icebergs over the sill and thereby evacuates ice from the deglaciated submarine trough.

Figure 2.4 shows stages in the disintegration of an ice shelf as its grounding line retreats on an upsloping bed and as the snowline rises at the end of a glaciation cycle. During the retreat, calving dynamics change as the pulling force (a) plucks icebergs along top and bottom tidal-flexure crevasses that form along the edge of the continental shelf and are widened by the pulling force until they meet at the freely-floating calving front beyond the continental shelf, (b) plucks icebergs along a calving front between pinning points on the continental shelf where the ice shelf is locally grounded on islands or shoals and is fractured by block faulting in the lee of the pinning points, (c) plucks icebergs along the grounding line after the calving front migrates around pinning points on the continental shelf and retreats into shallower water, (d) plucks ice slabs above and below a wave-cut groove along an ice wall standing in water too shallow to float the ice, (e) plucks ice slabs above a wave-cut groove at the base of an ice wall that ends at the shoreline of a beach, and (f) vanishes as surface melting causes the ice margin to retreat from the beach and converts the ice wall into an ice ramp. Calving events dislodge smaller blocks of ice as calving dynamics change for the changing boundary conditions (a) through (f) in Figure 2.4, and a temporary stillstand of the calving front is possible for each boundary condition. Therefore, ice calving rates exert critical rate controls on disintegrating ice shelves, retreating ice streams, and collapsing ice sheets, and these calving mechanisms need to be examined.

8

ICE CALVING

Disintegration of Ice Shelves

Boundary conditions change radically where ice streams cross the grounding lines of ice shelves. These changes produce longitudinal and transverse crevasses that become lines of weakness dividing the ice shelf into weakly coupled plates that are carried passively to the calving front. Bending forces at the calving front reactivate the intersecting lines of weakness, causing the ice shelf to disintegrate into tabular icebergs along the calving front. Since an ice shelf has an almost flat top surface, a slight increase in snowline elevation may suddenly convert the top surface from an accumulation zone to an ablation zone. In this event, surface meltwater drains toward the lines of weakness, where ice is slightly thinner, and deepens crevasses there that become filled with this meltwater. An ice shelf can then disintegrate in situ into plates separated by these lines of weakness.

On a macroscopic scale, migration of surface meltwater from plates on an ice shelf to lines of weakness between plates is analogous to migration of dislocations within grains to grain boundaries in polycrystalline ice, on a microscopic scale. Dislocation pileups increase strain energy in grain boundaries, thereby augmenting the thermal energy needed to initiate melting of polycrystalline ice, which can therefore disintegrate into individual grains by grain-boundary melting. Similarly, surface meltwater accumulating along lines of weakness provides thermal energy that augments the mechanical energy needed to deepen crevasses and thereby disintegrate the ice shelf into individual plates by melting and fracturing the lines of weakness between plates.

Mercer (1978) noted that ice shelves form along the western side of the Antarctic Peninsula when the summer surface air temperature falls below the melting point. Rising snowlines have moved this boundary south along the peninsula and have converted the top surface of the northern-most ice shelf from an accumulation zone into an ablation zone. This is Wordie Ice Shelf, and it is now disintegrating in situ along clearly defined lines of weakness, seen in Figure 8.1 (Doake and Vaughan, 1991). Lines of weakness exist in other Antarctic ice shelves even though their top surfaces remain accumulation zones, as illustrated for the Ross Ice Shelf in Figure 8.2. These lines of weakness are often obscured by the snow cover.

Fracture Criteria

If ice shelves disintegrate along lines of weakness where crevasses accumulate, a fracture criterion is needed to specify the fracture stress that opens a crevasse. No viscoplastic fracture criterion has been established for ice. However, a plastic yield criterion emerges when Equation (4.136) is plotted in Figure 8.3 after it is written in the form:

$$\frac{\dot{\varepsilon}_v}{\dot{\varepsilon}_o} = \left(\frac{\sigma_v}{\sigma_o}\right)^n \tag{8.1}$$

where σ_v is the viscoplastic creep stress and $\dot{\varepsilon}_v$ is the viscoplastic creep rate. An infinite $\dot{\varepsilon}_v$ results when σ_v only slightly exceeds plastic yield stress σ_o for $n = \infty$ at the plastic end of the viscoplastic creep spectrum. Therefore, a plastic fracture criterion is:

$$\frac{\sigma_v}{\sigma_o} = 1 \tag{8.2}$$

because fracture also produces an infinite $\dot{\varepsilon}_v$.

By analogy, it seems reasonable to postulate fracture criteria for $1 < n < \infty$, which includes $n \approx 3$ for ice, in which fracture occurs when a small increase in σ_v causes a large increase in $\dot{\varepsilon}_v$ that culminates in fracture. Figure 8.3 illustrates two fracture criteria that are compatible with this assumption. A "knee" develops on curves of $\dot{\varepsilon}_v/\dot{\varepsilon}_o$ versus σ_v/σ_o as n increases. The crack nucleation fracture criterion postulates that fracture stress σ_v' occurs at critical strain rate $\dot{\varepsilon}_v'$ at the knee because microcracks nucleate at that strain rate and produce the knee. The crack propagation fracture criterion postulates that fracture stress σ_v'' occurs at critical strain rate $\dot{\varepsilon}_v''$ just beyond the knee because microcracks propagate at that strain rate and lead to fracture. The two

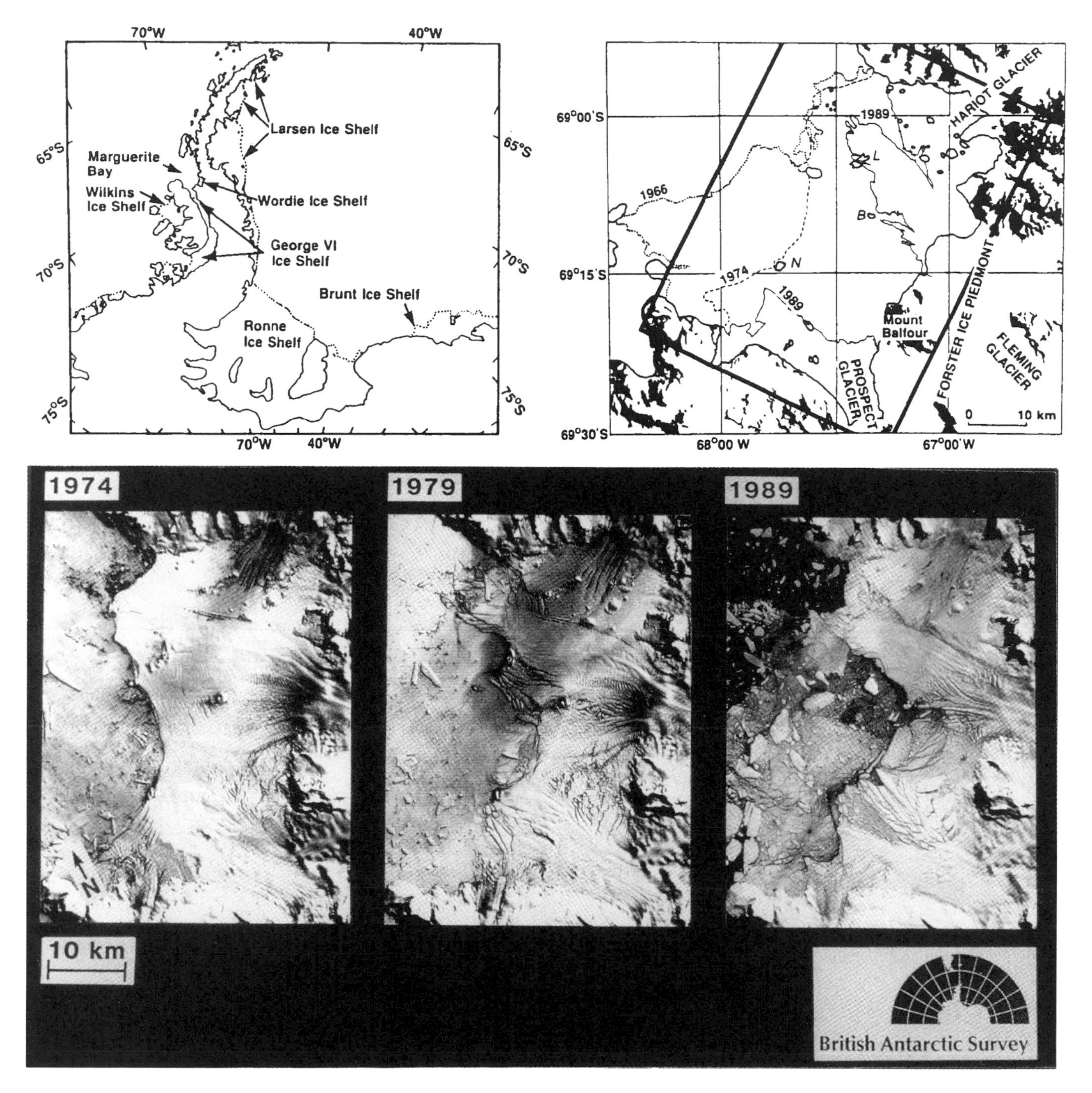

Figure 8.1: Disintegration of the Wordie Ice Shelf along lines of weakness. The Wordie Ice Shelf is at 67.5°W and 69.3°S on the Antarctic Peninsula. Satellite imagery provided by Christopher Doake. *Top right:* Location map. *Top left*: Image area. *Bottom*: Images

fracture criteria converge as n increases because s_v' and s_v'' both approach $\sigma_v = \sigma_o$, while $\dot{\varepsilon}_v'$ and $\dot{\varepsilon}_v''$ both approach $\dot{\varepsilon}_v = 0$. Both criteria become the plastic yield criterion at $n = \infty$, $\sigma_v = \sigma_o$, and $\dot{\varepsilon}_v = 0$. Both criteria become poorly defined for $n < 3$, and both are meaningless for $n = 1$ at the viscous end of the viscoplastic creep spectrum. Both criteria are linked to viscoplastic viscosity η_v.

Bends in the curves shown in Figure 8.3 have a dramatic effect on viscoplastic viscosity η_v of ice, which de-pends strongly on n, because η_v is the slope of curves in the viscoplastic creep spectrum. Differentiating Equation (8.1):

$$\eta_v = \frac{d\sigma_v}{d\dot{\varepsilon}_v} = \frac{\sigma_o^{\,n}}{n\,\dot{\varepsilon}_o\,\sigma_v^{\,n-1}} = \frac{\sigma_v}{n\,\dot{\varepsilon}_v} \tag{8.3}$$

where $\eta_v = \sigma_v / \dot{\varepsilon}_v$ is the fluid viscosity for viscous flow

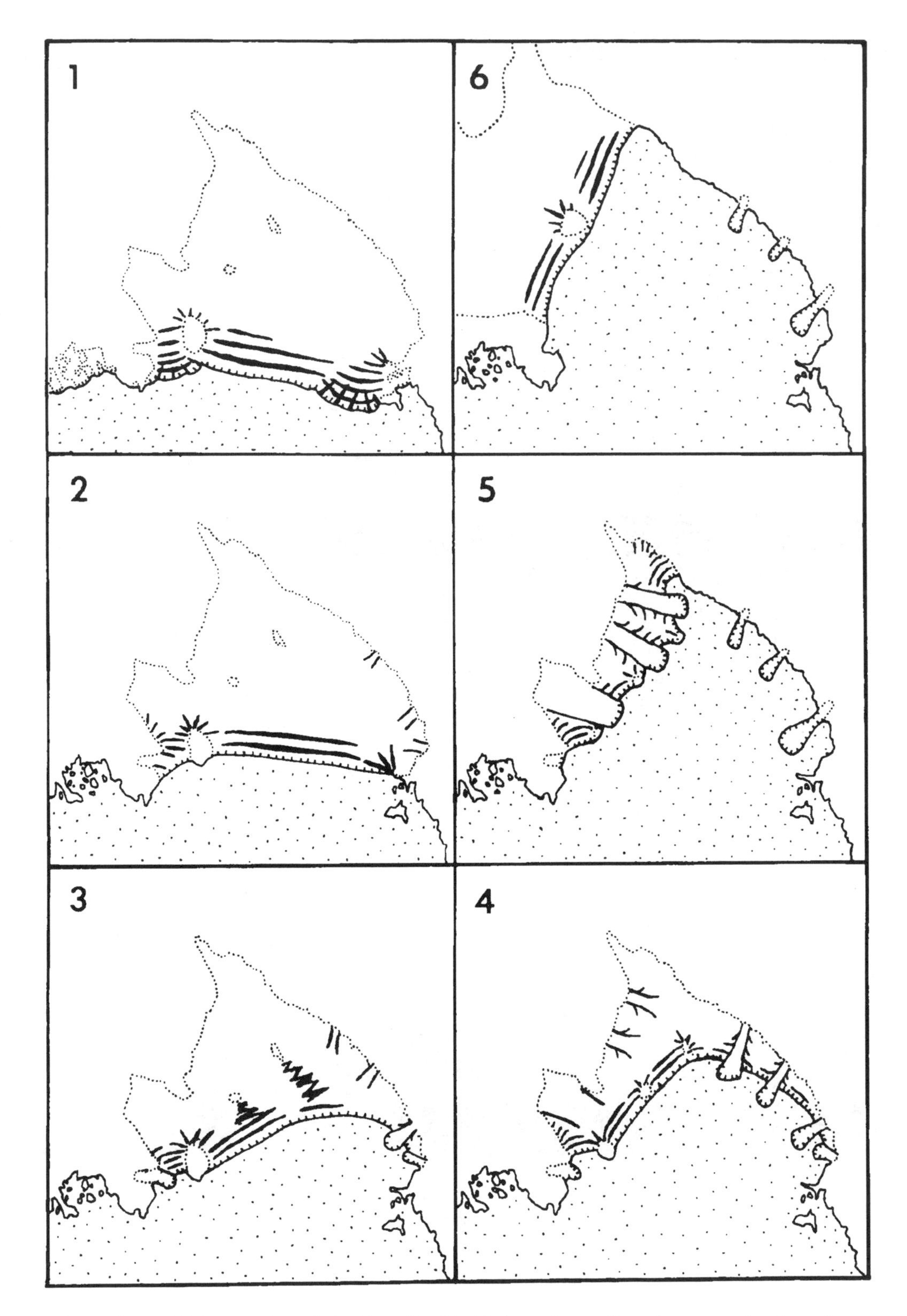

Figure 8.2: Disintegration of the Ross Ice Shelf along lines of weakness on the Ross Ice Shelf. The ice shelf is grounded along dotted lines and calves along hatchured lines. Longitudinal lines of weakness are rifts alongside ice streams and in the lee of ice rises. Transverse lines of weakness are transverse crevasses in ice streams and tidal flexure crevasses in ice crossing grounding lines. Progressive disintegration is shown in sequential stages 1 through 6. Compare with Figure 7.5.

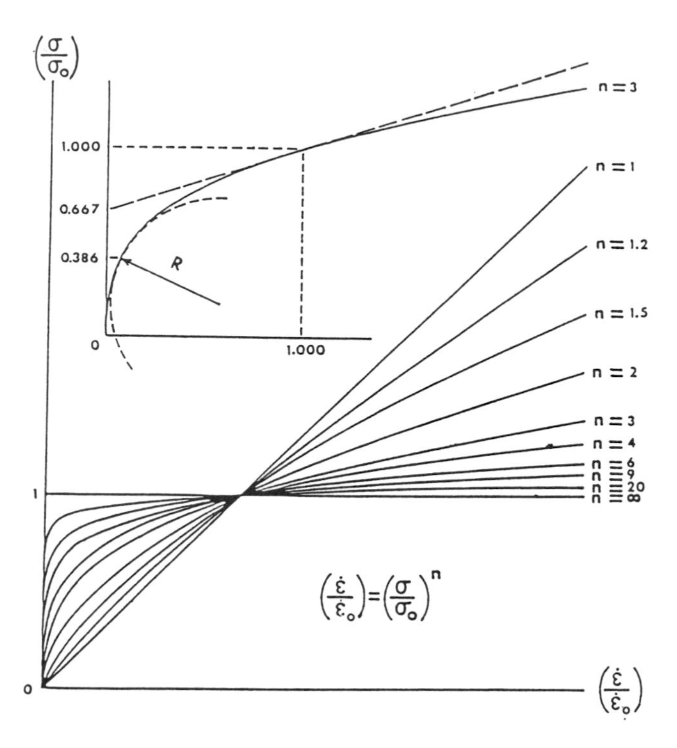

Figure 8.3: The viscoplastic spectrum for steady-state creep and two criteria for viscoplastic yielding. A sharp knee develops in stress-strain rate curves when the viscoplastic exponent n increases, where $\dot{\varepsilon}_m$ is the strain rate at the maximum shear stress τ_m, $\dot{\varepsilon}_o$ is the strain rate at the plastic yield stress σ_o and $\dot{\varepsilon}_m / \dot{\varepsilon}_o = (\tau_m / \sigma_o)^n$. Viscoplastic yielding occurs at the knee in the crack nucleation fracture criterion and at the stress intercept of the tangent line at $\dot{\varepsilon}_o$ in the crack propagation fracture criterion. For n = 3, $\sigma_v = 0.386\ \sigma_o$ at the knee and $\sigma_v = 0.667\sigma_o$ at the stress intercept (Hughes, 1983, Figure 4).

when n = 1 and $\eta_v = 0$ is the inviscid viscosity for plastic flow when $n = \infty$. Knees in the bends of curves in Figure 8.3 occur at stresses σ_v' where the curvature of these curves is greatest. The radius of creep curvature R_c obtained from Equation (8.1) is:

$$R_c = \frac{\left[1 + \left(\dfrac{d\left(\dot{\varepsilon}_v / \dot{\varepsilon}_o\right)}{d\left(\sigma_v / \sigma_o\right)}\right)^2\right]^{3/2}}{\dfrac{d^2\left(\dot{\varepsilon}_v / \dot{\varepsilon}_o\right)}{d\left(\sigma_v / \sigma_o\right)^2}} = \frac{\left[1 + n^2\left(\sigma_v / \sigma_o\right)^{2n-2}\right]^{3/2}}{n\,(n-1)\left(\sigma_v / \sigma_o\right)^{n-2}} \tag{8.4}$$

Setting $dR_c / d(\sigma_v / \sigma_o) = 0$ at the maximum curvature where $\sigma_v = \sigma_v'$ gives the crack nucleation viscoplastic fracture criterion:

$$\frac{\sigma_v'}{\sigma_o} = \left[\frac{n-2}{2\,(n-1)\,n^2}\right]^{\frac{1}{2n-2}} \tag{8.5}$$

Equation (8.5) gives $\sigma_v' = 0$ at n = 2 and $\sigma_v' = \sigma_o$ at $n = \infty$.

Critical strain rate $\dot{\varepsilon}_v'$ at σ_v' is obtained by combining Equations (8.1) and (8.5):

$$\frac{\dot{\varepsilon}_v'}{\dot{\varepsilon}_o} = \left[\frac{n-2}{2\,(n-1)\,n^2}\right]^{\frac{n}{2n-2}} \tag{8.6}$$

Equation (8.6) gives $\dot{\varepsilon}_v' = 0$ at n = 2 and at $n = \infty$.

Viscosity η_o is the slope of straight lines that are tangent to curves in Figure 8.3 at point σ_o, $\dot{\varepsilon}_o$ and gives viscoplastic stress σ_v'' at $\dot{\varepsilon}_v = 0$. The equation for these straight lines is:

$$\frac{\sigma_v}{\sigma_o} = \left[\frac{d\left(\sigma_v / \sigma_o\right)}{d\left(\dot{\varepsilon}_v / \dot{\varepsilon}_o\right)}\right]\frac{\dot{\varepsilon}_v}{\dot{\varepsilon}_o} + \frac{\sigma_v''}{\sigma_o} = \left(\frac{d\sigma_v}{d\dot{\varepsilon}_v}\right)\frac{\dot{\varepsilon}_v}{\sigma_o} + \frac{\sigma_v''}{\sigma_o} \tag{8.7}$$

Setting $\sigma_v = \sigma_o$ at $\dot{\varepsilon}_v = \dot{\varepsilon}_o$ and differentiating Equation (8.1) to obtain $\eta_o = d\,\sigma_v / d\dot{\varepsilon}_v = \sigma_v / n\,\dot{\varepsilon}_o$ gives the crack propagation viscoplastic yield criterion:

$$\frac{\sigma_v''}{\sigma_o} = \frac{n-1}{n} \tag{8.8}$$

Equation (8.8) gives $\sigma_v'' = 0$ at n = 1 and $\sigma_v'' = \sigma_o$ at $n = \infty$. Critical strain rate $\dot{\varepsilon}_v''$ at σ_v'' is obtained by combining Equations (8.1) and (8.8):

$$\frac{\dot{\varepsilon}_v''}{\dot{\varepsilon}_o} = \left(\frac{n-1}{n}\right)^n \tag{8.9}$$

Equation (8.9) gives $\dot{\varepsilon}_v'' = 0$ at n = 1 and at $n = \infty$.

Equations (8.3), (8.5), and (8.8) are plotted over the viscoplastic creep spectrum of n in Figure 8.4. The strong dependence of η_v, σ_v', and σ_v'' on n illustrates how superplasticity, like recrystallization, increases the creep rate. As n increases, η_v for viscoplastic flow falls rapidly below η_N for Newtonian flow, and both σ_v' and σ_v'' for viscoplastic yielding rise rapidly toward σ_o for plastic yielding.

Whether fracture begins at $\dot{\varepsilon}_v'$ or at $\dot{\varepsilon}_v''$ depends on the philosophical question of whether fracture begins when microcracks form at stress σ_v' or when microcracks propagate to open a crevasse at stress σ_v''. Fracture undergoes a transition from ductile to brittle for (1) a triaxial state of stress, (2) a low temperature, and (3) a high strain rate (Dieter, 1961, p. 216). Since $\dot{\varepsilon}_v' < \dot{\varepsilon}_v''$, it may be that microcrack formation constitutes a ductile fracture criterion in which σ_v' is the ductile fracture stress and that microcrack propagation constitutes a brittle fracture criterion in which σ_v'' is the brittle fracture stress.

Both ductile fracture and brittle fracture exists in ice, so the ductile-to-brittle transition may occur at a fracture stress that is the average of σ_v' and σ_v''. Also, fracture is usually predictable only in a statistical sense within a limiting range of fracture stresses that may be bracketed by σ_v' and σ_v''

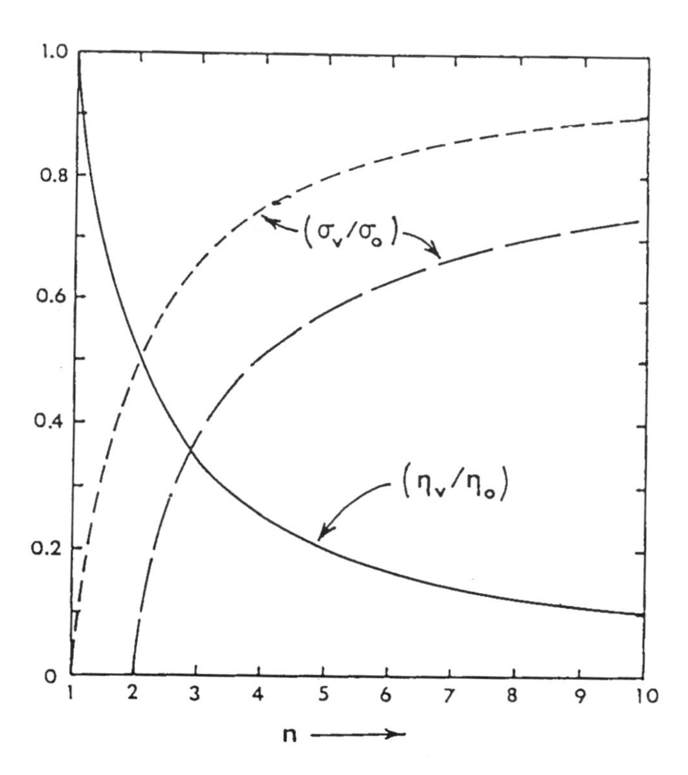

Figure 8.4: Variations of yield stress and viscosity across the viscoplastic spectrum of steady-state creep. Ratios of viscoplastic yield stress σ_v and plastic yield stress σ_0 increase with the viscoplastic exponent n according to Equation (8) for the maximum stress–curvature yield criterion (long dashed line) and according to Equation (10) for the critical strain-rate yield criterion (short dashed line). The ratio of effective viscosity η_v and fluid viscosity η_0 decreases with increasing n (solid curve) (Hughes, 1983, Figure 5).

(Dieter, 1961, pp. 215–216). If so, fracture stress $\bar{\sigma}_v$ can be defined as an average of σ_v' and σ_v'' that is bracketed by σ_v' and σ_v'':

$$\bar{\sigma}_v = \frac{1}{2}(\sigma_v' + \sigma_v'') = \frac{\sigma_o}{2}\left[\left(\frac{n-2}{2(n-1)n^2}\right)^{\frac{1}{2n-2}} + \left(\frac{n-1}{n}\right)\right] \qquad (8.10)$$

In this case the critical strain rate $\dot{\varepsilon}_v$ for fracture is also a statistical average of $\dot{\varepsilon}_v'$ and $\dot{\varepsilon}_v''$:

$$\bar{\dot{\varepsilon}}_v = \frac{1}{2}\left(\dot{\varepsilon}_v' + \dot{\varepsilon}_v''\right) = \frac{\dot{\varepsilon}_o}{2}\left[\left(\frac{n-2}{2(n-1)n^2}\right)^{\frac{n}{2n-2}} + \left(\frac{n-1}{n}\right)^n\right] \qquad (8.11)$$

An average viscoplastic viscosity $\bar{\eta}_v$ for fracture is therefore the average of η_v' and η_v'':

$$\bar{\eta}_v = \frac{1}{2}\left(\frac{\sigma_v'}{n\,\dot{\varepsilon}_v'} + \frac{\sigma_v''}{n\,\dot{\varepsilon}_v''}\right) = \eta_o\left[\frac{1}{2}\left(\frac{2n-2}{n-2}\right)^{1/2} + \frac{1}{2n}\left(\frac{n-1}{n-1}\right)^{n-1}\right] \qquad (8.12)$$

where $\eta_o = \sigma_o/\dot{\varepsilon}_o$. Taking n = 3 for ice, $\bar{\sigma}_v = (0.54 \pm 0.13)\,\sigma_o$, $\bar{\dot{\varepsilon}}_o = (0.132 \pm 0.114)\,\dot{\varepsilon}_o$, and $\bar{\eta}_v = (1.375 \pm 0.625)\,\eta_o$. Note that $\bar{\sigma}_v$ and $\bar{\dot{\varepsilon}}_v$ have values only for $n \geq 2$, and $\bar{\eta}_v$ has values only for $n > 2$.

If $\bar{\sigma}_v$ is the fracture stress, it can be computed from the stress field on an ice shelf to determine calving rates along lines of weakness. Fracture produces crevasses on the top and bottom surfaces, so $\bar{\sigma}_v$ is computed for plane stress conditions for which the Mohr circle construction relates principal stresses σ_1 and σ_2 to stresses σ_{xx}, σ_{yy}, and σ_{xy} (Dieter, 1961):

$$\sigma_1 = \frac{1}{2}\left(\sigma_{xx} + \sigma_{yy}\right) + \left[\frac{1}{4}\left(\sigma_{xx} + \sigma_{yy}\right)^2 + \sigma_{xy}^2\right]^{1/2} \qquad (8.13a)$$

$$\sigma_2 = \frac{1}{2}\left(\sigma_{xx} + \sigma_{yy}\right) - \left[\frac{1}{4}\left(\sigma_{xx} + \sigma_{yy}\right)^2 + \sigma_{xy}^2\right]^{1/2} \qquad (8.13b)$$

where x is parallel to and y is transverse to the flow direction. Since $\sigma_1 > \sigma_2$ and is tensile at the top surface, top crevasses open when $\sigma_1 = \bar{\sigma}_v$ on the ice shelf.

Primary Transverse Crevasses

Primary transverse crevasses on an ice shelf can begin as bottom crevasses at the heads of ice streams where a frozen bed becomes thawed. As these crevasses open, water and water-saturated till or sediments are injected under high basal pressures into the crevasses, where refreezing occurs. Expansion as water becomes ice exerts additional pressure against the crevasse walls, and forces the crevasse tip upward. These transverse bottom crevasses become primary lines of weakness that are transported relatively passively along the ice stream until they become reactivated by tidal flexure when the ice stream becomes afloat and merges with the ice shelf. They are then transported relatively passively from the grounding line to the calving front, where they again become reactivated by bending caused by the imbalance of opposing lithostatic and hydrostatic longitudinal forces along the calving front. Crevasses on the top surface should also form above bottom crevasses, because bottom crevasses reduce the transverse cross section, and thereby increase tensile stress σ_{xx}' for a given pulling force F_x. These top crevasses will be empty if they open in the accumulation zone, but they may fill with water if they move across an ablation zone on the way to the calving front. Figure 8.5 shows the deviator, lithostatic, and hydrostatic stresses that determine the depth of a top or bottom crevasse.

The longitudinal tensile deviator stress along an ice stream of length L_S is given by Equations (6.7) and (6.18):

$$\sigma_{xx}' = \frac{\rho_I\, g_z\, h_I}{4}\left(1 - \frac{\rho_I}{\rho_W}\right)\left(\frac{P_W}{P_I}\right)^2$$

$$= \frac{\rho_I\, g_z\, h_I}{4}\left(1 - \frac{\rho_I}{\rho_W}\right)\left(1 - \frac{x}{L_S}\right)^{2c} \qquad (8.14)$$

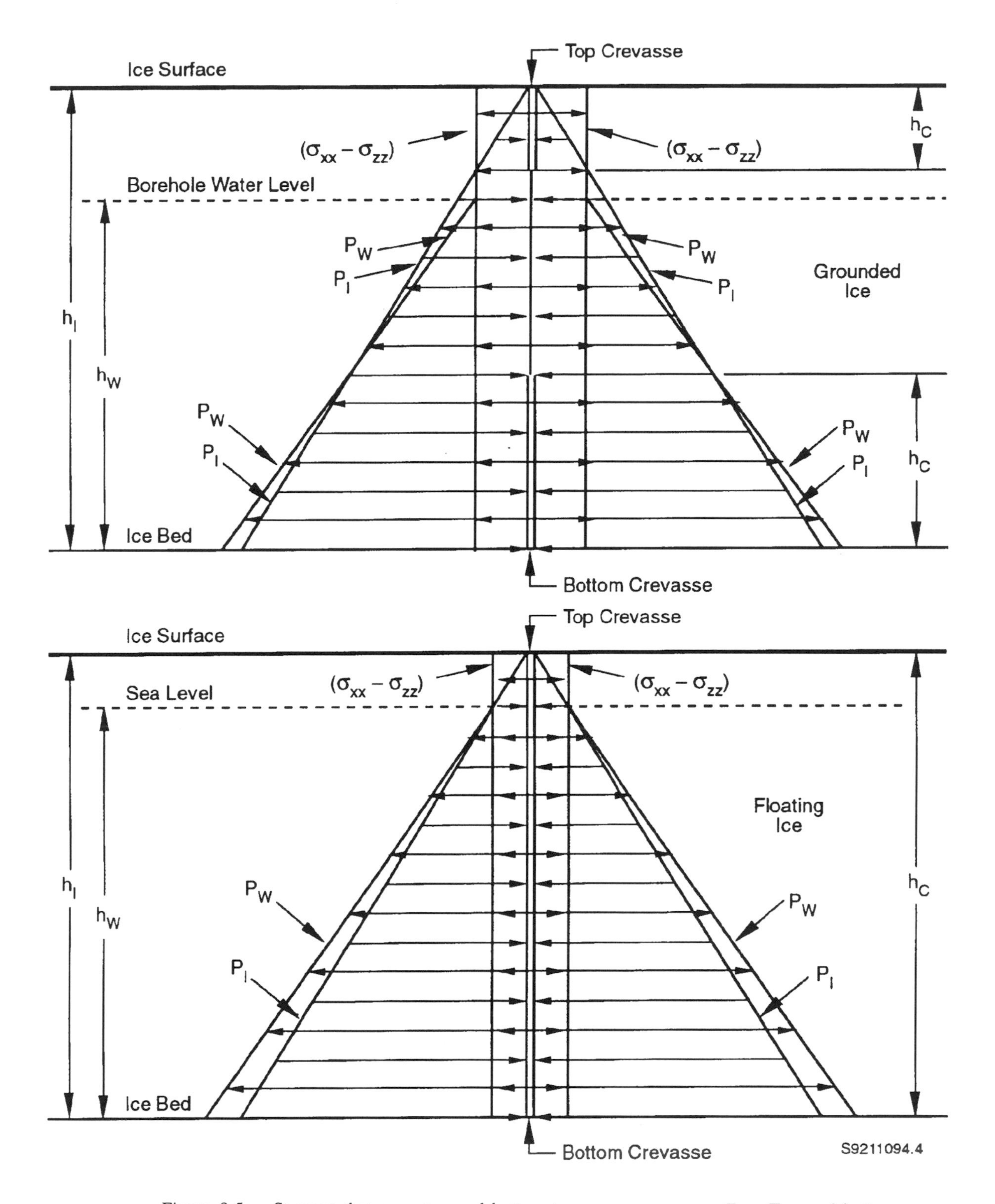

Figure 8.5: Stresses that open top and bottom transverse crevasses. *Top*: Top and bottom crevasses do not meet in an ice stream because ice is grounded enough to prevent basal water from rising to the flotation height in a borehole. *Bottom*: Top and bottom crevasses meet in an ice shelf because basal water rises to sea level in a borehole. Stresses shown are ice pressure P_I and water pressure P_W that increase linearly with depth, and longitudinal tensile stress $(\sigma_{xx} - \sigma_{zz})$ that is constant with depth. Crevasses open until $P_I - P_W = \sigma_{xx} - \sigma_{zz}$. Heights are h_I for ice, h_W for water, and h_C for crevasses.

Therefore, $\sigma_{xx}'/h_I = 0$ for $x = L_S$ where $P_W/P_I = 0$ for full basal coupling, and σ_{xx}'/h_I increases for $x < L_S$ to a maximum at the ice-shelf grounding line where $x = 0$ and $P_W/P_I = 1$.

Primary transverse crevasses are opened when longitudinal stress σ_{xx} is tensile at and just below the top ice surface and is less compressive than vertical stress σ_{zz} at all deeper depths. Therefore, the stress difference $(\sigma_{xx} - \sigma_{zz})$ is always tensile at all depths, and a crevasse will open if $(\sigma_{xx} - \sigma_{zz})$ equals or exceeds fracture stress $\bar{\sigma}_v$ in Equation (8.10). For an ice stream, where primary transverse crevasses form, Equation (4.42) gives:

$$(\sigma_{xx} - \sigma_{zz}) = 2\,\sigma_{xx}' \tag{8.15}$$

Nye (1952) proposed that a top surface crevasse would extend to depth h_C at which tensile stress $2\,\sigma_{xx}'$ that was constant through total ice thickness h_I would be canceled by compressive lithostatic pressure P_I that increased linearly with depth according to Equation (4.32) when z was measured vertically upward from the bottom surface. The lithostatic pressure at depth h_C is P_C given by:

$$P_C = \rho_I\, g_z\, h_C \tag{8.16}$$

Setting $(\sigma_{xx} - \sigma_{zz})$ equal to P_C and solving for h_C:

$$h_C = \frac{2\,\sigma_{xx}'}{\rho_I\, g_z} \tag{8.17}$$

If the top crevasse fills with water, ρ_I in Equation (8.17) is replaced by $\rho_I - \rho_W$, and P_C becomes ΔP_C such that:

$$\Delta P_C = (\rho_I - \rho_W)\, g_z\, h_C \tag{8.18}$$

Since $\rho_I < \rho_W$, ΔP_C is of opposite sign from P_C and causes extension instead of compression. Hence, ΔP_C adds to $(\sigma_{xx} - \sigma_{zz})$, and the maximum tensile stress at the crevasse tip is attained when:

$$h_C = h_I \tag{8.19}$$

Bottom crevasses fill with water or water–saturated muck, depending on whether the bed is hard or soft, that extends upward to height h_C, where the total extensional stress is zero:

$$2\,\sigma_{xx}' + \Delta P_C = 0 \tag{8.20}$$

Substituting Equation (8.18) into Equation (8.20) and solving for h_C:

$$h_C = \frac{2\,\sigma_{xx}'}{(\rho_W - \rho_I)\, g_z} \tag{8.21}$$

where ρ_W is the density of either water or of water-saturated muck.

Primary transverse crevasses first open when longitudinal gravitational pulling stress $2\,\sigma_{xx}'$ reaches fracture stress $\bar{\sigma}_v$ for ice. Gravitational pulling force F_G depends on the gravitational potential energy provided by an ice stream having surface height h_I above its bed, so F_G does not change when crevasses of depth h_C above the bed reduce the unfractured ice thickness. However, pulling stress $2\,\sigma_{xx}'$ that equals fracture stress $\bar{\sigma}_v$ for ice of unfractured thickness h_I must increase in ice of unfractured thickness $h_I - h_C$, since the longitudinal force balance requires that for an ice stream of width w_I:

$$(F_G)_x = h_I\, w_I\, \bar{\sigma}_v = 2\,(h_I - h_C)\, w_I\, \sigma_{xx}' \tag{8.22}$$

where $2\,\sigma_{xx}' = \bar{\sigma}_v$ when $h_C = 0$. Solving Equation (8.22) for h_C:

$$h_C = h_I\,(1 - \bar{\sigma}_v / 2\,\sigma_{xx}') \tag{8.23}$$

Since σ_{xx}' increases as h_C increases for a bottom crevasse, a top crevasse should open above the bottom crevasse. Alternatively, as h_C increases for a top crevasse, a bottom crevasse should open beneath the top crevasse. Therefore, top and bottom crevasses should open simultaneously, with top crevasses directly above bottom crevasses.

Weertman (1973) analyzed crevasses filled with water of arbitrary depth, using dislocation theory to compute the concentration of stresses at the crevasse tip. He concluded that $h_C = \pi\,\sigma_{xx}'/\rho_I\, g_z$ for empty top crevasses, $h_C = h_I$ for water-filled top crevasses, and $h_C = \pi\,\sigma_{xx}'/(\rho_W - \rho_I)\, g_z$ for water-filled bottom crevasses. He also concluded that closely spaced top crevasses would extend to depth $h_C = 2\,\sigma_{xx}'/\rho_I\, g_z$, whether they were empty or filled with water. Weertman (1980) also calculated how high a water-filled bottom crevasse could extend up into cold ice before it froze shut. He concluded that water-filled bottom crevasses would remain open in unconfined ice shelves thicker than about 400 m. His approach can be used to compute the spacing beneath transverse bottom crevasses in ice streams, and, since top crevasses should form above bottom crevasses for a given σ_{xx}', the spacing between top transverse crevasses as well.

Advance of the tip for water-filled top or bottom crevasses can be halted if water freezes onto the crevasse walls as rapidly as the crevasse widens. Fracture opens a transverse crevasse when $2\,\sigma_{xx}' \geq \bar{\sigma}_v$. A crevasse of width w_o has an opening rate (Weertman, 1980):

$$\begin{aligned} \partial w_o/\partial t &= h_C\,(\partial u_x/\partial z)_C = h_C\,(2\,\dot{\varepsilon}_{xz})_C \\ &= 2\,h_C\,\dot{\varepsilon}_v = 2\,h_C(\bar{\sigma}_v/n\,\bar{\eta}_v) \end{aligned} \tag{8.24}$$

where $\dot{\varepsilon}_v = (\dot{\varepsilon}_{xz})_C = 1/2\,(\partial u_x/\partial z)_C$ is the creep rate for viscoplastic fracture caused by viscoplastic fracture stress $\bar{\sigma}_v$ at the crevasse tip. Viscoplastic viscosity η_v for fracture is $\bar{\eta}$ related to σ_v and $\dot{\varepsilon}_v$ in Equation (8.3) by setting $\sigma_v = \bar{\sigma}_v$, $\dot{\varepsilon}_v = \dot{\varepsilon}_v$, and $\eta_v = \bar{\eta}_v$ in Equations (8.10), (8.11), and (8.12), respectively.

The law of thermal diffusion states that heat diffuses into the ice from a crevasse wall at the melting point to eliminate the temperature drop ΔT across ice layer of width w_r on the crevasse wall in time t such that (Weertman, 1980):

$$w_r \approx (4\,\kappa\,t)^{1/2} \tag{8.25}$$

where κ is the thermal diffusion coefficient of ice. Layer w_r is warmed to the melting point by removing latent heat of melting H_M from water and converting it into sensible heat until an ice layer of width w_c at freezing temperature T_M freezes onto area A_c of the crevasse wall. Removing H_M freezes volume $A_c\,w_c$ of water onto the two crevasse walls, so the heat exchange is:

$$A_c\,w_c\,H_M = 2\,A_c\,w_r\,C_H\,\Delta T \tag{8.26}$$

where C_H is the heat capacity for ice. Combining Equations (8.25) and (8.26):

$$w_c = 2\,w_r\,C_H\,DT\,/H_M = 4\,(\kappa\,t)^{1/2}\,C_H\,\Delta T\,/H_M \tag{8.27}$$

Differentiating Equation (8.27) gives the freezing rate of water onto the crevasse walls to produce a closing rate:

$$\frac{\partial w_c}{\partial t} = \frac{2\,C_H\,\kappa^{1/2}\,\Delta T}{H_M\,t^{1/2}} = \frac{2\,K\,\Delta T}{H_M\,\rho_I\,\kappa^{1/2}\,t^{1/2}} \tag{8.28}$$

where $C_H = K/\rho_I\,\kappa$ and K is the thermal conductivity coefficient for ice.

Net crevasse width w_t at time t is obtained by subtracting the integrated opening rate from the integrated closing rate respectively given by Equations (8.24) and (8.28):

$$w_t = w_o - w_c = \int_0^t \left(\partial w_o/\partial t\right) dt - \int_0^t \left(\partial w_c/\partial t\right) dt$$

$$= \left(2\,h_C\,\bar{\sigma}_v\,/n\,\bar{\eta}_v\right) t - \left(4\,C_H\,\kappa^{1/2}\,\Delta T\,/H_M\right) t^{1/2} \tag{8.29}$$

The crevasse becomes closed when $w_t = 0$ at time $t = t_C$ in Equation (8.29), which then gives:

$$t_C = \left(\frac{2\,n\,\bar{\eta}_v\,C_H\,\kappa^{1/2}\,\Delta T}{h_C\,\bar{\sigma}_v\,H_M}\right)^2 = \left(\frac{2\,n\,\eta_v\,K\,\Delta T}{h_C\,\bar{\sigma}_v\,H_M\,\rho_I\,\kappa^{1/2}}\right)^2 \tag{8.30}$$

The spacing C_x of primary transverse bottom crevasses depends on the time t_C needed for water-filled crevasses to freeze shut, because that is the time needed for longitudinal gravitational pulling force $(F_G)_x$ to switch from being applied to unfractured ice thickness $h_I - h_C$ to being applied to total ice thickness h_I. Therefore, if u_x is the mean longitudinal ice velocity:

$$C_x = t_C\,u_x \approx \left(\frac{2\,n\,\bar{\eta}_v\,C_H\,\Delta T}{h_C\,\bar{\sigma}_v\,H_M}\right)^2 \kappa\,u_x$$

$$= \left[\frac{2\,n\,\bar{\eta}_v\,C_H\,\Delta T}{h_I\,\bar{\sigma}_v\left(1 - \bar{\sigma}_v\,/2\,\sigma_{xx}'\right) H_M}\right]^2 \kappa\,u_x \tag{8.31}$$

where t_C is given by Equation (8.30) and h_C is given by Equation (8.23). For ice streams, u_x does not vary greatly with depth through the ice, and it is approximately equal to the mass-balance ice velocity.

For transverse bottom crevasses to open, $2\,\sigma_{xx}'$ must equal or exceed fracture stress $\bar{\sigma}_v$. Taking n = 3 for ice and σ_o = 1 bar = 10^5 Pa for the plastic yield stress (Paterson, 1981, p. 41), the average fracture stress given by Equation (8.10) has the range $\bar{\sigma}_v = (5.4 \pm 1.3) \times 10^4$ Pa.

Table 4.1 gives values of κ, C_H, K, and H_M for ice. In computing t_C from Equation (8.30), ΔT is obtained from the vertical temperature profile through an ice stream. Apart from ice accumulating on the top and bottom surfaces, most ice supplying ice streams comes from converging sheet flow advected from higher elevations at colder temperatures. Therefore, as a bottom crevasse propagates upward, ΔT at the crevasse tip increases as it penetrates colder advected ice, but ΔT may decrease somewhat if it nears the top surface. Therefore, if $T_b = T_M$ is the bottom-surface melting temperature and T_t is the top surface temperature obtained from the formula by Fastook and Prentice (1994):

$$\Delta T \geq T_b - T_t = h_I\left(\frac{\partial T_M}{\partial h_I}\right) - h\left(\frac{\partial T}{\partial h}\right) - L_E\left(\frac{\partial T}{\partial L_E}\right) - T_o \tag{8.32}$$

where $(\partial T_M/\partial h_I) = (\partial P/\partial h_I)\,(\partial T_M/\partial P) = \rho_I\,g_z\,(\partial T_M/\partial P) = -6.67 \times 10^{-4}$ °C/m is the depression of the melting point with increasing ice thickness for $(\partial T_M/\partial P) = 7.42 \times 10^{-8}$ °C/Pa, $(\partial T/\partial h) = -9.6 \times 10^{-3}$ °C/m is the adiabatic lapse rate, $(\partial T/\partial L_E) = -0.55$ °C/°L_E is the gradient of mean annual sea level air temperature with Earth latitude L_E, and T_o = 25°C is the equatorial mean annual sea level air temperature.

Primary transverse top and bottom crevasses appear in ice streams, which then transport these crevasses onto ice shelves as transverse lines of weakness. As an example, Equation (8.31) can be solved for the initial spacing of these crevasses when they first open along an ice stream. Take n = 3, $\eta_v = 10^{15}$ Pa s, $C_H = 2 \times 10^3$ J kg^{-1} °C^{-1}, $\kappa = 1.15 \times 10^{-6}$ m^2 s^{-1}, $H_M = 3.3 \times 10^5$ J kg^{-1}, and $\bar{\sigma}_v = 5.4 \times 10^4$ Pa for the ice stream, and take u_x = 500 m a^{-1} and h_I = 2000 m where crevasses first open on an ice stream at L_E = 80°N. Equation (8.32) then gives ΔT = 27.3°C. If C_x = 1 km is the initial crevasse spacing, Equation (8.31) can then be solved to give $\sigma_{xx}' = 2.8 \times 10^4$ Pa, for which x = 0.944 L_S when c = 0.5 in Equation (8.14). Equation (8.23) gives $h_C = 0.04\,h_I$ = 80 m for the combined depth and height of top and bottom crevasses. Equation (8.17) gives h_C = 6 m for top crevasses, and Equation (8.21) gives h_C = 57 m for bottom crevasses. In this example, crevasses first open just below the head of the ice stream. Crevasses first open farther downstream for larger values of c

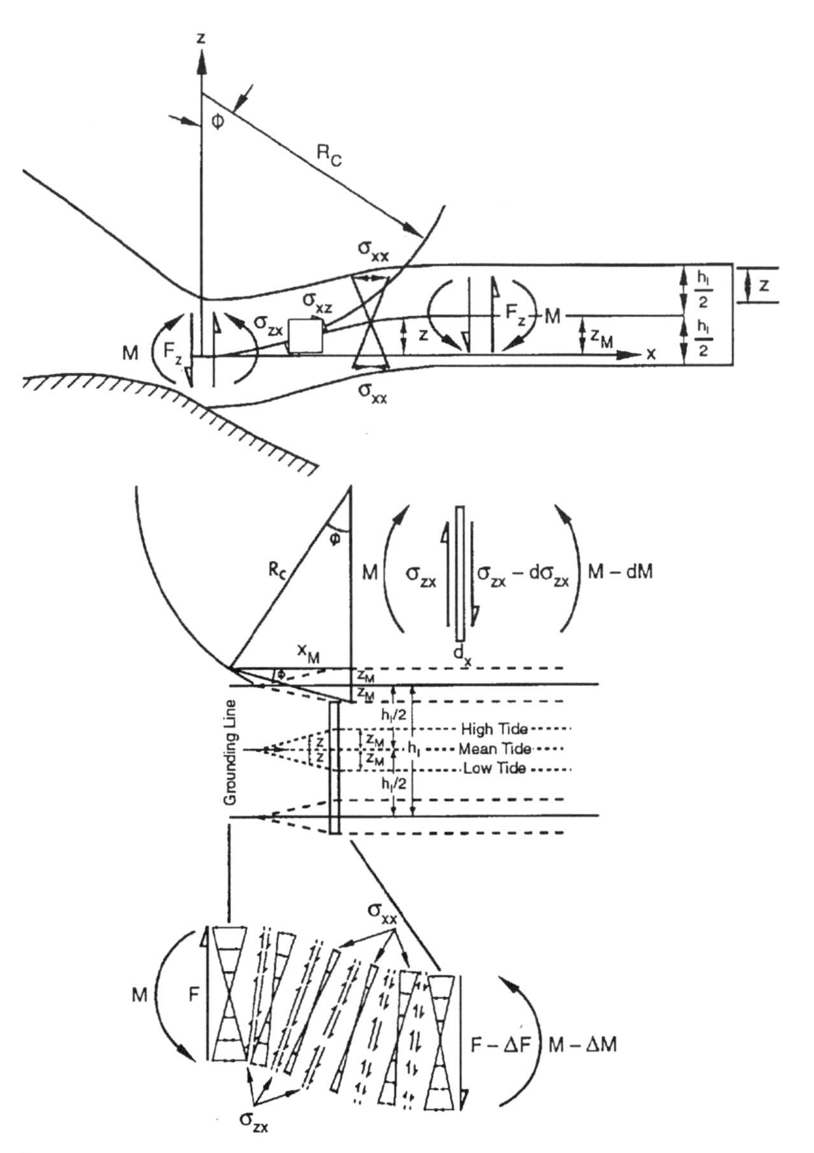

Figure 8.6: A schematic longitudinal cross-section of a floating ice shelf bent by tidal flexure. Tidal displacements are referred to reference coordinates x and z, whose origin is where the neutral axis (dotted lines) of the floating ice meets grounded ice. Tidal displacements z of the floating ice surfaces (dashed lines) from their mean tide positions (solid lines) vary along x from zero at the grounding line to z_f when the ice becomes freely floating. Tidal flexure is exaggerated to show the average radius of curvature r and angle of curvature ϕ over the longitudinal distance x_f experiencing tidal flexure. Actually, $h_I >> z_f$, so $\phi \approx \sin\phi \approx \tan\phi$. *Top*: The inset shows forces, stresses, and bending moments acting on an element of height h_I, length dx, and unit width according to (1), (2), and (3). *Bottom*: The enlargement shows stresses along the length $0 < x < x_f$. These stresses conform with an exponentially damped sinusoidal tidal flexure curve and consist of the shearing stress σ_{zx} due to the rise and fall of tide, which induces a shearing force F_z, and the longitudinal stress σ_{xx} due to nonrotation of the ice slab for $x < 0$ and $x > x_f$, which induces a bending moment M. From the ice surfaces to the center line, σ_{zx} varies parabolically from zero to a maximum, and σ_{xx} varies linearly from a maximum to zero, for elastic flexure. The maximum moment M_m along the length x_f where the shearing force is closest to F_z can be very crudely approximated by $M_m \approx \int_0^{x_f} F_z\, dx = F_z x_f = h_I \sigma_{zx} x_f \approx \int_{-h_I/2}^{+h_I/2} (h_I/2)\, \sigma_{xx}\, dz = \left(h_I^2/2\right) \sigma_{xx}$. Therefore $\sigma_{zx} \approx \sigma_{xx}$ when $x_f \approx h_I/2$, where σ_{zx} and σ_{xx} are the center line and surface values at the grounding line (Hughes, 1977, Figure 13).

in Equation (8.14). If c = 1 were used, crevasses would first open at $x = 0.767\ L_S$.

Grounding Line Crevasses

A floating ice shelf is raised and lowered by the tide, so the ice shelf undergoes tidal flexure along its grounding line. As it bends upward during high tide, bending stresses will be tensile along its bottom surface. As it bends downward during low tide, bending stresses will be tensile along its top surface. These tensile stresses can open grounding line crevasses if they equal or exceed the fracture stress given by Equation (8.31). This is usually the case. Swithinbank (1955) reported not only that top surface crevasses are common along grounding lines of Antarctic ice shelves, but also that the top surface also has a slight depression along the grounding line, perhaps because repeated bending during repeated tidal cycles causes localized "necking" of the ice shelf. Cyclic bending, in conjunction with pulling stress σ_{xx}' normal to ice-stream grounding lines, can indeed cause necking. The reason is that necking reduces transverse cross-sectional area $A_x = w_I h_I$ along the grounding line as flexural bending increases σ_{xx}', so pulling force $F_x = \sigma_{xx}' A_x$ remains constant as h_I decreases. As shown by Equation (4.111), during necking of a cylindrical specimen in uniaxial tension, this phenomenon allows strength exponent s in the strain hardening law given by Equation (4.109) to be equated with the reciprocal strain e_v during viscoplastic instability in Equation (4.109). The necking phenomenon allows an analysis of tidal flexure crevasses using the flow curve for ice described by Equation (4.109).

For ice flowing straight across the grounding line, x is normal to the grounding line and positive in the flow direction, y parallels the grounding line, and z is vertical and positive upward. As shown in Figure 8.6, the coordinate origin in this analysis is halfway through the ice thickness at the grounding line, which is the case for constant ice density. Actually, firnification below the top surface reduces the upper ice-shelf density, so the neutral axis of the ice shelf is displaced somewhat below the midpoint. Using these axes, Equation (4.109) becomes:

$$\varepsilon_{xx} = (\sigma_{xx} / \sigma_s)^s = z / R_c \tag{8.33}$$

where $z \leq h_I / 2$ is ice height above the neutral axis and R_c is the radius of bending curvature of the ice shelf at the grounding line. For elastic strain, $s = 1$ and $\sigma_s = \sigma_e$ is the elastic modulus. For viscoplastic strain, $s = 2$ and $\sigma_s = \sqrt{2}\, \sigma_v$ is the viscoplastic modulus given by Equation (4.113).

Tidal flexure of ice shelves begins with elastic strain. The radius of curvature R_c of the ice shelf is given by the elastic curve, for which:

$$\frac{1}{R_c} = \frac{d^2 z / dx^2}{\left[1 + (dz/dx)^2\right]^{3/2}} \approx \frac{d^2 z}{dx^2} \tag{8.34}$$

The boundary condition at the grounding line is $z = dz/dx = 0$ at $x = 0$ and $z = z_m$ as $x \rightarrow \infty$, where z_m is the maximum vertical displacement at high or low tide. Rising or falling tide produces a vertical shear stress σ_{zx} caused by gravitational force F_z when the ice shelf departs from hydrostatic equilibrium at mean tide along the grounding line:

$$\sigma_{zx} = dF_x / dx = d^2 M / dx^2 = -\rho_W\, g_z\, (z_m - z) \tag{8.35}$$

where M is the bending moment and, by definition, $\sigma_{zx} = dF_x / dx$ and $F_x = dM / dx$.

Equations (8.33) and (8.34) give:

$$M = \int_{-h_I/2}^{+h_I/2} \sigma_{xz}\, z\, dz = \int_{-h_I/2}^{+h_I/2} \sigma_s\, (\varepsilon_{xx})^{1/s}\, z\, dz$$

$$= \int_0^{h_I} \sigma_s\, (z / R_c)^{1/s}\, z\, dz = \frac{\sigma_s\, h_I^{2+1/s}}{(2 + 1/s)\, R_c^{1/s}} \tag{8.36}$$

Substituting R_c from Equation (8.36) into Equation (8.34) gives the differential equation of the flexure curve:

$$\frac{d^2 z}{dx^2} = \left[\frac{(2 + 1/s)\, M}{\sigma_s\, h_I^{2+1/s}}\right]^s \tag{8.37}$$

Differentiating twice more:

$$\frac{d^4 z}{dx^4} = \left[\frac{2 + 1/s}{\sigma_s\, h_I^{2+1/s}}\right]\left[(s-1)\, s\, M^{s-2} \left(\frac{dM}{dx}\right)^2 + s\, M^{s-1}\left(\frac{d^2 M}{dx^2}\right)\right] \tag{8.38}$$

Solutions of Equation (8.38) exist for elastic flexure and viscoplastic flexure, which can be added to obtain a solution for elastic-viscoplastic flexure.

The elastic solution is obtained for $s = 1$ when $\sigma_s = \sigma_e$ is the elastic modulus. The variation of vertical displacement z_e for elastic bending is (Robin, 1958):

$$z_e = z_m - [C_1 \cos(\lambda_e x) + C_2 \sin(\lambda_e x)] \exp(\lambda_e x)$$

$$+ [C_3 \cos(\lambda_e x) + C_4 \sin(\lambda_e x)] \exp(-\lambda_e x) \tag{8.39}$$

where C_1, C_2, C_3, and C_4 are constants determined by the boundary conditions and:

$$\lambda_e = (3\, \rho_W\, g_z / \sigma_e\, h_I^3)^{1/4} \tag{8.40}$$

is the elastic damping factor.

The viscoplastic solution is obtained for $s = 2$ when $\sigma_s = s^{1/s}\, \sigma_v = s^{\varepsilon_v}\, \sigma_v = \sqrt{2}\, \sigma_v$ is the viscoplastic modulus given by Equation (4.113) and (4.114). The variation of vertical displacement z_v for viscoplastic bending is (Hughes, 1977):

$$z_v = (K_v / 8\, \lambda_v^2)\, [2 \cos(2\, \lambda_v x)] \exp(-2\, \lambda_v x) + C_5 x + C_6 \tag{8.41}$$

where C_5 and C_6 are constants and

$$K_v = (6\, \rho_W\, g_z\, z_m^2\, \sigma_e / h_I^2\, \sigma_v) \tag{8.42}$$

is the viscoplastic damping factor.

Since elastic and viscoplastic strains are additive, the total vertical bending displacement is the sum of Equations (8.39) and (8.41) obtained by Lingle and others (1981):

$$z = z_e + z_v \approx z_m - [(z_m + K_v / 8\, \lambda_v^2) \cos(\lambda_e x)] \exp(-\lambda_e x)$$

$$- [(z_m - K_v / 8\, \lambda_v^2) \sin(\lambda_e x)] \exp(-\lambda_e x)$$

$$+ (K_v / 8\, \lambda_v^2)\, [2 - \cos(2\, \lambda_v x)] \exp(-\lambda_e x) \tag{8.43}$$

where $C_1 = C_2 = C_5 = 0$ because z is finite, $C_6 = 0$ because $z = z_m$ as $x \rightarrow \infty$, $C_3 = z_m + (K_v / 8\, \lambda_v^2)$ since $z = 0$ at $x = 0$, and $C_4 = z_m - (K_v / 8\, \lambda_v^2)$ since $dz / dx = 0$ at $x = 0$.

Lingle and others (1981) obtained a good fit to tidal flexure data for the lateral grounding lines of the ice-shelf terminus of Jakobshavns Isbræ floating in Jakobshavns Isfjord on the west coast of Greenland (69°10'N, 50°10'W) by assuming that $\lambda_v = \lambda_e$ and $z_m = (K_v / 8\, \lambda_v^2)$ in Equation (8.43), which then reduces to:

$$z = z_m\, \{1 - 2 \cos(\lambda_e x) \exp(-\lambda_e x)$$

$$+ [2 - \cos(\lambda_e x)] \exp(-2\, \lambda_v x)\} \tag{8.44}$$

Along a transverse transect about midway between the rear grounding line and the calving front, where $z_m = \pm\, 0.5$ m is the maximum vertical tidal displacement and $h_I \approx 800$ m is the ice thickness, a primary maximum bending stress of $s_m \approx 5 \times 10^5$ Pa at $x = 0$ on the side grounding line and a secondary maximum bending stress of $\sigma_m \approx 1 \times 10^5$ Pa at $x \approx 1700$ m from the side grounding line, both for an uncrevassed ice thickness of $h_I - h_C \approx 316$ m, were computed.

No ice moved across the side grounding line of Jakobshavns Isbræ, so the same ice experienced repeated tidal flexures and intense lateral shear. Ice crossing the rear grounding line would experience many fewer tidal flexures and no lateral shear. This explains why crevasses along the side grounding lines were much more prominent than crevasses along the rear grounding line. Ice crossing the rear grounding line came from two ice streams and an ice fall between the two ice streams. Therefore, the grounding line was fixed at the base of the ice fall, but the grounding line could migrate up and down the ice streams from high to low tide, because the ice fall had a high top surface slope and the ice streams had low top surface slopes. In addition, ice crossing the rear grounding line was much slower at the foot of the ice fall than at the foot of the ice streams. These features would produce tidal flexure crevasses along the rear grounding line that were much larger at the foot of the ice fall than at the foot of the ice streams.

Figure 8.7 illustrates four kinds of grounding lines for floating ice shelves, three of which exist for the floating terminus of Jakobshavns Isbræ. They exist for ice shelves in general. An ice shelf that abuts a fjord side wall has the most stationary hinge line during tidal cycles, and no ice crosses the grounding line, so large tidal flexure crevasses are produced. An ice shelf grounds on a steeply curving bed beneath an ice fall, so the hinge line migrates slightly during tidal cycles, and tidal flexure enlarges bottom crevasses formed by reverse bending at the foot of the ice fall. Staircase crevasses formed on the top surface as ice bends over the curving bed are pinched shut by reverse bending as ice becomes afloat. An ice shelf grounds on a gently sloping bed at the foot of an ice stream, so tidal cycles cause the grounding line to migrate back and forth as the tide rises and falls. This migration distributes tidal flexure over the grounding-line migration zone, thereby reducing the actual bending produced by the tide. This makes z_m in Equation (8.44) much less than the tidal amplitude, so bending stresses may be less than the fracture stress and tidal cycles may not create grounding-line crevasses. Then primary transverse crevasses formed along the ice stream will be the only transverse crevasses transported onto the ice shelf. However, if bending stresses reach the fracture stress, then spacing C_x of primary transverse crevasses may be the longitudinal tidal sweep of the grounding line. The fourth kind of grounding line is associated with Ice Stream C in West Antarctica, where a series of "pseudo ice shelves" above subglacial lakes or brine pockets along the ice stream produce step-like terraces on its surface (Robin and others, 1970; Hughes, 1973). The pseudo ice shelves are separated by basal sticky spots that form boudins along the ice stream.

Primary and secondary tidal flexure crevasses along both side and rear grounding lines of an ice shelf floating in a fjord divide the ice shelf into plates separated by longitudinal and transverse lines of weakness. When the ice shelf lies in a surface ablation zone, the ice shelf disintegrates into tabular icebergs along these lines of weakness before the floating ice are primary longitudinal crevasses that first develop in the advances very far. This is the case for Jakobshavns Isbræ. As seen in Figure 8.8, the calving front of Jakobshavns Isbræ is only about 10 km beyond its rear grounding line. Figure 8.8 also shows deep longitudinal crevasses along the centerline of Jakobshavns Isbræ after it becomes afloat.

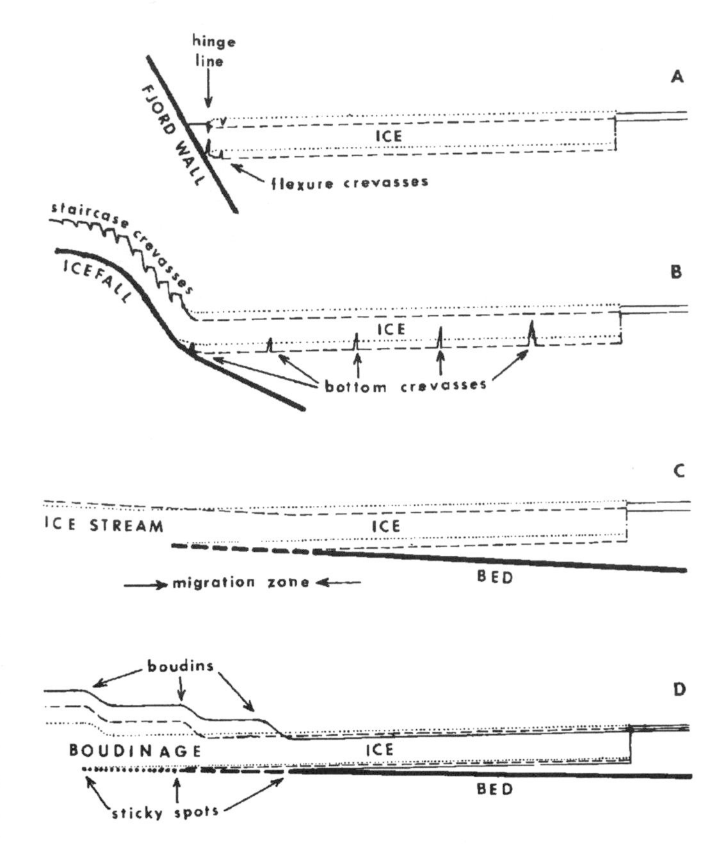

Figure 8.7: Four types of grounding lines for the ice shelves. A: Ice abutting the fjord sidewall has a grounding line that experiences maximum tidal flexure, minimum tidal migration, and no crossing ice flow. B: Ice at the base of the ice fall has a grounding line that experiences substantial tidal flexure, some tidal migration, and slowly crossing ice flow. C: Ice at the base of the ice stream has a grounding line that experiences minimal tidal flexure, maximum tidal migration, and rapidly crossing ice flow. D: Grounding lines with ice streams linked in boudinage have maximum tidal flexure within and maximum tidal migration between boudins. Types A, B, and C exist where Jakobshavns Isbræ becomes afloat in Jakobshavns Isfjord, Greenland (69°10'N, 50°10'W).

These crevasses, collectively called "the zipper," form at the base of the ice fall where the two ice streams merge. They lateral shear zones of the ice streams.

Primary Longitudinal Crevasses

Primary longitudinal crevasses begin as shear crevasses alongside ice streams. Shear crevasses form where σ_{xy} dominates σ_{xx}' and σ_{yy}' in an ice stream, which is the case in its lateral shear zones. From Equation (4.28a), with $y = 0$ at the centerline and $y = w_I/2$ on each side of the ice stream, $g_z = g \cos \alpha$, and $g_x = g \sin \alpha = g_z \tan \alpha \approx \alpha$:

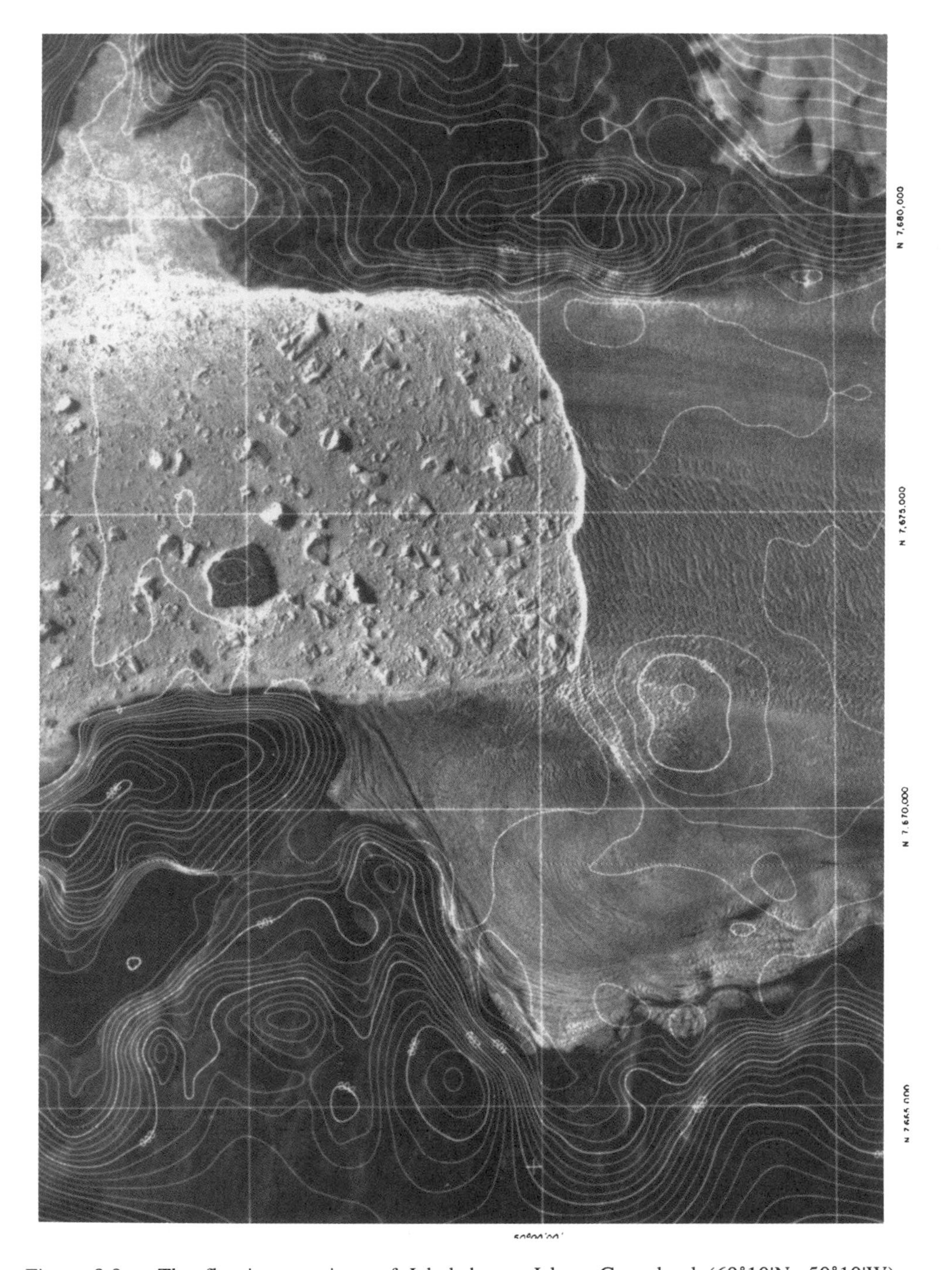

Figure 8.8: The floating terminus of Jakobshavns Isbræ, Greenland (69°10'N, 50°10'W). Primary transverse crevasses open in the main ice stream before it becomes afloat. Primary longitudinal crevasses open at the junction of the two ice streams, forming "The Zipper," and in the lateral shear zones of the floating terminus. Beyond the southern shear zone, Jakobshavns Isbræ spills over the southern wall of Jakobshavns Isfjord and ends as a grounded ice lobe.

$$\sigma_{xy} = \int_0^{\sigma_{xy}} d\sigma_{xy} = \int_0^{w_I/2} \rho_I \, g_x \, dy \approx \frac{1}{2} \rho_I \, g_z \, w_I \, \alpha \qquad (8.45)$$

where α is the top surface slope, $\partial \sigma_{xx}' / \partial x << \partial \sigma_{xy} / \partial y$ when σ_{xx}' is minimized by buttressing stress σ_B, and $\partial \sigma_{xz} / \partial z << \partial \sigma_{xy} / \partial y$ when side traction dominates basal traction.

As seen in Equation (8.14), longitudinal deviator stress σ_{xx}' that opens primary transverse crevasses in ice streams is reduced by buttressing stress σ_B. As seen in Equation (6.28), one component of σ_B is lateral traction alongside ice streams caused by mean side shear stress $\bar{\tau}_v$ averaged over distance x to the ice shelf. Equation (6.30) shows that $\bar{\tau}_v$ depends on $(\tau_v)_S$ at the head of an ice stream where $x = L_S$ and on $(\tau_v)_0$ at the foot of the ice stream where $x = 0$. If $(\tau_o)_0 = 0$ due to shear rupture that produces lateral rifts beyond the grounding line, as shown in Figure 7.5 for Byrd Glacier, then top and bottom shear crevasses meet. Ice streams become imbedded in a buttressing ice shelf when these lateral rifts freeze shut, but they continue as longitudinal lines of weakness all the way to the calving front.

For a plane state of stress in the xy plane, in which σ_{xy} dominates σ_{xx} and σ_{yy} on the top surface in the lateral shear zones, Equations (8.13) give:

$$\sigma_1 \approx \sigma_{xy} \tag{8.46a}$$

$$\sigma_2 \approx -\sigma_{xy} \tag{8.46b}$$

$$\sigma_3 = 0 \tag{8.46c}$$

This is pure shear, with $\sigma_{xy} = \tau_v$ for viscoplastic yield stress σ_v in the lateral shear zones, plastic yield stress σ_o according to the strain energy of distortion yield criterion is given by entering Equations (8.46) into Equation (4.14):

$$\sigma_o = \frac{1}{\sqrt{2}}\left[\left(\sigma_1-\sigma_2\right)^2+\left(\sigma_2-\sigma_3\right)^2+\left(\sigma_3-\sigma_1\right)^2\right]^{1/2}$$

$$= \sqrt{3}\,\sigma_{xy} = \sqrt{3}\,\tau_v \tag{8.47}$$

and σ_o according to the maximum shear stress yield criterion is given by entering Equations (8.46) into Equation (4.143):

$$\sigma_o = (\sigma_1 - \sigma_2) = 2\,\sigma_{xy} = 2\,\tau_v \tag{8.48}$$

The average plastic yield stress is:

$$\bar{\sigma}_o = \tfrac{1}{2}\left(\sqrt{3}+2\right)\tau_v \pm \tfrac{1}{2}\left(\sqrt{3}-2\right)\tau_v \tag{8.49}$$

Therefore, viscoplastic yielding occurs before plastic yielding.

Replacing σ_o with $\bar{\sigma}_o$ in Equation (8.10) gives mean fracture stress $\bar{\tau}_v$ for shear rupture that opens shear crevasses:

$$\bar{\tau}_v = \tau_v\left(\bar{\sigma}_v / \bar{\sigma}_o\right)$$

$$= \frac{\sigma_o}{\left(\sqrt{3}+2\right)}\left[\left(\frac{n-2}{2\,(n-1)\,n^2}\right)^{\frac{1}{2n-2}} + \left(\frac{n-1}{n}\right)\right] \tag{8.50}$$

Taking $\sigma_o = 10^5$ Pa, Equation (8.50) gives $\bar{\tau}_v = 2.9 \times 10^4$ Pa.

Shear rupture that rifts the whole ice thickness as an ice stream crosses the grounding line of an ice shelf, greatly reducing σ_{xy} in Equation (6.45), produces bottom crevasses that can freeze shut at height:

$$h_C = h_W = h_I\,(\rho_I/\rho_W) \tag{8.51}$$

Replacing $\bar{\sigma}_v$ with $\bar{\tau}_v$ and h_C with $h_I\,(\rho_I/\rho_W)$ in Equation (8.30) gives the time t_c needed to freeze the lateral rifts shut:

$$t_c = \left(\frac{2\,n\,\bar{\eta}_v\,\rho_W\,C_H\,\kappa^{1/2}\,\Delta T}{h_I\,\rho_I\,\bar{\tau}_v\,H_M}\right)^2 \tag{8.52}$$

Taking u_x as the ice velocity averaged over length L_C of primary longitudinal crevasses:

$$L_C \approx t_c\,u_x = \left(\frac{2\,n\,\bar{\eta}_v\,\rho_W\,C_H\,\Delta T}{h_I\,\rho_I\,\bar{\tau}_v\,H_M}\right)^2 \kappa\,u_x \tag{8.53}$$

Computing values of L_C from Equation (8.53) requires making a distinction between the linear Newtonian viscosity η_N for creep in fluids and the nonlinear viscoplastic viscosity η_v for creep in crystalline solids like ice. By definition, referring to Equations (4.6), (4.15), and (8.3):

$$\eta_N = \frac{\sigma_{ij}'}{\dot{e}_{ij}} = \frac{\sigma_{ij}'}{2\,\dot{\varepsilon}_{ij}} \tag{8.54a}$$

$$\eta_v = \frac{\sigma_v}{n\,\dot{\varepsilon}_v} = \frac{\sigma_c}{n\,\dot{\varepsilon}_c} = \frac{\left(\frac{1}{2}\,\sigma_{ij}'\,\sigma_{ij}'\right)^{1/2}}{n\left(\frac{1}{2}\,\dot{\varepsilon}_{ij}\,\dot{\varepsilon}_{ij}\right)^{1/2}}$$

$$= \frac{1}{n}\left[\frac{\sigma_{xx}'^2 + \sigma_{yy}'^2 + \sigma_{zz}'^2 + 2\,\sigma_{xy}^2 + 2\,\sigma_{yz}^2 + 2\,\sigma_{zx}^2}{\dot{\varepsilon}_{xx}^2 + \dot{\varepsilon}_{yy}^2 + \dot{\varepsilon}_{zz}^2 + 2\,\dot{\varepsilon}_{xy}^2 + 2\,\dot{\varepsilon}_{yz}^2 + 2\,\dot{\varepsilon}_{zx}^2}\right]^{1/2} \tag{8.54b}$$

For uniaxial tension, $\eta_N = \sigma_{xx}'/2\,\dot{\varepsilon}_{xx} = (n/2)\,\eta_v$. For simple shear, $\eta_N = \sigma_{xy}/2\,\dot{\varepsilon}_{xy} = (n/2)\,\eta_v$. Note that $\sigma_c = \sigma_{xx}'$ and $\dot{\varepsilon}_c = \dot{\varepsilon}_{xx}$ for uniaxial tension because $\sigma_{xx}' = -\sigma_{zz}'$ and $\dot{\varepsilon}_{xx} = -\dot{\varepsilon}_{zz}$ for $\sigma_{yy}' = 0$ and $\dot{\varepsilon}_{yy} = 0$ in ice streams and that $\dot{\varepsilon}_{xy} = 1/2\,(\partial u_x/\partial y + \partial u_y/\partial x) = 1/2\,(\partial u_x/\partial u_y)$ for simple shear because $1/2\,(\partial u_y/\partial x) = 0$ in the lateral shear zones alongside ice streams.

Data from Byrd Glacier, a major Antarctic ice stream supplying the Ross Ice Shelf, can be used to compute L_C for the primary longitudinal crevasses observed as lateral rifts in Figure 7.5 (Hughes, 1977). Measured from the grounding line, primary longitudinal crevasses formed lateral rifts 98 km long and 73 km long on the north and south sides of Byrd Glacier. About halfway along this distance, $h_I \approx 750$ m, $u_x = 800$ m a^{-1} and $\dot{\varepsilon}_v = \dot{\varepsilon}_{xy} = 1/2\,(\partial u_x/\partial y) = 7.2 \times 10^{-12}$ s^{-1} across width $w_I = 20$ km between the rifts, which is low because rifting almost eliminates side traction. Along the floating rifted length of Byrd Glacier, $\alpha = 6.5 \times 10^{-4}$ is the mean surface slope and $\Delta T = 19.3$°C from Equation (8.32). These data give $\sigma_{xy} = 5.86 \times 10^4$ Pa in Equation (8.45), which exceeds $\bar{\tau}_v = 2.9 \times 10^4$ Pa in Equation (8.50), so primary longitudinal crevasses should form. Equation (8.54) then gives $\bar{\eta}_v = \bar{\tau}_v / n\, \dot{\varepsilon}_v = 1.4 \times 10^{15}$ Pa s using n = 3 for ice. A primary longitudinal crevasse would freeze shut in length L_C = 67 km according to Equation (8.53). This is comparable to the observed lengths of Byrd Glacier rifts. Byrd Glacier advects cold East Antarctic ice onto the Ross Ice Shelf, so ΔT = 25°C may be more appropriate in Equation (8.53), which then gives L_C = 113 km, only 14 km longer than the longest rift.

Calving Ice Shelves

As longitudinal and transverse lines of weakness in an ice shelf are transported toward the calving front, they are exposed to bending stresses caused by the asymmetric imbalance of lithostatic and hydrostatic longitudinal forces at the calving front, as shown in Figure 2.2. This can reopen transverse crevasses along transverse lines of weakness, with tabular icebergs being released along intersecting longitudinal and transverse lines of weakness.

Figure 8.9 illustrates bending of an ice shelf toward its calving front in relation to bending forces and the bending moment, following the plane strain analysis by Reeh (1968). The origin of axes x, y, z is at the midpoint of the ice-shelf thickness and width at the grounding line, with x directed horizontally toward the calving front, y directed horizontally along the grounding line, and z directed vertically upward. Constant ice thickness, density, and temperature are assumed. Bending displaces the neutral axis of the ice shelf by vertical distances $\pm\, d_z$ that vary along the horizontal x axis. Stresses σ_{xx} and σ_{xz} are induced by bending and, as shown by the ice elements in Figure 8.9, these stresses give rise to bending moment M, longitudinal force per unit width F_x, and vertical force per unit width F_z defined as:

$$M = \int_{-h_I/2}^{+h_I/2} \sigma_{xx}\, z\, dz \qquad (8.55a)$$

$$F_x = \int_{-h_I/2}^{+h_I/2} \sigma_{xx}\, dz \qquad (8.55b)$$

$$F_z = \int_{-h_I/2}^{+h_I/2} \sigma_{xz}\, dz \qquad (8.55c)$$

The equilibrium equations are:

$$\frac{\partial M}{\partial x} = F_z \qquad (8.56a)$$

$$\frac{\partial F_z}{\partial x} = -\rho_W\, g_z\, d_z \qquad (8.56b)$$

where d_z is the vertical ice displacement due to bending over time t.

Solutions of Equations (8.56) require introducing dimensionless variables $\hat{t}$, $\hat{d}_z$, $\hat{x}$, $\hat{M}$, $\hat{F}_x$, and $\hat{F}_z$. Dimensionality is removed for M by introducing moment $\rho_W\, g_z\, I_M$, for F_x

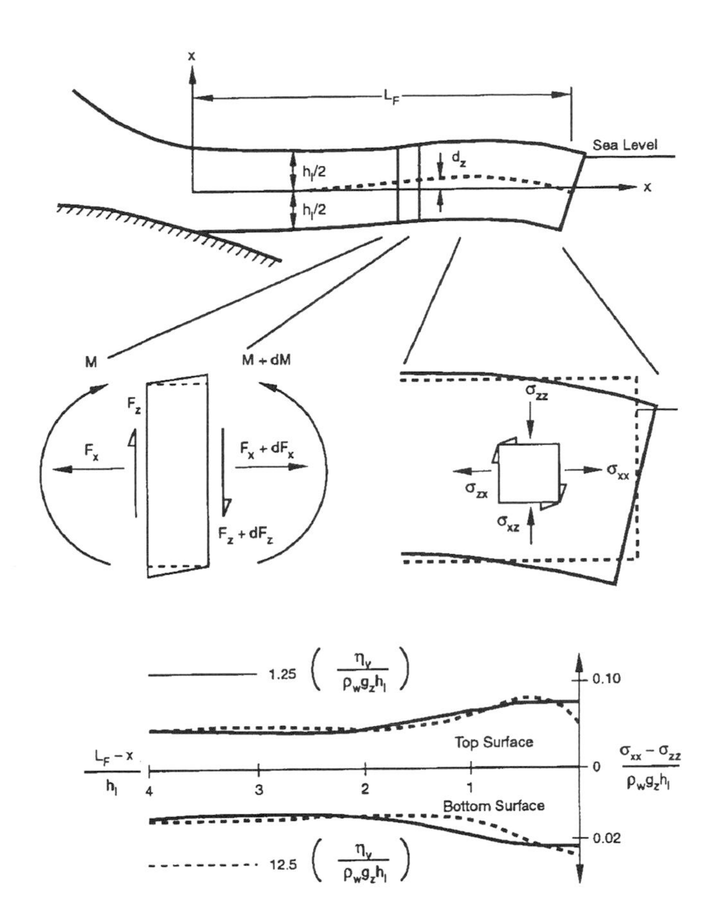

Figure 8.9: Bending near the calving front of an ice shelf. *Top*: Bending is due to the asymmetrical misbalance between the longitudinal lithostatic force in ice and the longitudinal hydrostatic force in water at the calving front, as illustrated in Figure 2.2. *Center left*: The normal force, the shear force, and the bending moment acting on an ice column near the calving front. *Center right*: Shear and axial stresses induced by bending at the calving front. *Bottom*: The variation of dimensionless longitudinal stress at the top and bottom surface with dimensionless distance from the calving front at dimensionless times 1.25 (solid lines) and 12.5 (dashed lines). Dimensionless stress is $(\sigma_{xx} - \sigma_{zz})/\rho_W\, g_z\, h_I$, dimensionless distance is $(L_F - x)/h_I$, and dimensionless time is $\eta_v / \rho_W\, g_z\, h_I$. Adopted from Reeh (1968).

by introducing average longitudinal force $1/2\ \rho_W\, g_z\, h_I^2$, and for F_z by introducing maximum vertical force $\rho_W\, g_z\, h_I^2$. By definition:

$$\hat{t} = t / t_c \qquad (8.57a)$$

$$\hat{d}_z = d_z / h_I \qquad (8.57b)$$

$$\hat{x} = x / L_F \qquad (8.57c)$$

$$\hat{M} = M / \rho_W\, g_z\, I_M = 12\, M / \rho_W\, g_z\, h_I^3 \qquad (8.57d)$$

$$\hat{F}_x = 2\, F_x / \rho_W\, g_z\, h_I^2 \qquad (8.57e)$$

$$\hat{F}_z = F_z / \rho_W g_z h_I^2 \qquad (8.57f)$$

where t_c is a time constant, η_v is the viscoplastic viscosity of ice, h_I is constant floating ice thickness, L_F is floating ice length from the grounding line at $\hat{x} = 0$ to the calving front at $\hat{x} = 1$, and $I_M = h_I^3/12$ is the moment of inertia for a rectangular cross section of unit width. Equations (8.56a), (8.56b), and (8.57f) now become:

$$\frac{\partial \hat{M}}{\partial \hat{x}} = \frac{12 L_F \hat{F}_z}{h_I} \qquad (8.58a)$$

$$\frac{\partial \hat{F}_z}{\partial \hat{x}} = -\frac{L_F \hat{d}_z}{h_I} \qquad (8.58b)$$

$$\hat{F}_x \approx -(\rho_I/\rho_W)^2 + 2(\rho_I/\rho_W)\hat{d}_z \qquad (8.58c)$$

Differentiating Equation (8.58a) with respect to $\hat{x}$ and substituting $d\hat{F}_z/d\hat{x}$ into Equation (8.58b) gives:

$$\frac{\partial^2 \hat{M}}{\partial \hat{x}^2} = -\frac{12 L_F^2 \hat{d}_z}{h_I^2} \qquad (8.59)$$

The normalized radius of curvature $\hat{R}_c$ of the bending ice shelf is defined by the equation:

$$\frac{1}{\hat{R}_c} = \frac{\dfrac{\partial^2 \hat{d}_z}{\partial \hat{x}^2}}{\left[1 + \left(\dfrac{\partial \hat{d}_z}{\partial \hat{x}}\right)^2\right]^{3/2}} \approx \frac{\partial^2 \hat{d}_z}{\partial \hat{x}^2} \qquad (8.60)$$

For elastic bending at time $t = 0$ controlled by elastic modulus σ_e:

$$\frac{\partial^2 d_z}{\partial x^2} = \frac{12 M}{\sigma_e h_I^3} \qquad (8.61)$$

For viscoplastic bending at time $t > 0$ controlled by viscoplastic viscosity h_v:

$$\frac{\partial}{\partial \hat{t}}\left(\frac{\partial^2 \hat{d}_z}{\partial \hat{x}^2}\right) = \frac{\rho_W g_z L_F^2 t_c}{4 \eta_v h_I}\left(\hat{M} + \frac{\rho_I}{\rho_W}\right) \qquad (8.62)$$

Differentiating Equation (8.62) twice with respect to $\hat{x}$ and substituting Equation (8.59) for $\partial^2 \hat{M}/\partial \hat{x}^2$ gives:

$$\frac{\partial}{\partial \hat{x}}\left(\frac{\partial^4 \hat{d}_z}{\partial \hat{x}^4}\right) = -\left(\frac{3 \rho_W g_z L_F^4 \hat{d}_z t_c}{\eta_v h_I^3}\right) = -3\hat{C}\left(\frac{L_F}{h_I}\right)^4 \hat{d}_z \qquad (8.63)$$

where $\hat{C}$ is the dimensionless constant:

$$\hat{C} = \left(\rho_W g_z h_I t_c / \eta_v\right) \qquad (8.64)$$

The solution of Equation (8.63) has the form:

$$\hat{d}_z = \hat{d}_0 + \hat{d}_1 \hat{t} + \hat{d}_2 \hat{t}^2 + \hat{d}_3 \hat{t}^3 + \cdots + \hat{d}_i \hat{t}^i \qquad (8.65)$$

where displacement $\hat{d}_z$ varies with $\hat{x}$ and $\hat{t}$, displacement function $\hat{d}_0$ gives the elastic displacement along $\hat{x}$ at $t = 0$, displacement functions $\hat{d}_i$ give viscoplastic displacements along $\hat{x}$ at $t > 0$, and the i series is infinite.

Displacements $\hat{d}_i$ in Equation (8.65) are evaluated from boundary conditions. The boundary condition at the grounding line, where $\hat{x} = 0$, is:

$$(\hat{d}_z)_0 = (\partial d_z / \partial \hat{x})_0 = 0 \qquad (8.66)$$

The boundary conditions at the calving front, where $\hat{x} = 1$, are illustrated in Figure 8.6 and give:

$$(\hat{F}_x)_1 \approx -(\rho_I/\rho_W)^2 + 2(\rho_I/\rho_W)(\hat{d}_z)_1 \qquad (8.67a)$$

$$(\hat{F}_z)_1 \approx 0 \qquad (8.67b)$$

$$(\hat{M})_1 \approx -3(\rho_I/\rho_W)^2 + 2(\rho_I/\rho_W)^3 + 6[(\rho_I/\rho_W) - (\rho_I/\rho_W)^2](\hat{d}_z)_1 \qquad (8.67c)$$

Substituting Equation (8.67c) into Equation (8.62) gives:

$$\left[\frac{\partial}{\partial \hat{t}}\left(\frac{\partial^2 \hat{d}_z}{\partial x^2}\right)\right]_1 = \frac{\hat{C} L_F^2}{4 h_I^2}\left[\left(\frac{\rho_I}{\rho_W}\right) - 3\left(\frac{\rho_I}{\rho_W}\right)^2 + 2\left(\frac{\rho_I}{\rho_W}\right)^3 + 6\left\{\left(\frac{\rho_I}{\rho_W}\right) - \left(\frac{\rho_I}{\rho_W}\right)^2\right\}(\hat{d}_z)_1\right] \qquad (8.68)$$

Differentiating Equation (8.62) with respect to $\hat{x}_1$ and substituting $\partial \hat{M}/\partial \hat{x}_1$ from Equation (8.58a) gives:

$$\frac{4 \eta_v h_I^2}{\rho_W g_z L_F^2 t_c} \frac{\partial}{\partial \hat{t}}\left(\frac{\partial^3 \hat{d}_z}{\partial \hat{x}_1^3}\right) = 12 \hat{F}_z \qquad (8.69)$$

Setting $\hat{x} = 1$ at $x = L_F$, so that $\hat{F}_z = 0$ from Equation (8.67b), reduces Equation (8.69) to:

$$\left[\frac{\partial}{\partial \hat{t}}\left(\frac{\partial^3 \hat{d}_z}{\partial \hat{x}_1^3}\right)\right]_1 = 0 \qquad (8.70)$$

Substituting Equation (8.65) into Equation (8.63) leads to a set of differential equations for displacement functions $\hat{d}_i$. If $\hat{d}_0$ is known, then all other $\hat{d}_i$ functions can be determined in succession by integrating Equation (8.63), with the constants of integration being determined from boundary conditions specified by Equations (8.66), (8.68), and (8.70).

Compared to viscoplastic displacements at t >> 0, elastic displacement is nil, and $\hat{d}_0 \approx 0$ can be used.

Once the curve for vertical viscoplastic displacements $\hat{d}_z$ along $\hat{x}$ has been determined from Equation (8.65), bending moment $\hat{M}$, longitudinal force $\hat{F}_x$, and vertical force $\hat{F}_z$ can be computed along $\hat{x}$ from Equations (8.58). Beam theory can then be used to compute dimensionless stresses $\hat{\sigma}_{xx}$, $\hat{\sigma}_{zz}$, and $\hat{\sigma}_{xz}$ from $\hat{M}$, $\hat{F}_x$, and $\hat{F}_z$. These stresses, made dimensionless by dividing by $\rho_W\, g_z\, h_I$, are:

$$\hat{\sigma}_{xx} = \sigma_{xx}/\rho_W\, g_z\, h_I = \tfrac{1}{2}\hat{F}_x + \hat{M}\hat{z} \tag{8.71a}$$

$$\hat{\sigma}_{zz} = \sigma_{zz}/\rho_W\, g_z\, h_I = -\left[(\rho_I/\rho_W) - \hat{d}_z\right]\left[\tfrac{1}{2} + \hat{z}\right] \tag{8.71b}$$

$$\hat{\sigma}_{xz} = \sigma_{xz}/\rho_W\, g_z\, h_I = \tfrac{3}{2}\hat{F}_z\left(1 - 4\hat{z}^2\right) \tag{8.71c}$$

Bending cracks open when longitudinal deviator stress $2\,\sigma_{xx}' = (\sigma_{xx} - \sigma_{zz})$ is a maximum tensile stress on the top or bottom surface of the ice shelf. In dimensionless form, from Equations (8.71):

$$\begin{aligned}\hat{\sigma}_{xx} - \hat{\sigma}_{zz} &= \frac{\sigma_{xx} - \sigma_{zz}}{\rho_W\, g_z\, h_I} \\ &= \tfrac{1}{2}\left[(\rho_I/\rho_W) - (\rho_I/\rho_W)^2 + \left\{2(\rho_I/\rho_W)^2 - 1\right\}\hat{d}_z\right] \\ &\quad + \hat{z}\left[\hat{M} + (\rho_I/\rho_W) - \hat{d}_z\right]\end{aligned} \tag{8.72}$$

Hence, $(\hat{\sigma}_{xx} - \hat{\sigma}_{zz})$ consists of one term for tension along the neutral axis of an ice shelf, and one term for the change in this tension toward the top and bottom surfaces of the ice shelf.

Figure 8.9 shows plots of Equation (8.57a) for $\hat{d}_z$ along the neutral axis and of Equation (8.72) for $(\hat{\sigma}_{xx} - \hat{\sigma}_{zz})$ along the top and bottom surfaces of an ice shelf at positions x/h_I and times t. Equations (8.57a) and (8.64) can be combined to give:

$$t = t_c\,\hat{t} = \frac{\eta_v\,\hat{C}\,\hat{t}}{\rho_W\, g_z\, h_I} = \frac{\eta_v}{\rho_W\, g_z\, h_I} \tag{8.73}$$

where $\hat{C}\ \hat{t} = 1$ makes t a time scaling factor introduced by Reeh (1968) to compare bending results for ice shelves with different average ice temperatures that control h_v. At t >> 0, the maximum $(\hat{\sigma}_{xx} - \hat{\sigma}_{zz})$ is at $x \approx 0.2\, h_I$ on the top surface and $x \approx 3.0\, h_I$ on the bottom surface. This indicates that narrow ice slabs will calve along top crevasses and wide tabular icebergs will calve along bottom crevasses.

Figure 8.9 shows that maxima in $(\hat{\sigma}_{xx} - \hat{\sigma}_{zz})$ are minor perturbations of a high tensile $(\hat{\sigma}_{xx} - \hat{\sigma}_{zz})$ on both top and bottom surfaces of the ice shelf, except at distances x/h_I within one ice thickness of the calving front. Therefore, these maxima are not likely to determine where crevasses will open. Instead, crevasses on both surfaces are likely to open along preexisting lines of weakness that began as primary transverse crevasses that opened in ice streams supplying the ice shelf, with bottom transverse crevasses becoming partly frozen shut as they were transported to the calving front.

The temperature dependence of η_v is given by Equations (4.108), (6.7), and (8.54):

$$\eta_v = \frac{\sigma_v}{n\,\dot{\varepsilon}_v} = \frac{A_o^n\, exp\left(C\,T_M/T\right)}{n\,\sigma_v^{\,n-1}} = \frac{A_o^n\, exp\left(C\,T_M/T\right)}{n\left[\frac{1}{2}\rho_I\, g_z\, h_I\left(1 - \rho_I/\rho_W\right)\right]^{n-1}} \tag{8.74}$$

where $\sigma_v = (\sigma_{xx} - \sigma_{zz}) = 2\,\sigma_{xx}'$ and $\phi = 0$ at the calving front of an ice shelf. Equation (8.74) allows t in Equation (8.73) to be computed for various values of η_v and h_I near the calving front of an ice shelf. For typical floating west Greenland outlet glaciers, $\rho_I \approx 900$ kg/m^3, $h_I \approx 600$ m, and T ≈ 269°K = – 4°C, giving t = 0.3 a. For typical Antarctic ice shelves, $\rho_I \approx 800$ kg/m^3, $h_I \approx 200$ m, and T ≈ 261°K = – 12°C, giving t = 19 a.

Calving rate u_c is the product of the spacing between transverse lines of weakness and the time needed for $(\sigma_{xx} - \sigma_{zz})$ to attain mean fracture stress $\bar{\sigma}_v$. Spacing C_x of primary transverse crevasses increases from initial spacing C_x' at their time of formation to final spacing C_x'' at the time of calving, after they have moved distance x in time t_x. If ε_{xx} is the average extending strain over this distance:

$$\varepsilon_{xx} = \int_{C_x'}^{C_x''} \frac{dC_x}{C_x} = ln\left(C_x''/C_x'\right) \tag{8.75}$$

Therefore, setting $\varepsilon_{xx} = \dot{\varepsilon}_{xx}\, t_x$ and using Equation (6.19):

$$\begin{aligned}C_x'' &= C_x'\, exp\,(\varepsilon_{xx}) = C_x'\, exp\,(\dot{\varepsilon}_{xx}\, t_x) \\ &= C_x'\, exp\,[(\hat{R}_{xx}^{\,n-1}/A^n)\,\sigma_{xx}'\, x/\bar{u}_x]\end{aligned} \tag{8.76}$$

where R_{xx} is given by Equation (6.18), σ_{xx}' is given by Equation (6.7), A is given by Equation (4.135), $\bar{u}_x$ is the average ice velocity over length x, and $t_x = x/\bar{u}_x$.

Let transverse lines of weakness have spacing C_x'' near the calving front, let $(\sigma_{xx} - \sigma_{zz}) = \bar{\sigma}_v$ be the fracture stress at time t given by Equation (8.73) along transverse lines of weakness having initial spacing C_x' given by Equation (8.31), and let $\bar{u}_x = 1/2\ (u_x' + u_x'')$ be averaged over $x = x'' - x'$ for velocity u_x' at x' and velocity u_x'' at x''. Calving velocity u_c for tabular icebergs is then:

$$u_c = \frac{C_x''}{t} = \frac{C_x'\, exp\left[\left(R_{xx}^{\,n-1}/A^n\right)\sigma_{xx}'\, x/\bar{u}_x\right]}{\eta_v/\rho_W\, g_z\, h_I}$$

$$= \frac{\rho_W g_z h_I'' \kappa u_x'}{\eta_v''} \left(\frac{2 n \eta_v' C_H \Delta T'}{h_C' (\sigma_{xx}')' H_M} \right)^2 exp \left(\frac{R_{xx}^{n-1} \sigma_{xx}' x}{A^n \bar{u}_x} \right) \tag{8.77}$$

where u_x', η_v', $\Delta T'$, h_C', and $(\sigma_{xx}')'$ are measured where transverse crevasses form, h_I'' and η_v'' are measured where tabular icebergs form, and R_{xx}, σ_{xx}', A, and $\bar{u}_x$ are averaged over longitudinal distance x between these sites. Equation (8.21) gives h_C', Equation (8.30) gives $\Delta T'$, and Equation (8.74) gives η_v' and η_v''.

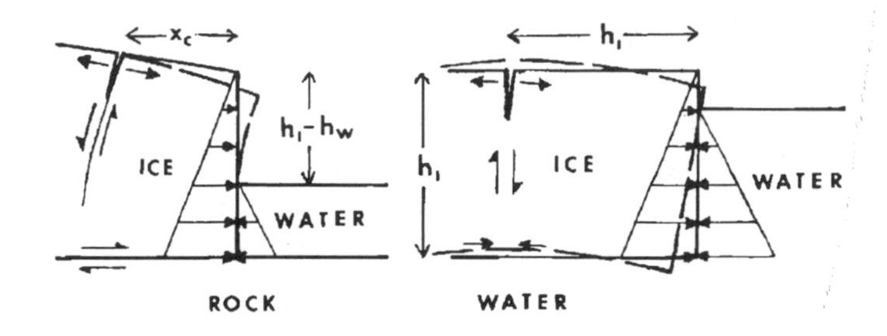

Figure 8.10: A comparison between forward bending of an ice wall and upward arching of an ice shelf due to asymmetry between lithostatic pressure in ice and hydrostatic pressure in water at the calving front (Hughes, 1992c, Figure 11).

Calving Ice Walls

An ice wall is an ice cliff grounded in water of variable depth or on dry land. The ice sheets in Greenland and Antarctica frequently end as ice walls calving in marine water of variable depth. The southern perimeter of the former ice sheets in North America and Eurasia included ice walls calving in lacustrine water of variable depth, especially as these ice margins retreated across isostatically depressed deglaciated landscapes after the last glacial maximum. Figure 8.10 compares calving from ice shelves and ice walls that results from forward bending in the two cases, height $h_I - h_W$ above water usually varies between 30 m and 70 m, regardless of water depth h_W.

By definition, ice walls are grounded, so bed traction resists their forward motion, unlike freely floating ice shelves. Consequently, ice walls lean forward and ice behind the walls is deformed by shear bands that rise vertically from basal ice and curve forward toward the ice surface, with shear offset within individual bands increasing from zero at the base to a maximum at the surface. Slip along these shear bands is analogous to slip between the pages of a book that is bent about its binding. This mechanism for creep deformation is illustrated in Figure 8.11. Shear rupture near the top of these shear bands allows crevasses to open where shear bands intersect the top surface behind the ice wall. Ice slabs calve from the overhanging ice wall along these crevasses. If x_c is the distance from the ice wall to the first crevasse and the ice wall leans forward at angle θ, the slab will calve by dropping downward if x_c and θ are small because the crushing stress of ice will be exceeded at the base of the slab, but the slab will calve by toppling forward if x_c and θ are large enough for the bending stress to peel the slab away from the ice wall. Figure 8.12 shows a slab about to calve from an ice wall for $x_c \approx 0.4\ h_I$ and $\theta \approx 18°$, conditions that appear to be transitional from downward to forward calving.

The mechanics of slab calving follow from the theory for bending beams. Popov (1952, p. 89) lists the steps in beam analysis that are needed to determine the shear force, the axial force, and the bending moment at any cross-section of a beam. An initially vertical beam of ice resting on the bed will be considered here. The beam has length x_c, width w_c, and mean height $\bar{h}_I$. In general, an ice wall stands in wa-

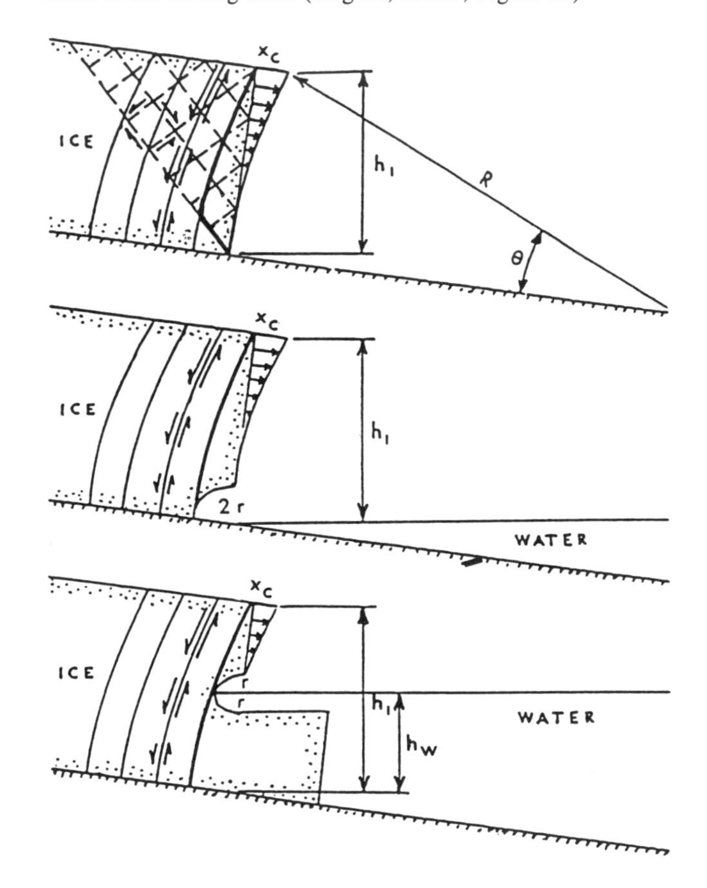

Figure 8.11: The bending creep mechanism for shear rupture in calving ice walls. Ice walls are shown grounded on dry land (top), at the shoreline of a beach (middle), and in deep water (bottom). The slip lines of maximum shear stress for homogeneous creep intersect the ice wall at 45° (straight dashed lines), shear bands rising vertically from the bed (curving solid lines) are produced by bending creep, and calving surfaces for shear rupture are also shown (heavy lines), (Hughes and Nakagawa, 1989, Figure 1).

ter, so distributed loads of ice and water act on opposite sides of the beam, causing the beam to bend forward. As shown in Figure 8.13, when water of height h_W rises against a wall of height h_I increasing buoyancy decreases basal shear stress

Figure 8.12: A slab about to calve from the ice wall of an Antarctic glacier (Hughes, 1992c, Figure 1, photographed by D. Allan).

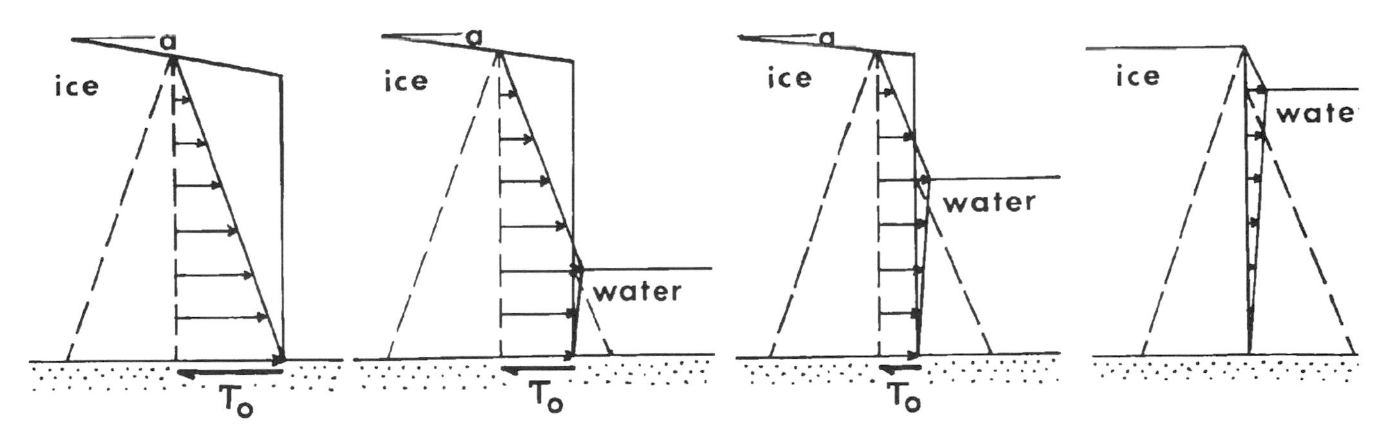

Figure 8.13: The effect of water depth on an ice wall. The difference between the lithostatic pressure in ice and the hydrostatic pressure in water (sloping dashed lines) is shown by arrows. This pressure difference, the basal shear stress τ_o, the surface slope α, and perhaps the slab-size decrease as water deepens, with $\tau_o = \alpha = 0$ at flotation depth, when tabular calving replaces slab calving (Hughes, 1992c, Figure 3). The greatest pressure difference pulls ice forward to sandwich it between a less dense medium above (air) and a more dense medium below (land or water). The height of the ice wall above water is usually 50 m ± 20 m, regardless of water depth.

τ_o, thereby causing a reduction of surface slope α, such that τ_o and α approach zero when the ice wall begins to float at $h_W = (\rho_I / \rho_W)\, h_I$, where ρ_I and ρ_W are ice and water densities, respectively, and are assumed to be constant. Questions to be examined are whether θ and x_c depend on h_W and h_I and how calving rate u_c depends on θ, x_c, h_W, and h_I. Specifically, an approximate calving rate that depends on these variables will be derived.

Figure 8.14 is a free-body diagram in which all forces acting on the slab, treated as a beam, are referenced with respect to the orthogonal coordinates originating on the bed at calving distance x_c behind the ice wall, with x horizontal and positive forward, y horizontal and parallel to the wall, and z vertical and positive upward. Gravitational forces acting on the slab are the distributed horizontal lithostatic force of ice F_I on the left side of the slab, the distributed horizontal hydrostatic force of water F_W on the right side of the slab, and the vertical body force F_B of the slab itself. Taking g_z as gravity acceleration, good approximations for these gravita-tional forces are:

$$F_I = \left(\tfrac{1}{2}\,\rho_I\, g_z\, h_c\right) w_c\, h_c = \tfrac{1}{2}\,\rho_I\, g_z\, w_c\, h_c^2 \tag{8.78}$$

where $1/2\ \rho_I\ g_z\ h_c$ is the mean lithostatic pressure pushing on area $w_c\ h_c$ for ice thickness h_c at calving distance x_c,

$$F_W = \left(\tfrac{1}{2}\,\rho_W\, g_z\, h_W\right) w_c\, h_W = \tfrac{1}{2}\,\rho_W\, g_z\, w_c\, h_W^2 \tag{8.79}$$

where $1/2\ \rho_I\ g_z\ h_W$ is the mean hydrostatic pressure pushing on area $w_C\ h_W$ for ice thickness h_I at the ice wall, and

$$F_B = w\, x_c\, \bar{h}_I\, \rho_I\, g_z - w_c\, x_c\, h_W\, \rho_W\, g_z$$

$$= \rho_I\, g_z\, w_c\, x_c\, \bar{h}_I\,(1 - \rho_W\, h_W/\rho_I\, \bar{h}_I) \tag{8.80}$$

where $\bar{h}_I = 1/2\ (h_I + h_c)$ is the average slab height, $w_c\ x_c\ \bar{h}_I$ is the slab volume, and $w_c\ x_c\ h_W$ is the water volume it displaces.

Gravitational action forces create stresses that produce opposing reaction forces. In the free–body diagram in Figure 8.14, these are a vertical basal normal force F_N, a horizontal basal traction force F_0, a vertical shear force F_S, and a horizontal tensile force F_C that opens the crevasse at x_c. The normal force is:

$$F_N = \sigma_{zz}\, w_c\, x_c \tag{8.81}$$

where horizontally averaged basal axial stress σ_{zz} acts through the substrate normal to basal area $w_c\ x_c$. The traction force is:

$$F_0 = \tau_o\, w_c\, x_c \tag{8.82}$$

where horizontally averaged basal shear stress τ_o acts parallel to basal area $w_c\ x_c$. The shear force is:

$$F_S = \tau_s\, w_c\, h_c \tag{8.83}$$

where vertically averaged bending shear stress τ_s induced by F_S acts parallel to calving area $w_c\ h_c$. The tensile force is:

$$F_C = \sigma_{xx}\, w_c\, h_c \tag{8.84}$$

where vertically averaged axial stress σ_{xx} acts normal to calving area $w_c\ h_c$. Reaction forces F_0, F_S, and F_C are kinematic forces because they result from bending motion of the slab.

Moment lever arms L_I, L_W, L_B, L_N, L_O, L_S, and L_C parallel to axes x, z for forces F_I, F_W, F_B, F_N, F_O F_S, and F_C, respectively, for bending moment M, are:

$$L_I = 1/3\ h_c \tag{8.85a}$$

$$L_W = 1/3\ h_W \tag{8.85b}$$

$$L_B = L_N = 1/2\ x_c \tag{8.85c}$$

$$L_0 = L_S = 0 \tag{8.85d}$$

$$L_C = 1/2\ h_c \tag{8.85e}$$

These are all shown in Figure 8.14. Equation (8.85c) is exact only if the surface and bed are parallel.

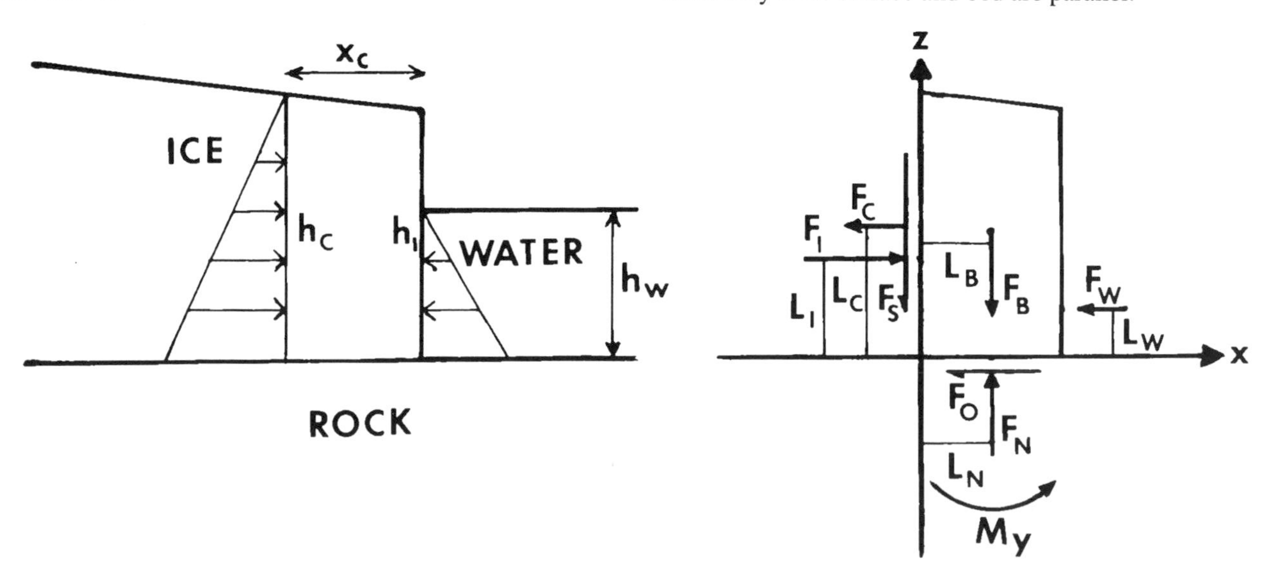

Figure 8.14: Free-body diagrams for a slab at an ice wall before shear rupture detaches the slab: (*left*): boundary conditions: (*right*): forces and moments (Hughes, 1992c, Figure 4).

The equations of statics are now applied to compute unknown reaction forces, assuming no inertial forces. Force equilibrium along x requires that:

$$F_I - F_W - F_O - F_C = 0 \qquad (8.86)$$

from which, up until a tensile crack opens the shear band :

$$\sigma_{xx} = \tfrac{1}{2}\rho_I g_z h_c - \tfrac{1}{2}\rho_W g_z h_W^2/h_c - \tau_o x_c/h_c$$

$$= \tfrac{1}{2}\rho_I g_z h_c\left(1 - \rho_W h_W^2/\rho_I h_c^2\right) - \tau_o\left(x_c/h_c\right) \qquad (8.87)$$

Force equilibrium along z requires that:

$$F_B - F_S - F_N = 0 \qquad (8.88)$$

from which:

$$\sigma_{zz} = \rho_I g_z \bar{h}_I - \rho_W g_z h_W - \tau_s h_c/x_c$$

$$= \rho_I g_z \bar{h}_I\left(1 - \rho_W h_W/\rho_I \bar{h}_I\right) - \tau_s\left(h_c/x_c\right) \qquad (8.89)$$

Moment equilibrium about the y axis requires that, until a tensile crack opens the shear band:

$$M + F_C L_C + F_N L_N + F_W L_W - F_B L_B - F_I L_I = 0 \qquad (8.90)$$

from which:

$$M = 1/2\, w_c c\, h_c (\tau_o + \tau_s) - 1/4\, \rho_I g_z w_c h_c^3 (1 - \rho_W h_W^2/\rho_I h_c^2)$$

$$+ 1/6\, \rho_I g_z h_c^3 (1 - \rho_W h_W^3/\rho_I h_c^3) \qquad (8.91)$$

where bending moment M causes bending creep at cross-section $w_c h_c$ that leads to shear rupture and crevassing in the shear band at calving distance x_c from the ice wall.

Forces can now be specified at cross-section $w_c h_c$. Axial force F_C is obtained from Equations (8.84) and (8.89):

$$F_C = 1/2\, \rho_I g_z w_c h_c^2 (1 - \rho_W h_W^2/\rho_I h_c^2) - \tau_o w_c x_c \qquad (8.92)$$

Shear force F_S is obtained from Equation (8.83) and (8.89):

$$F_S = \rho_I g_z w_c x_c \bar{h}_I (1 - \rho_W h_W/\rho_I \bar{h}_I) - \sigma_{zz} w_c x_c \qquad (8.93)$$

Force F_S causes shear rupture in the shear band, and force F_C causes the shear band to open into a crevasse.

The slab becomes isolated after the crevasse opens. As seen in Figure 8.12, the crevasse can separate most of the slab from the ice wall before it actually calves. Figure 8.15 is a free-body diagram for the ice slab at this stage of bending creep, showing the forces and bending moments acting on it immediately prior to calving.

Major changes in the action and reaction forces occur if some assumptions are made. First, assume a thawed bed allows water to fill the crevasse to height h_W so that:

$$F_I = F_W = 1/2\, \rho_W g_z w_c h_W^2 \qquad (8.94)$$

Second, downward penetration of the crevasse stops when $\sigma_{xx} = 0$ at the crevasse tip so that:

$$F_C = h_c w_c \sigma_{xx} = 0 \qquad (8.95)$$

Third, bending creep in numerous shear bands through length x_c has inclined the ice slab an angle q from the vertical, with θ increasing along z. Figure 8.16 shows similar triangles for which θ is the angle formed by the z axis and the distance vector to point x, z on the bending curve, and by velocity vectors u_x and u_z of point x, z. Therefore:

$$\theta = x/z = u_z/u_x \qquad (8.96)$$

where forward displacement x of the slab increases along z, with point x, z on the slab having forward velocity u_x and downward velocity u_z due to bending creep.

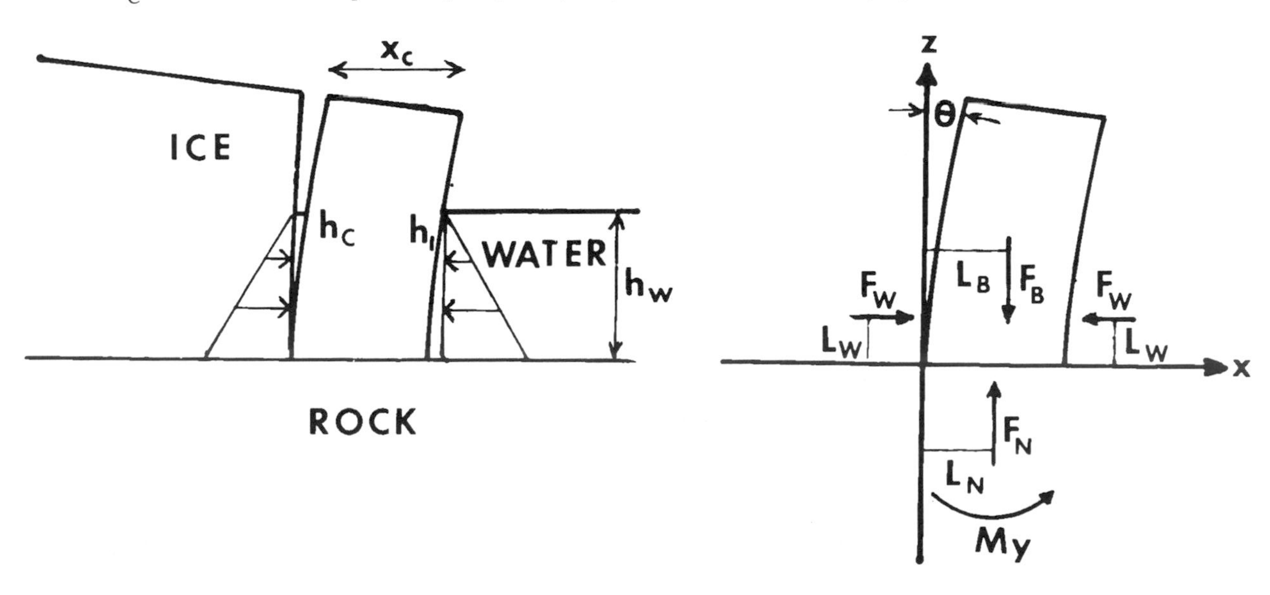

Figure 8.15: Free-body diagrams for a slab at an ice wall after shear rupture detaches the slab: (*left*) boundary conditions: (*right*) forces and moments (Hughes, 1992b, Figure 5).

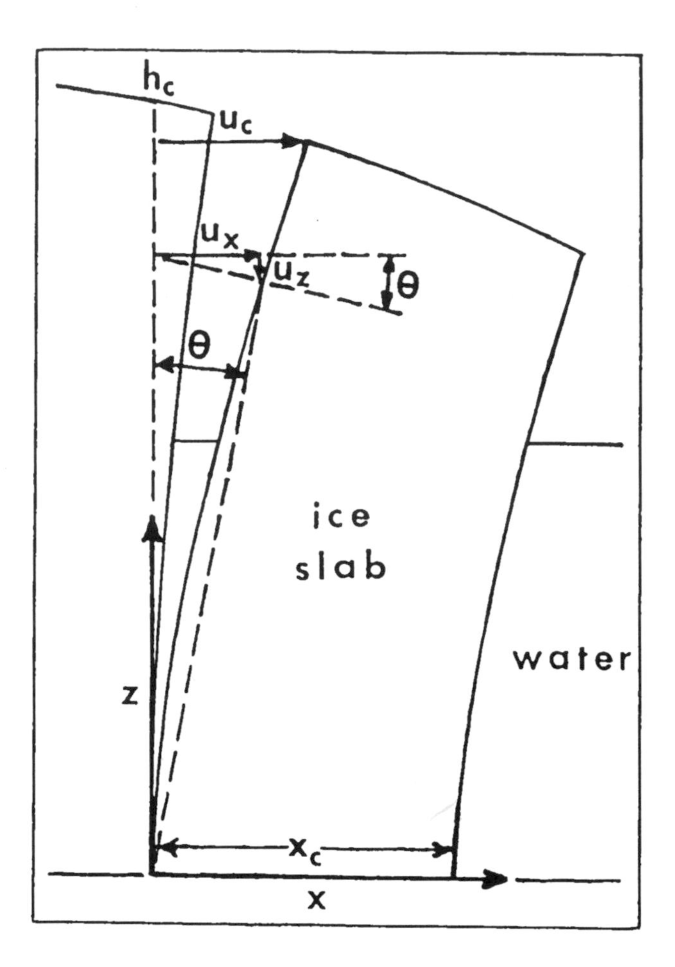

Figure 8.16: A bending ice slab after τ_S has caused shear rupture in a shear band and σ_{xx} has opened a crevasse along the ruptured shear band. The relationship between the angle θ in the crevasses, points x and z on the slab side of the crevasse and velocities u_x and u_z at points x and z are shown geometrically and are given by Equation (8.96). Note that $u_x = u_c$ at the top of the crevasse (Hughes, 1992c, Figure 6).

The equations of static equilibrium for the free body in Figure 8.15 can now be solved. No net horizontal forces are present. Summing vertical forces:

$$F_B - F_N = 0 \tag{8.97}$$

Summing bending moments:

$$M + F_N\,(1/2\ c) - F_B\,(1/2\ x_c + s) = 0 \tag{8.98}$$

where s is horizontal bending displacement x at height 1/2 $\bar{h}_I$ of the centroid of the slab. Therefore, from Equation (8.96):

$$s \approx 1/2\ \bar{h}_I\,\theta \tag{8.99}$$

Entering Equations (8.80) and (8.81) into Equation (8.97) gives:

$$\sigma_{zz} = F_B / w_c\,x_c = \rho_I\,g_z\,\bar{h}_I\,(1 - \rho_W\,h_W / \rho_I\,\bar{h}_I) \tag{8.100}$$

Combining Equations (8.97) through (8.100) gives:

$$M = F_B\,s \approx -\,1/2\ \rho_I\,g_z\,w_c\,x_c\,\bar{h}_I^2\,\theta\,(1 - \rho_W\,h_W / \rho_I\,\bar{h}_I) \tag{8.101}$$

A calving criterion can be developed for the slab by entering M into the flexure formula (Popov, 1952, pp. 98–102) and then deriving the differential equation for bending (Popov, 1952, pp. 269–273).

In deriving the flexure formula, bending stresses normal to the beam cross-section, tensile on the convex side and compressive on the concave side, create an internal bending moment that exactly resists the external bending moment (Popov, 1952). The important concepts used to derive the flexure formula are (1) beam geometry, which establishes the variation of strain in the beam cross-section, (2) beam rheology, which relates strain to stress, and (3) the laws of statics, which locate the neutral axis and resisting moment at the beam cross-section. The flexure formula usually assumes elastic bending only, so Hooke's law can be used to relate stress to strain. In this case, the flexure formula can be used to derive the differential equation for the elastic bending curve for a beam in static equilibrium (Popov, 1952, pp. 269–273). When beam height h_I along z is large compared to beam length x_c in bending direction x and beam width w_c, a good approximation is:

$$\frac{d^2 x}{d z^2} = \frac{M}{E I} \tag{8.102}$$

where E is the elastic modulus and $I = 1/12\ w_c\,x_c^3$ is the rectangular moment of inertial for basal cross-sectional area $w_c\,x_c$ of the bending beam.

The counterpart to Equation (8.102) for creep bending when the beam is an ice slab in dynamic equilibrium is:

$$\frac{d^2 u_x}{d z^2} = \frac{M}{\eta_v\,I} \tag{8.103}$$

where h_v is the viscoplastic viscosity of creep. Instead of relating stress to strain using Hooke's law, as was done in deriving Equation (8.102), Equation (8.103) is derived by relating stress to strain rate using Glen's law (Glen, 1955).

At the start of bending creep, Equation (8.91) is substituted into Equation (8.103), which is then integrated twice to give:

$$u_x = \frac{h_c}{2\,\eta_v\,x_c^3}\Big[6\,x_c\,(\tau_o + \tau_s) - 3\,\rho_I\,g_z\,h_c^2\left(1 - \rho_W\,h_W^2/\rho_I\,h_c^2\right) + 2\,\rho_I\,g_z\,h_c^2\left(1 - \rho_W\,h_W^3/\rho_I\,h_c^3\right)\Big]\,z^2 \tag{8.104}$$

where $I = 1/12\ w_c\,x_c^3$ is the rectangular moment of inertia for basal area $w_c\,x_c$ of the ice slab. At the top of the ice slab, $z = h_c$ and $u_x = u_c$ is the maximum velocity of bending creep, so:

$$u_c = \frac{h_c^3}{2\,\eta_v\,x_c^3}\Big[6\,x_c\,(\tau_o + \tau_s) - 3\,\rho_I\,g_z\,h_c^2\left(1 - \rho_W\,h_W^2/\rho_I\,h_c^2\right)$$

$$+ 2 \rho_I g_z h_c^2 \left(1 - \rho_W h_W^3 / \rho_I h_c^3\right)\Big] \qquad (8.105)$$

When bending creep is ended by shear rupture so that a fracture crevasse opens the shear band, Equation (8.101) is substituted into Equation (8.103) so that:

$$u_x = \frac{3 \rho_I g_z \bar{h}_I^2 \theta}{\eta_v x_c^2} \left(1 - \frac{\rho_W h_W}{\rho_I \bar{h}_I}\right) z^2 \qquad (8.106)$$

The slab calves when $u_x = u_c$ measured at $z = h_c \approx \bar{h}_I$:

$$u_c = \frac{3 \rho_I g_z \bar{h}_I^4 \theta}{\eta_v x_c^2} \left(1 - \frac{\rho_W h_W}{\rho_I \bar{h}_I}\right) = \frac{3 \rho_I g_z \bar{h}_I^4 \theta}{\eta_v (x_c / \bar{h}_I)} \left(1 - \frac{\rho_W h_W}{\rho_I \bar{h}_I}\right) \qquad (8.107)$$

where u_c is the calving velocity and $x_c / \bar{h}_I$ is the calving ratio. The smaller the calving ratio, the better Equation (8.107) gives the calving rate.

In applying Equation (8.107) to calving ice walls, such as the one in Figure 8.12, a question arises. Should θ be measured when the crevasse first opens, in which case θ is about 6° (0.1 radians), as measured by the forward tilt of the left side of the crevasse, or should θ be measured when the slab actually calves, in which case θ is about 18° (0.3 radians) as measured by the forward tilt of the right side of the crevasse? As seen in Figure 8.17, other crevasses open behind the primary crevasse before calving, so θ should apply to the calving event itself.

A second question concerns the calving ratio x_c / h_I. For convenience, let $h_I = \bar{h}_I = h_c$. Is x_c / h_I a function of total ice thickness h_I or just ice thickness $h_I - h_W$ above water? Both possibilities exist in Figure 8.12. Also, does x_c / h_I depend on water depth h_W through flotation ratio $(\rho_W / \rho_I) h_W / h_I$ for the ice wall? Figure 8.12 is for $h_W = 0$, so it provides no answers to these questions.

A third question concerns the value of viscoplastic viscosity η_v to be used in Equation (8.107). If creep in the bending slab is primarily in shear bands, not between shear bands, then creep by easy glide, not hard glide, dominates. Newtonian viscosity $\eta_N = n \eta_v$ in Equation (8.3). Paterson (1981, p. 41) gives $\eta_N = 8 \times 10^{13}$ kg m^{-1} s^{-1} for the minimum creep rate (hard glide) and $\sigma_o = 1$ bar as the plastic yield stress. Hughes and Nakagawa (1989) report that the creep rate increases tenfold in shear bands (easy glide), for which $\eta_N = 8 \times 10^{12}$ kg m^{-1} s^{-1}.

These questions can be addressed using data from twelve calving tidewater glaciers in Alaska (Brown and others, 1982). Table 8.1 lists the glaciers and values of h_I, $h_I - h_W$, h_W, and $\bar{u}_c$ for each one, where $\bar{u}_c$ is the mean annual calving rate that can be compared with u_c in Equation (8.107). Figure 8.12 is consistent with four expressions for x_c in Equation (8.107): a linear dependence on h_I:

$$x_c / h_I = 0.4 \qquad (8.108)$$

a linear dependence on $h_I - h_W$:

Figure 8.17: The transverse crevasses behind the ice wall of a tidewater outlet glacier in Greenland (Hughes, 1992c, Figure 8).

$$x_c / (h_I - h_W) = 0.4 \qquad (8.109)$$

a linear dependence on $(\rho_W / \rho_I) h_W / h_I$:

$$x_c / h_I = 0.4 - 0.35 (\rho_W / \rho_I) h_W / h_I \qquad (8.110)$$

$$(x_c / h_I)^2 = (0.22) / [(\rho_W / \rho_I) h_W / h_I] \qquad (8.111)$$

Values of x_c / h_I in Equations (8.110) and (8.111) are obtained by setting $u_c = \bar{u}_c$ in Equation (8.107) and solving for x_c / h_I:

$$\left(\frac{x_c}{h_I}\right)^2 = \frac{3 \rho_I g_z h_I^2 \theta}{\eta_v \bar{u}_c} \left(1 - \frac{\rho_W h_W}{\rho_I h_I}\right) \qquad (8.112)$$

Table 8.1: Average ice thickness h_I, ice height above water $h_I - h_W$, water depth h_W and mean annual calving rate $\bar{u}_c$ at the ice walls of 12 Alaskan tide-water glaciers (Brown and others, 1982) and a parabolic dependence on $(\rho_W / \rho_I) h_W / h_I$:

Glacier	h_I (m)	$h_I - h_W$ (m)	h_W (m)	$\bar{u}_c$ (m a^{-1})
McCarty	42	30	12	600
Harvard	104	68	36	1080
Yale	222	69	153	3500
Meares	90	59	31	1010
Columbia	161	86	75	2185
Tyndall	113	49	64	1740
Hubbard	172	92	80	2630
Grand Pacific	62	44	18	220
Margerie	75	60	15	463
Johns Hopkins	126	70	56	2290
Muir	160	60	100	3700
South Sawyer	234	48	186	3200

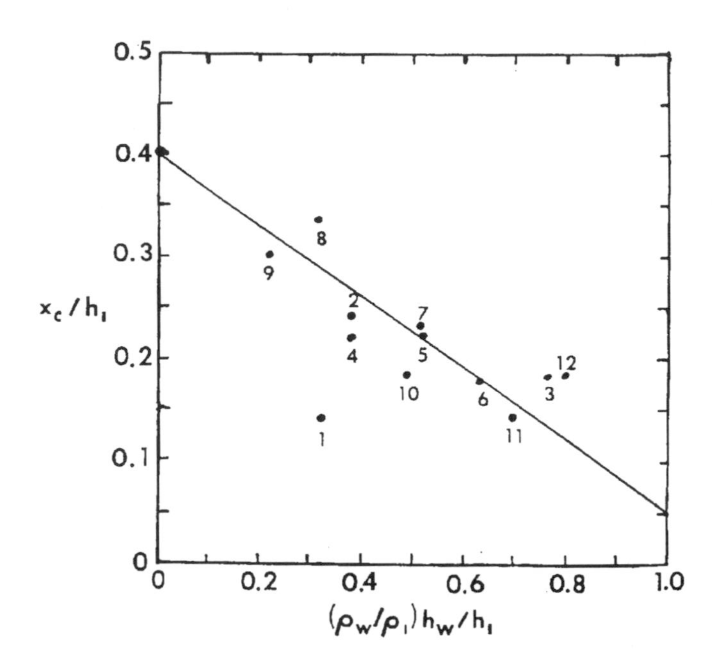

Figure 8.18: A linear dependence of calving ratio c/h on buoyancy ratio (ρ_W/ρ_I) d/h for the ice wall in Figure 8.12 and the 12 tidewater glaciers in Tables 8.1 and 8.2 (Hughes, 1992c, Figure 9).

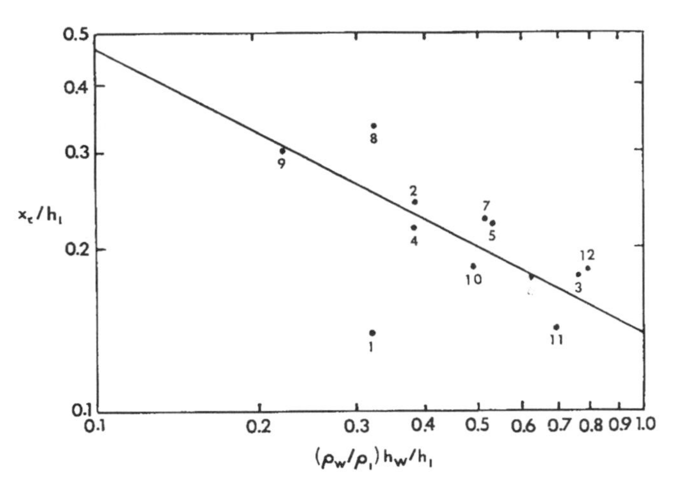

Figure 8.19: A logarithmic dependence of calving ratio x_C/h_I on buoyancy ratio (ρ_W/ρ_I) h_W/h_I for the ice wall in Figure 8.12 and the 12 tidewater glaciers in Tables 8.1 and 8.2 (Hughes, 1992c, Figure 10).

Figures 8.18 and 8.19 show respective linear and logarithmic plots of x_c/h_I versus (ρ_W/ρ_I) h_W/h_I obtained from Equation (8.112), for the purpose of comparing Equations (8.110) and (8.111). The linear plot is preferred because it can include $x_c/h_I = 0.4$ at $h_W = 0$, from Figure 8.12.

Table 8.2 shows results from entering Equations (8.108) through (8.111) into Equation (8.107). Ideally, $u_c/\bar{u}_c = 1.00$. Reasonable approximations of this ideal are at-tained for most of the twelve glaciers if $\theta = 0.3$ radian and bending creep of the ice wall is controlled by $\eta_N = 8 \times 10^{12}$ kg m^{-1} s^{-1} for easy glide within shear bands, $\eta_N = 8 \times 10^{13}$ kg m^{-1} s^{-1} for hard glide between shear bands, or $\eta_N = 4 \times 10^{13}$ kg m^{-1} s^{-1} for a combination of easy and hard glide within and between shear bands. In fact, $u_c/\bar{u}_c = 1.03 \pm 0.01$ averaged for all twelve glaciers, except when u_c is computed using Equation (8.111), and $u_c/\bar{u}_c = 0.95$ averaged for all the u_c values in Table 8.2. These results are summarized in Table 8.3. The same results are obtained for $\theta = 0.1$ radian and $\eta_v = 1/3\ \eta_N$ for n = 3 in Equation (8.3). However, Equation (8.103), and therefore Equation (8.107), are exact only for n = 1 in Equation (8.3) (Popov, 1952, pp. 109–113).

Results in Table 8.2 are consistent with both easy glide within shear bands and hard glide between shear bands contributing to bending creep in the ice slab. This was confirmed by relating bending creep to ice fabrics in and between shear bands behind the calving ice wall on Deception Island (Hughes and Nakagawa, 1989). The calving ratio was $x_c/h_I \approx 0.3$, but the ends of crevasses at calving distance x_c behind the ice wall experienced compression from converging ice flow. This violated the assumption that end effects did not exist or could be ignored. The assumption is valid for the ice wall in Figure 8.10, where $x_c/h_I = 0.4$. Field studies on tidewater glaciers are needed to determine the dependence, if any, of x_c/h_I on water buoyancy at the ice wall, as measured by $P_W/P_I = \rho_W h_W/\rho_I h_I$.

Table 8.2: Theoretical calving rates u_C and their ratios to the observed mean annual calving rates $\bar{u}_c$ from the ice walls of Alaskan tidewater glaciers

Glacier	Equation (8.108) u_c(m/a)	$u_c/\bar{u}_c$	Equation (8.109) u_c(m/a)	$u_c/\bar{u}_c$	Equation (8.110) u_c(m/a)	$u_c/\bar{u}_c$	Equation (8.111) u_c(m/a)	$u_c/\bar{u}_c$
McCarty	241	0.40	94	0.16	137	0.23	120	0.20
Harvard	1333	1.23	624	0.58	895	0.83	808	0.70
Yale	2316	0.66	4830	1.38	6274	1.79	2800	0.80
Meares	1001	0.99	468	0.46	670	0.66	604	0.60
Columbia	2504	1.15	1753	0.80	2242	1.03	2046	0.94
Tyndall	947	0.64	1018	0.58	1386	0.94	942	0.64
Hubbard	2858	1.09	2000	0.76	2822	1.07	2334	0.88
Grand Pacific	521	2.37	208	0.94	299	1.36	266	1.20
Margerie	875	1.89	273	0.60	400	0.86	308	0.66
Johns Hopkins	1610	0.70	1043	0.46	1479	0.65	1254	0.54
Muir	1567	0.42	2234	0.60	3007	0.81	1718	0.46
South Sawyer	2245	0.70	10690	3.34	7190	2.25	2820	0.88

Table 8.3: Average calving velocity ratios $u_c/\bar{u}_c$ computed for combinations of wall bending angles θ and ice viscosities η_N and η_v

	Equation (8.108)	Equation (8.109)	Equation (8.110)	Equation (8.111)
average ($u_c/\bar{u}_c$)	1.02	1.04	1.04	0.71
$\theta = 18°$ = 0.3 radian				
η_N(kg m^{-1}s^{-1})	8×10^{12}	4×10^{13}	8×10^{13}	4×10^{13}
$\theta = 6°$ = 0.1 radian				
η_v(kg m^{-1}s^{-1})	$8/3\times10^{12}$	$4/3\times10^{13}$	$8/3\times10^{13}$	$4/3\times10^{13}$

During rapid calving retreat of Columbia Glacier from 1976 to 1994, x_c/h_I decreased as height above buoyancy, $h_I - h_W(\rho_W/\rho_I)$, decreased from 100 m to 20 m (Van der Veen, 1995). Given the scatter of data in Figure 8.18, it is reasonable to replace Equation (8.110) with:

$$x_c/h_I = 0.4\,(1 - \rho_W h_W/\rho_I h_I)$$
$$= 0.4\,(1 - P_W/P_I) \qquad (8.113)$$

Substituting Equation (8.113) in-to Equation (8.107) gives:

$$u_c = \frac{19\,\rho_I g_z \bar{h}_I^{\,2}\,\theta}{\eta_v\,(1 - \rho_W h_W/\rho_I h_I)} \qquad (8.114)$$

This substitution preserves the close match between u_c and $\bar{u}_c$ in Table 8.3 and supports the expectation that $\bar{h} \rightarrow h_I$ and $\theta \rightarrow 0$ when $h_W \rightarrow (\rho_I/\rho_W)\,h_I$ as the ice wall becomes afloat. If forward tilt θ from bending shear vanish when ice floats, then u_c is indeterminate in Equation (8.114), which was derived for grounded ice. In fact, θ does not vanish as the ice wall becomes afloat, because arching of the floating ice causes the calving front to lean forward. However, crevasses open where $x_c/h_I \approx 1$, as this is the top of the arch where σ_{xx}' is maximized at the ice surface (Reeh, 1968).

As illustrated by the Greenland tidewater glacier in Figure 8.17, a series of transverse crevasses, about equally spaced, open and widen toward the calving ice wall. If this crevassed section of the glacier were to become afloat, columnar slabs would calve if $x_C/h_I < 1$ and tabular icebergs would calve if $x_c/h_I > 1$. Water would fill the crevasses up to sea level. When water in crevasses between slabs reaches the flotation depth, the crevassed section becomes afloat if water continues to deepen, and the forward bending creep of a grounded ice wall also includes the upward flexural creep of a floating ice shelf, as shown in Figure 8.20.

A nearly linear relationship was observed between the calving rate and the ice velocity during calving retreat of Columbia Glacier (van der Veen, 1995). This relationship

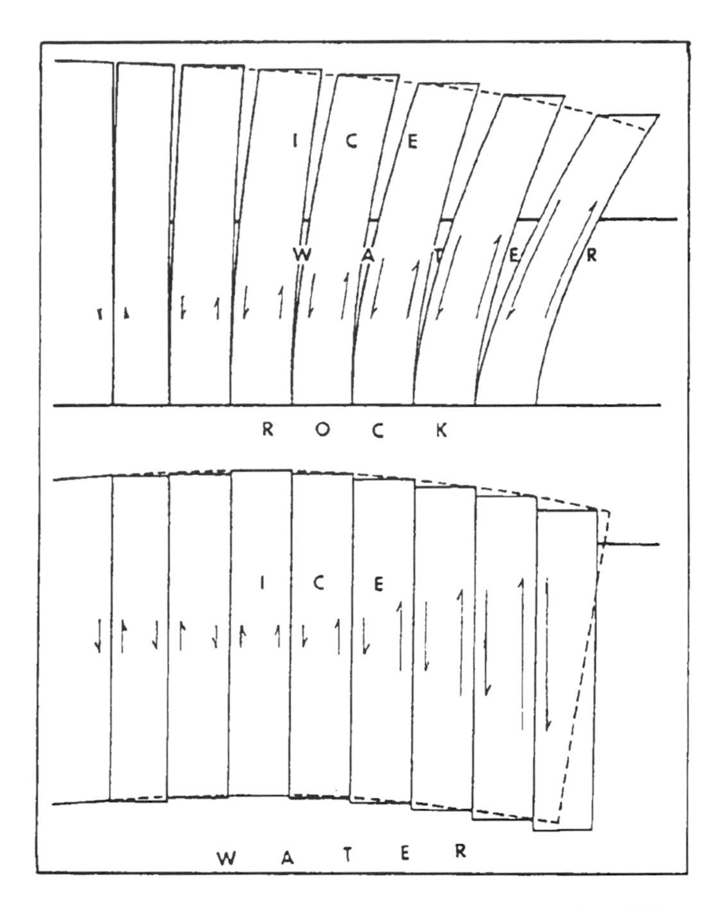

Figure 8.20 The dependence of calving rate u_c (m/a), sliding velocity u_s (m/a), basal water pressure P_W (bars), and ice height h (m) above water at the calving termini of ice streams according to Equation (8.125). Note that in the mid range of P_W, equivalent to water depths of order 500 m, $u_c > u_s$ when u_s is low and $u_c < u_s$ when u_s is high. The peak u_c is for h = 100 m, and decreases from 20 km/a to 13 km/a as u_s increases from 0.5 km/a to 32 km/a.

can be obtained from Equation (8.107) if forward bending angle θ is equated with surface slope a over distance x_c behind the calving front, when $0 < x_c < h_I$. This is reasonable in terms of the geometry in Figure 8.10 if a over this distance is determined largely by forward bending, especially after basal traction is reduced when P_W/P_I exceeds a critical value. In this case θ is given by $a = \Delta h/\Delta x$ in Equation (6.19), $P_W/P_I = \rho_W h_W/\rho_I h_I$ from Equation (6.3), and Equation (8.107) becomes:

$$u_c = \frac{3\,\rho_I g_z \bar{h}_I^{\,4}}{\eta_v x_c^2}\left(1 - \frac{\rho_W h_W}{\rho_I h_I}\right)\theta = \frac{3\,\rho_I g_z \bar{h}_I^{\,4}}{\eta_v x_c^2}\left(1 - \frac{P_W}{P_I}\right)\frac{\Delta h}{\Delta x}$$

$$= \frac{3\,\rho_I g_z \bar{h}_I^{\,4}}{\eta_v x_c^2}\left(1 - \frac{\rho_I}{\rho_W}\right)\left(1 - \frac{P_W}{P_I}\right)\left(\frac{P_W}{P_I}\right)^2$$

$$\left\{\frac{\bar{h}_I\,a}{a\,x + h_0 u_0} - \frac{\bar{h}_I^{\,2} R_{xx}^{\;n-1}}{a\,x + h_0 u_0}\left[\frac{\rho_I g_z h_I}{4A}\left(1 - \frac{\rho_I}{\rho_W}\right)\left(\frac{P_W}{P_I}\right)^2\right]^n\right\}$$

$$+\frac{3\,\rho_I\,g_z\,\bar{h}_I^{\,5}}{\eta_v\,x_c^{\,2}}\left(1-\frac{\rho_I}{\rho_W}\right)\left(1-\frac{P_W}{P_I}\right)\frac{P_W}{P_I}\frac{\Delta}{\Delta x}\left(\frac{P_W}{P_I}\right)$$

$$+\frac{3\,\bar{h}_I^{\,4}}{\eta_v\,x_c^{\,2}}\left(1-\frac{P_W}{P_I}\right)\left(\frac{2\,\tau_s}{w_I}+\frac{\tau_o}{h_I}\right) \tag{8.115}$$

When ice calves, lateral shear rupture requires that $\sigma_{xy} = \tau_s = 0$ along length x_c on both sides of the ice stream. Then Equation (8.115) includes only σ_{xx}', $\partial\sigma_{xx}'/\partial x$, and σ_{xz}. When $\theta \approx \alpha$, and noting that $2\,\dot{\varepsilon}_{xz} = (\partial u_x/\partial z) + (\partial u_z/\partial x)$ for basal shear $\partial u_x/\partial z$ and bending shear $\partial u_z/\partial x$:

$$\eta_v \approx \frac{d\sigma_{zx}}{d(\partial u_z/\partial x)} \approx \frac{d\sigma_{xz}}{d(\partial u_x/\partial z)} \approx \frac{d\sigma_{xz}}{d\dot{\varepsilon}_{xz}}$$

$$= \frac{d\sigma_{xz}}{d(\sigma_{xz}/A)} = \frac{A^{\,n}}{n\,\sigma_{xz}^{\,n-1}} \approx \frac{A^{\,n}}{n\,\tau_o^{\,n-1}} \tag{8.116}$$

where $\dot{\varepsilon}_{xz} = (\sigma_{xz}/A)^n$ is the flow law of ice and $\tau_o \approx \sigma_{xz}$ is the basal shear stress.

Substituting Equation (8.116) into Equation (8.115):

$$u_c = \frac{3\,\rho_I\,g_z\,\bar{h}_I^{\,4}\,n\,\tau_o^{\,n-1}}{x_c^{\,2}\,A^{\,n}}\left(1-\frac{\rho_I}{\rho_W}\right)\left(1-\frac{P_W}{P_I}\right)\left(\frac{P_W}{P_I}\right)^2$$

$$\left\{\frac{\bar{h}_I\,a}{a\,x+h_0\,u_0}-\frac{\bar{h}_I^{\,2}\,R_{xx}^{\,n-1}}{a\,x+h_0\,u_0}\left[\frac{\rho_I\,g_z\,\bar{h}_I}{4\,A}\left(1-\frac{\rho_I}{\rho_W}\right)\left(\frac{P_W}{P_I}\right)^2\right]^n\right\}$$

$$+\frac{3\,\rho_I\,g_z\,\bar{h}_I^{\,5}\,n\,\tau_o^{\,n-1}}{x_c^{\,2}\,A^{\,n}}\left(1-\frac{\rho_I}{\rho_W}\right)\left(1-\frac{P_W}{P_I}\right)\frac{P_W}{P_I}\frac{\Delta}{\Delta x}\left(\frac{P_W}{P_I}\right)$$

$$+\frac{3\,\bar{h}_I^{\,3}\,n\,\tau_o^{\,n-1}}{x_c^{\,2}\,A^{\,n}}\left(1-\frac{P_W}{P_I}\right) \tag{8.117}$$

Assuming that ice velocity u is almost entirely due to basal sliding velocity u_s given by Equation (4.26):

$$\tau_o = B\,(1 - P_W/P_I)^2\,u_s^{1/m} \tag{8.118}$$

Substituting Equation (8.118) into Equation (8.117) and setting $P_I = \rho_I\,g_z\,\bar{h}_I$:

$$u_c = \frac{3\,P_W\,\bar{h}_I^{\,3}\,n\,B^{\,n-1}}{x_c^{\,2}\,A^{\,n}}\left(1-\frac{\rho_I}{\rho_W}\right)\left(1-\frac{P_W}{P_I}\right)^{2n-1}\left(\frac{P_W}{P_I}\right)$$

$$\left\{\frac{\bar{h}_I\,a}{a\,x+h_0\,u_0}-\frac{\bar{h}_I^{\,2}\,R_{xx}^{\,n-1}}{a\,x+h_0\,u_0}\left[\frac{\rho_I\,g_z\,\bar{h}_I}{4\,A}\left(1-\frac{\rho_I}{\rho_W}\right)\left(\frac{P_W}{P_I}\right)^2\right]^n\right\}u_s^{\frac{n-1}{m}}$$

$$+\frac{3\,P_W\,\bar{h}_I^{\,4}\,n\,B^{\,n-1}}{x_c^{\,2}\,A^{\,n}}\left(1-\frac{\rho_I}{\rho_W}\right)\left(1-\frac{P_W}{P_I}\right)^{2n-1}\frac{\Delta}{\Delta x}\left(\frac{P_W}{P_I}\right)u_s^{\frac{n-1}{m}}$$

$$+\frac{3\,\bar{h}_I^{\,3}\,n\,B^{\,n}}{x_c^{\,2}\,A^{\,n}}\left(1-\frac{P_W}{P_I}\right)^{2n+1}u_s^{n/m} \tag{8.119}$$

In the Weertman (1957a) sliding theory, $m = (n + 1)/2$. Taking $n = 3$, exponent $(n - 1)/m = 1$ and exponent $n/m = 3/2$. Therefore, the first two terms give $u_c \propto u_s$ in Equation (8.119), as observed during fast retreat of Columbia Glacier when stream flow dominates (Van der Veen, 1995), and the third term becomes unimportant as $P_W \rightarrow P_I$. The third term dominates when $P_W \rightarrow 0$ for sheet flow and gives $u_c \propto u_s^{3/2}$. Which term dominates depends on P_W/P_I and $\Delta(P_W/P_I)/\Delta x$. Since $P_W = \rho_I\,g_z\,h_W$, the first two terms dominate the third term as water depth increases, so $u_c \propto h_W$ as a first approximation. This was reported by Van der Veen (1995) for Columbia Glacier and by Brown and others (1982) and Pelto and Warren (1991) for calving tidewater glaciers in general. In Equation (8.119), P_W/P_I is large enough so that $\alpha = \theta$, for which $\bar{h}_I \approx h_c \approx h_0 \approx h_I$, where h_c is h_I at $x = x_c$ and h_0 is h_I at $x = 0$. Since basal sliding dominates internal creep in this case, $\dot{\varepsilon}_{xx}$ is the dominant strain rate. Therefore, $u_s \approx -u_0$ and $R_{xx} \approx 1$, according to Equation (6.9). Also, $a\,x \ll h_0\,u_0$ over distance x_c.

In the absence of data on how P_W/P_I varies with distance x behind the calving ice wall, Equation (6.37) can be used to evaluate Equation (8.119) by setting $x = x_c$ for $(P_W/P_I)_c$ so that:

$$x_c = L_S\,[1 - (P_W/P_I)_c^{\,1/c}] \tag{8.120}$$

from which:

$$\left[\frac{\Delta(P_W/P_I)}{\Delta x}\right]_c = -\frac{c}{L_S}\left(1-\frac{x_c}{L_S}\right)^{c-1} = -\frac{c}{x_c}\left[1-\left(\frac{P_W}{P_I}\right)_c^{1/c}\right]\left(\frac{P_W}{P_I}\right)_c^{\frac{c-1}{c}} \tag{8.121}$$

Setting $P_I = \rho_I\,g_z\,h_I$ and making all of the above substitutions, Equation (8.119) becomes:

$$u_c = \frac{3\,P_W\,h_I^{\,3}\,n\,B^{\,n-1}}{x_c^{\,2}\,A^{\,n}}\left(1-\frac{\rho_I}{\rho_W}\right)\left(1-\frac{P_W}{P_I}\right)_c^{2n-1}\left(\frac{P_W}{P_I}\right)_c$$

$$\left\{h_I^{\,2}\left[\frac{P_I}{A}\left(1-\frac{\rho_I}{\rho_W}\right)\left(\frac{P_W}{P_I}\right)_c^2\right]^n - a\right\}u_s^{\frac{n-m-1}{m}}$$

$$-\left\{\frac{3\,c\,P_W\,h_I^{\,4}\,n\,B^{\,n-1}}{x_c^{\,3}\,A^{\,n}}\left(1-\frac{\rho_I}{\rho_W}\right)\left(1-\frac{P_W}{P_I}\right)_c^{2n-1}\right.$$

$$\left.\left[1-\left(\frac{P_W}{P_I}\right)_c^{1/c}\right]\left(\frac{P_W}{P_I}\right)_c^{\frac{c-1}{c}}\right\}u_s^{\frac{n-1}{m}}$$

$$+\left\{\frac{3\,h_I^{\,3}\,n\,B^{\,n}}{x_c^{\,2}\,A^{\,n}}\left(1-\frac{P_W}{P_I}\right)_c^{2n}\right\}u_s^{n/m} \tag{8.122}$$

Two empirical observations can be applied to Equation (8.112). First, x_c is given by Equation (8.117) to a good approximation. This suggests that x_c is determined primarily by the height of the ice wall above the buoyancy height. Second, as seen in Table 8.1, the height h above sea level of the ice wall is rarely more than 100 m, where:

$$h_I = h + h_W = h + P_W / \rho_W \, g_z \qquad (8.123)$$

Then:

$$P_I = \rho_I \, g_z \, h_I = \rho_I \, g_z \, h + (\rho_I / \rho_W) \, P_W \qquad (8.124)$$

Substituting Equations (8.117), (8.123), and (8.124) for x_c, h_I, and P_I in Equation (8.122) gives:

$$u_c = \frac{19 \, P_W \, n \, B^{\,n-1}}{A^{\,n}} \left(h + \frac{P_W}{\rho_W \, g_z}\right)\left(1 - \frac{\rho_I}{\rho_W}\right)\left(1 - \frac{P_W}{P_I}\right)_c^{2n-3} \left(\frac{P_W}{P_I}\right)_c$$

$$\left\{\left(h + \frac{P_W}{\rho_W \, g_z}\right)^2\right.$$

$$\left.\left[\left(\frac{\rho_I \, g_z \, h + (\rho_I / \rho_W) \, P_W}{A}\right)\left(1 - \frac{\rho_I}{\rho_W}\right)\left(\frac{P_W}{P_I}\right)_c^2\right]^n - a\right\} u_s^{\frac{n-2}{m}}$$

$$-\left\{\frac{19 \, c \, P_W \, n \, B^{\,n-1}}{A^{\,n}} \left(h + \frac{P_W}{\rho_W \, g_z}\right)\left(1 - \frac{\rho_I}{\rho_W}\right)\left(1 - \frac{P_W}{P_I}\right)_c^{2n-3}\right.$$

$$\left.\left[1 - \left(\frac{P_W}{P_I}\right)_c^{1/c}\right]\left(\frac{P_W}{P_I}\right)_c^{\frac{c-1}{c}}\right\} u_s^{\frac{n-1}{m}}$$

$$+\left\{\frac{19 \, n \, B^{\,n}}{A^{\,n}} \left(h + \frac{P_W}{\rho_W \, g_z}\right)\left(1 - \frac{P_W}{P_I}\right)_c^{2n}\right\} u_s^{n/m} \qquad (8.125)$$

where $h = 100$ m is an upper limit for calving ice walls.

Figure 8.21 is a plot of u_c / u_s versus P_W using Equation (8.125), when $A = 0.65$ bar $a^{1/n}$ at $n = 3$ for ice at the pressure melting point when a calving ice wall is standing in water, which gives $B = 0.02$ bar $a^{1/m}$ $m^{-1/m}$ at $m = 2$ (Hughes, 1981), $\rho_I = 917$ kg/m^3, $\rho_W = 1000$ kg/m^3 for lake water and $\rho_W = 1020$ kg/m^3 for sea water, $g_z = 9.8$ m/s^2 $= 9.8 \times 10^{-5}$ bar m^2/kg, $a = -1$ m/a for surface melting along x_C, 500 m/a $\leq u_s \leq$ 32,000 m/a, 50 m $\leq h \leq$ 100 m, and $c = 1$. Equation (8.125) is insensitive to c for $0.1 \leq c \leq 10$. Note that u_c approaches u_s in both the low and the high ranges of P_W, depending on h. This is what Van der Veen (1995) reported for Columbia Glacier as u_s increased from 1 km/a to 7 km/a. Equation (8.125) breaks down at the highest and lowest values of P_W for a given h, beyond which u_c is negative. The assumption that $\alpha \approx \theta$ needed to derive Equation (8.125) is invalid under these conditions. When P_W is too low, basal shear stress τ_o is great enough so that surface slope a is substantially greater than forward bending angle θ of the ice wall. When P_W is too high, the ice wall becomes afloat and τ_o vanishes.

The conclusion from Chapter 8 is that calving rate u_c is given by Equation (8.114) when $P_W < 20$ bars, by Equation (8.125) when 20 bars $< P_W < P_I$, and by Equation (8.77) when $P_W = P_I$, over the usual range 50 m $\leq h \leq$ 100 m for calving ice walls. A real problem with using Equations (8.114) and (8.125) is knowing what value of A to use, because A for temperate ice is used to determine B. Creep experiments conducted on ice samples taken from a calving ice wall standing on dry land on Deception Island, Antarctica, using the mean annual air temperature on Deception Island (– 3°C), gave $n = 2.07$ for $A = 86.3$ k Pa $a^{1/n}$ between natural shear bands and $A = 25.2$ k Pa $a^{1/n}$ within a shear band (Hughes and Nakagawa, 1989). These are equivalent to $A = 0.91$ bar $a^{1/n}$ and $A = 0.40$ bar $a^{1/n}$ for $n = 3$, respectively. Equation (8.125), as plotted in Figure 8.21, gives a maximum value of u_c / u_s in the middle range of P_W, with $u_c > u_s$ when u_s is in the low velocity range for ice streams and $u_c < u_s$ when u_s is in the high velocity range for ice streams. For any given h, the maximum in u_c / u_s increases nearly linearly as u_s decreases; for example, from $u_c / u_s = 0.4$ when $u_s = 32{,}000$ m/a to $u_c / u_s = 40$ when $u_s = 500$ m/a for $h = 100$ m. Antarctic ice streams buttressed by ice shelves are in the 500 m/a range. Unbuttressed Ant-arctic ice streams are in the 2000 m/a range. Unbuttressed Greenland ice streams having high surface melting rates are in the 8000 m/a range. Peak surge velocities of temperate glaciers are in the 32,000 m/a range. The most significant result is that the calving rate surpasses the sliding velocity for the most common water depths at marine and lacustrine ice-sheet margins, with the calving rate in the range 13 km/a to 20 km/a for all sliding velocities, whereas the sliding velocity surpasses the calving rate for both shallower and deeper water, especially when the sliding velocity is high. Therefore, when sliding velocities are high at the beginning of an ice-stream life cycle, it is unlikely that calving can prevent ice-sheet advance by stream flow; but when sliding velocities are low at the end of a life cycle, it is unlikely that sliding can prevent ice-sheet retreat by calving. Water depth $h_W = P_W / \rho_W \, g_z$ at the calving ice wall depends on the depth of a glacioisostatically depressed bed below contemporary sea level at marine margins and below the lowest spillway at lacustrine margins. Equation (5.31) gives the glacioisostatically depressed bed depth or elevation with respect to contemporary sea level, with r given by Equation (5.25) for an advancing ice margin and by Equation (5.26) for a retreating ice margin.

Equation (8.113) predicts that x_c / h_I decreases as P_W / P_I increases, so that calving slabs become thinnest as ice becomes afloat. This is seen in Figure 8.22, where a tabular iceberg calving from Miles Glacier consists of a sequence of thin ice slabs that are also calving, indicating that slab calving has caused the calving front to retreat into deeper water which shows Miles Glacier that can float and release tabular icebergs. The transition from slab calving given by Equation (8.125) to tabular calving given by Equation (8.77) seems to exhibit this behavior. Recent workshops on the calving dy–namics of glaciers have been conducted by Reeh (1994) and by Van der Veen (1997).

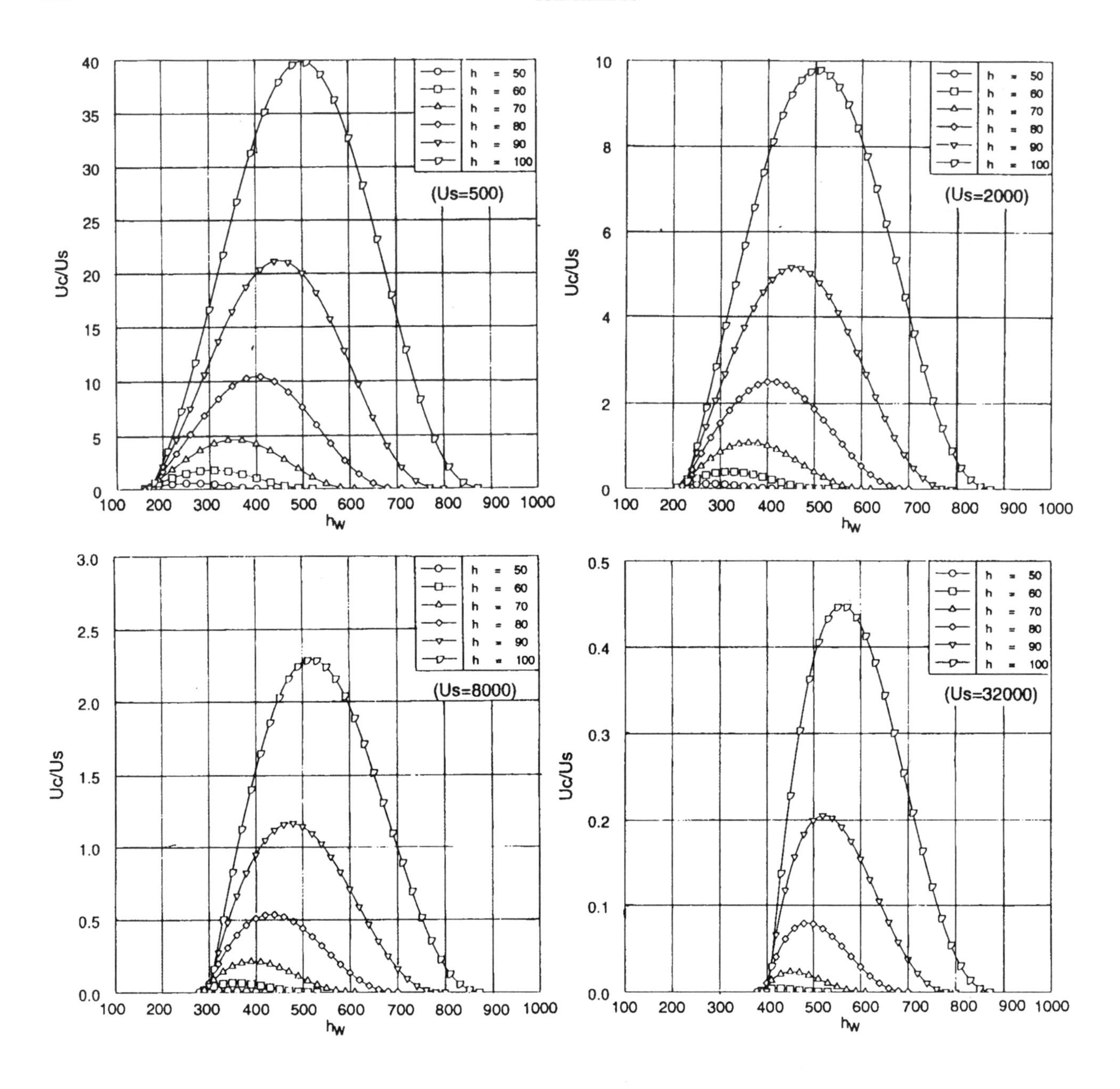

Figure 8.21: The dependence of calving rate u_c (m/a), sliding velocity u_s (m/a), basal water depth h_W (m) P_W (bars), and ice height h (m) above water at the calving termini of ice streams according to Equation (8.125). Note that in the mid range of P_W, equivalent to water depths of order 500 m, $u_c > u_s$ when u_s is low and $u_c < u_s$ when u_s is high. The peak u_c is for h = 100 m and decreases from 20 km/a to 13 km/a as u_s increases from 0.5 km/a to 32 km/a. Plot by James L. Fastook.

Figure 8.22: Calving from Miles Glacier, Alaska (U.S. Geological Survey photograph 87V2-058, supplied by Austin Post). The righthand part of Miles Glacier shows an apparent transition from slab calving to tabular calving as the glacier becomes afloat.

9

THE LAST GLACIATION

A Diagnostic Approach to Modeling the Last Glaciation

Traditionally, climate modelers needed ice-sheet modelers to provide three essential boundary conditions for climate models: (1) the areal extent of ice sheets, with surface accumulation and ablation zones denoted, in order to specify global surface albedos that reflected insolation, (2) the volume of ice sheets, with marine and terrestrial portions denoted, in order to compute the global ocean-atmosphere heat exchange for the reduced sea-surface area as increasing ice volume lowered global sea level, and (3) the elevation of ice sheets, with ice-surface topography denoted, in order to compute the surface pressure and wind fields for atmospheric circulation. Albedo is higher in accumulation zones than in ablation zones of ice sheets because granular snow has a higher albedo than bubbly ice. Marine ice sheets lower sea level less than terrestrial ice sheets because part of a marine ice sheet lies below sea level. Ice surface topography directs surface winds over ice sheets and focuses katabatic winds toward the downdrawn slopes of ice streams. These features of ice sheets allow ice sheets to impose boundary conditions on climate models that are largely passive, thereby encouraging the belief that ice sheets are merely passive components of Earth's climate machine. However, ice sheets may initiate climatic change and glacial geology may reveal the history of ice sheets that accompanies climatic change. This is demonstrated by a diagnostic approach to modeling the last glaciation cycle.

The view that ice sheets are passive components of Earth's climate machine is changing as a new paradigm of ice-sheet dynamics emerges. The outline of the emerging paradigm took shape during the Chapman Conference on Fast Glacier Flow, sponsored in 1986 by the American Geophysical Union. Proceedings of this Chapman Conference were published in 1987 in volume 32 (B9) of the *Journal of Geophysical Research*. Glaciologists presented field studies of three kinds of fast-moving glaciers: (1) surging mountain glaciers having regular surge cycles, (2) tidewater glaciers that enter calving bays, and (3) ice streams that drain present-day ice sheets. These studies showed that all three kinds of glaciers had episodes of fast flow that could begin and end abruptly and that were independent of forcing from other components of Earth's climatic machine. Moreover, the dynamics controlling fast flow in each kind of glacier seemed to be explainable by underlying physical processes common to all three kinds of glaciers. In particular, ice streams seem to have life cycles that are analogous to surge cycles of much smaller mountain glaciers but are much longer in duration, and marine ice streams ending as calving ice tongues or ice shelves have similarities with tidewater glaciers that calve into embayments called calving bays. Since the 1986 Chapman Conference, leading glaciologists have been studying these underlying physical processes for major ice streams draining present and former ice sheets. Internal mechanisms of rapid glaciation and deglaciation have been discovered that are largely independent of external climatic forcing and that may significantly regulate ocean and atmosphere circulation. In short, ice sheets are increasingly viewed as active, and perhaps even controlling, components of Earth's climate machine.

Inverse diagnostic ice sheet models

The most accurate models for computing the area, volume, and elevation of former ice sheets are inverse diagnostic models that use the glacial geologic record as direct model input to diagnose the size, shape, and flow of former ice sheets. This modeling strategy depends upon accurately mapping and dating the glacial geology. The necessary accuracy is generally available only for the last glacial maximum and for the last deglaciation. However, glacial geologists are beginning to map geomorphological landforms and glacial geological features that were produced prior to the last glacial maximum and that survived deglaciation. Interpretations of field evidence documenting the inception and growth stages of the last glaciation cycle include those by Kleman (1990, 1992), Kleman and Borgström (1990, 1994, 1996), Kleman and Stroeven (1997), Kleman and others (1997), Lagerbäck (1988a, 1988b), Lunqvist (1992), Fastook and

Holmund (1994) and Holmund and Fastook (1993, 1995) for the Scandinavian Ice Sheet and by Sugden and Watts (1977), Dredge and Thorleifson (1987), Vincent and Prest (1987), and Boulton and Clark (1990) and Boulton (1996) for the Laurentide Ice Sheet.

Dated glacial geology and associated geomorphology reveal the precise areal extent of former ice sheets during the last glacial maximum, and during the major stadials and interstadials during the last deglaciation, as well as the pattern and distribution of basal thermal zones beneath these ice sheets and flowlines of the ice sheet. Basal thermal zones include zones where the bed is frozen, thawed, freezing, and melting. Freezing and melting zones lie between frozen and thawed zones and consist of a mosaic of frozen and thawed patches. Frozen patches in a thawed matrix grade into thawed patches in a frozen matrix across a freezing zone, and thawed patches in a frozen matrix grade into frozen patches in a thawed matrix across a melting zone. As shown in Chapter 5, a thawed bed and thawed patches give less basal traction to resist ice flow than do a frozen bed and frozen patches. Therefore, an ice sheet has greater thickness and elevation above a frozen bed than above a thawed bed, with intermediate thicknesses and elevations above freezing and melting beds. Since glacial geology and related geomorphology allow basal thermal zones to be mapped accurately, ice-sheet models which use these field data directly as model input give the most accurate calculations of ice thickness and elevation as model output. Moreover, glacial geology and associated geomorphology include longitudinal and transverse lineations that give the direction of ice flow, including changing flow directions over time. All these data are primary model input to the initial-value finite-difference recursive equations for reconstructing flowline profiles that were derived in Chapter 5 for sheet flow and in Chapter 6 for stream flow. Numerical solutions of these equations constitute output in inverse diagnostic models for reconstructing former ice sheets.

Since they depend so heavily on dated glacial geology and related geomorphology, inverse diagnostic models are best employed to reconstruct former ice sheets at snapshots in time where these dated landforms are mapped accurately. The models are ill suited for continually simulating advance and retreat of former ice sheets during a glaciation cycle, and they generally do not incorporate the equations of heat flow in their solutions. Therefore, inverse diagnostic models are simple to construct but difficult to use, because labor saved by excluding time dependence and heat flow in the models must be expended in conducting field studies to map changing basal thermal zones and ice-flow directions. Nothing comes cheaply. Important milestones for the inverse diagnostic approach to ice-sheet modeling are the initial development and application of an inverse diagnostic flowline model to reconstruct global ice sheets from glacial geology at the last glacial maximum (Hughes, 1981), constructing flowlines as output rather than input in the models (Reeh, 1982; Fisher and others, 1985), allowing surface mass balance and converging or diverging flow to vary along flowlines in the models (Fastook, 1984; Hughes, 1986), incorporating deforming beds and heat flow into the models (Clark and others, 1996) and making the models two-dimensional and time-dependent (Fastook and Chapman, 1989).

Inverse diagnostic models appeal to glacial geologists because these models can be reduced to simplified forms that need only a hand calculator for computing accurate flowline profiles in the field (e.g., Hughes, 1995). Glacial geologists are often interested in former ice sheets at a snapshot in time, and they specialize in obtaining the geological field data that are input to inverse diagnostic models. Diagnosis of former glacial conditions is a primary goal of glacial geologists. Glaciologists, on the other hand, are primarily interested in the dynamics of ice sheets, both their response to external forcing and their capability to generate internal forcing that can change external conditions. These dynamics allow a prognosis of future behavior for present-day ice sheets. Therefore, glaciologists tend to favor forward prognostic ice-sheet models.

Forward prognostic ice-sheet models

The most accurate models for simulating the internal capacity of ice sheets to act independently from external climatic forcing, and perhaps to initiate climatic change, are forward prognostic models. These models use the time-dependent continuity equation and the equations of heat flow to predict the distribution of basal thermal zones and surface flowlines over time from the distribution of surface temperature and accumulation or ablation rates over time, perhaps as specified by climate models. In forward prognostic models, glacial geology and associated geomorphology are used primarily as a reliability check of model output at specific times during a glaciation cycle. Hence, great effort is expended in developing the model in order to reduce the effort needed to apply the model without having to map, date, and interpret field geology showing the history, extent, flowlines, and basal thermal zones of former ice sheets. When a known climate forcing is assumed, such as Milankovitch insolation variations, forward prognostic models give as output a continuous simulation of the advance and retreat of ice sheets over time. Since insolation variations can be computed into the future, future inception, growth, retreat, and collapse of ice sheets can be simulated by forward prognostic models, if Milankovitch forcing drives climate change. Unlike inverse diagnostic models, however, forward prognostic models are ill suited for replicating the precise ice extent, flow pattern, and distribution of basal thermal zones at a given snapshot in time for which these features have been mapped and dated.

Since forward prognostic models incorporate more physics than do inverse diagnostic models, they are best suited to show how ice-sheet dynamics may exert some control on climate dynamics. In particular, since ice streams drain at least 90 percent of present-day ice sheets and probably of former and future ice sheets, and ice streams are the most dynamic and variable components of an ice sheet, it is likely that ice sheets exert some control over climate dynamics primarily through the dynamics of their ice streams. Understanding and simulating ice-stream dynamics, therefore, is the primary goal of forward prognostic ice-sheet models.

Forward prognostic models cannot simulate a glaciation cycle precisely because (1) the critical basal thermal conditions are computed indirectly from remote ice surface conditions, (2) surface thermal conditions of temperature and mass balance change over time in response to a complex interplay of external climatic forcing mechanisms, (3) rates of basal heat generation and transport change as a frozen bed becomes thawed and vice versa, and (4) surface and basal boundary conditions change during the substantial time needed to transport basal heat to the surface. Despite these disadvantages, forward prognostic models can (1) produce abrupt changes arising from the temporal instability mechanisms inherent in the dynamics of ice sheets, (2) generate life cycles for ice streams as temporal variations in heat flow cause alternating melting and freezing conditions beneath the ice sheet, (3) construct surface flowlines that change as an ice sheet advances and retreats and as ice streams pass through successive life cycles, and (4) simulate the migration and proliferation of domes and saddles along interior ice divides and the creation of local ice domes between ice streams as ice streams pull ice out of an ice sheet.

The full capabilities of forward prognostic models have not yet been attained, especially regarding the critical role of ice streams in the dynamics of ice sheets. Important milestones toward these goals were achieved by Budd and Smith (1981) for simulating a full glaciation cycle, by Budd and others (1984) for modeling the role of ice streams in controlling the stability of the marine West Antarctic Ice Sheet, by Peltier (1987), for incorporating a rheological model for glacial isostasy, by Lingle and Brown (1987) for including isostasy and deformable bed hydrology in ice-stream dynamics, by Clarke (1987a, 1987b), Alley (1989a, 1989b) and Kamb (1991) for including creep of a deformable bed in ice-stream dynamics, by MacAyeal (1989) for linking sheet flow, stream flow, and shelf flow with a deforming bed under an ice stream, by Lindstrom (1990) for combining heat and mass transport in modeling a full glaciation cycle, by Huybrechts (1990) for modeling both mass balance and ice dynamics in three dimensions, and by Fastook (1993) for linking surface topography and velocity to the distribution of internal creep and basal sliding in the Antarctic Ice Sheet.

Diagnostic-prognostic models

The ideal strategy for modeling the last glaciation cycle makes use of both inverse diagnostic models and forward prognostic models in a way that maximizes the strengths and minimizes the weaknesses in each kind of model. A main strength of inverse diagnostic models is their ability to produce the most accurate reconstructions of ice sheets at snapshots in time for which the glacial geology is mapped and dated, and a main weakness is their inability to reconstruct with accuracy a full glaciation cycle because the changing glacial geology needed to control the models is unknown. A main strength of forward prognostic models is their ability to use all known physical processes to simulate the continuous advance and retreat of ice sheets for a full glaciation cycle, and a main weakness is their inability to specify the precise size and behavior of an ice sheet at particular times during the cycle.

Inverse diagnostic models place ice dynamics in the service of glacial geology in order to make the most accurate reconstructions of ice sheets at a particular time during a glaciation cycle. Forward prognostic models place glacial geology in the service of ice dynamics in order to reproduce the most accurate history of advance and retreat of ice sheets during a glaciation cycle. Diagnostic-prognostic models place ice dynamics and glacial geology in partnership, in order to understand how ice dynamics produces glacial geology which can alter ice dynamics, with the goal of both reconstructing ice sheets and reproducing their advance and retreat as accurately as possible during a glaciation cycle. This goal is ambitious and largely unrealized, but it is the prize that will inspire the most creative ice-sheet modelers.

Diagnostic-prognostic models maximize advantages and minimize disadvantages of diagnostic and prognostic models. Fastook and Hughes (1991) combined the advantages of using glacial geology and the time-dependent continuity equation to produce a two-dimensional finite-element model that simulates a full glaciation cycle for ice sheets. The model includes ice-stream and ice-shelf dynamics in parameterized form. Tushingham and Peltier (1991) combined the advantages of using the global glacioisostatic record of changing relative sea level and the time-dependent continuity equation to produce a two-dimensional finite-element model that simulates a full glaciation cycle for ice sheets coupled to Earth's mantle rheology. These two models are the first milestones in diagnostic-prognostic modeling of ice sheets. The models have been applied to the last glaciation cycle (Fastook, 1990, 1993; deBlonde and others, 1992). Future diagnostic-prognostic models need to incorporate the equations of heat flow to become three-dimensional, to combine ice-stream dynamics with glacioisostatic dynamics beneath ice streams to produce time-dependent abrupt changes in ice-sheet dynamics, and to include the basal erosion and deposition processes that produce glacial geology. MacAyeal (1993a, 1993b) and Verbitsky and Saltz-man (1994, 1995) have pioneered work on these models.

Modeling Stages of the Last Glaciation Cycle

This book presents the inverse diagnostic approach to modeling former ice sheets. Earlier chapters derived the equations needed in this approach, and this chapter applies the approach to reconstructing Northern Hemisphere ice sheets at critical stages during the last glaciation cycle. A rich and growing literature illustrates the capabilities of forward prognostic models, and nearly all glaciological modeling research is geared toward improving these models. In contrast, no comparable body of literature and research activity is devoted to developing and applying the advantages of inverse diagnostic models. This book is an attempt to promote research geared toward improving and applying these models. Advances in understanding how ice sheets produce glacial geology and advances in mapping and dating the glacial geological record are needed to develop the inverse diagnostic modeling approach. It is therefore an approach that uniquely combines the expertise of both glaciologists and glacial geologists. In this chapter, an accurate but greatly simplified version of the inverse diagnostic model developed in earlier chapters will be presented and used to show glacial geologists how the model can take advantage of their field data to reconstruct ice sheets at critical stages of the last glaciation cycle.

External and internal forcing mechanisms allow a Northern Hemisphere glaciation cycle to be treated as consisting of five hypothetical stages (see Chapter 1).

1. Initiation of the glaciation cycle is assumed to result from a rapid, widespread albedo increase in the Northern Hemisphere caused by thickening and grounding of sea ice on broad Arctic continental shelves and from expansion of snowfields on Arctic plateaus, the latter in response to regional lowering of snowlines during cooling periods of the Milankovitch insolation cycles. Rapid lowering of sea level began when grounded sea ice sealed the Arctic coastline, so all rivers draining into the Arctic became dammed by ice. This produced a vast "White Hole" in the Northern Hemisphere, from which no precipitation could escape.

2. Advance of Northern Hemisphere ice sheets is assumed to result from a positive mass balance within the White Hole. Once sea ice grounded on Arctic continental shelves, continued surface precipitation caused grounded sea ice to dome, thereby becoming marine ice sheets that transgressed onto Arctic coasts and moved inland to eventually coalesce with terrestrial ice sheets that grew from precipitation over plateau snowfields. Landward advance of both marine and terrestrial ice was over a frozen bed between ice-dammed lakes and over a thawed bed beneath ice-dammed lakes. Reduced basal coupling on lake beds allowed ice streams to advance up the river valleys of ice-dammed lakes. Seaward advance followed marine ice-grounding lines, especially in deep water.

3. The glaciation cycle is assumed to reach maturity when a stable core developed in the central portion of ice sheets. This core roughly corresponded to crystalline shields for terrestrial ice, such as the Canadian, Scandinavian, and Angaran shields, and to shallow polar or subpolar embayments and seas for marine ice, such as Hudson Bay, Foxe Basin, the Gulf of Bothnia, and the Barents, Kara, Laptev, East Siberian, and Chukchi Seas. Stable sheet flow in the core areas became unstable stream flow beyond core areas, with transient marine ice streams occupying straits and inter-island channels on the outer continental shelves and transient terrestrial ice streams occupying linear lakes and river valleys in continental interiors. For all cases, stream flow began when ice-cemented sediments thawed and ended when these sediments refroze.

4. Retreat of Northern Hemisphere ice sheets is assumed to have resulted from a negative mass balance in the outer peripheral zone of ice streams that disintegrated the ice shelves buttressing marine ice streams and melted the ice lobes buttressing terrestrial ice streams. Disintegrating ice shelves and melting ice lobes were a response to rising snowlines during warming periods of the Milankovitch insolation cycles. Since ice shelves, and ice lobes to a lesser degree, are essentially flat, a small increase in snowline elevation can convert most or all of the ice-shelf or ice-lobe surface from an accumulation zone to an ablation zone. Ice shelves generally occupied high latitudes, so they disintegrated in response to the warming hemicycle of Earth's 41,000-year cycle of axial tilt. Ice lobes generally occupied mid-latitudes, so they melted in response to the warming hemicycle of Earth's 23,000-year cycle of axial precession. Warming in both high and mid-atitudes overlapped significantly twice in a 100,000-year cycle of Pleistocene glaciation. These were all times when the unstable ice-sheet periphery could retreat to the stable ice-sheet core.

5. Termination of a glaciation cycle occurred when Northern Hemisphere ice sheets collapsed over their core areas. Collapse could have been triggered by continuing isostatic depression beneath the core areas, thereby increasing the depth and extent of water around the perimeter of the core areas. This would have exposed core areas to the unstable, irreversible deglaciation mechanisms inherent in floating ice shelves (Weertman, 1974; Thomas, 1977; Thomas and Bentley, 1978; Stuiver and others, 1981), and at calving ice walls (see Chapters 7 and 8). Collapse could also have been triggered along terrestrial margins if flow converging on terrestrial ice streams converted ice ridges between ice streams into local ice domes. A small increase in snowline elevation could then have given these domes a negative mass balance that made them shrink rapidly and disappear (Weertman, 1961; Hughes, 1987; Hughes, 1992a). Ice streams lying between former ice ridges would then have become ice lobes exposed to rapid surface melting that, in turn, unbuttressed ice streams so they could move faster and pull out more interior ice. This produces new local domes between ice streams and allows mass wastage to propagate irreversibly into the core area of the ice sheet.

These five stages of the last glaciation cycle allow irregular intervals of abrupt change to punctuate the more regular and gradual periodicities of the Milankovitch insolation cycle. Although Milankovitch forcing is thought to be the "pacemaker" for Pleistocene glaciation cycles (Hays and others, 1976; Imbrie and others, 1984; Imbrie, 1985), it does not account for abrupt forcing that is superimposed on gradual insolation variations (Schnitker, 1979; Broecker and Denton, 1989). Moreover, abrupt forcing produced records of climatic change in ice cores (Paterson and Hammer, 1987) and ocean cores (Heinrich, 1988) that do not correlate well with Milankovitch insolation forcing. In fact, as shown in Figure 1.1, the spikiness of abrupt change is so pronounced and so irregular that it all but masks the more subdued and regular Milankovitch forcing (Johnsen and others, 1992).

Because of this, inception of a glaciation cycle may be as abrupt as its termination.

An ice-sheet model, whether of the inverse diagnostic or forward prognostic variety, must be capable of responding to both external and internal forcing and of simulating both regular and irregular responses to forcing. The only known external forcing that is regular enough to be quantified mathematically is the Milankovitch insolation variation (Berger, 1978; Berger and others, 1992). Other possible external forcings, such as variations in atmospheric carbon dioxide, can be used to simulate a glaciation cycle (Lindstrom and MacAyeal, 1989; Lindstrom, 1990; Verbitsky, 1992), but they are not understood well enough to be quantified mathematically. Whatever forcing is prescribed, however, MacAyeal (1993a, 1993b) and Verbitsky and Saltzman (1994, 1995) showed that ice-sheet models are capable of producing irregular and abrupt responses. However, the responses cannot be predicted mathematically by the models with any degree of confidence because the physics of the internal mechanisms is poorly understood and subject to too many variables. This limits their utility to diagnostic, rather than prognostic, studies.

With these limitations in mind, a modeling strategy employing an inverse diagnostic model will be illustrated for reconstructing Northern Hemisphere ice sheets for each of the five stages of the last glaciation cycle. These stages are correlated with changing mass balance over ice sheets during the glaciation cycle. Mass-balance changes are correlated with changing equilibrium-line altitudes on glaciers found in present-day climate regimes.

Mass Balance and Climate Regimes

As ice moves from the ice divides to the ice margins of an ice sheet, basal traction beneath sheet flow that is pulled into ice streams provides the primary resistance to the gravitational pulling force in ice streams. Without basal traction, an ice sheet would be pulled down until it became as flat and thin as a floating ice shelf. This pulling down or downdraw by ice streams converts gravitational potential energy, stored as ice elevation above the flotation thickness of ice, into kinetic energy of motion, particularly fast motion during the life cycles of ice streams when basal traction is reduced greatly. Gravitational potential energy is restored by the mass balance and increases when accumulation over the ice-sheet interior equals or exceeds ablation around the perimeter of the ice sheet. Specifying the mass balance correctly during a glaciation cycle is therefore a primary goal of ice-sheet modeling. Accumulation and ablation zones on an ice sheet are separated by the equilibrium line, which moves toward the ice divide or the ice margin as snowlines rise or fall on the time scale of climatic change, not the much shorter seasonal time scale. Precipitation falls as snow above snowlines and as rain below snowlines. Snowlines intersecting an ice sheet from different directions define a surface across which precipitation changes from snow to rain, as shown in Figure 2.5. The only climatic time scale that can be quantified as known changing input to ice-sheet models is the time scale of Milankovitch insolation variations (Milankovitch, 1941), particularly the 41,000-year and 23,000-year cycles of Earth's axial tilt and precession (Berger, 1978), as illustrated in Figure 2.8.

At 75°N latitude, where Northern Hemisphere marine ice sheets originated on polar continental shelves, insolation is dominated by the tilt cycle. At 50°N latitude, which was an average limit of Northern Hemisphere terrestrial ice sheets at Pleistocene glacial maxima, insolation is dominated by the precession cycle. Although the insolation at any latitude at any time can be calculated (Berger, 1978), the equations are not simple. The glacial mass balance in these latitudes depends on many variables in complex ways that have been circumvented by computing the mass balance from the energy balance (Gallée and others, 1991; Huybrechts and others, 1991; deBlonde and others, 1992). However, the simplest assumption is that mass balance depends on snowline elevation, which depends on insolation. Figure 2.8 shows that insolation cycles are approximately sinusoidal, with a period of 41,000 years at 75°N and a period of 23,000 years at 50°N. An adequate assumption in using inverse diagnostic ice-sheet models is that insolation variations are purely sinusoidal, and therefore the snowline elevation and the mass balance are also sinusoidal.

These two simplifications are justified in inverse diagnostic models because ice-sheet elevation h is computed directly from basal shear stress τ_o, see Equation (5.56), and when accumulation rate a is constant, $\tau_o \propto a^{1/4}$ for n = 3 and a frozen bed, see Equation (5.95), and $\tau_o \propto a^{2/5}$ for m = 2 and a thawed bed, see Equation (5.96). Therefore, h is insensitive to a. Figure 2.9 shows intersections between the snowline and the surface of a Northern Hemisphere ice sheet in latitudinal cross-section. The range of intersections as snowline elevation and slope change defines the sweep of the equilibrium line during a glaciation cycle. Changes in snowline elevation and slope are caused by a 41,000-year cycle of snowline elevation at the north pole and a 23,000-year cycle of snowline elevation at the equator. Both cycles are sinusoidal.

Assuming that Milankovitch insolation cycles control the snowline, mass balance can be quantified in the one-dimensional flowline finite-difference model (FDM) of ice-sheet dynamics presented in Chapters 5 and 6 by allowing snowline altitude to vary with latitude as follows:

$$H = H_P + R\, S_R \qquad (9.1)$$

where H is snowline height at radial distance R from the north pole, H_P is H above or below sea level at the pole, and S_R is snowline slope measured along R. For snowline variations controlled by insolation cycles, ΔH_P is the elevation change at the pole during the 41,000-year period t_p of axial tilt, and ΔH_e is the elevation change at the equator during the 23,000-year period t_e of axial precession. Taking $R_e = 10^7$ m as the latitude distance from the north pole to the equator, and letting δH_p and δH_e be the fractions of ΔH_p and ΔH_e at time t after the start of a tilt cycle (which initiates a glaciation cycle):

$$S_R = (H_e - H_p)/R_e = (H_e^\circ - \delta H_e)/R_e - (H_p^\circ - \delta H_p)/R_e$$
$$= (H_e^\circ/R_e) - (\Delta H_e/R_e)\,sin^2(\pi t/t_e) - (H_p^\circ/R_e)$$
$$+ (\Delta H_p/R_e)\,sin^2(\pi t/t_p) \quad (9.2)$$

where H_e° and H_p° are H_e and H_p at $t = 0$. The rate of accumulation or ablation a at distance x from the ice-sheet margin, measured along a flowline, is given by Equation (5.35):

$$a = a_o'\,exp\,(-\lambda'\,x^k) + a_o''\,exp\,(-\lambda''\,x^k) \quad (9.3)$$

where a_o' is the (negative) ablation rate and a_o'' is the (positive) accumulation rate at the ice margin, λ' and λ'' are damping factors such that a peak in a occurs at some point between the equilibrium line and the ice divide, as observed for present-day ice sheets, and k is a constant such that $k \approx 1.0$ for the convex surface of sheet flow and $k \approx 1.5$ for the concave surface of stream flow. The variation of a with ice elevation is required in order to relate Equation (9.3) to the equilibrium-line altitude (ELA) method for specifying mass balance (Pelto and others, 1990). Using Equation (5.5) for ice elevation h above ice-margin elevation h_0, keeping basal shear stress τ_o, ice density ρ_I, and gravity acceleration g_z constant:

$$h = h_0 + (2\,\tau_o/\rho_I\,g_z)^{1/2}\,x^{1/2} \quad (9.4)$$

Substituting this expression for x in Equation (9.3):

$$a = a_o'\,exp\,[-\Gamma'\,(h - h_o)^{2k}] + a_o''\,exp\,[-\Gamma''\,(h - h_o)^{2k}] \quad (9.5)$$

where Γ' and Γ'' are adjusted until Equation (9.5) gives a least-squares fit to mass-balance data for present-day glaciers and ice sheets. By definition:

$$\Gamma' = \lambda'\,(\rho_I\,g_z/2\,\tau_o)^k \quad (9.6a)$$

$$\Gamma'' = \lambda''\,(\rho_I\,g_z/2\,\tau_o)^k \quad (9.6b)$$

Take h_E as the height of the equilibrium line, formed by connecting points where snowlines intersect flowlines on an ice sheet, and located at distance x that gives $a = 0$ in Equation (9.3) for a given flowline. The mass-balance curve defined by Equation (9.5) can then be displaced to higher or lower altitudes to ensure that $h_E = H$ at $a = 0$ as the snowline elevation changes through time. This is accomplished by subtracting δH from $h - h_0$ in Equation (9.5), where δH is the change in H in time t, so that :

$$a = a_o'\,exp\,[-\Gamma'\,(h - h_0 - \delta H)^{2k}]$$
$$+ a_o''\,exp\,[-\Gamma''\,(h - h_0 - \delta H)^{2k}] \quad (9.7)$$

Figure 9.1 shows how δH affects a in Equation (9.7). Note that a_o' and a_o'' change as δH changes in Figure 9.1 over the range $0 \leq \delta H \leq \Delta H$ during insolation cycles t_e and t_p.

The effect of Milankovitch insolation variations on rates of advance and retreat of landward margins of Northern Hemisphere ice sheets can be calculated from Equations (9.1), (9.2), (9.4), and (9.7). After time t since the coming

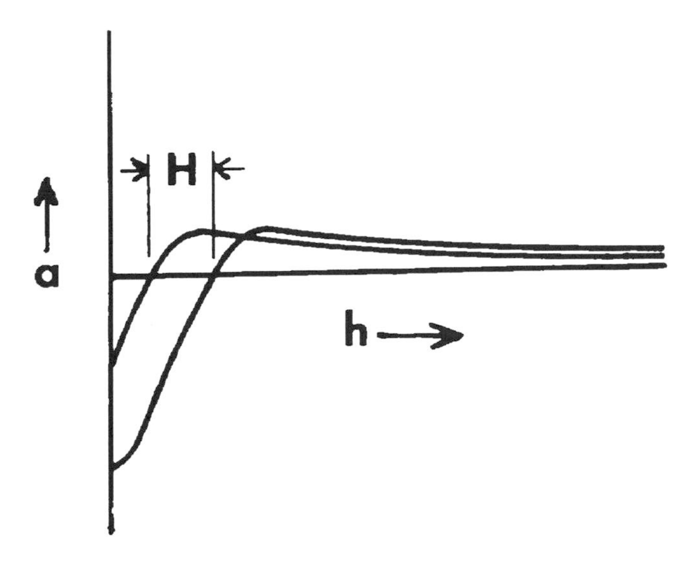

Figure 9.1: Migration of the equilibrium line on an ice sheet in response to orbital variations in insolation. The snowline slope varies over the 23,000-yr cycle of axial precession and over the 41,000-yr cycle of axial tilt. Maximum and minimum snowline elevations occur when tilt and precession are in phase. Maximum and minimum snowline slopes occur when tilt and precession are out of phase. Compare with Figure 2.8.

glaciation cycle begins, present-day snowline elevations H lower amounts dH. From Equation (9.1):

$$\delta H = \delta H_p + R\,\delta S_R \quad (9.8)$$

where δH_P is the difference between H_p at $t = 0$ and H_p at $t > 0$:

$$\delta H_p = \Delta H_p\,sin^2(\pi t/t_p) \quad (9.9)$$

and δS_R is the difference between S_R at $t = 0$ and S_R at $t > 0$ in Equation (9.2):

$$\delta S_R = (\Delta H_e/R_e)\,sin^2(\pi t/t_e) - (\Delta H_p/R_e)\,sin^2(\pi t/t_p) \quad (9.10)$$

As in Equation (9.2), ΔH_p and ΔH_e are the maximum changes in snowline elevation H_p and H_e at the pole and the equator in time periods $t_p = 41{,}000$ a and $t_e = 23{,}000$ a for Earth's axial tilt and axial precession cycles, respectively. Equation (9.4) gives ice flowline elevations h at distances x in from the ice-sheet margin for the first Δx step in Equation (5.56), and $h - h_0 = 0$ at $x = 0$. Setting $h - h_0 = 0$ at $x = 0$ in Equation (9.7) and using $k = 1$ for sheet flow, the change δa_0 in net ablation or accumulation rate a_0 at the ice margin for change δH in snowline elevation H is:

$$\delta a_0 = (a_0' + a_0'') - (a_0'\,exp\,[-\Gamma'\,(\delta H)^2] + a_0''\,exp\,[-\Gamma''\,(\delta H)^2])$$
$$= a_0'\,(1 - exp\,[-\Gamma'\,(\delta H)^2]\,) + a_0''(1 - exp\,[-\Gamma'\,(\delta H)^2]) \quad (9.11)$$

Early in the glaciation cycle, continental margins of the advancing ice sheets will occupy high latitudes so polar continental values of a_0', a_0'', Γ', and Γ'' should be used in Equation (9.11). As seen in Equations (9.8) through (9.10), $\delta H = 0$ when $t = 0$ at the beginning of a glaciation cycle when $\delta H_p = \delta S_R = 0$, and changes as H changes over time t. Therefore, $\delta a_0 = 0$ at $t = 0$ and changes with t in Equation (9.11).

According to the Weertman (1961) ice-sheet stability criteria illustrated in Figures 2.6 through 2.8, the rate and extent of advance or retreat depends on both the elevation and the slope of the snowlines, which vary over time with latitudinal variations of Milankovitch insolation, as illustrated in Figure 9.2. Milankovitch insolation is insolation at the top of Earth's atmosphere, as distinct from insolation at Earth's surface after filtering by carbon dioxide, clouds, dust, and so on, in the atmosphere. Milankovitch insolation is the only external variable boundary condition that can so far be specified for any time and latitude during the former and future glaciations.

As an external control on a glaciation cycle, snowlines will be highest for a given latitude at times $t \approx i\, t_e \approx j\, t_p$, where i, j = 0, 1, 2, 3, etc., so that substituting Equations (9.9) and (9.10) into Equations (9.1) and (9.2) gives:

$$H = (H_p^\circ - \delta H_p) + R\,(S_R^\circ - \delta S_R) = H_p^\circ + (R/R_e)\,(H_e^\circ - H_p^\circ) \tag{9.12}$$

Equation (9.12) applies at time $t = 0$ during interglaciations when Milankovitch insolation is a maximum at all Northern Hemisphere latitudes, and at times $t > 0$ when this condition exists during glaciations, so ice-sheet retreat rates are fastest.

Snowlines are lowest for a given latitude during glaciations at times $t \approx (i/2)\, t_e \approx (j/2)\, t_p$, where i, j = 0, 1, 2, 3, etc., so that substituting Equations (9.9) and (9.10) into Equations (9.1) and (9.2) gives:

$$H = (H_p^\circ - \delta H_p) + R\,(S_R^\circ - \delta S_R)$$

$$= (H_p^\circ - \Delta H_p) + (R/R_e)\,[(H_e - \Delta H_e) - (H_p - \Delta H_p)] \tag{9.13}$$

Ice-sheet advance rates were fastest at times $t > 0$ during the last glaciation when Equation (9.13) is satisfied.

Snowlines have the lowest slope during glaciations at times $t \approx (i/2)\, t_e \approx (j/2)\, t_p$, where i, j = 0, 1, 2, 3, etc., so that Equations (9.2) and (9.10) give:

$$S_R = (H_e^\circ - H_p^\circ)/R_e - \delta S_R = (H_e^\circ - H_p^\circ)/R_e - \Delta H_e/R_e \tag{9.14}$$

Ice sheets tend toward unstable equilibrium at times $t > 0$ during glaciations when Equation (9.14) is satisfied, so the advance of low-latitude ice margins and the retreat of high-latitude ice margins are of greatest extent because unstable equilibrium tends to be irreversible.

Snowlines have the highest slope during glaciations at times $t \approx (i/2)\, t_e \approx (j/2)\, t_p$, where i, j = 0, 1, 2, 3, etc., so Equations (9.2) and (9.10) give:

$$S_R = (H_e^\circ - H_p^\circ)/R_e + \delta S_R = (H_e^\circ - H_p^\circ)/R_e + \Delta H_p/R_e \tag{9.15}$$

Ice sheets tend toward stable equilibrium at times $t > 0$ during glaciations when Equation (9.15) is satisfied, so the advance of high-latitude ice margins and the retreat of low-latitude ice margins are of least extent because stable equilibrium tends to be reversible.

Values of a_o', a_o'', Γ', Γ'', and k can be specified for pre

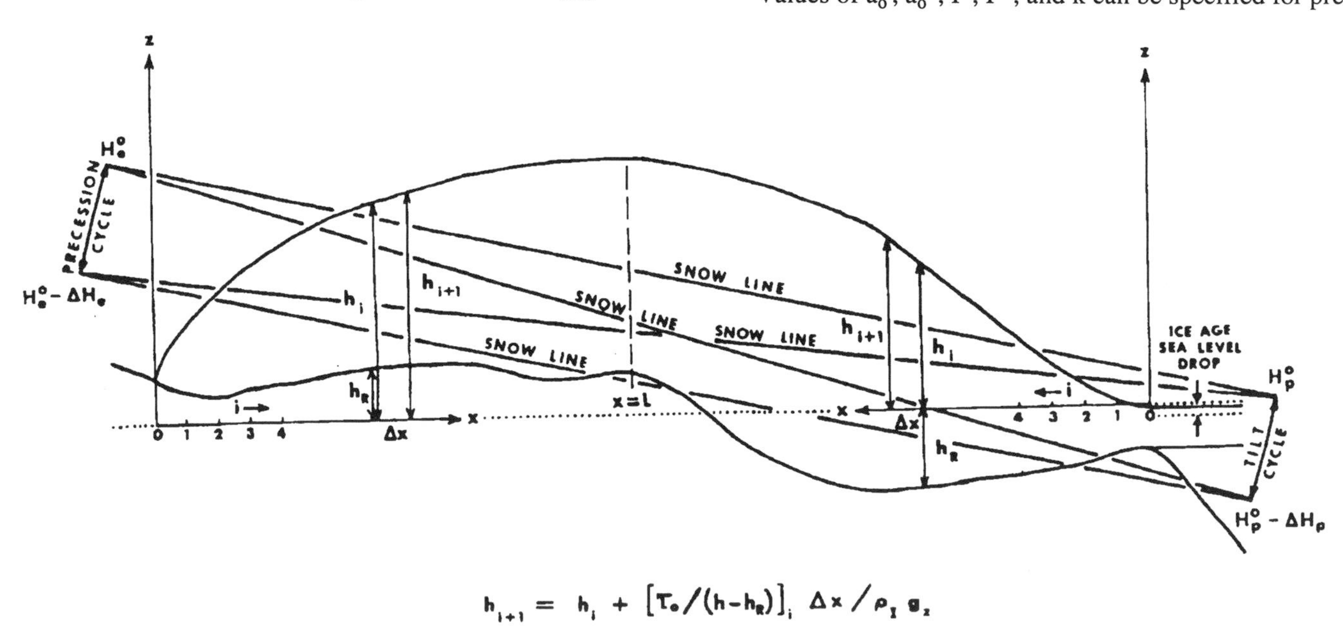

Figure 9.2: The effect of snowline elevation H on the mass-balance curve of net accumulation rates (positive a) or ablation rates (negative a) versus ice-sheet elevation h in Equation 7 (Hughes, 1992a, Figure 4).

Table 9.1: Climatic conditions associated with each of five climate regimes

Climate Regime	Winter temp.(C°)	Summer temp. (C°)	Annual precip. (m)
Polar continental	< –20	< 8	< 0.4
Polar mix	< –15	< 6	0.4 – 1.0
Sub-polar mix	–10 to –15	8–12	0.6 – 1.2
Sub-polar maritime	–6 to –12	6–10	> 1.2
Temperate maritime	0 to –6	10–16	> 1.5

Table 9.2: Linear mass-balance gradients

Climate Regime	Ablation Zone	Peak Accumula-tion Zone	Interior Accumula-tion Zone
Polar Continental (PC)			
*Range of accumulation (+) or ablation (–) in m/yr	0 to –3.5	0.3 to 0.6	0.033/100
*Slope of balance gradient in m/km	2.0/100	0.25/100	
Polar Mix (PM)			
*Range of accumulation (+) or ablation (–) in m/yr	–1.0 to –4.0	0.5 to 0.7	0.033/100
*Slope of balance gradient in m/km	3.0/100	.35/100	
Subpolar Continental (SC)			
*Range of accumulation (+) or ablation (–) in m/yr	–4.0 to –6.0	0.6 to 0.8	0.033/100
*Slope of balance gradient in m/km	4.0/100	0.5/100	
Subpolar Mix (SX)			
*Range of accumulation (+) or ablation (–) in m/yr	–5.0 to –8.0	0.8 to 1.2	0.033/100
*Slope of balance gradient in m/km	5.0/100	0.6/100	
Subpolar Maritime (SM)			
*Range of accumulation (+) or ablation (–) in m/yr	–9.0 to –12	1.2 to 3.0	0.033/100
*Slope of balance gradient in m/km	6.0/100	0.9/100	

sent-day glaciers and ice sheets that occupy distinct climate regimes. Five of the eight climate regimes that have been proposed by Pelto and others (1990), based on distinct ranges of summer temperature, winter temperature, and annual precipitation, are shown in Table 9.1. Pelto calls these regimes polar mix (PX), polar continental (PC), subpolar maritme (SM), subpolar mix (SX), subpolar continental (SC), temperate maritime (TM), temperate mix (TX), and temperate continental (TC) climates. Table 9.2 lists mass-balance gradient B(n) data for present-day glaciers and ice caps that can be grouped into the SM, TM, PX, PC, and SX climate regimes, as shown in Figure 9.3 (Pelto and others, 1990). Mauri Pelto and James Fastook listed and plotted mass-balance data for glaciers and ice caps in these climate regimes, with a least-square fit of Equation (9.5) to these data for $\Delta H = 0$ and $k = 1$, to obtain the values of a_0', a_0'', Γ', and Γ'' for each climate regime in Table 9.3. Using Equations (9.6), the corresponding values of λ' and λ'' for $\tau_0 = 100$ kPa are listed in Table 9.4. The resulting mass-balance curves obtained from Equation (9.3) with $k = 1$ are plotted in Figure 9.4.

The geographical distribution of climate regimes proposed by Mauri S. Pelto (Global Ice Sheet Modeling, 1989 Annual Report, prepared by the University of Maine for Battelle, Pacific Northwest Laboratories, unpublished) during a glaciation cycle is depicted in Figure 9.5. Changes in the pattern of these climate regimes during the last glaciation cycle are based on published results from general atmospheric circulation models and biological indicators of climate change, notably pollen and marine microfossils. Changes in the geographical distribution of these climate regimes are accompanied by changes in δH in Equation (9.7), with δH given by Equations (9.1) and (9.2) for $\Delta H = 0$ at $t = 0$ during the present-day interglacial and at the beginning of the last glaciation cycle, and $\delta H > 0$ at $t > 0$ during the glaciation cycle. In solving Equations (9.1) and (9.2), maximum and minimum values of H_e and H_p are obtained from Figure 9.6 for the PC, PX, SX, SM, and TM climate regimes of the North American Cordillera (Pelto, 1992). Extrapolating the ELA slopes for cirque glaciers in Figure 9.6 for present-day and glacial-maximum climate regimes in Figure 9.5 gives H_e at the equator and H_p at the north pole for the PC, PX, SX, SM, and TM climate regimes. Tangent lines to smooth curves through the data between 40°N and 70°N give H_e and H_p for these climate regimes today and during the last glacial maximum. These data allow H to be computed at any Northern Hemisphere latitude during a glaciation cycle, using Equations (9.1) and (9.2), which then allows a to be computed from Equation (9.7) for each climate regime during the glaciation cycle, with δH being H at $t = 0$ minus H at $t > 0$ and values of a_0', a_0'', Γ', and Γ'' for flowlines of ice sheets occupying these climate regimes. The variation of a with x along flowlines is then calculated from Equation (9.3) using values of λ' and λ'' obtained from Γ' and Γ'' in Equations (9.6). This method for computing the mass balance can be applied to snapshots in time during the glaciation cycle.

Modeling Stadial and Interstadial Ice Sheets

Abrupt climatic changes depicted by oxygen isotope variations in Figure 1.1 consist of Dansgaard–Oeschger events that change in a few centuries and last a few millennia and that occur in "sawtooth" clusters of irregular but prolonged cooling lasting 7000 to 15,000 years, called Bond cycles, separated by abrupt warming. Cold minima of Dansgaard–Oeschger events and Bond cycles are often associated with layers of ice-rafted sediments in the North Atlantic and, secondarily, in the North Pacific. A developing consensus links Dansgaard–Oeschger events to episodes of abrupt climatic cooling and warming, Bond cycles to episodes of irregular

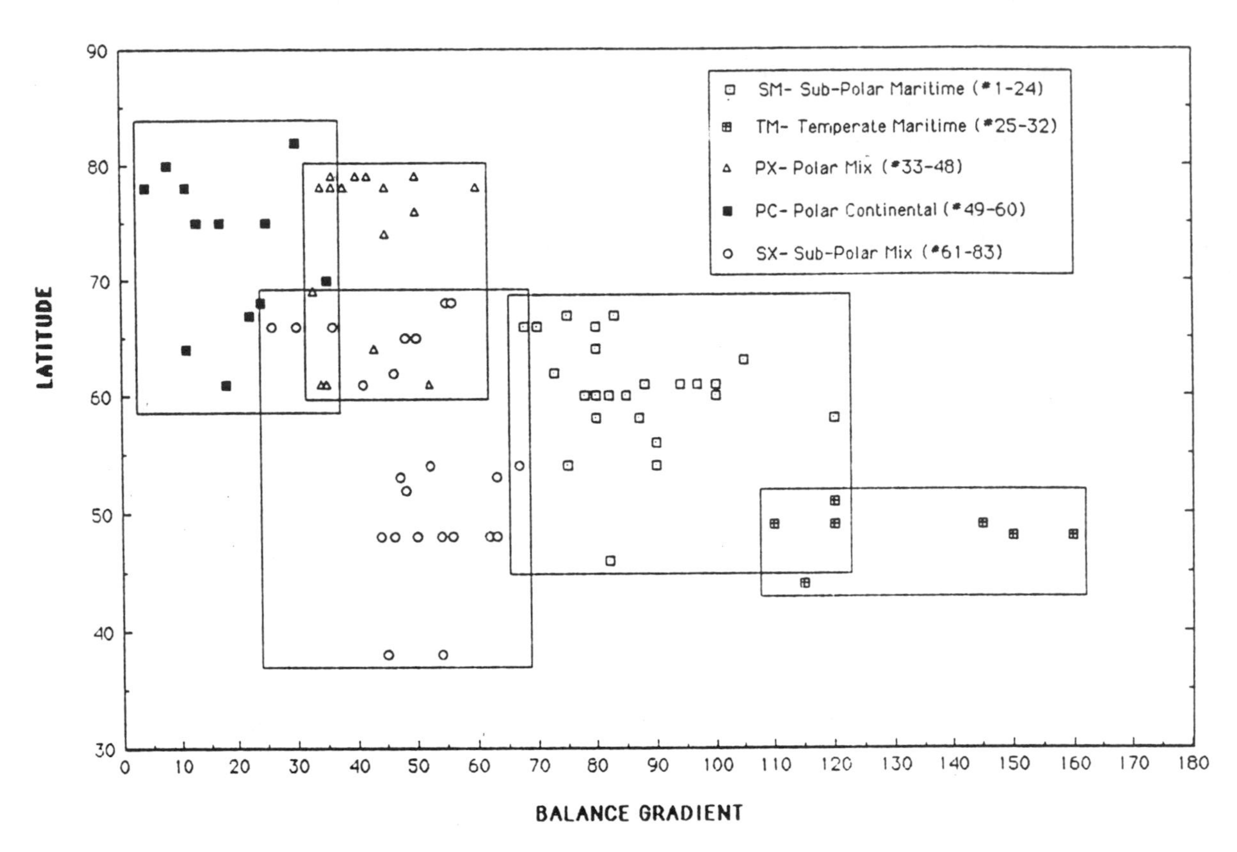

Figure 9.3: The balance gradient of present-day glaciers versus latitude. Five distinct populations result, indicative of five climatic settings for the glaciers Reproduced with permission of Pelto and others (1990).

Table 9.3: Mass-balance constants used in Equation (9.5) to construct the balance gradient for each climate zone: a_o' and a_o'' are the ablation and accumulation rates at the ice margin, respectively; Γ' and Γ" determine the changes in ablation and accumulation with elevation

Climate Regime	a_o'	a_o''	Γ'	Γ"	ELA
PC	−1.2936	0.2998	1.5023×10^{-6}	3.6619×10^{-8}	312
PX	−2.8490	0.8378	8.5345×10^{-6}	3.4053×10^{-8}	379
SM	−7.5766	2.5689	2.7733×10^{-6}	2.9162×10^{-7}	660
SX	−5.8563	0.8575	1.3228×10^{-6}	9.4618×0^{-8}	125[illegible]
TM	−11.6877	3.67407	1.0838×10^{-6}	3.6411×10^{-8}	105[illegible]

Table 9.4: Constants used to generate the mass balance curves for each flowline using Equation (9.3) when k = 1

Climate Regime	a_o' (cm/yr)	a_o'' (cm/yr)	λ' (km^{-1})	λ" (km^{-1})
Temperate Maritime (TM)	−500	300	0.006	0.0012
Subpolar Maritime (SM)	−350	200	0.007	0.0010
Subpolar Mix (SX)	−400	150	0.006	0.0009
Subpolar Continental (SC)	−450	150	0.003	0.0007
Polar Mix (PM)	−250	150	0.005	0.0008
Polar Continental (PC)	−100	80	0.004	0.0010

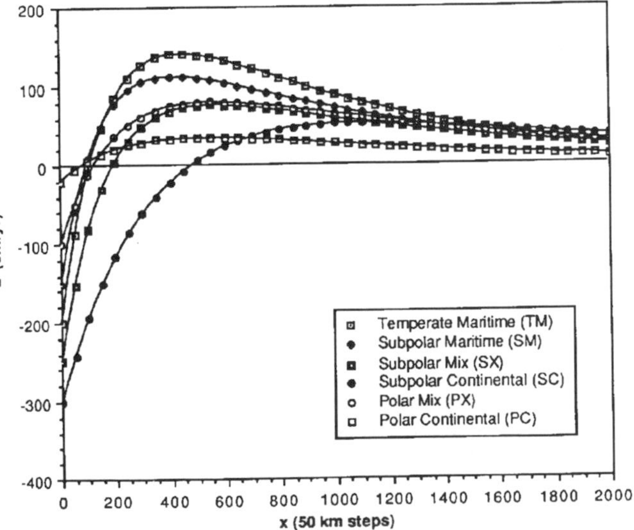

Figure 9.4: A plot of Equation (9.3) for k = 1 and the data in Table 9.4.

but prolonged lowering of sea level followed by short intervals of rapidly rising sea level, and ice-rafted sediments to outbursts of icebergs from disintegrating ice shelves and stagnating ice streams. In the emerging paradigm, stadials occur at the end of Bond cycles, after ice sheets thicken over

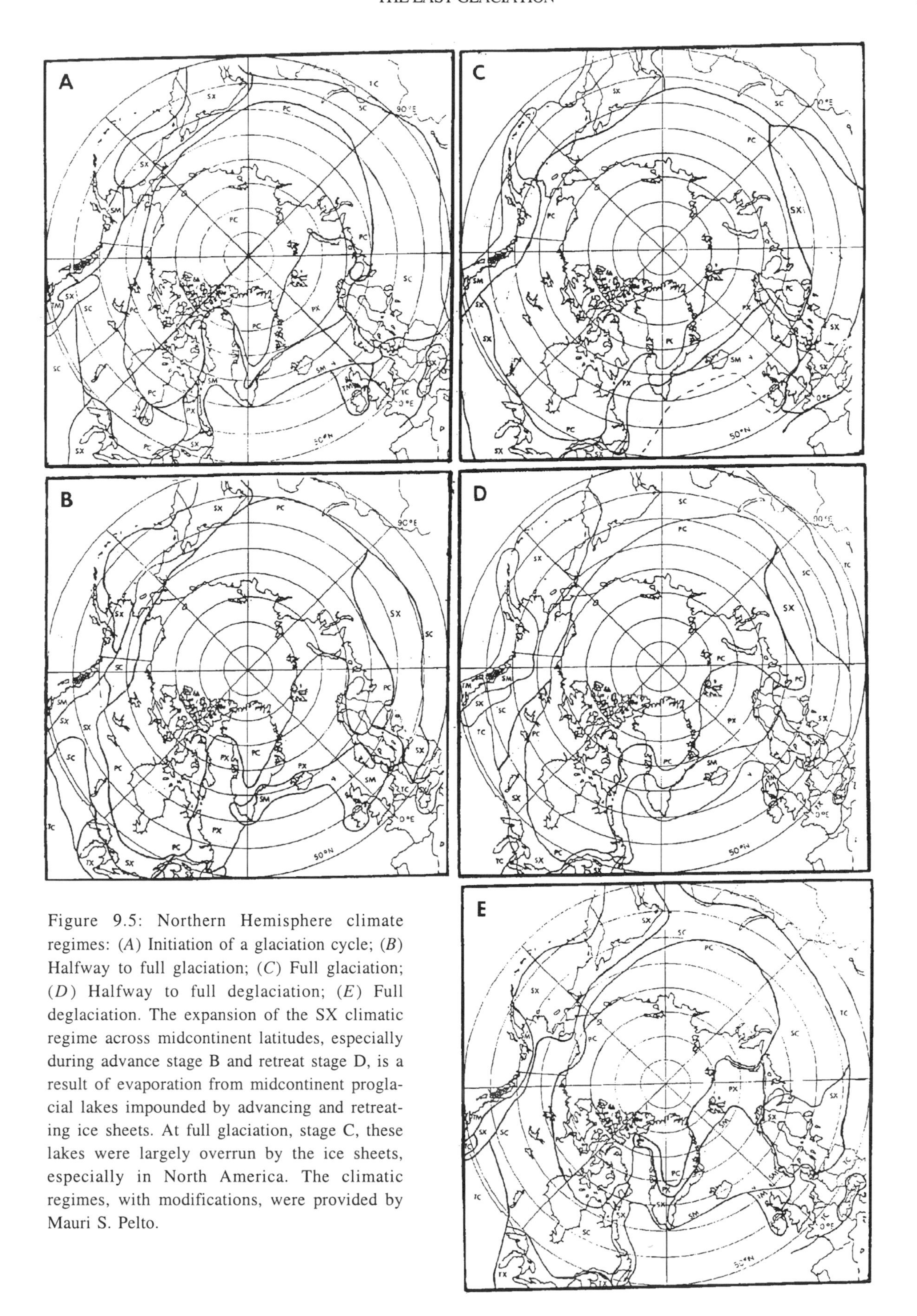

Figure 9.5: Northern Hemisphere climate regimes: (*A*) Initiation of a glaciation cycle; (*B*) Halfway to full glaciation; (*C*) Full glaciation; (*D*) Halfway to full deglaciation; (*E*) Full deglaciation. The expansion of the SX climatic regime across midcontinent latitudes, especially during advance stage B and retreat stage D, is a result of evaporation from midcontinent proglacial lakes impounded by advancing and retreating ice sheets. At full glaciation, stage C, these lakes were largely overrun by the ice sheets, especially in North America. The climatic regimes, with modifications, were provided by Mauri S. Pelto.

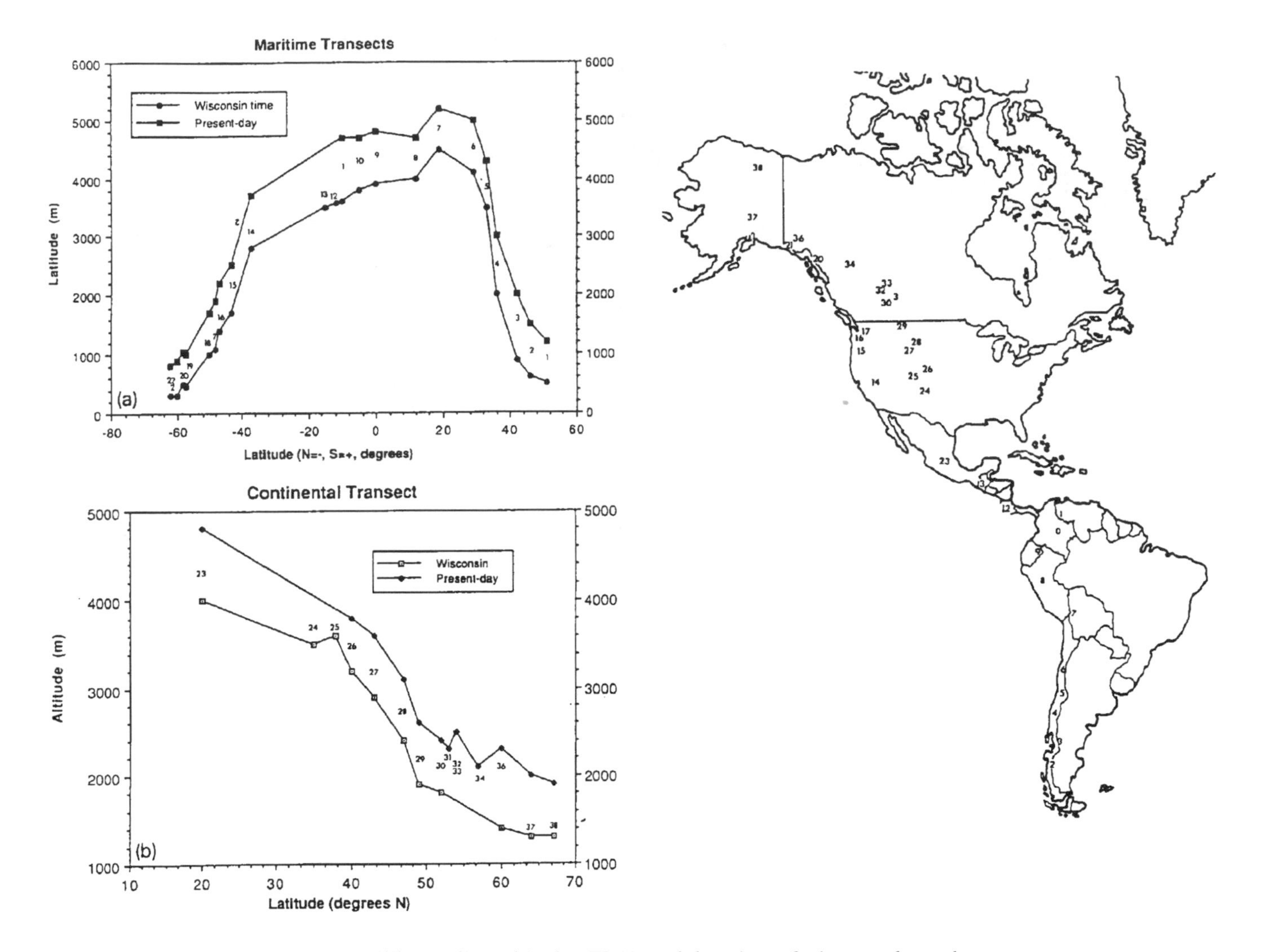

Figure 9.6: Equilibrium-line altitude (ELA) and location of cirques along the western cordillera of the Americas. Comparisons are between present and late Wisconsin cirque glacier ELA along a maritime climatic transect from Patagonia to Alaska (upper left) and along a continental transect from Mexico to Alaska (lower left) at the locations shown by numbers (right). The numbers are located vertically above or below the respective data points and refer to the data sources in Table 9.1. Reproduced by courtesy of M. S. Pelto and Elsevier Science, from *Palaeogeography, Palaeoclimatology, Palaeoecology, 95,* (1992), pp. 42–43, Figures. 1 and 2.

several millenia during interstadials that began with rapid retreat from overextended stadial ice-sheet margins. Advance during ice-stream life cycles and retreat between ice-stream life cycles cause stadial-interstadial-stadial transitions and cause rapid climatic changes recorded as Dansgaard-Oeschger events.

Representative stadial and interstadial ice sheets during the last glaciation are reconstructed from Equation (6.19), but ignoring second-order effects on flowline profiles. Second-order effects include variations of R_{xx} and $\Delta(P_W/P_I)/\Delta x$. The term containing $\Delta(P_W/P_I)/\Delta x$ has a first-order effect primarily for calving ice walls, as shown in Figure 6.8 and Equation (8.119). Let $R_{xx} = 1$ and $\Delta(P_W/P_I)/\Delta x = 0$ for convenience and replace a with a in Equation (6.28), where:

$$a = \mathrm{a} - \partial h_I/\partial t \qquad (9.16)$$

Therefore, a is an effective accumulation (positive) or ablation (negative) rate that is defined as the actual rate a, as given by Equation (9.7) for example, minus the ice thickening rate (or plus the ice thinning rate) $\partial h_I/\partial t$ over time t. Using a instead of a allows effective steady-state flowline profiles to be reconstructed from Equation (6.19). With these modifications, Equation (6.19) becomes:

$$\frac{\Delta h}{\Delta x} = \left(1 - \frac{\rho_I}{\rho_W}\right)\left(\frac{P_W}{P_I}\right)^2 \left\{\frac{a\, h_I}{a\, x + h_0 u_0} - \frac{h_I^2}{a\, x + h_0 u_0}\right.$$

$$\left[\frac{\rho_I g_z h_I}{4A}\left(1-\frac{\rho_I}{\rho_W}\right)\left(\frac{P_W}{P_I}\right)^2\right]^n\Bigg\} + \frac{2\,\tau_s}{\rho_I g_z w_I} + \frac{\tau_o}{\rho_I g_z h_I} \qquad (9.17)$$

Equation (9.17) describes sheet flow when $P_W/P_I \to 0$, stream flow when $0 < P_W/P_I < 1$, and shelf flow when $P_W/P_I = (1 - \sigma_B/\sigma_P)^{1/2}$, where $\sigma_P = 1/4\, \rho_I\, g_z\, h_I\, (1 - \rho_I/\rho_W)$ is the gravitational pulling stress for a flowband of constant width and σ_B is the buttressing stress when an ice shelf is partially confined, see Equation (7.19). This flexibility allows Equation (9.17) to be a universal expression for generating surface profiles along flowlines of ice sheets, after modifications are made for accumulation and ablation rates, convergence and divergence of flow, frozen and thawed bed conditions, and glacioisostatic adjustments, all of which can vary along a flowline.

For shelf flow, $\tau_o = 0$ and $\Delta h/\Delta x = (1 - \rho_I/\rho_W)\, \Delta h_I/\Delta x$ for floating ice. Equation (7.19) gives:

$$\frac{P_W}{P_I} = \left(1-\frac{\sigma_B}{\sigma_P}\right)^{1/2} = \left[1 - \frac{4\,\sigma_B}{\rho_I g_z h_I}\left(\frac{\rho_W}{\rho_W-\rho_I}\right)\right]^{1/2} \qquad (9.18)$$

Equation (6.17) then gives as the ice surface slope along flowlines:

$$\frac{\Delta h}{\Delta x} = \left\{1 - \frac{4\,\sigma_B}{\rho_I g_z h_I}\left(\frac{\rho_W}{\rho_I-\rho_W}\right)\right\}$$
$$\left\{\frac{a}{a\,x + h_0 u_0} - \frac{h_I^2}{a\,x+h_0 u_0}\left[\frac{\rho_I g_z h_I}{4A}\left(1-\frac{\rho_I}{\rho_W}\right) - \frac{\sigma_B}{A}\right]^n\right\}$$
$$+ \frac{2\,\tau_s}{\rho_I g_z w_I} \qquad (9.19)$$

In Equation (9.19), $x = 0$ at the grounding line of the ice shelf and is positive along flowlines toward the calving front. Therefore, u_0 is the positive ice velocity across the grounding line. Buttressing stress σ_B decreases from a maximum at the grounding line to zero at the calving front in a way that depends on the number and distribution of pinning points beneath the ice shelf and the geometry of the embayment that confines the ice shelf. Side shear stress τ_s is greatest alongside ice streams as they enter the ice shelf, but as stream flow becomes shelf flow, τ_s is transferred to the sides of pinning points and the confined sides of the ice shelf. For modeling purposes, σ_B can be taken as decreasing linearly from the grounding line to the calving front and τ_s can be taken as constant over this distance.

For sheet flow and stream flow, $\tau_o/h_I >> 2\,\tau_s/w_I$ is a reasonable assumption along a flowline. Now Equation (7.19) gives:

$$\frac{\Delta h}{\Delta x} = \left(1-\frac{\rho_I}{\rho_W}\right)\left(\frac{P_W}{P_I}\right)^2$$
$$\left\{\frac{a\,h_I}{a\,x + h_0 u_0} - \frac{h_I^2}{a\,x+h_0 u_0}\left[\frac{\rho_I g_z h_I}{4A}\left(1-\frac{\rho_I}{\rho_W}\right)\left(\frac{P_W}{P_I}\right)^2\right]^n\right\}$$
$$+ \frac{\tau_o}{\rho_I g_z h_I} \qquad (9.20)$$

where x is now positive toward the ice divide and u_0 is therefore negative.

Specifying conditions in the ablation zone

The ablation zone of an ice sheet includes the melting surface from the equilibrium line to the ice margin, calving ice walls at the equilibrium line, and floating ice shelves which ablate by calving at their grounding lines. For the ablation zone of sheet flow and ablating terminal lobes of ice streams, both having $0 \le P_W << P_I$, a nearly horizontal bed, $\tau_s = 0$, and a nearly constant τ_o, Equation (9.17) reduces to:

$$\frac{\Delta h}{\Delta x} \approx \frac{\tau_o}{\rho_I g_z h_I} = \frac{\tau_o}{\rho_I g_z (h - h_R)} \qquad (9.21)$$

where $h_I = h - h_R$. Equation (9.21) can be integrated to give parabolic flowline profiles for the ablation zone, with x increasing from the ice margin to the equilibrium line:

$$h = h_R + \left(\frac{2\,\tau_o\, x}{\rho_I g_z}\right)^{1/2} \qquad (9.22)$$

where τ_o ranges from 150 k Pa when sheet flow ends on rough frozen bedrock to 20 k Pa when terminal lobes of ice streams lie on soft wet sediments or till.

For sheet flow, including ice lobes, when calving removes the zone of surface melting that had a flowline profile given by Equation (9.22), setting $x = E$ at the equilibrium line gives $h = h_E$ as the height of the calving ice wall:

$$h_E = h_R + \left(\frac{2\,\tau_o\, E}{\rho_I g_z}\right)^{1/2} \qquad (9.23)$$

For stream flow ending as a calving ice wall, any surface meltwater refreezes in crevasses so height h_E of the equilibrium line is the height of the calving ice wall. Taking $\Delta h/\Delta x = \tau_o/\rho_I\, g_z\, h_I$ for ice streams in bending creep near the ice wall, solving Equation (9.20) for $h_I = h_E$ gives:

$$h_E = \left[\frac{a}{\left[\frac{\rho_I g_z}{4A}\left(1-\frac{\rho_I}{\rho_W}\right)\left(\frac{P_W}{P_I}\right)^2\right]^n}\right]^{\frac{1}{n+1}} \qquad (9.24)$$

Since $P_W/P_I = \rho_W\, h_W/\rho_I\, h_I$, Equation (9.24) for $h_I = h_E$ is also:

$$h_E = \left[\frac{\left[\frac{\rho_W g_z}{4A}\left(\frac{\rho_W}{\rho_I} - 1\right)\right]^n}{a}\right]^{\frac{1}{n-1}} h_W^{\frac{2n}{n-1}} \qquad (9.25)$$

For ice streams ending as a floating ice tongue, h_E for ablation by calving of the floating ice tongue is given by Equation (9.24) when $P_W/P_I = 1$:

$$h_E = \left[\frac{a}{\left[\frac{\rho_I g_z}{4A}\left(1-\frac{\rho_I}{\rho_W}\right)\right]^n}\right]^{\frac{1}{n+1}} \qquad (9.26)$$

Specifying conditions in the accumulation zone

Basal shear stress τ_o in Equation (9.20) depends on frozen fraction f_f and thawed fraction f_t of the bed, on P_W/P_I when the bed is thawed, on the divergence and convergence of flow along a flowline, and on the distribution of a along the flowine. In general, sheet flow diverges from ice domes and converges on ice streams, with a increasing downslope from high domes which convective storms cannot cross and decreasing downslope from low saddles which convective storms can cross, as shown in Figure 5.10. Therefore, flowlines lie in flowbands of variable width w along which a varies.

Creep in ice, basal sliding, and bed deformation can contribute to mass-balance velocity u, as shown in Chapter 3. For creep in ice of height h:

$$u_C = \frac{2\,R_{xz}\,h}{n+2}\left(\frac{\tau_o}{A}\right)^n \tag{9.27}$$

For sliding over the bed at the base of the ice:

$$u_S = \left[\frac{\tau_o}{B\,(1-P_W/P_I)^2}\right]^m \tag{9.28}$$

For deformation of bed thickness $\lambda_D << h$ beneath the ice:

$$u_S = 2\,\lambda_D\left[\frac{\tau_o}{C\,(1-P_W\,P_I)}\right]^c \tag{9.29}$$

For mass-balance continuity:

$$\begin{aligned}\Gamma(x) &= \int_x^L a(x)\,w(x)\,dx = u(x)\,w(x)\,h(x)\\ &= (f_C\,u_C + f_S\,u_S + f_D\,u_D)\,w\,h\\ &= (f_C\,u_C + f_B\,u_B)\,w\,h\end{aligned} \tag{9.30}$$

where w is the flowband width, f_C, f_S, and f_D are fractions of mass-balance velocity u due to creep in ice, basal sliding, and bed deformation, respectively, and $f_B = f_S + f_D$ is the fraction of u that occurs at the ice-bed interface where $u_B = u_S + u_D$ is the velocity of basal ice.

In reconstructing former ice sheets using Equation (9.20) to obtain ice elevations along specified flowlines, τ_o must be specified separately for fraction f_f of the bed that is frozen and fraction f_t of the bed that is thawed, such that:

$$\begin{aligned}\tau_o &= f_f\,(\tau_o)_C + f_t\,(\tau_o)_S + f_t\,(\tau_o)_D\\ &= (1-f_t)\,(\tau_o)_C + f_t\,(\tau_o)_B\end{aligned} \tag{9.31}$$

where $(\tau_o)_C$, $(\tau_o)_S$, and $(\tau_o)_D$ are values of τ_o when ice velocity is due entirely to ice creep, basal sliding, and bed deformation, respectively, and $(\tau_o)_B = (\tau_o)_S + (\tau_o)_D$. When glacial geology was not produced, the bed was frozen and $\tau_o = (\tau_o)_C$. When glacial geology was produced, $\tau_o = (\tau_o)_B$ is used because glacial geology is usually unable to distinguish how much of u_B was due to u_S and due to u_D. It is therefore useful to define u_C as:

$$u_C = h\left(\frac{\tau_o}{A_C}\right)^n \tag{9.32}$$

and to define u_B as:

$$u_B = \left[\frac{\tau_o}{D_B\,(1-P_W/P_I)^v}\right]^m \tag{9.33}$$

where A_C is a parameter for creep of ice above the bed defined as:

$$A_C = A\left[\frac{n+2}{2\,R_{xz}}\right]^{1/n} \tag{9.34}$$

and D_B is a parameter for displacement of basal ice defined as:

$$D_B = f_S\,B + f_D\,\frac{C}{(2\,\lambda_D)^{1/c}} \tag{9.35}$$

with $D_B = B$, $v = 2$, and $m = (n+1)/2$ when $f_S = 1$ and $D_B = C/(2\lambda_D)^{1/c}$, $v = 1$, and $m = c$ when $f_D = 1$.

For a frozen bed, $f_t = 0$, $f_C = 1$ and Equation (9.30) reduces to:

$$\Gamma(x) = u_C\,w\,h = (\tau_o/A_C)^n\,w\,h^2 = (\rho_I\,g_z\,\alpha/A_C)^n\,w\,h^{n+2} \tag{9.36}$$

where $\tau_o = \rho_I\,g_z\,h\,\alpha$ using $h = h_I$ in Equation (5.4) for sheet flow. Surface slope α obtained from Equation (9.36) is:

$$\alpha = \frac{dh}{dx} = \frac{A_C}{\rho_I\,g_z}\left[\frac{\Gamma(x)}{w(x)}\right]^{\frac{1}{n}}\frac{1}{h^{\frac{n+2}{n}}} \tag{9.37}$$

Integrating Equation (9.37) from h_E to h and solving for h:

$$h = \left[h_E^{\frac{2n+2}{n}} + \left(\frac{2\,n+2}{n}\right)\frac{A_C}{\rho_I\,g_z}\int_E^x\left(\frac{\Gamma(x)}{w(x)}\right)^{\frac{1}{n}}dx\right]^{\frac{n}{2n+2}} \tag{9.38}$$

Combining Equations (9.37) and (9.38) gives $(\tau_o)_C$:

$$(\tau_o)_C = \rho_I\,g_z\,h\,\alpha = \frac{A_C\left(\frac{\Gamma(x)}{w(x)}\right)^{\frac{1}{n}}}{\left[h_E^{\frac{2n+2}{n}} + \left(\frac{2\,n+2}{n}\right)\frac{A_C}{\rho_I\,g_z}\int_E^x\left(\frac{\Gamma(x)}{w(x)}\right)^{\frac{1}{n}}dx\right]^{\frac{1}{n+1}}} \tag{9.39}$$

Define $\Pi_C(x)$ as:

$$\Pi_C(x) = \int_E^x\left(\frac{\Gamma(x)}{w(x)}\right)^{\frac{1}{n}}dx = \int_E^x\left[\frac{1}{w(x)}\int_x^L a(x)\,w(x)\,dx\right]^{\frac{1}{n}}dx \tag{9.40}$$

For a thawed bed, $f_t = 1$, $f_B \rightarrow 1$, and Equation (9.30) reduces to:

$$\begin{aligned}\Gamma(x) &\approx u_B\,w\,h = \left[\tau_o/D_B\,(1-P_W/P_I)^v\right]^m w\,h\\ &\approx \left[\frac{\rho_I\,g_z\,\alpha}{D_B\,(1-P_W/P_I)^v}\right]^m w\,h^{m+1}\end{aligned} \tag{9.41}$$

where $\tau_o = \rho_I g_z h \alpha$ using $h = h_I$ in Equation (5.4) for sheet flow. Surface slope α obtained from Equation (9.41) is:

$$\alpha = \frac{dh}{dx} = \frac{D_B (1 - P_W/P_I)^v}{\rho_I g_z} \left[\frac{\Gamma(x)}{w(x)}\right]^{\frac{1}{m}} \frac{1}{h^{\frac{m+1}{m}}} \tag{9.42}$$

Integrating Equation (9.42) from h_E to h and solving for h:

$$h = \left[h_E^{\frac{2m+1}{m}} + \left(\frac{2m+1}{m}\right)\frac{D_B}{\rho_I g_z}\left(1 - \frac{P_W}{P_I}\right)^v \int_E^x \left(\frac{\Gamma(x)}{w(x)}\right)^{\frac{1}{m}} dx\right]^{\frac{m}{2m+1}} \tag{9.43}$$

Combining Equations (9.42) and (9.43) gives $(\tau_o)_B$:

$$(\tau_o)_B = \rho_I g_z h \alpha$$

$$= \frac{D_B\left(1 - \frac{P_W}{P_I}\right)^v \left(\frac{\Gamma(x)}{w(x)}\right)^{\frac{1}{m}}}{\left[h_E^{\frac{2m+1}{m}} + \left(\frac{2m+1}{m}\right)\frac{D_B}{\rho_I g_z}\left(1 - \frac{P_W}{P_I}\right)^v \int_E^x \left(\frac{\Gamma(x)}{w(x)}\right)^{\frac{1}{m}} dx\right]^{\frac{m}{2m+1}}} \tag{9.44}$$

Define $\Pi_D(x)$ as follows:

$$\Pi_D(x) = \int_E^x \left(\frac{\Gamma(x)}{w(x)}\right)^{\frac{1}{m}} dx = \int_E^x \left[\frac{1}{w(x)}\int_x^L a(x)\, w(x)\, dx\right]^{\frac{1}{m}} dx \tag{9.45}$$

Evaluations of $\Pi_C(x)$ and $\Pi_D(x)$ for average values of w and a are obtained by solving Equations (9.39) and (9.44) for average accumulation rate $\bar{a}$ along a flowline. In this case, Equation (9.30) with $\bar{w}$ replacing w(x) for average flowband width w along the flowline is written:

$$\Gamma(x)/\bar{w} = \int_x^L \bar{a}\, dx = \bar{a}\,(L - x) \tag{9.46}$$

For which Equation (9.40) for a frozen bed becomes:

$$\Pi_C(x) = \int_E^x \left[\frac{\Gamma(x)}{\bar{w}}\right]^{1/n} dx = \int_E^x [\bar{a}\,(L - x)]^{1/n} dx$$

$$= \frac{n\,\bar{a}^{1/n}}{n+1}\left[(L-E)^{\frac{n+1}{n}} - (L-x)^{\frac{n+1}{n}}\right] \tag{9.47}$$

and Equation (9.45) for a thawed bed becomes:

$$\Pi_D(x) = \int_E^x \left[\frac{\Gamma(x)}{\bar{w}}\right]^{1/m} dx = \int_E^x [\bar{a}\,(L - x)]^{1/m} dx$$

$$= \frac{m\,\bar{a}^{1/m}}{m+1}\left[(L-E)^{\frac{m+1}{m}} - (L-x)^{\frac{m+1}{m}}\right] \tag{9.48}$$

Equation (9.39) now reduces to:

$$(\tau_o)_C = \frac{A_C[\Gamma(x)/w(x)]^{1/n}}{\left[h_E^{\frac{2n+2}{n}} + \left(\frac{2n+2}{n}\right)\frac{A_C}{\rho_I g_z}\Pi_C(x)\right]^{\frac{1}{n+1}}}$$

$$= \frac{A_C[\bar{a}\,(L-x)]^{1/n}}{\left\{h_E^{\frac{2n+2}{n}} + \frac{2\,\bar{a}^{1/n} A_C}{\rho_I g_z}\left[(L-E)^{\frac{n+1}{n}} - (L-x)^{\frac{n+1}{n}}\right]\right\}^{\frac{1}{n+1}}} \tag{9.49}$$

and Equation (9.44) now reduces to:

$$(\tau_o)_B = \frac{D_B\left(1 - \frac{P_W}{P_I}\right)^v \left(\frac{\Gamma(x)}{w(x)}\right)^{\frac{1}{m}}}{\left[h_E^{\frac{2m+1}{m}} + \left(\frac{2m+1}{m}\right)\frac{D_B}{\rho_I g_z}\left(1 - \frac{P_W}{P_I}\right)^v \Pi_D(x)\right]^{\frac{1}{2m+1}}}$$

$$= \frac{D_B\left(1 - \frac{P_W}{P_I}\right)^v [\bar{a}\,(L-x)]^{1/m}}{\left\{h_E^{\frac{2m+2}{m}} + \left(\frac{2m+1}{m+1}\right)\frac{\bar{a}^{-1/m} D_B}{\rho_I g_z}\left(1 - \frac{P_W}{P_I}\right)^v \left[(L-E)^{\frac{m+1}{m}} - (L-x)^{\frac{m+1}{m}}\right]\right.} \tag{9.50}$$

Equations (9.49) and (9.50) depend on w(x) and $a(x)$, or on $\bar{a}$ only, and on E and L for a given flowline, using n when the bed is frozen and m when the bed is thawed.

Specifying glacial isostasy

Equation (5.29) gives height h of an ice sheet at distance x along a flowline from x = 0 at the grounded ice margin in relation to height above or depth below sea level h_0 at x = 0. Isostatically depressed height h* is needed when $0 \leq r \leq r_o$ specifies the degree of glacioisostatic depression, with $r_o \approx 1/3$ according to Equation (5.24). Solving Equation (5.29) for h:

$$h = h_0 + (1 + r)^{1/2}(h^* - h_0) \tag{9.51}$$

Setting $\Delta h = h_{i+1} - h_i$ at distance x from the grounded ice margin and applying Equation (9.51) to both h_{i+1} and h_i, Δh in Equation (9.20) becomes:

$$\Delta h = h_{i+1} - h_i = \left[h_0 + (1 + r_{i+1})^{1/2}(h_{i+1}^* - h_0)\right]$$
$$- \left[h_0 + (1 + r_i)^{1/2}(h_i^* - h_0)\right]$$
$$\approx (1 + r)^{1/2}(h_{i+1}^* - h_i^*) \tag{9.52}$$

where $r \approx r_{i+1} \approx r_i$ because r varies slowly along x.

Let $h_I = h - h_R$ in Equation (9.20), with h and Δh being specified by Equations (9.51) and (9.52). Writing Equation (9.20) for $\Delta h^* = h_{i+1}^* - h_i^*$ then gives:

$$h_{i+1}^* = h_i^* + \frac{(1 - \rho_I/\rho_W)}{(1+r)^{1/2}}\left(\frac{P_W}{P_I}\right)^2$$

$$\left\{ \frac{a\left[h_0 + (1+r)^{1/2}(h^* - h_0) - h_R\right]_i}{a\,x + h_0 u_0} \right.$$

$$- \frac{\left[h_0 + (1+r)^{1/2}(h^* - h_0) - h_R\right]_i^{n+2}}{a\,x + h_0 u_0}$$

$$\left. \left[\frac{\rho_I g_z}{4A}\left(1 - \frac{\rho_I}{\rho_W}\right)\left(\frac{P_W}{P_I}\right)^2\right] \right\} \Delta x$$

$$+ \left\{ \frac{(1 - f_t)(\tau_o)_C}{\rho_I g_z \left[h_0 + (1+r)^{1/2}(h - h_0) - h_R\right]_i} \right\} \Delta x$$

$$+ \left\{ \frac{f_t(\tau_o)_B}{\rho_I g_z \left[h_0 + (1+r)^{1/2}(h - h_0) - h_R\right]_i} \right\} \Delta x \quad (9.53)$$

where Equations (9.23) through (9.26) for h_E give $h_i = h_0$ for $i = 0$ and Equation (9.31) is substituted for τ_o. Variations of $(\tau_o)_C$ and $(\tau_o)_B$ with distance x along flowlines are given by Equations (9.49) and (9.50), respectively. For sheet flow, $P_W/P_I = 0$ when the bed is frozen and $P_W/P_I << 1$ when the bed is thawed. For stream flow, $0 < P_W/P_I < 1$. Measuring along flowlines from 0.2 L to 0.8 L, $(\tau_o)_C$ decreases almost linearly from 80 k Pa to 20 k Pa and $(\tau_o)_B$ decreases almost linearly from 50 k Pa to 10 k Pa, with both decreasing rapidly to 0 k Pa at the ice divide, when $\bar{a} = 0.2$ m/a along a flowband of constant width (Hughes, 1981).

For an ice sheet advancing over time t:

$$r = r_o\,[1 - exp\,(-t/t_o)] \quad (9.54)$$

where t = 0 at the beginning of advance and $r = r_a$ at $t = t_a$ when advance ends and retreat beings. For an ice sheet retreating over time t, with retreat beginning at t = 0:

$$r = r_a\,exp\,(-t/t_o) \quad (9.55)$$

Time constant t_o depend on the size of an ice sheet at time t_a, being larger for larger ice sheets, up to $t_o = 7800$ a for the Laurentide Ice Sheet (Andrews, 1970). Values of r are calculated along flowlines for the maximum ice–sheet extent, with Equation (9.54) being used for the time t when a given point on the bed was covered by ice and Equation (9.55) being used for the time t when that point was not covered by ice, both times beginning with the beginning of glaciation. In this scheme, $r \rightarrow r_o$ over time in the core area of an ice sheet that remains intact during stadial advances and interstadial retreats, but r decreases from $r = r_o$ in the core to $r = 0$ at the maximum ice margin. This area beyond the core is subjected to repeated ice-sheet advances during stadials and retreats during interstadials. Equation (5.31) describes how height h_R* (or depth, both with respect to present-day sea level) of the landscape changes with glacioisostatic depression expressed by Equations (9.54) and (9.55).

Specifying stadial and interstadial conditions

In the prognostic modeling approach, ice thickness change $\delta h_I/\delta t$ during stadial-interstadial transitions is obtained along flowlines by subtracting thicknesses obtained from Equation (9.53) in time δt, using a obtained from Equations (9.1) through (9.15) or by some other method. Then a can be calculated along flowlines using Equation (9.16) and is used to calculate $(\tau_o)_C$, and $(\tau_o)_B$ in an interative calculation. Ideally, f_t and P_W/P_I are calculated by applying the equations of heat transport to internal creep and basal sliding along the flowline, including creep in wet deforming sediments or till beneath ice streams. The most important time interval δt in a Pleistocene glaciation cycle is the time needed for an ice sheet to fluctuate between its stadial and its interstadial configurations. Advance to stadial positions and retreat to interstadial positions occurred over distance ranging from 400 km to 800 km for most Pleistocene ice sheets. Advance was primarily by ice streams, and retreat was primarily by ice calving. Based on ice-stream and ice-calving velocities ranging from 500 m/a to 7000 m/a for present-day ice sheets, $\delta t = 160$ a is realistic for Pleistocene ice sheets.

In the diagnostic modeling approach, $(\tau_o)_C$, $(\tau_o)_B$, f_t, and P_W/P_I are specified from glacial geology. This allows a to be calculated from Equations (9.39) and (9.44) or $\bar{a}$ to be calculated from Equations (9.49) and (9.50) when $\Gamma(x)$ and $\Pi(x)$ are obtained from variable flowband widths w deduced from glacial geology and variable accumulation rates a, as calculated using Equations (9.1) through (9.15) or some other mass balance formulation. As in prognostic modeling, a is related to a in Equation (9.16) by obtaining ice thickness change δh_I along flowlines in time change δt as an ice sheet thins while advancing during stadials and thickens while retreating during interstadials. Values of $(\tau_o)_C$, $(\tau_o)_B$, f_t, P_W/P_I, and w are deduced from first order glacial geology produced during interstadials and from second order glacial geology produced during stadials.

Mass balance is useful in ice-sheet models primarily in calculating how rapidly an ice sheet thins as it advances or thickens as it retreats. In forward prognostic models, the surface mass balance is also used to calculate where the bed is frozen, thawed, melting, and freezing, using the equations of heat and mass transport. In inverse diagnostic models, these basal thermal conditions are determined by mapping and interpreting the glacial geology. Basal thermal conditions are crucial in reconstructing former ice sheets accurately because they control basal traction that resists ice motion and therefore determine ice elevation for a given distribution of accumulation and ablation on the ice surface. Although stream flow discharges over 90 percent of the ice, sheet flow occurs over 90 percent of the bed. Sheet flow is slower over a frozen bed than over a thawed bed, so ice elevation will be higher. Sheet flow slows down across a freezing bed and speeds up across a melting bed, so the ice surface slope will be less over a freezing bed than over a melting bed. Glacial geology reveals basal frozen, thawed, freezing, and melting zones directly and therefore allows surface elevations and slopes to be calculated directly. The surface mass balance determines these surface conditions indirectly by influencing bed conditions. Hence, the surface mass balance does not have the dominant role in inverse diagnostic models that it has in forward prognostic models.

The inverse diagnostic approach to ice-sheet modeling takes advantage of the fact that glacial geology is more reliable than mass balance as a guide in reconstructing former ice sheets. Using the inverse diagnostic approach, however, requires that the glacial geology be not only mapped and dated for a reconstruction at a particular snapshot in time but also interpreted correctly. The key to correct interpretation is distinguishing between first-order and second-order glacial geology. First-order glacial geology is reinforced by each glaciation cycle and therefore places an increasingly dominant glacial imprint on the landscape. This implies that ice sheets have a nearly steady-state core within which glacial erosion and deposition processes are constantly active. Second-order glacial geology is erased, reoriented, or overprinted by advances and retreats during a glaciation cycle and during successive cycles, so its strongest imprint was produced during the most recent retreat. This implies that ice sheets have a transient periphery where successive advances and retreats produce different patterns of glacial erosion and deposition. Advance from the stable core produces stadials and retreat to the stable core produces interstadials in the glaciation cycle. Second-order glacial geology produced during the initial advance and final retreat of a glaciation cycle will overprint first-order glacial geology in the stable core, but it will be too transient to obliterate the steady state imprint of first-order glacial geology.

First-Order Glacial Geology

Mapping first-order glacial geology is like observing the forest instead of the trees. Major features related to first-order glacial geology are the depth and extent of permafrost, the size and distribution of glacially eroded lakes, selective linear glacial erosion along channels radiating from a common center, glacial erosion of bedrock by areal scouring, glacial deposition of moraine belts, and postglacial rebounding basins. In turn, they identify conditions beneath Pleistocene ice sheets during initial advance, during interstadials, during advances from interstadial to stadial positions, during stadials, during retreat from stadial to interstadial positions, and during final retreat. Therefore, these features identify concentric zones that expand and contract as an ice sheet advances and retreats, so the zonation becomes the dominant imprint on the glaciated landscape over time. The zones therefore constitute the first-order glacial geology.

Permafrost

The distribution of permafrost reveals transitions from frozen to thawed landscapes beneath Pleistocene ice sheets during their initial advance. The thickness and extent of permafrost is intimately associated with whether the core area beneath former ice sheets was frozen or thawed and, therefore, whether it was surrounded by a melting or freezing zone. Figure 9.7 shows the limits of continuous, discontinuous, and sporadic permafrost in the Arctic, which are also zones of relative permafrost thickness. Permafrost is thickest and continuous in northeastern Siberia, where it extends far out onto the continental shelf in the Laptev and East Siberian Seas. Marine ice sheets that formed when sea ice thickened and grounded on these continental shelves initially would have a frozen bed. Where permafrost is thinner and discontinuous, such as on the mainlands and islands bordering the Kara and the Barents Seas in Eurasia and bordering Foxe Basin and Hudson Bay in Canada, ice sheets would cover a mostly frozen bed. Where permafrost is sporadic or absent, such as on the floors of the Kara Sea, the Barents Sea, Foxe Basin, Hudson Bay, in interisland channels, and in subpolar latitudes on land, ice sheets would cover a partly or wholly thawed bed. In all cases thickening ice would thin underlying permafrost and could eventually thaw an initially frozen bed because the ice cover is an insulating blanket. In polar and subpolar lake studded areas, permafrost would generally be absent on lake floors, so these would be thawed patches under an advancing ice sheet.

Lakes

Glacially eroded lakes were thawed patches on bedrock beneath Pleistocene ice sheets during interstadials. The lakes tend to be clustered on a slightly elevated landscape that forms broad arcs beyond the isostatically rebounding core area and the low peripheral zone eroded by ice streams. These lakes tend to get larger and more elongated with increasing distance from the rebounding core, and the long axes of these lakes tends to align along lines radiating from the core. This elongation of lakes coincides with a tendency for hills to become drumlinized and for drumlins to become flutes with increasing radial distance from the core area. Figure 9.8 identifies these arcs of lakes as lying between the Hudson Bay lowlands and the edge of the Canadian Shield in North America, between the Gulf of Bothnia lowlands and the edge of the Scandinavian Shield in Europe, and along the arctic coastal plains of Siberia and Alaska. In Canada, most lakes occupy a broad arc more than 300 m (1000 feet) above present-day sea level. All lake-studded arcs have relatively high elevations in the rebounding core areas. The arcuate distribution of lakes about rebounding core areas and the relatively high elevations of these arcs are consistent with zones of either basal melting beyond an interior frozen zone beneath the central ice divides of former ice sheets or basal freezing beyond an interior thawed zone beneath these ice divides. Basal melting and freezing zones consist of frozen and thawed patches, with glacial erosion quarrying and abrading the thawed patches so that they become lakes on the deglaciated landscape. Frozen patches do not erode, so the overall elevation of freezing and melting zones is lowered less by glacial erosion than is the case in thawed zones. This

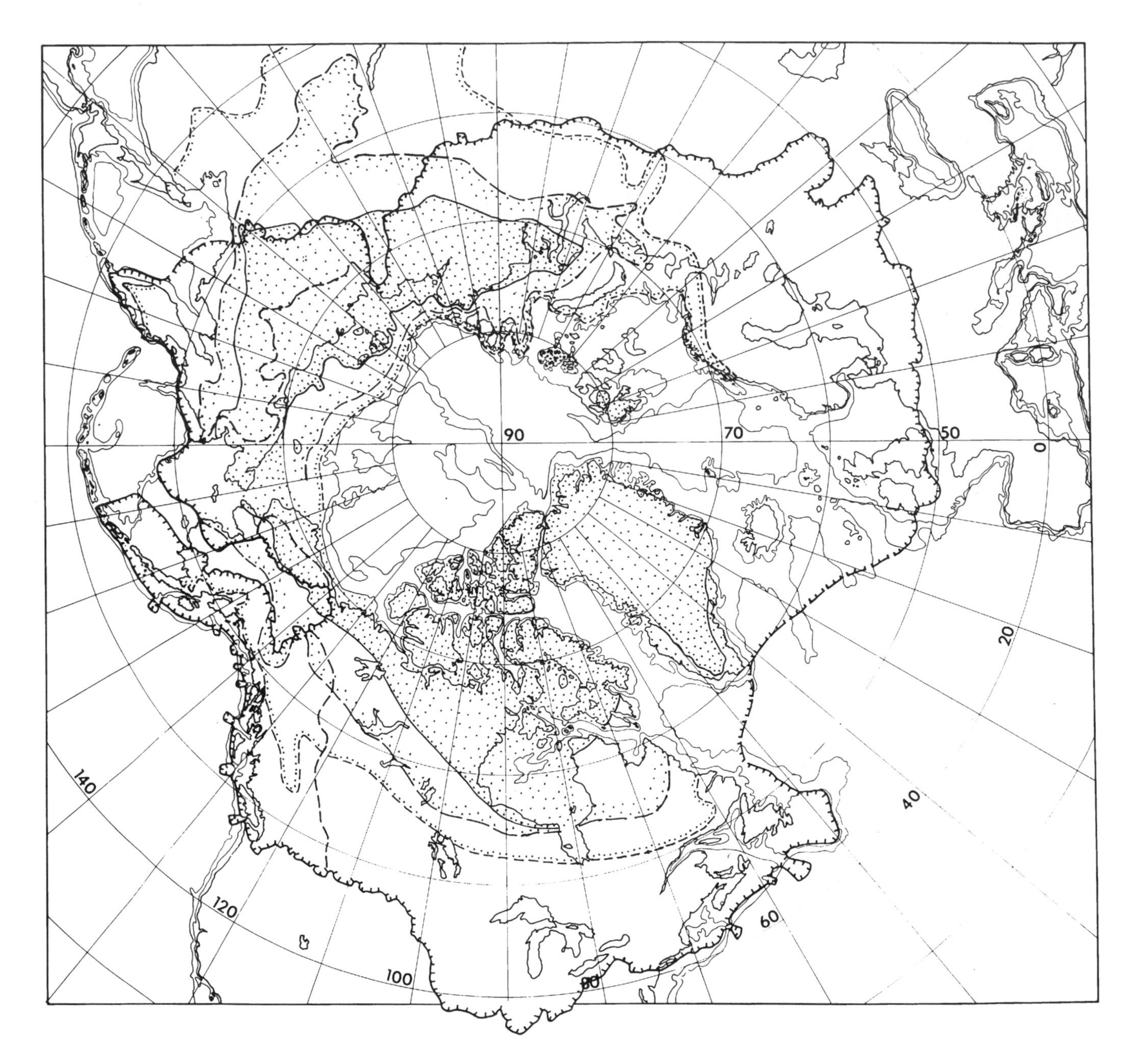

Figure 9.7: Evidence for transitions from frozen to thawed landscapes beneath Pleistocene ice sheets during their initial advance: present-day limit of continuous permafrost (solid lines bordering dotted areas); present-day limit of discontinuous permafrost (broken lines bordering dotted bands); present-day limit of sporadic permafrost (dashed lines bordering dotted lines); the outer limit of Pleistocene glaciation (hatchured line); and separation of marine and mountain glaciation (broken line)

accounts for the relatively higher elevation of these zones of long-term basal melting or freezing conditions.

Channels

Glacially modified channels were occupied by ice streams when Pleistocene ice sheets advanced during transitions from interstadials to stadials. Glacially modified channels often radiate from formerly glaciated highland plateaus and marine basins. Glacial modification consists of longitudinal fore-deepening and transverse U-shaping of channels by selective linear erosion (Sugden and John, 1976). Selective linear erosion of these channels occurred along fjords and their through valleys transecting mountainous polar and subpolar coasts, along straits and interisland channels crossing polar

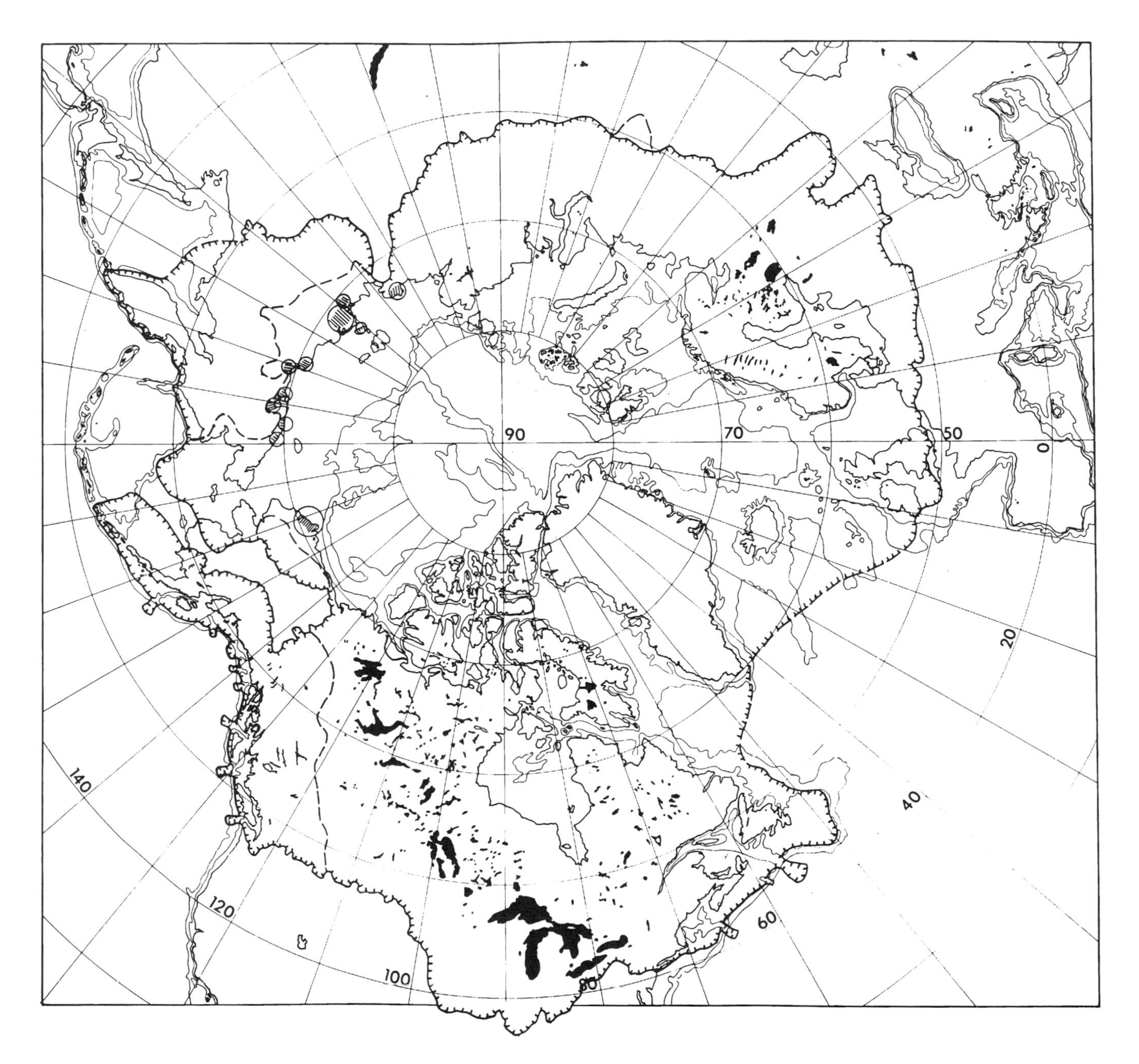

Figure 9.8: Evidence for subglacial freezing and melting zones beneath Pleistocene ice sheets during interstadials: lakes surrounding centers of postglacial isostatic rebound (black areas); oriented lakes on Arctic coastal plains (circled areas); lake orientations (parallel lines); the outer limit of Pleistocene glaciation (hatchured line); and separation of marine and mountain glaciation (broken line).

continental shelves, along linear lakes bordering continental shields, and along river valleys beyond continental shields. These channels are often aligned to form continuous troughs, and they are all submerged linear depressions floored by water-saturated marine, lacustrine, or fluvial sediments that provide virtually no traction resisting the sliding of overlying ice. Marine and terrestrial ice streams therefore, naturally flow along these channels during a glaciation cycle. That this was the case is confirmed by submarine deposits that form sills at fjord entrances and form fans on continental slopes beyond troughs on continental shelves and by lobate moraines beyond linear lakes and across river valleys on the continents. Ice streams reoccupied these channels during successive glaciation cycles, deepening, widening, and lengthening them, thereby making them first-order glacial geological features. Large-scale glacial channels are indicated by arrows that radiate from the major domes of former ice sheets in Figure 9.9. On a smaller scale, glacial

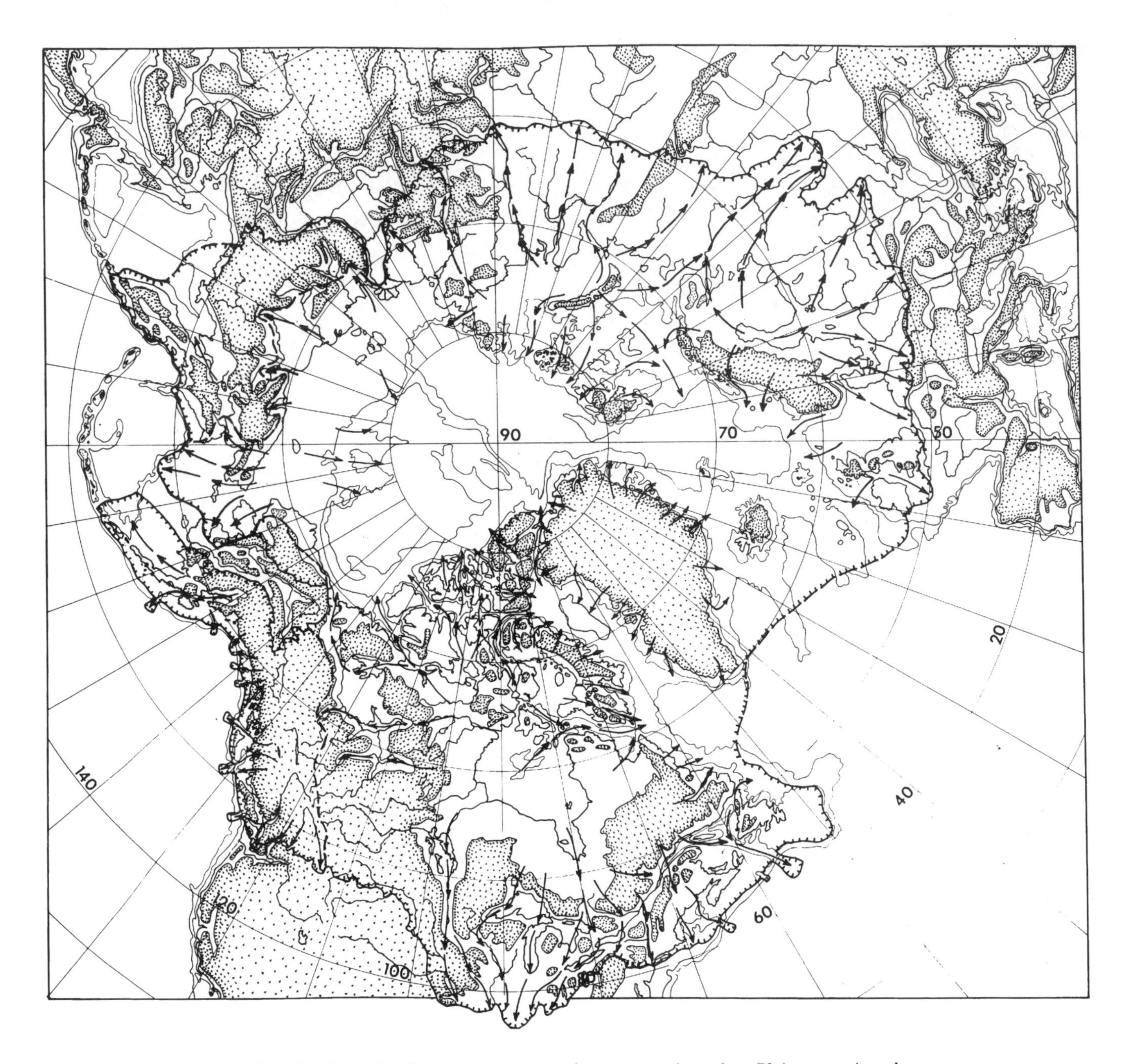

Figure 9.9: Evidence for former ice streams that were active when Pleistocene ice sheets advanced from interstadial to stadial positions: ice streams that followed submarine troughs, interisland channels, major fjords, glacial lakes, and river valleys (black arrows); land elevations over 300 m (dotted areas); the outer limit of Pleistocene glaciation (hatched line); and separation of marine and mountain glaciation (broken line). Major rivers are shown.

channels also radiate from former large ice caps, such as those over Newfoundland, Iceland, and Spitzbergen.

Erosion

Glacial erosion by areal scouring occurred beneath Pleistocene ice sheets during stadials. Figure 9.10 shows the distribution of crystalline and sedimentary surficial rocks that were covered by Northern Hemisphere ice sheets during the last glaciation cycle. These are regions of long-term thawed bed conditions beneath the ice sheet, so basal sliding caused areal scouring that exposed the bedrock. Areal scouring was most pronounced when it stripped away sedimentary bedrock and exposed the underlying crystalline shield.

Bedrock beneath the former Laurentide Ice Sheet in North America consists of a Paleozoic cap of sedimentary bedrock over most of Hudson Bay and the Hudson Bay low-

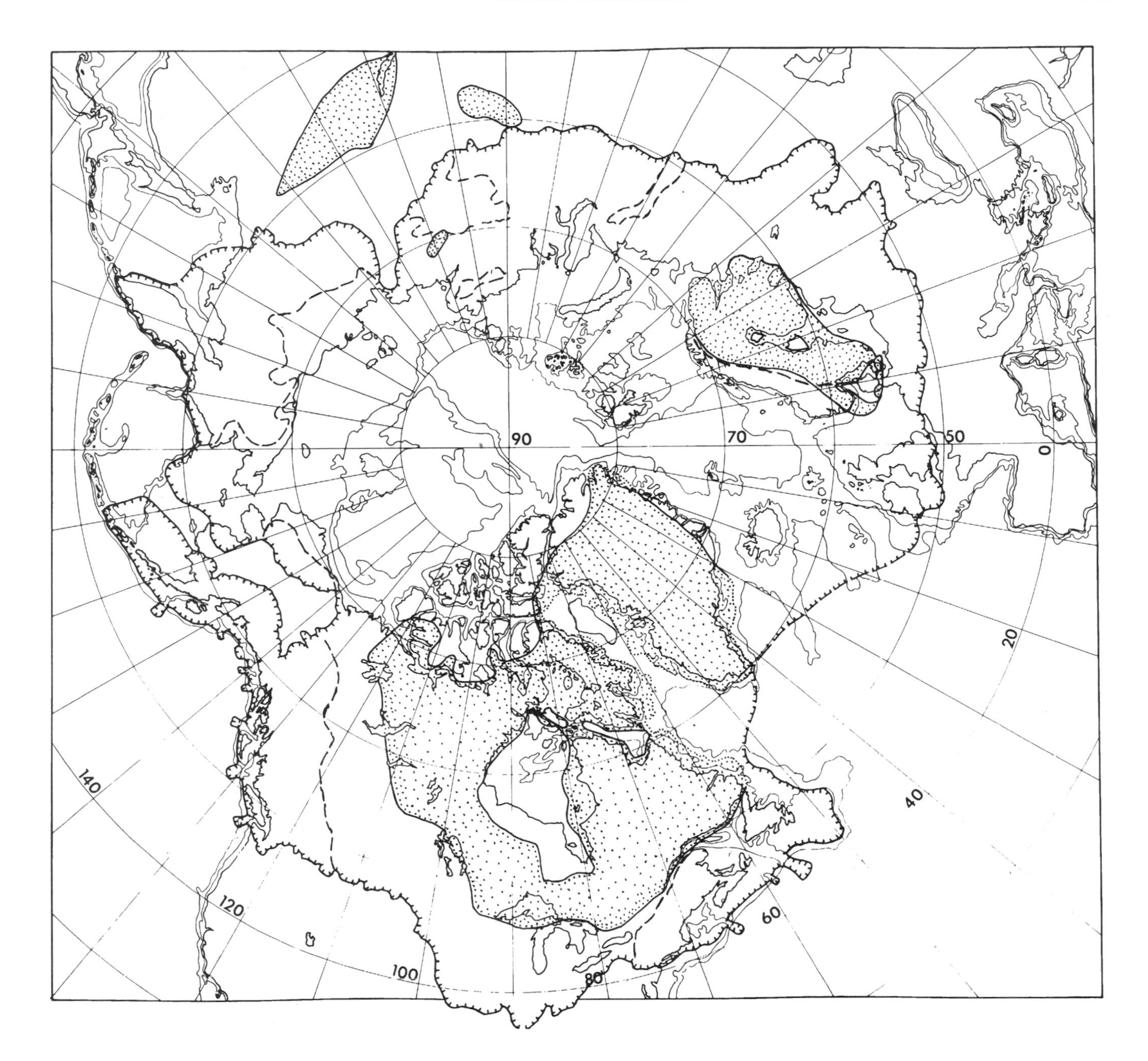

Figure 9.10: Evidence for glacial erosion by areal scouring beneath Pleistocene ice sheets during stadials: exposed crystalline shield (dotted areas);. the outer limit of Pleistocene glaciation (hatched line); and separation of marine and mountain glaciation (broken line).

lands at the center of the ice sheet, a Precambrian crystalline shield surrounding this cap and extending to the outer limit of the steady state core of the ice sheet, and a Paleozoic sedimentary bedrock platform beneath the transient margin of the ice sheet. This distribution is consistent with slowly moving ice under a central ice dome causing little glacial erosion, so the Paleozoic cap was not removed, with ice moving more rapidly away from the ice dome and causing more glacial erosion that stripped away the Paleozoic bedrock to expose the Precambrian shield, and with more slowly moving ice in the ablation zone along the ice margin that was incapable of eroding the Paleozoic cover.

Bedrock beneath the Scandinavian Ice Sheet in Europe consists of two small Paleozoic rock caps in the Gulf of Bothnia, surrounded by a Precambrian crystalline shield that extended to the equilibrium line of the ice sheet, with a Paleozoic platform beneath the ablation zone beyond the equilbrium line. This is consistent with little glacial erosion by relatively stagnant ice under the central ice dome, more rapid erosion by faster ice out to the equilibrium line, and slower erosion by slower ice beyond the equilibrium line.

The major ice dome in Siberia was located over the Kara

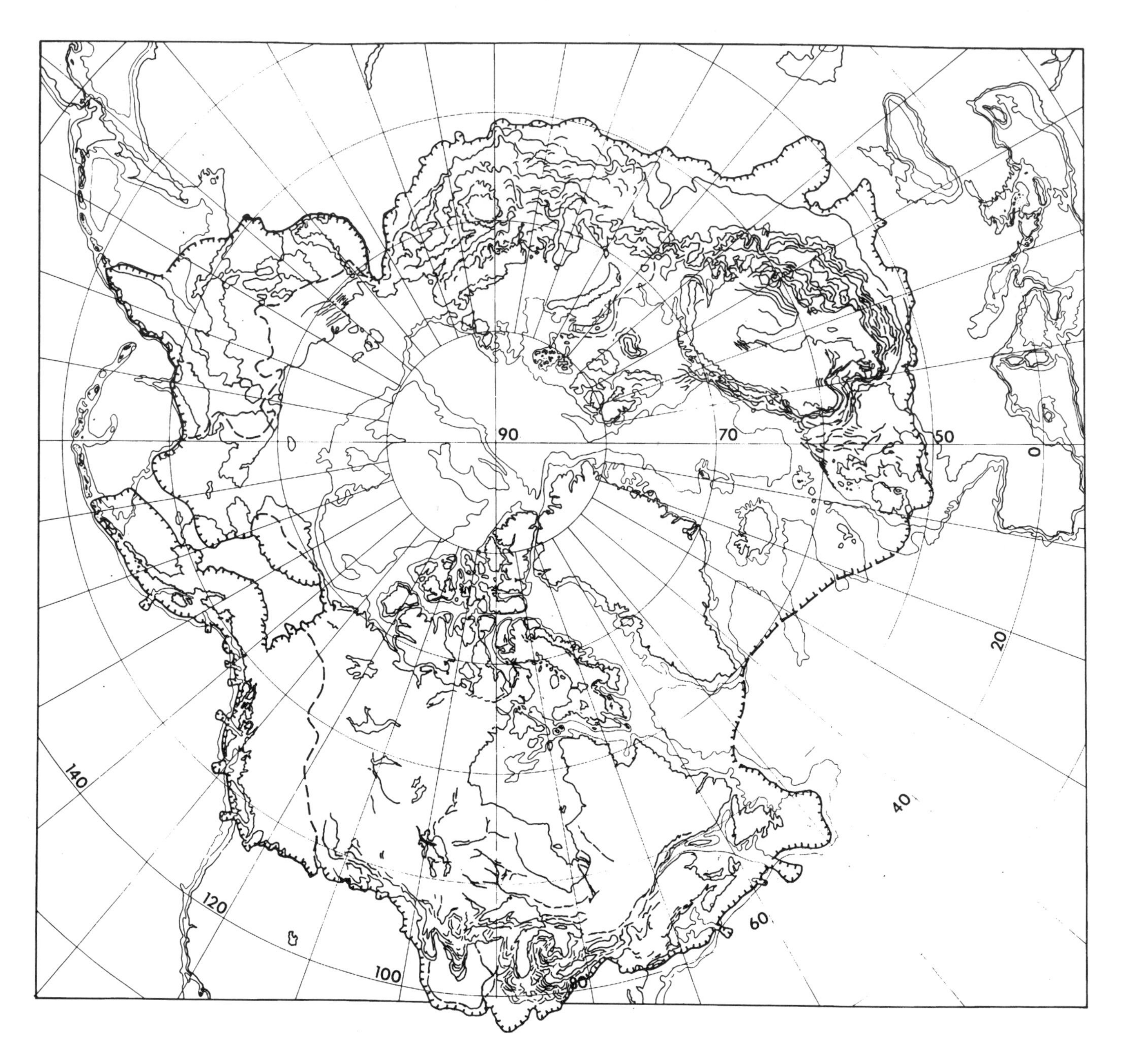

Figure 9.11: Evidence for glacial deposition beneath Pleistocene ice sheets during retreat from stadial to interstadial positions: moraine belts, mostly recessional moraines deposited since the last glacial maximum, beyond which are terminal moraines from all Pleistocene glaciations (wiggly lines); the outer limit of Pleistocene glaciation (hatched line); and separation of marine and mountain glaciation (broken line).

Sea. Paleozoic rock on the Kara Sea floor is consistent with slow erosion by nearly stagnant ice beneath the ice divide. The Paleozoic platform south of the Kara Sea is consistent with slow erosion by ice moving slowly beneath the southern ablation zone. The exposed Precambrian crystalline shield on the mainland to the south is consistent with local highland glaciation that stripped off the Paleozoic cover.

Other important ice domes existed in North America over the Queen Elizabeth Islands in Arctic Canada and along the western cordillera and in Eurasia over Ireland and Britain, Spitzbergen, the New Siberian Islands, and Wrangel Island. However, these ice domes were not broad enough to extend far beyond a bed that was either frozen or mountainous, and therefore they did not permit the kind of differentiation of erosion into concentric zones that existed under the larger Laurentide, Scandinavian, and Kara ice domes.

Deposition

Glacial deposition occurred along the margins of Pleistocene ice sheets during final retreat. Glacial deposition zones are shown in Figure 9.11. A broad zone of glacial deposition exists in ice flow directions beyond the Canadian, Scandinavian, and Angaran Precambrian shields of North America, Europe, and Asia; narrower zones of glacial deposition exist along the Arctic coast of Russia from Finland to Alaska, and on Arctic continental slopes. Soil and rock removed by glacial erosion exposed these crystalline shields and was deposited on the Paleozoic platforms beyond the shields. Much of this soil and rock was reduced to glacial flour during abrasion of the shield by sliding regelation ice in which imbedded rock fragments acted as cutting tools. Since glacial erosion is mainly beneath sheet flow and over ninety percent of sheet flow became stream flow toward the ice margin, more than 90 percent of glacial deposition formed sills or debris flows at and beyond grounding lines of marine ice streams and moraines or outwash fans at and beyond terminal lobes of terrestrial ice streams. First-order depositional features are therefore glacial depositional piles on continental slopes beyond foredeepened troughs associated with interisland channels that radiate from rebounding marine embayments, moraine sills on outer continental shelves at the ends of these channels or channels beyond coastal fjords, and lobate end moraines beyond elongated lakes or river valleys that radiate from centers of former glaciation in marine embayments or on highland plateaus. Especially diagnostic are glacial depositional piles beyond marine troughs, coastal fjords, and through valleys that form continuous channels radiating from centers of isostatic rebound, such as the huge Egga Moraines on the outer continental shelf of Norway. They are at the ends of foredeepened marine troughs that extend beyond major coastal fjords at the ends of glacial through valleys that cross the Scandinavian mountains and radiate from the rebounding Gulf of Bothnia in the Baltic Sea (Andersen, 1981).

Basins

Isostatically rebounding basins after glacial terminations are present where the core areas of Pleistocene ice sheets underwent final collapse. The core area of former ice sheets is revealed by isostatically rebounding basins, which produce negative gravity anomalies, raised marine beaches, and tilted lake paleoshorelines. Rebounding basins exist over Hudson Bay, Foxe Basin, and the Queen Elizabeth Islands in North America and over the Baltic Sea, the Barents Sea, the Kara Sea, the Laptev Sea, the East Siberian Sea, and the Chukchi Sea in Eurasia. These rebounding basins are shown in Figure 9.12. Evidence for rebound consists of negative gravity anomalies and raised beaches in North America and Europe and negative gravity anomalies in Asia. The reason why raised beaches did not form along Asian arctic coasts may be because fast ice along coastal Siberia prevents the wave action that produces beaches. Also marine ice sheets in the Laptev, East Siberian, and Chukchi Seas may have been buttressed by ice shelves along their northern margins and ablated slowly along their southern margins, so they may have thinned slowly and lasted well into the Holocene, thereby reducing postglacial isostatic rebound. Isostasy contour lines produced by mapping negative gravity anomalies and mapping and dating raised beaches indicate changing ice thickness areally at the last glacial maximum and temporally during deglaciation. When nonisostatic subglacial topography is also taken into account, isostasy contours can reveal domes and saddles along former ice divides. For example, maxima over the Queen Elizabeth Islands of Arctic Canada, over Hudson Bay, over the Gulf of Bothnia, and over Spitzbergen allow ice domes to be located at these sites, and minima between these maxima allow ice saddles to be located in Barrow Strait for Arctic Canada and in the Barents Sea for Arctic Europe.

Zonation

Figure 9.13 presents the analysis by Robin (1955) that showed how decreasing surface temperature and precipitation as ice elevation increases cause basal ice temperature to increase under the central dome of an ice sheet, because basal geothermal heat is conducted upward more slowly, but that advection of cold ice at high interior elevations to lower elevations near the ice margin tends to keep basal temperatures cold away from the central ice dome because basal geothermal heat is conducted upward more rapidly. Therefore, the bed becomes thawed only near the ice margin. Countering this tendency, the melting-point isotherm at the base of permafrost rises beneath a thickening ice sheet that increasingly depresses the bed, so the bed first becomes thawed beneath the thick ice dome, as seen in Figure 9.14. The pressure melting temperature increases as ice thickness decreases, so if basal ice is at or near its melting point under ice domes, it can remain well below its melting point near ice margins. In addition to these effects, basal frictional heat increases with increasing ice velocity from ice domes to the surface equilibrium line, beyond which surface ablation causes ice velocity to decrease toward the ice margin, except in ice streams. These competing effects are illustrated in Figure 9.15.

Zonation is different for terrestrial and marine sectors of an ice sheet, as illustrated in Figure 9.16. Permafrost is thick on polar uplands and thin on subpolar lowlands. Figure 9.16 shows an ice sheet at the last glacial maximum with terrestrial flowlines that extended from frozen uplands onto thawed lowlands and marine flowlines that extended from a thawed marine embayment onto a frozen continental plain.

Terrestrial flowlines from the upland to the lowland had a frozen bed under the ice divide and a thawed bed under the ice margin with a melting bed in between. As shown in Figure 9.16, melting would begin where the melting-point isotherm at the base of the permafrost came into contact with low places on the bed, such as the floors of former

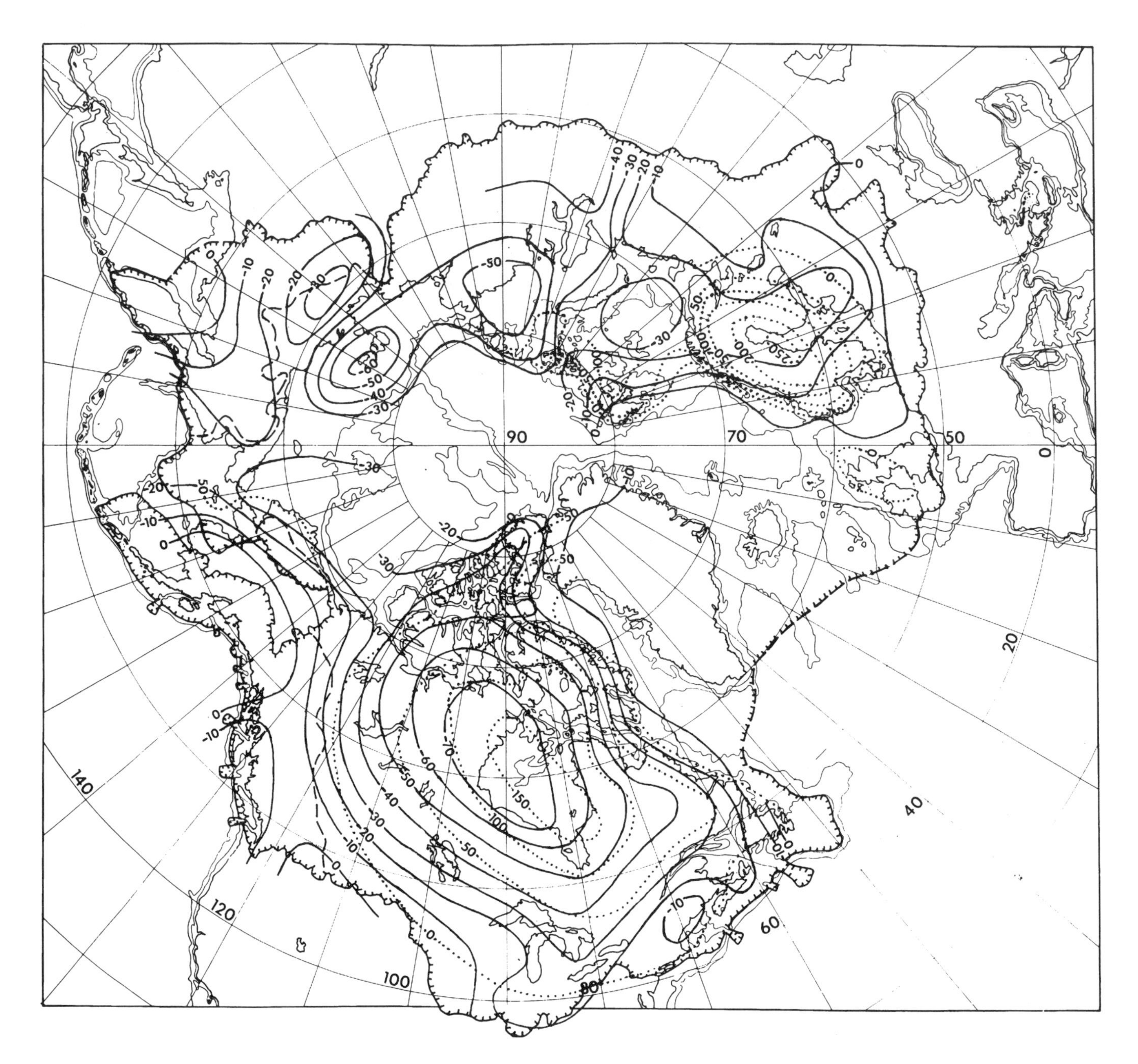

Figure 9.12: Evidence for postglacial isostatic rebound beneath Pleistocene ice sheets after their final retreat: negative gravity anomalies (m gal), from Tarakanov and others (1987) (solid lines); raised beaches (m), raised beaches date from 7 ka BP in North America (Peltier and Andrews, 1983), from 13 to 10 ka BP, in Europe (Andersen and Borns, 1994), and are beyond radio-carbon dating in Siberia and Alaska (dashed lines); the outer limit of Pleistocene glaciation (hatched line); and separation of marine and mountain glaciation (broken line).

lakes. Glacial sliding would be possible in these thawed patches, so glacial erosion would deepen them and lengthen them in the direction of flow. Under steady-state conditions, the number of ice flowlines entering a given patch due to basal melting and leaving the patch due to basal freezing would be identical, so the patches would deepen but not expand over time. However, the patches would be larger and more numerous across the melting zone until the entire bed was thawed. As the ice sheet advanced or retreated, the melting zone would advance or retreat with it. During advance, more ice flowlines would enter a given patch than leave, so the patch would expand. During retreat, more ice flowlines would leave a given patch than enter, so the patch would contract. The melting-point isotherm ascends during advance if ice thickens and descends during retreat if ice thins. This compels advance and retreat of the melting zone.

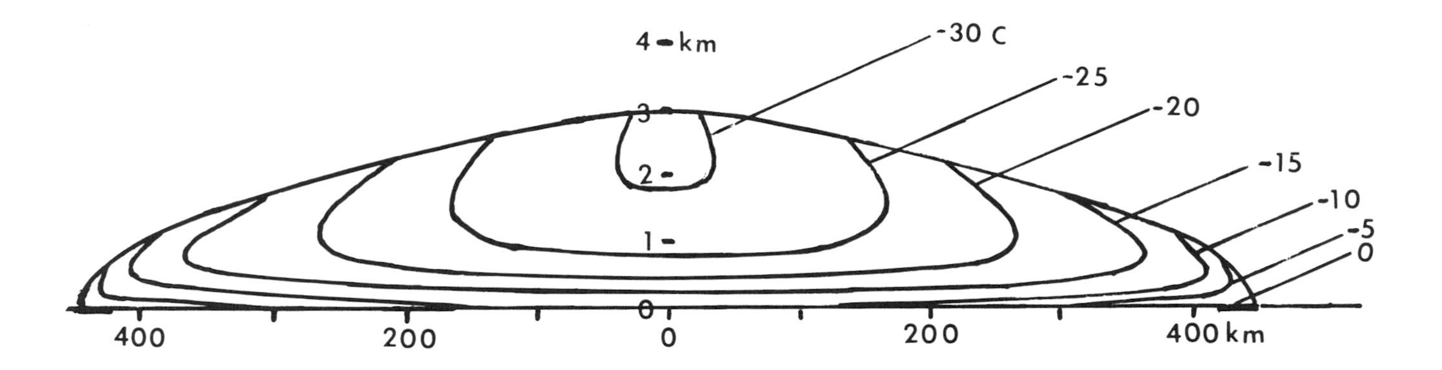

Figure 9.13: Schematic temperature isotherms for a steady-state ice sheet resulting from ice precipitation rates and mean surface temperature that decrease with ice elevation and ice advection rates that increase from the ice divide to the ice margin. Temperatures, elevations, and lengths are chosen for comparison with the Greenland Ice Sheet. Reproduced by courtesy of G. deQ. Robin and the International Glaciological Society, from the *Journal of Glaciology, 2*(18) (1955), p. 531, Figure. 5.

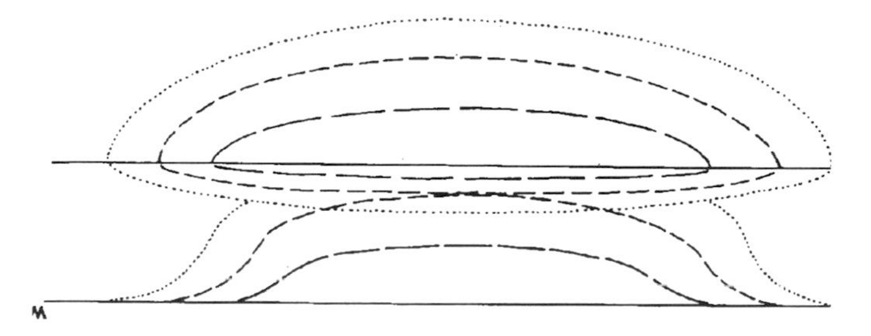

Figure 9.14: Ascent of melting point isotherm T_M at the base of permafrost beneath a growing ice sheet. The bed is thawed when T_M coincides with the ice-bed interface: the ice-sheet surface, bed, and T_M are shown at successive stages of growth (solid, broken, dashed, and dotted lines).

On the other hand, marine flowlines from the marine embayment onto a high interior plain would probably have a thawed bed under the ice divide and a frozen bed under the ice margin, with a freezing bed in between. As shown in Figure 9.16, freezing would begin where the melting-point isotherm at the base of the permafrost was below the ice-bed interface, notably the tops of hills beneath the ice sheet. These frozen hills would be larger and more numerous across the freezing zone, and the hills would become elongated in the direction of ice flow due to glacial erosion by sliding over the thawed areas between hills. The number of ice flowlines entering and leaving the thawed areas would be identical under steady-state conditions, so the frozen hills would remain unchanged. However, more flowlines would enter than leave if the ice sheet advanced, and more would leave than enter if the ice sheet retreated, causing the frozen hills to become smaller as the melting-point isotherm ascended when ice thickened during advance and to become larger as the isotherm descended when ice thinned during retreat. This compels advance and retreat of the freezing zone. The hills

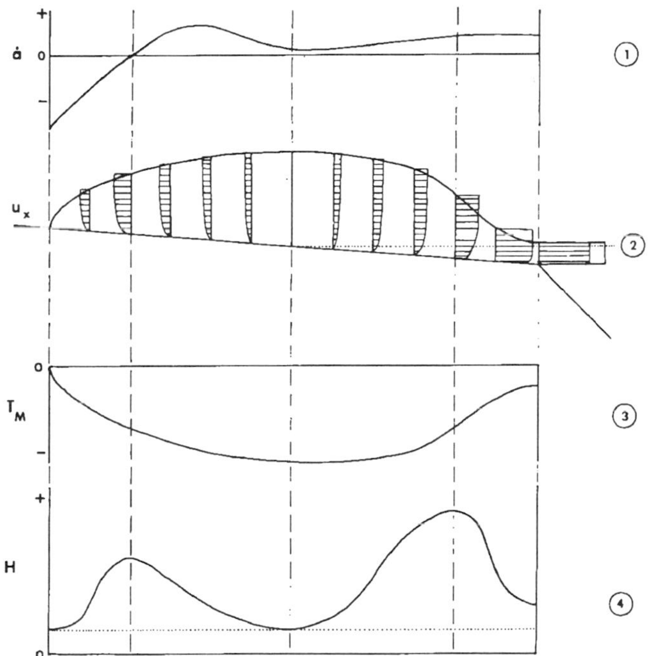

Figure 9.15: The relation between (1) surface accumulation or ablation rates a, (2) horizontal ice velocity u_x, (3) basal melting temperature T_M, and (4) basal heat H for an ice sheet having terrestrial (left) and marine (right) margins. A constant geothermal heat flux (dotted line) is assumed in (4).

would become drumlinized in the process, with the drumlins becoming longer and more sculptured from the ice divide to the ice margin.

Glacial erosion for sheet flow and stream flow is illustrated in Figure 9.17. As an ice sheet thickens and expands, permafrost beneath the ice sheet thins most under the central ice dome, and the bed thaws when the melting-point iso-

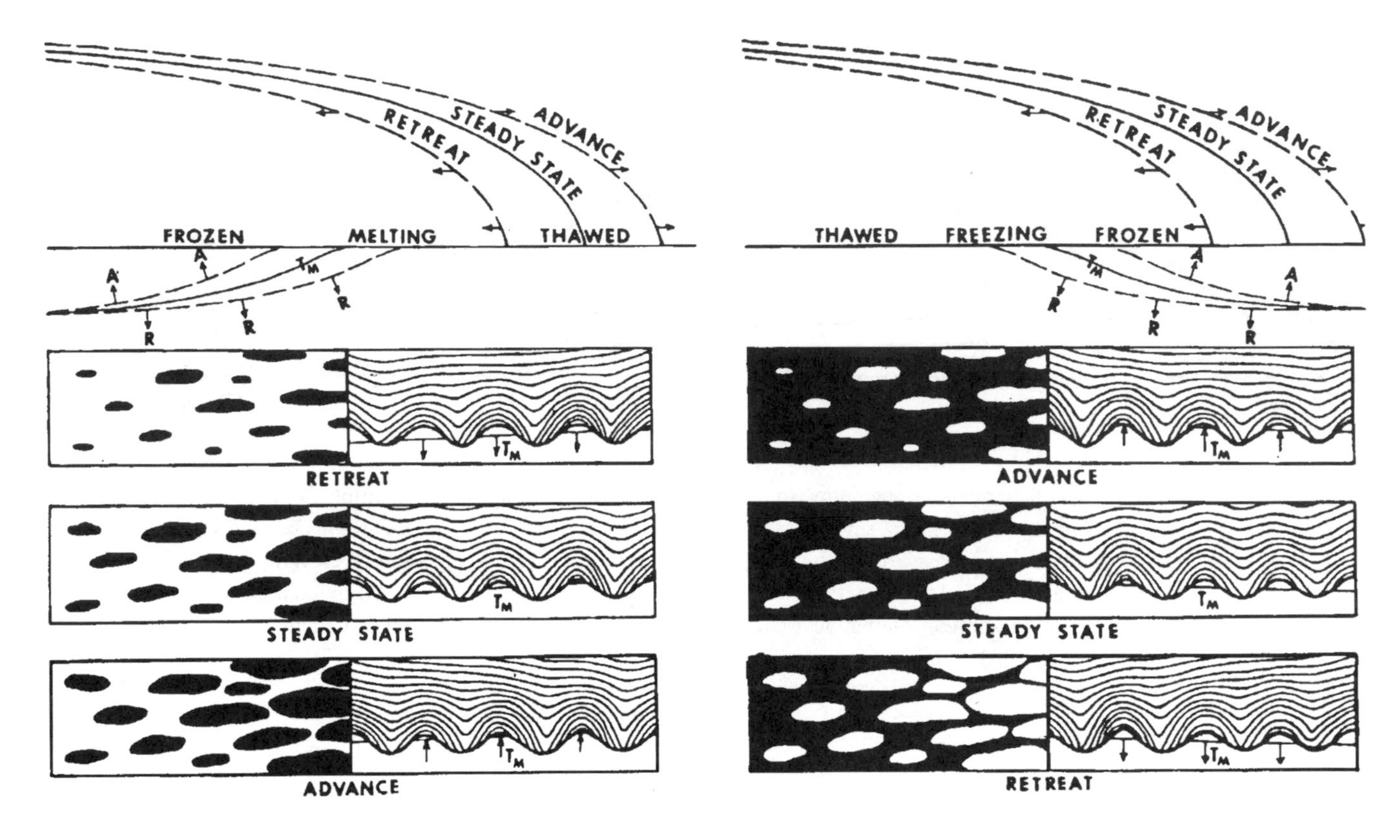

Figure 9.16: Advance and retreat of basal melting and freezing zones with advance and retreat of an ice sheet. *Left*: Profile and plan views along a terrestrial flowline from a frozen center to a thawed margin. *Right*: Profile and plan views along a marine flowline from a thawed center to a frozen margin. The ice-sheet profile and the basal melting temperature isotherm T_M are shown during steady-state, advance A and retreat R (solid and dashed lines), thawed (black) and frozen (white) basal areas in plan view, and ice flowlines entering and leaving thawed basal areas in longitudinal cross section.

therm coincides with the ice-bed interface. Since this interface has topography, the melting-point isotherm meets the bed first in the low places, which then become thawed patches on the bed that can be further deepened by glacial erosion. If the ice sheet is centered over an isostatically depressed marine embayment, as was the Laurentide Ice Sheet, and has a wholly thawed bed under a thick ice dome, a freezing bed consisting of frozen and thawed patches will surround the thawed bed because only bed hollows touches the melting-point isotherm. An ice sheet centered over mountains, such as the Cordilleran Ice Sheet, may have a frozen bed under ice divides where thin ice covered a mountain range but could have a melting bed consisting of frozen patches along ridges and thawed patches along valleys that flank the mountain range. A "trimline" separating serrated upper ridge crests from smooth lower ridge crests will mark the thermal boundary, with glacial sliding smoothing the lower ridges. These trimlines can be misinterpreted as the maximum ice elevation instead of the maximum elevation of the basal melting point isotherm. For both marine and highland ice sheets, ice streams develop near ice-sheet margins. Glacial erosion causes foredeepening at their heads, notably quarrying at fjord head-walls. Glacial deposition occurs at their feet, producing end moraines at the terminal lobes of terrestrial ice streams and at the grounding lines of marine ice streams.

Basal frozen, thawed, and melting zones can account for apparent weathering zones on headlands between fjords on Baffin Island (Pheasant and Andrews, 1972; Boyer and Pheasant, 1974), Labrador (Ives, 1957; Andrews, 1963), Newfoundland (Grant, 1977; Brookes, 1982), and Nova Scotia (Grant, 1971, 1976). If the bed was frozen beneath thin ice domes on these headlands, preglacial tors and felsenmeer would be undisturbed; thick ice streams would slide over a thawed bed in fjords between these headlands, with selective linear erosion polishing and striating the bedrock; and ice flowing into these ice streams from the local ice domes would pass over a melting bed in which hollows were thawed patches that could be eroded to produce a deglaciated rolling landscape of lakes and hills spattered with erratics (Hughes, 1987). Figure 9.18 illustrates the transition from a frozen to a melting to a thawed bed as ice moves from local ice domes on headlands to ice streams in fjords. Iverson (1991) has analyzed quarrying in thawed patches

Figures 9.16 through 9.18 show ice melting in the back part of thawed patches and meltwater refreezing in the front

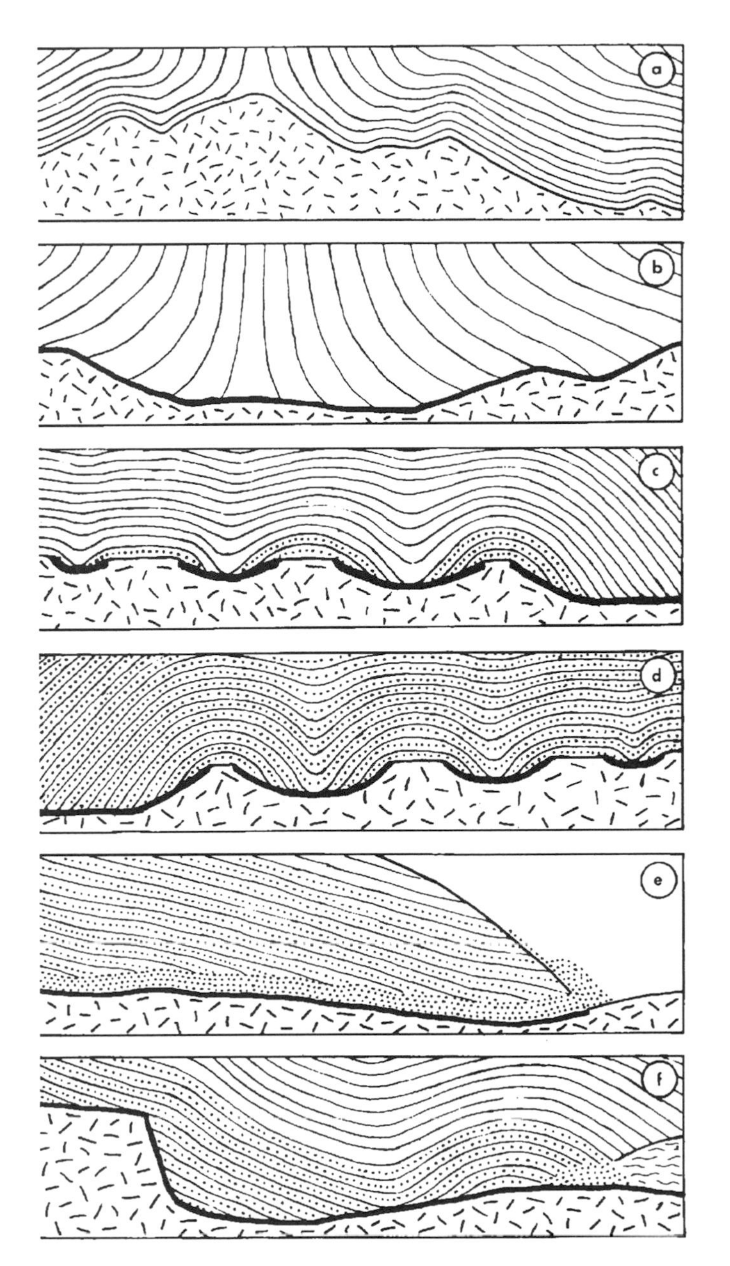

The sequence of basal thermal zones associated with Figure 9.17: Erosion and deposition beneath an ice sheet: flowline trajectories in vertical longitudinal cross-section (thin lines); eroded debris entrained in basal regelation ice (dotted lines); steady-state deposition of eroded debris as regelation ice melts (dotted areas); bedrock (stippled areas); a frozen bed (light line between bedrock and ice); a thawed bed (heavy line between bedrock and ice). (a) Flow over frozen highlands beneath the central ice divide. Ice flowlines descend to the bed and diverge parallel to the bed, with no motion at the ice-bed interface. (b) Flow over a thawed basin beneath the central ice divide. Ice flowlines descend and intersect the bed, with sliding at the ice-bed interface. (c) Flow at the inner part of a basal melting zone. Ice melts in the upstream end of isolated thawed patches and the meltwater freezes in the downstream end, creating bands of regelation ice that eventually melt in the outer part of the basal melting zone. (d) Flow at the outer part of a basal freezing zone. Regelation ice from the inner part of the basal freezing zone melts in the upstream end of isolated thawed patches and the meltwater freezes in the downstream and, creating new bands of regelation ice that lie beneath the older regelation ice. Active steady-state quarrying takes placer in the thawed patches shown in (c) and (d), creating pits that become the sites of lakes after deglaciation. (e) Flow in the outer deposition zone of basal melting near the ice-sheet margin. Debris entrained in regelation ice melts out to create a ground moraine across the zone and a terminal moraine along the ice margin. (f) Flow in a fjord occupied by an ice stream. Regelation ice melts above the headwall, and the meltwater refreezes on the headwall to create new regelation ice that melts, refreezes, and melts on the fjord floor, leaving an end moraine as a sill.

part. This is purely for purposes of illustration. Within a thawed patch, areas of melting and freezing change continually as glacial erosion within the patch continually modifies the ice-rock interface. Therefore, abrasion where ice melts and quarrying where the meltwater refreezes do not necessarily reveal the direction of ice sliding. Basal water freezes on the part of thawed patches that will be quarried, with rock fragments becoming imbedded in basal regelation ice. Basal ice melts on the part of thawed patches that will be abraded by rock fragments melting out of basal regelation ice. For steady-state sliding, the patches neither expand nor contract, they only deepen. However, the patches expand if the ice thickens and contract if the ice thins, making the melting or freezing zone expand or contract. Thawed patches become larger and closer together as ice crosses a basal melting zone and become smaller and farther apart as ice crosses a basal freezing zone. The size and spacing of lakes with distance from an ice dome indicate whether the bed under the dome was frozen and surrounded by a melting zone or thawed and surrounded by a freezing zone.headlands and fjords along the margins of an ice sheet, illustrated in Figure 9.18, can also exist in the central core area of the ice sheet. Beneath the Laurentide ice sheet, a cap of Paleozoic rock over Hudson Bay and the Hudson Bay low-lands would have been preserved if the bed had been either frozen or thawed, because nearly stagnant ice under the ice divide would have produced little glacial erosion. The distribution of exposed bedrock and unconsolidated rock in Figure 9.19 suggests that the Laurentide ice divide had a thawed bed on its seaward side and a frozen bed on its landward side, if the distribution is a first-order feature produced by glacial erosion and deposition processes in the steady-state core of the Laurentide Ice Sheet. Unconsolidated rock covers the cap of Paleozoic rock. Exposed bedrock is concentrated along the eastern and

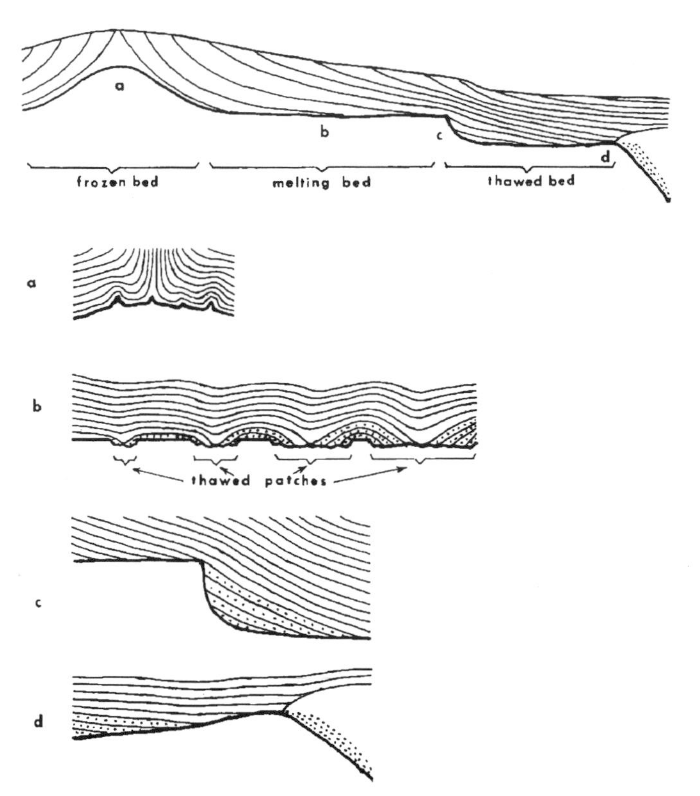

Figure 9.18: A geomorphic interpretation of weathering zones. *Top*: Generalized ice trajectories along a flowline for an ice sheet with flow diverging from a local ice dome over the frozen bed of a highland summit (left), spreading over the melting bed of a highland plateau (center), and converging over the thawed bed of a coastal fjord (right). *Bottom*: Details of flow, (a) flowlines over and around tors on the highland summit; (b) flowlines entering and leaving thawed patches on the melting bed of the highland plateau, with dotted areas being debris-filled refrozen ice formed in the downstream part of thawed patches, which get bigger and more numerous in the downstream direction; (c) flowlines intersecting the melting bed upstream from the fjord headwall, originating at the freezing bed of the fjord headwall, or intersecting the melting bed downstream from the fjord headwall, with active quarrying at the fjord headwall to produce debris-filled refrozen ice shown as dotted area; (d) flowlines intersecting the melting bed upstream from the grounding line and the melting underside of the floating ice shelf downstream from the grounding line as the ice stream leaves the fjord, with dotted areas showing refrozen ice (left) and a submarine terminal moraine (right).

northwestern shores of Hudson Bay, along the arc of large intracontinental lakes bordering the outer edge of the Precambrian shield, and in the fjordlands of Newfoundland, Labrador, and Baffin Island. When considered along with the distribution of isostasy and lakes, this distribution is consistent with a thawed bed on the seaward side of the ice divide, where ice converged primarily on a major marine ice stream in Hudson Strait and secondarily on smaller ice streams in coastal fjords, and with a frozen bed on the landward side of the ice divide, where ice converged on terrestrial ice streams occupying the large linear lakes along the edge of the Precambrian shield. On the seaward side of the ice divide, the thawed bed became a freezing bed as ice crossed the plateaus of Quebec. Basal freezing in the thawed patches of the freezing zone removed the unconsolidated rock and quarried pits that became lakes after deglaciation, whereas unconsolidated rock remained in the frozen patches. The freezing bed became a melting bed at the heads of Hudson Strait and coastal fjords, where ice converged on ice streams sliding on a thawed bed in Hudson Strait and in the fjords. As with the inner freezing zone, unconsolidated rock would have been removed by glacial erosion in thawed patches and would remain in frozen patches of the outer melting zone. On the landward side of the ice divide, the frozen bed became a melting bed beyond the Hudson Bay lowlands, with glacial erosion removing unconsolidated rock, producing pits now occupied by lakes in the thawed patches, and leaving unconsolidated rock in the frozen patches. Thawed patches became larger and coalesced as ice converged on terrestrial ice streams at the outer edge of the Precambrian shield. These ice streams had a thawed bed, and they occupied and eroded the large linear lakes at the edge of the shield, notably Great Bear Lake, Great Slave Lake, Lake Athabasca, Lake Manitoba, Lake Winnipeg, Lake Superior, Lake Michigan, Lake Huron, Lake Erie, Lake Ontario, and Lake Champlain.

Second-Order Glacial Geology

Mapping second-order glacial geology is like observing the trees instead of the forest. Second-order glacial geology

→ Figure 9.19: Glacial erosion and deposition beneath the Laurentide Ice Sheet. *Top*: Surface materials map of Canada showing mostly exposed bedrock (light areas), bedrock mostly covered with unconsolidated material (gray areas) and bedrock about half exposed and half covered (dark areas). Bedrock is exposed by areal scouring of a thawed bed (light areas), but not by ice moving over a frozen bed or by deposition over a thawed bed (gray areas), with quarrying concentrated in basal melting and freezing zones that migrate (dark areas). This is a glaciological interpretation of the surface materials map on pages 37–38 in the *National Atlas of Canada*, Fourth Edition (1974), Her Majesty the Queen in Right of Canada with permission of Natural Resources Canada. *Bottom*: Basal thermal zones in the erosion zone beneath the Laurentide Ice Sheet are correlated with surface materials (top) and consist of frozen areas, freezing areas, melting areas, and thawed areas. The erosion zone is separated from the deposition zone by a heavy line. When the bed becomes mostly thawed in the erosion zone, partial gravitational collapse causes the ice sheet to spread from the erosion zone to the deposition zone, primarily as ice streams (dotted flowbands). This spreading is shown for the last glacial maximum. Ice flowlines are shown in the erosion zone (*solid lines*) and in the deposition zone (*broken lines*). Areas of mountain glaciation are enclosed by solid lines and ice-shelf grounding lines are dashed. →

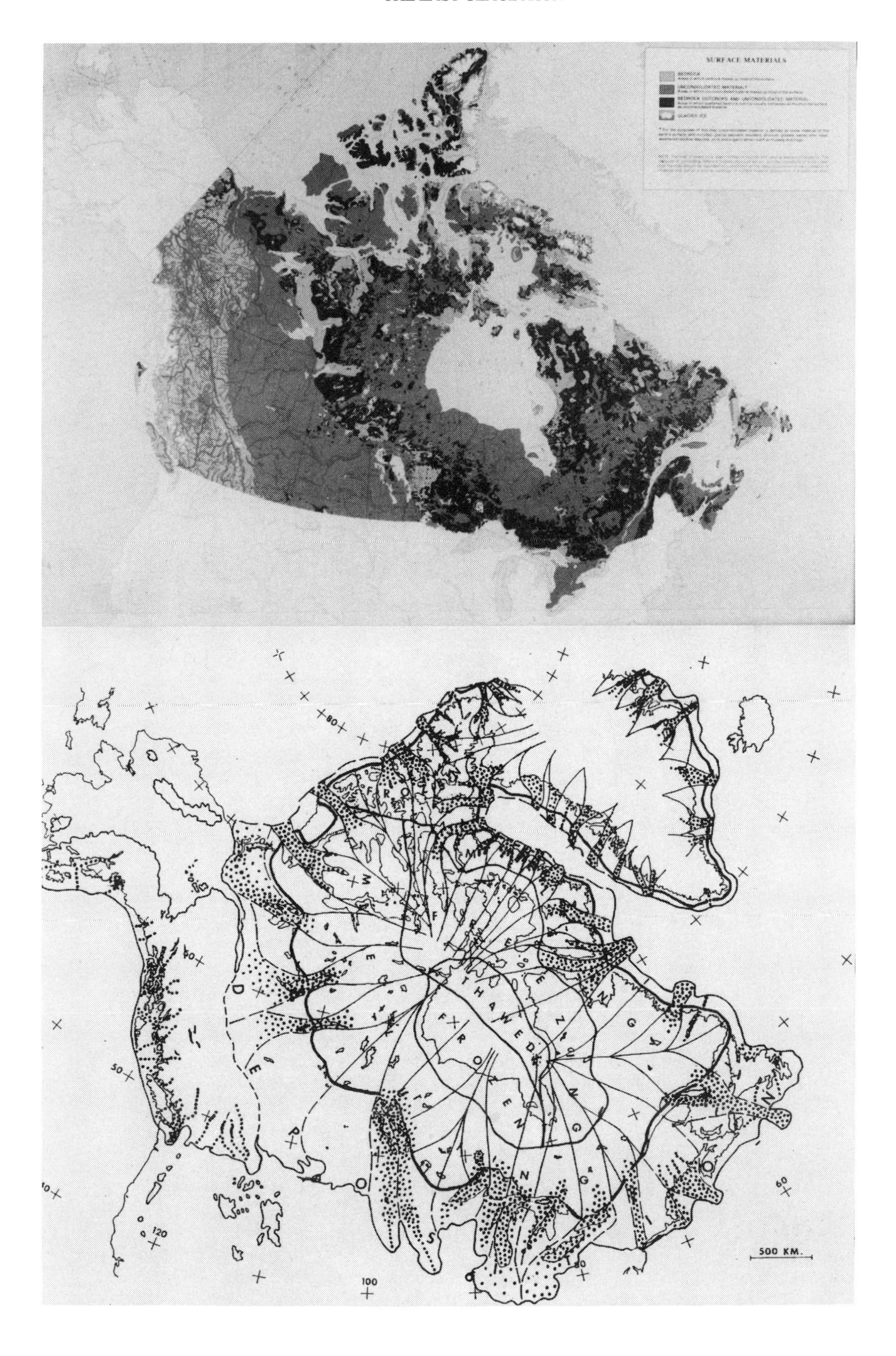
SURFACE MATERIALS
80
60
50
120
100
60
500 KM.

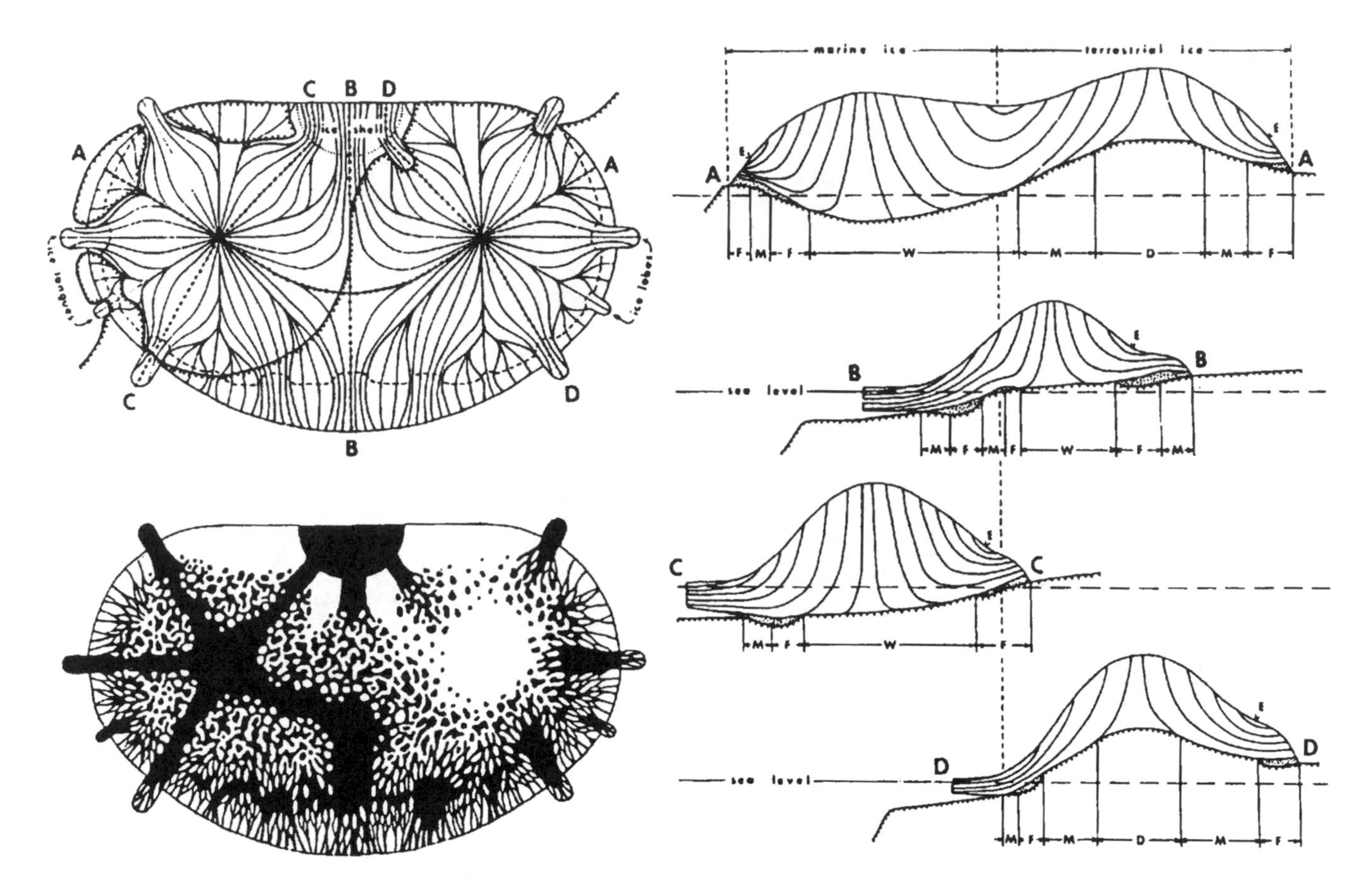

Figure 9.20: Idealized ice-sheet flow regime. *Left*: Flow in plan view for surface ice (top) and basal water (bottom). Shown at top are surface flowlines (solid lines) radiating from a terrestrial dome (inside hatched line) and a marine dome (outside hatched line); a surface equilibrium line (dashed line) that is highest on the equatorward flank of the ice sheet; and an ice-shelf grounding line (dotted line) on the poleward flank of the ice sheet. Shown at bottom are thawed patches (isolated black areas), where quarrying creates lakes, frozen patches (isolated white areas), where lodgement till creates drumlins; the arc of exhumation (black areas beneath the surface equilibrium line), where quarrying and regelation occur along a basal equilibrium that separates an inner melting regime from an outer freezing regime; the arc of deposition (between the arc of exhumation and the ice margin), where regelation ice melts; selective linear erosion in ice-stream channels (broad black bands radiating from ice domes); and selective linear deposition in eskers (narrow black lines between broad black bands). *Right*: Flow of ice in vertical cross-section along dotted surface flowlines A–A' through D–D' shown on the left. Shown at top are longitudinal flowline profiles along the crest of the ice divide (section A–A'), along opposite flanks of a saddle on the ice divide (section B–B'), along opposite flanks of the marine ice dome (section C–C'), and along opposite flanks of the terrestrial ice dome (section D–D'). At the base (hatched line), a dry bed (D) is frozen everywhere, a wet bed (W) is thawed everywhere, melting beds (M) and freezing beds (F) have a mix of frozen and thawed areas, and regelation ice (dotted areas) forms at ice-stream headwalls and over the thawed parts of a freezing bed.

weakly overprints first-order glacial geology, and consists of lineations that form near ice-sheet margins, with longitudinal lineations being parallel to ice-flow directions and transverse lineations being parallel to ice margins. Longitudinal lineations include striations and grooves on bedrock, flutes on till blanketing bedrock, drumlins and roches moutinnes, beaded chains of linear glacial lakes, hole and hill pairs, tunnel valleys, and eskers. Transverse lineations include push moraines, terminal moraines, recessional moraines, dump moraines, ribbed or washboard moraines, belts of stagnation terrain, ice marginal rivers, and ice-contact fluviatile accumulations. Concentric thawed, freezing, frozen, and melting basal zones expand and contract as an ice sheet advances and retreats, so these zones produce first-order glacial geology that records "breathing" by the ice sheet, during which second-order glacial geology, whether areal scouring of a thawed zone or quarrying and abrasion within thawed patches of freezing and melting zones, is produced. Each time the ice sheet exhales (advances) and inhales (retreats), the second-order glacial geology is overprinted because the zones of first-order glacial geology migrate outward and inward.

Figure 9.20 illustrates the relationship between first-order and second-order glacial geology. Surface flowlines diverging from domes and converging from saddles along the ice divide converge on ice streams or diverge from ice ridges between ice streams along the ice margin. This pattern of surface flowlines produces first-order and second-order glacial geology on the bed. First-order glacial geology consists of a thawed bed surrounded by a freezing zone beneath a thick marine ice dome over a marine embayment, a frozen bed surrounded by a melting zone beneath a thin terrestrial ice dome over a continental plateau, and linear depressions that radiate from these domes and that were occupied by ice streams. These depressions are typically straits and interisland channels for marine ice streams, linear lakes and river valleys for terrestrial ice streams, and through valleys into fjords that meet troughs across the continental shelf for ice streams that cross coastal mountain ranges. Individual flowlines can cross a dry bed (D) that is frozen, a wet bed (W) that is thawed, a melting bed (M) that begins as isolated thawed patches and ends as isolated frozen patches, and a freezing bed (F) that begins as isolated frozen patches and ends as isolated thawed patches. After deglaciation, with increasing distance from former ice domes in directions of former ice flow, isolated thawed patches become increasingly elongated lakes and isolated frozen patches become increas-ingly elongated hills, both as a result of increasing glacial erosion. These trends are illustrated in Figure 9.20 as black patches for thawed patches and white patches for frozen patches. The wet bed tends to consist of subglacial melt-water channels that often form eskers where sheet flow extends to the ice margin under ice ridges between ice streams.

Figure 9.20 also relates ice flowlines to basal thermal zones that are dry (D), wet (W), melting (M), and freezing (F) in longitudinal cross-sections along selected flowlines. Section AA shows flowline profiles along the central ice divide, which includes the marine dome, the terrestrial dome, and the intervening saddle. Flow is strongly divergent all along this ice divide. The embayment beneath the marine ice dome is wet, and the plateau beneath the terrestrial ice dome is dry, primarily because the preglacial temperature is higher at lower elevations and the pressure melting temperature is lower under thicker ice. The ice divide bows toward the terrestrial ice margin because the grounded ice margin is higher on land than in the sea. Section BB shows flowline profiles down the flanks of the central ice saddle, to the calving front of a floating ice shelf on the marine side and to the terminal moraine of a grounded ice piedmont on the terrestrial side. The ice shelf is formed from the confluence of three marine ice streams. The ice piedmont is formed from the confluence of five terrestrial ice streams. The major ice streams in both cases form on opposite flanks of the ice saddle. Beneath these ice streams, a freezing bed becomes a melting bed, with glacial erosion producing a foredeepened trough along the marine ice stream and an elongated depression along the terrestrial ice stream. The trough becomes a strait and the depression becomes a lake after deglaciation. Section CC shows flowline profiles on opposite flanks of the marine ice dome, one along a flowline to a floating marine ice tongue imbedded in the ice shelf and one along a flowline to a grounded terrestrial ice lobe. Section DD shows flowline profiles on opposite flanks of the terrestrial ice dome, becoming the respective centerlines of marine and terrestrial ice streams, as in Section CC. The surface equilibrium line is a dashed line in the plan view of Figure 9.20, and crosses flowlines at point E in the flowline cross sections. It shows that the surface ablation zone widens toward the equator.

Properly interpreted, crystalline shields glaciated by Quaternary ice sheets can be gigantic Rosetta stones for understanding Quaternary glacial history and processes. Kleman (1994) has developed criteria for linking the steady-state basal thermal zones in Figure 9.20, which produce a first-order glacial landscape if the ice sheet is undisturbed, with the migrations of these basal thermal zones that accompany advance or retreat of the ice sheet. Migrations of these zones produce the second-order glacial landscape of overprinted lineations analyzed by Boulton and Clark (1990) for the Laurentide Ice Sheet. In his model of glacial geomorphic systems, Kleman (1994) recognizes:

1. A dry bed system in which no new landforms are created so long as the bed under an ice sheet remains frozen
2. A wet bed system in which basal sliding produces landforms aligned in the ice-flow direction and which may reveal multiple flow directions
3. A marginal meltwater system consisting of delicate landforms, such as eskers and meltwater channels, that preserve only the last ice-flow direction
4. A supraglacial drift depositional system that is most active where debris-charged regelation ice outcrops along the ice-sheet margin at its maximum extent
5. An interstadial subaerial system consisting of periglacial landforms

Figure 9.21 illustrates the first three geomorphic systems and shows their time-transgressive nature as basal thermal zones expand and contract with advance and retreat of the ice sheet. Following Hooke (1977), this model assumes sheet flow without stream flow and assumes that the ice sheet forms over land-based permafrost, such as on a plateau, rather than in a marine embayment. A more realistic representation would place a melting zone of frozen and thawed patches between the inner frozen zone and the outer thawed zone, and the thawed zone would penetrate the frozen zone along ice streams, as shown in Figure 9.20.

The main consequence of taking these and other additional features into account would be to modify the circular symmetry of the geomorphic systems displayed in Figure

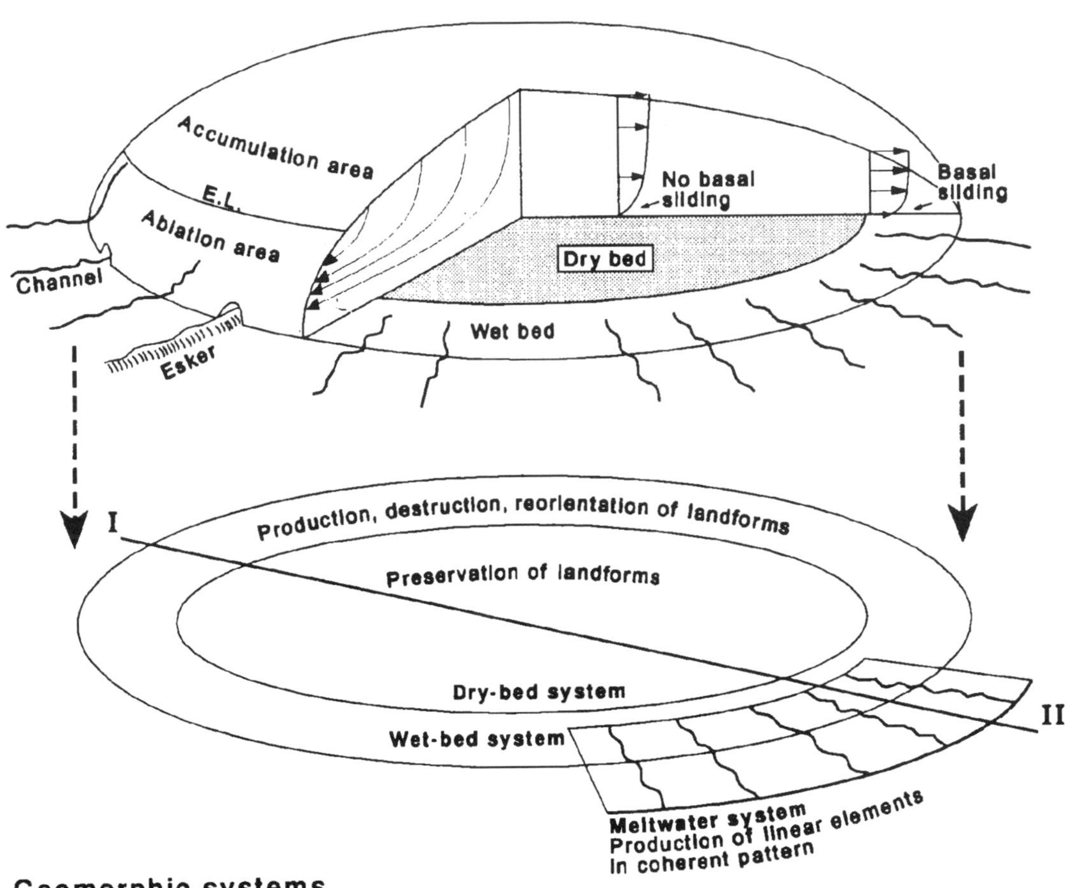
Ice sheet
Accumulation area
E.L.
Ablation area
Channel
Esker
No basal sliding
Basal sliding
Dry bed
Wet bed
I
II
Production, destruction, reorientation of landforms
Preservation of landforms
Dry-bed system
Wet-bed system
Meltwater system
Production of linear elements
in coherent pattern

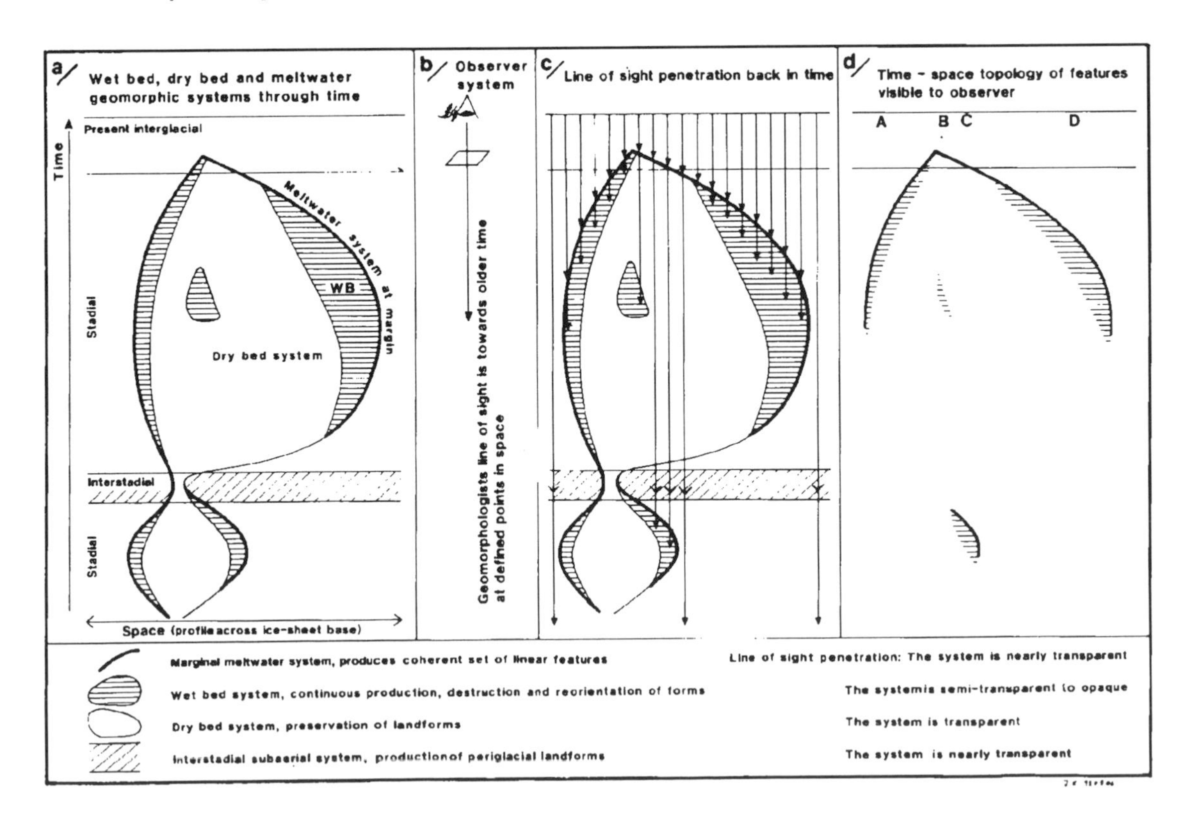
Geomorphic systems
a/ Wet bed, dry bed and meltwater geomorphic systems through time
Present interglacial
Time
Stadial
Interstadial
Stadial
Meltwater system at margin
WB
Dry bed system
Space (profile across ice-sheet base)
b/ Observer system
Geomorphologists line of sight is towards older time at defined points in space
c/ Line of sight penetration back in time
d/ Time - space topology of features visible to observer
A
B C
D
Marginal meltwater system, produces coherent set of linear features
Wet bed system, continuous production, destruction and reorientation of forms
Dry bed system, preservation of landforms
Interstadial subaerial system, productionof periglacial landforms
Line of sight penetration: The system is nearly transparent
The systemis semi-transparent to opaque
The system is transparent
The system is nearly transparent

← Figure 9.21: Glacial geomorphic systems proposed by Kleman (1994). *Top*: An ice sheet is sectioned to show velocity profiles for creep over a frozen (dry) bed and sliding over a thawed (wet) bed, flowline trajectories, the surface mass-balance regime, and the basal thermal regime. Dry bed, wet bed, and marginal meltwater geomorphic systems are shown for one-dimensional symmetry. *Bottom*: A demonstration of how time-space histories can be deduced by observing geomorphic landforms along transects such as I-II. The actual history is shown in (a), the line-of-sight along a given transect is shown in (b), what can be seen along all possible lines-of-sight is shown in (c), and what can be seen in sectors A, B, C, and D is shown in (d). Reproduced courtesy of J. Kleman and Elsevier Science, from *Geomorphology, 9,* (1994), p. 28, Figure 7.

9.21 so that a geomorphologist walking along a transect such as I-II would gain a different look into the past history of the ice sheet for each transect he followed. These different looks along different transects are represented by one-dimensional space-time relationships in Figure 9.21. The actual relationship is shown in (a). The ice sheet begins to grow over a frozen bed that remains a dry bed system under the ice sheet except along the margin, where a wet bed system lies inside a marginal meltwater system. The ice sheet grows and shrinks through a short stadial, becomes a small ice cap during a shorter interstadial, and expands into a much larger ice sheet during the much longer stadial that follows before shrinking and vanishing during the present interglacial. At the last glacial maximum, a wet bed system developed under the ice dome within the dry bed system, and the outer wet bed system and marginal meltwater system reappear beyond the dry bed system.

The geomorphologist's view into this glacial history along various transects is shown in (b). Lines of sight along various transects are shown in (c), and these reveal how much of the glacial history in (a) can be reconstructed by the geomorphologist if he follows all of the transects. Lines of sight with two or more arrowheads show that the marginal meltwater system is nearly transparent, the wet bed system is transparent for some distance into the system but not through the system, the dry bed system is wholly transparent, and the interstadial subaerial system is nearly transparent.

The information available along transects in various sectors is displayed in (d). In sector A, transects reveal the times and locations of the marginal meltwater system, the wet bed system, and, where the wet bed system is narrow, the interstadial subaerial system. In sector B, transects reveal the meltwater marginal system and the interior wet bed system, because no outer wet bed system exists to block the line of sight. In sector C, the interior wet bed system is absent, so the line of sight reveals the marginal meltwater system, the interstadial subaerial system, and the marginal meltwater and dry bed systems of the first stadial. In sector D, transects reveal the same information revealed by transects in sector A.

Reconstructing Interstadial and Stadial Ice Sheets

The model of geomorphic systems developed by Kleman (1994) is very useful in deducing the history of the last glaciation. In applying the model, first-order and second-order glacial geology can be used for reconstructing ice sheets during interstadials and stadials, respectively. The centers of glaciation to which the model can be applied fall into distinct types:

1. An Innuitian Ice Sheet that develops over an island archipelago in the high Arctic
2. A Cordilleran Ice Sheet that develops on a broad plateau
3. A Laurentide Ice Sheet that develops over a marine embayment in a continental interior
4. Maritime ice sheets that develop on islands surrounding the North Atlantic
5. A Scandinavian Ice Sheet that develops on a large peninsula incised by fjords and embayments
6. A Siberian Ice Sheet that develops on a broad Arctic continental shelf

Flowline profiles for these ice sheets are reconstructed using Equation (9.53) as modified for sheet, stream, and shelf flow, ablation by both melting and calving, variable flowband widths and accumulation rates, and changing glacio-isostatic adjustments through time.

Innuitian Ice

North American glaciation probably began with nucleation and growth of a marine ice sheet over the Queen Elizabeth Islands of Arctic Canada. Blake (1970) called this the Innuitian Ice Sheet, and, by mapping and dating raised beaches, he showed that it had an ice divide extending northeast from a saddle in eastern Viscount Melville Sound across a dome in Norwegian Bay to a saddle in Kane Basin at the last glacial maximum, and its final collapse came in the early Holocene when calving bays carved away ice streams occupying interisland channels. England (1987) maintained that a late Quaternary Innuitian Ice Sheet did not exist, that the islands and interisland channels in the Queen Elizabeth Islands are horsts and grabens produced by late Tertiary block faulting, after which ice caps formed only on the larger and higher eastern islands, primarily Ellesmere Island, Axel-Heiberg Island, and Devon Island, which are partly glaciated today. These opposing viewpoints can be reconciled by making a distinction between first-order and second-order glacial geology, based on the discussion of field evidence by Mayewski and others (1981).

First-order glacial geology consists of the pattern of dated raised beaches, which leaves no doubt that the Innuitian Ice Sheet existed during the last glacial maximum, and the pattern of interisland channels and through valleys across Ellesmere Island that all radiate from the ice dome over Norwegian Bay. Reeh (1984) reconstructed the Innuitian Ice Sheet at the last glacial maximum, using a computer model that plotted ice flowlines as model output. Flowlines showed ice converging on ice streams in these interisland channels. The role of ice streams in glacial erosion is primarily to concentrate material eroded beneath an ice sheet, as ice streams discharge more than 90 percent of ice from the present-day Greenland and Antarctic Ice Sheets. Because basal shear stress is low beneath ice streams and because glacial erosion increases as the product of basal shear stress, which is a measure of the ice grip on the bed, and basal sliding velocity that detaches and transports basal material, glacial erosion by the low basal shear stress and high sliding velocity of stream flow need not be enormously greater than glacial erosion by the high basal shear stress and low sliding velocity of sheet flow. Therefore, it is quite likely that the interisland channels and through valleys in the Queen Elizabeth Islands originated from Tertiary block faulting and were only modified by Quaternary ice streams. What cannot be disputed is that ice streams would have occupied the interisland channels and through valleys that ended in coastal fjords and would have caused foredeepening of the channels and fjords by inner glacial erosion and outer glacial deposition.

Second-order glacial geology consists mainly of glacial erratics scattered over headlands between fjords and plateaus behind fjords and recessional moraines on the presently glaciated larger islands (England and Bradley, 1978) and of freshly striated pavements and erratics at lower elevations on eastern Ellesmere Island and on smaller offshore islands (Blake, 1977; Hudson, 1983). Widespread second-order glacial geology is largely absent on the Queen Elizabeth Islands, however, especially the smaller and lower western islands. This is consistent with an Innuitian Ice Sheet that had a frozen bed over all of the lower islands and over the higher elevations of the larger islands but had a thawed bed in the interisland channels and fjords (Hättestrand and Stroeven, 1996). This ice sheet isostatically depressed all of the islands, hence the isostatically rebounding raised beaches mapped and dated by Blake (1970). Glacial erratics scattered over the rolling upland landscape described by England (1987) is consistent with a melting bed between the frozen bed under highland ice caps and the thawed bed in coastal fjords, as illustrated in Figure 9.18. Glacial erosion in thawed patches of the rolling uplands would produce both the depressions in the landscape and the scattered erratics. However, most of the rock eroded in the uplands would be funneled into ice streams, where they would become cutting tools that foredeepened the fjords and later accumulated at ice-stream grounding lines to produce depositional sills at fjord entrances or submarine moraines beyond fjords.

Making a distinction between first-order and second-order glacial geology leads to a probable history of the Innuitian Ice Sheet during the last glaciation cycle. Snowlines began lowering rapidly in the Queen Elizabeth Islands after the high latitude insolation maximum at 125 ka BP, as seen in Figure 1.10, as cooler summers and warmer winters allowed greater winter precipitation that did not completely melt during the summer (Berger, 1978). This lowered snowlines on ice caps on the larger eastern islands and thickened sea ice in interior marine basins. Ice caps merged on Ellesmere Island and on Axel-Heiberg Island. Sea ice grounded in the shallow water of Hecla and Griper Bay, Prince Philip Basin, and Norwegian Bay. The beds of these interior marine basins may have been frozen upon grounding, because sea ice would freeze at a freezing point of salt water that became lower as its salinity was increased by salt expelled at the freezing interface. Perennial snowfields would have developed on islands surrounding these interior marine basins. Virtually the whole area of the Queen Elizabeth Islands would then be covered with terrestrial ice on islands and marine ice in interisland channels, with the bed largely frozen everywhere.

Ice thickened to produce the Innuitian Ice Sheet at a rate that was almost the surface accumulation rate, because ice would be discharged slowly through the outer interisland channels if channel beds were frozen. Ice probably accumulated at a rate between 50 mm/a and 100 mm/a, allowing the Innuitian Ice Sheet to attain a steady-state thickness of 3 km under its central ice dome over Norwegian Bay in 30,000 a to 60,000 a. That ice thickness and ice accumulation rate would allow the marine bed beneath the ice dome to become thawed, according to the heat flow analysis by Robin (1955), if the mean annual surface temperature was about –30°C on the central ice dome, which would be the case for a mean annual sea surface temperature of –15°C and an adiabatic lapse rate of –6°C/km to an ice dome 2.5 km above sea level. This conclusion is supported by the thinner permafrost in interisland channels, which is consistent with the present–day marine environment (Taylor, 1988). The thawed bed in interior basins would expand into the outer interisland channels, where converging ice flow would concentrate basal frictional heat. Ice streams would develop in these channels when the thawed bed reached the ice margin. Major ice streams flowed down the flanks of the ice saddles that separated the Innuitian Ice Sheet from the Laurentide Ice Sheet to the south and the Greenland Ice Sheet to the east. Ice streams flowed east in Lancaster Sound (Mayewski and others, 1981), west in M'Clure Strait (Vincent, 1978) from the southern ice saddle over Barrow Strait, and north (Hudson, 1983; deFreitas, 1990) and south (Blake, 1977, 1992) in Nares Strait from the eastern ice saddle over Kane Basin. The Innuitian Ice Sheet probably maintained this steady-state maximum from the high-latitude insolation minimum at 25 ka BP to as late as 14 ka BP, when sea level was rising rapidly.

Retreat and collapse of the Innuitian Ice Sheet occurred during the Holocene (Blake, 1972). The high-latitude insolation maximum at 12 ka BP shown in Figure 2.8, combined with rising sea level, allowed warm Pacific water to breach Bering Strait and flow into the Canadian Basin of the Arctic Ocean. This would have disintegrated an ice shelf that floated over the Canadian Basin and buttressed the northwest margin of the Innuitian Ice Sheet. Marine ice streams along this margin occupied M'Clure Strait, Prince Gustav Adolf Sea, Peary Channel, and Nansen Sound, all of which were foredeepened by glacial erosion that left depositional sills on the outer continental shelf. Once rising sea level floated ice-stream grounding lines over these sills, irreversible grounding-line retreat would have continued at least to where the rear of the foredeepened troughs were as high as the sills, as illustrated in Figure 2.2. Since the troughs were also isostatically depressed by the Innuitian Ice Sheet, with maximum depression under the ice dome over Norwegian Bay, it is likely that the grounding lines of these ice streams continued to retreat until the dome collapsed and only a disintegrating ice shelf existed in Norwegian Bay. M'Clure Strait was ice-free by 12.6 ka BP, and the other northwest inter-island channels were deglaciated by 10.0 ka BP. An open sea in M'Clure Strait, Viscount Melville Sound, Barrow Strait, and Lancaster Sound separated the Innuitian Ice Sheet from the Laurentide Ice Sheet between 9.5 ka BP and 9.2 ka BP. The ice shelf in Norwegian Bay had disintegrated by 9.0 ka BP. The ice saddle over Kane Basin between Ellesmere Island, and Greenland did not collapse until 5.0 ka BP (Blake, 1992; England, 1997), and ice caps still remain on Ellesmere Island, Axel-Heiberg Island, and Devon Island.

Cordilleran Ice

The last glaciation of the North American cordillera illustrates the advantage of inverse diagnostic ice-sheet models that rely on glacial geology over forward prognostic ice-sheet models that rely on mass balance. Using a forward prognostic model to simulate the last glaciation cycle, Budd and Smith (1981) discovered that the mass-balance pattern of accumulation and ablation needed to simulate advance and retreat of the Laurentide Ice Sheet reasonably also simulated an unreasonable advance and retreat of the Cordilleran Ice Sheet. Contrary to the glacial geological record, Cordilleran ice accumulated faster than Laurentide ice, buried all of Alaska under 2 km of ice and extended as far south as New Mexico during the last glacial maximum, and remained longer than Laurentide ice in the model simulation. This contradiction developed because the mass-balance input to the model was overly dependent on changing snowline elevations in response to Milankovitch insolation variations, and the gridpoint spacing in the model did not represent lower elevations adequately in western North America, notably interior Alaska between the Alaska Range and the Brooks Range. All forward diagnostic models have a tendency to overestimate Cordilleran ice in order to reproduce Laurentide ice, unless the mass-balance input is rigged in arbitrary ways. These problems are avoided by inverse diagnostic models, because glacial geology rather than mass balance is the controlling input, and ice flowlines used as input to the model can follow glaciated mountain valleys, thereby using lower elevation regions of western North America to prevent Cordilleran ice from becoming unreasonably thick.

In broad terms, the cordillera of western North America consists of a broad plateau bordered by a continental mountain belt along interior North America and a maritime mountain belt along the Pacific coast. From Alaska to Mexico, the continental mountain belt includes the Brooks Range and the various ranges of the Rocky Mountains. From Alaska to Mexico, the maritime mountain belt includes the Alaska Range and the various ranges of the Coast Mountains. The general trend of the cordillera in this region is to the northwest. Figure 9.6 shows higher glacier equilibrium lines in the continental mountain belt than in the maritime mountain belt, both today and during the last glacial maximum. Cordilleran climate regimes shown in Figure 9.5 for the last glaciation cycle have a pattern that can be represented schematically by Figure 9.22, which shows a north-south transect across Alaska and an east-west transect across British Columbia.

For a north-south transect across Alaska, the rise in glacier equilibrium lines from the Pacific maritime to the continental interior can be understood as the combined effect of stacked snowlines, one snowline for each climate regime. From the highest to the lowest snowline, the polar continental (PC) snowline defines the equilibrium line for glaciers on the north slope of the Brooks Range, the polar mixed (PX) snowline defines the equilibrium line for glaciers on the south slope of the Brooks Range, the subpolar mixed (SX) snowline defines the equilibrium line for glaciers on the north slope of the Alaska Range, and the subpolar maritime (SM) snowline defines the equilibrium line for glaciers on the south slope of the Alaska Range. A line connecting these equilibrium lines creates the illusion of a snowline that slopes up from south to north, whereas the actual PC, PX, SX, and SM snowlines slope down from

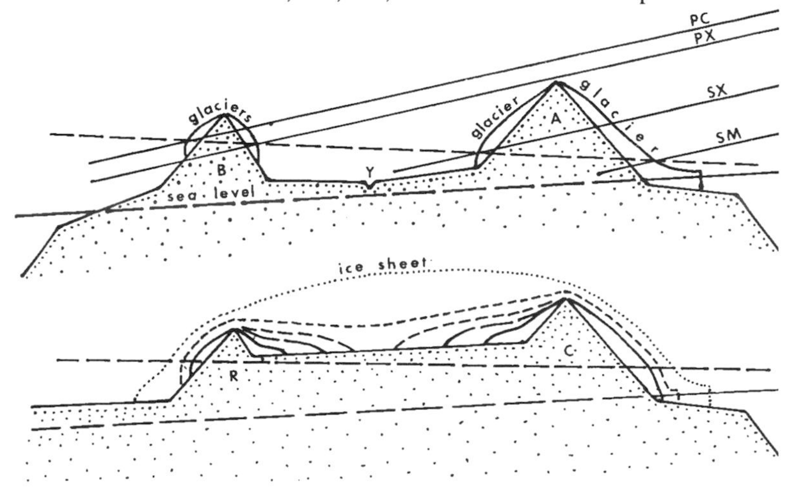

Figure 9.22: Growth of the Cordilleran Ice Sheet. *Top*: The relationship between the equilibrium line (dashed) and snowlines (solid) for Alaskan glaciers. Snowlines are shown for subpolar maritime (SM), subpolar mixed (SX), polar mixed (PX), and polar continental (PC) climates when moving from south to north across the Alaska Range (A), the Yukon River (Y), and the Brooks Range (B). *Bottom*: Irreversible growth of the Cordilleran Ice Sheet in Canada when the equilibrium line lies below the elevation of the plateau between the Rocky Mountains (R) and the Coast Range (C). This would also apply to an ice sheet over Tibet.

south to north. This distinction must be made in order to express Milankovitch insolation variations in terms of snowline slope variations during a glaciation cycle, as depicted in Figure 9.2, for the purpose of modeling advance and retreat of ice sheets.

For east-west transects across the cordillera in British Columbia, snowline slopes are much less, as shown in Figure 2.5. Therefore, when mountain glaciers descend to the intermontane plateau between the continental and maritime mountain belts in British Columbia, they advance over a relatively horizontal plateau, so the ablation zone has a relatively constant width but the accumulation zone continues to widen. This situation, also shown in Figure 9.22 and analyzed by Weertman (1961), allows irreversible advance of glaciers across the intermontane plateau. Ice over the plateau must thicken to at least the height of the flanking continental and maritime mountain belts, thereby producing the Cordilleran Ice Sheet. In Alaska, the intermontane plateau was lowered by the Yukon River drainage system, so mountain glaciers in the Brooks Range and the Alaska Range never descended to the intermontane landscape, preventing the Cordilleran Ice Sheet from glaciating interior Alaska. These considerations provide a framework for understanding the history of the Cordilleran Ice Sheet during the last glaciation cycle.

Cordilleran glaciation was fed by moisture-bearing westerly winds that precipitated snow preferentially on the Pacific slopes of the maritime mountain belt, and mountain glaciers extended farther and lower on these slopes, in accord with the highland-origin, windward-growth hypothesis for nucleating ice sheets (Flint, 1971). This is the case today. It is therefore likely that the Cordilleran Ice Sheet began with increased glaciation of the maritime mountain belt, which includes the Alaska Range, the Chugach Mountains, the Wrangel Mountains, the Coast Mountains, and Vancouver Island. Icefields in these mountains would have grown and merged to become an ice sheet by 115 ka BP, when insolation at both mid- and high north latitudes reached a minimum, as shown in Figure 2.8. Seaward advance of this ice sheet could have extended only 100 km to 300 km to the edge of the North American continental shelf, where ice would have calved into the Pacific Ocean, primarily from ice streams that occupied coastal fjords. Landward ice advance would have been much less restricted, and if the snowline reached the elevation of the intermontane plateau, the landward ice margin would have advanced across the plateau until it met glaciers advancing west from the continental mountain belt. This would have compelled the ice divide of the Cordilleran Ice Sheet to migrate east from an original axis along the Coast Mountain chain.

By the time of the last glacial maximum at 25 ka BP, when insolation again was minimal in mid- and high northern latitudes, the Cordilleran ice divide in British Columbia was an arc that bowed west over the intermontane plateau, with ice domes over the Selkirk Mountains in the southeast, the Cascade Range in the southwest, and the Stikine Mountains in the north. Ice flowing down the inner flank of this arc flooded east along the Frazer River valley and merged with the Laurentide Ice Sheet. Ice flowing down the outer flank of the arc also followed river valleys, becoming terrestrial ice streams that ended as ice lobes to the south, notably in the Rocky Mountain Trench and in Puget Sound, and becoming marine ice streams occupying troughs across the continental shelf, notably in the Strait of Juan de Fuca, Queen Charlotte Sound, and Dixon Entrance (Booth, 1987; Pelto and Roberts, 1993). In the Yukon, ice domes existed over the Selwyn Mountains and over the Saint Elias Mountains, with a major terrestrial ice stream moving northwest down the Yukon River valley from an ice saddle between these ice domes. In Alaska, ice domes existed over the Alaska Range, the Wrangel Mountains, and the Chugach Mountains, all supplying a major ice stream in Cook Inlet. Although insolation increased after 25 ka BP, the Cordilleran Ice Sheet attained its maximum extent at 14 ka BP. Figure 9.23 presents a reconstruction of the Cordilleran Ice Sheet for 14 ka BP, according to glacial geology.

Retreat of Cordilleran ice was rapid after 14 ka BP. An ice-free corridor appeared between the Cordilleran and Laurentide Ice Sheets by 13.6 ka BP (St. Onge, 1972). The Puget Sound lobe was gone by 11.5 ka BP (Armstrong and others, 1965). Only the Coast Mountains were glaciated by 9.5 ka BP (Fulton, 1971). The Yukon was deglaciated by 10 ka BP, except for an ice cap over the Saint Elias Mountains (Denton, 1974). In Alaska, Cook Inlet was ice free by about 12 ka BP (Schmoll and others, 1972). As shown in Figure 2.8, retreat took place during a time of maximum insolation in mid and high northern latitudes. Therefore, Cordilleran deglaciation was probably driven by high snowline elevations and rapid ablation rates at ice margins.

Laurentide Ice

Figure 9.24 shows the major longitudinal and transverse lineations comprising second-order glacial geology produced by the Laurentide Ice Sheet during the last glaciation cycle. Most of these lineations were produced during its final retreat, and the major stillstands are dated. Boulton and Clark (1990) mapped successively overprinted patterns of second-order lineations and grouped these patterns into time transgressive sequences in an effort to map the advance and retreat history of the Laurentide Ice Sheet during the last glaciation cycle. Figure 9.25 presents their lineation patterns. These overprinted lineation patterns, called palimpsests, were not superimposed over the entire glaciated area in a way that allows an unambiguous glacial history to be determined. The authors chose an interpretation in which Hudson Bay was deglaciated about halfway through the glaciation cycle, according to amino acid dating of marine sequences in the Hudson Bay lowlands by Andrews and others (1983). However, amino acid dating is strongly dependent on temperature, which was unknown when the sequences were emplaced, and Andrews and others (1983) proposed three deglaciations of Hudson Bay on the basis of these dates, not the single deglaciation allowed by Boulton and Clark (1990). Other investigators favor two deglaciations, one 80,000 years ago and one 40,000 years ago (Forman and others,

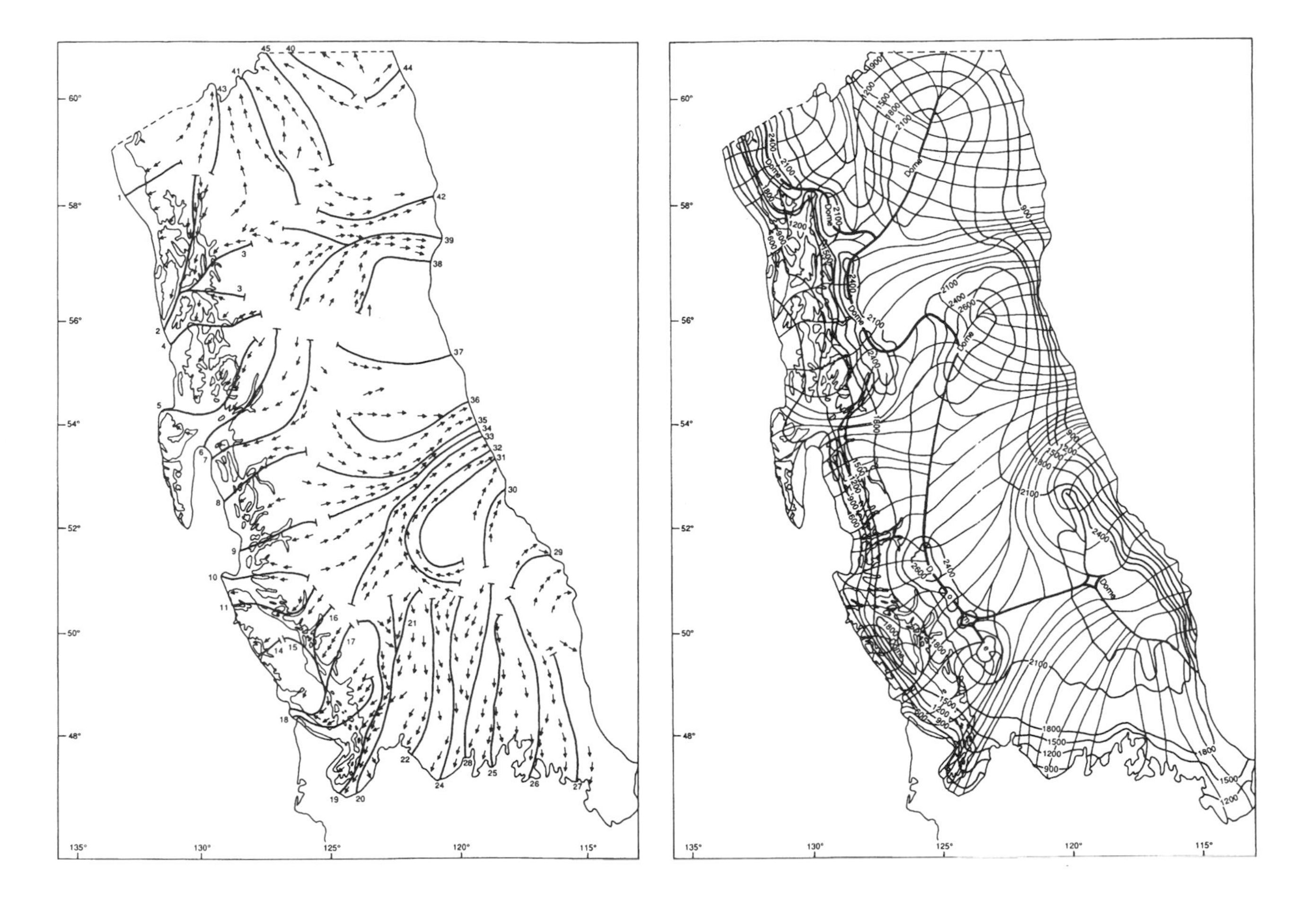

Figure 9.23: The Cordilleran Ice Sheet in Canada. *Left*: Ice-flow directions determined by glacial geology. *Right*: Ice-elevation contours (m) and flowlines at the last glacial maximum. Ice elevations were computed along the numbered flowlines. Heavy lines are ice divides. Reconstruction by Mauri S. Pelto.

1987; Berger and Nielsen, 1991; Laymon, 1991). Since the proposed multiple deglaciations were about 40,000 years apart, it is tempting to attribute them to warm hemicycles in the 41,000 year cycle of Earth's axial tilt. However, deglaciation of Hudson Bay is not supported by the sea-level data in Figure 1.1.

Palimpsests mapped by Boulton and Clark (1990) also allow a glacial history in which a nearly steady-state core persisted over Hudson Bay throughout the last glaciation cycle. Figure 9.26 presents discrete lineation patterns in relative age relationships that allow this interpretation. A glacial history that is compatible with these patterns is presented in Figure 9.26, which postulates the following stages:

1. Ice flowed over Keewatin from a marine ice dome in Foxe Basin.
2. A terrestrial ice dome developed over Keewatin, with possible ice divides extending to the south and to the west.
3. A large central marine ice dome grew over Hudson Bay, from which ice radiated to the east over Quebec, to the northwest over Keewatin, and to the west, southwest, south, and southeast over the Hudson Bay lowlands, as far as the outer edge of the Precambrian shield.
4. A high terrestrial ice dome developed over Quebec, with extensive flow to the southwest, and a terrestrial ice dome developed again over Keewatin, with strong flow into Hudson Bay from both domes, indicating that the marine ice dome over Hudson Bay had become a saddle.
5. Strong westward and southwestward flow from the Keewatin ice dome, and strong radial flow from the Quebec ice dome, except to the north, indicate the saddle over Hudson Bay was much higher and an ice stream in Hudson Strait had shut down, possibly because the bed had become frozen in these regions. This may have been the last glacial maximum.
6. Diverging flow to the west and northwest from Keewatin required a high Keewatin ice dome; strongly converging northward flow into Ungava Bay suggests that an ice stream from a high Quebec ice dome crossed Hudson Strait and ended as a terminal ice lobe on southeastern Baffin Island; and strong southern flow from Hudson Bay implies

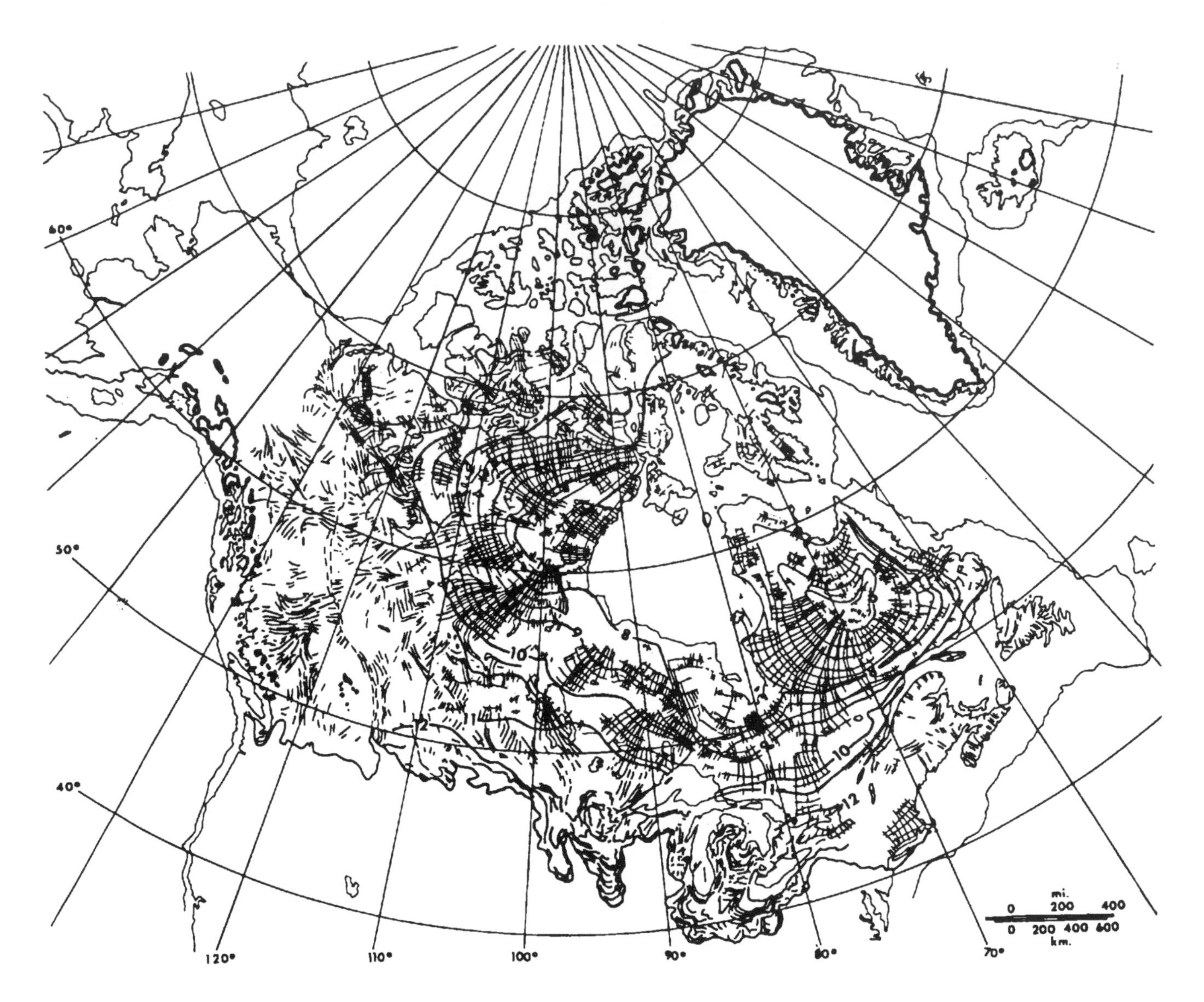

Figure 9.24: Second-order glacial geology of North America. Trends of two types of glacial lineations believed to be formed predominantly near or at the retreating ice margin are shown. Longitudinal lineations aligned predominantly with ice flow include eskers, drumlins, flutes, and striations. Transverse lineations aligned predominantly normal to ice flow include various kinds of recessional moraines, belts of stagnation terrain, and ice-contact fluviatile accumulations. Numbers are ages for ice-margin positions for prominent still-stands in millenia. After reconstructing the last glacial and deglacial ice sheets, the *National Atlas of Canada* (1974), Her Majesty the Queen in Right of Canada with permission of Natural Resources Canada and Boulton and others (1985). From Hughes (1987, Figure 17).

that the ice saddle over Hudson Bay was being downdrawn to sea level by a major terrestrial ice stream, possibly in James Bay.

7. Strong flow east into Hudson Bay indicates collapse of the Keewatin ice dome, whereas the Quebec ice dome may have blocked Hudson Strait, according to Kleman and others (1994).

Dating the sequence of discrete lineation patterns in Figure 9.26 may be possible using the time scale in Figure 1.1 obtained from sea-level data and ocean-floor oxygen isotope stratigraphy for microfossil sediments, which provides first-order proxy data for global ice volume changes. This possibility requires that ice volume changes in Figure 9.26 are synchronous with global ice volume changes. The Laurentide Ice Sheet would then begin as a marine ice dome in Foxe Basin, with a terrestrial southwestward advance into Keewatin to eventually produce the Keewatin ice dome and a marine southward advance into Hudson Bay to eventually produce the central Hudson Bay ice dome of the Laurentide

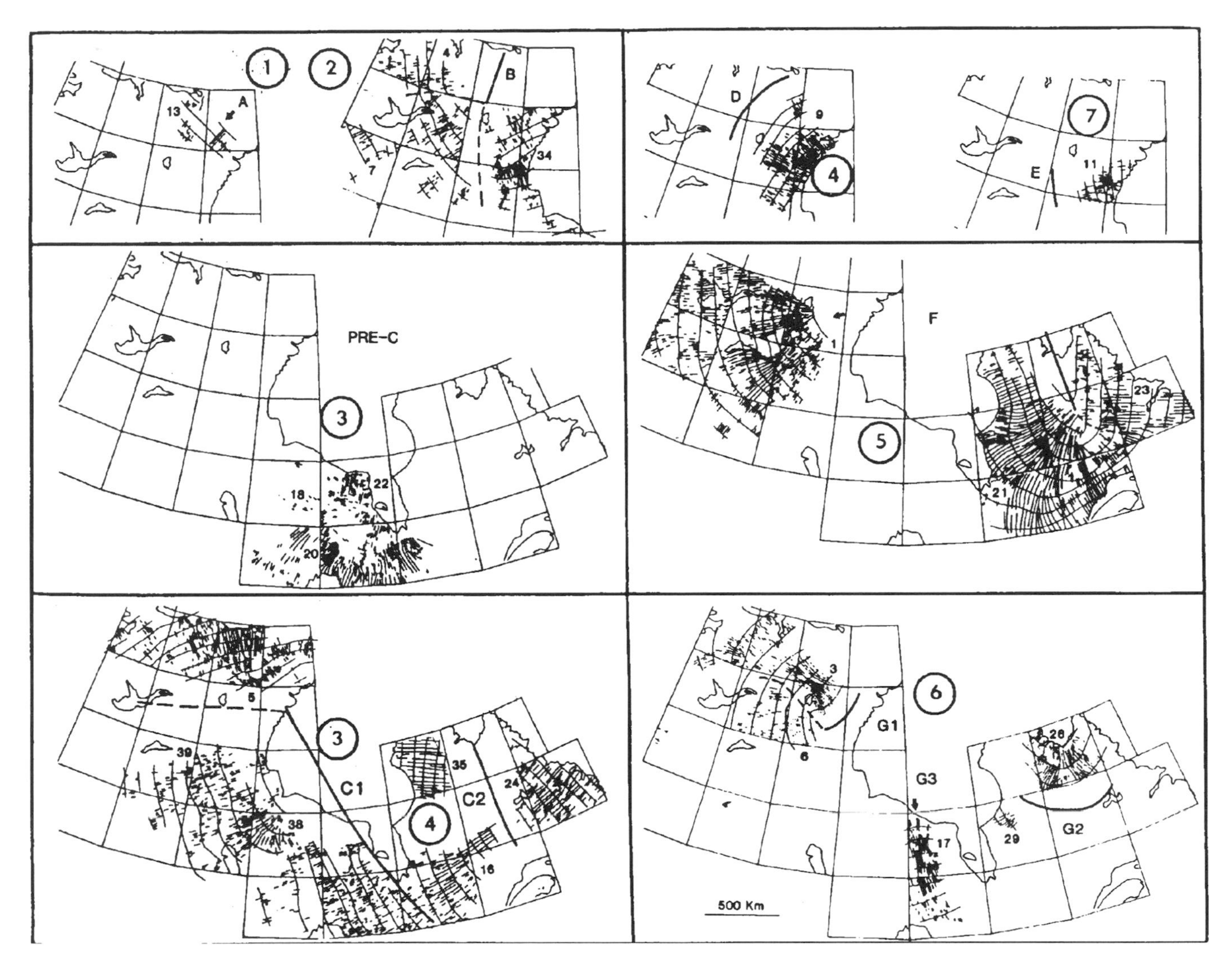

Figure 9.25: Flow stages reconstructed from discrete flow sets, and their relative age relationships. Two possible time sequences are presented. They are lettered A through G in the sequence by Boulton and Clark (1990) and numbered 1 through 7 in an alternative sequence. The postulated locations of ice divides in sequence A–G are shown by solid and dashed lines. In flow stage C, an early ice divide (C1) and a later ice divide C2 are distinguished. Hudson Bay is deglaciated halfway through the last glaciation cycle in sequence A–G but not in sequence 1–7. Sea-level data in Figure 1.1 supports sequence 1–7. Sequence A–G is reproduced by permission of the Royal Society of Edinburgh, G. S. Boulton, and C. D. Clark, from *Transactions of the Royal Society of Edinburgh: Earth Sciences, 81*(4), (1990), pp. 327–347.

Ice Sheet. This expansion would occur quickly, from 115 ka to 112 ka before present (BP), when sea level fell about 20 m. As the Hudson Bay dome became dominant, the Laurentide Ice Sheet expanded to the outer edge of the Canadian Shield by 90 ka BP, when sea level was some 45 m below present-day sea level. This established the stable core of the Laurentide Ice Sheet. Sea level rose about 32 m by 84 ka BP, fell about 64 m by 70 ka BP, and fluctuated around 60 m below present-day sea level between 70 ka BP and 34 ka BP, before falling to 120 m below present-day sea level at 22 ka BP. These ice volume fluctuations, expressed as sea level changes, are approximately in phase with maximum and minimum insolation changes related to hemicycles of Earth's axial precession presented in Figure 9.2. The ice volume changes may have been in large part fluctuations of the southern Laurentide ice margin, where ice ablation rates were most sensitive to insolation variations controlled by the 23 ka cycle of axial precession. The maximum ice volume at 70 ka BP, when sea level had dropped 76 m, resulted from insolation minima linked to both axial precession and axial tilt. Sea level rose about 22 m by 50 ka BP, with increasing insolation, before beginning an unsteady drop to 125 m below present-day sea level at 23 ka BP, when both axial precession and axial tilt contributed to an insolation

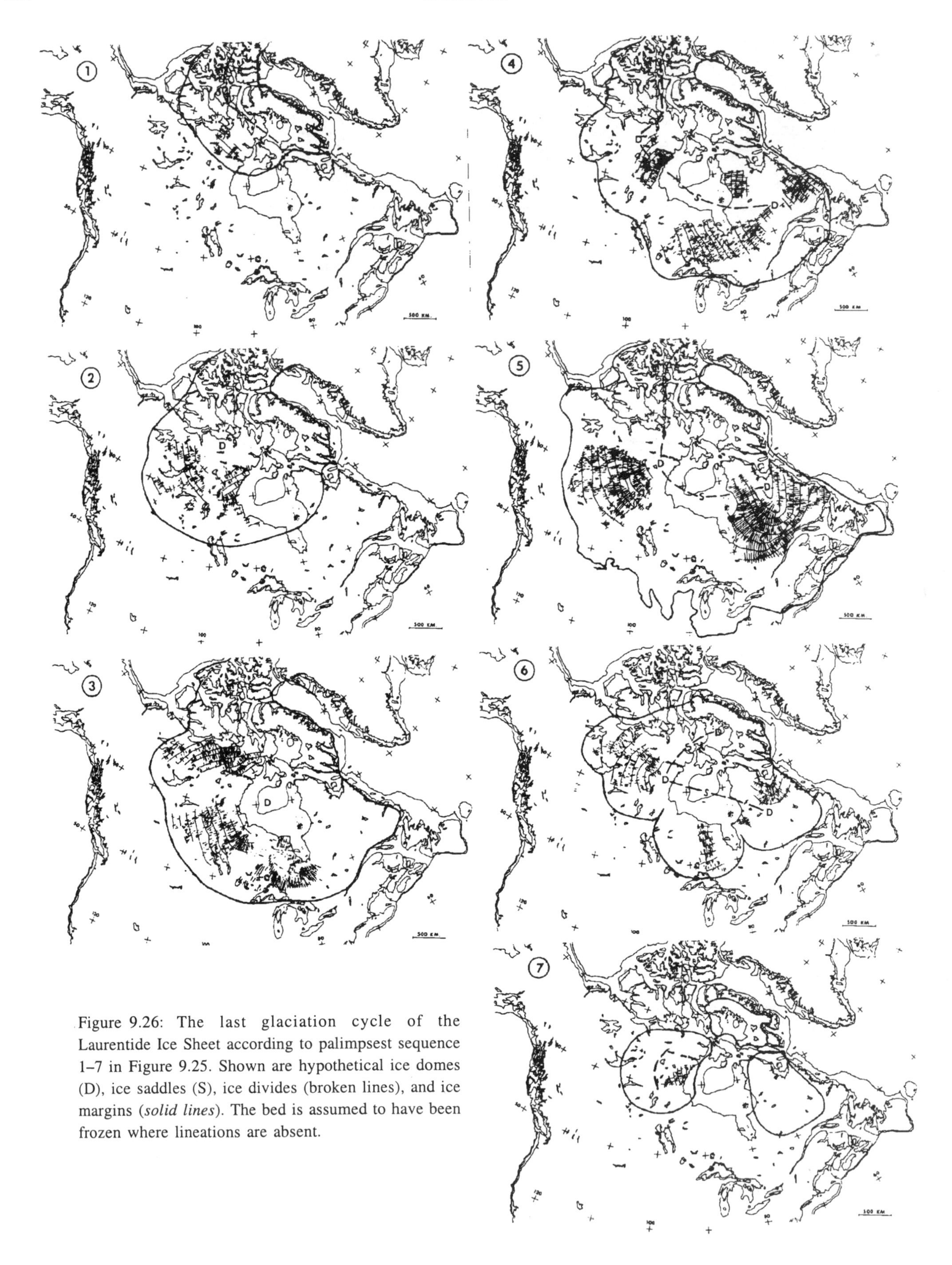

Figure 9.26: The last glaciation cycle of the Laurentide Ice Sheet according to palimpsest sequence 1–7 in Figure 9.25. Shown are hypothetical ice domes (D), ice saddles (S), ice divides (broken lines), and ice margins (*solid lines*). The bed is assumed to have been frozen where lineations are absent.

minimum. The Laurentide Ice Sheet attained its maximum extent at this time, and held close to its maximum until 14 ka BP. The ice volume records in Figure 1.1, along with the ice lineation patterns in Figure 9.26, do not support complete collapse of the Hudson Bay ice dome, followed by a marine incursion into Hudson Bay, at any time from 112 ka BP to 9 ka BP during the last glaciation cycle. At most, the Hudson Bay dome may have become a saddle during insolation maxima.

The only clear evidence that an ice saddle developed over Hudson Bay is the pattern of glacial lineations showing radial spreading from separate Keewatin and Quebec ice domes after 14 ka BP, during the rapid collapse of the Laurentide Ice Sheet, when sea level rose rapidly according to the ice volume records in Figures 1.1 and 2.8. Final collapse of the Laurentide Ice Sheet occurred after 11 ka BP, when ice streams collapsed the ice-divide saddle between Laurentide and Innuitian ice domes (Hodgson, 1994), and after 8.4 ka BP when the Hudson Bay ice-divide saddle was downdrawn to sea level by a south-flowing terrestrial ice stream in James Bay that spread over the Hudson Bay lowlands as the Cochrane ice lobe (Dyke and Prest, 1987), and by a northeast-flowing marine ice stream from Hudson Bay that joined a southeast-flowing marine ice stream from Foxe Basin to become an east-flowing marine ice stream in Hudson Strait (Laymon, 1992).

Separate Keewatin and Quebec Ice Sheets existed after Laurentide ice collapsed over Hudson Bay. The Keewatin and the Quebec Ice Sheets wasted away along their terrestrial ice margins. However, the Keewatin ice divide arced around a marine ice stream in Chesterfield Inlet that carried sediments from the Dubawnt red beds into Hudson Bay (Shilts, 1982), and the Quebec ice divide arced around a marine ice stream entering Ungava Bay that transported carbonates onto southeastern Baffin Island (Miller and Kaufman, 1990). Ice downdrawn into these marine ice streams collapsed the Keewatin and the Quebec Ice Sheets. These three ice streams, one terrestrial and two marine, prevented recovery of the Laurentide Ice Sheet. Applying the model of geomorphic systems presented in Figure 9.21, Kleman and others (1994) found that ice flowing north into Ungava Bay after the last life cycle of the ice stream in Hudson Strait was sheet flow from the Quebec Ice Sheet. Subsequently, the Quebec Ice Sheet retreated slowly over a largely frozen bed. This allowed a calving bay to migrate up Hudson Strait and to carve out Ungava Bay.

Figure 9.19 uses the distribution of exposed bedrock as a first-order glaciological feature that reveals the distribution of relatively stable basal thermal zones in the nearly steady-state core of the Laurentide Ice Sheet. Figure 9.26 presents lineations as a second-order glaciological feature that overprints the first-order glacial geology, and that reveals fluctuations of the ice margins and the ice divides. Together, these figures illustrate stadials and interstadials of the Laurentide Ice Sheet during the last glaciation cycle. During stadials, the bed thawed beneath a high ice dome over Hudson Bay, resulting in partial gravitational collapse that converted the dome into a saddle, advanced landward ice margins across proglacial lakes, and caused iceberg outbursts from ice streams along marine ice margins. Nearly all of the bed thawed as the dome became a saddle, so glacial sliding produced the glacial lineations in Figure 9.26. During interstadials, much of the bed became frozen, allowing interior ice to thicken and to convert the low saddle into a high dome, causing the marine ice streams to shut down, and causing stagnation along terrestrial ice margins, which then retreated rapidly by ablation, first by melting and then by calving as proglacial lakes reformed. Retreat halted at the edge of the Canadian Shield because basal traction on the shield was high, even where the bed remained thawed.

Figure 9.27 is a map showing how much of the bed was thawed, contoured at 25 percent intervals, during the interstadial preceding the last glacial maximum. This distribution of thawed basal conditions is assumed to have remained relatively unchanged during interstadials of the last glaciation cycle, and therefore to have produced the distribution of exposed and covered bedrock shown in the map of surface materials on pages 37 and 38 of *The National Atlas of Canada*, Fourth Edition (1974). This map is reproduced in Figure 9.19. Fulton (1995) has updated the distribution of surficial materials. Assuming that glacial erosion by Laurentide ice sliding over a thawed bed exposes bedrock, 75 to 100 percent of the bed is thawed where bedrock is exposed over 75 to 100 percent of the area, 25 to 75 percent of the bed is thawed where bedrock is exposed over 25 to 75 percent of the area, and 0 to 25 percent of the bed is thawed where bedrock is exposed over 0 to 25 percent of the area. Glacial deposition is assumed to blanket bedrock beyond the region mapped in Figure 9.27. These outer areas were covered by advancing Laurentide ice during stadials, with deposition occurring over these areas during the subsequent ice retreat to the interstadial core area.

Figure 9.28 is a map showing how much of the bed was thawed, contoured at 50 percent intervals, during the stadial at the last glacial maximum. Advance of Laurentide ice during stadials was relatively brief, so time was insufficient for glacial erosion to expose bedrock beyond what is shown in Figure 9.19. Therefore, bed relief was used instead of exposed bedrock as the criterion for determining the distribution of thawed basal conditions. The bed was assumed to be mostly frozen in areas of high bed topography, especially near ice margins where basal heat is conducted rapidly to the surface because ice is thinnest, and ice in high latitudes because the ice surface is coldest. The map of subaerial and submarine relief on pages 1 and 2 of *The National Atlas of Canada*, Fourth Edition (1974), was used for this purpose. For the Canadian mainland, the subglacial bed was assumed to be thawed below the present-day 305 m (1000 ft) elevation contour, from 100 percent to 50 percent thawed between the 305 m and 610 m (2000 ft) elevation contours, and from 50 percent to 0 percent thawed between the 610 m and 914 m (3000 ft) elevation contours. For the Canadian arctic islands, the bed was assumed to be 100 percent thawed in troughs below the 305 m (1000 ft) bathymetric contour, from 100 percent to 50 percent thawed from the 305 m bathymetric contour to present-day sea level, from 50 per-

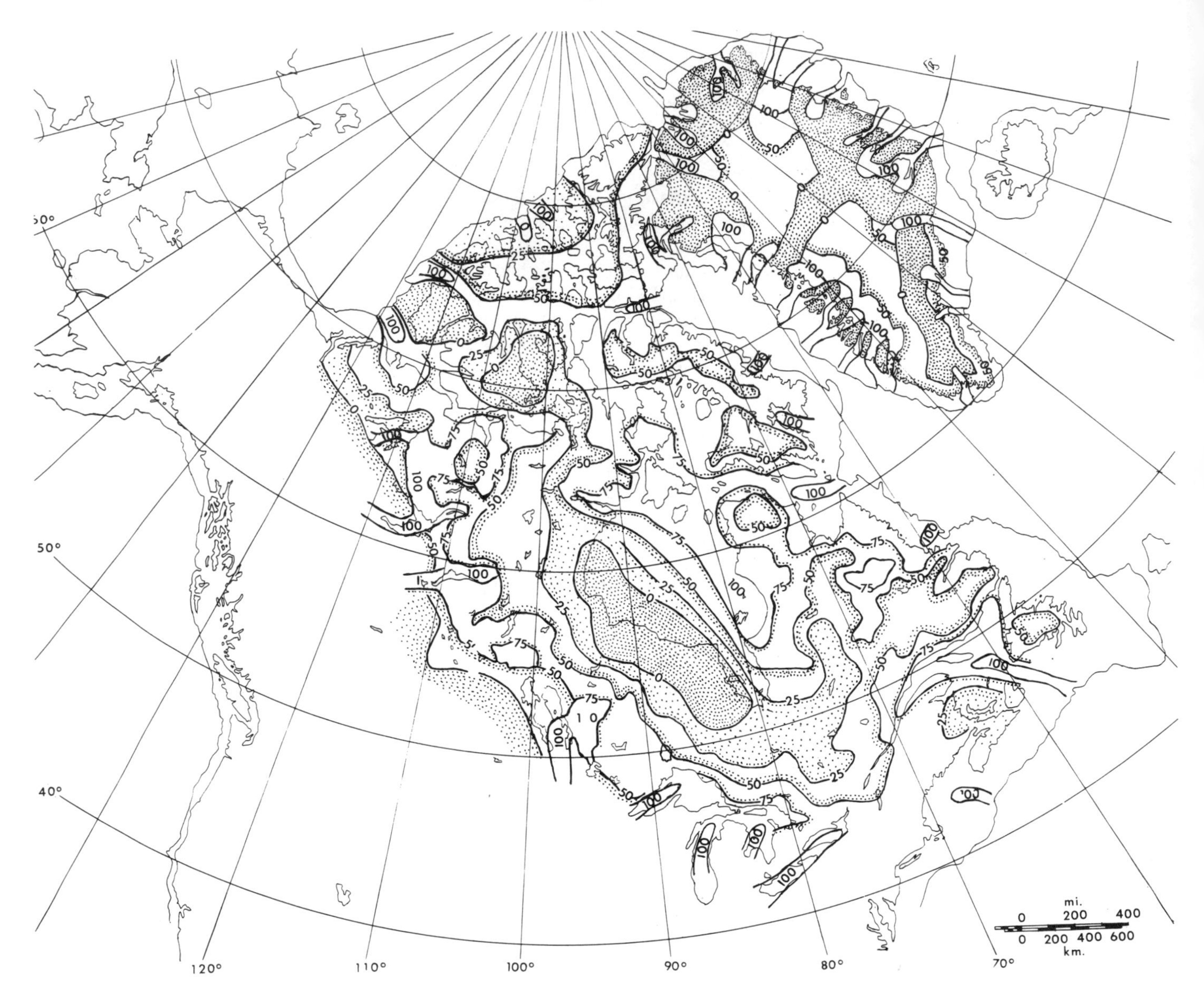

Figure 9.27: The thawed fraction of the bed beneath the Laurentide Ice Sheet during interstadials of the last glaciation. Contours of 100, 75, 25, and 0 percent thawed bed are based on the surface materials map on pages 37 and 38 of *The National Atlas of Canada*, Fourth Edition (1974).

cent thawed to 0 percent thawed from present-day sea level to the 305 m elevation contour, and frozen above the 305 m elevation contour. The relationship between bed topography and percent thawed bed in Figure 9.28 beneath former Laurentide ice should be compared with that shown in Figures 3.14 through 3.17 beneath present-day Antarctic ice. The relative merits of using exposed bedrock and bed topography as criteria for mapping high-traction sticky spots beneath former ice sheets is still a matter of debate (e.g., Clark and Walder, 1994; Fulton, 1995; Marshall and others, 1996). However, both approaches give a largely frozen bed in the high plains of western Canada, which was the case (Moran and others, 1980; Hooke, 1970, 1998).

Figure 9.29 is a map of Laurentide ice elevations, contoured at 200 m intervals, for the interstadial preceding the last glacial maximum. It is based on the distribution of proglacial lakes, based on the relief map on pages 1 and 2 of *The National Atlas of Canada*, Fourth Edition (1974), assuming that the landscape beyond the ice margin had rebounded after previous ice advances. Ice elevations were computed along the flowlines shown in Figure 9.29, with ice elevation contours drawn for equal elevations along flowlines. In reconstructing the interstadial ice sheet using Equation 9.53, ice streams were kept short, calving ice walls were 50 m above water along lacustrine and marine ice margins, and Laurentide ice was generally confined to areas where bedrock was often exposed. Thee constraints were assigned in order to maximize interstadial ice elevations, since

Figure 9.28: The thawed fraction of the bed beneath the Laurentide Ice Sheet during stadials of the last glaciation. Contours of 100, 50, and 0 percent thawed bed are based on the relief map on pages 1 and 2 of *The National Atlas of Canada*, Fourth Edition (1974). The thick solid line is the ice margin at the last glacial maximum, with thick dashed lines showing where Laurentide ice merged with Cordilleran ice in the west and with Greenland ice in the northeast. Thin dashed lines show the maximum Quaternary limits of Laurentde ice.

little ice was downdrawn by ice streams, ablation was mainy by calving along calving ice walls, and bed traction was high over bedrock, especially the crystalline bedrock of the Canadian Shield. With these constraints, the interstadial Laurentide Ice Sheet had an ice volume that lowered sea level by 50 m and had a single elongated ice dome over 3600 m high above Hudson Bay. This allowed Laurentide ice to spill over the Queen Elizabeth Islands and to merge with Greenland ice in Nares Strait. Whether the interstadial Laurentide Ice Sheet really had this configuration is unknown.

Figure 9.30 is a map of Laurentide ice elevations, contoured at 200 m intervals, for the major stadial that produced the last glacial maximum. It is based on the distribution of thawed bed conditions mapped in Figure 9.28. A substantially larger fraction of the bed is thawed in Figure 9.28, compared to Figure 9.27. This additional decoupling of ice from the bed by basal meltwater causes partial gravitational collapse of Laurentide ice, so the ice sheet is lower and broader in Figure 9.30 compared to Figure 9.29. Whether basal decoupling is by basal meltwater that promotes basal sliding or till deformation or both, as examined by Boulton (1996), is immaterial in calculating ice elevations along the flowlines shown using Equation (9.53), since parameters D and m in Equation (9.50) used to calculate $(\tau_o)_B$ encompass both mechanisms for specifying basal ice velocity u_B. Lowering interior ice elevations was accomplished by increasing

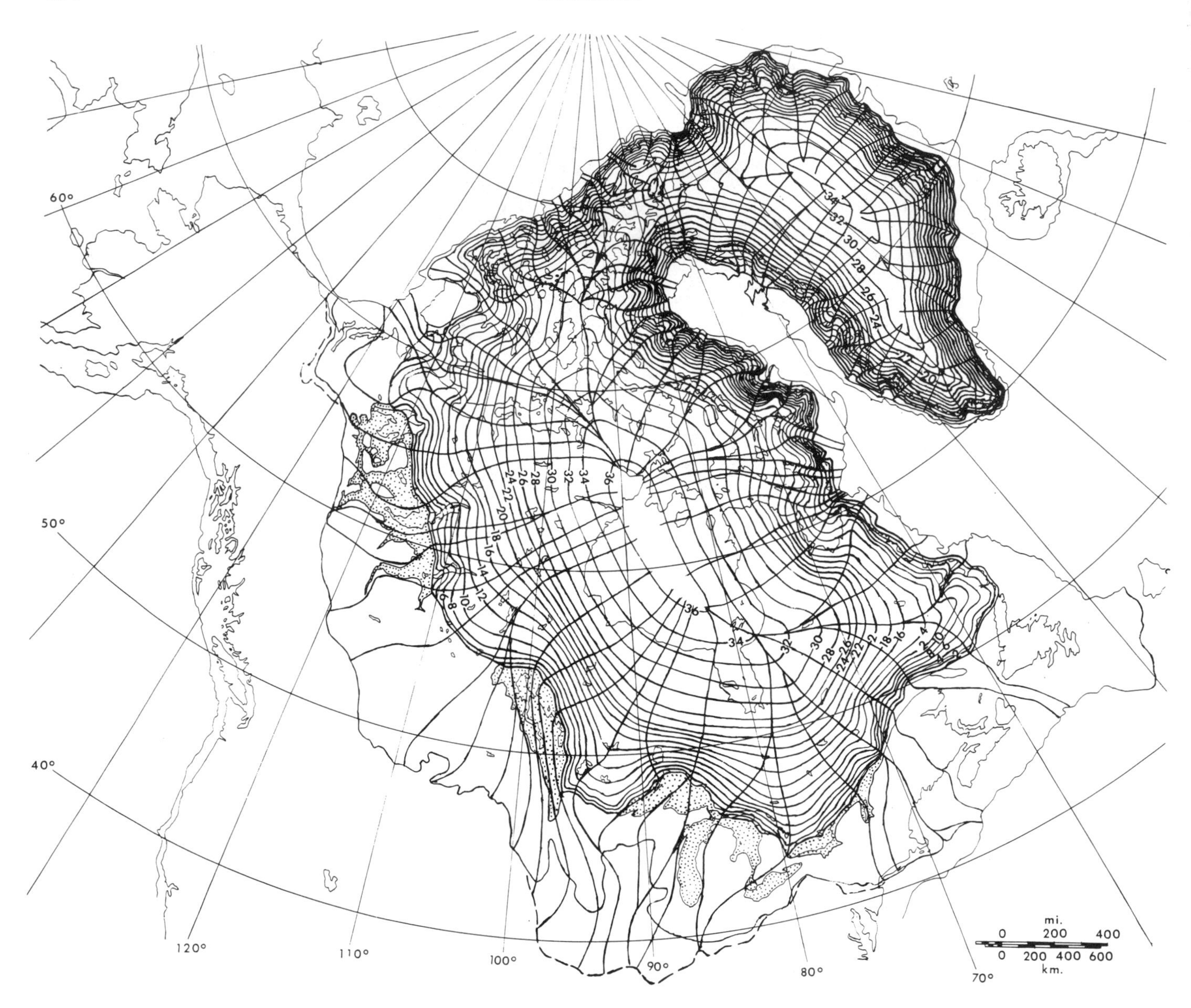

Figure 9.29: A reconstruction of the Laurentide and Greenland Ice Sheets during a glacial interstadial. Surface ice flowlines and contour lines (200 m intervals) are computed for first-order glacial geology described in Figures 9.7 through 9.12, with particular attention to Figure 9.27. Ablation is primarily by calving from ice walls grounded in ice-dammed lakes (dotted areas) and ice-free seas surrounding the ice sheet. A substantially frozen bed, short ice streams, and ablation mainly by calving maximizes the height-to-area ration of the ice sheets. Laurentide flowlines are shown extending to ice margins at the last glacial maximum (solid lines) and to the Quaternary glacial maximum (dashed lines).

f_t, so that $(\tau_o)_B$ replaces $(\tau_o)_C$ in controlling basal traction. Extending the ice margin was accomplished by extending ice streams, along which P_W / P_I in-creases, from the crystalline Canadian Shield to the ice margin that has been mapped for the last glacial maximum (e.g., Mayewski and others, 1981; Dyke and Prest, 1987; Lundqvist and Saarnisto, 1995).

Ice streams flowling east and west in Parry Channel diverted Laurentide ice from flowing north over the Queen Elizabeth Islands, which were covered by an Innuitian Ice Sheet that also supplied these ice streams (Blake, 1970; Hättestrand and Stroeven, 1996), and ice streams that flowed north and south in Nares Strait and were also supplied by Greenland Ice (England, 1997). Laurentide ice lowered 800 m over Hudson Bay, converting the dome to a saddle, but ice volume increased to contribute another 24 m to lowered sea level because the lower and broader Laurentide Ice Sheet was a more effective precipitation sink. Its accumulation area was greatly enlarged and convective storm systems could traverse farther over the lower ice. Laurentide ice flowing west ran into the Cordilleran Ice Sheet, which advanced onto the

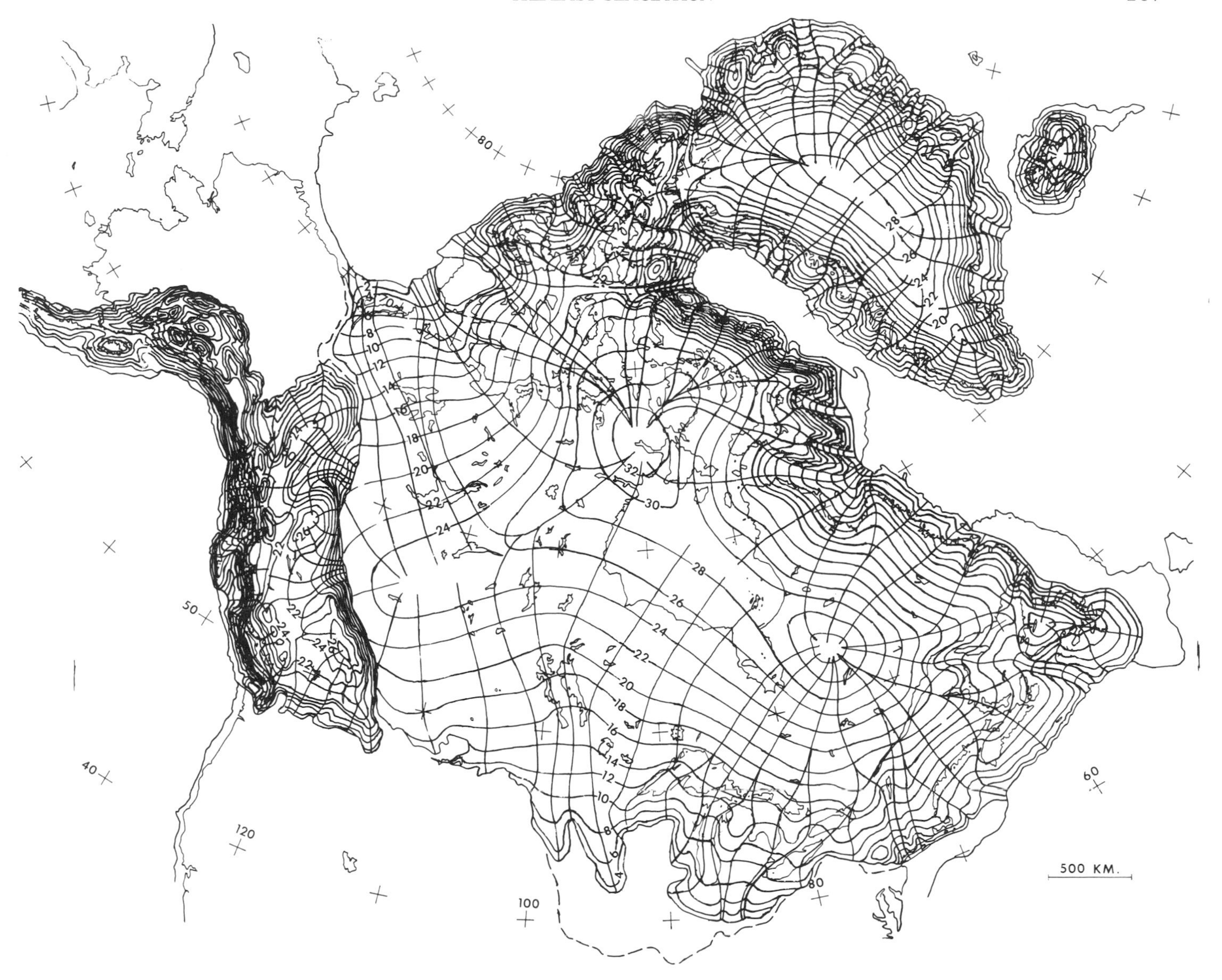

Figure 9.30: A reconstruction of the Laurentide, Cordilleran, Innuitian, and Greenland Ice Sheets during a glacial stadial. Thc stadial associated with the last glacial maximum is depicted. Surface flowlines and contour lines (200 m intervals) are computed for second-order glacial geology described in Figure 9.24 and in the map of glacial geology on pages 33 and 34 of *The National Atlas of Canada,* Fourth Edition (1974), with particular attention to Figure 9.28. Ablation is by melting of ice lobes on land and by calving from ice walls and ice shelves in the sea. A substantially thawed bed, long ice streams, and a wide ablation zone along southern ice margins minimizes the height-to-area ratio of the ice sheets. The Quaternary limit of Laurentide ice is also shown (*dashed lines*).

high planes of western Canada toward the end of the last glacial maximum. By 14 ka BP, this blockage of westward Laurentide Ice had produced a high ice ridge that extended west-southwest to the Cordilleran Ice Sheet. Glacial lineations in the map of glacial geology on pages 33 and 34 of *The National Atlas of Canada*, Fourth Edition (1974), show that ice flowed south-southeast from this ridge along Lake Winnipeg and Lake Manitoba, where it became an ice stream down the valley of the Red River (of the North) to end as twin ice lobes in the valleys of the James River and the Des Moines River, and ice flowed north-northwest from the ridge to become an ice stream in the Mackenzie River valley. These glacial lineations crosscut older lineations produced by ice flowing radially from the collapsing Hudson Bay ice dome to the northwest, west, and southwest. Preservation of the older lineations is evidence that these parts of the bed became frozen when the younger lineations formed.

The Cordilleran Ice Sheet is not shown in Figure 9.29

because its extent during the interstadial preceding the last glacial maximum is unknown. The extent of the Cordilleran Ice Sheet shown in Figure 9.30 during the last glacial maximum is based on the map of glacier retreat on pages 31 and 32 of *The National Atlas of Canada*, Fourth Edition (1974).

Maritime Ice

Islands bordering the North Atlantic Ocean had a largely maritime ice cover during the last glaciation cycle. The ice sheet over Greenland expanded, large ice caps grew on Newfoundland, Iceland, Ireland, Great Britain, and Spitzbergen, and small ice caps covered the Faeroe, Shetland, and Orkney Islands and Jan Mayen Island (Andersen, 1981; Mayewski and others, 1981). A dominant feature of maritime glaciation is the heavy precipitation on slopes facing moisture-bearing winds. Flint (1971) noted this characteristic in developing his highland-origin, windward-growth hypothesis for initiating a glaciation cycle. All islands bordering the North Atlantic have highlands that were exposed to moisture-laden maritime winds during much of the last glaciation cycle, especially at the beginning, so their glaciation history is explained best by this hypothesis.

Newfoundland glacial geology has been interpreted in two ways, both examined by Brookes (1982). An interpretation by Grant (1977) assumed that the glacial geology represented weathering zones, in which case glaciation barely reached present-day sea level during the last glaciation cycle. The high elevations on headlands between coastal fjords were covered by heavily weathered frost-shattered rocks called felsenmeer, so they were assumed to lie above the ice limit. At lower elevations, a pitted and rolling landscape was dotted with glacial erratics and was less uniformly weathered, so it was assumed to have been ice-covered early in the glaciation cycle or during a previous glaciation cycle. The lowest elevations showed pervasive evidence for recent glaciation, notably freshly polished and striated bedrock on fjord walls and young end moraines on coastal plains beyond glacially sculptured valleys. This interpretation has been applied to similar maritime glacial geology in Labrador by Ives (1978), on Baffin Island by Pheasant and Andrews (1972, 1973) on Ellesmere Island by England and Bradley (1978), in Greenland by Funder and Hjort (1973), in Scandinavia by E. Dahl (1955), and in Spitzbergen by Boulton (1979). The alternative explanation, by R. Dahl (1967) for Scandinavia, by Sugden (1974) for Greenland, and by Sugden and John (1976), Hughes and others (1981), and Hughes (1987) for maritime glacial geology in general, postulate subglacial thermal zones during the last glaciation cycle instead of weathering zones. Ice sliding on a thawed bed converged on fjord headwalls, glacially polishing and striating bedrock on fjord walls and isolating headlands between fjords, which were therefore covered by local ice caps on a frozen bed, so weathered felsenmeer was left undisturbed. A melting basal zone between the headlands and the fjords consisted of frozen and thawed patches, glacially eroded in thawed patches, so a rolling and pitted landscaped scattered with erratics was produced. Ice was much thicker in this interpretation. In Newfoundland, foredeepened submarine troughs extending from the larger fjords to the edge of the continental shelf radiate from the central highlands. This first-order glacial geology is evidence that ice streams radiated from a high central ice dome over Newfoundland and calved into deep water beyond the continental shelf. In this interpretation, Newfoundland and the Grand Banks were covered by a large ice cap that merged with the Laurentide Ice Sheet during the last glacial maximum, and glacial sliding in basal thermal zones produced the second-order glacial geology.

Greenland glacial geology, like that in Newfoundland, can be interpreted as reflecting either time-transgressive weathering zones or subglacial thermal zones. In eastern Greenland, Funder and Hjort (1973) interpreted weathering zones as requiring unglaciated headlands and outlet glaciers confined to fjords, notably in Scoresby Sound. In western Greenland, Weideck (1968) and Ten Brink and Weideck (1974) interpreted areal scouring and offshore submerged moraines as requiring a limited extension of the present-day ice sheet margin. In northern Greenland, Weideck (1972, 1976) limited glaciation mainly to expansion of the present-day outlet glaciers. However, these interpretations can be questioned. Moraines in Greenland are often associated with high raised beaches that require substantial glacioisostatic rebound after removal of a heavy ice load, and dates are often of Holocene age or are based on shells of mixed ages (Sugden, 1972; Blake, 1975; Andersen, 1981; Hughes and others, 1981). In northern Greenland, no moraines are older than Holocene and Peary Land is a center of isostatic uplift (Andersen, 1981). Holocene ice recession progressed from south to north, so uplift is relatively young in northern Greenland. Widespread isostatic rebound in eastern, western, and northern Greenland requires much thicker ice during the last glaciation and points to subglacial thermal zones as an explanation for weathering horizons. Thicker ice would allow the Greenland Ice Sheet to expand far onto the surrounding continental shelf and to merge with the Innuitian Ice Sheet along Nares Strait (Blake, 1992; England, 1997). Many foredeepened troughs extend to the edge of the continental shelf from the larger coastal fjords and present-day outlet glaciers, which implies glacial erosion by former ice streams that occupied these troughs during the last glaciation cycle.

Iceland glacial geology displays the same pattern found in Newfoundland. First-order glacial geology consists of submarine troughs across the continental shelf that often continue into coastal fjords and radiate from interior highlands and raised beaches that are concentric around the island and date from 12.5 ka BP on the southern coast. Second-order glacial geology consists of end moraines of this and younger ages on the south coast (Einarsson, 1968; Andersen, 1981) and a prominent submarine end moraine 100 km off the west coast (Olafsdottir, 1975). These glacial features are compatible with a large ice cap over Iceland that was drained by ice streams extending to the edge of the surrounding continental shelf during the last glacial maximum. These ice streams probably calved into the North Atlantic on the south, but they may have merged with a thick ice shelf floating in the Greenland and Norwegian Seas to the north (Lindstrom and MacAyeal, 1986). This would explain why raised beaches and end moraines are rare on the north coast, since features would not form until after the ice shelf had disintegrated.

Irish and British glacial geology is consistent with the highland origin, windward-growth hypothesis of Flint (1971) for initiating the last glaciation cycle. In the North Atlantic, moisture-bearing westerly winds beat against the rocky highlands of Ireland and Scotland, where ice caps formed and advanced west across the continental shelf toward the Porcupine and Rockall Banks. Ice then flowed east into Connemara in Ireland (Borns and Warren, 1990) and onto the Outer Hebrides in Scotland (Peacock, 1991). This ice and ice from the Scottish Highlands merged over North Channel to form a saddle on the ice divide. A marine ice stream flowed into the North Atlantic and a terrestrial ice stream flowed through the Irish Sea from opposite flanks of this ice divide (Eyles and McCabe, 1991). This situation existed on both sides of Great Britain (Charlesworth 1957; Andersen, 1981; Balson and Jeffreys, 1991). The terrestrial ice dome over the Scottish Highlands merged with Scandinavian ice advancing west across the North Sea, producing a saddle on the ice divide. From this ice saddle, a marine ice stream moved north between the Orkney and the Shetland Islands toward the North Atlantic and a terrestrial ice stream moved south to produce an ice lobe in the southwestern North Sea, which was then on dry land (Boulton and others, 1977; Catt, 1991). During deglaciation, Irish ice retreated to the Connemara highlands and the north-central lowlands (Warren, 1991), and British ice retreated to the Scottish Highlands and the Outer Hebrides (Sutherland, 1991).

Spitzbergen was glaciated by moisture-bearing winds from the Norwegian Sea, which sustain highland icefields and the Nordaustlandet ice cap today. During the last glaciation cycle, the icefields and the ice cap advanced onto the surrounding continental shelf as a single large ice cap that became a marine ice dome when it merged with Siberian ice at the last glacial maximum (Schytt and others, 1968). Ice streams drained the Spitzbergen ice dome through the major fjords, interisland channels, and marine troughs (Isaksson, 1992). An arcuate pattern of raised beaches is evidence for the marine ice dome, and the major fjords, interisland channels, and marine troughs generally radiate from the center of greatest uplift. Mangerud and others (1992) mapped and dated second-order glacial geology showing that the outer continental shelf was deglaciated between 15 and 12.5 ka BP, with ice streams still occupying the inner fjords during Younger Dryas cooling from 11 to 10.2 ka BP, and subsequent warming from 10 to 5 ka BP causing Holocene deglaciation.

Islands in the North Atlantic that were covered by maritime ice caps and ice sheets during the last glaciation are shown in Figures 9.27 through 9.30 for the western Atlantic and in Figures 9.31 through 9.33 for eastern Atlantic. Ice caps on Newfoundland, Ireland, and Britain during the interstadial preceding the last glacial maximum merged with the respec-tive continental Laurentide and Scandinavian Ice Sheets at the last glacial maximum. These island ice caps then became peripheral ice domes of these large ice sheets. The Greenland Ice Sheet may have merged with the Innuitian Ice Sheet across Nares Strait during the interstadial preceding the last glacial maximum and certainly did during the glacial maximum, as shown in Figures 9.29 and 9.30. Iceland was covered by an ice cap during the last glacial maximum, remnants of which survive today. Maritime ice on these islands during the last glaciation is reconstructed in Figures 9.29, 9.30, 9.32, and 9.33.

Maritime ice on Greenland is of particular interest since it was the only North Hemisphere ice sheet to survive termination of the last glaciation. Although much of its subglacial landscape has been mapped from radio-echo radar sounding (Letréguilly and others, 1991a, 1991b), its glacial geology is known only beyond the present-day ice margin (e.g., Andersen, 1981). Therefore reconstructing the Greenland Ice Sheet for the interstadial preceding the last glacial maximum and for the stadial at the last glacial maximum must rely on mapped bed topography and on areas of basal thawing generated by Huybrechts (1996) using his three-dimensional thermomechanical ice sheet model.

Figure 9.29 includes a map of Greenland ice elevations, contoured at 200 m intervals, for the interstadial preceding the last glacial maximum. It is based on the distribution of thawed bed conditions shown in Figure 9.27. Ice margins generally followed the 152 m (500 ft) bathymetric contour, along which calving ice walls rose 50 m above sea level, except where major ice streams occupied the deeper submarine troughs across the Greenland continental shelf. Many Greenland ice streams today probably occupy continuations of these troughs beneath the Greenland Ice Sheet, but resolution of the radio-echo data is not sufficient to map them. Ice elevations were computed along the flowlines in Figure 9.29, using Equation (9.53), but ice streams were kept short and much of the bed was kept frozen, as shown in Figure 9.27. This maximized ice elevations during the interglacial preceding the last glacial maximum, with a central ice dome over 3400 m high. Figures 9.27 and 9.29 should be compared with those by Huybrechts (1994, 1996) for similar basal thermal conditions.

Figure 9.30 includes a map of Greenland ice elevations, contoured at 200 m intervals, for the major stadial that produced the last glacial maximum. It is based on the distribution of thawed bed conditions shown in Figure 9.28. Ice margins generally followed the 610 m (2000 ft) bathymetric contour, which lies along the edge of the Greenland continental shelf. Calving ice walls formed along this ice margin in the south, but in the north it was the grounding line of floating ice shelves, for the purpose of applying Equation (9.53) to the ice flowlines in Figure 9.30. Ice streams occupied submarine troughs across the Greenland continental shelf and, unlike the interstadial ice streams in Figure 9.29, a thawed bed extended continuously from these ice streams to the central ice dome, as shown in Figure 9.28. This permitted ice downdrawn into the ice streams to lower the central ice dome by nearly 800 m. As with Lau rentide ice, partial gravitational collapse over the interior and rapid advance of the ice margin was accomplished using Equation (9.53) by employing high values of f_t for sheet flow over the ice-sheet interior and high values of P_W/P_I for stream flow toward the ice-sheet margin. The transformation of basal thermal zones from interstadial to stadial conditions, shown in Figures 9.27 and 9.28, is illustrated by the contrast between basal thermal conditions for terrestrial and marine domes of the ice sheet shown as a cartoon in Figure 9.21. Partial gravitational collapse and accompanying spreading from the Greenland Ice Sheet in Figure 9.27 to the one in Figure 9.28 is similar to that depicted in Figure 1.6, which might trigger a new glaciation cycle by discharging icebergs with a volume and rate that are adequate to slow or halt production of North Atlantic Deep Water in the Greenland Sea and the Labrador Sea. This transformation can be quite abrupt, perhaps within years or decades, according to climate records in the GISP 2 glaciochemical series from the summit corehole of the Greenland Ice Sheet (Mayewski and others, 1997; Meeker and others, 1997), and the rapid response of ocean circulation in the North Atlantic on a decadal time scale (Lehman and Keigwin, 1992; Hughen and others, 1996; Sy and others, 1997; Dickson, 1997; McCartney, 1997; Sutton and Allen, 1997; Howard, 1997).

Depositional fans on the Greenland continental slope at the ends of the submarine troughs across the continental shelf are an indication that Greenland ice streams once extended this far before either calving or becoming afloat (Reeh, 1994). This result is quite different from the one by Huybrechts (1994, 1996), who obtained this extended ice margin by thickening interior ice.

Scandinavian Ice

Figures 9.31 through 9.33 show the extent of Scandinavian ice. Kleman and others (1997) produced a study of second-order glacial geology for the Scandinavian Ice Sheet, using the inversion model of glacial landscapes by Kleman (1994) and Kleman and Borgström (1996), that is comparable to the Boulton and Clark (1990) study for the Laurentide Ice Sheet. In several respects, Scandinavian geography is a smaller-scale version of North American geography. The Scandinavian mountains are analogous to the North American cordillera. The Scandinavian Shield is analogous to the Canadian Shield. Centered in these Precambrian shields, the Baltic Sea is analogous to Hudson Bay. The Baltic Sea has a single outlet through the Norwegian Channel into the North Sea, and Hudson Bay has a single outlet through Hudson Strait into the Labrador Sea. Perhaps these geographical similarities accommodated similar glaciation histories. If so, glaciation of the Scandinavian mountains was like glaciation of the North American cordillera, with precipitation from moisture bearing westerly winds and the primary direction of glacier advance toward the western fjordlands in both cases, according to the highland origin, windward growth hypothesis of Flint (1971). This opens the possibility that the steady-state core of the Scandinavian Ice Sheet developed from a marine ice dome in the Gulf of Bothnia where sea ice thickened and grounded. This would be analogous to the steady-state core of the Laurentide Ice Sheet, which developed from sea ice that thickened and grounded in Hudson Bay to produce a marine ice dome. A present-day analogue for initial glaciation of Scandinavia may be Spitzbergen, which has icefields in the maritime climate of the western highlands and an ice cap in the continental climate of Nordaustlandet, separated by Hinlopenstretet, an ice-free corridor. These ice centers merged to become a Spitzbergen ice dome during the last glacial maximum (Schytt and others, 1968; Isaksson, 1992; Mangerud and others, 1992). These similarities are speculative. What is not speculative is that the central ice divide of the Scandinavian Ice Sheet was downdrawn by a Baltic Ice Stream, just as the central ice divide of the Laurentide Ice Sheet was downdrawn by a Hudson Strait Ice Stream (Boulton and Jones, 1979; Fastook and Holmlund, 1994; Holmlund and Fastook, 1993, 1995). Whether second-order glacial geology can reveal the entire glaciation history is a question now being addressed.

Kaitanen (1969), Kujansuu (1975), Lundqvist (1986), Lagerbäck (1988a, 1988b), Lagerbäck and Robertsson (1988), Rodhe (1988), Andersen and Mangerud (1989), Kleman (1990, 1992), Kleman and Borgström (1990), and Kleman and others (1992, 1997) have undertaken the task of determining the history of the last Scandinavian glaciation cycle by mapping second-order glacial geology. The bed under the Scandinavian Ice Sheet was a mosaic of frozen and thawed patches, allowing a palimpsest glacial landscape to develop. Lineations and other glacial features produced in thawed patches during the earlier stages of the last glaciation are preserved in frozen patches of its later stages. The earliest lineations are lateral moraines and channels alongside large outlet glaciers moving east from an ice divide approximately along the mountain divide coinciding with the Norwegian-Swedish frontier (Kleman, 1992). Kleman (1992) concluded that landforms along its eastern ice margin are compatible with two ice recessions which may correlate with the insolation maxima and ice volume minima around 105 ka BP and 82 ka BP shown in Figure 1.1. If the Scandinavian Ice Sheet formed from a merger of a marine ice sheet in the Gulf of Bothnia and a terrestrial ice sheet over the Scandinavian mountains, this retreating ice margin may have represented two partings of marine and terrestrial ice in Scandinavia that was analogous to, and perhaps synchronous with, partings of marine Laurentide ice and terrestrial Cordilleran ice in North American, producing ice-free corridors in both cases. Since there is no field evidence for a marine ice sheet in the Gulf of Bothnia early in the last glaciation cycle, it would have advanced and retreated over a frozen landscape if it existed. Early mountain glaciation is confirmed (Kleman and Stroeven, 1997).

Large through valleys across the Scandinavian mountains radiate from the Gulf of Bothnia and continue as large coastal fjords that become wide submarine troughs ending at the edge of the Norwegian continental shelf. These first-order glacial geological features are proof that a marine ice dome in the Gulf of Bothnia was the primary spreading center for

Figure 9.31: Second-order glacial geology of northern Europe. This interpretation is based on boulder dispersal patterns (Salonen, 1987, Figure 3), the general map of Quaternary deposits in Finland (Kujansuu and Niemelä, 1984), and recessional moraines of the last glaciation in northern Europe (Andersen, 1981). Successive positions 1–4 indicate the shifting of ice-flow centers during deglaciation. The areas of more rapidly flowing ice lobes and the areas of passive ice which prevailed during the deglaciation were determined by Punkari (1980). The calving ice front indicates the area with major glaciolacustrine sedimentation (area of waterlain tills). The Scandinavian Ice Sheet has been observed to have divided into several ice lobes during deglaciation. The lobes have their own flow regimes, and long belts of ice divides formed only in Sweden during the last phases of deglaciation (Lundqvist, 1986). From Andersen (1981) and Bouchard and Salonen (1990).

marine ice streams draining the Scandinavian Ice Sheet. This conclusion is reinforced by the pattern of isostatic rebound revealed by raised beaches and negative gravity anomalies around the Gulf of Bothnia. Both transverse ice marginal lineations and longitudinal ice dispersal lineations in Figure 9.31 indicate widespread stagnation of Scandinavian ice in Finland during the final deglaciation. Salonen (1987) and Bouchard and Salonen (1990) postulate from erratic dispersal trains that ice-spreading centers radiated across Finland from an ice dome in the Gulf of Bothnia, and spreading produced a highly lobate retreating ice margin. Lobate recessional moraines looping around the Gulf of Riga in the Baltic Sea, Lake Ladoga, Lake Onega, and the Gulf of Onega in the White Sea are aligned with the erratic dispersal trains. These alignments are evidence that the Gulf of Bothnia was the primary spreading center for terrestrial ice streams draining the Scandinavian Ice Sheet. The largest Scandinavian ice stream was the Baltic Ice Stream, which also retreated to the Gulf of Bothnia and which produced the lobate recessional moraines looping around Kiel Bay in the western Baltic Sea (Fastook and Holmlund, 1994; Holmlund and Fastook, 1993, 1995).

By 14 ka BP, rapidly increasing insolation was causing rapid melting along southern margins of Northern Hemisphere ice sheets and a corresponding rapid rise in sea level, as shown in Figure 1.1. Melting Scandinavian ice produced a Baltic Ice Lake, which was prevented from draining into the North Sea because the terminal lobe of the Baltic Ice Stream still occupied Kiel Bay. Rising sea level caused Scandinavian ice to calve back along the isostatically depressed Norwegian Trough in the North Sea, probably by mechanisms illustrated in Figures 7.5 and 7.6. Calving from the north and melting from the south deglaciated the North Sea and separated Irish-British maritime ice from Scandinavian ice between 17 ka BP and 16 ka BP (Andersen, 1981).

Glacial geological lineations produced by the final collapse of the Scandinavian Ice Sheet can be related to life cycles of the Baltic Ice Stream. The maximum southern advance of the Scandinavian Ice Sheet coincided with the 23 ka BP insolation minimum and ice volume maximum in Figure 1.1. Terminal moraines in Figure 9.31 were not markedly lobate at that time, indicating that sheet flow rather than stream flow characterized the advance. By 14 ka BP, however, terminal moraines were strongly lobate and associated with Baltic Sea embayments, elongated lakes, and river valleys. This implies stream flow. Unlike other lobes, the lobe of the Baltic Ice Stream held close to its most advanced position during the intervening 9000 years. This would be possible if the Baltic Ice Stream was becoming increasingly dominant by expanding its ice drainage basin at the expense of smaller ice streams. However, from 14 ka BP to 11 ka BP, the Baltic Ice Stream apparently retreated 600 km almost to the Gulf of Bothnia before readvancing during the Younger Dryas from 11 ka BP to 10 ka BP, followed by a final retreat into the Gulf of Bothnia from 10 ka BP to 9 ka BP. These events are consistent with a life cycle of the Baltic Ice Stream from 23 ka BP to 14 ka BP, followed by a quiescent stage from 14 ka BP to 11 ka BP that allowed the Gulf of Bothnia ice dome to recover and sustain a shorter, second life cycle from 11 ka BP to 10 ka BP. This second life cycle lowered the Gulf of Bothnia ice dome enough that it could not recover, so the spent ice stream was carved away by a calving bay in the Baltic Sea that continued into the Gulf of Bothnia. Carving out the marine heart of the Scandinavian Ice Sheet left terrestrial Scandinavian ice in Sweden and Finland to waste away slowly over a largely frozen bed, producing ribbed moraines as the bed thawed along the retreating ice margin (Hättestrand, 1997). There is no evidence from the glacial geology that an ice-free corridor developed between a marine ice dome in the Gulf of Bothnia and a terrestrial ice divide along the Scandinavian Mountains, as might have happened during early interstadials of the glaciation cycle under conditions that allowed the marine ice dome to recover. This deglaciation history is summarized in Figure 9.31, and was developed primarily by Lundqvist (1986) and Salonen (1987). As the marine Hudson Strait Ice Stream caused final collapse of the Laurentide Ice Sheet, so the marine Baltic Ice Stream caused the final collapse of the Scandinavian Ice Sheet.

As with the Laurentide Ice Sheet, the glacial history of the Scandinavian Ice Sheet during the last glaciation cycle can be related to stadials and interstadials. An ice sheet confined to the Scandinavian Shield and the Scandinavian Mountains will thicken when the bed is frozen or has many frozen patches, because basal traction will be high. Ice calving rates will be high in the Baltic, North, Norwegian, and Barents Seas, which surround the ice sheet, thereby preventing the ice margin from advancing as the ice interior thickens. This is the interstadial stage reconstructed in Figure 9.32. When ice thickened enough to thaw the interior bed, partial gravitational collapse caused ice margins to advance rapidly, primarily as terrestrial ice streams along river valleys south of the Baltic Sea and as marine ice streams along submarine troughs in the Norwegian Sea. The terrestrial ice streams produced lobate terminal moraines and the marine ice streams produced iceberg outbursts. This is the stadial stage reconstructed in Figure 9.33. Advancing Scandinavian ice merged with British ice in the North Sea and with Siberian ice in the Barents Sea. Partial gravitational collapse produced a saddle in the ice divide between the southern Scandinavian Mountains and the Gulf of Bothnia. Refreezing of the bed beneath this ice divide restored the high central ice dome and the overextended ice margins retreated by melting and calving during the subsequent interstadial. Holmlund and Fastook (1993, 1995) and Fastook and Holmlund (1994) show how glacial geology can be used to model the last Scandinavian glaciation cycle, with particular attention to the Baltic Ice Stream and the Younger Dryas stadial.

Siberian Ice

Figures 9.34 through 9.37 show the extent of Siberian ice. Figure 9.34 presents second-order glacial geology produced by Siberian ice, according to Grosswald (1993a, 1993b) and Grosswald and Spector (1993). During the last glaciation cycle, the Siberian Ice Sheet probably began with expansion of permanent ice caps over the islands of Novaya Zemlya, Franz Josef Land, and Severnaya Zemlya and with formation of ice caps over the New Siberian Islands and Wrangel Island after 125 ka BP. These island ice caps would then be nuclei for marine ice domes in the Kara Sea and the East Siberian Sea. By the insolation minimum at 115 ka BP in Figure 1.1, Kara Sea ice had advanced north to the edge of the continental shelf, had advanced west across the Barents Sea to merge with a marine ice cap over the islands of Spitzbergen to the north and with the Scandinavian Ice Sheet to the south, had advanced southwest across the Yamal Peninsula to merge with a terrestrial ice cap in the northern Ural Mountains, had advanced southeast to merge with terrestrial ice caps on the Putorana Plateau and the Anabar Plateau, and had advanced eastward to merge with a marine ice sheet over the New Siberian Islands, according to modeling experiments by Lindstrom (1990) and by Fastook and Hughes (1991). Fluctuations during Milankovitch insolation oscillations, produced interstadials at 105, 82, and 50 ka BP and stadials at 95 and 69 ka BP that ratcheted toward increasing ice volume until the last glacial maximum at 25 ka BP, when a Eurasian Ice Sheet extended from Ireland to Alaska.

Siberian ice from the Kara Sea spilled west across Novaya Zemlya into the Barents Sea, where it formed an ice saddle triple junction revealed by isostatic uplift contours that dip toward it from Scandinavia and Spitzbergen (see Figure 9.12). Three ice divides met at this saddle, a northwest ice divide to the high marine ice dome over Spitzbergen, a southeast ice divide to the higher marine ice dome over the Gulf of Bothnia, and an eastern ice divide to the highest marine ice dome in the Kara Sea. From this ice sad-

Figure 9.32: A reconstruction of the Scandinavian Ice Sheet during a glacial interstadial. Surface ice flowlines and contour lines (200-m intervals) are computed for first-order glacial geology desribed in Figures 9.7 through 9.12 and a largely frozen bed. Basal traction is maximized by ice frozen to bedrock or to ice-cemented lacustrine and marine sediments. Ablation is primarily by calving from ice walls grounded in ice-dammed lakes and ice-free seas surrounding the ice sheet. High bed traction and ablation by calving maximizes the height-to-area ratio of the ice sheet.

dle and between the ice divides, a large ice stream flowed south between the Kola Peninsula and the Kanin Peninsula to end as an ice lobe in the Mezen River valley, a larger ice stream flowcd northward through thc Franz-Victorla Trough between Spitzbergen and Franz Josef Land, and the largest ice stream flowed east through Bjørnøyrenna between Spitzbergen and northern Scandinavia. Marine ice in the Barents Sea extended to the edge of the continental shelf, and the ice stream in Bjørnøyrenna dumped a huge submarine sediment fan off the continental slope (Everhøi and Solheim, 1983; Solheim and Kristoffersen, 1984; Vorren and others, 1988; Elverhøi and others, 1990; Sættem and others, 1992; Gataullin and others, 1993). Many interisland channels in Franz Josef Land are branches of a broad trough across the continental shelf to the north. This suggests that Franz Josef Land was a region of converging flow into an ice stream occupying the trough, rather than a local ice dome, at the last glacial maximum. The Kara Sea ice dome would then be the source of ice flowing across Franz Josef Land. Most large submarine troughs in the Barents Sea are foredeepened, so they are first-order glacial geological features that were occupied by marine ice streams, even if they were originally grabens produced by block faulting.

Numerous first-order glacial geological features radiate from the Kara Sea, as would be the case if a major marine ice dome were grounded there during the last glacial maximum. These radial features include the submarine St. Anna Trough and Voronin Trough between Franz Josef Land and Severnaya Zemlya to the north, interisland channels continuing as submarine troughs in Severnaya Zemlya to the northeast, glacial through valleys across highlands of the Taimyr Peninsula to the southeast and highlands of Novaya Zemlya to the northwest, and numerous linear gulfs to the south. Most of these gulfs are estuaries of rivers flowing northward into the Kara Sea, notably the estuaries of the Ob River and the Yenisei River. If the various linear depressions radiating from the Kara Sea were occupied by ice streams draining a marine ice dome, the marine ice sheet would have extended to the edge of the continental shelf in the north, would have spread into the Laptev Sea to the east, would have occupied the estuarine gulfs to the south, and would have spread into the Barents Sea to the west.

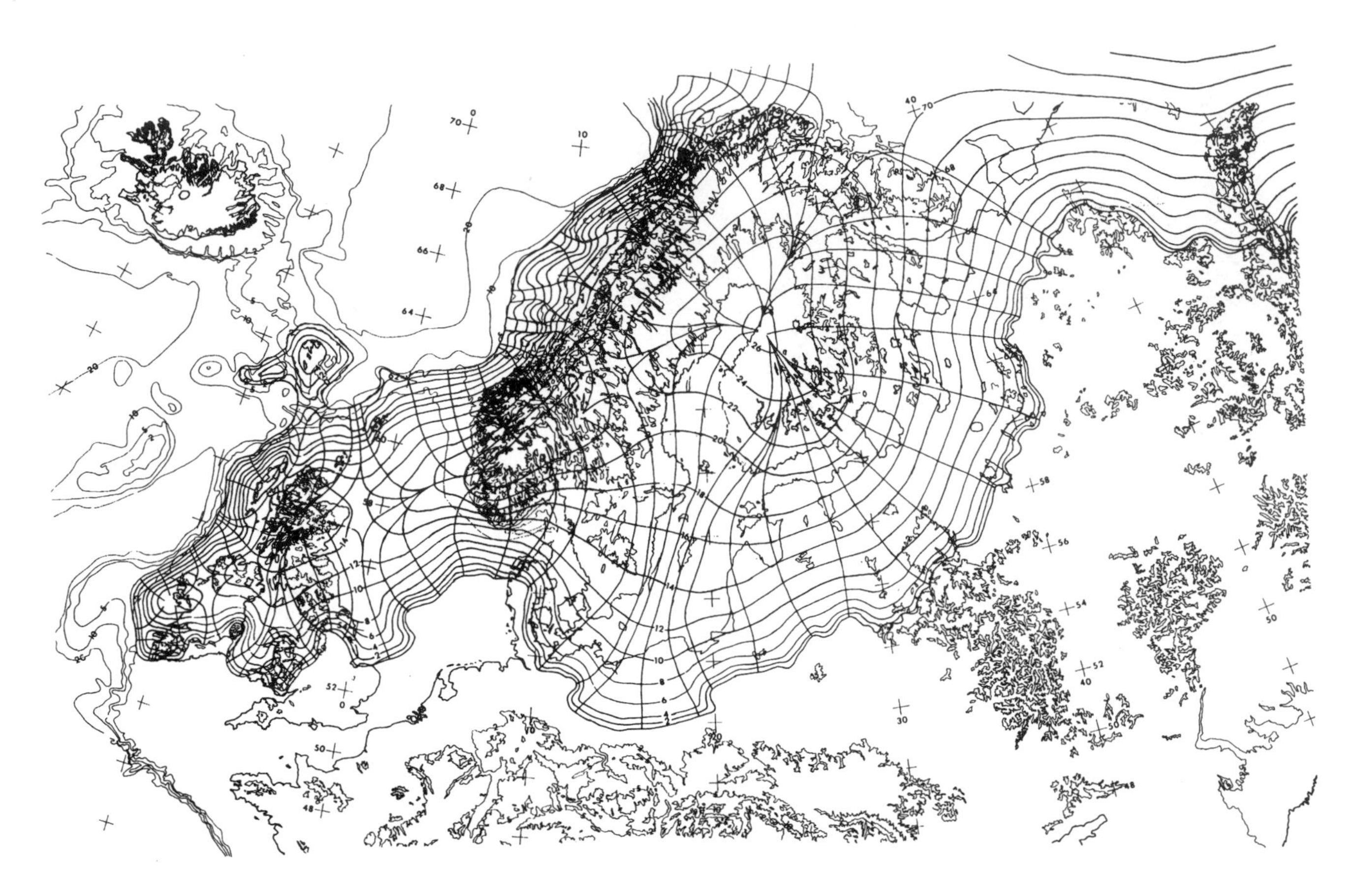

Figure 9.33: A reconstruction of the Scandinavian Ice Sheet during a glacial stadial. Surface ice flowlines and contour lines (200 m intervals) are computed for second-order glacial geology described in Figure 9.31 and a largely thawed bed. Basal traction is minimized by ice sliding over water-saturated lacustrine and marine sediments. Ablation is by melting of ice lobes on land and by calving from ice shelves in the sea. Low bed traction, long ice streams, and a wide ablation zone minimize the height-to-area ratio of the ice sheet.

Ice streams entering estuarine gulfs would have continued as terrestrial ice streams following the river valleys. This would have created ice-dammed lakes ahead of the advancing terminal lobes of the ice streams. By the time of the last glacial maximum, Mansi Lake flooded the vast West Siberian Lowland drained by the Ob River, and Yenisei Lake was a much smaller dendritic lake in the Central Siberian Upland drained by the Yenisei River. Spillways connected these lakes to each other, and to much enlarged Aral and Caspain Seas that spilled over into the Black Sea and ultimately drained into the Mediterranean Sea (Grosswald, 1988, 1993a). West of the Ural Mountains, Siberian ice from the Kara Sea marine ice dome dammed the Pecora River and the Mezen River, producing ice-dammed lakes that spilled over into the headwaters of the Volga River system and entered the Black Sea by that route.

Second-order glacial geological directional indicators, notably ice-shoved hill-hole pairs, glaciotectonic folds, glacial grooves and striae, drumlins and flutes, and terminal moraines, allowed Grosswald (1993a) to map the outer limit of Siberian ice at the last glacial maximum. From the North Russian Plain across the West Siberian Lowland to the Central Siberian Upland, these directional indicators show ice radiating to the southwest, south, and southeast from a marine ice dome in the Kara Sea. This ice reached as far south as 61°N in the central Ural Mountains and 62°N for the terminal lobes of terrestrial ice streams in the Ob and Yenisei River valleys. Between these lobes, Kara Sea ice extended to about 63°N and often ended as calving ice walls in Mansi Lake. In the Central Siberian Upland, Kara Sea ice overrode ice caps on the Putorana Plateau and the Anabar Plateau and may have reached its southeastern limit along the Olenek River valley if the Olenek River was an ice-marginal meltwater stream along most of its present length.

Kara Sea ice spilled over Taimyr Peninsula into the Laptev Sea, where an ice-free corridor may have initially existed between marine ice domes in the Kara Sea and the East Siberian Sea. The broad Arctic continental shelf of Siberia is relatively narrow in the Laptev Sea, and a huge sediment pile on the reentrant continental slope is suggested by the bathymetric contours. A sediment pile would develop in this location if the Lena River had passed through an ice-

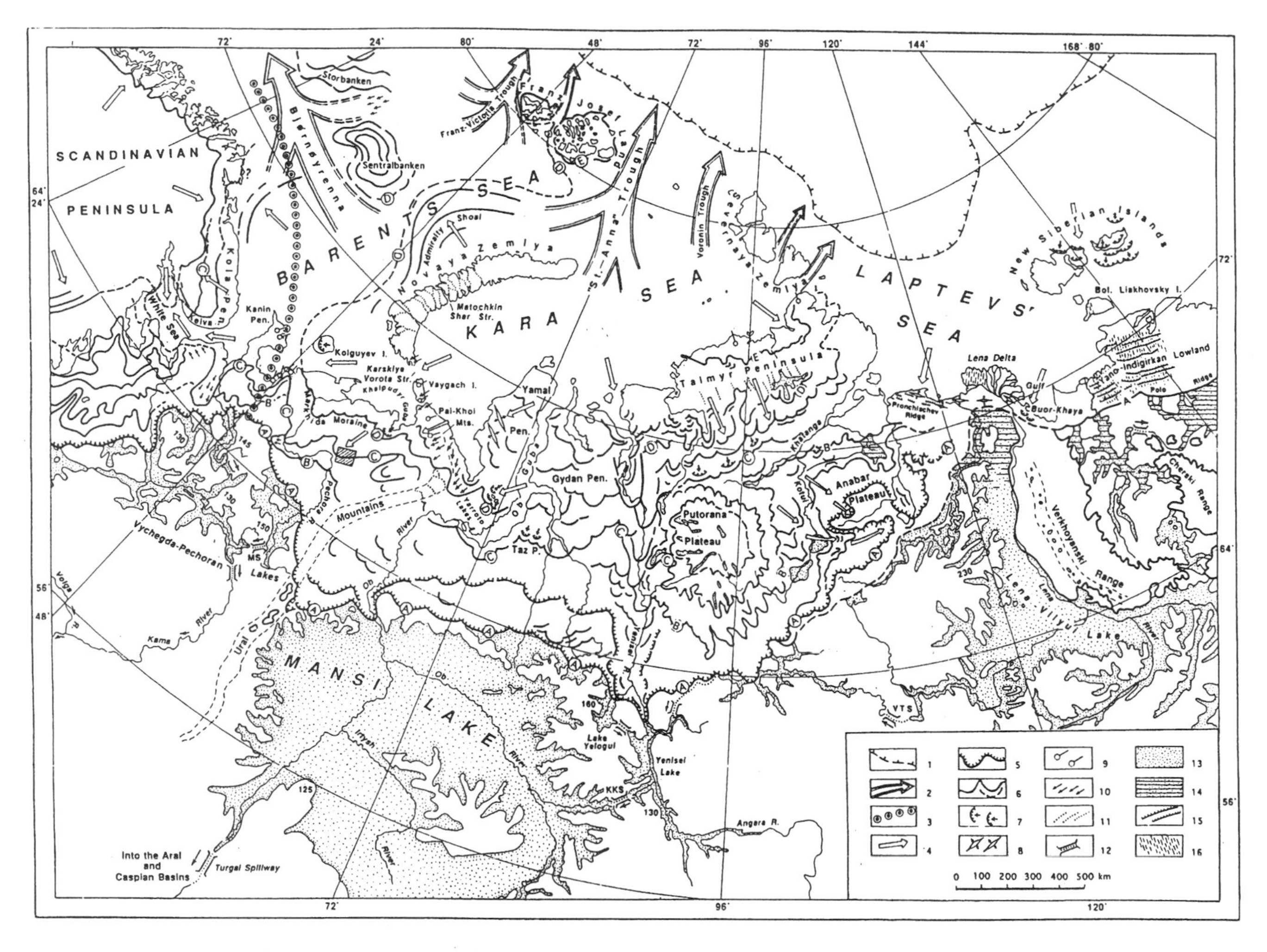

Figure 9.34: Evidence for glaciation of Arctic Russia by a marine ice sheet according to Grosswald (1993a, 1993b) and Grosswald and Spector (1993). Symbols on the map are: 1 = Brink of the continental shelf, 2 = Directions and relative size of submarine troughs (channels), 3 = Boundary of the Scandinavian and Kara Ice Sheets during the last glacial maximum, 4 = Inferred directions of ice flow, 5 = Outer limit of the Scandinavian and Kara Ice Sheets during the last glacial maximum, (moraine belt A), 6 = Ice limits, late glacial and Holocene stages (moraines B, C, D, E, and intermittent), 7 = Hill and hole pairs, 8 = Ice-shoved folding in unconsolidated sediments, 9 = Glacial grooves and striations, 10 = Drumlins, drumlinoids, fluting, and other lineations, 11 = Glacial breaches (through valleys), 12 = Major spillways, meltwater overflow channels, 13 = Ice-dammed lakes—their maximum extent at the last glacial maximum, 14 = "Yedoma" areas (accumulations of ice-rich silt and sand), 15 = "Alas" valleys (parallel valleys created by melting of ground ice), 16 = Direction of long axes of "oriented lakes."

free corridor and formed a delta on the continental slope or if northward converging flow from a saddle on an ice divide in the Laptev Sea between ice domes in the Kara Sea and the Siberian Sea produced a marine ice stream that became afloat at the edge of the continental shelf, causing most glacially eroded material to be dumped on the continental slope. In the latter case, southward converging flow would have produced a terrestrial ice stream that ended as an ice lobe in the Lena River valley. Lakes in the western half of the present-day Lena River delta are beaded and elongated in a direction that is due south of the presumed sediment pile on the continental slope. This is suggestive of stream flow, and Grosswald and Lasca (1993) argue that these elongated beaded lakes are the surface expression of an underlying glacially fluted landscape. Their model disputes the traditional explanations. Thermokarst processes produce lakes on Arctic coastal lowlands, and Washburn (1980) discussed other mechanisms proposed to orient these lakes under present-day conditions.

In the East Siberian Sea, marine ice probably spread initially from an ice cap that formed over the New Siberian Islands, which have glaciotectonic ice thrust features (Grosswald, 1988). Westward flow into the Laptev Sea would have met ice moving eastward from the Kara Sea, creating a saddle on the ice divide of the Eurasian Ice Sheet at the last glacial maximum. Southward flow transgressed onto the Siberian mainland, and produced ice-shoved hill-hole pairs and glaciotectonic thrust features in the Tiksi lowlands south of Buorkhaya Bay (Grosswald and others, 1992). It is tempting to put the southern limit of marine ice transgression against the northern foothills of the Cherskiy Mountains and the Gydan Mountains, in which case the lowlands of the Indigirka River and the Kolyma River would have been covered by marine ice from the East Siberian Sea. In particular, the Kolyma River runs northeast along the Gydan foothills, instead of flowing directly north across the lowlands, as if this was an ice-marginal channel it cut when the East Siberian Ice Sheet abutted the foothills. Lakes in parts of the coastal lowlands have a north-south orientation that Grosswald and Lasca (1993) ascribe to underlying glacial fluting caused by marine ice flowing south from the East Siberian Sea.

Siberian ice spread east into the Chukchi Sea, where it may have transgressed onto coastal Alaska as far as the DeLong Mountains and the Colville River, and poured south through Bering Strait into the Bering Sea, at least as far as St. Lawrence Island and possibly to the southwest edge of the continental shelf. Colville River flows east before turning north into the Beaufort Sea, as if it had cut an ice marginal channel along that course. Elongated lakes west of Point Barrow are suggestive of underlying glacial fluting, and they point northwest to an ice-spreading center in the Chukchi Sea (Grosswald and Lasca, 1993). North-south through valleys cut across the DeLong Mountains and may be glacial valleys. North-south through valleys also cut across the mountains of Chukchi Peninsula in Siberia and Seward Peninsula in Alaska, and coastal fjords in western Bering Strait have north-south trends (Grosswald, 1993a). Raised beaches from Seward Peninsula almost to Point Barrow are suggestive of Holocene glacioisostatic rebound, although they have been assigned to a much earlier time (Hamilton, 1991; Hamilton and Brigham-Grette, 1991). Glaciotectonic ice thrust features and glacial erratics on the north side of St. Lawrence Island have been linked to Siberian ice moving southward (Brigham-Grette and others, 1992; Heiser and others, 1992). The Yukon River now loops far south before turning north into Norton Sound, whereas it once entered Norton Sound much more directly from an abandoned channel. Perhaps it became an ice marginal river rerouted around an ice lobe from the Chukchi Sea that crossed Seward Peninsula and Norton Sound.

Two major stages of deglaciation were preserved by the second-order glacial geology for Siberian ice (Grosswald, 1993a). The last glacial maximum was attained between 28 ka BP and 14 ka BP according to this record. As shown in Figure 1.1, rapidly increasing insolation caused rapid melting along the southern margins of Northern Hemisphere ice sheets and a corresponding rapid rise in sea level between 25 ka BP and 12 ka BP. Rates of rising sea level peaked at 12.0 ka BP and 9.5 ka BP (Fairbanks, 1989), with minimum rates during the Younger Dryas cooling from 11.0 ka BP to 10.2 ka BP and the Boreal cooling from 8.5 ka BP to 8.0 ka BP, when Siberian ice readvanced or maintained stillstands (Grosswald, 1993a). The southern ice limits at 24, 14, 10, and 8 ka BP are shown in Figure 9.34, following Grosswald (1993a). Rising sea level by 14.5 ka BP had probably caused the grounding line of Siberian ice to retreat as far as the White Sea in the western Barents Sea, where Bjørnøyrenna Ice Stream was downdrawing ice that produced a strong meltwater spike in the Norwegian Sea at that time (Jones and Keigwin, 1988; Lehman and others, 1991; Weinelt and others, 1991).

The Younger Dryas advance brought Siberian ice from the Kara Sea over most of the Kola Peninsula from the northeast after Scandinavian ice retreated, inundated the White Sea with a marine ice stream from the east, and invaded the Mezen River valley with a terrestrial ice stream from the north (Grosswald, 1993a). Drainage of proglacial lakes into the Barents Sea by way of the White Sea was temporarily halted, once again causing these lakes to drain into the Mediterranean Sea by way of the Black Sea during the Younger Dryas. This apparently coincided with a readvance of the Baltic Ice Stream that redammed the Baltic Ice Lake (Lundqvist, 1987). The southern margin of Siberian ice during the Younger Dryas produced moraine sequences that Grosswald (1993a) was able to trace from the terminal ice lobe in the Mezen River valley, continuing east along the North Russian Plain, crossing the northern Ural Mountains, forming lobes south of Ob Bay, looping south around the Putorana Plateau, and following the Khatanga River valley to the Laptev Sea. Kara Sea ice no longer overrode the Putorana Plateau, which was covered by a local ice dome.

The Boreal stillstand was marked by the Markhida moraine on the North Russian Plain, which has been linked to submarine moraines associated with Admiralty Shoal west of Novaya Zemlya and present-day ice caps in Franz Josef Land to the north and to moraines in the Pai-Khoi Mountains of the northernmost Urals, a submarine moraine across Ob Gulf, and lobate moraines at the ends of through valleys across highlands in the Taimyr Peninsula to the east (Grosswald, 1993a). The ice dome on the Putorana Plateau had become an isolated ice cap that was separated from the Kara Sea ice domes by an ice-free corridor from the Barents Sea to the Laptev Sea. This corridor was rebounding isostatically and was a marine channel, so the Kara Sea ice dome became a Kara Ice Sheet surrounded by polar seas. Major marine ice streams in St. Anne Trough and Voronin Trough between Franz Josef Land and Severnaya Zemlya drew down the Kara Ice Sheet during the Holocene, leaving residual ice caps on islands surrounding the Kara Sea in Novaya Zemlya, Franz Josef Land, and Severnaya Zemlya. These ice caps remain today. Deglaciation of southern Novaya Zemlya revealed second-order glacial geology consisting of drumlins, flutes, giant glacial grooves, and striae, all showing ice moving from the Kara Sea to the Barents Sea. This flow

Figure 9.35: Second-order glacial geology of Arctic Eurasia and Alaska. Topographic and bathymetric contours are in hundreds of kilometers. Glacial erosional and depositional features are identified in Figure 9.34. From Grosswald and Hughes (1995).

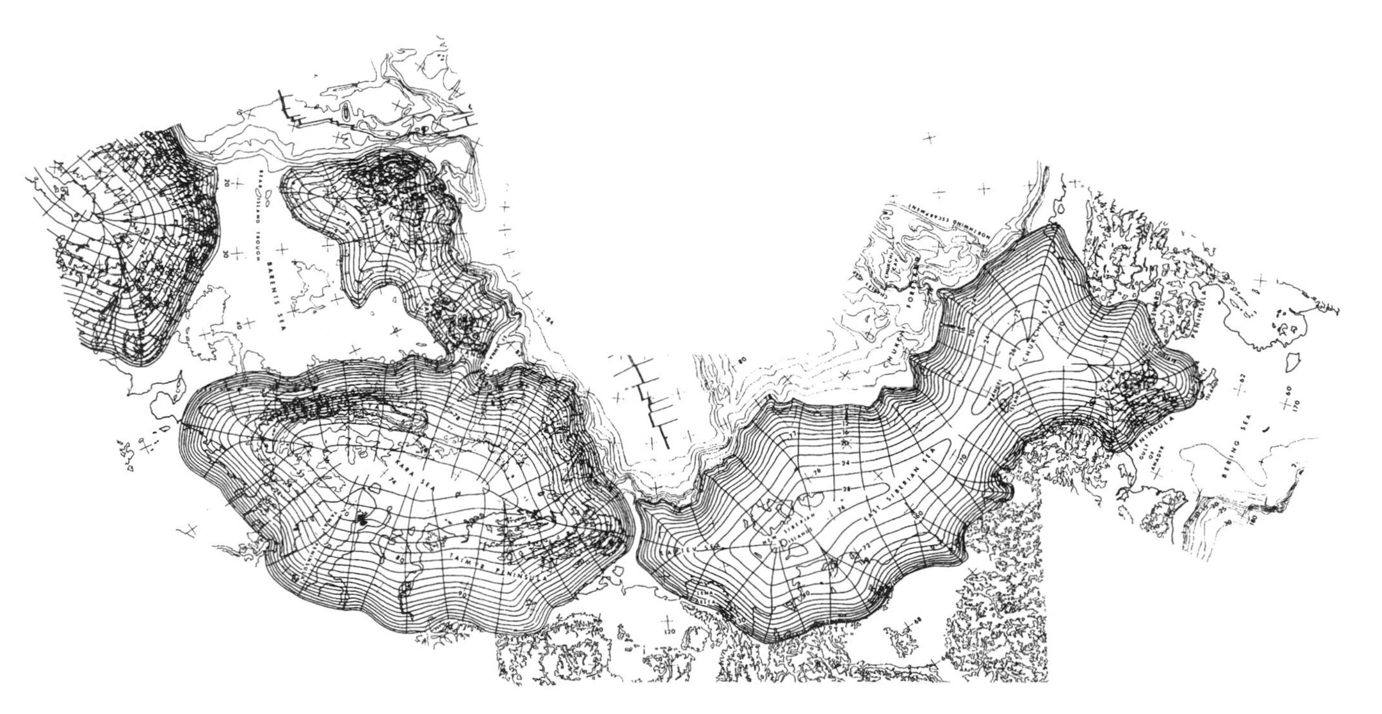

Figure 9.36: A reconstruction of the Siberian Ice Sheet as it separated into three ice sheets during a glacial interstadial. Surface ice flowlines and contour lines (200-m intervals) are computed for first-order glacial geology described in Figures 9.7 through 9.12 and a largely frozen bed. Ablation is primarily by calving from ice walls grounded in ice-dammed lakes and ice-free seas bordering the ice sheet, and from an ice shelf in the Arctic Ocean. High bed traction and ablation by calving maximizes the height-to-area ratio of the ice sheet.

Figure 9.37: A reconstruction of the Siberian Ice Sheet as it joined the Scandinavian Ice Sheet to produce a single Eurasian Ice Sheet during a glacial stadial. Surface ice flowlines and contour lines (200 m intervals) are computed for second-order glacial geology described in Figure 9.35 and a largely thawed bed. Basal traction is minimized by ice sliding over water-saturated lacustrine and marine sediments. Ablation is by calving in ice-dammed lakes and melting of ice lobes on land and by calving from ice walls and ice shelves in the sea. Low bed traction and a wide ablation zone minimizes the height-to-area ratio of the ice sheet. A lobe of the Eurasian Ice Sheet crosses the Bering Sea continental shelf and calves directly into the Pacific Ocean, thereby blocking the Bering land bridge between Siberia and Alaska.

direction is also revealed by east-west through valleys across Novaya Zemlya, which constitute first-order glacial geology.

Siberian ice retreated north in the East Siberian Sea, if glaciotectonic folds and ice-shoved features on the New Siberian Islands were produced during deglaciation (Grosswald, 1988). These structures were produced by ice moving south over a frozen bed. Deglaciation also extended "yedomas" north. Yedomas are thick accumulations of ice-rich silt and fine sand. They formed under very cold and dry conditions in proglacial river deltas dammed by the East Siberian Ice Sheet. Summer meltwater in these deltas spread out in a thin layer over the Arctic coastal lowlands and froze completely during the following winter. Therefore, yedomas accumulated in annual layers and advanced north against the retreating margin of the East Siberian Ice Sheet. Ground till may underlie yedomas where the retreating ice sheet had a thawed bed, along terrestrial ice streams, for example. Rising sea level invaded Bering Strait during deglaciation and caused Siberian ice to retreat in Alaska's Seward Peninsula and Siberia's Chukchi Peninsula, leaving local ice caps. Wastage of these ice caps ended with valley glaciers retreating into highlands and superimposed radial second-or-der glacial geology on first-order glacial geology that consisted of north-south through valleys and fjords cut by overriding ice during the glacial maximum.

First-order glacial geology in Figures 9.7 through 9.12 can be combined with second-order glacial geology in Figure 9.34 to reconstruct stadial and interstadial stages of a Siberian Ice Sheet, as was done for the Laurentide and Scandinavian Ice Sheets. Figure 9.35 combines the glacial landforms in these figures and includes sites of oriented lakes and glacial through valleys in Siberia and Alaska that may not be related to ice-flow directions. The interstadial stage of Siberian glaciation is shown in Figure 9.36. An East Siberian Ice Sheet was separated from a West Siberian Ice Sheet by the Lena River, which crossed the Laptev Sea (Fütterer, 1993, 1994). Alternatively, the Lena River channels that were cut into the Laptev Sea continental shelf, the depositional wedge on the continental slope, and the numerous iceberg plough marks that Fütterer (1993, 1994) described could have resulted from one or more outbursts of an ice-dammed lake in the Lena River valley, when an ice-divide saddle of the marine Siberian Ice Sheet collapsed over the Laptev Sea after the stadial stage in Figure 9.37. One such outburst may have occurred 15.7 ka BP (Stein and others, 1994), or later as shown in Figure 1.16.

The interstadial ice sheet in Figure 9.36 was reconstructed over a largely frozen bed, and equilibrium ice thicknesses are shown. It is likely that time was insufficient for equilibrium ice thicknesses to be attained for the low accumulation rates that prevailed during stadials and interstadials of the last glaciation. Therefore, the reconstruction in Figure 9.36 must be considered to represent one extreme of possible interstadials during the last glaciation cycle. For that extreme, thawing the bed would produce the stadial stage of Siberian glaciation shown in Figure 9.37. The actual late Quaternary glacial history of central Arctic Siberia was investigated by a German-Russian project from 1994 to 1997. Initial results have been published, but they are tentative (Melles, 1994; Fütterer, 1994).

The East Siberian and the West Siberian Ice Sheets would merge in the Laptev Sea to become a single Siberian Ice Sheet that poured through Bering Strait to calve into the Pacific Ocean in the Bering Sea, and poured into the Barents Sea to merge with the Scandinavian and Spitzbergen Ice Sheets to produce a single Eurasian Ice Sheet. The Eurasian Ice Sheet had Chukchi, East Siberian, Kara, Spitzbergen, and Scandinavian ice domes. At its maximum Pleistocene extents, it would have blocked migration across the Beringian land bridge between Siberia and Alaska, would have extended as far as 60°S in the West Siberian Lowlands, and would have nearly reached the Black Sea in Europe. This did not happen during the last glacial maximum, but it did attain this extent during other Quaternary glaciations. Collapse of the Siberian Ice Sheet would occur first at the major saddles along the ice divide, notably saddles in the Barents and the Laptev Seas that separated major ice domes in Scandinavia, the Kara Sea, and the East Siberian Sea. Collapse of these ice saddles, followed by collapse of the ice domes, would be accompanied by outbursts of ice-dammed lakes into the North Atlantic Ocean and into the Arctic Ocean during the last deglaciation (Stein and others, 1994).

The prevailing view is that the Eurasian Ice Sheet occupied only the Barents and the Kara Seas, and adjacent island and coastal plains, during the last glacial maximum. Sher (1991, 1992) has emphasized the importance of faunal and floral records in the Siberian Arctic in placing limits on a Eurasian Ice Sheet during the last glaciation cycle. For example, mammoths may have lived on Wrangel Island during the last glacial maximum, and they definitely survived there in isolation well into the Holocene (Vartanyan and others, 1993). This would limit the extent of a marine Eurasian Ice Sheet in the East Siberian and Chukchi Seas at that time. Fütterer (1993, 1994) believes that the Laptev Sea was unglaciated at the last glacial maximum because seismic profiling revealed that several river channels filled with acoustically transparent sediments had incised the continental shelf when it was above sea level, and these channels link the Lena River and smaller Siberian rivers to sediment piles on the continental slope. However, as seen in Figure 9.38, present-day outflow from the Lena River produces a wedge of freshwater ice in Tiksi Bay, and fast ice extends from the coastline far into the Laptev Sea, even beyond the New Siberian Islands (Dethleff, 1992). When temperature and sea level fell during the last glaciation, this freshwater ice and fast ice could have thickened and grounded on the continental shelf. It would then have become part of a marine Eurasian Ice Sheet. In that case, the incised, sediment-filled channels and sediment wedges record outbursts of ice-dammed lakes during interstadials when a saddle on the ice divide collapsed in the Laptev Sea.

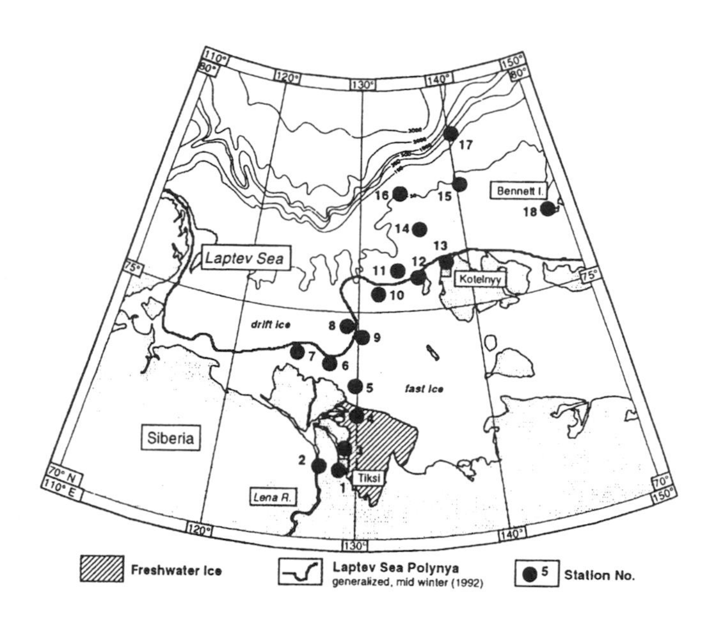

Figure 9.38: Ice conditions and sampling locations in the Laptev Sea during the 1992 East Siberian Arctic Regions Expedition. Reproduced with permission of Joint Oceanographic Institutions, and D. Dethleff, *The Nansen Icebreaker*, newsletter 3, (1992), p. 4, Fig. 1.

Problematic Ice Sheets

Owing to lack of comprehensive, dated, glacial geological field studies comparable to those in North America and Europe, all Quaternary ice sheets in Asia are problematic. Even the Siberian ice-sheet reconstructions in Figures 9.36 and 9.37 are disputed, especially during the last glaciation when wooly mammoths and other extinct species inhabited the region (e.g., Sher, 1991, 1992, 1995; Vartanyan and others, 1993; Rutter, 1995; Velichko, 1995). However, undated glacial landforms are consistent with glaciation by ice sheets. What is mainly in dispute is when they existed. If it is granted that marine ice sheets did occupy the broad Arctic continental shelf of Siberia at the last glacial maximum, then mass-balance formulations like the one employed in this chapter and illustrated by Figures 9.2 through 9.5, or by climate models (e.g., Verbitsky and Oglesby, 1992; Marsiat, 1994; Huybrechts and T'siobel, 1995), invariably lower snowlines enough to glaciate mountain highlands in Asia if they lower snowlines enough to glaciate the Arctic continental shelf. Ice sheets form over two highland areas in particular, the mountains of the northwest Pacific rim and the Tibetan Plateau.

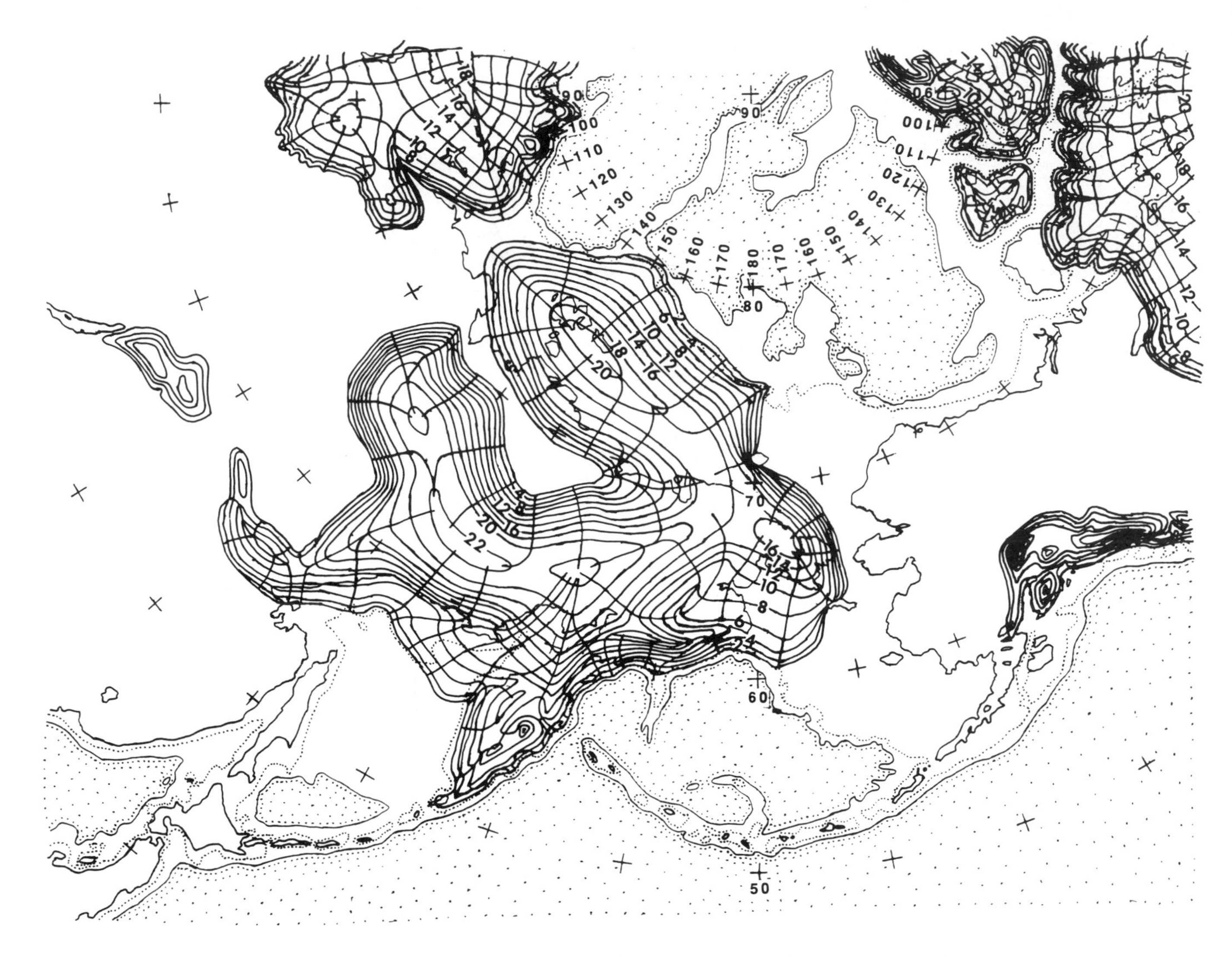

Figure 9.39: Problematic glaciation of the northwest Pacific rim during an interstadial of the last glaciation. Surface ice flowlines and contour lines (*200 m intervals*) are computed from first-order glacial geology and ice margins described by Grosswald and Hughes (1995, 1998) and Grosswald (1997). Bathymetric contours are shown at 100 fathoms (*dotted line*), which approximates the lowest late Quaternary sea level, and 1000 fathoms (*solid line*), which separates continental crust (*white areas*) from oceanic crust (*dotted areas*).

Glaciation of the northwest Pacific rim

Glaciation of the northwest Pacific rim has been examined by Grosswald and Hughes (1995, 1998) and Grosswald (1997) as part of discon-tinuous glaciation around the entire Pacific rim that includes Antarctic glaciation to the south, South American Andean glaciation to the southeast, North American Cordilleran glaciation to the northeast, Beringian glaciation to the north, glaciation in and around the Sea of Okhotsk to the northwest, and glaciation of the Southern Alps in New Zealand to the southwest. Denton and Hughes (1981) documented former Pacific rim ice sheets in the Andes and Cordillera of the Americas, and New Zealand glaciation. Grosswald and Hughes (1995) added a marine ice sheet in Beringia that spilled over the Chukchi Peninsula into the Bering Sea from the Chukchi Sea. Grosswald and Hughes (1998) added a highland Pacific rim ice sheet that spilled into the Sea of Okhotsk from the surrounding mountains of northeast Siberia, including Kamchatka Peninsula.

Figures 9.39 and 9.40 show interstadial and stadial constructions, respectively, of the northwest Pacific rim ice sheet and associated ice sheets during the last glaciation. The interstadial northwest Pacific rim ice sheet over the highlands of northeast Siberia descended to the Arctic lowlands of northeast Siberia on the north and into the Sea of Okhotsk on the south from an elongated ice dome that paralleled the Cherskiy Range. Its eastern limit was a lower ice dome over Chuckchi Peninsula. Ice from this may have reached Bering

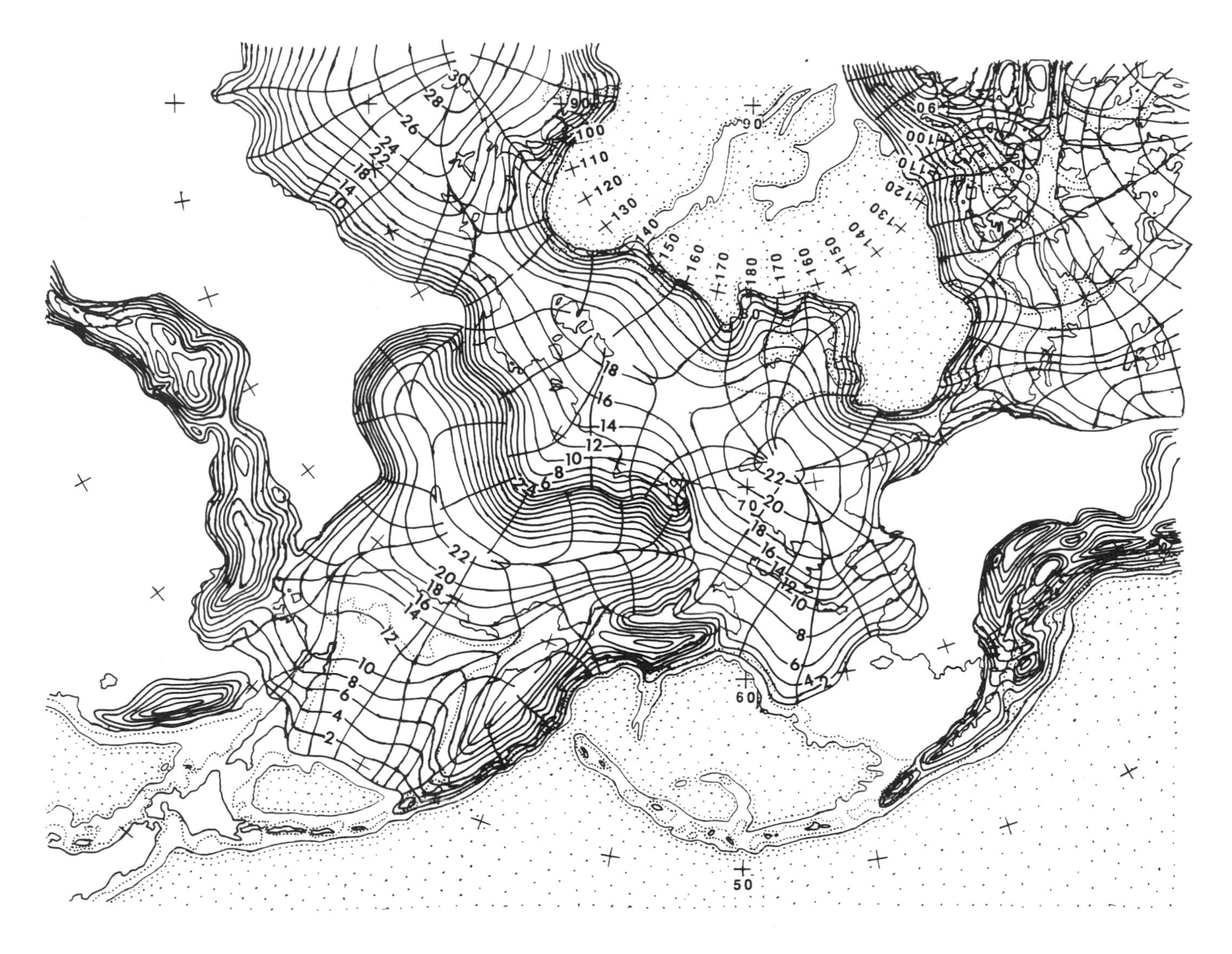

Figure 9.40: Problematic glaciation of the northwest Pacific rim during a stadial of the last glaciation. Surface ice flowlines and contour lines (*200 m intervals*) are computed from second-order glacial geology and ice margins described by Grosswald and Hughes (1995, 1998) and Grosswald (1997). Bathymetric contours are shown at 100 fathoms (*dotted line*), which approximates the lowest late Quaternary sea level, and 1000 fathoms (*solid line*), which separates continental crust (*white areas*) from oceanic crust (*dotted areas*).

Strait to the east (Elias and others, 1996), southern Wrangel Island to the north (Vartanyan and others, 1993), Saint Lawrence Island to the south (Heiser and others, 1992), and the lowlands of the Anadyr River to the southwest. The ice-marginal Lena River was the western limit of the northeast Pacific rim ice sheet. Partial deglaciation during the interstadial was accomplished by melting along terrestrial margins and calving along marine margins. In particular, calving bays migrated up stagnating ice streams occupying submarine troughs in the Sea of Okhotsk, Bering Strait, and interisland channels of Arctic Canada. During the subsequent stadial, the highland northwest Pacific rim ice sheet merged with the marine Siberian ice sheet that advanced southward across the Arctic coastal plain, while it advanced to the southern edge of the continental shelf in the Sea of Okhotsk. There it either formed a calving ice wall or continued as an ice shelf that calved along the Kuril Islands, which would have been ice-shelf pinning points. Its western ice margins remained relatively unchanged from the interstadial position,but on its eastern margin, marine Siberian ice overrode Chukchi Peninsula and poured through Bering Strait into the Bering Sea (Grosswald and Hughes, 1995).

Evidence for mountain glaciation is widespread in northeast Siberia, but it is largely undated. What led Grosswald and Hughes (1998) to conclude that an ice sheet produced the mountain glaciation was ubiquitous evidence that the Sea of Okhotsk had also been glaciated, mainly by an ice sheet that overrode mountains to the north, but that also included local ice domes in the Koryak Mountains and on Kamchatka Peninsula to the east. This required an ice sheet on the northwest Pacific rim. Erosional features predominate on the shallow continental shelf and deposi-tional features predominate on the deep ocean floor in the respective northern and southern parts of the Sea of Okhotsk. In particular, evidence was strong that an ice stream began at the head of Shelekhov Gulf at the northern end of the Sea of Okhotsk and extended south in a foredeepened trough to the edge of the continental shelf. Such widespread and large-scale glacial features on the sea floor are consistent with an overriding ice sheet, but not with mountain glaciers that ringed the Sea of Okhotsk and ended near the present-day shoreline. Ice-rafted detritus in the northwest Pacific from the last glacial maximum is also more consistent with calving from the marine margins of an ice sheet on the northwest Pacific rim than with tidewater calving from mountain glaciers on the rim (Kotilainen and Shackleton, 1995).

Mammoth remains on Wrangel Island (Vartanyan and others, 1993) and in the Arctic lowlands of northeast Sibera, as reported by Sher (1991, 1992, 1995), are consistent with migrations from Alaska and through the interstadial ice-free corridor between the marine Siberian ice sheet and the terrestrial northeast Pacific rim ice sheet constructed in Figure 9.39, just as migrations took place through the interstadial ice-free corridor between the Laurentide and Cordilleran Ice Sheets of North America. In addition, the marine Beringian ice sheet and the terrestrial northeast Pacific rim ice sheet reconstructed in Figure 9.40 for stadials would have prevented migrations from Siberia to Alaska dur-ing the last glacial maximum. This may explain the problematic archeological record for human habitation in the Americas before 12,000 years ago, after which that record is undisputed (e.g., Morell, 1990; Bahn, 1993; Hughes and Hughes, 1994). As noted in Chapter 1, knowing that Asians crossed a stormy ocean passage to Australia 80,000 years ago, it strains the mind to believe that Asians could not cross a land bridge 1000 km wide and teeming with game, and with the west wind at their backs, until 12,000 years ago. Perhaps the "land bridge" was blocked by a Circum-Pacific Ice Sheet for most of the 80,000 years.

Glaciation of the Tibetan Plateau

The prevailing view is that glaciation of the Tibetan Plateau during the last glacial maximum consisted primarily of limited expansions of present-day mountain glaciers (Li Binyauan and Li Jijun, 1991). However, Kuhle (1987, 1988a, 1988b, 1988c, 1989, 1990, 1991, 1994, 1995, 1996) postulated a Tibetan Ice Sheet, as shown in Figures 9.41 and 9.42, after mapping the full suite of glacial land-

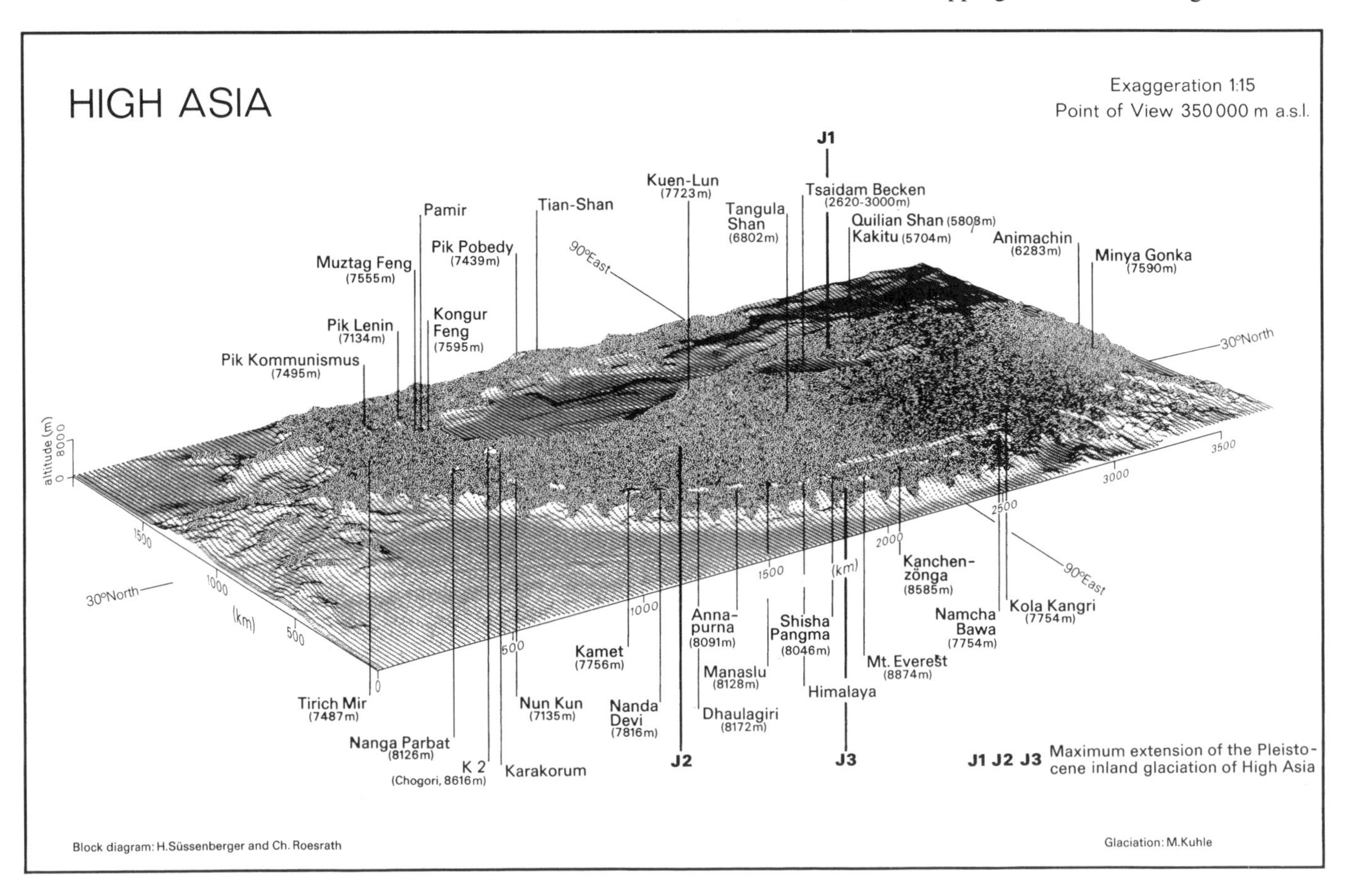

Figure 9.41: The Tibetan Ice Sheet according to Kuhle (1989). Reproduced with permission.

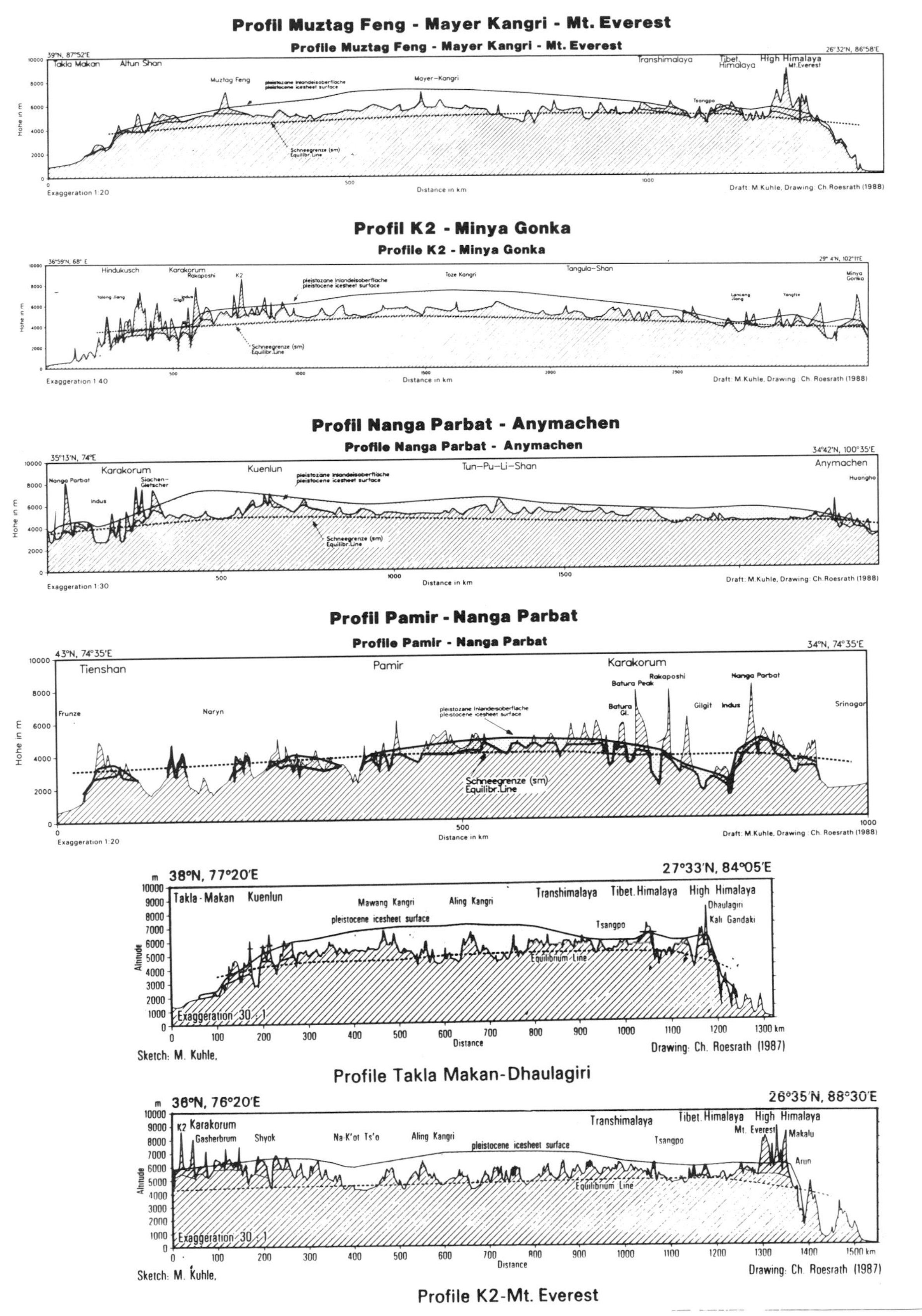

Figure 9.42: Cross-sectional profiles through the Tibetan Ice Sheet in Figure 9.41, according to Kuhle (1989). Reproduced with permission.

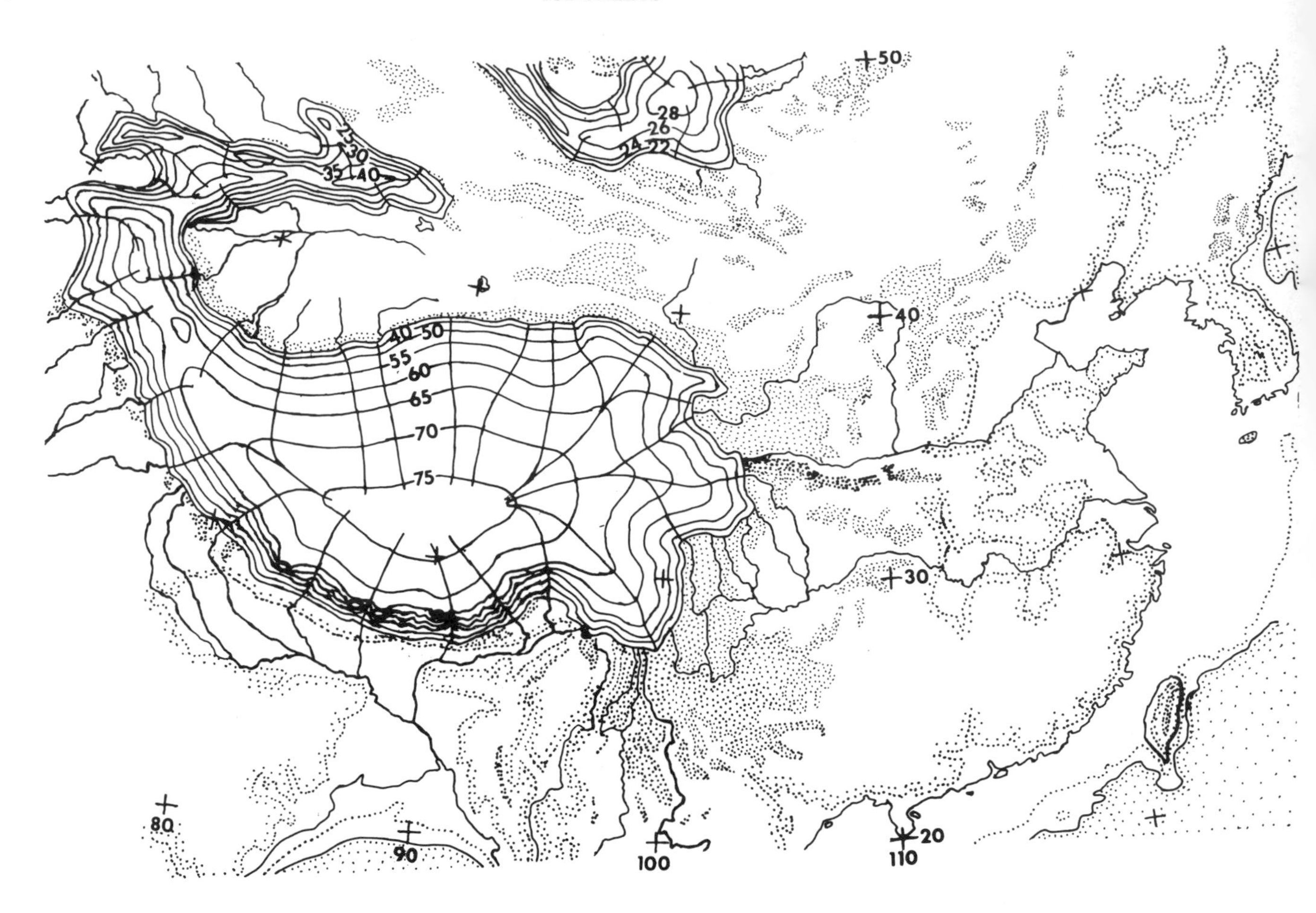

Figure 9.43: Problematic Tibetan Ice Sheet and mountain glaciation of central Asia. Surface ice flowlines and contour lines (*elevations in hectometers*) are based on second-order glacial geolgy. Rivers fed by melting ice margins are shown. Land elevations are shown by the double-dotted border at 500 feet (152 m) and by dotted areas above 5000 feet (1525 m). Bathymetric contours are shown at 500 feet (*dotted line*), which approximates the lowest late Quaternary sea level, and 5000 feet (*solid line*) at the deep ocean floor (*dotted area*).

forms over vast regions of Tibet, including elevations below these ice limits. His work included formulating a more accurate calculation of the equilibrium-line altitude on glaciers that took glacial geometry into account (Kuhle, 1988b), using information theory to identify ice-marginal ramps as glacial landforms (Kuhle, 1990), and determining rates of postglacial isostatic rebound (Kuhle, 1995). Although Yu Kin and others (1994) ascribed widespread negative gravity anomalies in the –500 m Gal range to crustal thickening under Tibet, these high anomalies could mask those in the –50 m Gal range expected for glacioisostatic depression of Tibet by Kuhle's ice sheet.

A Tibetan Ice Sheet during the last glaciation would have been upwind from the thick Loess deposits of central China studied by Kukla (1987), and by Porter and others (1992). Katabatic winds from the ice sheet may have entrained silt from Tarim Basin into the prevailing westerly winds across China. Wind erosion would be intensified during stadials, as was Loess deposition (Mayewski and others, 1994).

Kuhle (1988c) noted that, during the present interglacial, Tibet receives four times as much solar energy as do the latitudes between 60° and 70° north or south, where Quaternary ice sheets developed, because of the low latitudes and high elevations of the Tibetan Plateau. An ice sheet over this plateau during the last glaciation would have reflected a similar amount of solar energy, owing to its high albedo. Therefore, the timing and extent of Tibetan glaciation, including advances and retreats of a hypothetical Tibetan Ice Sheet during stadials and interstadials of the last glaciation, along with Quaternary uplift of the Tibetan Plateau, are problems that must be addressed in any attempt to understand the relationship between glaciation and Quaternary climatic change.

Figure 9.41 shows the areal extent of the Tibetan Ice Sheet mapped by Kuhle (1995), based on the glacial geology of the Tibetan Plateau that he mapped during twenty field seasons. Figure 9.42 shows six cross-sectional profiles through the Tibetan Ice Sheet that Kuhle (1995) produced, based on his field studies. They show equilibrium-line altitudes, ice elevations, ice thicknesses, and bed topography. The ice elevations are minimal, being based on the highest elevations of glacial erosional and depositional features. Figure 9.43 is a recon-struction of the Tibetan Ice Sheet based on Equation (9.53). Although the ice sheet was drained by ice streams that occupied valleys descending from the Tibetan Plateau, these are too short and narrow to appear in Figure 9.43. However, most of the flowlines used in this reconstruction followed these valleys. There is reasonable agreement between this theory-based reconstruction and the field-based reconstruction by Kuhle (1995). The major difference is that the theoretical reconstruction glaciates the Tsaidam Basin region of Tibet, but the Kuhle (1995) reconstruction does not.

High resolution imagery from Earth-orbiting satellites should be able to distinguish between evidence for limited extensions of interglacial mountain glaciers and a large ice sheet on the Tibetan Plateau during the last Quaternary glacial cycle. In addition, surficial dating by cosmogenic nuclides should be able to establish a Quaternary glacial chronology. This strategy also applies to other parts of Asia where problematic ice sheets have been postulated. Therefore, it is likely that the question of problematic Asian ice sheets can be resolved within a few years. If their existence is confirmed, however, a major effort will be required to determine their history, their dynamics, and their contribution to Quaternary climate change.

10

THE COMING GLACIATION

A Prognostic Approach to the Coming Glaciation Cycle

As noted in Chapter 1, another glaciation looms over our northern horizon. Although the approach to modeling ice sheets presented in this book was developed for diagnostic modeling of earlier glaciations, it can also be used for prognostic modeling of future glaciations. Initiation, advance, mature, retreat, and termination stages of an ice sheet are linked externally to the mass balance of distinct climatic regimes that expand and shrink in their geographical extent during a glaciation cycle. These slowly varying external boundary conditions are paced by Milankovitch insolation cycles, especially by a predictable rhythm controlled by known slow variations in tilt and precession of Earth's rotation axis (cycles of minor changes in orbital eccentricity are ignored). This rhythm is punctuated by relatively unpredictable abrupt changes in internal boundary conditions that are inherent in ice-sheet dynamics. These abrupt changes begin with widespread basal melting, causing ice to thin and advance during stadials, and end with widespread basal freezing, causing ice to thicken and retreat during interstadials. In the record of the last glaciation, extended positions produce stadials and retracted positions produce interstadials, as recorded by Dansgaard-Oeschger events bundled into Bond cycles in the stratigraphy from ice-sheet and ocean-floor cores. It is this dated stratigraphic record, together with the known insolation variation with latitude through time, that provide the internal and external boundary conditions that constrained the diagnostic model of the last glaciation presented in Chapter 9.

In Chapter 10, a prognostic model of the coming glaciation utilizes known future insolation variations as the primary external boundary conditions, but internal boundary conditions must be deduced from the mechanisms that produced the stratigraphic record of stadials and interstadials during the last glaciation, since that record for the coming glaciation has yet to be written. The mechanisms that will produce this future record can be grouped into four "machines" that operate during a glaciation. As described in Chapter 1, these are the ice machine, the dirt machine, the dust machine, and the cloud machine. To the extent that the effect of these machines on internal ice-sheet dynamics can be quantified, a prognostic model of the coming glaciation can be advanced. A prognostic simulation of the coming glaciation, which immediately follows "greenhouse" warming, concludes Chapter 10.

After inception of ice sheets for the coming glaciation, particularly inception of a hypothetical Arctic Ice Sheet, abrupt changes in their size and shape depend on turning the ice, dirt, dust, and cloud machines on and off, according to the new paradigm being promoted in this book. Inception of an Arctic Ice Sheet may itself be an abrupt change from our present-day world. In addition, abrupt changes in internal ice-sheet boundary conditions may trigger abrupt changes in external boundary conditions. Milankovitch insolation variations at the top of the atmosphere are the only external boundary conditions that are known to be immune to abrupt changes of internal ice-sheet boundary conditions. Other external boundary conditions, such as atmospheric carbon dioxide and cloudiness, might be a consequence of internal ice-sheet dynamics.

Modeling External Boundary Conditions

Modeling inception of the Arctic Ice Sheet must include mechanisms for increasing precipitation, lowering snowlines, and thickening sea ice. The mass balance at gridpoints within the area in Figure 1.10 can be computed from Equation (7.6) for thickening of floating sea ice by snow accumulation on the top surface and water freezing on the bottom surface, assuming that little heat is imported to this area by rivers or ocean currents so that setting $\dot{Q}_H = 0$ is acceptable. In addition, Equation (4.58) shows that $\dot{\varepsilon}_{zz} = 0$ is reasonable when $h_I < 100$ m, where $\dot{\varepsilon}_{zz} = -2\,\dot{\varepsilon}_{xx}$ by Equation (4.51). In any case, $\dot{\varepsilon}_{zz} = 0$ when thickening sea ice is grounded around its entire perimeter. Snow accumulation rates on the top of sea ice and on Arctic islands and mainlands can be computed at each gridpoint using the approach by Fastook and Prentice (1994). Mean annual surface temperature $\bar{T}_a$ in °C is:

$$\bar{T}_a = T_s + T_c + (\partial T/\partial h)_c\, h + (\partial T/\partial N)_c\, N \quad (10.1)$$

where T_s is the sea-surface temperature, T_c is a constant tuning temperature for a given climate regime, $(\partial T/\partial h)_c$ is the constant adiabatic lapse rate of temperature T with surface elevation h and $(\partial T/\partial N)_c$ is the constant change of temperature with north latitude N expressed in degrees. The absolute free atmosphere isothermal layer temperature T_f in °A computed from $\bar{T}_a$ is:

$$T_f = 0.67\,(\bar{T}_a + 273°C) + 88.9°A \quad (10.2)$$

where T_f is used to calculated the following constants:

$$C_1 = -9.09718\,[(273.16°A/T_f) - 1] \quad (10.3a)$$

$$C_2 = -3.56654\,\log\,(273.16°A/T_f) \quad (10.3b)$$

$$C_3 = 0.876793\,[1 - (T_f/273.16°A)] + 0.785835 \quad (10.3c)$$

Equations (10.2) and (10.3) are standard expressions. Saturation water vapor pressure P_v is:

$$P_v = 10^{(C_1 + C_2 + C_3)} \quad (10.4)$$

Surface accumulation rate a" is:

$$a'' = (\partial a''/\partial P_v)_c\, P_v + a_1''\,(\partial h/\partial x) + a_2'' \quad (10.5)$$

where constant gradient $(\partial a''/\partial P_v)_c$ and constant accumulation rates a_1'' and a_2'' are all fitting parameters to make present-day observed accumulation rates a" fit data for variable water vapor pressure P_v and variable surface slope $\partial h/\partial x$. Fastook and Prentice (1994) used $(\partial a''/\partial P_v)_c = 0.1914$ m/a mbar, $a_1'' = 0.009228$ m/a, and $a_2'' = -0.007389$ m/a for Antarctica. They computed surface ablation rates a' from the annual number of positive degree days, days when the surface temperature is above the melting temperature by some specified °C. Mean monthly insolation $\bar{Q}_m$ at north latitude N has a peak $(\bar{Q}_m)_p$ at north latitude N_p and a relatively constant latitudinal gradient $(\partial \bar{Q}_m/\partial N)_c$ such that:

$$\bar{Q}_m = (\bar{Q}_m)_p + (\partial \bar{Q}_m/\partial N)_c\, N \quad (10.6)$$

Mean annual insolation $\bar{Q}_a$ at the top of the atmosphere is $\bar{Q}_m$ for each month averaged over twelve months:

$$\bar{Q}_a = \frac{1}{12}\sum_{i=1}^{i=12} \left(\bar{Q}_m\right)_i \quad (10.7)$$

Mean monthly temperature $\bar{T}_m$ in °C is the sum of mean annual temperature $\bar{T}_a$ and a monthly temperature variation that is proportional to the difference between mean monthly insolation $\bar{Q}_m$ and mean annual insolation $\bar{Q}_a$:

$$\bar{T}_m = \bar{T}_a + (\partial T/\partial Q)_c\,(\bar{Q}_m - \bar{Q}_a) \quad (10.8)$$

where gradient $(\partial T/\partial Q)_c$ is the proportionality constant. The annual sum of positive degree days is temperature T_d defined as $\bar{T}_m$ for each of thirty-day months summed over twelve months:

$$\bar{T}_d = 30\sum_{i=1}^{i=12} \left(\bar{T}_m\right)_i \quad (10.9)$$

The surface ablation rate is:

$$a' = (\partial a'/\partial T_d)_c\, T_d \quad (10.10)$$

where constant gradient $(\partial a'/\partial T_d)_c = 0.006$ m/a°C for West Greenland (Braithwaite and Olesen, 1989). The net mass balance at each gridpoint in the Arctic is:

$$a = a' + a'' \quad (10.11)$$

Fastook and Prentice (1994) give values for constants in Equations (10.1) through (10.10). The effect of albedo feedback on a is not included in their analysis.

Turning the Ice Machine On and Off

The ice-machine hypothesis postulates twelve mechanisms that link positive and negative feedbacks during a glaciation cycle.

1. Marine ice sheets are grounded on polar continental shelves far enough below sea level for their seaward perimeter to form floating ice shelves (Mercer, 1968). The only present-day marine ice sheet is in West Antarctica, but marine ice was a major component of Quaternary ice sheets in the Northern Hemisphere (Grosswald, 1988). The grounding lines of ice shelves are shallower than the isostatically depressed continental shelf beneath the marine ice sheet. This tends to make marine ice sheets inherently unstable; they either advance to the edge of the continental shelf or they collapse (Weertman, 1974). The first analytical demonstration of irreversible grounding retreat was for a Laurentide ice stream in the Gulf of Saint Lawrence (Thomas, 1977).

2. Collapse is triggered when the ice-shelf grounding line retreats on the downsloping bed beneath the marine ice sheet, making retreat irreversible until the grounding line stabilizes on an upsloping bed at a shallower depth. Initial retreat is caused (a) by ice thinning during surge-like life cycles of ice streams crossing the grounding line, (b) by a lag in basal isostatic depression along the grounding line, (c) by rising sea level at the grounding line, (d) by climatic warming that melts ice along the grounding line, and (e) by calving that removes ice above the grounding line (see Figure 2.3). In particular, irreversible advance to and retreat from the continental-shelf edge is possible for ice streams entering the North Atlantic from Europe to the east and from North America to the west, because the snowline is relatively horizontal in east-west directions, making this an inherently unstable orientation for ice-sheet margins. Tabular icebergs calve when top and bottom crevasses meet, as illustrated in Figure 2.4.

3. An ice shelf is sustained by its own surface and basal mass balance and by the mass balance between calving along its terminal ice cliff and ice streams crossing its grounding line. Ice shelves are small if the net surface/basal mass balance becomes negative, that is, if summed over both top and bottom surfaces, a net loss of ice occurs (Oerlemans and Van der Veen, 1984). Ice shelves vanish if grounding-line advance rates overtake calving rates or if calving rates exceed grounding-line retreat rates. These various rates are very sensitive to environmental constraints, notably (a) ice-stream dynamics, (b) climatic changes, and (c) changing sea level in relation to grounded parts of the ice-shelf perimeter and internal pinning points at islands, ice rises, and ice rumples where the ice shelf is locally grounded.

4. Grounding lines create lines of weakness in an ice shelf. Grounding lines can produce (a) lateral rifts where lateral ungrounding occurs when an ice stream enters the ice shelf, (b) tidal flexure cracks between ice streams, (c) horst and graben gashes in the lee of pinning points, and (d) shear-rupture crevasses (Hughes, 1983). Ice flowing across grounding lines brings these lines of weakness onto the ice shelf, where they form a network of weak links that allows rapid disintegration of the ice shelf if climatic warming produces ubiquitous surface melting. A present-day example of rapid disintegration is discerpation of Wordie Ice Shelf on the Antarctic Peninsula (Doake and Vaughan, 1991).

5. Marine ice streams develop in the straits and interisland channels near the grounded margin of a marine ice sheet. They supply most of the ice to ice shelves, which formed by a confluence of floating ice-stream tongues. Ice streams are therefore the dynamic links between grounded and floating parts of a marine ice sheet. It can be argued that ice shelves are produced by ice streams during a life cycle consisting of inception, growth, mature, decline and terminal stages (Hughes, 1992a, 1992b). During its life cycle, an ice stream has fast flow akin to surges of certain mountain glaciers (Kamb and others, 1985). However, a large ice stream drains an area some 1000 times larger than a typical surging glacier, so its life cycle is probably 1000 times longer than the one- to three-year duration of a surge.

6. Inception of stream flow occurs when permafrost thaws in unconsolidated sediments on the floor of straits and interisland channels beneath a marine ice sheet or on river bottoms overrun by terrestrial ice sheets. Thawing converts a rigid bed into a deformable bed in which shear deformation mobilizes interstitial water, reducing basal shear traction drastically and allowing rapid stream flow and downdraw in the overlying ice. Stream flow can probably be initiated in other ways and may even have a steady-state mode, but these possibilities are not involved in the proposed life cycle classification.

7. Growth of stream flow occurs when ice streams lengthen, producing floating ice tongues that coalesce into ice shelves at their seaward terminus, grounded ice lobes that coalesce into ice piedmonts at their landward terminus, and downdrawn interior ice drainage basins behind them. Lengthening occurs when basal frictional heat produced by converging flow at the heads of ice streams causes the thawing front of basal permafrost to retreat rapidly into marine embayments behind islands on the outer continental shelf. Hudson Bay and the Barents Sea are marine embayments that were sites of Pleistocene marine ice sheets.

8. Mature stream flow occurs when glacial erosion strips away enough of the thawed permafrost beneath ice streams to expose bedrock outcrops that act as sticky spots retarding basal sliding and when ice shelves become sufficiently confined in marine embayments and anchored at pinning points to effectively buttress ice streams. West Antarctic ice streams supplying the Ross and the Ronne Ice Shelves seem to be in a mature stage.

9. Declining stream flow occurs when ice streams have downdrawn so much interior ice that further lowering does not release enough gravitational potential energy to sustain the pulling power of ice streams at the time when proliferating sticky spots increase ice-bed coupling beneath ice streams and proliferating pinning points increase ice-shelf buttressing beyond ice streams. Glacial erosion of sediments under the ice stream uncovers more bedrock sticky spots, and glacioisostatic rebound under the ice shelf pro-duces more bedrock pinning points. Pulling power in ice streams is also reduced when retreating ice-shelf grounding lines begin moving upslope out of marine embayments, because the pulling force is proportional to ice thickness squared at the grounding line (see Figure 2.2).

10. Termination of stream flow occurs when ice streams become so thin that the pulling force is negligible at grounding lines and basal heat is conducted to the surface rapidly enough to transform the bed from sticky frozen spots in a slippery thawed matrix to slippery thawed spots in a sticky frozen matrix as a result of basal freezing. In this case, the forward ice velocity lags far behind the calving rate, and the marine embayment becomes a calving bay within which the ice shelf disintegrates rapidly, leaving a calving ice wall that continues to retreat. Pauses in the calving retreat rate are possible (a) along lines of ice-shelf pinning points in the embayment, (b) when the calving front approaches the grounding line to produce an ice shelf that calves along tidal flexure cracks, (c) when the calving front meets the ground-ing line so that an ice wall replaces the ice shelf and tidal-flexure calving is replaced by subaerial and submarine calving along the wall, (d) when the ice wall retreats to a beach so that submarine calving no longer occurs, and (e) when melting converts the ice wall to an ice ramp so subareal calving no longer occurs (see Figure 2.4).

11. Rejuvenated stream flow is possible when ice-bed coupling by basal freezing allows grounded ice to thicken and sheet flow to replace stream flow and when disintegration of a floating ice shelf removes ice-shelf buttressing. This restores conditions before ice streams began their life cycle, so a new life cycle becomes possible if environmental conditions allow the marine ice sheet to become sufficiently large. This permits stadials and interstadials in the glaciation cycle. If rejuvenated stream flow is not possible because collapse of marine ice sheets has released too much gravitational potential energy, then the glaciation cycle ends in an abrupt termination.

12. The ice machine is turned on when the life cycle of an ice stream begins, and it is turned off when the life cycle ends. Preliminary work on Antarctic ice streams suggests that their life cycles can last a millennium or more, but their inception and terminal stages may take place in less than a century (Thomas and Bentley, 1978; Fastook, 1984, 1987; Whillans and others, 1987; Shabtaie and Bentley, 1988; MacAyeal, 1989). Life cycles are strongly dependent on water depths and bed slopes at grounding lines and on the dimensions of marine embayments. The great variety of these conditions for the Pleistocene marine ice sheets, especially those concentrated on continental shelves around the North Atlantic, might explain the spikiness of oxygen isotope stratigraphy in the Dye 3 core that records both ice volume and climatic changes (Hammer and others, 1985) and "Heinrich events" that record episodes of ice-rafted deposition in the North Atlantic (Heinrich, 1988; Broecker and others, 1992; Andrews and Tedesco, 1992). This spikiness is superimposed on the more gradual stadials and interstadials and the glacial terminations that are correlated with Milankovitch insolation variations (Hays and others, 1976).

Temperature field for thickening ice domes

The first task in applying the ice-machine hypothesis to modeling the coming glaciation is to determine what ice thickness is needed to thaw a frozen bed, both for marine ice grounding on shallow continental shelves or in shallow marine embayments and for terrestrial ice formed from thickening icefields on highland plateaus or from a thickening snow cover at lower elevations. During initial stages of ice thickening, spreading flow will be small, so the heat-flow analysis by Robin (1955) can be employed to compute approximate temperature profiles through the ice thickness. Ignoring compression of snow into ice in the upper ten meters, a good approximation for relating ice compression rate dz/dt to ice accumulation rate a is:

$$dz/dt = -(z/h_I)\, a \tag{10.12}$$

where z is vertical distance above the bed, t is time, and h_I is ice thickness. Since horizontal temperature gradients and advection rates are small, the equation of heat flow reduces to vertical heat transport:

$$dT/dt = \kappa\, d^2 T/dz^2 \tag{10.13}$$

where T is temperature and κ is thermal diffusivity for ice. Combining Equations (10.12) and (10.13):

$$\frac{dT}{dt} = \frac{dT}{dz}\frac{dz}{dt} = \frac{dT}{dz}\left(-\frac{a\,z}{h_I}\right) = \kappa\frac{d^2T}{dz^2} = \kappa\frac{d(dT/dz)}{dz} \tag{10.14}$$

Writing Equation (10.14) in integral form:

$$\int_{(dT/dz)_0}^{dT/dz} \frac{d(dT/dz)}{(dT/dz)} = -\left(\frac{a}{\kappa\, h_I}\right)\int_0^z z\, dz \tag{10.15}$$

Integrating Equation (10.15):

$$dT/dz = (dT/dz)_0\, exp\,(-a\, z^2/2\,\kappa\, h_I) \tag{10.16}$$

where basal ice temperature gradient $(dT/dz)_0$ is the ratio of the basal geothermal heat flux Q_G and the thermal conductivity K of basal ice:

$$(dT/dz)_0 = Q_G/K \tag{10.17}$$

Integrating Equation (10.16) gives temperature T_z at height z above the bed in terms of mean annual surface temperature $\bar{T}_a$ in Equation (10.1):

$$T_z = \bar{T}_a + (dT/dz)_0 \int_0^z exp\left(-a\, z^2/2\,\kappa\, h_I\right) dz$$

$$= \bar{T}_a + \left(2\,\kappa\, h_I\, Q_G/a\, K\right)\left[erf\left(2\,\kappa\, h_I\, z/a\right) - erf\left(2\,\kappa\, h_I^2/a\right)\right] \tag{10.18}$$

Abramowitz and Stegun (1965) have tabulated the error function:

$$erf\left(2\,\kappa\, h_I\, z/a\right) = \int_0^{(2\kappa h_I z/a)} exp\,(-\lambda^2)\, d\lambda \tag{10.19}$$

Figure 10.1 shows steady-state temperature profiles that

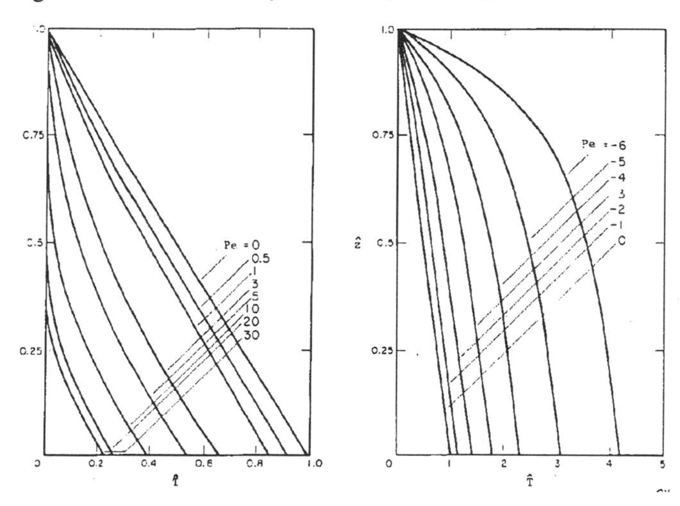

Figure 10.1: Vertical temperature profiles through ice sheets when ice accumulation and ablation rates greatly exceed ice spreading rates. Profiles in the accumulation (left) and ablation (right) zones are shown for dimensionless internal ice height above the bed $\hat{z} = z/h$ versus dimensionless internal ice temperature $\hat{T} = K\,(T - T_s)/Q_G\, h$ for various Peclet numbers $Pe = a_t\, h/\kappa$ in relation to ice thickness h, top surface temperature T_s, top surface accumulation rate a_t, Geothermal heat flux Q_G, thermal conductivity of ice K, and thermal diffusivity of ice κ. Reproduced with permission from Clarke and others (1977).

Clarke and others (1977) obtained from Equation (10.18). MacAyeal (1993a, 1993b) numerically calculated temperature profiles when $\bar{T}_a$ changes as ice sheets thicken and thin.

Glacial lineations mapped by Boulton and Clark (1990) for the last Laurentide glaciation were produced when basal ice temperature T_0 rose to the basal melting temperature T_M, which happpens when surface ice temperature $\bar{T}_a$ and ice thickness h_I give $T_z = T_M$ at $z = 0$ in Equation (10.18). As interpreted in Figure 9.26, this lineation pattern is consistent with a marine ice dome growing first in Foxe Basin and then migrating into Hudson Bay as marine ice spreads over the Hudson Bay lowlands. A similar history of marine ice transgression could be postulated for marine ice domes in the Gulf of Bothnia for Scandinavia and the Kara Sea and the East Siberian Sea for Siberia. It is reasonable to assume that the coming glaciation will repeat this pattern, since it is the pattern that produces first-order glacial geology that makes a deeper imprint on the landscape with successive glaciation cycles.

In computing ice-sheet flowline profiles for this growth stage of the coming glaciation, inception of the Arctic Ice Sheet creates the White Hole shown in Figure 1.11 so that ice accumulates on both the embryonic Arctic Ice Sheet and the surface of shallow lakes in broad river valleys impounded by the Arctic Ice Sheet. Winter lake ice thickens to the bed in shallow water and melts from the top down in the summer. When the annual energy budget is in deficit, not all winter lake ice melts in the summer, and giant "icings" can form along the margin of the Arctic Ice Sheet where shallow lakes are impounded. Similar giant icings exist today in the Moma Depression of northeastern Siberia, where rivers are frozen to their beds so surface melt-water in the summer overflows the river channels and spreads over the lowlands as a thin water blanket that freezes completely during the subsequent winter. This process is repeated every year, and giant icings result. A similar process produced the yedomas of northeastern Siberia when shallow ice-dammed lakes formed along the southern margin of the marine ice sheet that transgressed onto the Arctic coastal plain of northeastern Siberia during the last glaciation (Grosswald and Lasca, 1993).

Icings or yedomas along the landward margins of the growing Arctic Ice Sheet allow the ice-covered part of the White Hole in Figure 1.11 to expand faster by growth of shallow ice-impounded lakes than by forward creep of the Arctic Ice Sheet. Therefore, the thickness increase of the ice cover lags far behind the increase of its areal extent. Elevation profiles of thickening grounded ice can be calculated along surface flowlines using Equation (5.56), provided that r is given by Equation (5.27) for $r_0 = 1/3$ and $t_0 = 7800$ a, with $f_t = 0$ for an initially frozen bed. For the first Δx step in Equation (5.56), basal shear stress $\tau_o = (\tau_o)_C$ is constant along x but should increase over time t in the same way that the increasing ice thickness increases r, namely:

$$\tau_o = \tau_v \left(1 - e^{-t/t_o}\right) \tag{10.20}$$

where $\tau_v \approx 1$ bar $= 100$ k Pa is the viscoplastic yield stress for ice and $t_o = 7800$ a is from Andrews (1970, p. 38).

Spreading rates for thickening ice domes

Ice elevation changes dh/dt at the ice margin are changes δa_0 in net ice ablation and accumulation rate a_0, so setting $dh/dt = \delta a_0$ at a given t and differentiating Equation (9.4) with respect to t gives as the advance or retreat rate of the ice-sheet margin at that time t:

$$v_x = \frac{dx}{dt} = \frac{d}{dt}\left[\frac{(h - h_0)^2}{2\,\tau_o/\rho_I\, g_z}\right] = \left[\frac{\rho_I\, g_z (h - h_0)}{\tau_o}\right]\frac{dh}{dt} = \left(\frac{P_o}{\tau_o}\right)\delta a_0 \approx \frac{\delta a_0}{\alpha_0} \tag{10.21}$$

In Equation (10.21), Milankovitch insolation variations cause δa_0 to change as δH changes, $P_o = \rho_I\, g_z\, (h - h_0)$ is ice pressure at the base of a terminal ice wall of height $(h - h_0)$ that calves into shallow proglacial lakes, and $\tau_o \approx P_o\, \alpha_0$ for surface slope α_0 at the top of the ice wall. Note that $P_o = 0$ when $h - h_0 = 0$ and $P_o = 1$ bar when $h - h_0 = 10$ m. Equation (10.21) predicts that the advance or retreat rate of the ice margin decreases as ice-surface slope increases at the ice margin. The initial landward advance rate of the Arctic Ice Sheet is very great because α_0 is initially very small.

The component of snowline elevation change δH that responds to Milankovitch insolation variations during the coming glaciation has four distinct effects on the rate and extent of ice-sheet advance or retreat. When Milankovitch insolation is a maximum in all Northern Hemisphere latitudes, $\delta H = 0$ and is obtained from Equation (9.8) through (9.10) by setting $t \approx i\, t_e \approx j\, t_p$, where i, j = 0, 1, 2, 3, etc. Entering this δH in Equation (9.11) gives for Equation (10.21):

$$v_x = \delta a_0/\alpha_0 = a_0'\,[1 - \exp(0)] + a_0''\,[1 - \exp(0)]/\alpha_0 = 0 \tag{10.22}$$

Therefore, the ice-sheet margins are nearly stationary, as is true for the Greenland Ice Sheet today. When Milankovitch insolation is a minimum, so that $t \approx (i/2)\, t_e \approx (j/2)\, t_p$ in these questions, where i, j = 1, 3, 5, 7, etc., then:

$$v_x = \delta a_0/\alpha_0 = a_0'\,[1 - \exp(-\Gamma'\{\Delta H_P + (R/R_e)(\Delta H_e - \Delta H_P)\}^2)]/\alpha_0 + a_0''\,[1 - \exp(-\Gamma''\{\Delta H_P + (R/R_e)(\Delta H_e - \Delta H_P)\}^2)]/\alpha_0 \tag{10.23}$$

This is the maximum advance rate of ice-sheet margins. When Milankovitch insolation is high in high latitudes and low in mid latitudes, the snowline slope is a minimum so that $t \approx i\, t_e$ for i = 0, 1, 3, 5, 7, etc., and $t \approx (j/2)\, t_p$ for j = 1, 3, 5, 7, etc., in these equations:

$$v_x = \delta a_0/\alpha_0 = a_0'\,[1 - \exp(-\Gamma'\{(R/R_e)\,\Delta H_e\}^2)]/\alpha_0 + a_0''\,[1 - \exp(-\Gamma''\{(R/R_e)\,\Delta H_e\}^2)]/\alpha_0 \tag{10.24}$$

Ice-sheet margins undergo major advances in mid latitudes and major retreats in high latitudes, because the low snowline slope is the closest approach to unstable equilibrium. This favors irreversible advance and retreat, as seen by comparing Figure 9.2 with Figure 2.6. When Milankovitch insolation is low in high latitudes and high in mid-latitudes, the snowline slope is a maximum so that $t \approx (i/2)\, t_e$ for $i = 0, 1, 3, 5, 7$, etc., and $t \approx j\, t_p$ for $j = 0, 1, 2, 3$, etc., in these equations, then:

$$v_x = \delta a_0/\alpha_0 = a_0'\,[1 - \exp(-\Gamma'\{\Delta H_P - (R/R_e)\,\Delta H_P\}^2)]/\alpha_0 + a_0''\,[1 - \exp(-\Gamma''\{\Delta H_P - (R/R_e)\,\Delta H_P\}^2)]/\alpha_0 \qquad (10.25)$$

Ice-sheet margins undergo minor retreats in mid latitudes and minor advances in high latitudes, because the high snowline slope is the closest approach to stable equilibrium. This favors reversible advance and retreat, as seen by comparing Figure 9.2 with Figure 2.6.

For the first Δx step in Equation (5.56), the mean surface slope $\bar{\alpha}_0$ obtained by differentiating Equation (9.4) with respect to x is:

$$\bar{\alpha}_0 = \frac{\Delta h}{\Delta x} \approx \frac{dh}{dx} = \frac{\tau_o}{\rho_I\, g_z\,(h - h_0)} \qquad (10.26)$$

where τ_o is given by Equation (10.20), h_0 is ice elevation at the ice margin, and h is ice elevation at the first Δx step in Equation (5.56). In computing rates of advance or retreat for the ice margin, $\bar{\alpha}_0$ in Equation (10.26) is used for α_0 in Equations (10.23) through (10.25), with $\bar{\alpha}_0 \to \alpha_0$ as $\Delta x \to 0$. Near the ice margin, if $h \to h_0$ a stagnant ice ramp of small slope exists, for which $\tau_o \to 0$. At the ice margin, when an ice wall stands on dry land windswept from drifting snow or stands in proglacial lakes impounded by the ice sheet, $h = h_0$ and Equation (10.26) gives $\alpha_0 = \infty$, which is the infinite slope of the vertical ice wall. Hooke (1970) describes ice-marginal conditions conducive to forming an ice wall, an ice ramp, or an ice-cored moraine.

Episodic discharge of icebergs into the North Atlantic during life cycles of major marine ice streams is the major mechanism by which an Arctic Ice Sheet might cause abrupt climatic change, so long as a frozen bed separates ice streams from interior ice during the growth stage of the ice sheet. Eventually interior ice will become thick enough to allow widespread thawing beneath grounded interior ice, as predicted by Equation (10.18). When glacial erosion becomes possible over a substantial part of the interior bed, due to glacial sliding in the thawed patches, and this eroded material becomes funneled into ice streams, a dirt machine can be turned on during life cycles of ice streams, and turned off between life cycles. This is a second mechanism that might cause abrupt climate change.

Turning the Dirt Machine On and Off

The dirt machine hypothesis postulates five mechanisms that act in sequence during a glaciation cycle. All of these mechanisms are hypothetical.

1. Northern Hemisphere ice sheets originate as marine ice sheets on Arctic continental shelves, where reduction of insolation caused by Earth's 41,000–year axial tilt cycle depresses the snowline enough to allow sea ice to thicken and ground in water less than 200 m deep in a few centuries. Marine ice domes then develop from surface precipitation because the snowline falls below sea level on these continental shelves (Fastook and Hughes, 1991). Summer meltwater soaks into the winter snowpack and refreezes, so there is no net ablation.

2. Marine ice sheets on Arctic continental shelves transgress onto the Eurasian and North American continents as the snowline continues to fall during the low-insolation hemicycle of axial tilt. The advancing continental ice sheets move south across permafrost, so basal traction is high. High basal traction and a narrow summer ablation zone along the southern ice-sheet margins give these margins a steep surface slope (Fastook and Hughes, 1991).

3. As the continental ice sheets pass from the exposed bedrock of Precambrian granitic shields onto softer sedimentary rocks and their thicker soil cover, permafrost beneath the ice sheets begins to thaw as increasingly thick ice moves over a given area, causing increasing depression of the melting point at the bed and increasing insulation and frictional heating from overlying moving ice. For ice moving across the southern borders of the shields, the subglacial bed becomes a mix of sticky frozen spots and slippery thawed spots in which thawed permafrost is slippery because melting interstitial ice produces a high pore-water pressure that mobilizes the unconsolidated sediments being sheared by overlying ice (Fastook and Hughes, 1991). A gradation from slippery thawed spots in a sticky frozen matrix to sticky frozen spots in a slippery thawed matrix is the condition for a basal melting zone, and it converts sheet flow into stream flow along river valleys where this change is complete, so the bed is completely thawed water-soaked silt.

4. Ice thins and slows toward the southern ice margin, in the terminal lobes of terrestrial ice streams, and in sheet flow between ice streams. This raises the basal-pressure melting point, increases heat flow to the surface, and reduces basal frictional heating, thereby producing a southward gradation from sticky frozen spots in a slippery thawed matrix to slippery thawed spots in a sticky frozen matrix beneath the ice sheet, the condition for a basal freezing zone. With net basal freezing, a thickening layer of debris-charged regelation (refrozen) ice forms at the base of the ice sheet, as illustrated in Figure 9.17. This layer outcrops in the surface ablation zone along the southern ice-sheet margin, which is still advancing over permafrost where there are no ice-dammed lakes. This situation is especially likely where ice streams terminate as ice lobes, because eroded debris from large areas is funneled into ice streams (Hughes, 1987).

5. Glacial flour is a major constituent of regelation ice. It is first produced in the basal melting zone where the ice sheets are sliding over exposed bedrock on the granitic shields, and it is transported into the basal freezing zone further south. This glacial flour is carried in suspension within subglacial drainage channels and in water from

melting surface regelation ice exposed in the ablation zone, to accumulate on outwash plains along the southern ice-sheet margin. Beneath regelation ice, thawed permafrost and basal till are also transported to the ice margin by laminar flow in these substrates (Boulton and Jones, 1979; Boulton and others, 1985; Boulton, 1996). This transport of eroded basal till and sediment to the ice-sheet margin in continuous conveyor-belt fashion has been called the "dirt machine" (Kotef, 1974). The dirt machine turns on and off with the life cycles of ice streams, on when the bed is thawed and off when the bed is frozen.

Heat flow during stream flow

The life cycle of an ice stream depends on (1) regulation of the thawed thickness of water-saturated basal sediment or till, (2) the basal mass balance of regelation ice which controls this thawed thickness, (3) the rate of heat flow from the bed to the surface which controls basal freezing and melting rates, (4) the surface mass balance which allows heat flow to be higher to a colder surface accumulation zone than to a warmer surface ablation zone, and (5) climatic cooling that expands the surface accumulation zone or climatic warming that expands the surface ablation zone. The only predictable mechanism for climatic change is insolation variations linked to cycles of precession and tilt of Earth's rotation axis and eccentricity in Earth's orbit, which lowers snowlines during cooling hemicycles and raises snowlines during warming hemicycles. Other mechanisms for climatic change, such as biological or anthropogenic production of greenhouse gases, meteorological production of atmospheric dust, or oceanic production of North Atlantic Deep Water, are not understood well enough to be predictable.

Climatic warming or cooling, by whatever mechanism, exerts indirect control on the life cycle of ice streams by regulating heat flow from the bed to the surface. The equation of heat flow has a nearly analytical solution for pure stream flow, because the ice stream has nearly constant thickness and almost no basal traction, except near its head, so it creeps much like a slab on a frictionless bed.

Stream flow develops when basal melting, either of basal ice or of the ice fraction in thawing subglacial permafrost, produces a low-traction bed along a given flowband that allows ice to creep rapidly. This converts the convex surface of sheet flow into the concave surface of stream flow, thereby reducing ice thickness and extending ice length along the flowband. The floating ice tongue of a marine ice stream has no basal traction, but it can provide buttressing if it merges with other ice tongues to form a floating ice shelf confined in a marine embayment. The grounded ice lobe of a terrestrial ice stream extends the ice-sheet margin, allowing basal meltwater to drain away around the perimeter of the ice lobe, thereby reestablishing basal traction that converts the concave surface of stream flow back to the convex surface of sheet flow, so ice thins rapidly toward the lobe terminus. The thinning ice toward the lobe terminus allows basal heat to be conducted more rapidly to the surface, causing basal freezing and a further increase in basal traction. The increase of basal traction when basal meltwater either drains away or refreezes converts extending stream flow into compressive sheet flow toward the lobe terminus, so the ice lobe buttresses the ice stream.

A first approximation to these conditions is provided by a nearly analytical solution for an ice stream treated as a sliding ice slab of uniform thickness which begins with suface accumulation and basal melting, the conditions at the head of an ice stream, and ends with surface ablation and basal freezing, the conditions at the foot of an ice stream. Sediment or till in the basal melting zone is dragged by sliding ice at the ice-bed interface into the basal freezing zone, where it is incorporated with regelation ice into the sliding ice slab. Surface accumulation and ablation rates are typically much greater than basal melting and freezing rates, so extending flow exists toward the head of the ice slab and compressive flow exists toward its foot. Creep in the slab will be controlled by the flow law for glacial ice given by Equation (4.43), with $\dot{\varepsilon}_{xx}$ given by Equation (4.50).

Flowlines for pure stream flow

Velocity components parallel to the x, y, and z rectilinear axes are u_x, u_y, and u_z, respectively, where x and y are parallel and transverse to flow and z is vertical, u_o is the basal velocity of the ice stream at $z = 0$, u_c is the creep velocity within ice thickness h, u_s is the shear velocity at the top of a deforming layer of sediment or till beneath the ice, u_y is negligible, a_t is the top accumulation ($-a_t$) or ablation ($+a_t$) rate for the ice stream at $z = h$, and a_b is the bottom freezing ($+a_b$) or thawing ($-a_b$) rate for the ice stream at $z = 0$. Figure 10.2 shows ice flowlines in the slab for various combinations of a_t and a_b. For a low-traction bed, u_x is nearly constant with z and u_z changes linearly with z in the range $0 \leq z \leq h$. These conditions are approached when u_c is mostly due to longitudinal extension or compression in the ice. Therefore:

$$u_x \approx u_o \approx u_c \tag{10.27a}$$

$$u_y \approx 0 \tag{10.27b}$$

$$u_z \approx a_b + (a_t - a_b)\, z/h \tag{10.27c}$$

Assume that a is constant with x. Since conservation of volume requires that $(\partial u_x/\partial x) + (\partial u_y/\partial y) + (\partial u_z/\partial z) = 0$, Equations (10.27) require that:

$$du_x /dx = -\, du_z /dz = -\,(a_t - a_b)\,/h \tag{10.28}$$

where the sediment or till becomes thawed at $x = 0$, thereby initiating the creep condition $u_c \approx u_o$ in the ice stream. The ice stream moves as a rigid body when $a_t = a_b$, moves in extending flow when $a_t < a_b$, and moves in compressive flow when $a_t > a_b$.

Particle velocities along an ice flowline are always in a

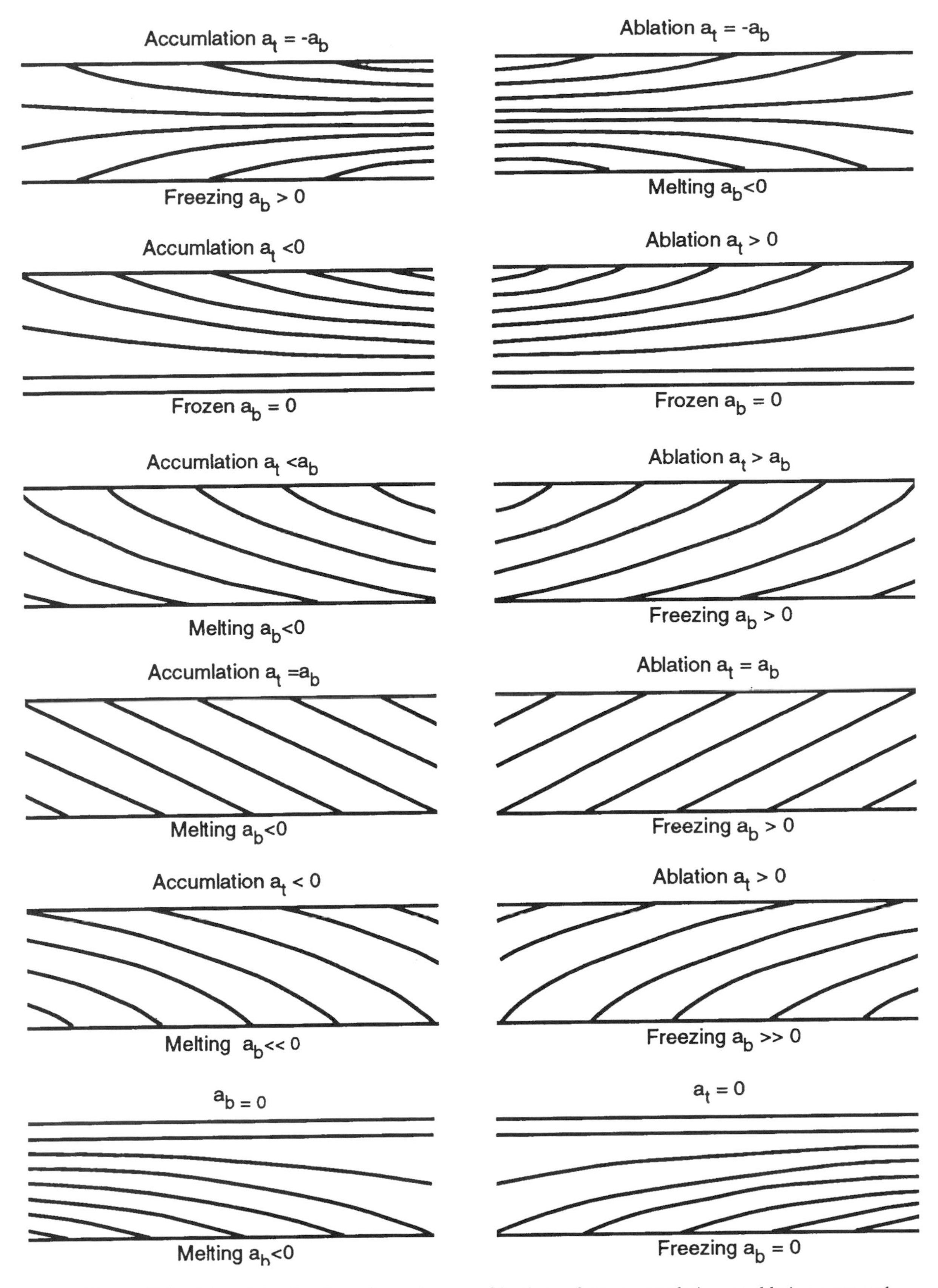

Figure 10.2: Internal ice flowlines for various combinations of top accumulation or ablation rates and bottom melting or freezing rates on an ice stream represented by an ice slab. From Equation (10.30).

direction tangent to the flowline. Hence, the equation of a flowline has the form $dz/dx = u_z/u_x$, and using Equations (10.27c) and (10.28):

$$\frac{dz}{dx} = \frac{a}{u} = \frac{h\,a_b + (a_t - a_b)\,z}{h\,u_o - (a_t - a_b)\,x} \tag{10.29}$$

The flowline curves are then obtained by integrating Equation (10.29):

$$z = \frac{h\,k}{h\,u_o - (a_t - a_b)\,x} - \frac{h\,a_b}{(a_t - a_b)} \tag{10.30}$$

where k is the constant of integration. As seen in Figure 10.2, these curves are hyperbolas which degenerate into straight lines of slope u_z/u_o when $a_t = a_b$.

Temperature field for pure stream flow

Temperatures inside an ice stream creeping in the xz plane on a low-traction bed must obey the heat balance equation (Paterson, 1981):

$$\partial T/\partial t - \kappa\,(\partial^2 T/\partial x^2) - \kappa\,(\partial^2 T/\partial z^2) + u_x\,(\partial T/\partial x) + u_z\,(\partial T/\partial z) - (\kappa/K)\,\dot{\varepsilon}_c\,\sigma_c = 0 \tag{10.31}$$

where t is time, T is temperature, κ is thermal diffusivity, K is thermal conductivity, σ_c is effective creep stress, and $\dot{\varepsilon}_c$ is effective creep rate. Heat balance in Equation (10.31) includes heat transfer due to climate change (first term), longitudinal conduction (second term), vertical conduction (third term), longitudinal ice motion (fourth term), vertical ice motion (fifth term), and internal frictional heating (sixth term). Paterson (1994) and Hooke (1988) show how Equation (10.31) is derived from the equation of energy conservation for deformation within an ice sheet.

Steady-state heat transfer occurs when $(dT/dt) = 0$. Longitudinal conduction can be ignored in a long slab since $(d^2T/dx^2) << (d^2T/dz^2)$. Internal frictional heating is confined to shear in the basal sediment or till layer, above which $\dot{\varepsilon}_c$ is small enough to be ignored. If the coordinate origin moves downslope at velocity u_x, the effect of the adiabatic lapse rate of ice-surface temperature on the longitudinal heat flow for ice advecting from higher to lower surface elevations will be experienced as surface ice warming even in steady-state flow. In this case the substitution $u_x = dx/dt$ is made in Equation (10.31), which then becomes:

$$(dT/dt)_c + (dT/dt)_a - \kappa\,(d^2T/dz^2) + u_z\,(dT/dz) = 0 \tag{10.32}$$

where $(dT/dt)_c$ is from climatic change and $(dT/dt)_a$ is from ice advection.

In the range $0 \leq z \leq h$ and in the absence of climatic change or significant advection, Equation (10.32) becomes:

$$\frac{d^2T}{dz^2} = \frac{u_z}{\kappa}\frac{dT}{dz} \tag{10.33}$$

Integrating Equation (10.33) from 0 to z:

$$\frac{dT}{dz} = \frac{(T_t - T_b)\int_0^z (u_z/\kappa)\,dz}{\int_0^h \left[exp\int_0^z (u_z/\kappa)\,dz\right]dz} \tag{10.34}$$

where T_t and T_b are temperatures at $z = h$ and $z = 0$, respectively. Integrating Equation (10.34) from 0 to z:

$$\hat{\theta} = \frac{T - T_b}{T_t - T_b} = \frac{\int_0^z \left[exp\int_0^z (u_z/\kappa)\,dz\right]dz}{\int_0^h \left[exp\int_0^z (u_z/\kappa)\,dz\right]dz} \tag{10.35}$$

where $\hat{\theta}$ is the normalized temperature.

Solutions of Equation (10.35) need the variation of u_z with z. Zotikov (1961, 1963, 1986) solved this equation by using the dimensionless ratios $\hat{a} = a_b/a_t$, $\hat{z} = z/h$, and the Peclet number $Pe = a_t\,h/\kappa$. Solutions for six combinations will be presented: (1) $a_t < a_b$ for extending flow, (2) $a_t > a_b$ for extending flow, (3) $a_t < a_b$ for compressive flow, (4) $a_t > a_b$ for compressive flow, (5) $a_t = a_b > 0$ for plug flow, and (6) $a_t = a_b = 0$ for no flow.

Temperature profiles for extending flow in ice streams

Surface accumulation greater than basal melting, or basal freezing greater than surface ablation, is required to maintain constant thickness for an ice stream creeping in extending flow. This condition is satisfied for many combinations of a_t and a_b. Each combination produces a characteristic temperature profile.

The ice stream cools from the top down when ice accumulates on the top faster than ice melts from the bottom, so that $\hat{a} < 1$. In this case

$$u_z = -a_b + (a_t - a_b)\,z/h = -a_t\,[\hat{a} - (1 - \hat{a})\,\hat{z}] \tag{10.36}$$

Substituting Equation (10.36) into Equation (10.35) gives:

$$\hat{\theta} = \frac{\int_0^z exp\left\{(-a_t z/2\kappa)\left[\hat{z}(1-\hat{a}) + 2\hat{a}\right]\right\}dz}{\int_0^h exp\left\{(-a_t z/2\kappa)\left[\hat{z}(1-\hat{a}) + 2\hat{a}\right]\right\}dz} \tag{10.37}$$

Using the function:

$$f = (Pe/2)^{1/2}\,[\hat{z}\,(1 - \hat{a})^{1/2} + \hat{a}/(1 - \hat{a})^{1/2}] \tag{10.38}$$

allows the substitution:

$$(a_t\,z/2\,\kappa)\,[\hat{z}\,(1 - \hat{a}) + 2\,\hat{a}] = f^2 - (Pe/2)^{1/2}\,\hat{a}^2/(1 - \hat{a}) \tag{10.39}$$

so that Equation (10.37) becomes:

$$\hat{\theta} = \frac{\int_{\zeta}^{\xi(z)} exp\left(-f^2\right) df}{\int_{\zeta}^{\xi(h)} exp\left(-f^2\right) df} \tag{10.40}$$

where ζ replaces 0 and ξ replaces z or h as the limits of integration. Equation (10.40) can now be evaluated using the tabulated function (Abramowitz and Stegun, 1965):

$$erf(f) = \left(2/\sqrt{\pi}\right) \int_0^f exp\left(-\lambda^2\right) d\lambda \tag{10.41}$$

The solution is:

$$\hat{\theta} = \frac{erf\left\{(Pe/2)^{1/2}\left[\hat{z}\left(1-\hat{a}\right)^{1/2} + \hat{a}/\left(1-\hat{a}\right)^{1/2}\right]\right\} - erf\left\{(Pe/2)^{1/2}\,\hat{a}/\left(1-\hat{a}\right)^{1/2}\right\}}{erf\left\{(Pe/2)^{1/2}\left[\left(1-\hat{a}\right)^{1/2} + \hat{a}/\left(1-\hat{a}\right)^{1/2}\right]\right\} - erf\left\{(Pe/2)^{1/2}\,\hat{a}/\left(1-\hat{a}\right)^{1/2}\right\}} \tag{10.42}$$

The ice stream warms from the bottom up when water freezes onto its bottom faster than ice ablates from its top, so that $\hat{a} > 1$. In this case:

$$u_z = a_b - (a_b - a_t)\, z/h = a_t\, [\hat{a} - (\hat{a} - 1)\, \hat{z}] \tag{10.43}$$

Substituting Equation (10.43) into Equation (10.35) gives:

$$\hat{\theta} = \frac{\int_0^z exp\left\{\left(+a_t z/2\kappa\right)\left[\hat{z}\left(\hat{a}-1\right) - 2\hat{a}\right]\right\} dz}{\int_0^h exp\left\{\left(+a_t z/2\kappa\right)\left[\hat{z}\left(\hat{a}-1\right) - 2\hat{a}\right]\right\} dz} \tag{10.44}$$

Using the function:

$$f = (Pe/2)^{1/2}\, [\hat{z}\,(\hat{a} - 1)^{1/2} - \hat{a}/(\hat{a} - 1)^{1/2}] \tag{10.45}$$

allows the substitution:

$$(a_t\, z/2\kappa)\, [\hat{z}\,(\hat{a} - 1) - 2\,\hat{a}] = f^2 - (Pe/2)^{1/2}\, \hat{a}^2/(\hat{a} - 1) \tag{10.46}$$

so that Equation (10.44) becomes:

$$\hat{\theta} = \frac{\int_{\zeta}^{\xi(z)} exp\left(+f^2\right) df}{\int_{\zeta}^{\xi(h)} exp\left(+f^2\right) df} \tag{10.47}$$

where ζ replaces 0 and ξ replaces z or h as the limits of integration. Equation (10.47) can now be evaluated using the tabulated function (Abramowitz and Stegun, 1965):

$$int(f) = \int_0^f exp\left(+\lambda^2\right) d\lambda \tag{10.48}$$

The solution is:

$$\hat{\theta} = \frac{erf\left\{(Pe/2)^{1/2}\left[\hat{z}\left(\hat{a}-1\right)^{1/2} - \hat{a}/\left(\hat{a}-1\right)^{1/2}\right]\right\} + erf\left\{(Pe/2)^{1/2}\,\hat{a}/\left(\hat{a}-1\right)^{1/2}\right\}}{erf\left\{(Pe/2)^{1/2}\left[\left(\hat{a}-1\right)^{1/2} - \hat{a}/\left(\hat{a}-1\right)^{1/2}\right]\right\} + erf\left\{(Pe/2)^{1/2}\,\hat{a}/\left(\hat{a}-1\right)^{1/2}\right\}} \tag{10.49}$$

Transformation to Equations (10.40) and (10.47) is via the formula:

$$\int_{\zeta}^{\xi} F(f)\, df = \int_M^N F[\phi(z)]\, \phi'(z)\, dz \tag{10.50}$$

where $\zeta = \phi(M)$, $\xi = \phi(N)$, and $f = \phi(z)$.

The ice stream cools from the top down and warms from the bottom up when ice accumulates on the top and water freezes on the bottom. Ice moves down through thickness $h/(1 + \hat{a})$ and moves up through thickness $\hat{a}\, h/(1 + \hat{a})$. The level where downward and upward flow converges has velocity $u_z = 0$ and temperature $T = T_o$.

Substituting $\hat{a}\, h/(1 + \hat{a})$ for z = h and setting $a_t = 0$ in Equation (10.49) gives the temperature profile through ice thickness $0 \le z \le \hat{a}\, h/(1 + \hat{a})$ when T_o replaces T_t:

$$\frac{T - T_b}{T_o - T_b} = \frac{erf\left\{(Pe/2)^{1/2}\left(\frac{\hat{a}}{1+\hat{a}}\right)^{1/2}\left[\hat{z}\left(\frac{1+\hat{a}}{\hat{a}}\right) - 1\right]\right\} + erf\left\{(Pe/2)^{1/2}\left(\frac{\hat{a}}{1+\hat{a}}\right)^{1/2}\right\}}{erf\left\{(Pe/2)^{1/2}\left(\frac{\hat{a}}{1+\hat{a}}\right)^{1/2}\right\}} \tag{10.51}$$

Substituting $\hat{a}\, h/(1 + \hat{a})$ for z = 0 and setting $a_b = 0$ in Equation (10.43) gives the temperature profile through ice thickness $\hat{a}\, h/(1 + \hat{a}) \le z \le h$ when T_o replaces T_b:

$$\frac{T - T_o}{T_t - T_o} = \frac{erf\left\{(Pe/2)^{1/2}\left(\frac{\hat{a}}{1+\hat{a}}\right)^{1/2}\left[\hat{z} - \left(\frac{\hat{a}}{1+\hat{a}}\right)^{1/2}\right]\right\}}{erf\left\{(Pe/2)^{1/2}\left(\frac{\hat{a}}{1+\hat{a}}\right)^{1/2}\right\}} \tag{10.52}$$

Differentiating Equations (10.51) and (10.52) and equating the resulting temperature gradients at $T = T_o$ gives a smooth, continuous temperature profile for which:

$$\frac{T - T_o}{T_t - T_o} = \frac{erf\left\{(Pe/2)^{1/2}\left(\frac{\hat{a}}{1+\hat{a}}\right)^{1/2}\left[\hat{z} - \left(\frac{\hat{a}}{1+\hat{a}}\right)^{1/2}\right]\right\}}{erf\left\{(Pe/2)^{1/2}\left(\frac{\hat{a}}{1+\hat{a}}\right)^{1/2}\right\}} \tag{10.53}$$

Temperature profiles for compressive flow in ice streams

Surface ablation greater than basal freezing, or basal melting greater than surface accumulation, is required to maintain constant thickness for an ice stream creeping in compressive flow. This condition is satisfied for many combinations of a_t and a_b. Each combination produces a characteristic temperature profile.

The ice stream warms from the top down when ice ablates from the top faster than water freezes onto the bottom, so that $\hat{a} < 1$. In this case:

$$u_z = a_b - (a_t + a_b)\, z/h = a_t\,[\hat{a} + (1 - \hat{a})\,\hat{z}] \tag{10.54}$$

Substituting Equation (10.54) into Equation (10.35) gives:

$$\hat{\theta} = \frac{int\left\{(Pe/2)^{1/2}\left[\hat{z}\,(1-\hat{a})^{1/2} + \hat{a}/(1-\hat{a})^{1/2}\right]\right\} - int\left\{(Pe/2)^{1/2}\hat{a}/(1-\hat{a})^{1/2}\right\}}{int\left\{(Pe/2)^{1/2}\left[(1-\hat{a})^{1/2} + \hat{a}/(1-\hat{a})^{1/2}\right]\right\} - int\left\{(Pe/2)^{1/2}\hat{a}/(1-\hat{a})^{1/2}\right\}} \tag{10.55}$$

The ice stream cools from the bottom up when ice melts on the bottom faster than it accumulates on the top, so that $\hat{a} > 1$. In this case:

$$u_z = (a_b + a_t)\, z/h - a_b = a_t\,[(r - 1)\,\hat{z} - \hat{a}] \tag{10.56}$$

Substituting Equation (10.56) into Equation (10.35) gives:

$$\hat{\theta} = \frac{int\left\{(Pe/2)^{1/2}\left[\hat{z}\,(\hat{a}-1)^{1/2} - \hat{a}/(\hat{a}-1)^{1/2}\right]\right\} + int\left\{(Pe/2)^{1/2}\hat{a}/(\hat{a}-1)^{1/2}\right\}}{int\left\{(Pe/2)^{1/2}\left[(\hat{a}-1)^{1/2} - \hat{a}/(\hat{a}-1)^{1/2}\right]\right\} + int\left\{(Pe/2)^{1/2}\hat{a}/(\hat{a}-1)^{1/2}\right\}} \tag{10.57}$$

The ice stream warms from the top down and cools from the bottom up when ice ablates on the top and melts on the bottom. Ice moves up through thickness $h/(1 + \hat{a})$ and moves down through thickness $\hat{a}\,h/(1 + \hat{a})$. The level where upward and downward flow diverges has velocity $u_z = 0$ and temperature $T = T_o$.

Substituting $\hat{a}\,h/(1 + \hat{a})$ for $z = h$ and setting $a_t = 0$ in Equation (10.58) gives the temperature profile through ice thickness $0 \le z \le \hat{a}\,h/(1 + \hat{a})$ when T_o replaces T_t:

$$\frac{T - T_b}{T_o - T_b} = \frac{int\left\{(Pe/2)^{1/2}\left(\frac{\hat{a}}{1+\hat{a}}\right)^{1/2}\left[\hat{z}\left(\frac{\hat{a}}{1+\hat{a}}\right)^{1/2} + 1\right]\right\} + int\left\{(Pe/2)^{1/2}\left(\frac{\hat{a}}{1+\hat{a}}\right)^{1/2}\right\}}{int\left\{(Pe/2)^{1/2}\left(\frac{\hat{a}}{1+\hat{a}}\right)^{1/2}\right\}} \tag{10.58}$$

Substituting $\hat{a}\,h/(1 + \hat{a})$ for $z = 0$ and setting $a_b = 0$ in Equation (10.55) gives the temperature profile through ice thickness $\hat{a}\,h/(1 + \hat{a}) \le z \le h$ when T_o replaces T_b:

$$\frac{T - T_t}{T_t - T_o} = \frac{int\left\{(Pe/2)^{1/2}\left(\frac{\hat{a}}{1+\hat{a}}\right)^{1/2}\left[\hat{z}\left(\frac{\hat{a}}{1+\hat{a}}\right)^{1/2} + 1\right]\right\}}{int\left\{(Pe/2)^{1/2}\left(\frac{\hat{a}}{1+\hat{a}}\right)^{1/2}\right\}} \tag{10.59}$$

Differentiating Equations (10.58) and (10.59) and equating the resulting temperature gradients at $T = T_o$ gives a smooth continuous temperature profile for which:

$$T_o = T_t\,\frac{int\left\{(Pe/2)^{1/2}\left(\frac{\hat{a}}{1+\hat{a}}\right)^{1/2}\right\} + T_b\, int\left\{(Pe/2)^{1/2}\right\}}{int\left\{(Pe/2)^{1/2}\left(\frac{\hat{a}}{1+\hat{a}}\right)^{1/2}\right\} + int\left\{(Pe/2)^{1/2}\right\}} \tag{10.60}$$

Table 10.1 gives the reduced forms of Equations (10.42), (10.49), (10.51), (10.52), (10.55), (10.57), and (10.59) when top surface mass flux is zero ($\hat{a} = \infty$), bottom

Table 10.1: Limiting equations for steady-state temperature profiles through an ice stream represented by an ice slab experiencing various creep conditions[a].

Creep condition	Boundary condition	Normalized temperature
Extending flow	$a_t = 0$	$\hat{\theta} = \dfrac{erf\left\{(Pe_b/2)^{1/2}(\hat{z}-1)\right\} + erf\left\{(Pe_b/2)^{1/2}\right\}}{erf\left\{(Pe_b/2)^{1/2}\right\}}$
Extending flow	$a_b = 0$	$\hat{\theta} = \dfrac{erf\left\{(Pe_t/2)^{1/2}\hat{z}\right\}}{erf\left\{(Pe_t/2)^{1/2}\right\}}$
Compressive flow	$a_t = 0$	$\hat{\theta} = \dfrac{int\left\{(Pe_b/2)^{1/2}(\hat{z}-1)\right\} + int\left\{(Pe_b/2)^{1/2}\right\}}{int\left\{(Pe_b/2)^{1/2}\right\}}$
Compressive flow	$a_b = 0$	$\hat{\theta} = \dfrac{int\left\{(Pe_t/2)^{1/2}\hat{z}\right\}}{int\left\{(Pe_t/2)^{1/2}\right\}}$
Plug flow	$a_t = a_b > 0$	$\hat{\theta} = \dfrac{exp(-Pe_t) - 1}{exp(-Pe_t) - 1}$
No flow	$a_t = a_b = 0$	$\hat{\theta} = \hat{z}$

[a] $Pe_t = a_t\,h/\kappa$ and $Pe_b = a_b\,h/\kappa$

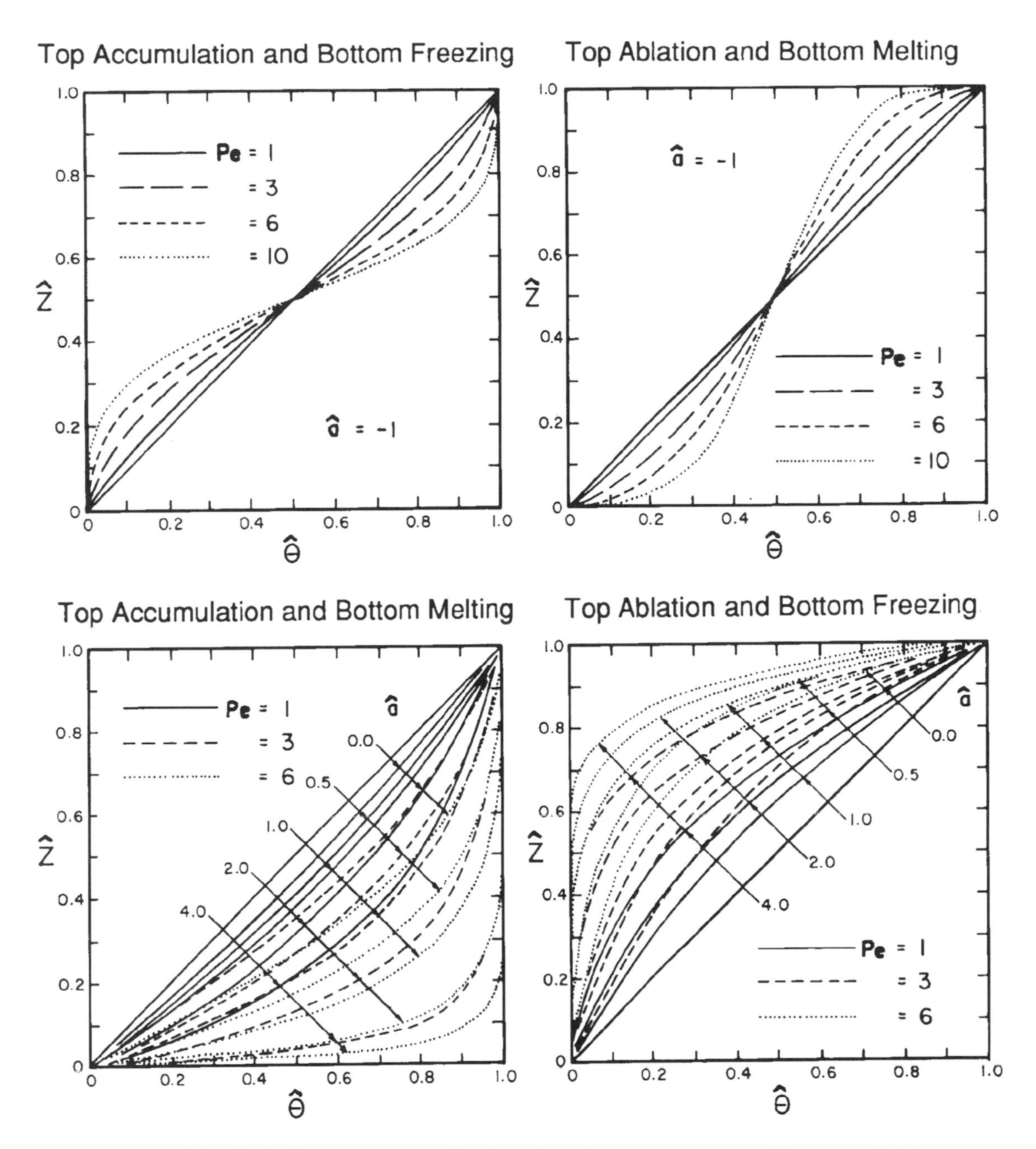

Figure 10.3: Vertical temperature profiles through an ice stream represented by an ice slab for various surface accumulation or ablation rates, basal freezing or melting rates, and surface temperatures. Dimen-sionless internal ice temperatures $\hat{\theta}$ are given by Equations (10.42), (10.49), (10.55), and (10.57) for dimensionless values $\hat{z} = z/h$, $\hat{a} = a_b / a_t$, and $Pe = a_t h/\kappa$. Reproduced by courtesy of the International Association of Hydrological Sciences and I. Zotikov, From *Bulletin of the International Association of Scientific Hydrology, 8*(1), (1963), pp. 38 and 41, Figures. 1 and 3.

surface mass flux is zero ($\hat{a} = 0$), top surface mass flux equals bottom surface mass flux ($\hat{a} = 1$), and when top surface and bottom surface mass fluxes are both zero. Figure 10.3 plots $\hat{\theta}$ versus $\hat{z}$ through an ice stream for specified values of $\hat{a}$, Pe, and T_0 in these equations. Creep can obviously significantly modify the steady-state linear temperature profile through an ice stream.

Basal heat flow for an ice stream in extending flow

Geothermal heat and frictional heat from shear deformation of basal sediments or till are supplied to unit area of basal ice at the rate:

$$Q_b = \rho_I H_M a_b - K (\partial T / \partial z)_b \qquad (10.61)$$

where $\rho_I H_M a_b$ is the rate at which basal heat melts unit area at the base of the ice stream, $K (\partial T / \partial z)_b$ is the rate per unit area at which basal heat is conducted to the surface of the ice stream, ρ_I is ice density, H_M is latent heat of melting for ice, and $(\partial T / \partial z)_b$ is the basal temperature gradient, which is always negative in an ice sheet.

An expression for $(\partial T / \partial z)_b$ is obtained by differentiating Equation (10.60) and then setting z = 0:

$$\left(\frac{\partial T}{\partial z}\right)_b = \left(\frac{T_t - T_b}{h}\right) \frac{(2/\pi)^{1/2} Pe^{1/2} \left(1-\hat{a}\right)^{1/2} exp\left[-(Pe/2)\,\hat{a}^2/\left(1-\hat{a}^2\right)\right]}{erf\left\{Pe^{1/2}\left[\left(1-\hat{a}\right)^{1/2} + \hat{a}/\left(1-\hat{a}\right)^{1/2}\right]\right\} - erf\left\{Pe^{1/2}\,\hat{a}/\left(1-\hat{a}\right)^{1/2}\right\}} \tag{10.62}$$

If $a_b = 0$, then $\hat{a} = 0$, and T_b can be below the basal melting temperature T_M. When $\hat{a} = 0$, Equation (10.62) reduces to:

$$\left(\frac{\partial T}{\partial z}\right)_b = \left(\frac{T_t - T_b}{h}\right) \frac{(2/\pi)^{1/2} Pe^{1/2}}{erf\left\{Pe^{1/2}\right\}} \tag{10.63}$$

where $T_b = T_M$ when $\hat{a} \neq 0$ in Equation (10.62), with both basal freezing (positive a_b) and basal melting (negative a_b) being possible.

The requirement that $T_b = T_M$ at $a_b = 0$, and therefore at $\hat{a} = 0$, is that stream thickness $h = h_M$. Solving Equation (10.61) for $h = h_M$ at $T_b = T_M$ and $\hat{a} = 0$ in Equation (10.62) gives:

$$h_M = \frac{T_t - T_M}{(\partial T / \partial z)_M} \frac{(2/\pi)^{1/2} (Pe)_M^{1/2}}{erf\left\{(Pe)_M^{1/2}\right\}}$$

$$= \left(\frac{T_M - T_t}{Q_b / K}\right) \frac{(2/\pi)^{1/2} (Pe)_M^{1/2}}{erf\left\{(Pe)_M^{1/2}\right\}} \tag{10.64}$$

where $(Pe)_M = a_t h_M / \kappa$. Basal melting takes place when $h > h_M$. An expression for Q_b is:

$$Q_b = Q_G + J\, u_o \tau_o = Q_G + u \rho_I g h \alpha \tag{10.65}$$

where Q_G is the geothermal heat flux per unit basal area, J is the mechanical equivalent of heat, and $J u_o \tau_o$ is the rate per unit basal area at which frictional heat is generated. In ice streams, u_o increases as τ_o decreases, but the product $u_o \tau_o$ is relatively stable, such that $Q_G \approx J u_o \tau_o$.

The geothermal heat flux around the Great Lakes of North America, where ice lobes formed along the southern margin of the Laurentide Ice Sheet, is $Q_G \approx 1.25 \times 10^{-6}$ cal/cm^2 sec (Sugden, 1978). Frictional heat from shear deformation of subglacial till beneath the ice lobes would no

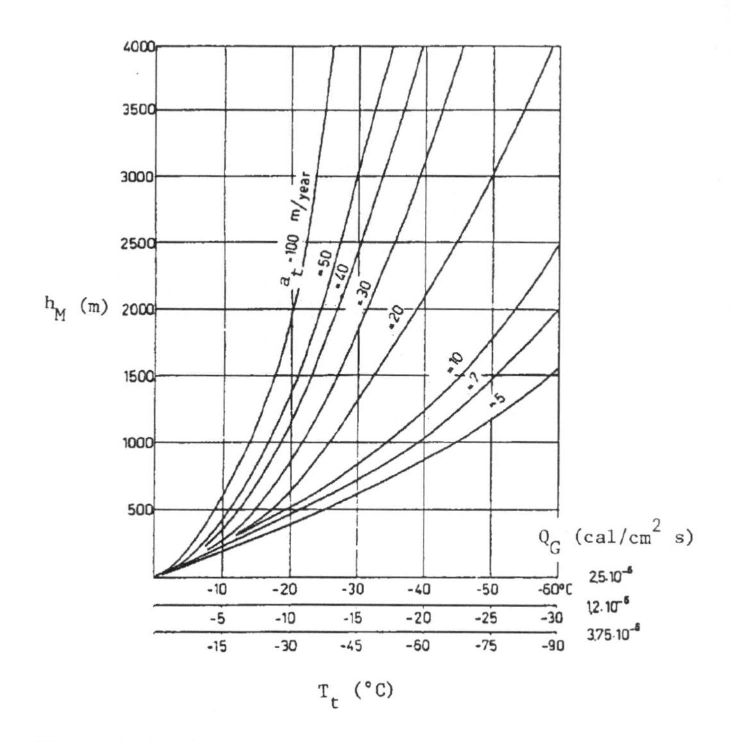

Figure 10.4: Conditions for which an ice stream represented by an ice slab will freeze to its bed. Critical ice thicknesses h_M for basal freezing vary with mean annual top surface temperature T_t, top surface ice accumulation rates a_t, and basal geothermal heat fluxes Q_G, according to Equation (10.64). Reproduced with permission from Zotikov (1963).

more than double or triple this amount for the range of τ_o and u_o values associated with these lobes (J = 25 x 10^3 cal/bar m^3, 0.1 bar < τ < 1.0 bar, 500 m/a > u_o > 50 m/a). Figure 10.4 shows the variation of h_M with T_t and a_t for three values of Q_b. For the accumulation zone of the ice stream, 5 m/a < a_t < 10 m/a and –20°C < T_t < –10°C is the probable range of accumulation rate and mean annual temperature on the top surface. For the range of Q_b in Figure 10.4, this gives 150 m < h_M < 1250 m as the range of ice thickness below which the bed would be frozen.

Basal heat flow for an ice stream in compressive flow

Compressive flow exists in the ablation zone of the ice stream, where a basal freezing zone exists if the bed is not frozen. For the ablation zone of the ice stream, h_M cannot be computed from Equation (10.64) because a_t in *Pe* is negative. In any case, Equation (10.64) shows that $h_M \rightarrow 0$ as $a_t \rightarrow 0$, regardless of a_t and Q_b. This is compatible with retaining a frozen bed at the ice margin so long as the ice sheet is advancing over a frozen bed.

If surface accumulation and basal melting rates for extending flow in the upslope part of the ice stream had not already modified the temperature profile, temperature profiles and the resulting basal heat flow in the downslope part of the ice stream could be computed from Equation (10.55) for surface ablation and basal freezing rates that cause compressive flow. This complication requires a steady-state solution in which changing temperature profiles are computed in a column of ice as it moves from the accumulation zone to the ablation zone of an ice stream. There is no analytical solution for this case, but a numerical solution has been developed by Budd and others (1971) and applied to flowlines of the Antarctic Ice Sheet. A major assumption is that u_x does not vary with z through the ice column, which is the case for creep in an ice stream sliding on a low-traction bed.

Basal boundary conditions

Zotikov (1986) has analyzed the heat balance and has computed the resulting temperature profiles for the various boundary conditions at the ice-bed interface. These include (1) ice frozen to bedrock, (2) ice separated from bedrock by a thin water layer, (3) ice frozen to ice-cemented permafrost that warms with depth and blankets frozen bedrock, (4) ice frozen to isothermal ice-cemented permafrost that is at its melting point but remains frozen and that blankets thawed bedrock, (5) ice in contact with permafrost that warms and thaws with depth and that blankets thawed bedrock, (6) ice in contact with thawed permafrost that warms with depth and blankets thawed bedrock, and (7) ice with a temperate basal layer over thawed permafrost which warms with depth and which blankets thawed bedrock. In all these cases, ice-cemented permafrost consists of sediment or till, bedrock warms with depth from geothermal heat, and frozen or thawed bedrock is colder or warmer than the pressure melting temperature of water. Over time, all of these conditions can occur at the ice-bed interface. Ice initially frozen to ice-cemented permafrost that blankets frozen bedrock can warm, causing the melting-point isotherm to ascend first through the bedrock, then through the permafrost until the ice fraction melts and allows the sediment or till to warm, and then through the basal ice to produce a temperate basal ice layer. In the meantime, the thawed basal sediment or till layer can be sheared away by basal sliding, leaving a film of meltwater between basal ice and bedrock. This water film may subsequently freeze, leaving basal ice frozen to bedrock.

Temperature profiles through time

Basal ice temperatures change through time primarily because surface ice temperatures change through time. Climate changes and ice-sheet elevation changes through time, and both change the surface temperature. For cyclic climatic change, a first approximation for the change in surface temperature T_t through time t might be:

$$T_t = T_\alpha \sin \omega t \tag{10.66}$$

where T_α is the amplitude and $\omega/2\pi$ in the frequency of climatic change. If heat is transported downward only by thermal conduction, the surface temperature change should decrease exponentially downward into the ice. At depth z below the surface, temperature T would then be:

$$T = T_\alpha \exp\left(-z\sqrt{\omega/2\kappa}\right)\sin\left(\omega t - z\sqrt{\omega/2\kappa}\right) \tag{10.67}$$

where the amplitude of the temperature wave propagates downward at velocity $\sqrt{\omega/2\kappa}$ and decreases downward exponentially as $exp\,(-z\sqrt{\omega/2\kappa})$.

The vertical temperature gradient obtained from Equation (10.67) is:

$$\begin{aligned}\partial T/\partial z = {} & T_\alpha\left(-\sqrt{\omega/2\kappa}\right)\exp\left(-z\sqrt{\omega/2\kappa}\right)\sin\left(\omega t - z\sqrt{\omega/2\kappa}\right)\\ & + T_\alpha\left(-\sqrt{\omega/2\kappa}\right)\exp\left(-z\sqrt{\omega/2\kappa}\right)\cos\left(\omega t - z\sqrt{\omega/2\kappa}\right)\end{aligned} \tag{10.68}$$

Differentiating Equation (10.68) to obtain $\partial(\partial T/\partial z)/\partial z$ gives:

$$\partial^2 T/\partial z^2 = T_\alpha(\omega/\kappa)\exp\left(-z\sqrt{\omega/2\kappa}\right)\cos\left(\omega t - z\sqrt{\omega/2\kappa}\right) \tag{10.69}$$

The rate of temperature change given by Equation (10.67) is:

$$\partial T/\partial t = T_\alpha\,\omega\exp\left(-z\sqrt{\omega/2\kappa}\right)\cos\left(\omega t - z\sqrt{\omega/2\kappa}\right) \tag{10.70}$$

Combining Equations (10.69) and (10.70) yields the partial differential equation:

$$\partial T/\partial t = \kappa\,(\partial^2 T/\partial z^2) \tag{10.71}$$

Equation (10.31) reduces to Equation (10.71) when longitudinal conduction is small compared to vertical conduction, so that $(\partial^2 T/\partial x^2) << (\partial^2 T/\partial z^2)$; when heat transported by advection is small compared to heat transported by conduction, so $u_x\,(\partial T/\partial x)$ and $u_z\,(\partial T/\partial z)$ can be ignored; and when ice flow is sluggish, so $\dot{\varepsilon}_c\,\sigma_c$ is nil. These conditions are approximated at high domes of an ice sheet where surface accumulation rates are small and at terminal lobes of ice streams when ice is stagnating. Under these conditions, Equation (10.67) is a reasonable solution to the general equation of heat flow for a cyclic surface temperature change.

Three periods $(2\pi/\omega)$ of surface temperature change are of particular interest for studying the interaction between ice sheets and climatic change, $(2\pi/\omega) = 1$ a for annual cycles, $(2\pi/\omega) = 2500$ a for Dansgaard-Oeschger events and the life cycle of ice streams, and $(2\pi/\omega) = 100{,}000$ a for Quaternary glaciation cycles and life cycles of earlier Quaternary ice sheets. The respective depths z at which the amplitude is 5 percent of the surface amplitude are 10.2 m, 509 m, and 3220 m, respectively; this amplitude propagates downward at velocities v_z of 21.3 m/a, 0.427 m/a, and 0.067 m/a, respectively; and the time lags z/v_z are 0.48 a, 1192 a, and 48,000 a, respectively (Paterson, 1994). These calculations indicate that the summer-to-winter temperature change damps out at about 10 m in depth, to become the mean annual temperature. If the surface temperature cooling is caused by a Dansgaard-Oeschger event, the cold wave is much more likely to propagate to the bed of a thin ice lobe than of a thick ice dome, and a thawed bed would freeze about 1200 years later. On the other hand, if a transition from interglacial to glacial conditions causes the surface cooling, the thawed base of an ice dome about 3200 m thick would freeze about 48,000 years later. Of course, all of this assumes that heat is transported downward only by conduction. As advective heat transport becomes more important, these numbers lose validity. The Péclet number measures the relative importance of advection and conduction.

Paterson (1994) discusses in detail the various factors that influence temperature profiles through ice sheets. In general, advection shortens the time when the bed responds to a temperature change at the surface. Paterson (1994) gives examples of the resulting departures from linearity in temperature profiles down boreholes through present-day ice sheets and ice caps. The departure is not great in the polar desert environment at Vostok, near the ice divide of the East Antarctic Ice Sheet, and at Camp Up B on Ice Stream B, which drains the West Antarctic Ice Sheet.

The dynamics of dirt

The ice machine can be instrumental in securing the coming glaciation through the Jakobshavns Effect, if it floods the Labrador, Greenland, and Norwegian Seas with Greenland icebergs fast enough and long enough to suppress formation of North Atlantic Deep Water, thereby throwing global climate into a glacial mode from the present interglacial mode. Snowlines will then lower in high Northern Hemisphere latitudes and allow marine ice grounded on Arctic continental shelves to transgress onto the Arctic mainland and merge with ice caps growing on Arctic highland plateaus. These events could occur as soon as the coming century, and they should raise the high-latitude albedo enough to allow slow but prolonged thickening of the ice cover over rapidly expanding areas of the Subarctic mainlands of Canada and Siberia. Greenland ice streams will pass through surge-like life cycles ranging from 200 a to 1000 a during this time, after which they will be dormant for perhaps 4000 a while the Greenland Ice Sheet thickens and restores its gravitational potential energy. Toward the end of that dormant period, Laurentide ice spreading from Hudson Bay will have reached the edge of the Canadian Shield, and Siberian ice spreading from the Kara Sea will have merged with a large ice cap on the Putorana Plateau, according to the ice-sheet advance rate predicted by Equation (10.21). Terrestrial ice streams will then develop in the large linear lakes at the edge of the Canadian Shield and in the valleys of large rivers flowing toward the Siberian Arctic. Marine ice streams will occupy straits and interisland channels on outer Arctic continental shelves. The dirt machine will be turned on first in these ice streams.

The ice streams form because permafrost is usually absent on sea floors, lake floors, and river bottoms, so channelized glacial sliding will take place. The bed beneath the ice sheet initially will be frozen from central ice divides to these ice streams, so growth of ice streams requires retreat of a basal melting-point isotherm at the head of each ice stream. The only source of dirt for the dirt machine will be sediments on the lake floors and river bottoms and thawing permafrost at the heads of ice streams that occupy these channels. However, thickening of interior ice will eventually lower the basal melting temperature enough to allow basal thawing beneath interior ice divides, as calculated from Equation (10.18). Ongoing ice thickening will expand the zone of interior basal thawing, which will eventually meet the retreating thawed bed at the heads of ice streams. Before that meeting, ice streams are able to collect material that has eroded from the inner thawed zone and frozen into regelation ice as it crossed an intermediate freezing zone before it crosses the outer frozen zone to then melt out beneath the ice streams. After that meeting, the ice streams will be able to collect more material eroded beneath interior ice by adding material dragged along at the sliding ice-bed interface to material frozen into basal regelation ice. Ice streams are therefore able to collect material eroded from most of the basal area beneath the ice sheet and deposit it as end moraines or outwash fans at or just beyond the terminal lobes of terrestrial ice streams and the grounding lines of marine ice streams. This ability of ice streams to collect material eroded over vast internal areas and to dump it at specific sites along the perimeter is the essence of the dirt machine. Since ice streams have life cycles, dumping is rapid during life cycles and slow in dormant periods between life cycles.

Rates of collecting and dumping the eroded material are predicted from rates of basal freezing and melting calculated from Equations (10.36) through (10.60), in particular, basal freezing rates ($+ a_b$) and basal melting rates ($- a_b$) of basal regelation ice. These rates are high during ice-stream life cycles and low between life cycles. The rates are sensitive to changing ice thickness (h_I), mean annual ice-surface temperature ($\bar{T}_a$), accumulation rates ($- a_t$), or ablation rates ($+ a_t$) that respond to Milankovitch insolation cycles, and other climate forcing mechanisms. Life cycles of ice streams end when Equation (10.64) predicts a frozen bed for specified combinations of these variables.

Dirt dumping rates at the ends of terrestrial and marine ice streams depend on the concentration of eroded material entrained in regelation ice as well as on freezing and melting rates of regelation ice, and on the volume of eroded material being dragged along at the ice-bed interface. The volume of material entrained or dragged will be low when the bed is bedrock and high when the bed is sediment or till. Therefore, the dirt machine will produce little dirt when the ice margin is on a crystalline shield and will produce increasingly more dirt as the ice margin moves over unlithified material beyond the shield. The maximum productivity of the dirt machine will generally occur during glacial maxima, when the maximum area is subjected to glacial erosion and when ice streams have moved the maximum distance beyond crystalline shields. Glacial flour in dirt deposited in end moraines and outwash plains is entrained by the dust machine.

Turning the Dust Machine On and Off

The dust machine hypothesis postulates seven mechanisms that regulate conversion of glacially eroded dirt into atmospheric dust.

1. Katabatic winds, possibly in excess of 100 km/hr for weeks on end, probably flow down margins of Quaternary ice sheets, being strongest above the downdrawn surfaces of ice streams, beneath which the dirt machine is most rapidly transporting most of the glacially eroded material to ice-stream termini. This convergence of katabatic winds along ice streams is observed and has been modeled for the Antarctic Ice Sheet today (Parish and Bromwich, 1987; Bromwich, 1989; Bromwich and others, 1990). During the life cycles of ice streams, katabatic winds may produce intermittent dust storms along ice-sheet margins by entraining ice crystals on the ice streams and glacial flour and silty till deposited in end moraines and outwash fans. This can be called the "dust machine."

2. The dust machine may first be turned on when Arctic marine ice sheets transgress onto dry land and where the ice sheet is high enough to generate katabatic winds, provided that the marine ice sheets are grounded on continental-shelf sediments that were either thawed initially or became thawed after grounding and dry land is in the permafrost condition. Marine ice moving landward from a thawed bed beneath the ice divide onto a frozen bed beyond the ice margin must then pass over a freezing bed, above which a thickening layer of regelation ice carrying glacially eroded sediments will develop. Hence, the dirt machine will be turned on as soon as regelation ice forms, and the dust machine will be turned on as soon as regelation ice outcrops along the advancing ice margin and katabatic winds become strong enough to entrain surface ice crystals and sediments or till released by ablation. However, abrupt changes in dust production will require ice streams, which may form later in the coming glaciation cycle. As shown in Figure 1.1, dust spikes became prominent late in the last glaciation cycle at the GRIP corehole site in Greenland (Hammer and others, 1985).

3. Insolation in mid-latitudes controlled by the precession cycle of Earth's rotation axis during the last glaciation began to increase by 22,000 BP and peaked at 11,000 BP in the Northern Hemisphere, yet the Laurentide (Mayewski and others, 1981) and Scandinavian (Andersen, 1981) ice-sheet margins fluctuated close to their 20,000-BP maximum southern extents until 14,000 BP. A glaciological explanation for this is that more rapid southern ablation was offset by a more rapid southward velocity of ice, so the ice margin neither advanced nor retreated by great amounts (Fastook and Hughes, 1991). The more rapid ice ablation occurred as the snowline rose in response to both lower latitudes and the warming hemicycle of precession-controlled insolation. The more rapid ice velocity occurred as the ice sheet advanced over permafrost that became less continuous and more sporadic toward warming lower latitudes so that the bed became more slippery. Dust storms cooled the ice-sheet margin by blocking insolation, making ablation rates less than they would otherwise have been, provided that surface katabatic winds entrained dust high enough to be drawn back over the ablation zone by inflowing air above the katabatic wind field, as illustrated in Figure 1.8c. This allowed a certain fraction of the dust to be recycled continually, thereby making dust storms increasingly prominent later in the glaciation cycle. These processes will repeat during the coming glaciation.

4. Decreasing bed traction and increasing surface ablation will transform the southern ice margin from sheet flow to stream flow, especially in Laurentide ice lobes south of the Great Lakes (Hughes, 1987), which will approach shelf flow if ice lobes merge. This may have been a common condition along southern margins of Northern Hemisphere ice sheets just prior to the last termination, and it would produce a broad zone of thin, fast-moving ice on a slippery bed (Boulton and Jones, 1979; Boulton and others, 1985; Boulton, 1996). Decreased bed traction along the southern margin of the Eurasian Ice Sheet would be linked to ice-dammed lakes in the valleys of rivers flowing northward in Siberia and on the North European Plain (Grosswald, 1988). In the coming glaciation, the dust machine will be most active at the lobes of ice streams between their life cycles and at the end of stadials, when retreating ice will produce outwash fans and expose ground moraines, but before proglacial lakes form. At these times, katabatic winds will dry and entrain glacial flour.

5. Termination of the coming glaciation will occur when the broad zone of fast-moving ice becomes thin enough to conduct basal heat to the surface rapidly enough to extract the latent heat of freezing from the bed. At this critical ice thickness, the bed will freeze suddenly and provide traction too great to be overcome by the gravitational pulling force in the thin ice. The greatly reduced ice velocity along the southern ice-sheet margin will no longer offset the increasing melting rate, and the ice sheet will retreat rapidly (Fastook and Hughes, 1991). Curtailed glacial erosion at the frozen bed may stop dirt production, but dust may still be entrained from dirt in basal moraines exposed by retreating ice lobes.

6. Retreat of southern terrestrial ice-sheet margins in North America will be irreversible when the retreating margins expose a bed that is sufficiently isostatically depressed by the former overburden of ice to impound proglacial lakes. Ablation meltwater will accumulate in these lakes, where ablation rates will be accelerated by ice calving, and where glacial flour and silt become suspended in water instead of being entrained into the atmosphere by katabatic winds. Dust storms will stop, allowing insolation to warm the ice-sheet margins. Dust flushed out of the atmosphere by precipitation will no longer be replaced because the dust machine will be turned off.

7. Once stream flow develops along advancing ice-sheet margins, ice downdrawn into ice streams can convert ice ridges between ice streams into local ice domes. The terminal lobes of ice streams may also become local ice domes after ice streams shut down, if the snowline is sufficiently low. Being small, low, and fed by local accumulation only, these new ice domes will be extremely unstable (Hughes, 1987; Weertman, 1961). A slight fall in snowline elevation will make these ice domes expand rapidly, thereby restoring sheet flow all along the advancing ice margin. A slight rise in snowline elevations will make the new ice domes shrink rapidly, leaving exposed ice-stream lobes that will retreat rapidly by ablation. This will also restore sheet flow at the ice-sheet margin, allowing stream flow to abruptly and drastically destabilize the entire margin of the ice sheet by producing local ice domes. Dirt and dust production will change from heavy and discontinuous to light and continuous as stream flow changes to sheet flow and a barren outwash periglacial environment is replaced by a lacustrine environment.

The dust-creating component of the dust machine is katabatic winds strong enough to scour ice crystals from the surface of ice streams and strong enough to entrain glacial flour and silt discharged from the termini of ice streams. In his pioneering study of surface winds on the Antarctic Ice Sheet, Parish (1982, 1984) described and modeled the meteorological processes illustrated in Figure 3.3 that determine the surface wind field. Surface winds are a response to (1) long-wave radiation from the ice surface, especially during the polar night, which produces a temperature inversion in air above the ice surface, (2) the sloping ice surface, which allows the cold air to convert its density-induced gravitational potential energy into kinetic energy by flowing downslope, (3) Earth's rotation, which introduces a Coriolis "force" that diverts surface airflow to the left of the downslope direction in the Southern Hemisphere, and (4) frictional energy dissipation caused by vertical wind shear between inflowing geostrophic winds and outflowing surface winds. These processes tend to produce nearly steady-state meteorological conditions over the ice-sheet interior, with surface winds being persistent in both direction and speed. These stable surface winds are called "inversion winds" because they are sustained by the temperature inversion in air above the surface (Parish, 1982). The dependence of inversion winds on ice-surface slopes causes surface winds to converge and pool at the heads of ice streams, where outbursts of katabatic winds rush down the ice streams until the pool of air is depleted (Parish, 1984). These katabatic winds are transient, but they are intense enough to scour ice crystals from the steep upper slope of an ice stream and to entrain the fine fraction of dirt deposited at the ice-stream terminus. Entrained ice crystals and dirt particles then become the dust in the dust machine.

Parish and Bromwich (1986, 1987) modeled the transformation of Antarctic surface winds from steady-state inversion winds converging on ice streams to time-dependent katabatic winds rushing down ice streams. Their model produced surface winds with directions and velocities that are in good agreement with observations of Antarctic surface winds, as shown in Figure 2.20. A particularly good test of the model was its accurate simulation of inversion winds converging on David Glacier and becoming katabatic winds that produce a polynya in Terra Nova Bay (Parish and Bromwich, 1989; Bromwich and others, 1990). David Glacier, a major East Antarctic ice stream, is located in Figure 2.15. These results justify using the model to map the distribution and intensity of surface winds over Northern Hemisphere ice sheets during the coming glaciation, with the goal of predicting how the dust machine will be turned on and off by the dirt-producing life cycles of ice streams and the dust-producing katabatic winds flowing down ice streams.

In the analysis by Parish and Bromwich (1986), velocity u of surface winds is the vector sum of fall-line wind velocity u_x perpendicular to ice elevation contour lines and thermal wind velocity u_y parallel to these contours, such that:

$$u_x = u \cos \gamma \tag{10.72a}$$

$$u_y = u \sin \gamma \tag{10.72b}$$

$$u = (u_x^2 + u_y^2)^{1/2} \tag{10.72c}$$

where γ is the angle subtended by the surface wind direction and the fall line direction.

The equations for steady-state flow of surface winds resolved in the x and y directions are:

$$g_z \frac{\Delta T}{\overline{T}} \frac{\partial h}{\partial x} + V \frac{\partial P}{\partial x} - f_C u_y + \frac{k u u_x}{H} = 0 \tag{10.73a}$$

$$g_z \frac{\Delta T}{\overline{T}} \frac{\partial h}{\partial y} + V \frac{\partial P}{\partial y} + f_C u_x + \frac{k u u_y}{H} = 0 \tag{10.73b}$$

where g_z is gravity acceleration, $\overline{T}$ is mean air temperature in an inversion layer of thickness H, ΔT is the vertical temperature decrease through H, h is ice-surface elevation, $\partial h/\partial x$ and $\partial h/\partial y$ are ice-surface slopes along and normal to the fall line, V is the specific volume of air, P is atmospheric pressure, $\partial P/\partial x$ and $\partial P/\partial y$ are horizontal atmospheric pressure gradients along and normal to the fall line, f_C is the Coriolis parameter that depends on latitude N and Earth's rotation rate Ω, and k is a frictional constant. Typically, $H \approx 300$ m and $k \approx 5 \times 10^{-3}$. Each term in Equations (10.73) is a force in the x and y directions, respectively. The first term is the gravitational force caused by the ice-surface slope along x and y, with $\partial h/\partial y = 0$ by definition. The second term is the horizontal force induced by pressure gradients in the free atmosphere along x and y. The third term is the Coriolis force along x and y. The fourth term is a quadratic drag formulation for the frictional force along x and y.

The synoptic pressure gradient force in Equations (10.73) is small compared to the other forces and can be ignored, allowing Parish (1981) to obtain the following solution to these equations:

$$u = (h F_x \cos \gamma / k)^{1/2} = (F_x / f_C) \sin \gamma \tag{10.74a}$$

$$\gamma = \cos^{-1} [(1 + J^2)^{1/2} - J] \tag{10.74b}$$

where:

$$J = h f_C^2 / 2 k F_x \tag{10.75a}$$

$$F_x = - g_z (\Delta T / \overline{T}) \partial h / \partial x \tag{10.75b}$$

$$f_C = 2 \Omega \sin N \tag{10.75c}$$

Cloudy conditions interrupt long-wave radiation from the ice surface, causing the inversion layer to vanish, but it appears within twenty-four hours of cloud dissipation (Cerni and Parish, 1984). Surface winds over an ice sheet are therefore a persistent feature.

Figure 10.5 shows the velocity u and direction γ of surface winds obtained from Equations (10.74) by Parish and Bromwich (1986) with respect to ice-sheet flowline profiles obtained from Equation (5.5), with ΔT decreasing from 7.0°C at the ice divide to 2.5°C at the ice margin. Along most of the flowline, $u_x \approx 7$ m/s and $\gamma \approx 45°$, but over the last 50 km u_x increases sharply and γ decreases sharply to $u_x \approx 20$ m/s and $\gamma \approx 5°$ at the ice margin. These results allow u and γ to be estimated from Equations (10.74) for surface winds on Quaternary ice sheets during the coming glacial maximum. The computed present-day surface wind field over the Antarctic Ice Sheet is shown in Figure 2.20. Surface winds begin at ice divides, converge on ice streams, attain a high velocity above the maximum surface slope at the heads of ice streams, maintain high velocities along ice streams, and attain peak velocities above the steep slopes of terminal ice lobes. Convergent inversion winds become katabatic winds strong enough to scour ice crystals from the surface of ice streams and to entrain the fine fraction of dirt melted out from regelation ice or discharged by subglacial streams around the perimeter of ice lobes. If entrained ice and dust particles rise high enough to be transported back over the ice sheet by inflowing geostropic winds, they may become cloud condensation nuclei that enhance precipitation over the ice sheet and lower the snowline. If these particles are entrained long enough, easterly anticyclonic storms may carry them along the ice-sheet margin, which may then be shrouded in clouds that increase snow precipitation, reduce ice ablation, and lower the snowline. These are features of a hypothetical cloud machine.

Turning the Cloud Machine On and Off

The cloud machine hypothesis postulates eight mechanisms by which the ice, dirt, and dust machines may regulate planetary albedo by controlling cloudiness in the North Atlantic and, perhaps, globally.

1. During the last glacial termination, Northern Hemisphere ice sheets discharged some 5 x 10^7 km^3 of meltwater into the North Atlantic in about 7000 years. Since marine components of these ice sheets bordered the North Atlantic, most of the meltwater could have been linked to outbursts of icebergs from the ice machine. Up to 4 x 10^{23} cal of latent heat would then have been extracted from sensible heat in the North Atlantic, with abrupt cooling episodes caused by icebergs produced during the life cycles of marine ice streams. Life cycles could have been triggered by disintegration of buttressing ice shelves. Cooling surface water by 5°C may have halted production of North Atlantic Deep Water (NADW) and thereby triggered the Younger Dryas stadial (Broecker and Denton, 1989). Cooling by 5°C does not densify surface water enough to make up for the density reduction caused by lower salinity resulting from reduced

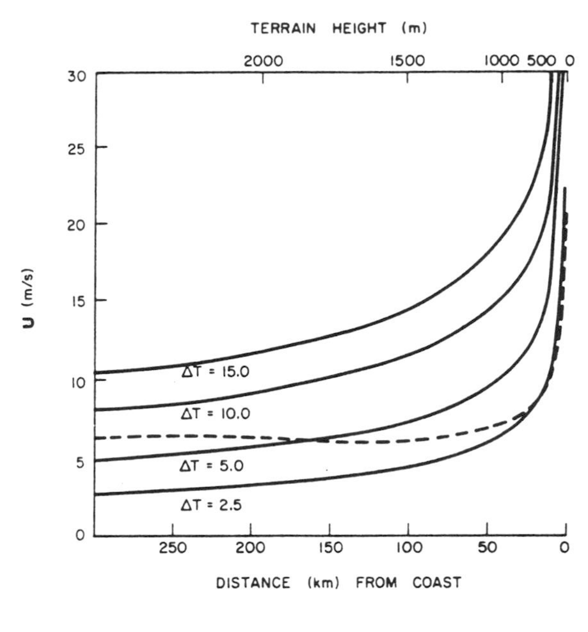

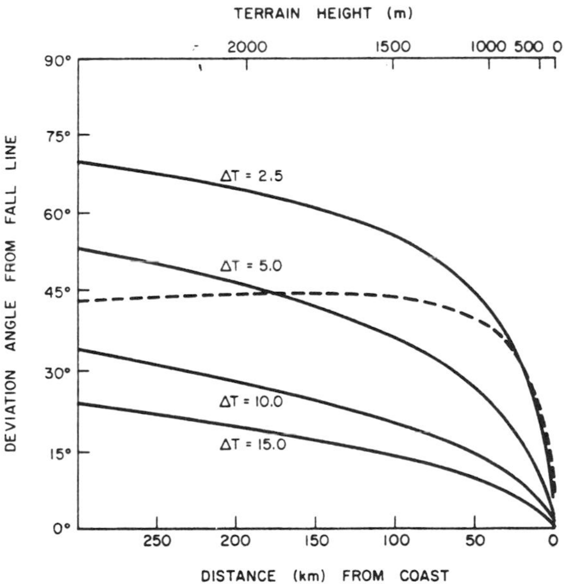

Figure 10.5: Surface winds over ice sheets computed from Equations (10.74). Wind speed u (*top*) increases rapidly and angular departure γ of winds from the downslope direction (*bottom*) decreases rapidly closest to the edge of the ice sheet for various temperature decreases ΔT that are constant through the inversion layer for all surface slopes (solid curves). Actual inversion strength decreases with increasing slope. If ΔT = 2.5°C at the coast and ΔT = 7.0°C over the interior, then u and γ are more nearly constant over the interior and change more rapidly near the coast (dashed curves). Reproduced by courtesy of R. Parish, D. Bromwich, and the American Meteorological Society, from *Monthly Weather Review, 114*(5), (1986), p. 852, Figure. 2.

evaporation. Iceberg outbursts have been proposed for the last deglaciation on both sides of the North Atlantic (Mercer, 1969; Miller and Kaufman, 1990; Andrews and others, 1990; Lehman and others, 1991) and also in the North Pacific (Kotilainen and Shackleton, 1995; Thunell and Mortyn, 1995; Kennett and Ingram, 1995; Keigwin, 1995). Life cycles of ice streams bordering the North Atlantic could have the same role in causing climatic changes associated with termination of the coming glaciation. Shutting down NADW production increases the thermal gradient between the equator and the poles, so stronger desert winds produce more atmospheric dust (Mayewski and others, 1994). By providing more cloud-condensation nuclei, increased dust levels produce increased cloudiness.

2. Iceberg and meltwater outbursts from marine ice streams would also decrease North Atlantic salinity. It has been argued that 3.5 x 10^5 m^3/s of incoming water vapor reduces salinity of surface water by 0.5 ‰, enough to halt NADW production (Broecker and Denton, 1989). An average of 2 × 10^5 m^3/s of incoming water was provided by Northern Hemisphere ice sheets during the last deglaciation. This input could easily attain 3.5 × 10^5 m^3/s during the life cycles of major marine ice streams. Hence, marine ice streams may control shutdowns of NADW production, with a concurrent increase in global cloudiness linked to a windier and dustier climate during stadials of a glaciation cycle.

3. Meltwater outbursts beneath present-day marine ice streams are highly turbid (Domack and Williams, 1990). Katabatic winds rushing down marine ice streams disperse the turbid water and entrain supermicrometer cloud-condensation nuclei (dust and sea salt) by bubble-film bursting, whitecapping, and spin-drift during the life cycles of these ice streams. Entrainment of super-micrometer nuclei varies logarithmically with wind speed over the North Atlantic today, such that doubling present-day wind speed from 10 m/s to 20 m/s increases the large condensation nuclei for marine stratus clouds tenfold, from ~ 5/cm^3 to ~ 50/cm^3, thereby increasing albedo enough to compensate for projected CO_2-induced climatic warming (Latham and Smith, 1990). Katabatic winds could increase the wind speed to 30 m/s (about 8 km/hr) over dirty water produced by former marine ice streams during their life cycles, thereby increasing cloud-condensation nuclei another tenfold to ~ 500/cm^3 for ice streams calving into the North Atlantic during the coming glaciation.

4. Katabatic winds rushing down marine ice streams can entrain cloud condensation nuclei directly by scouring ice crystals and their condensation nuclei from the surface of the ice streams. Wind ablation of Antarctic ice streams today is clearly observed in satellite imagery, and it is a major meteorological process (Bromwich, 1989; Bromwich and others, 1990). Katabatic winds that scoured marine ice streams draining the eastern Laurentide Ice Sheet may have entrained cloud condensation nuclei that subsequently entered the northern branch of the jet stream, as it flowed down Baffin Bay and into the Labrador Sea during the last glacial maximum (Manabe and Broccoli, 1985; Kutzbach, 1987; COHMAP Members, 1988). Then, a continuous cloud bank may have existed where the jet stream crossed the North Atlantic. This situation may repeat during the coming glaciation.

5. Dust entrained by katabatic winds flowing down terrestrial ice streams along the southern Laurentide ice margin was much more voluminous than dust entrained above marine ice streams along the eastern ice margin, because the extensive outwash fans of terrestrial ice streams were dry and largely unvegetated (Wright, 1987). These were primary sources for loess during the last glacial maximum, with loess deposits being a measure of cloud-producing atmospheric dust. Even more than clouds produced by katabatic winds above marine ice streams (Harvey, 1988), terrestrial dustiness increased global cloudiness and planetary albedo enough to cause cooling during the last glacial maximum. The amount of cooling depended on the relative efficiency and abundance of supermicrometer and submicrometer cloud-condensation nuclei (Anderson and Charlson, 1990). The dust machine associated with terrestrial ice streams may produce the larger nuclei in great abundance during the coming glaciation.

6. The dirt machine was especially productive beneath southern Laurentide terrestrial ice streams during their fast-flow life cycles. The dust machine would therefore also have been unusually productive at these times, and as terminal ice lobes retreated between life cycles, and perhaps vigorous enough to mix dust entrained in the anticyclonic easterly winds along the southern margin with the much stronger westerlies to the south and with the jet stream overhead. Regional and global cloudiness would have been enhanced by these mixing processes, with stratocumulus clouds being carried across the North Atlantic by the westerlies and the jet stream (Albrecht and others, 1988). Today, insolation delivers about 10^5 cal/cm^2 to the North Atlantic, providing 2 x 10^{22} cal/a over an area of 2 x 10^{17} cm^2 above 35°N, an amount of heat increased some 25 percent by incoming warm ocean water (Broecker and Denton, 1989). If the dust machine sustained a continuous belt of high-albedo stratocumulus clouds across the North Atlantic that reduced the heat input to 3 x 10^{21} cal/a, North Atlantic surface water would cool by about 5°C. This cooling could have halted NADW production at the last glacial maximum and sustained Northern Hemisphere ice sheets, even though insolation was about the same as now. The coming glaciation may also be sustained by this hypothetical cloud belt.

7. The dirt machine has the potential for directly changing planetary albedo through dust-regulated upper atmosphere cloudiness (Starr, 1987; Cox and others, 1987; LaBrecque, 1990) and possibly by directly providing essential trace nutrients to phytoplankton whose enhanced metabolism may have reduced atmospheric CO_2 to the 200-ppm levels that existed during the last glacial maximum (Martin, 1990; Peng and Broecker, 1991; Mortlock and others, 1991). Such direct linkages to global climatic change are independent of intermediate processes, such as proposed flip-flops in ocean circulation that must be propagated along oceanic conveyor belts requiring some 1000 years per circuit (Jones, 1991; Birchfield and Broecker, 1990; Mortlock and others, 1991). Direct aerosol-induced cloudiness registers a climatic response in decades or less (Mayewski and others, 1990; 1997). The most abrupt climatic changes may result from direct link-ages among the dirt, dust, and cloud machines during the coming glaciation.

8. The dust-induced cloud belt along southern margins of Northern Hemisphere ice sheets would depress the equilibrium line by increasing snow precipitation over the ice and by reflecting insolation above the clouds, as illustrated in Figure 1.8d. Inflowing air above outflowing katabatic winds would draw these clouds over the ice sheet, thereby expanding the accumulation zone and shrinking the ablation zone. By this mechanism, ice sheets may control much of their own mass balance, allowing former ice sheets to remain close to their maximum southern limit as late as 14 ka BP, when insolation was high.

The major significance of the cloud machine during the coming glaciation is its ability to destabilize the southern terrestrial margin of an Arctic Ice Sheet. Local ice domes form between ice streams as they pass through their life cycles and may develop from their stagnating ice lobes at the end of life cycles. These ice domes are vulnerable to minor changes in snowline elevation. This vulnerability has received little attention, but it may exert a major influence on the overall stability of an ice sheet. As illustrated in Figure 2.6, a small change in the size of a local ice dome is reversible if the snowline depends only on latitude, irreversible if it depends only on altitude, and either reversible or irreversible if it depends on both latitude and altitude. Latitude-only dependence is depicted by the vertical snowline, so expansion of the ice dome increases the ablation area and forces contraction, but contraction of the ice dome decreases the ablation area and forces expansion. These perturbations are reversible. Altitude-only dependence is depicted by the horizontal snowline. Expansion of the ice dome now increases the accumulation area and causes more expansion, whereas contraction decreases the accumulation area and causes more contraction. These perturbations are irreversible. A sloping snowline depends on both latitude and altitude, with expansions or contractions of local ice domes being reversible for high snowline slopes and irreversible for low snowline slopes.

If the snowline slope depends only on Milankovitch insolation variations, Figure 9.2 shows that for north-south transects across Northern Hemisphere ice sheets, the snowline slope will be highest when insolation is a minimum for the tilt cycle and a maximum for the precession cycle and lowest when insolation is a maximum for the tilt cycle and a minimum for the precession cycle. These effects on the stability of an ice sheet as a whole are magnified greatly for local ice domes, because relatively small changes in snowline slope can cause large relative changes in their ratios of accumulation area to ablation area.

The Weertman (1961) analysis of the stability of ice-age ice sheets can be applied to local ice domes between terrestrial ice streams and to their terminal ice lobes along the southern margin of an Arctic Ice Sheet. The origin of coordinates x, z is on the bed beneath the center of the local ice dome, with x horizontal along flowlines from the ice dome to the ice margin and z vertical from the ice bed to the ice surface. In relation to these coordinates, the snowline height H_S above the bed is assumed to be linear:

$$H_S = H_0 + S\,x \tag{10.76}$$

where H_0 is the snowline height at $x = 0$ and S is the snowline slope along x for a given flowline. Height H_0 and slope S in Equation (10.76) can be determined by computing H from Equation (9.1) at points $x = 0$ and $x > 0$ on the flowline.

For a flowband of constant unit width, mass-balance continuity requires that:

$$u\,h = \int_0^x a\,dx \tag{10.77}$$

where u is the mass-balance ice velocity, h is ice-surface height above a horizontal bed, and a is the net rate of ice accumulation or ablation at a given point along x. Ignoring longitudinal stresses, the force balance requires that:

$$\tau_o = -\rho_I\,g_z\,h\,dh/dx \tag{10.78}$$

where τ_o is the basal shear stress, ρ_I is ice density, g_z is gravity acceleration, and dh/dx is ice-surface slope. The relationship between u and τ_o used by Weertman (1961) is a generalized flow or sliding law given by:

$$u = C\,\tau_o^{\,c} \tag{10.79}$$

Equations (10.77) through (10.79) combine to give:

$$C\,\rho_I^2\,g_z^2\,h^3\,(dh/dx)^2 = \int_0^x a\,dx \tag{10.80}$$

where $C = 3 \times 10^{-16}\ m^3\,s^3\,kg^{-2}$ and $c = 2$ were obtained by fitting Equation (10.80) to flowline profiles of present-day ice caps. The solution of Equation (10.80) is the equilibrium profile of flowlines for an ice dome or an ice cap.

Separate solutions of Equation (10.80) exist for the accumulation zone and the ablation zone of the ice dome. If E and L are the respective distances along x from the center of the ice dome to the equilibrium line and to the ice margin, the average accumulation rate $\bar{a}''$ is given by:

$$\bar{a}''\,E = \int_0^E a\,dx \tag{10.81}$$

and the average ablation rate $\bar{a}'$ is given by:

$$\bar{a}'\,(L - E) = -\int_E^L a\,dx \tag{10.82}$$

Subtracting Equation (10.82) from Equation (10.81):

$$\bar{a}''\,E - \bar{a}'\,(L - E) = \int_0^E a\,dx + \int_E^L a\,dx = \int_0^L a\,dx = 0 \tag{10.83}$$

Equation (10.83) is the requirement for mass-balance equilibrium.

Based on Equations (10.81) through (10.83), the integral form of Equation (10.80) for accumulation over $0 \le x \le E$ is:

$$\int_{h_0}^{h} h^{3/2}\, dh = \left(\frac{\bar{a}''}{C \rho_I^2 g_z^2}\right)^{1/2} \int_0^x x^{1/2}\, dx \qquad (10.84)$$

and for ablation over $E \le x \le L$ is:

$$\int_{h_E}^{h} h^{3/2}\, dh = \left(\frac{\bar{a}'}{C \rho_I^2 g_z^2}\right)^{1/2} \int_{\bar{a}'' E}^{\bar{a}'' E - \bar{a}' (x-E)} x^{1/2}\, dx \qquad (10.85)$$

where h_0 and h_E are flowline heights at $x = 0$ and $x = E$, respectively. The solution of Equation (10.84) for the accumulation zone gives the expression for the flowline profile when $0 \le x \le E$:

$$h^{5/2} = h_0^{5/2} - C^* \bar{a}''^{1/2} x^{3/2} \qquad (10.86)$$

and the solution of Equation (10.85) for the ablation zone gives the expression for the flowline profile when $E \le x \le L$:

$$h^{5/2} = h_E^{5/2} + C^* \bar{a}'^{1/2} \{[\bar{a}'' E - \bar{a}' (x - E)]^{3/2} - [\bar{a}'' E]^{3/2}\} \qquad (10.87)$$

where $C^* = 2\, m^{1/2} a^{1/2}$ is a constant defined as:

$$C^* = 5/(3\, C^{1/2} \rho_I g_z) \qquad (10.88)$$

Equilibrium-line elevation h_E is obtained from Equation (10.87) by setting $h = 0$ at $x = L$, so that $\bar{a}'' E - \bar{a}' (L - E) = 0$ from Equation (10.83):

$$h_E = (\bar{a}'' E)^{3/5} (C^*/\bar{a}')^{2/5} \qquad (10.89)$$

Figure 2.7 gives the variation of h_E with E according to Equation (10.89) for the equilibrium line.

Setting $H_S = h_E$ at $x = E$ in Equation (10.76) gives the snowline elevation at the equilibrium line:

$$H_S = h_E = H_0 + S\, E \qquad (10.90)$$

Figure 2.7 gives the variation of h_E with E according to Equation (10.90) for the snowline.

Let E be small, so that $h_E \approx H_0$. Equation (10.89) can be solved for E:

$$E = h_E^{5/3}\, \bar{a}'^{2/3}/\bar{a}''\, C^{*2/3} \qquad (10.91)$$

Equation (10.86) can be solved for h_0 when $h = h_E$ at $x = E$, with E given by Equation (10.91):

$$h_0 = h_E (1 + \bar{a}'/\bar{a}'')^{2/5} \qquad (10.92)$$

Equation (10.83) can be solved for L, with E given by Equation (10.91) and h_E given by Equation (10.92):

$$\begin{aligned} L &= (1 + \bar{a}'/\bar{a}'')\, E \\ &= (1 + \bar{a}'/\bar{a}'')\, (h_E^{5/3}\, \bar{a}'^{1/3}\, C^{*2/3}) \\ &= [(1 + \bar{a}'/\bar{a}'')\, h_0^5/\bar{a}\, C^{*2}]^{1/3} \end{aligned} \qquad (10.93)$$

Note that both h_0 and L decrease when h_E decreases or $\bar{a}''$ increases. Since the volume of the ice dome decreases when h_0 and L decrease but volume increases when h_E decreases and $\bar{a}''$ increases, Equations (10.91) through (10.93) describe an ice dome in unstable equilibrium. The ice dome will therefore grow or shrink irreversibly with small changes in snowline elevation or accumulation rate, both of which are controlled by the cloud machine.

Let E be large, so that $h_E \approx S\, E$. Equation (10.89) can be solved for E:

$$E = C^*\, \bar{a}''^{3/2}/\bar{a}'\, S^{5/2} \qquad (10.94)$$

Equation (10.92), with S E substituted for h_E and E given by Equation (10.94), becomes:

$$h_0 = (1 + \bar{a}''/\bar{a}')^{2/5} (\bar{a}''/\bar{a}')^{3/5}\, \bar{a}''^{1/2}\, C^*/S^{3/2} \qquad (10.95)$$

Equation (10.93), with Equation (10.95) substituted for h_0, becomes:

$$L = (1 + \bar{a}''/\bar{a}')\, C^*\, \bar{a}''^{3/2}/\bar{a}'\, S^{5/2} \qquad (10.96)$$

Note that both h_0 and L increase as $\bar{a}''$ increases and $\bar{a}'$ decreases. Since the volume of the ice dome increases when H_0 and L increase and volume increases when $\bar{a}''$ increases and $\bar{a}'$ decreases, Equation (10.94) through (10.96) describe an ice dome in stable equilibrium. The ice dome will then grow or shrink reversibly with small change in accumulation or ablation rates.

Figure 2.7 plots Equations (10.89) and (10.90) for a range of snowline elevations H_0 and snowline slopes S. The H_S line is above the h_E curve for all E values, is tangent to the h_E curve at one E value, or intersects the h_E curve at two E values. For two intersections, the ice dome is unstable when E is small and stable when E is large. For a tangent point, the ice dome is metastable. The significance of Figure 2.7 is that E is always small for local ice domes between ice streams that drain an ice sheet and for the terminal lobes of stagnating ice streams. Therefore, these ice domes and ice lobes grow irreversibly when the snowline lowers and shrink irreversibly when the snowline rises. This is demonstrated in Figure 10.6, which shows that growth or shrinkage changes from irreversible to reversible when ice domes and ice lobes exceed a critical size in response to small changes in snowline slope and in rates of accumulation or ablation. When the irreversible response is expansion, the ice dome eventually becomes large enough to allow reversible responses tolowering the snowline, increasing accumulation rates and reducing ablation rates. The growing local ice domes cause a substantial advance of the southern terrestrial margin of an Arctic Ice Sheet until this stabilization occur. By pinching off ice streams between local ice domes, growth of these domes turns off the cloud machine by shutting down the dirt

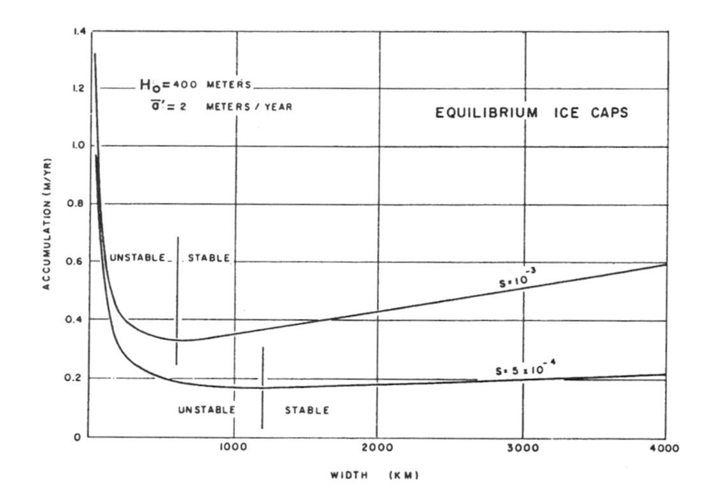

Figure 10.6: Stable and unstable equilibrium widths 2 L of ice domes or ice lobes in relation to mean surface accumulation rates $\bar{a}''$ above snowlines for two snowline slopes S. The snowline at the North Pole is H_0 = 400 m above sea level and the mean surface ablation rate below the snowline is $\bar{a}'$ = 2 m/a. Note that ice domes and ice lobes below the critical size grow or shrink irreversibly for respective increases or decreases in the mean accumulation rate, and reversibly for respective increases or decreases in these rates. The critical sizes are the minima in the curves of $\bar{a}''$ versus 2 L. These sizes become larger as S decreases. Reproduced from J. Weertman, *Journal of Geophysical Research, 66* (1961) p. 3790, Copyright American Geophysical Union.

and dust machines. The snowline rises and snow precipitation decreases when the cloud belt along southern ice margins dissipates. Stagnating lobes of terrestrial ice streams can also behave like local ice domes, experiencing irreversible advance as the snowline lowers and irreversible retreat as the snowline rises. Advance eventually stabilizes the ice lobe; retreat makes the ice lobe vanish. Figure 10.7 shows local ice domes and terminal ice lobes produced by stagnating ice streams along the southern terrestrial margin of the Laurentide Ice Sheet at 14 ka BP (Hughes, 1987).

At higher latitudes, the snowline lowered by the cloud machine allowed central Alaska and northeast Siberia to be glaciated, as demonstrated by applying the Weertman (1961) analysis to mountain glaciers on a sloping bed that becomes nearly horizontal at the base of Alaskan and northeast Siberian mountain ranges. If α is the surface slope, β is the bed slope, x is parallel to the bed slope and positive downslope, and z is normal to the bed slope and positive upward, basal shear stress τ_o is:

$$\tau_o = -\rho_I g_z h (\alpha + \beta) \qquad (10.97)$$

Equation (10.80) for the surface profile of the ice flowline now becomes:

$$C \rho_I^2 g_z^2 h^3 (\beta + dh/dx)^2 = u_0 + \int_0^x a\, dx \qquad (10.98)$$

where u_0 is the mean ice velocity at the origin of coordinates x, z, and α = dh/dx. If the snowline is horizontal, $S = -\beta$ in Equation (10.76) for coordinates x, z on a sloping bed, so that:

$$H_S = H_0 - \beta x \qquad (10.99)$$

where H_0 depends on where the coordinate origin is located on the sloping bed. Length L of the flowline from this origin is given by Equation (10.96) with β substituted for S:

$$L = (1 - \bar{a}' / \bar{a}'')\, C^* \bar{a}''^{3/2} / \bar{a}' \beta^{5/2} \qquad (10.100)$$

Since β is in the denominator and raised to the 5/2 power, L becomes very large when β becomes small. Therefore, glaciers moving down mountain slopes become very long if they reach valleys or plains between mountain ranges because β becomes small. So long as the snowline remains depressed, these valleys or plains will fill with glacial ice up to and even above the summits of the flanking mountain ranges. This is probably how the x forms during glaciation cycles. In Canada, the major Cordilleran mountain ranges are the Rocky Mountains and the Coast Mountains, which trend north to northwest and are separated by a highland plateau. Snowlines normal to these ranges have low slopes that dip from the continental Rocky Mountains to the maritime Coast Mountains. These are conditions under which Equation (10.94) predicts that moun-tain glaciers moving down the western flank of the Rocky Mountains and the eastern flank of the Coast Mountains produce a Cordilleran Ice Sheet over the highland plateau.

Initiating the Coming Glaciation

Figures 10.8 through 10.14 preconstruct ice sheets at various stages during the coming glaciation cycle. Glaciation is initiated when area ice thickens and grounds on Arctic continental shelves. Converting thickening sea ice into an ice shelf that spreads under its own weight requires an initial thickening of about 5 m to prevent leads from opening and isolating floes of sea ice, assuming that wind and current tensile stresses do not exceed 50 k Pa in perennial pack ice. Suppressing leads fundamentally changes the mechanics (Parkinson and Washington, 1979; Hibler, 1979) and the thermodynamics (Perovich and Maykut, 1990) of sea ice. Figure 10.8 shows the advance of terrestrial ice margins and of grounding lines for Arctic sea ice during initiation of the coming glaciation cycle. Equation (10.21) gives the advance rate of terrestrial ice margins. Floating ice grounds at ice thicknesses h_I on Arctic continental shelves calculated from Equation (7.6) for $\dot{Q}_H = \dot{\varepsilon}_{zz} = 0$ and for a calculated from Equation (10.11). For both grounded and floating ice, present-day top surface accu-mulation rates are augmented by snow precipitation origi-nating from water evaporated over ice-free polar seas warmer than today, following Fastook and Prentice (1994), as summarized in Equations (10.1) through (10.11). In Canada, sea ice grounds first in the interisland channels of the Queen Elizabeth Islands, and these grounding

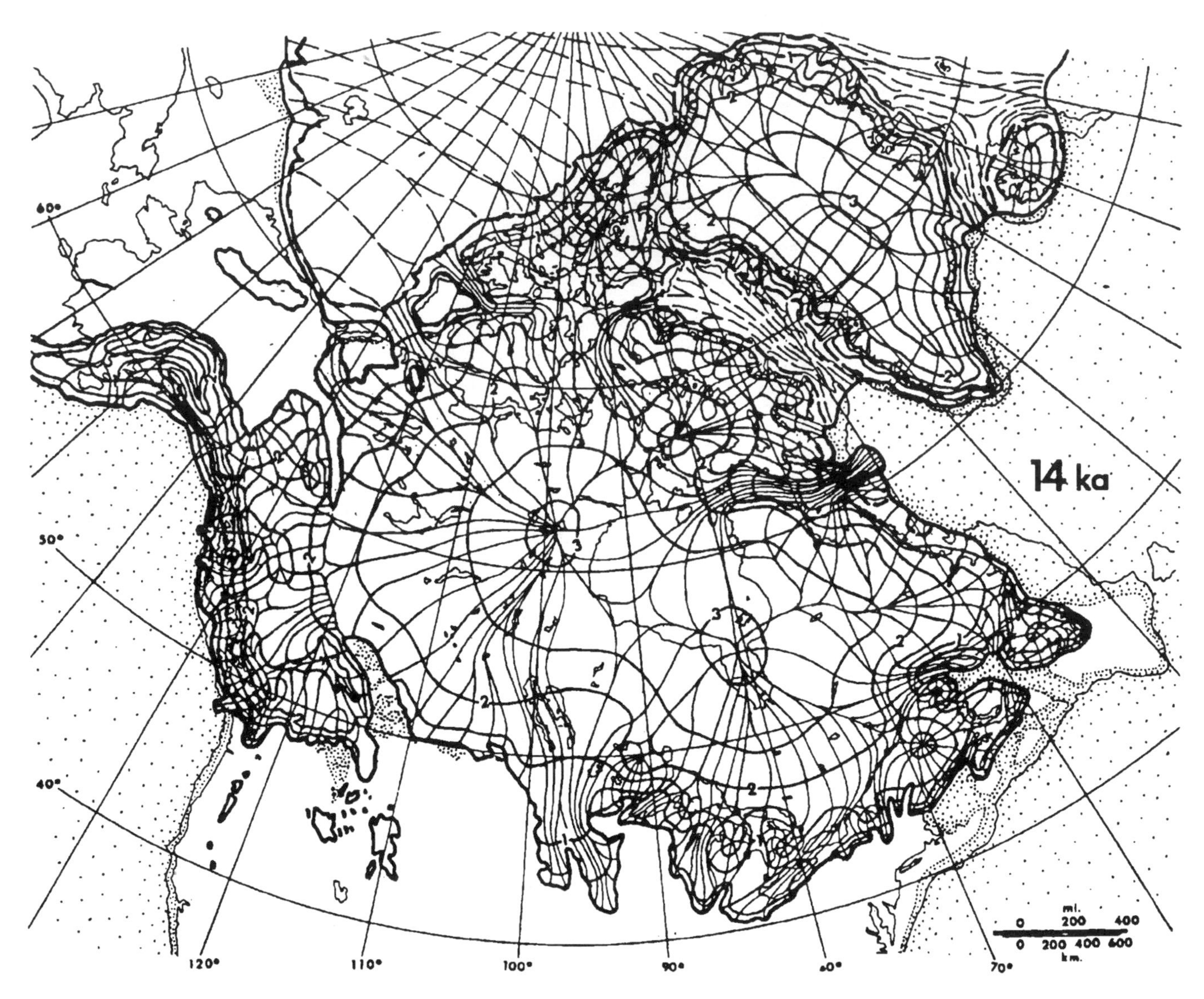

Figure 10.7: Local ice domes and ice lobes associated with ice streams along the southern margin of the Laurentide Ice Sheet 14,000 years ago. Ice domes and ice lobes are the beginnings and ends of ice surface flowlines drawn normal to ice elevation contours, which are spaced 0.5 km apart. Local ice domes lie between the heads of terrestrial ice streams, which end as ice lobes. Flowlines on ice shelves beyond marine ice streams are dashed. Heavy lines enclose glaciated areas. Dotted areas are oceans and lakes beyond the glaciated areas. Present-day lakes and coasts and the 500-m bathymetric contour are also shown. From Hughes (1987).

lines then migrate toward the Canadian mainland and across Foxe Basin to the entrance of Hudson Bay. Grounding lines also migrate from the shoreline to the center of Hudson Bay. A terrestrial ice cap develops on the central plateau of Quebec. In Siberia, sea ice grounds first around islands and along the Siberian coast in the East Siberian Sea and the Laptev Sea and then over the Kara Sea and the Chukchi Sea. This requires that fresh water from Siberian rivers spreads out over the top surface of fast ice frozen to Arctic coasts, where it freezes as thin annual layers instead of reaching the bottom surface through leads and tidal cracks. This is the situation in the Laptev Sea today beyond the Lena River delta as shown in Figure 9.36 (Dethleff, 1992). Freezing this water on the top surface augments a from snowfall and keeps $\dot{Q}_H \approx 0$ on the bottom surface.

Landward spreading is retarded by calving into proglacial lakes impounded along the advancing ice margin. Therefore, growth occurs primarily by ice thickening until enough gravitational potential energy accumulates and converts into kinetic energy of motion that drives the ice margin across the lakes during glacial stadials. As described in Chapter 1, ice thickening followed by spreading converts an interstadial ice sheet into a stadial ice sheet.

Figure 10.8 shows the marine ice domes at the end of the inception stage of the coming glaciation cycle, using Equation (9.53) to construct flowline profiles for sheet flow

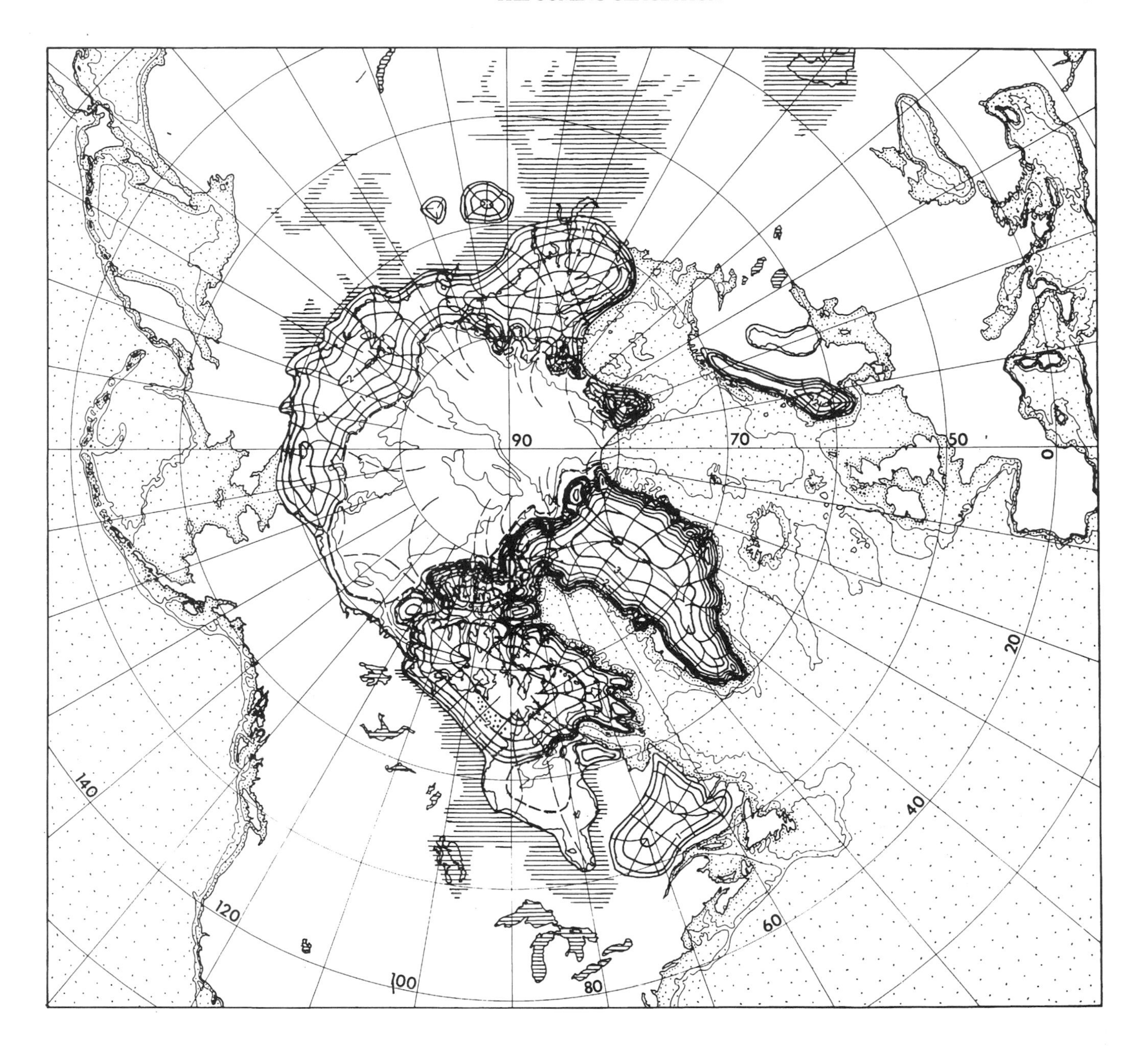

Figure 10.8: Initiation of an Arctic Ice Sheet during the coming glaciation. Shown in Figures 10.8 through 10.14 are grounded and floating ice margins (heavy lines), ice elevation contour lines every 0.5 km (solid lines), flowlines for grounded (light solid lines) and floating (light broken lines) ice, ice streams (heavily dotted flowbands), relative thawing of the bed beneath grounded ice sheets (the relative spacing of heavy dots), proglacial lakes (areas shaded by parallel lines), oceans (areas shaded by light dots), and present day shorelines.

over a mostly frozen bed and Equation (9.11) for snowline lowering. The northern Greenland Ice Sheet has thickened owing to increased evaporation from the Greenland Sea, made ice-free by "greenhouse" warming, and the resulting higher snow precipitation over the ice sheet (Huybrechts, 1994). Terrestrial ice caps over islands in the Queen Elizabeth Islands have merged to become the marine Innuitian Ice Sheet. The Laurentide Ice Sheet has begun as a marine ice dome in Foxe Basin that then spreads into Hudson Bay and across Keewatin, duplicating the pattern of oldest Laurentide lineations discussed by Boulton and Clark (1990) for the beginning of the last glaciation. Terrestrial ice caps on the New Siberian Islands have expanded over the shallow continental shelf of the East Siberian Sea to become the marine East Siberian Ice Sheet. Somewhat later, terrestrial ice caps on Severnaya Zemlya and Wrangel Island advance over shal-

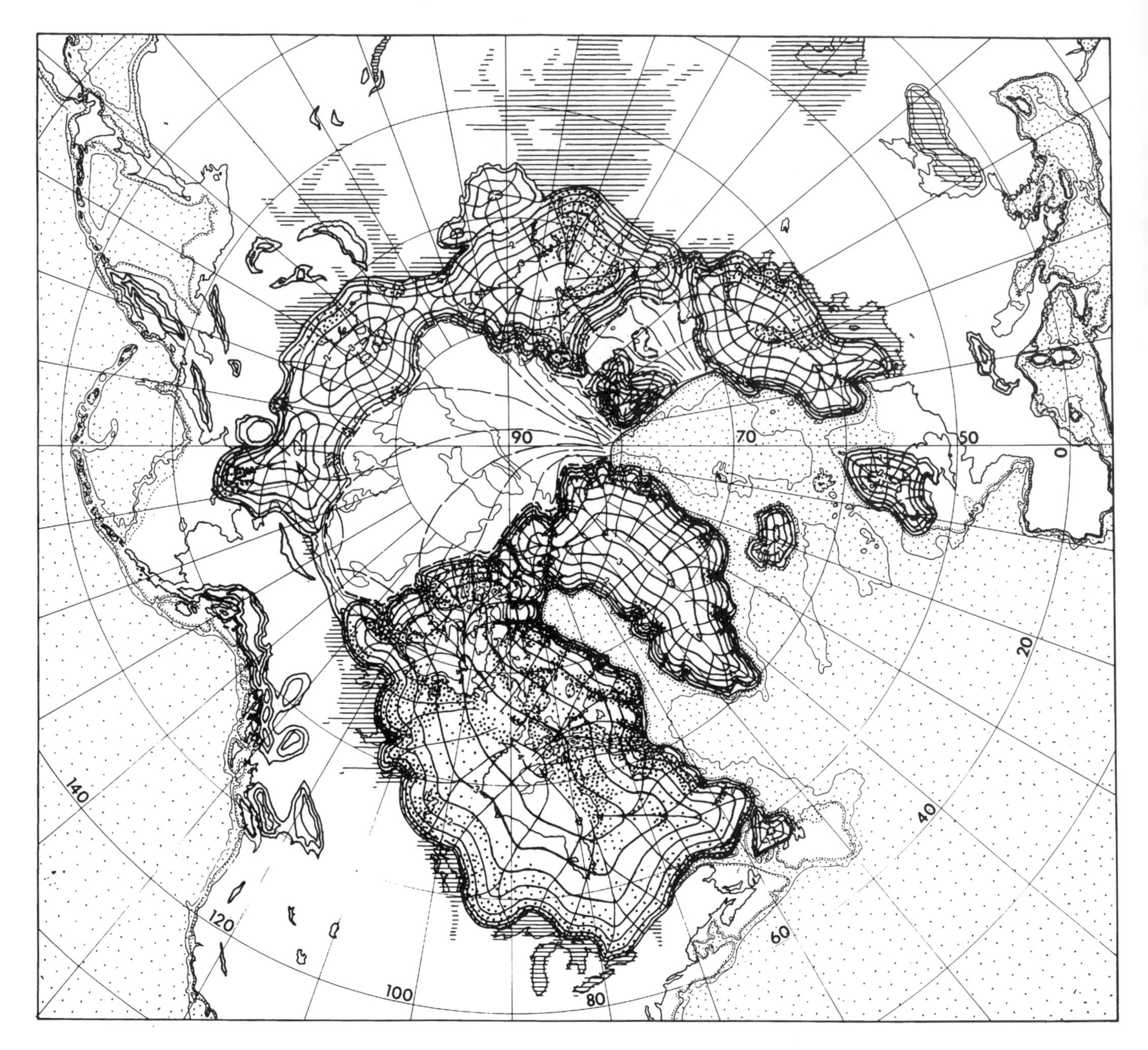

Figure 10.9: Thickening ice domes during growth of the coming glaciation. See the caption of Figure 10.8 for identification of features.

low continental shelves in the Kara Sea and Chukchi Sea to become the marine Kara Ice Sheet and Chukchi Ice Sheet, respectively. These incipient ice sheets in Canada and Siberia merge and are connected by floating ice some 200 m thick over the deep basins of the Arctic Ocean, so they behave as a unified dynamic system called the Arctic Ice Sheet.

The Arctic Ice Sheet impounds all rivers flowing into the Canadian and Siberian Arctic. All precipitation, whether rain or snow, falling over the Arctic Ice Sheet or within the watersheds of these rivers is trapped until spillways develop to the south. This combined area, shown in Figure 1.6, is therefore a White Hole analogous to the Black Holes in outer space in that nothing entering can escape. If mean annual precipitation of 250 mm/a of ice equivalent falls over just the 24×10^6 km^2 ice-covered area, 6×10^3 km^3/a of ice will accumulate and cause sea level to drop 11.6 mm/a. Evaporation from large proglacial lakes impounded along the southern margin of the Arctic Ice Sheet will create high snow precipitation rates akin to a subpolar maritime climatic regime, which replaces the present-day subpolar continental climatic regime and results in rapid ice-sheet growth.

Thickening of a Growing Ice Sheet

Figure 10.9 relates basal thermal conditions to the extent of an Arctic Ice Sheet at t = 20,000 years after inception of the

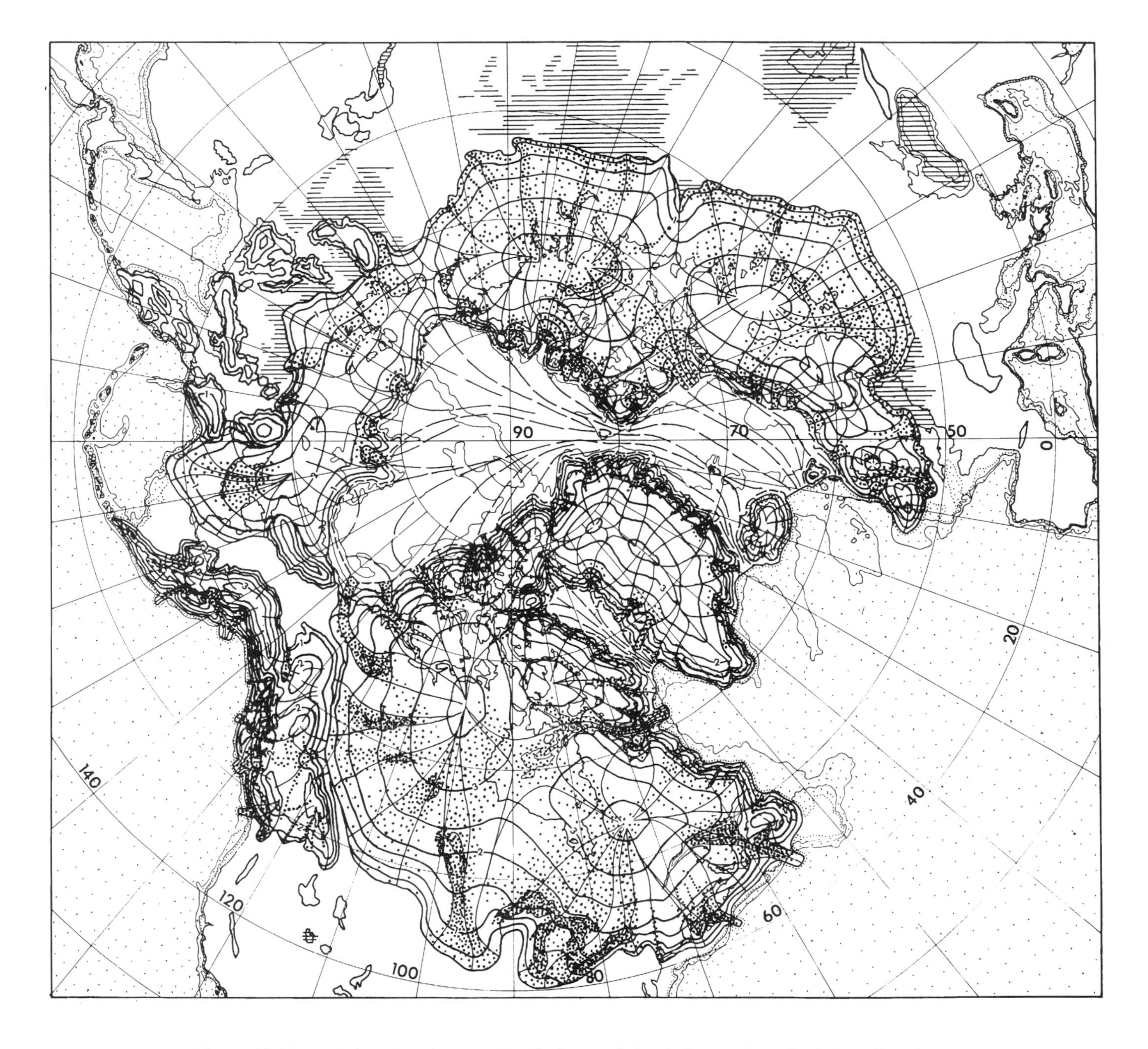

Figure 10.10: Advancing ice margins during stadials of the coming glaciation. See the caption of Figure 10.8 for identification of features.

coming glaciation, using Equations (9.8) through (9.11) and Equations (10.20), (10.21), and (10.26) to compute flowline profiles from Equation (9.53). Growth is primarily by thickening, the interstadial stage of an ice sheet, as discussed in Chapter 1. The thickness to area ratio of grounded ice during most of the 20,000 years of growth remains well below the steady-state ratio, as calculated by Paterson (1972), so creep rates of ice are low and Equation (10.18) gives reasonable temperature profiles through the ice thickness. After 20,000 years, however, the steady-state ratio is approached, because calving into deepening proglacial lakes replaces the advancing ice ramp with a more stationary calving ice wall (see Figure 2.4). Equation (8.107) gives the ice calving rate, and replaces Equation (10.21) because $v_x = 0$ when $\alpha_o = \infty$ for an ice wall. Using mean annual surface temperatures $\bar{T}_a$ in Equation (10.1) to compute basal temperature T_o in Equation (10.18) shows that $T_o < T_M$ for nearly all grounded portions of the Arctic Ice Sheet, especially on land where permafrost was initially thick (see Figure 9.10).

An outburst of Greenland ice into the North Atlantic, resulting from rejuvenation of Greenland ice streams when "greenhouse" warming triggers the Jakobshavns Effect, is assumed to have shut down production of North Atlantic Deep Water and thrown global climate into a glacial mode.

This reduces $\bar{T}_a$ sharply in high Northern Hemisphere latitudes and diverts the North Atlantic Current from the Arctic so that $Q_H \approx 0$ in Equation (7.1), and sea ice thickens rapidly to become an ice shelf floating in the deep ocean basins of the Arctic. An ice shelf floating in Baffin Bay forms during the partial collapse of the northern Greenland Ice Sheet, after the Jakobshavns Effect triggers surges of its northern ice streams (see Figure 1.10).

After 20,000 years, Laurentide ice has reached the edge of the crystalline Canadian Shield in North America and is connected to Greenland ice across Nares Strait. The ice shelf floating in Baffin Bay has disintegrated. Greenland ice has recovered from ice thinning during the 300 years of The Jakobshavns Effect. A Cordilleran Ice Sheet has formed in western North America. In Eurasia, Scandinavian ice has grounded in the Gulf of Bothnia, grounded ice in the Kara Sea has advanced across the Taimyr Peninsula to merge with a large ice cap on the Putorana Plateau, and has advanced into the shallow water of the eastern Barents Sea to become an ice shelf floating in the deep water of the western Barents Sea, and grounded ice in the Laptev, East Siberian, and Chukchi Seas has transgressed onto the Arctic coasts of Siberia and Alaska. An ice shelf floats over the Arctic Ocean and expands south in the Norwegian and Greenland Seas. Marine ice streams occupy the deeper marine troughs and terrestrial ice streams occupy the deeper lake basins, but they are sluggish with high c val-ues in Equation (6.37). Elsewhere, the bed remains frozen, and ice continues to thicken. This situation changes radically when the bed thaws in these regions.

Stadial Advances of a Mature Ice Sheet

Figure 10.10 relates basal thermal conditions to the extent of an Arctic Ice Sheet at $t = 25{,}000$ years after inception, and 5000 years after widespread basal melting beneath the ice sheet shown in Figure 10.9. Growth occurs primarily by thinning and spreading, as conversion of potential energy into kinetic energy thaws much of the bed beneath the ice sheet in Figure 10.9 and causes partial gravitational collapse when the frozen bed loses traction upon thawing. As MacAyeal (1993a, 1993b) showed, when basal temperature are calculated from Equation (10.18), gravitational collapse ends when the bed refreezes. Equation (10.64) gives the ice thickness for refreezing. Advance of the ice margin introduces the stadial stage of an ice sheet, as discussed in Chapter 1. Stadials in the coming glaciation can last a few millennia, depending on how many ice streams participate in the advance.

Spreading takes place in ice streams, which follow marine troughs, ice-dammed lakes, and rivers beyond the ice margin and which slide with little or no basal traction on the water-soaked marine, lacustrine, and fluvial sediments. Ice streams lengthen by downdrawing interior ice, which supplies terrestrial ice lobes and marine ice shelves beyond the ice streams. Ice downdrawn at the heads of ice streams converts ice ridges between ice streams into local ice domes and produces saddles on central ice divides. Figure 10.10 shows the larger ice domes between major saddles. Local ice domes between ice streams are in unstable equilibrium, because their mass balance is described by Equations (10.91) through (10.93). If the cloud machine is turned on, the higher precipitation rates and the lower snowlines cause the local ice domes to grow irreversibly; if the cloud machine is turned off, the lower precipitation rates and the higher snowlines cause the ice domes to shrink irreversibly. The same irreversible advance and retreat may apply to ice lobes at the foot of terrestrial ice streams. Growth leads to the most extensive stadials of the glaciation cycle. Shrinkage causes the ice sheet to retreat to the large stable ice domes described by Equations (10.94) through (10.96) and shown in Figure 10.9. These contracted positions are interstadials, and they allow the central ice domes to thicken on a newly frozen bed until Equation (10.18) predicts the bed will thaw, leading to renewed gravitational collapse, spreading, and another stadial in the glaciation cycle. Proglacial lakes are a moisture source for thickening ice.

Maximum Extent of the Coming Glaciation

Figure 10.11 depicts the maximum size of a future Arctic Ice Sheet that expands to the maximum limits of earlier Quaternary glaciations, as if these limits could be attained during a single glacial maximum. This is unlikely, but it is even more unlikely that the coming glaciation could expand beyond these limits. Therefore, Figure 10.11 represents the end-member in the spectrum of Quaternary glaciations. The Arctic Ice Sheet for this end-member is produced by allowing the stagnating terrestrial ice lobes in Figure 10.9 to expand irreversibly according to the Weertman (1961) mass-balance mechanism, described by Equations (10.91) through (10.93), until their size is large enough for expansion to become reversible, as described by Equations (10.94) through (10.96). The stagnating ice lobes form at the terminus of terrestrial ice streams, but ice-stream dynamics play a minor role in this maximum expansion. It is an expansion that takes place after gravitational collapse of the large interior ice domes has progressed too far to provide potential energy that converts to the kinetic energy that drives ice streams. The local ice domes grow because the cloud machine has depressed the snowline along southern margins of the Arctic Ice Sheet.

Another consequence of the cloud machine is depression of the snowline of mountain glaciers in highlands just beyond landward margins of the Arctic Ice Sheet. This is most striking in the western cordillera of North America and in the mountains of Beringia. The lowered snowline allows Laurentide ice to merge with a Cordilleran Ice Sheet in western Canada and perhaps to override Cordilleran ice in some east-west through valleys, so Laurentide ice calves directly into the Pacific Ocean. The lowered snowline in the Alaskan Brooks Range and in mountains on the Chukchi Peninsula of Siberia allows the marine ice dome in the Chukchi Sea to merge with these expanded highland ice caps

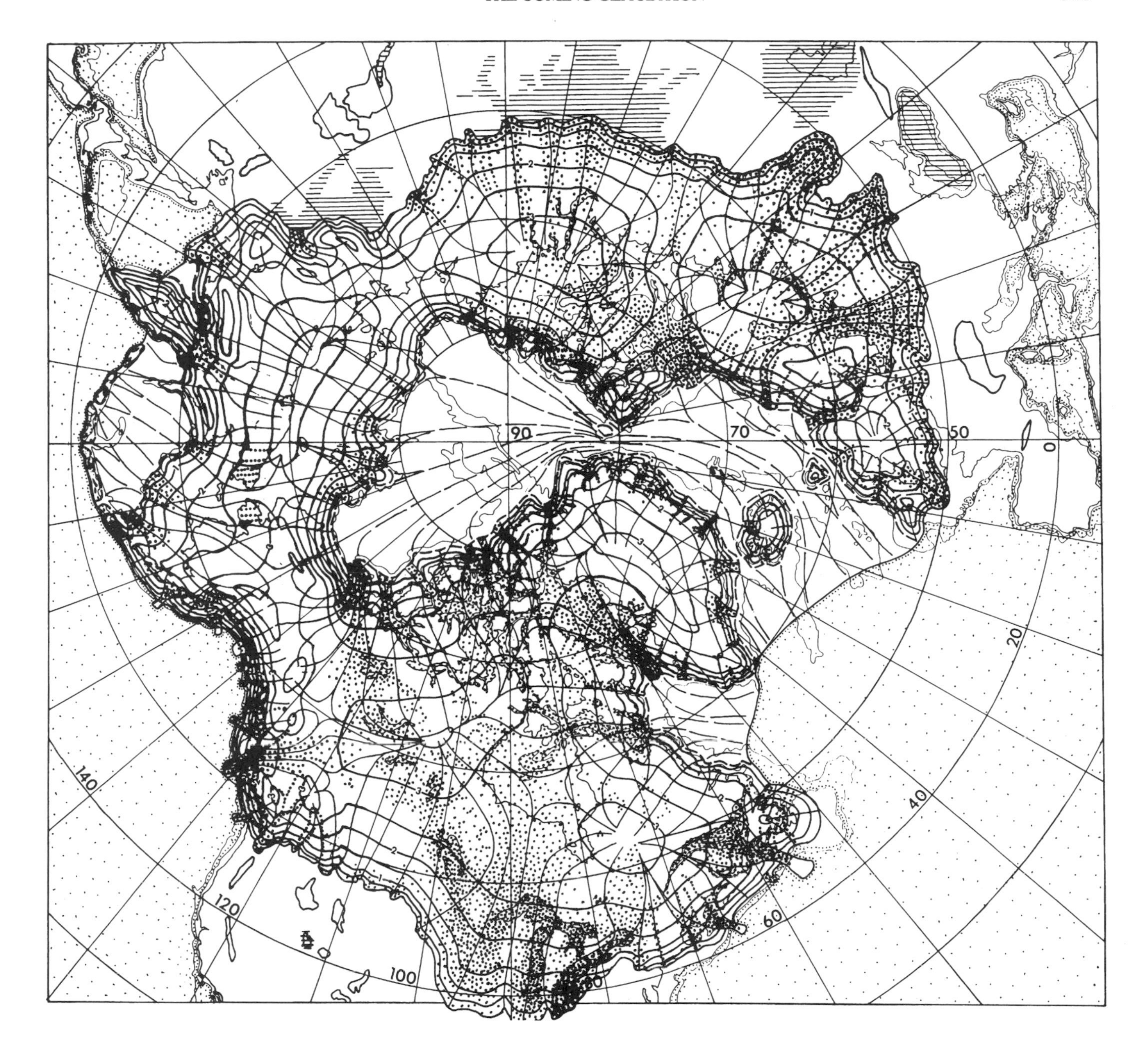

Figure 10.11: Maximum extent of the Arctic Ice Sheet during the coming glaciation. See the caption of Figure 10.8 for identification of features.

on opposite sides of Bering Strait, so a huge ice stream forms in Bering Strait and continues across the Beringian land bridge to either calve into the deep southwest Bering Sea or become a floating ice shelf that calves into the Pacific Ocean between island pinning points in the Aleutian chain. This ice gate closes the land bridge, as hypothesized for earlier Quaternary glaciations in Chapter 1.

The glacial maximum in Figure 10.11 is the most extensive stadial, after a series of stadial advances as interior ice thins and interstadial retreats as interior ice thickens. These cycles become more pronounced as glacioisostatic depression progresses, so more water is flushed through the southern water gates in Figure 1.14 when ever-deeper proglacial lakes are overrun by ice during successive stadials. Ice volume increases more with each stadial and decreases less with each interstadial because glacioisostatic depression increases and rebound decreases, so a given ice thickness coexists with a lower ice elevation, allowing the heavy precipitation from convective storm systems to extend over more of the ice sheet. Along lacustrine ice margins, ice streams advance across the soft beds of ever-deeper proglacial lakes, so landward ice margins increasingly acquire the marine

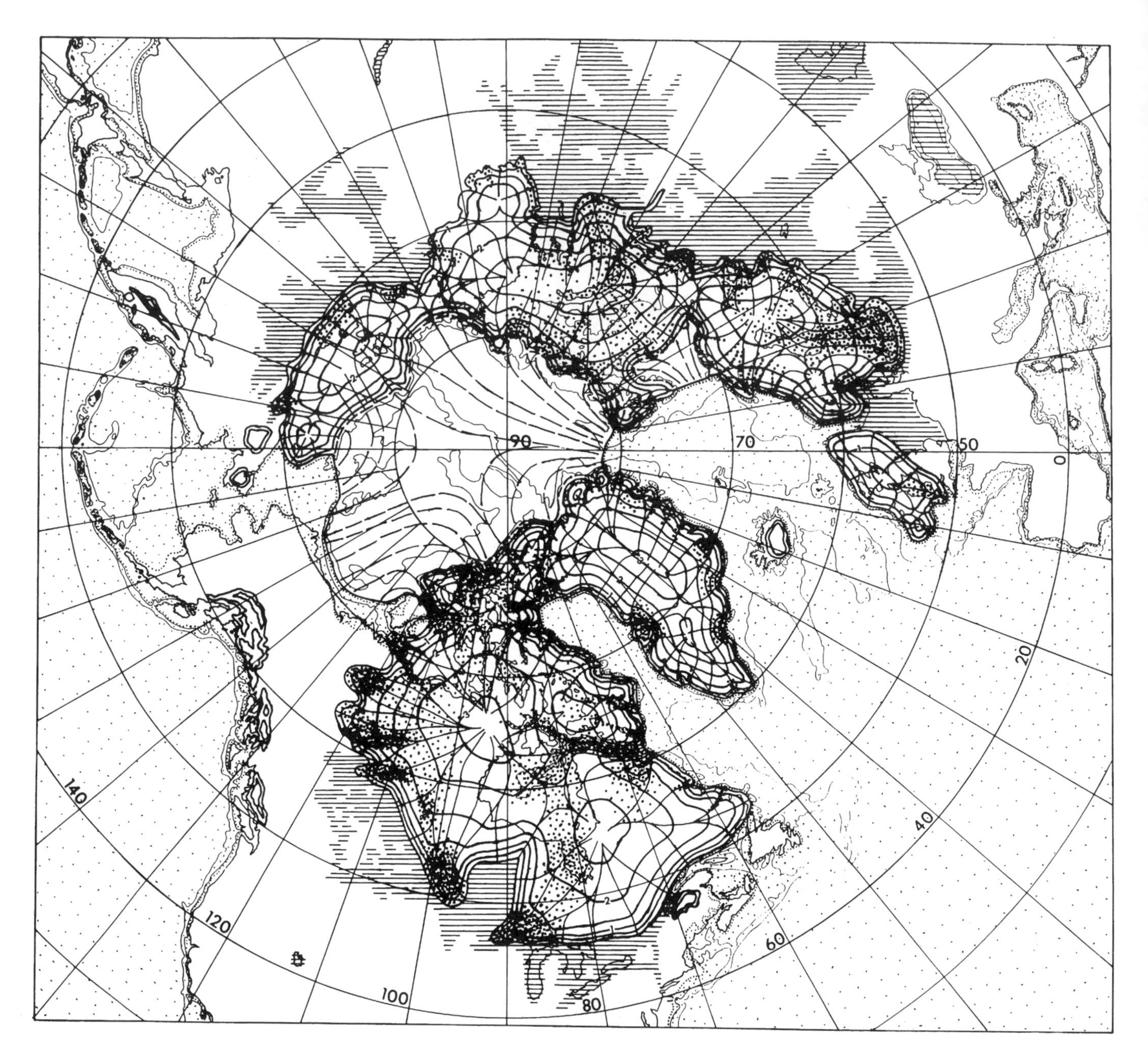

Figure 10.12: Retreating ice margins during interstadials of the coming glaciation. See the caption of Figure 10.8 for identification of features.

character of seaward ice margins. Advance and retreat of ice streams along both margins are then controlled by the irreversible grounding-line dynamics incorporated into Equation (7.50) and the ice-calving dynamics incorporated into Equation (8.125). This results in termination of the glaciation cycle when grounding-line retreat rates and ice calving rates dominate basal freezing rates after a stadial, so the stadial is followed by an interglacial instead of an interstadial.

Interstadial Retreats of a Mature Ice Sheet

Figure 10.12 depicts retreat of an Arctic Ice Sheet from its unstable glacial stadials, to its more stable interstadial core areas over the Canadian Shield, the Scandinavian Shield, and the Kara Sea. In these core areas, exposed crystalline bedrock in the shields provides substantial basal traction, and islands or highlands surrounding the Kara Sea provide substantial buttressing along grounded ice margins, even when the bed is thawed. Smaller core areas exist over island archipelagoes, notably over the Queen Elizabeth Islands in Arctic Canada, over Spitsbergen and Franz Josef Land in the Barents Sea, and over the New Siberian Islands in the East Siberian Sea. The bed would probably remain frozen over these islands and would also freeze in interisland channels after the life cycles of ice streams occupying these channels came to an end when Equation (10.64) was satisfied.

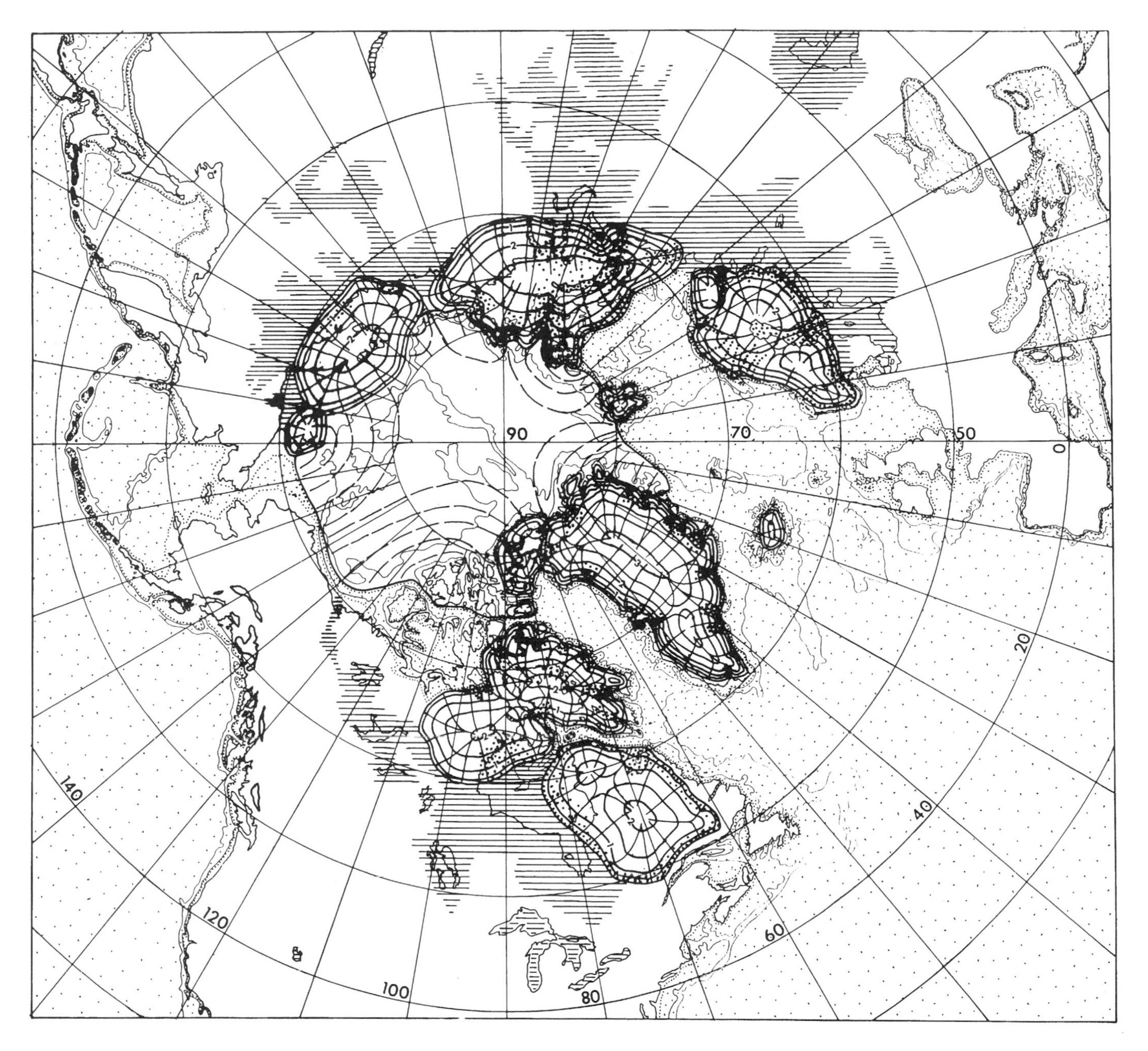

Figure 10.13: Collapsing ice domes during termination of the coming glaciation. See the caption of Figure 10.8 for identification of features.

Retreat opens the central water gates shown in Figures 1.15 and 10.12, allowing ice-dammed lakes at the edge of the Canadian Shield to discharge rapidly, first along the Hudson River valley and later through the Saint Lawrence Estuary, and discharge of the Baltic Ice Lake, first through the English Channel and later through the Norwegian Trough. If glacio-isostatic depression is insufficient to impart the irreversible grounding-line dynamics of buoyant marine ice streams to these lacustrine ice margins, the greater basal traction and perimeter buttressing in these core areas will reverse gravita-tional collapse and initiate thickening of these ice domes. The lowered elevation of these domes will allow greater precipitation from convective storm systems, so thickening can be rapid. Thickening stops when ice domes are again down-drawn by ice streams that renew their life cycles when the bed thaws in interisland channels and along the valleys of ice-dammed rivers.

Termination of the Coming Glaciation

Termination of the glaciation cycle opens the northern water gates shown in Figures 1.16 and 10.13. This occurs because ongoing, glacioisostatic depression beneath the core areas of the Arctic Ice Sheet will impose the irreversible grounding-

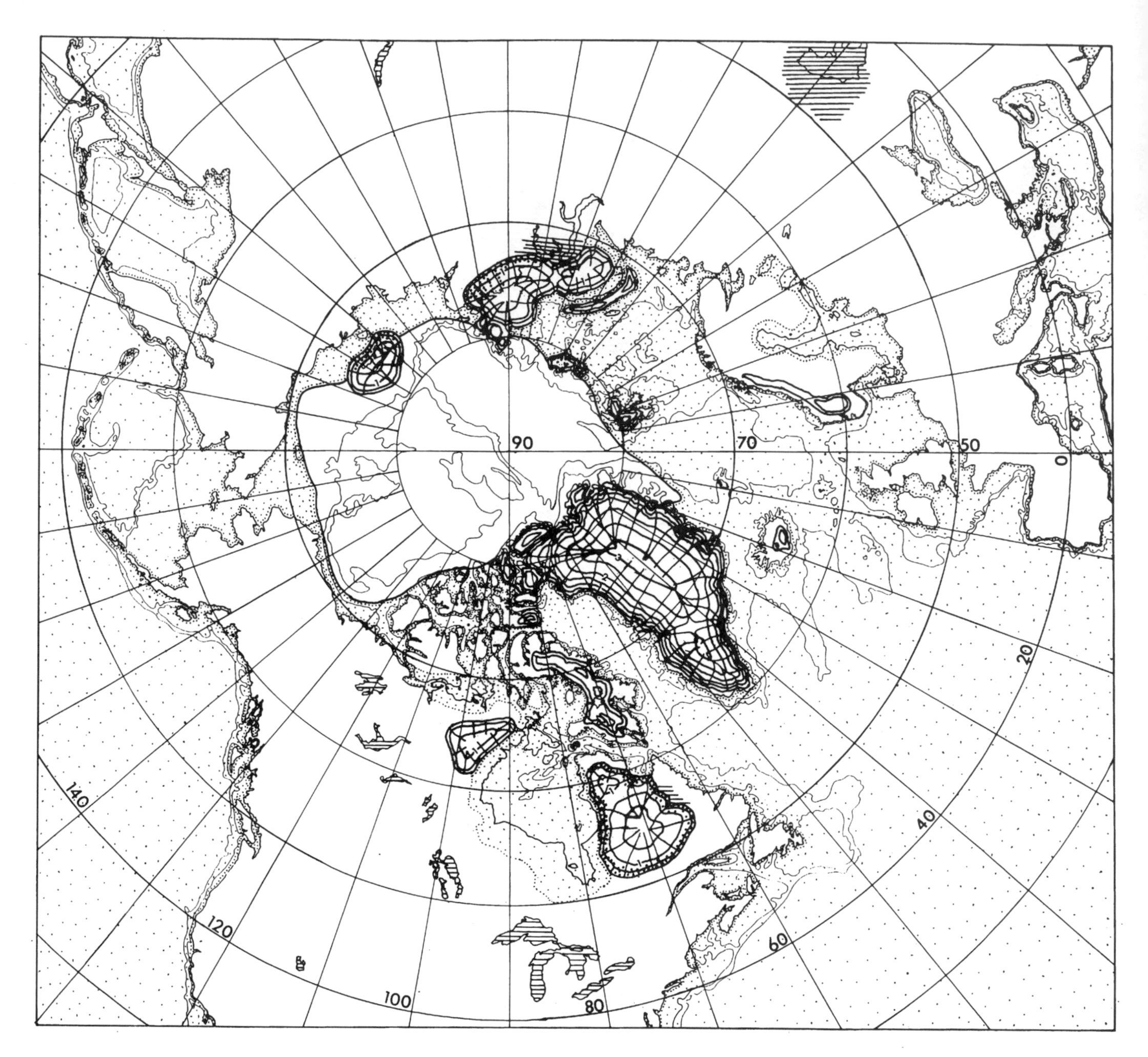

Figure 10.14: Residual ice domes after termination of the coming glaciation. See the caption of Figure 10.8 for identification of features.

line retreat rates and rapid ice-calving rates of nearly buoyant ice, given by Equations (7.50) and (8.125), respectively, along both the marine and the lacustrine margins of the interstadial ice domes. The final stadial advance will then be followed by a retreat that does not reverse when the interstadial ice domes remain, but continues until retreating ice-stream grounding lines completely collapse the saddles between these domes. The result is termination of the glaciation cycle. Ice domes of the Arctic Ice Sheet become separate ice sheets, and the interconnecting ice shelf thins to become perennial sea ice, as shown in Figures 10.13 and 10.14.

Proglacial lakes dammed by Laurentide ice discharge initially down the Mackenzie River,and discharge through Hudson Strait occurs when a saddle between the Quebec and Keewatin domes on the Laurentide ice divide is downdrawn to sea level over Hudson Bay by ice streams in Hudson Strait and James Bay, on opposite flanks of the ice divide. Discharge into the Barents Sea occurs when a saddle between the Kara Sea and Scandinavian domes on the Eurasian ice divide is downdrawn to sea level by ice streams in Bear Island Trough and the White Sea, on opposite flanks of the ice divide. Discharge into the Laptev Sea occurs when a saddle between the Kara Sea dome and the dome over the

New Siberian Islands on the Eurasian ice divide is downdrawn to sea level by ice streams on opposite flanks of the ice divide. Discharge into the Chukchi Sea oc-curs when the Wrangel Island ice dome shrinks enough to open the water gate in Long Strait. Discharge into the East Siberian Sea occurs when an ice saddle between ice domes over the New Siberian Islands and Wrangel Island is down-drawn to sea level by flanking ice streams. The remaining ice domes waste away by calving around their perimeters and melting over their surfaces. The retreat rate is given by Equation (8.114) for calving and by Equations (10.91) through (10.93) for melting. In the anthropomorphic analogy, hardening of the ice-stream "arteries" stops the stadial-interstadial-stadial "heartbeat" so that the ice sheet expires and decays.

Residual Ice Sheets

Figure 10.14 shows residual ice sheets after an ice shelf floating in the Arctic Ocean is transformed into perennial sea ice. In the Weertman (1961) analysis of ice-sheet stability, only the Greenland Ice Sheet is large enough to remain in stable equilibrium. All of the smaller ice sheets are in unstable equilibrium, and they will shrink irreversibly until they vanish. Remnant ice caps may remain on some Arctic islands and highlands, as is the case today. This is the interglacial condition. Note the similarity between Figure 10.14 and Figure 10.8 for the onset of a glaciation cycle. The crescendo of ice has come full circle, ending where it began, with an interlude that anticipates the next performance of thc Quaternary symphony of ice sheets.

Climatic changes that accompany volume changes in Northern Hemisphere ice sheets depicted in Figures 10.8 through 10.14 may cause changes in the Southern Hemisphere that are simultaneous within the ± 300 a accuracy of radiocarbon dating. Broecker (1994) asked whether this was possible for the last glaciation. The feed-back mechanisms allowing such climatic changes for typical interstadial-stadial-interstadial transitions are as follows. For the interstadial-to-stadial transition, ice-stream velocities calculated from Equation (6.24), when P_W / P_I goes from unity to zero during the life cycle of ice streams according to Equation (6.61), allow the interstadial lacustrine ice margins in Figure 10.9 to advance to the stadial terrestrial ice margins in Figure 10.10 or 10.11 in less than 300 years. A rise in sea level in 300 years occurs when proglacial lakes are flushed by terrestrial ice streams and icebergs flood into the North Atlantic from some marine ice streams. This rise is enough to initiate life cycles in other marine ice streams, causing grounding-line retreat according to Equation (7.50) and a further rise in sea level. This affects ice sheets in both hemispheres, and the cap of cold surface water should suppress formation of North Atlantic Deep Water and perhaps alter formation of Antarctic Bottom Water, such that an interstadial-to-stadial transition occurs within 300 years in both hemispheres. The stadial-to-interstadial transition depicted by retreat of Northern Hemisphere ice-sheet margins from the positions in Figures 10.10 or 10.11 to those in Figures 10.9 or 10.12 may also take place within 300 years, according to retreat rates calculated from Equations (8.77), (8.114), and (8.125) for ablation by calving and from Equation (10.21) for ablation by melting. The interstadial in Figure 10.9 is stable and leads to another stadial. The interstadial in Figure 10.12 is unstable and can lead to termination of the glaciation cycle. Theoretical ice stream velocities and iceberg calving rates calculated from these equations can duplicate the fastest known speeds, as observed on Jakobshavns Isbrae in Greenland.

Periodicities of Quaternary Glaciation Cycles

The ultimate goal of climate research is to reconstruct the past in order to preconstruct the future. This incudes the comings and goings of ice sheets. Although the coming glaciation presented in this chapter is fanciful, it is rooted in past Quaternary glaciation cycles. The best means for understanding the coming glaciation, however, is understanding long records of climatic change that are contained in ice cores through present-day ice sheets. The climate record of the last glaciation that was obtained from the GISP 2 ice core through the summit of the Greenland Ice Sheet, as interpreted by Mayewski and others (1993, 1994, 1997) and by Meeker and others (1997), is particularly instructive. They derived changes in Earth's climatic regime from a glaciochemical series of seasalts and terrestrial dusts that accumulated over Greenland with the surface precipitation, and that had diagnostic sources and timings linked to changing patterns of atmospheric circulation. This time series represented the average size and intensity of atmospheric circulation, the Polar Circulation Index (PCI) that mainly responds to changes in Earth's axial tilt, and the Mid-Low Latitude Circulation Index (MLCI) that mainly responds to changes in Earth's axial precession. Figure 10.15 shows the PCI over the last 110,000 years, and how 95 percent of its variability can be explained by band-pass components at power-spectral maxima having eight periodicities ranging from over 70,000 years to 510 years. The remaining 5 percent includes younger, higher resolution records that reveal decadal and seasonal periodicities.

Most of the periodicities in Figure 10.15 are strongly influenced by the presence and dynamics of ice sheets. The longer periodicities are climatic responses to Milankovitch insolation variations. Periodicities exceeding 70,000 years are linked to precessions in the eccentricity of Earth's elliptical orbit. Periodicities of 38,500 years are linked to precessions in the tilt of Earth's rotation axis. Periodicities of 22,500 + 11,100 years are linked to precession of the equinoxes caused by wobble of Earth's axis, for which insolation is out–out–phase across the equator. The climatic impact of these long-term gradual insolation variations is amplified by the presence of Quaternary ice sheets. They thicken and advance slowly in response to a positive surface mass balance, and thin and retreat slowly in response to a negative surface mass balance, with corresponding changes

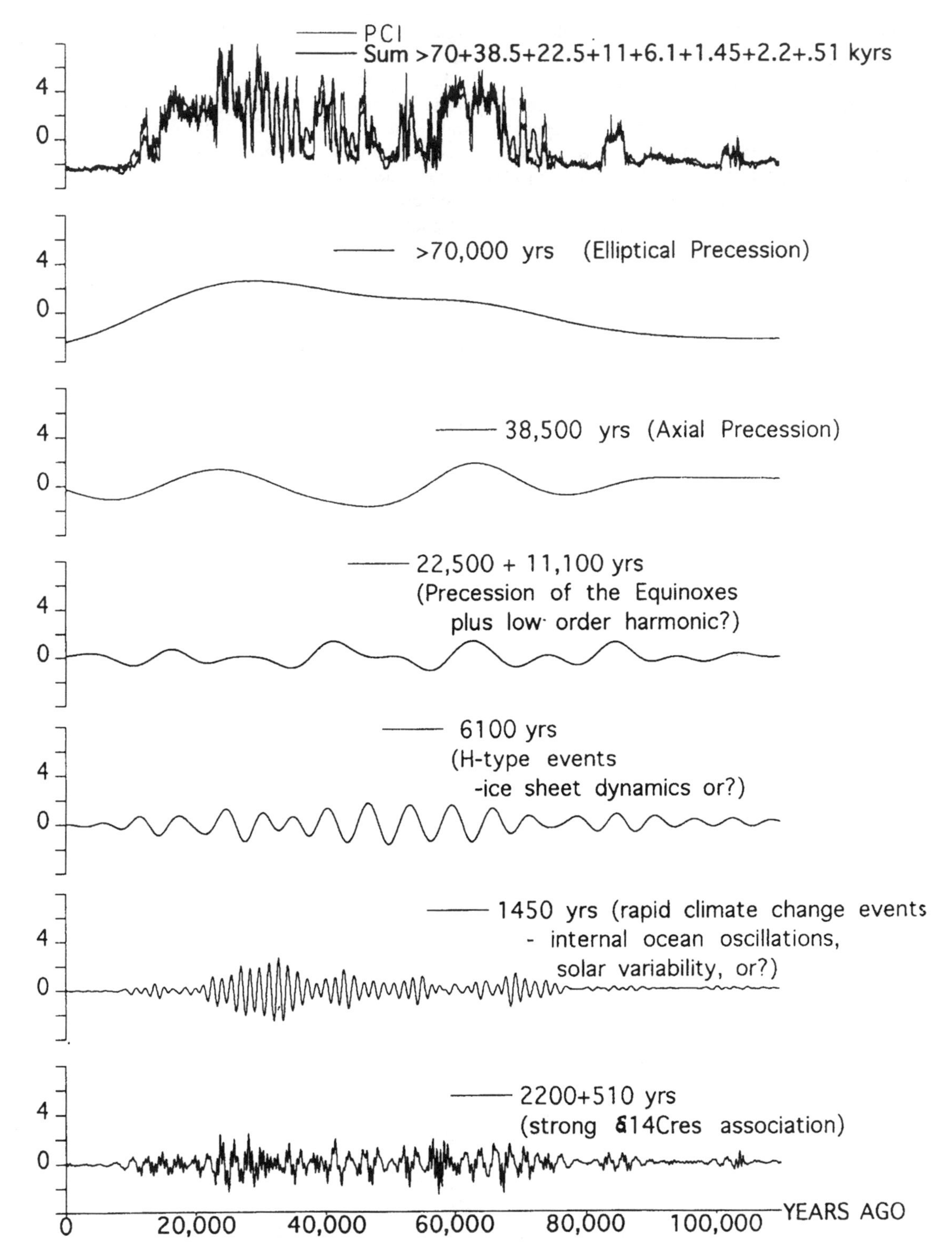

Figure 10.15: Periodicities of the last glaciation cycle from the glaciochemical series obtained from the Greenland GISP 2 corehole. The Polar Circulation Index, PCI, is expressed in normalized units at 50–year sampling intervals (top), consist of band-pass components, bpc, having preriodicities of > 70,000, 38,500, 22,500, 11,100, 6100, 2200, 1450, and 510 years. This figure was provided by Paul A. Mayewski and the Glacier Research Group in the Institute for the Study of Earth, Oceans, and Space at the University of New Hampshire. See Mayewski and others (1997) for a detailed analysis.

in the size and intensity of atmospheric circulation, as recorded by the Polar Circulation Index.

In contrast to the long-term passive response of ice sheets to gradual external insolation variations that cause gradual changes of accumulation and ablation rates in the surface mass balance, shorter periodicities in Figure 10.15 seem to be related to the internal dynamics of ice sheets, notably the basal mass balance that allows abrupt changes. Due to the phase transformations between ice and water that accompany melting and freezing rates in the basal mass balance, ice sheets lower and advance rapidly when their beds thaw, and thicken and retreat rapidly when their beds refreeze, with corresponding changes in the Polar Circulation Index. Periodicities of 6100 years are related to iceberg outbursts from the Laurentide Ice Sheet that produce Heinrich events in the climate record. Periodicities of 1450 and 2200 + 510 years may be related to iceberg outbursts from other Quaternary ice sheets, or to stadial-interstadial flushings of proglacial lakes by these ice sheets. The shorter periodicities include Dansgaard-Oeschger events in the climate record.

Gradual changes in the surface mass balance and temperature can cause abrupt changes in the basal mass balance of ice sheets. Thermal waves that propagate from the surface to the bed, even though greatly attenuated with increasing depth, need only supply the final increment of latent heat needed to thaw a frozen bed of ice-cemented sediments or till, thereby converting a rigid bed to mush, and then to remove the final increment of latent heat needed to refreeze the mush (MacAyeal, 1992b). These abrupt changes at the bed cause rapid advance and retreat of ice sheets, thereby producing abrupt changes in the climate record.

Mayewski and others (1997) favor changes in solar energy output to account for periodicities of 2200 years, and perhaps of 1450+510 years as well, with periodicities of 1450 years resonating with internal oscillations of the ocean-climate system to magnify the climatic impact of solar variability. Such solar-based external forcing also impacts on ice sheets, creating thermal waves that reach the bed with lags and attenuations that depend on the size and number of ice sheets. Therefore, climatic instabilities caused by ice-sheet instabilities will tend to cluster around maxima in Quaternary glaciation cycles, when ice sheets are most numerous and have the greatest range in sizes. In Figure 10.15, periodicities of 1450 years and of 2200+510 years to a lesser degree show this tendency to cluster around the last glacial maximum.

Meeker and others (1997) investigated periodicities on the decadal and centennial scales, which reveal two modes of atmospheric circulation, a warm interglacial mode similar to present-day patterns and a cold glacial mode during which today's largely zonal polar circulation extends deep into middle latitudes. Rapid transitions (200 years or less) between these modes are apparently associated with a critical volume (about 50 m in lowered sea level) and extent of Pleistocene ice sheets. Periodicities of 2200 years may reflect variations in solar energy output (Denton and Karlén, 1973). Other periodicities seem to be combination tones or low-order harmonics of the longer cycles. As the physical basis for these periodicities becomes understood, they will provide the constraints that replace fancy with fact in simulating the coming glaciation.

Taylor and others (1997) show that a major climatic change occurred stepwise within 40 years during the Younger Dryas to Holocene transition. Even Holocene ice sheets can have abrupt changes on the decadal to centennial time scales that are great enough to cause substantial changes in sea level, and to sustain these changes on a millenial time scale. Kindler and others (1997) show that interglacial ice sheets were responsible for abrupt rises in sea level of 5 m to 7 m lasting from 132.6 to 125.4 ka BP and 123.2 to 119.9 ka BP during the Eemian interglacial that preceded the Holocene interglacial. Sea level was high for both a wet climate as the Eemian began and a dry climate as the Eemian ended in the western Mediterranean Sea, indicating sea-level changes that were independent of Eemian climate changes in mid latitutudes. Hollin (1980) attributed such rapid rises to surges in sectors of interglacial ice sheets.

The restoration of interglacial ice sheets can be as rapid as their collapse, according to the Eemian sea-level data by Kindler and others (1997). Therefore, after sea level rose during substantial gravitational collapse in sectors of the interglacial Greenland and Antarctic Ice Sheets, a mechanism of enhanced precipitation over these collapsed sectors would be required to restore former ice elevations, thereby lowering sea level. Partial collapse due to surge-like life cycles of ice streams has been proposed to account for rapid rises in sea level. Enhanced precipitation of water evaporated from seas and lakes that became calving bays between life cycles and that were kept ice-free by katabatic winds coming off the ice sheets have been proposed to allow rapid restoration of ice elevations and the accompanying rapid fall in sea level.

Alternatively, sea level might fall rapidly if collapsed marine ice sheets on Arctic continental shelves could reappear as sea ice thickened and grounded, thereby damming North American and Eurasian rivers that empty into the Arctic during interglacials and interstadials. This produces a White Hole in the Arctic into which much precipitation enters but from which little precipitation escapes. Phillips and Gantz (1997) describe these conditions in the Amerasian Basin of the Arctic Ocean during Quaternary glaciations.

Figure 10.15 reveals the rhythms within Quaternary glaciation cycles that introduced the new paradigm for the ice age in Chapter 1 and now concludes Chapter 10 in anticipation of the coming glaciation. Understanding these rhythms in terms of glaciological process in past, present, and future ice sheets has been the major theme of this book. All mechanisms for abrupt changes in ice sheets that were presented here are largely untested and their timing to account for changing climate and sea level are unknown. This book is one step on a long road leading to an understanding of abrupt changes in ice sheets, sea level, and climate. It is one of the most exciting and important roads that scientists travel. Few other roads have signposts warning of more serious hazards to the social, political, and economical fabric of humanity.

REFERENCES

Abramowitz, M., and Stegun, I. A. (1965). *Handbook of mathematical functions*. New York: Dover.

Albrecht, B. A., Randall, D. A., and Nicholls, S. (1988). Observations of marine stratocumulus clouds during FIRE. *Bulletin of the American Meteorological Society*, 69(6), 618–626.

Alley, R. B. (1984). *A Non–Steady Ice–Sheet Model Incorporating Longitudinal Stresses*, No. 84. Institute of Polar Studies, The Ohio State University.

Alley, R. B. (1989a). Water–pressure coupling of sliding and bed deformation: I. Water system. *Journal of Glaciology*, 35(119), 108–118.

Alley, R. B. (1989b). Water–pressure coupling of sliding and bed deformation: II. Velocity–depth profiles. *Journal of Glaciology*, 35(119), 119–129.

Alley, R. B. (1990). Multiple steady states in ice–water–till systems. *Annals of Glaciology*, 14, 1–5.

Alley, R. B. (1991). Deforming–bed origin for southern Laurentide till sheets? *Journal of Glaciology*, 37(125), 67–76.

Alley, R. B. (1992). Flow–law hypotheses for ice–sheet modeling. *Journal of Glaciology*, 38(129), 245–256.

Alley, R. B., and Whillans, I. M. (1984). Response of the East Antarctic Ice Sheet to sea–level rise. *Journal of Geophysical Research*, 89(C4), 6487–6493.

Alley, R. B., and MacAyeal, D. R. (1994). Ice–rafted debris associated with binge/purge oscillations of the Laurentide Ice Sheet. *Paleoceanography*, 9(4), 503–511.

Alley, R. B., Blankenship, D. D., Rooney, S. T., and Bentley, C. R. (1987a). Till beneath ice stream B. 3. Till deformation: Evidence and implications. *Journal of Geophysical Research*, 92(B9), 8921–8929.

Alley, R. B. (1987b). Till beneath ice stream B. 4. A coupled ice–till flow model. *Journal of Geophysical Research*, 92(B9), 8931–8940.

Alley, R. B. (1989). Sedimentation beneath ice shelves – the view from ice stream B. *Marine Geology*, 85, 101–120.

Alley, R. B., Meese, D. A., Shuman, C. A., Gow, A. J., Taylor, K. C., Grootes, P. M., White, J. W. C., Ram, M., Waddington, E. D., Mayewski, P. A., and Zielinkski, G. A. (1993). Abrupt increase in Greenland snow accumulation at the end of the Younger Dryas event. *Nature*, 362(Letter), 527–529.

Andersen, B. G. (1981). Late Weichselian ice sheets in Eurasia and Greenland. In G. H. Denton and T. J. Hughes (Eds.), *The Last Great Ice Sheets* (pp. 1–65). New York: Wiley–Interscience.

Andersen, B. G., and Mangerud, J. (1989). The last interglacial–glacial cycle in Fennoscandia. *Quaternary International*, 3–4, 21–29.

Andersen, B. G., and Borns, H. W., Jr. (1994). *The Ice Age World*. Oslo: Scandinavian University Press. 208 pages.

Anderson, J. B., Kurtz, D. D., and Weaver, F. M. (1979). Sedimentation on the Antarctic continental slope. *Society of Economic Paleontologists and Mineralogists*, Special Publication 27, 265–283.

Anderson, J. B., Kurtz, D. D., Domack, E. W., and Balshaw, K. M. (1980). Glacial and glacial marine sediments of the Antarctic continental shelf. *Journal of Geology*, 88, 399–414.

Anderson, J. B., and Thomas, M. A. (1991). Marine ice–sheet decoupling as a mechanism for rapid, episodic sea–level change: The record of such events and their influence on sedimentation. In K. T. Biddle and W. Schlager (Eds.), *The Record of Sea–Level Fluctuations*, (Vol. 70, pp. 87–104). New York: Elsevier Science.

Anderson, J. B., Shipp, S. S., Bartek, L. R., and Reid, D. E. (1992). Evidence for grounded ice sheet on the Ross Sea continental shelf during the late Pleistocene and preliminary paleo–drainage reconstruction. *Antarctic Research Series*, 57, 39–62.

Anderson, T. L., and Charlson, R. J. (1990). Ice–age dust and sea salt. *Nature*, 345(6274), 393.

Andrews, J. T. (1963). End moraines and late–glacial chronology in the northern Nain–Okak section of the Labrador coast. *Geografiska Annaler*, 45, 645–665.

Andrews, J. T. 1970). *A geomorphological study of post–glacial uplift with particular reference to Arctic Canada*. London: Institute of British Geographers.

Andrews, J. T., and Tedesco, K. (1992). Detrital carbonate–rich sediments, northwestern Labrador Sea: Implications for ice–sheet dynamics and iceberg rafting (Heinrich) events in the North Atlantic. *Geology*, 20, 1087–1090.

Andrews, J. T., Shilts, W. W., and Miller, G. H. (1983). Multiple deglaciation of the Hudson Bay lowlands since deposition of the Missinaibi (last interglacial?) Formation. *Quaternary Research*, 19(1), 18–37.

Andrews, J. T., Evans, L. W., Williams, K. M., Briggs, W. M., Jull, A. J. T., Erlenkeuser, H., and Hardy, I. (1990). Cryosphere/ocean interactions at the margin of the Laurentide Ice Sheet during the Younger Dryas chron: S.E. Baffin shelf, Northwest Territories. *Paleoceanography*, 5(6), 921–935.

Andrews, J. T., Tedesco, K., and Jennings, A. E. (1993). Heinrich events: Chronology and processes, east–central Laurentide Ice Sheet and northwest Labrador Sea. In W. R. Peltier (Ed.), *Ice in the climate system*, (pp. 167–186). Berlin: Springer.

Andrews, J. T., Erlenkeuser, H., Tedesco, K., Aksu, A., and Jull, A. J. T. (1994). Late (Stage 2 and 3) meltwater and Heinrich Events, northwest Labrador Sea. *Quaternary Research*, 41,

26–34.

Anonymous (1994). Reconstructing the last glacial and deglacial ice sheets. *EOS*, 74(36):82–84.

Armstrong, J. E., Crandell, D. R., Easterbrook, D. J., and Noble, J. B. (1965). Late Pleistocene stratigraphy and chronology in southwestern British Columbia and northwestern Washington. *Geological Society of America Bulletin*, 76, 321–330.

Bader, H. (1961). The Greenland Ice Sheet. In *Army Cold Regions Science and Engineering Laboratories Monograph* (Vol. 1–B2, pp. 18). Hanover, New Hampshire: Army Cold Regions Science and Engineering Laboratories.

Baer, D. (1993). Team traces four trails from Asia: DNA suggests divisions in 'First Wave' Americans. *Mammoth Trumpet*, 8(3), 1, 4–5.

Bahn, P. G. (1993). 50,000–year–old Americans of Pedra Furada. *Nature*, 362, 114–115.

Balson, P. S., and Jeffrey, D. H. (1991). The glacial sequence of the southern North Sea. In J. Ehlers, P. L. Gibbard, and J. Rose (Eds.), *Glacial Deposits in Great Britain and Ireland* (pp. 245–253). Rotterdam: A.A. Balkema.

Bandy, O. L. (1973). Paleontology. *Antarctic Journal of the United States,* VIII(3), 86–92.

Bard, E., Hamelin, B., Fairbanks, R. G., and Zindler, A. (1990). Calibration of the ^{14}C timescale over the past 30,000 years using mass spectrometric U–Th ages from Barbados corals. *Nature*, 345, 405–410.

Barnhardt, W. A., W. R. Gehrels, D. F. Belknap, J. T. Kelley. (1995). Late Quaternary relative sea–level change in the western Gulf of Maine: Evidence for a migrating glacial forebulge. *Geology*, 23(4): 317–320.

Barry, R. G., and Chorley, R. J. (1971). Atmosphere, Weather and Climate. London: Metheun.

Bauer, A., Fontanel, A., and Grau, G. (1967). The application of optical filtering in coherent light to the study of aerial photographs of Greenland glaciers. *Journal of Glaciology*, 6, 781–793.

Belknap, D. F. and Kraft, J. C. (1977). Holocene relative sea–level changes and coastal stratigraphic units on the northwest flank of the Baltimore Canyon Trough geosyncline. *Journal of Sedimentary Petrology*, 47, 610–629.

Bender, M. L., Fairbanks, R. G., Taylor, F. W., Matthews, R. K., Goddard, J. G., and Broecker, W. S. (1979). Uranium–series dating of the Pleistocene reef tracts of Barbados, West Indies. *Geological Society of America Bulletin*, Part I, 90, 577–594.

Bentley, C. R. (1987). Antarctic ice streams: A review. *Journal of Geophysical Research*, 92(B9), 8843–8858.

Berger, A. (1978). Long term variations of daily insolation and Quaternary climatic changes. *Journal of Atmospheric Science*, 35, 2362–2367.

Berger, A., Fichefet, T., Gallée, H., Tricot, C., and van Ypersele, J. P. (1992). Entering the glaciation with a 2–D coupled climate model. *Quaternary Science Reviews*, 11, 481–493.

Berger, G. W., and Nielsen, E. (1991). Evidence from thermo-luminescence dating for middle Wisconsinan deglaciation in the Hudson Bay Lowland of Manitoba. *Canadian Journal of Earth Sciences*, 28, 240–249.

Bindschadler, R. A. (1984). Jakobshavns Glacier Drainage Basin: A Balance Assessment. *Journal of Geophysical Research*, 89(2066–2072).

Bindschadler, R. A. (Ed.). (1995). *West Antarctic Ice Sheet Initiative Science and Implementation Plan*. Arlington: United States National Science Foundation. 75 pages.

Bindschadler, R. A., and Gore, R. (1982). A time–dependent ice sheet model: Preliminary results. *Journal of Geophysical Research*, 87(C12), 9675–9685.

Bindschadler, R. A., Stephenson, S. N., MacAyeal, D. R., and Shabtaie, S. (1987). Ice dynamics at the mouth of ice stream B, Antarctica. *Journal of Geophysical Research*, 92(B9), 8885–8894.

Binyuan, L., and Jijun, L. (1991). *Quaternary glacial distribution map of Qinghai–Kizang (Tibet) Plateau.* Beijing: Science Press.

Birchfield, G. E., and Broecker, W. S. (1990). A salt oscillator in the glacial Atlantic. *Paleoceanography*, 5(6), 835–843.

Bischof, J. F., and Clark, D. L. (1992). Rhythmical dropstone discharge into the western Arctic Ocean – the direct link between the terrestrial and the deep sea glacial record. In A. F. Spilhaus Jr., S. Cole, and M. C. White (Ed.), *AGU 1992 Fall Meeting Abstracts*, (pp. 160). San Francisco: American Geo-physical Union.

Blake, W., Jr. (1970). Studies of glacial history in Arctic Canada, I: Pumice, radiocarbon dates, and differential postglacial uplift in the eastern Queen Elizabeth Islands. *Canadian Journal of Earth Sciences*, 6, 634–664.

Blake, W., Jr. (1972). Climatic implications of radiocarbon–dated driftwood in the Queen Elizabeth Islands, Arctic Canada. In Y. Vasaro and others (Eds.), *Climatic Changes in Arctic Areas During the Last Ten–Thousand Years* (pp. 77–104). Acta Universitatis Ouluensis.

Blake, W., Jr. (1975). Radiocarbon age determination and post–glacial emergence at Cape Storm, southern Ellesmere Island, Arctic Canada. *Geografiska Annaler*, 57(A), 1–71.

Blake, W., Jr. (1977). Glacial sculpture along the east–central coast of Ellesmere Island, Arctic Archipelago. *Geological Survey of Canada*, 77(C), 107–115.

Blake, W., Jr. (1992). Holocene emergence at Cape Herschel, east–central Ellesmere Island, Arctic Canada: implications for ice sheet configuration. *Canadian Journal of Earth Sciences*, 29(9), 1958–1980.

Blanchon, P., and Shaw, J. (1995). Reef drowning during the last deglaciation: Evidence for catastrophic sea–level rise and ice–sheet collapse. *Geology*, 23(1), 4–8.

Blankenship, D. D., Bentley, C. R., Rooney, S. T., and Alley, R. B. (1987). Till beneath Ice Stream B: I. Properties derived from seismic travel times. *Journal of Geophysical Research*, 92, 8903–8911.

Blankenship, D. D., Bell, R. E., Hodge, S. M., Brozena, J. M., Behrendt, J. C., and Finn, C. A. (1993). Active volcanism beneath the West Antarctic ice sheet and implications for ice–sheet stability. *Nature*, 361, 526–529.

Bloom, A. L. (1967). Pleistocene shorelines: a new test of isostasy. *Geological Society of America Bulletin*, 78, 1477–1494.

Bloom, A. L., and Yonekura, N. (1990). Graphic analysis of dislocated Quaternary shorelines. In Geophysics Study Com-mittee (Eds.), *Sea–level change,* (pp. 104–115). Washington, D.C.: National Academy Press.

Bond, G., and Lotti, R. (1995). Iceberg discharges into the North Atlantic on millennial time scales during the last glaciation. *Science*, 267, 1005–1010.

Bond, G., Heinrich, H., Broecker, W., Labeyrie, L., McManus, J., Andrews, J., Huon, S., Jantschik, R., Clasen, S., Simet, C., Tedesco, K., Klas, M., Bonani, G., and Ivy, S. (1992). Evidence for massive discharges of icebergs into the North Atlantic ocean during the last glacial period. *Nature*, 360, 245–249.

Bond, G., Broecker, W., Johnsen, S., McManus, J., Labeyrie, L., Jouzel, J., and Bonani, G. (1993). Correlations between climate records from North Atlantic sediments and Greenland ice. *Nature*, 365, 143–147.

Bonnichsen, R., and Turnmire, K. L. (Eds.). (1991). *Clovis: Origins and adaptations*. Corvallis: Center for the Study of the First Americans.

Booth, D. B. (1987). Timing and processes of deglaciation along the southern margin of the Cordilleran ice sheet. In W. F. Ruddiman and H. E. Wright, Jr. (Eds.), *North America and adjacent oceans during the last deglaciation*, (Vol. K–3, Chapter 4). Boulder: The Geological Society of America.

Borns, H. W., and Warren, W. P. (1990). New evidence for ice flow directions over Connemara, Ireland. *Connemara Discussion Meeting Abstracts,* 24–25 September 1990. Department of Geology, University College, Galway Ireland.

Bouchard, M. A., and Salonen, V.–P. (1990). Boulder transport in shield areas. In R. Kujansuu and M. Saarnisto (Eds.), *Glacial Indicator Tracing* (pp. 87–107). Rotterdam/Brookfield: A.A. Ballkema.

Boulton, G. S. (1979). Glacial history of the Spitsbergen archipelago and the probelm of a Barents Shelf ice sheet. *Boreas*, 8, 31–57.

Boulton, G. S. (1993). Two cores are better than one. *Nature*, 366, 507–508.

Boulton, G. S. (1996). Theory of glacial erosion, transport, and deposition as a consequence of subglacial sediment deformation. *Journal of Glaciology*, 42, 43–62.

Boulton, G. S., and Jones, A. S. (1979). Stability of temperate ice caps and ice sheets resting on beds of deformable sediment. *Journal of Glaciology*, 24, 29–43.

Boulton, G. S., and Hindmarsh, R. C. A. (1987). Sediment deformation beneath glaciers: Rheology and geological consequences. *Journal of Geophysical Research*, 92, 9059–9082.

Boulton, G. S., and Clark, C. D. (1990). The Laurentide ice sheet through the last glacial cycle: the topology of drift lineations as a key to the dynamic behaviour of former ice sheets. *Transactions of the Royal Society of Edinburgh: Earth Sciences*, 81, 327–347.

Boulton, G. S., Jones, A. S., Clayton, K. M., and Kenning, M. J. (1977). A British ice–sheet model and patterns of glacial erosion and deposition in Britain. In F. W. Shotton (Eds.), *British Quaternary Studies, Recent Advances* (pp. 231–246). Oxford: Clarendon Press.

Boulton, G. S., Smith, G. D., Jones, A. S., and Newsome, J. (1985). Glacial geology and glaciology of mid–latitude ice sheets. *Journal of the Geological Society of London*, 124, 447–474.

Boyer, S. J., and Pheasant, D. R. (1974). Delineation of weathering zones in the fiord areas of eastern Baffin Island, Canada. *Geological Society of America Bulletin*, 85, 805–810.

Braithwaite, R. J., and Olesen, O. B. (1989). Calculation of glacier ablation from air temperature, West Greenland. In J. Oerlemans (Ed.)^(Eds.), *Glacier Fluctuations and Climate Change*, (pp. 219–233). Dordrecht: Kluwer Academic Press.

Brecher, H. H. (1982). Photographic determination of surface velocities and elevations on Byrd Glacier. *Antarctic Journal of the United .States*, 17(5), 79–81.

Brecher, H. H. (1986). Surface velocity determination on large polar glaciers by aerial photogrammetry. *Annals of Glaciology*, 8, 22–26.

Brigham–Grette, J., Benson, S., Hopkins, D. M., Heiser, P. A., Ivanov, V. F., and Basilyan, A. (1992). Middle and Late Pleistocene Russian glacial ice extent in the Bering Strait region: Results of recent field work. In *Abstracts with Programs – Annual Meeting*, Abstracts. Cincinnati: Geological Society of America, A346.

Broecker, W. S. (1975). Floating ice cap on the Arctic Ocean. *Science*, 188, 1116–1118.

Broecker, W. S. (1994). Massive iceberg discharge as triggers for global climate change. *Nature*, 372, 421–424.

Broecker, W. S., and van Donk, J. (1970). Insolation changes, ice volumes, and the ^{18}O record in deep–sea cores. *Reviews of Geophysics and Space Physics*, 8(1), 169–198.

Broecker, W. S., and Denton, G. H. (1989). The role of ocean–atmosphere reorganizations in glacial cycles. *Geochimica et Cosmochimica Acta*, 53(10), 2465–2501.

Broecker, W. S., Bond, G., Klas, M., G., B., and Wolfli, W. (1990). A salt oscillator in the Glacial Atlantic? 1. The Concept. *Paleoceanography*, 5, 469–477.

Broecker, W. S., Bond, G., Mieczyslawa, K., Clark, E., and McManus, J. (1992). Origin of the northern Atlantic's Heinrich events. *Climate Dynamics*, 6, 265–273.

Bromwich, D. H. (1989). Satellite analysis of Antarctic katabatic wind behavior. *Bulletin of the American Meteorological Society*, 70(7), 738–749.

Bromwich, D. H., Parish, T. R., and Zorman, C. A. (1990). The confluence zone of the intense katabatic winds at Terra Nova Bay, Antarctica, as derived from airborne sastrugi surveys and mesoscale numerical modeling. *Journal of Geophysical Research*, 95(D5), 5495–5509.

Bromwich, D. H., Du, Y., and Parish, T. R. (1994). Numerical simulation of winter katabatic winds from West Antarctica crossing Siple Coast and the Ross Ice Shelf. *Monthly Weather Review*, 122, 1417–1435.

Brookes, I. A. (1982). Ice marks in Newfondland: A history of ideas. *Gèographie Physique et Quaternaire*, 36(1–2), 139–163.

Brown, C. S., Meier, M. F., and Post, A. (1982). The calving relation of Alaskan tidewater glaciers, with application to Columbia Glacier. In *United States Geological Survey Professional Paper* (pp. 13). Washington, D.C.: United States Government Printing Office.

Budd, W. F. (1969). *The Dynamics of Ice Masses* (Australian National Antarctic Research Expeditions (ANARE) Interim Reports, Series A (IV), Glaciology No. 108). Antarctic

Division, Department of Supply.

Budd, W. F., and Smith, I. N. (1981). The growth and retreat of ice sheets in response to orbital radiation changes. *International Association of Hydrologic Sciences Publication,* 131, 369–409.

Budd, W. F., and Jacka, T. H. (1989). A review of ice rheology for ice sheet modeling. *Cold Regions and Technology*, 16, 107–144.

Budd, W., Dingle, W. R. J., and Radok, U. (1966). The Byrd snow drift project: outline and basic results. In M. J. Rubin (Ed.), *Studies in Antarctic Meteorology*, (pp. 71–134). Washington, D.C.: American Geophysical Union.

Budd, W. F., Jensen, D., and Radok, U. (1970). The extent of basal melting in Antarctica. *Polarforschung*, 6, 293–306.

Budd, W. F., Jensen, D., and Radok, U. (1971). *Derived physical characteristics of the Antarctic Ice Sheet* (Australian National Antarctic Research Expeditions (ANARE) Interim Reports, Series A (IV), Glaciology No. 120). University of Melbourne, Meteorology Department.

Budd, W. F., Keage, P. L., and Blundy, N. A. (1979). Empirical studies of ice sliding. *Journal of Glaciology*, 23(89), 157–170.

Budd, W. F., Jenssen, D., and Smith, I. N. (1984). A three–dimensional time–dependent model of the West Antarctic ice sheet. *Annals of Glaciology*, 5, 29–36.

Budyko, M. (1974). *Climate and Life*. New York: Academic Press.

Carbonnell, M., and Bauer, A. (1968). Exploitation des couvertures photographiques aériennes répétées du front des glaciers vêlant dans Diske Bugt en Umanak Fjord, Juin–Juillet, 1964. *Meddelelser om Grønland*, 173(5), 1–78.

Carrara, P. (1979). Former extent of glacial ice in the Orville Coast region, Antarctic Peninsula. *Antarctic Journal of the United States*, 14, 45–46.

Carrara, P. (1981). Evidence for a former large ice sheet in the Orville coast–Ronne Ice Shelf area, Antarctica. *Journal of Glaciology*, 27, 487–491.

Cathless, L. M., III (1975). *The Viscosity of the Earth's Mantle*. Princeton: Princeton. 386 pages.

Catt, J. A. (1991). Late Devensian glacial deposits and glaciations in eastern England and the adjacent offshore region. In J. Ehlers, P. L. Gibbard, and J. Rose (Eds.), *Glacial Deposits in Great Britain and Ireland* (pp. 61–68). Rottersdam: A.A. Balkema.

Cerni, T. A., and Parish, T. R. (1984). A radiative mopdel of the stable nocturnal boundary layer with application to the polar night. *Journal of Climate and Applied Meteorology,* 23(11), 1563–1572.

Chappell, J., and Shackleton, N. J. (1986). Oxygen isotopes and sea level. *Nature*, 324, 137–140.

Charlesworth, J. K. (1957). *The Quaternary Era*. London: Edwin Arnold.

Chlachula, J. (1994). A Palaeo–American (Pre–Clovis) Settlement in Alberta. *Current Research in the Pleistocene*, 11, 21–23.

Chlachula, J., Drozdov, N. I., and Chekha, V. P. (1994). Early Palaeolithic in the Minusinsk Basin, Upper Yenisei River Region, Southern Siberia. *Current Research in the Pleistocene*, 11, 128–130.

Clark, J. A. (1980). The reconstruction of the Laurentide ice sheet of North America from sea level data: Method and preliminary results. *Journal of Geophysical Research*, 85(B8), 4307–4323.

Clark, P. U. (1992). Surface form of the southern Laurentide Ice Sheet and its implications to ice–sheet dynamics. *Geological Society of America Bulletin*, 104, 595–605.

Clark, P. U. (1994). Unstable behavior of the Laurentide Ice Sheet over deforming sediment and its implications for climate change. *Quaternary Research*, 41, 19–25.

Clark, P. U. (1995). Fast glacier flow over soft beds. *Science*, 267, 43–44.

Clark, P. U. and Walder, J. S. (1994). Subglacial drainage, eskers, and deforming beds beneath the Laurentide and Eurasian ice sheets. *Geological Society of America Bulletin*, 106, 304–314.

Clark, P. U., Liccardi, J. M., MacAyeal, D. R., Jensen, J. W., and Currans, K. (1996). Numerical reconstruction of a soft–bedded Laurentide Ice Sheet during the last glacial maximum. *Geology*, 24(8), 679–682.

Clarke, G. K. C. (1987a). Fast glacier flow: Ice streams, surging, and tidewater glaciers. *Journal of Geophysical Research*, 92(B9), 8835–8841.

Clarke, G. K. C. (1987b). Subglacial till: A physical framework for its properties and processes. *Journal of Geophysical Research*, 92(B9), 9023–9036.

Clarke, G. K. C., Nitsan, U., and Paterson, W. S. B. (1977). Strain heating and creep instability in glaciers and ice sheets. *Reviews of Geophysics and Space Physics*, 15, 235–247.

COHMAP Members (1988). Climate changes of the last 18,000 years: Observations and model simulations.*Science*, 241, 1043–1052.

Colinvaux, P. A. (1980). Vegetation of the Bering Land Bridge Revisited. *Quarterly Review of Anthropology*, 1, 2–15.

Cox, S. K., McDougal, D. S., Randall, D. A., and Schiffer, R. A. (1987). FIRE – The first ISCCP Regional Experiment. *Bulletin of the American Meteorological Society*, 68(2), 114–117.

Crary, A. P. (1960). Arctic ice island and ice shelf studies: Part 2. *Arctic*, 13, 32–50.

Cwynar, L. C., and Ritchie, J. C. (1982). Arctic Steppe–Tundra: A Yukon Perspective. *Science*, 208, 1375–1377.

Dahl, E. (1955). Biogeographic and geologic indications of unglaciated areas in Scandinavia during the glacial ages. *Geological Society of America Bulletin*, 66, 1499–1520.

Dahl, R. (1967). Post–glacial micro–weathering of bedrock surfaces in the Narvik district of Norway. *Geografiska Annaler*, 49(A), 155–166.

Dansgaard, W., Johnsen, S. J., Clausen, H. B., and Langway, C. C. (1971). A flow model and a time scale for the ice core from Camp Century, Greenland. *Journal of Glaciology*, 8, 215–223.

Dansgaard, W., Johnsen, S. J., Clausen, H. B., Dahl–Jensen, Gundestrup, N. S., Hammer, C. U., Hvidberg, C. S., Steffensen, J. P., Jouzel, J., and Bond, G. (1993). Evidence for general instability of past climate from a 250–kyr ice–core record. *Nature*, 364(6434), 218–220.

Davis, J. L., and Mitrovica, J. X. (1996). Glacial isostatic adjustment and the anomalous tide gauge record of eastern North America. *Nature*, 379, 331–333.

deBlonde, G., Peltier, W. R., and Hyde, W. T. (1992). Simulations of continental ice sheet growth over the last glacial–interglacial cycle: experiments with a one level seasonal energy balance model including seasonal ice albedo feedback. *Global and Planetary Change*, 6(1), 37–55.

deFreitas, T. A. (1990). Implications of glacial sculpture on Hans Island, between Greenland and Ellesmere Island (Nares Strait). *Journal of Glaciology*, 36(122), 129–130.

Denton, G. H. (1974). Quaternary glaciations of the White River Valley, Alaska, with a regional synthesis for the northern St. Elias Mountains, Alaska, and Yukon Territory. *Geological Society of America Bulletin*, 85, 871–892.

Denton, G. H., and Karlén, W. (1973). Holocene climatic variations – Their pattern and possible causes. *Quaternary Research*, 3, 155–205.

Denton, G. H., & Hughes, T. J. (Eds.). (1981). *The Last Great Ice Sheets*. New York: Wiley-Interscience.

Denton, G. H. and Hughes, T. (1983). Milankovitch theory of ice ages: hypothesis of ice–sheet linkages between regional insolation and global climate. *Quaternary Research*, 20, 125–144.

Denton, G. H., Bockheim, J. G., Wilson, S. L., and Schlüchter, C. (1986a). Late Cenozoic history of Rennick Glacier and Talos Dome, northern Victoria Land, Antarctic. In E. Stump (Ed.), *Geological investigations in northern Victoria Land*:, (Vol. 46, pp. 339–375): American Geophysical Union.

Denton, G. H., Andersen, B. G., and Conway, H. B. (1986b). Late Quaternary surface ice level fluctuations of the Beardmore Glacier, Antarctica. *Antarctic Journal of the United States*, 21, 90–92.

Denton, G. H., Bockheim, J. G., Wilson, S. C., and Stuiver, M. (1989). Late Wisconsin and Early Holocene Glacial History, Inner Ross Embayment, Antarctica. *Quaternary Research*, 31, 151–182.

Denton, G. H., Prentice, M. L., and Burckle, L. H. (1991). Cainozoic history of the Antarctic Ice Sheet. In R. J. Tingey (Ed.) *The Geology of Antarctica*, (pp. 365–433). Oxford: Oxford University Press.

Denton, G. H., Bockheim, J. G., Rutford, R. H., and Andersen, B. G. (1992). Glacial history of the Ellsworth Mountains, West Antarctica. In G. F. Webers, C. Craddock, and J. F. Splettstoesser (Eds.), *Geology and Paleontology of the Ellsworth Mountains, West Antarctica*, (Vol. Geological Society of America Memoir 170, pp. 403–431). Boulder.

Dethleff, D. (1992). GEOMAR's E.S.A.R.E. '92 expedition to the Laptev Sea. *The Nansen Ice Breaker(Newsletter from the Nansen Arctic Drilling Program* , Fall(3), 4.

Dickson, B. (1997). From the Labrador Sea to global change. *Nature*, 386, 649–650.

Dieter, G., E. Jr. (1961). *Mechanical Metallurgy*. New York: McGraw–Hill Book Company, Inc. 615 pages.

Dillehay, T. (1989). *Monte Verde: A Late Pleistocene Settlement in Chile*. Washington, D.C.: Smithsonian Institution.

Doake, C. S. M., and Vaughan, D. G. (1991). Rapid disintegration of the Wordie Ice Shelf in response to atmospheric warming. *Nature*, 350, 328–330.

Domack, E. W. (1982). Sedimentology of glacial and glacial marine deposits on the George V – Adelie continental shelf, East Antarctica. *Boreas*, 11, 79–97.

Domack, E. W., and Williams, C. R. (1990). Fine structure and suspended sediment transport in three Antarctic fjords. In *Antarctic Research I* (pp. 71–89). Washington, D.C.:

Dott, R. H., and Batten, R. L. (1971). *Evolution of the Earth*. New York: McGraw–Hill.

Dowdeswell, J. A., Maslin, M. A., Andres, J. T., and McCave, I. N. (1995). Iceberg production, debris rafting, and the extent and thickness of Heinrich layers (H–1, H–2) in North Atlantic sediments. *Geology*, 23(4), 301–304.

Dredge, L. A., and Thorleifson, L. H. (1987). The middle Wisconsinan history of the Laurentide Ice Sheet. *Gèographié Physique et Quaternairé*, XLI, 215–235.

Drewry, D. J. (1983). Antarctica: Glaciological and Geophysical Folio. Cambridge: Scott Polar Research Institute, University of Cambridge.

Drewry, D. J., Jordan, S. R., and Jankowski, E. (1982). Measured properties of the Antarctic Ice Sheet: surface configuration, ice thickness, volume, and bedrock characteristics. *Annals of Glaciology*, 3, 83–91.

Dupont, T. K. (1996, May). *Millenial scale climate variability in the sea–ice/ocean system: A low–order dynamical model*. Unpublished Master of Science Thesis in Quaternary Studies, University of Maine.

Duval, P. (1976). Lois du fluage transitoire on permanent de la glace polycristalline pour divers etats de contrainte. *Annals de Géophysique*, 32(4), 335–350.

Duxbury (1971). *The Earth and its Ocean*. Reading, MA: Addison–Wesley.

Dwyer, G. S., Cronin, T. M., Baker, P. A., Raymo, M. E., Buzas, J. S., and Correge, T. (1995). North Atlantic deepwater temperature change during late Pliocene and late Quaternary climatic cycles. *Science*, 263, 796–800.

Dyke, A. S., and Prest, V. K. (1987). Late Wisconsinan and Holocene history of the Laurentide Ice Sheet. *Gèographie Physique et Quaternaire*, 41, 237–263.

Echelmeyer, K., and Kamb, B. (1986). Stress–gradient coupling in glacial flow: II. Longitudinal averaging in the flow response to small perturbations in ice thickness and surface slope. *Journal of Glaciology*, 32(111).

Echelmeyer, K., and Zhongxiang, W. (1987). Direct observation of basal drift at sub–freezing temperatures. *Journal of Glaciology*, 33(113), 83–98.

Echelmeyer, K., and Harrison, W. (1990). Jakobshavns Isbrae, West Greenland: seasonal variations in velocity–or lack thereof. *Journal of Glaciology*, 36(122), 82–88.

Echelmeyer, K., Clarke, T. S., and Harrison, W. D. (1991). Surficial glaciology of Jakobshavns Isbræ, West Greenland: Part I. Surface morphology. *Journal of Glaciology*, 37(127), 368–382.

Echelmeyer, K., Harrison, W. D., Clarke, T. S., and Benson, C. (1992). Surficial glaciology of Jakobshavns Isbræ, West Greenland: Part II. Ablation, accumulation and temperature. *Journal of Glaciology*, 38(128), 169–181.

Edwards, R. L. (1995). Paleotopography of glacial–age ice sheets. *Science*, 267(Technical Comments), 536.

Eglinton, G., Bradshaw, S. A., Rosell, A., Sarnthein, M., Pflaumann, U., and Tiedemann, R. (1992). Molecular record of secular sea surface temperature changes on 100–year timescales for glacial terminations I, II, and IV. *Nature*, 356, 423–425.

Einarsson, T. (1968). *Jardfraedi–Saga bergs og lands*. Reykjavik: 335 pages.

Elias, S. A., Short, S. K., Nelson, C. H., and Birks, H. H. (1996). Life of times of the Beringian land bridge. *Nature*, 382, 60–63.

El–Sayed, S. Z. (1973). Biological oceanography. *Antarctic Journal of the United States*, VIII(3), 93–100.

Elverhøi, A. (1981). Evidence for a late Wisconsin glaciation of the Weddell Sea. *Nature*, 293, 292–293.

Elverhøi, A., and Solheim, A. (1983). The Barents Sea Ice Sheet – a sedimentological discussion. *Polar Research*, 1, 23–42.

Elverhøi, J., Nyland–Berg, M., Russwaurm, L., and Solheim, A. (1990). Late Weichselian ice recession in the Central Barents Sea. In U. Bleil and J. Thiede (Eds.), *Geological History of the Polar Oceans: Arctic versus Antarctic* (pp. 289–307). Dordrecht: Kluwer Academic Publishers.

Engelhardt, H., Humphrey, N., Kamb, B., and Fahnestock, M. (1990). Physical Conditions at the Base of a Fast Moving Antarctic Ice Stream. *Science*, 248(4951), 57–59.

England, J. H. (1976). Late Quaternary glaciation of the eastern Queen Elizabeth Islands, N.W.T., Canada: Alternative models. *Quaternary Research*, 6, 185–202.

England, J. H. (1987). Glaciation and the evolution of the Canadian high arctic landscape. *Geology*, 15, 419–424.

England, J. H. (1997). *The last glaciation of Nares Strait: Ice configuration, deglacial chronology, and regional implications*. Paper presented at the 27th Arctic Workshop, Ottawa.

England, J. H., and Bradley, R. S. (1978). Past glacial activity in the Canadian high Arctic. *Science*, 200, 265–270.

Epprecht, W. (1987). A major calving event of Jakobshavns Isbrae, West Greenland, on 9 August 1982. *Journal of Glaciology*, 33(114), 169–172.

Evans, D. A., Beukes, N. J., and Kirschvink, J. L. (1997). Low–latitude glaciation in the Palaeoproterozoic era. *Nature*, 386, 262–266.

Eyles, N., and McCabe, M. (1991). Glaciomarine deposits of the Irish Sea Basin: The role of glacio–isostatic disequilibrium. In J. Ehlers, P. L. Gibbard, and J. Rose (Eds.), *Glacial Deposits in Great Britain and Ireland* (pp. 311–331). Rotterdam: A.A. Balkema.

Fairbanks, R. G. (1989). A 17,000–year glacio–eustatic sea level record: Influence of glacial melting rates on the Younger Dryas event and deep–ocean circulation. *Nature*, 342, 637–642.

Fairbridge, R. W. (Ed.). (1967). *Encyclopedia of Atmospheric Sciences and Astrogeology*. New York: Reinhold.

Fahnestock, M., Bindschadler, R., Kwok, R., and Jezek, K. (1993). Greenland ice sheet surface properties and ice dynamics from ERS–1 SAR imagery. *Science*, 262, 1530–1534.

Fastook, J. L. (1984). West Antarctica, the sea–level controlled marine instability; Past and Future. In J. E. Hansen and T. Takahashi (Eds.), *Climate Processes and Climate Sensitivity* (pp. 275–287). Washington, D.C.: American Geophysical Union.

Fastook, J. L. (1987). Use of a new finite–element continuity model to study the transient behavior of ice stream C, and causes of its present low velocity. *Journal of Geophysical Research*, 92(B9), 8941–8949.

Fastook, J. L. (1990). A map–plane finite–element program for ice sheet reconstruction: A steady state calibration with Antarctica and a reconstruction of the Laurentide Ice Sheet for 18,000 B.P. In H. U. Brown III (Eds.), *Computer assisted analysis and modeling on the IBM 3090* (pp. 45–80). White Plains: IBM Scientific and Technical Computing Department.

Fastook, J. L. (1993). The finite–element method for solving conservation equations in glaciology. *Computer Magazine*, October issue.

Fastook, J. L., and Schmidt, W. F. (1982). Finite element analysis of calving from ice fronts. *Annals of Glaciology*, 3, 103–106.

Fastook, J. L., and Chapman, J. E. (1989). A map–plane finite–element model: three modeling experiments. *Journal of Glaciology*, 35(119), 48–50.

Fastook, J. L., and Hughes, T. J. (1991). Changing ice loads on Earth's surface during the last glaciation cycle. In R. Sabadini, K. Lamback, and E. Boschi (Eds.), *Glacial Isostasy, Sea Level, and Mantle Rheology* (pp. 165–201). Dordrecht: Kluwer Academic Publishers.

Fastook, J. L., and Holmlund, P. (1994). A glaciological model of the Younger Dryas event in Scandinavia. *Journal of Glaciology*, 40(134), 125–131.

Fastook, J. L., and Prentice, M. L. (1994). A finite–element model of Antarctica: Sensitivity test for meteorological mass balance relationship. *Journal of Glaciology*, 40(134), 167–175.

Ferrigno, J. G., Lucchitta, B. K., Mullins, K. F., Allison, A. L., Allen, R. J., and Gould, W. G. (1993). Velocity measurements and changes in position of Thwaites Glacier/iceberg tongue from aerial photography, Landsat images and NOAA AVRR data. *Annals of Glaciology*, 17, 239–244.

Fisher, D. A., Reeh, N., and Langley, K. (1985). Objective reconstructions of the Late Wisconsinan Laurentide Ice Sheet and the significance of deformable beds. *Gèographié physique et Quaternairé*, 39(3), 229–238.

Fletcher, J. O. (1969). *Ice Extent on the Southern Ocean and its Relation to World Climate*. (Memorandum–RM–5793–NSF). Santa Monica: The Rand Corporation.

Fletcher, J. O. (1972). *Rumination on climate dynamics and program management* (Memorandum): NSF Office of Polar Programs, Washington, D.C.

Flint, R. F. (1971). *Glacial and Quaternary geology*. New York: John Wiley.

Forman, S. L., Thorleifson, L. H., Wintle, A. G., and Wyatt, P. H. (1987). Thermoluminescence properties and age estimates for Quaternary raised marine sediments, Hudson Bay Lowland, Canada. *Canadian Journal of Earth Sciences*, 24, 2405–2411.

Fulton, R. J. (1971). Radiocarbon geochronology of southern British Columbia. *Geological Survey of Canada Paper*, 71(37), 28 p.

Fulton, R. J. (1995). Surficial materials of Canada. *Geological Survey of Canada*, Map 1880A, scale 1:5,000,000.

Funder, S., and Hjort, C. (1973). Aspects of the Weichselian chronology in central East Greenland. *Boreas*, 2, 69–84.

Funk, M., Echelmeyer, K., and Iken, K. (1994). Mechanisms of fast flow in Jakobshavns Isbrae, West Greenland: Part II. Modeling of englacial temperatures. *Journal of Glaciology*,

40(136) 569–585.

Fütterer, D. (1993). ARCTIC '93: A major breakthrough in Arctic Ocean research cooperation. *The Nansen Ice breaker (Newsletter)*, 5, 1 and 8.

Fütterer, D. (Ed.). (1994). *The Expedition ARCTIC '93*. (Vol. 149/94). Bremerhaven: Alfred–Wegener Institute for Polar and Marine Research, 244 pages.

Gallée, H., van Ypersele, J. P., Fichefet, T., Tricot, C., Marsiat, I., and Berger, A. (1991). Simulation of the last glacial cycle by a coupled sectorially averaged climate–ice–sheet model. I. The Climate Model. *Journal of Geophysical Research*, 96, 13,139–13,161.

Gallup, C. D., Edwards, R. L., and Johnson, R. G. (1994). Timing of high sea levels over the past 200,000 years. *Science*, 263, 796–800.

Gataullin, V. N., Polyak, L., Epstein, O., and Romanyuk, B. (1993). Glacigenic deposits of the Central Deep: A key to the Late Quaternary evolution of the eastern Barents Sea. *Boreas*, 22, 47–58.

Gibbons, A. (1993). Geneticists trace the DNA trail of the First Americans. *Science*, 259, 312–313.

Giovinetto, M. B., Bentley, C. R., and Bull, C. B. B. (1989). Choosing between some incompatible regional surface mass balance data sets in Antarctica. *Antarctic Journal of the United States*, 24, 7–13.

Glen, J. W. (1955). The creep of polycrystalline ice. *Proceedings of the Royal Society of London*, Series A(228), 519–538.

Glen, J. W. (1958). The flow law of ice. *International Association of Scientific Hydrology*, 47, 171–183.

Gordon, A. L. (1973). Physical oceanography. *Antarctic Journal of the United States*, VIII(3), 61–69.

Gow, A. J. (1970). Deep core studies of crystal structure and fabrics of Antarctic glacier ice. *United States Army Cold Regions Research and Engineering Laboratory (CRREL) Research Report 282*.

Grant, D. R. (1971). Glaciation of Cape Breton Island, Nova Scotia. *Geological Survey of Canada Paper*, 71(1B), 455–463.

Grant, D. R. (1976). Reconnaissance of early and middle Wisconsinan deposits along the Yarmouth–Digby coast of Nova Scotia. *Geological Survey of Canada Paper*, 76(1B), 363–369.

Grant, D. R. (1977). Altitudinal weathering zones and glacial limits in western Newfoundland with particular reference to Gros Morne National Park. *Geological Survey of Canada Paper*, 77(1A), 455–463.

Gray, J. M., and Coxon, P. (1991). The Loch Lomond Stadial glaciation in Britain and Ireland. In J. Ehlers, P. L. Gibbard, and J. Rose (Eds.), *Glacial Deposits in Great Britain and Ireland* (pp. 89–105). Rotterdam: A.A. Balkema.

Greenberg, J. (1987). *Language in the Americas*. Stanford: Stanford University Press.

GRIP, Greenland Ice–core Project members. (1993). Climate instability during the last interglacial period recorded in the GRIP ice core. *Nature*, 364(6434), 203–207.

Grootes, P. M., Stuiver, M., White, J. W. C., Johnsen, S., and Jouzel, J. (1993). Comparison of oxygen isotope records from the GISP2 and GRIP Greenland ice cores. *Nature*, 366, 552–554.

Grosswald, M. G. (1980). Late Weichselian ice sheet of northern Eurasia. *Quaternary Research*, 13, 1–32.

Grosswald, M. G. (1988). Antarctic–style ice sheet in the Northern Hemisphere: Toward the New global glacial theory. *Data of Glaciological Studies, in Russian, or Polar Geography and Geology*, in English (inadequate translation), 63R or 12E, 3–25R or 239–267E.

Grosswald, M. G. (1993a). Extent and melting history of the late Weichselian ice sheet, the Barents–Kara continental margin (unpublished abstract).

Grosswald, M. G. (1993b). Ice age environments of northern Eurasia (with special reference to Northeastern Siberia), (unpublished abstract).

Grosswald, M. G. (1997). Late Weichselian ice sheets in Arctic and Pacific Siberia. *Quaternary International*, in press.

Grosswald, M. G., and Lasca, N. P. (1993). Oriented lakes of the Arctic Coastal Lowlands: A glacialistic approach to the mystery (unpublished abstract).

Grosswald, M. G., and Spector, V. B. (1993). Glacial landforms of the Tiksi Region (western coast of the Buor–Khaya Bay, North Yakutia). *Geomorfologiya*, 1, 72–82 (In Russian).

Grosswald, M. G., and Hughes, T. (1995). Paleoglaciology's Grand Unsolved Problem. *Journal of Glaciology*, 41(138), 313–332.

Grosswald, M. G., and Hughes, T. (1998). Evidence for Quaternary glaciation of the Sea of Okhotsk. *International Project on Paleolimnology and Late Cenozoic Climate*, Newsletter No. 11.

Grosswald, M. G., Karlén, W., Shishorina, Z., and Bodin, A. (1992). Glacial landforms and the age of deglaciation in the Tiksi area, East Siberia. *Geografiska Annaler*, 74(A), 295–304.

Gruhn, R., and Bryan, A. L. (1984). The record of Pleistocene Meagafauna at Taima–taima, Venezuela. In P. S. Martin and R. G. Klein (Eds.), *Quaternary Extinctions: A Prehistoric Revolution* (pp. 128–137). Tuscon: University of Arizona Press.

Gundestrup, N. S., Dahl–Jensen, D., Johnsen, S. J., and Rossi, A. (1993). Bore–hole survey at dome GRIP 1991. *Cold Regions Science and Technology*, 21, 399–402.

Guthrie, R. D. (1989). Wooly arguments against the mammoth steppe – A new look at the Palynological data. *Review of Archaeology*, 10(1), 16–34.

Guthrie, R. D. (1990). *Frozen Fauna of the Mammoth Steppe: The Story of Blue Babe*. Chicago: The University of Chicago Press.

Hall, D. A. (1992). Siberian site defies theories on peopling: Pebble tools are dated to 3 million years. *Mammoth Trumpet*, 7(3), 1, 4–5.

Hamilton, T. D. (1991). Late Cenozoic glaciation of Alaska. In G. Plafker and H. C. Berg (Eds.), *The Geology of Alaska*, (Vol. G–1 [Chapter 32]). Boulder: Geological Society of America.

Hamilton, T. D., and Brigham–Grette, J. (1991). The last interglaciation in Alaska: Stratigraphy and paleoecology of potential sites. *Quaternary International*, 10–12, 49–71.

Hammer, C. U., Clausen, H. B., Dansgaard, W., Neftel, A., Kristinsdottir, P., and Johnson, E. (1985). Continuous impurity analysis along the Dye 3 deep core. In C. C. Langway, H.

Oeschger, and W. Dansgaard (Eds.), *Greenland Ice Core: Geophysics, Geochemistry, and the Environment* (pp. 90–94). Washington, D.C.: American Geophysical Union.

Harvey, L. D. D. (1988). Climatic impact of ice–age aerosols. *Nature*, 334(6180), 333–335.

Hättestrand, C. (1997). Ribbed moraines in Sweden—distribution pattern and palaeoglaciological implications. *Sedimentary Geology*, 111, 41–56.

Hättestrand, C., and Stroeven, A. P. (1996). Field evidence for wet–based ice sheet erosion from south–central Queen Elizabeth Islands, Northwest Territories, Canada. *Arctic and Alpine Research*, 28(4), 466–474.

Hays, J. D., Imbrie, J., and Shackleton, N. J. (1976). Variations in the Earth's orbit: Pacemaker of the ice ages. *Science*, 1974, 1121–1132.

Hedgpeth, J. W. (1973). Systematic zoology. *Antarctic Journal of the United States*, VIII(3), 106–108.

Heinrich, H. (1988). Origin and consequences of cyclic rafting in the northeast Atlantic Ocean during the past 130,000 years. *Quaternary Research*, 29, 142–152.

Heiser, P. A., Hopkins, D. M., Brigham–Grette, J., Benson, S., Ivanov, V. F., Lozhkin, A., and Svknii, M. (1992). Pleistocene glacial geology of St. Lawrence Island, Alaska. In *Abstracts with Programs – Annual Meeting*, Abstracts . Cincinnati: Geological Society of America, A345–A346.

Hibler, W. D. I. (1979). A dynamic/thermodynamic sea ice model. *Journal of Geophysical Oceanography*, 9, 815–846.

Hibler, W. D. I. (1985). *Numerical modeling of sea ice dynamics and ice thickness characteristics* (A final report No. 85–5). US Army Corps of Engineers – Cold Regions Research and Engineering Laboratory.

Hill, R. (1964). *The Mathematical Theory of Plasticity*. London, Oxford University Press.

Hodgson, D. A. (1994). Episodic ice streams and ice shelves during retreat of the northwesternmost sector of the late Wisconsinan Laurentide Ice Sheet over the central Canadian Arctic Archipelago. *Boreas*, 23, 14–28.

Holdsworth, G. (1977). Tidal interaction with ice shelves. *Annales de Geophysique*, 33(1/2), 133–146.

Hollin, J. T. (1980). Climate and sea level in isotope stage 5: an East Antarctic ice surge at ≈ 95,000 BP? *Nature (London)*, 283, 629-633.

Holmlund, P., and Fastook, J. L. (1993). Numerical modelling provides evidence of a Baltic Ice Stream during the Younger Dryas. *Boreas*, 22, 77–86.

Holmlund, P., and Fastook, J. L. (1995). A time dependent glaciological model of the Weichselian Ice Sheet. *Quaternary International*, 27(2), 53–58.

Hooke, R. LeB. (1970). Morphology of the ice–sheet margin near Thule, Greenland. *Journal of Glaciology*, 9(57), 303–324.

Hooke, R. LeB. (1977). Basal temperatures in polar ice sheets: A qualitative review. *Quaternary Research*, 7, 1–13.

Hooke, R. LeB. (1998). *Principles of Glacier Mechanics*. Upper Saddle River, NJ: Prentice Hall, 248 pages.

Hooke, R. LeB., Hanson, B., Iverson, N. R., Jansson, P., and Fischer, U. H. (1997). Rheology of till beneath Storglaiären, Sweden. *Journal of Glaciology*, 43(143), 172–179.

Horai, S., Kondo, R., Sonoda, S., and Tajima, K. (1993). The First Americans: Different waves of migration to the New World inferred from Mitochondrian DNA sequence polymorphisms. In T. Akazawa (Eds.), *Prehistoric Dispersal of Mongoloids*. New York: Oxford University Press.

Howard, W. R. (1997). A warm future in the past. *Nature*, 388, 418–419.

Hudson, R. D. (1983). Direction of glacial flow across Hans Island, Kennedy Channel, N.W.T., Canada (letter). *Journal of Glaciology*, 29(102), 353–354.

Hughen, K. A., Overpeck, J. T., Peterson, L. C., and Trumbore, S. (1996). Rapid climate changes in the tropical Atlantic region during the last deglaciation. *Nature*, 380, 51–57.

Hughes, B. A., and Hughes, T.J. (1994). Transgressions: Rethinking Beringian glaciation. *Palaeogeography, Palaeoclimatology, Palaeoecology*, 110(3–4), 275–294.

Hughes, T. (1973). Is the West Antarctic Ice Sheet disintegrating? *Journal of Geophysical Research*, 78, 7884–7910.

Hughes, T. (1977). West Antarctic ice streams. *Reviews of Geophysics and Space Physics*, 15(1), 1–46.

Hughes, T. (1981a). Numerical reconstructions of paleo ice sheets. In G. H. Denton and T. Hughes (Eds.), *The Last Great Ice Sheet* (pp. 221–261). New York: Wiley–Interscience.

Hughes, T. (1981b). The weak underbelly of the West Antarctic Ice Sheet (letter). *Journal of Glaciology*, 27(97), 518–525.

Hughes, T. (1981c). Lithosphere deformation by continental ice sheets. *Proceedings of the Royal Society of London*, Series A–378:507–527.

Hughes, T. (1983). On the disintegragtion of ice shelves: The role of fracture. *Journal of Glaciology*, 29(101), 98–117.

Hughes, T. (1985). The Great Cenozoic Ice Sheet. *Palaeogeography, Palaeoclimatology, Palaeoecology*, 50, 9–43.

Hughes, T. (1986a). The Jakobshavns Effect. *Geophysical Research Letters,* 13(1), 46–48.

Hughes, T. (1986b). The marine ice transgression hypothesis. *Geografiska Annaler*, 69A(2), 237–250.

Hughes, T. (1987). Ice dynamics and deglaciation models when ice sheets collapsed. In W. F. Ruddiman and H. E. Wright Jr. (Eds.), *North America and Adjacent Oceans During the Last Deglaciation* (pp. 183–220). Boulder: Geological Society of America.

Hughes, T. (1992a). Abrupt climatic change related to unstable ice–sheet dynamics: toward a new paradigm. *Palaeogeography, Palaeoclimatology, Palaeoecology (Global and Planetary Change Section)*, 97, 203–234.

Hughes, T. (1992b). On the pulling power of ice streams. *Journal of Glaciology*, 38(128), 125–151.

Hughes, T. (1992c). Theoretical calving rates from glaciers along ice walls grounded in water of variable depths. *Journal of Glaciology*, 38(129), 282–294.

Hughes, T. (1995). Ice sheet modelling and the reconstruction of former ice sheets from glacial geo(morpho)logical field data. In J. Menzies (ed.),.*Modern Glacial Environments – Processes, Dynamics, and Sediments* (pp. 77–99, Chapter 3). Oxford (Linacre House, Jordan Hill): Butterworth–Heinemann Ltd.

Hughes, T. J., and Fastook, J. L. (1981). Byrd Glacier: 1978–1979 field results. *Antarctic Journal of the United States*, 16(5), 86–89.

Hughes, T. J., and Nakagawa, M. (1989). Bending shear: The rate–controlling mechanism for calving ice walls. *Journal of Glaciology*, 35(120), 260–266.

Hughes, T. J., Denton, G. H., and Grosswald, M. G. (1977). Was there a late Würm Arctic Ice Sheet? *Nature*, 266, 596–602.

Hughes, T. J., Denton, G. H., Andersen, B. G., Schilling, D. H., Fastook, J. L., and Lingle, C. S. (1981). The last great ice sheets: A global view. In G. H. Denton and T. J. Hughes (Eds.), *The Last Great Ice Sheets* (pp. 263–317). New York: Wiley–Interscience.

Hutter, K. (1983). *Theoretical Glaciology: Material Science of Ice and the Mechanics of Glacier and Ice Sheets*. Rotterdam: D. Reidel.

Huybrechts, P. (1990). A 3–D model for the Antarctic Ice Sheet: a sensitivity study on the glacial–interglacial contrast. *Climate Dynamics*, 5, 79–92.

Huybrechts, P. (1994). The present evolution of the Greenland ice sheet: an assessment by modelling. *Global and Planetary Change*, 3(4), 399–412.

Huybrechts, P. (1996). Basal temperature conditions of the Greenland ice sheet during the glacial cycles. *Annals of Glaciology*, 23, 226–236.

Huybrechts, P., & T'siobel, S. (1995). Thermomechanical modelling of Northern Hemisphere ice sheets with a two-level mass-balance parameterization. *Annals of Glaciology*, 21, 111-116.

Huybrechts, P., Letreguilly, A., and Reeh, N. (1991). The Greenland ice sheet and greenhouse warming. *Global and Planetary Change*, 3(4), 399–412.

Imbrie, J. (1985). A theoretical framework for the Pleistocene ice ages. *Journal of the Geological Society of London*, 142, 417–432.

Imbrie, J., and Imbrie, J. Z. (1980). Modelling the climatic response to orbital variations. *Science*, 207, 943–953.

Imbrie, J., Hays, J. D., Martinson, D., McIntyre, A., Mix, A., Morley, J., Pisias, N., Prell, W., and Shackleton, N. J. (1984). The orbital theory of Pleistocene climate: support for a revised chronology of the marine $\delta^{18}O$ record. In A. L. Berger, I. Imbrie, A. Hays, G. Kukla, and B. Saltzman (Eds.), *Milankovitch and Climate, Part 1* (pp. 269–305). Dordrecht: D. Reidel.

Imbrie, J., Boyle, E. A., Clemens, S. C., Duffy, A., Howard, W. R., Kukla, G., Kutzbach, J., Martinson, D. G., McIntyre, A., Mix, A. C., Molfino, B., Morley, J. J., Peterson, L. C., Pisias, N. G., Prell, W. L., Raymo, M. E., Shackleton, N. J., and Toggweiler, J. R. (1992). On the structure and origin of major glaciation cycles 1. Linear Responses to Milankovitch forcing. *Paleoceanography*, 7(6), 701–738.

Imbrie, J., Berger, A., Boyle, A. E., Clemens, S. C., Duffy, A., Howard, W. R., Kukla, G., Kutzbach, J., Martinson, D. G., McIntyre, A., Mix, A. C., Molfino, B., Morley, J. J., Peterson, L. C., Pisias, N. G., Prell, W. L., Raymo, M. E., Shackleton, N. J., and Toggweiler, J. R. (1993). On the structure and origin of major glaciation cycles, 2, The 100,000–year cycle. *Paleoceanography*, 8(6), 699–735.

Isaksson, E. (1992). The western Barents Sea and the Svalbard archipelago 18,000 years ago – a finite–difference computer model reconstruction. *Journal of Glaciology*, 38(129), 295–301.

Iverson, N. R. (1991). Potential effects of subglacial water–pressure fluctuations on quarrying. *Journal of Glaciology*, 25(125), 27–36.

Iverson, N.R., Hanson, B., Hooke, R. LeB., and Jansson, P. (1995). Flow mechanism of glaciers on soft beds. *Science*, 267(5194), 80–81.

Iverson, R. M. (1985). A constitutive equation for mass movement behavior. *Journal of Geology*, 93, 143–160.

Iverson, R. M. (1986a). Unsteady, nonuniform landslide motion: 1. Theoretical dynamics and the steady datum state. *Journal of Geology*, 94(1), 1–15.

Iverson, R. M. (1986b). Unsteady, nonfunctional landslide motion: 2. Linearized theory and the kinematics of transient response. *Journal of Geology*, 94, 349–364.

Ives, J. D. (1957). Glaciation of the Torngat Mountains. *Geographical Bulletin*, 10, 67–87.

Ives, J. D. (1978). The maximum extent of the Laurentide ice sheet along the east coast of North America during the last glaciation. *Arctic*, 31, 24–53.

Ives, J. D., Andrews, J. T., and Barry, R. G. (1975). Growth and decay of the Laurentide Ice Sheet and comparison with Fenno–Scandinavia. *Die Naturwissenschaften*, 62, 118–125.

Jacobs, S. S., Amos, A. F., and Bruchhausen, P. M. (1970). Ross Sea oceanography and Antarctic bottom water formation. *Deep–Sea Research*, 17, 935–962.

Jacobs, S. S., Hellmer, H., and Jenkins, A. (1996). Antarctic Ice Sheet melting in the Southeast Pacific. *Geophysical Research Letters*, 23(9), 957–960.

Jenssen, D., Budd, W. F., Smith, I. N., and Radok, U. (1985). *On the surging potential of polar ice streams, Part II: Ice streams and physical characteristics of the Ross Sea drainage basin, West Antarctica* No. DE/ER/60197–3). Meteorology Department, University of Melbourne and Cooperative Institute for Research in Environmental Sciences, University of Colorado.

Jenkins, A., Vaughan, D. G., Jacobs, S. S., Hellmer, H. H., and Keys, J. R. (1997). Glaciological and oceanographic evidence of high melt rates beneath Pine Island Glacier, West Antarctica. *Journal of Glaciology*, 43(143), 114–121.

Jezek, K. C. (1984). A modified theory of bottom crevasses used as a means for measuring the buttressing effect of ice shelves on inland ice sheets. *Journal of Geophysical Research*, 89(B3), 1925–1931.

Jin, Y., McNutt, M. K., and Zhu, Y. (1994). Evidence from gravity and topography data for folding of Tibet. *Nature*, 371, 669–674.

Johnsen, S. J., Clausen, H. B., Dansgaard, W., Fuhrer, K., Gundestrup, N., Hammer, C. U., Iverson, P., Jouzel, J., Stauffer, B., and Steffensen, J. P. (1992). Irregular glacial interstadials recorded in a new Greenland ice core. *Nature*, 359, 311–313.

Jones, G. A. (1991). A start–stop ocean conveyor. *Nature*, 349, 364–365.

Jones, G. A. (1994). An abiotic central Arctic during the last glacial maximum: Evidence for an Arctic ice shelf? *EOS (Transactions, American Geophysical Union)*, 75, 226.

Jones, G. A., and Keigwin, L. D. (1988). Evidence from Fram Strait (78°N) for early deglaciation. *Nature*, 336, 56–59.

Jouzel, J., Barkov, N. I., Barnola, J. M., Bender, M., Chappellaz, J.,

Genthon, C., Kotlyakov, V. M., Lipenkov, V., Lorius, C., Petit, J. R., Raynaud, D., Raisbeck, G., Ritz, C., Sowers, T., Stievenard, M., Yiou, F., and P., Y. (1993). Extending the Vostok ice–core record of paleoclimate to the penultimate glacial period. *Nature*, 364, 407–412.

Kaitanen, V. (1969). A geographical study of the morphogenesis of northern Lapland. *Fennia*, 99(5), 1–85.

Kamb, B. (1970). Sliding motion of glaciers: theory and observation. *Reviews of Geophysics and Space Physics*, 8, 673–728.

Kamb, B. (1986). Stress–gradient coupling in glacial flow: III. Exact longitudinal equilibrium equation. *Journal of Glaciology*, 32(112), 335–341.

Kamb, B. (1987). Glacier surge mechanism based on linked cavity configuration of the basal water conduit system. *Journal of Geophysical Research*, 92(B9), 9083–9100.

Kamb, B. (1991). Rheological nonlinearity and flow instability in the deforming bed mechanism of ice stream motion. *Journal of Geophysical Research*, 96(B10), 16,585–16,959.

Kamb, B., and Echelmeyer, K. (1986a). Stress–gradient coupling in glacial flow: I. Longitudinal averaging of the influence of ice thickness and surface slope. *Journal of Glaciology*, 32(111), 267–283.

Kamb, B., and Echelmeyer, K. (1986b). Stress–gradient coupling in glacial flow: IV. Effects on the "T" term. *Journal of Glaciology*, 32(112), 342–349.

Kamb, B., Raymond, C. F., Harrison, W. D., Engelhardt, H., Brugman, M. M., and Pfeffer, T. (1985). Glacier surge mechanism: 1982–1983 surge of Variegated Glacier, Alaska. *Science*, 227(4686), 469–479.

Kapitsa, A. P., Ridley, J. K., Robin, G. d., Siegert, M. J., and Zotikov, I. A. (1996). A large deep fresh water lake beneath the ice of central East Antarctica. *Nature*, 381, 684–686.

Keigwin, L. D. (1995). The North Pacific through the millennia. *Nature*, 377(Letters), 485–486.

Kellogg, T. B., and Kellogg, D. E. (1987a). Late Quaternary deglaciation of the Amundsen Sea: implications for ice sheet modelling. *The Physical Basics of Ice Sheet Modelling*, (Proceedings of the Vancouver Symposium, August, 1987). IAHS Publication no. 170. 349–357.

Kellogg, T. B., and Kellogg, D. E. (1987b). Microfossil distributions in modern Amudnsen Sea sediments. *Marine Micropaleontology*, 12, 203–222.

Kellogg, T. B., and Kellogg, D. E. (1987c). Recent glacial history and rapid ice–stream retreat in the Amundsen Sea. *Journal of Geophysical Research*, 92(B9), 8859–8864.

Kellogg, T. B., Truesdale, R. S., and Osterman, L. E. (1979). Late Quaternary extent of the West Antarctic ice sheet: new evidence from Ross Sea cores. *Geology*, 7, 249–253.

Kellogg, T. B., Kellogg, D. E., and Hughes, T. (1985). Amundsen Sea sediment coring. *Antarctic Journal of the United States, 1985 Review*, 19(5), 79–81.

Kellogg, T. B., Hughes, T. J., and Kellogg, D. E. (1996). Late Pleistocene interactions of East and West Antarctic ice flow regimes: Evidence from the McMurdo Ice Shelf. *Journal of Glaciology*, 42(142), 486–500.

Kennett, J. P. and Ingram, B. L. (1995). A 20,000–year record of ocean circulation and climate change from the Santa Barbara basin. *Nature*, 377(Letters), 510–513.

Kindler, P., Davaud, E., & Strasser, A. (1997). Tyrrhenian coastal deposits from Sardinia (Italy): a petrographic record of high sea levels and shifting climate belts during the last interglacial (isotopic substage 5e). *Palaeogeography, Palaeoclimatology, Palaeoecology*, 133(1–2), 1–25.

Kleman, J. (1990). On the use of glacial striae for reconstruction of paleo–ice sheet flow patterns – With application to the Scandinavian ice sheet. *Geografiska Annaler*, 72A(3–4), 217–236.

Kleman, J. (1992). The palimpsest glacial landscape in northwestern Sweden – Late Weichselian deglaciation forms and traces of older west–centered ice sheets. *Geografiska Annaler*, 74A(4), 305–325.

Kleman, J. (1994). Preservation of landforms under ice sheets and ice caps. *Geomorphology*, 9, 19–32.

Kleman, J., and Borgström, I. (1990). The boulder fields of Mt. Fulufjället, west–centeral Sweden – Late Weichselian boulder blankets and interstadial periglacial phenomena. *Geografiska Annaler*, 72A(1), 63–78.

Kleman, J. (1994). Glacial land forms indicative of a partly frozen bed. *Journal of Glaciology*, 40(135), 255–264.

Kleman, J., and Borgström, I. (1996). Reconstruction of paleo–ice sheets: The use of geomorphological data. *Earth Surface Processes and Landforms*, 21, 893–909.

Kleman, J., and Stroeven, A. P. (1997). Preglacial surface remnants and Quaternary glacial regimes in northwestern Sweden. *Geomorphology*, 19, 35–54.

Kleman, J., Borgström, I., Robertsson, A.–M., and Lilliesköld, M. (1992). Morphology and stratigraphy from several deglaciations in the Transtrand mountains, western Sweden. *Journal of Quaternary Science*, 6, 1–17.

Kleman, J., Borgström, I., and Hättestrand, C. (1994). Evidence for a relict glacial landscape in Quebec–Labrador. *Palaeogeography, Palaeoclimatology, Palaeoecology*, 111(3–4), 217–2228.

Kleman, J., Hättestrand, C., Borgström, I., and Stroeven, A. P. (1997). Fennoscandian paleoglaciology reconstructed using a glacial geological inversion model. *Journal of Glaciology*, 43(145).

Knopoff, L. (1964). The convection current hypothesis. *Reviews of Geophysics*, 2(1), 89–122.

Kominz, M. A., Heath, G. R., Ku, T. L., and Pisias, N. G. (1979). Brunhes time scales and the interpretation of climatic change. *Earth and Planetary Science Letters*, 45, 394–410.

Kotef, C. (1974). The morphologic sequence concept and deglaciation of southern New England. In D. R. Coates (Eds.), *Glacial Geomorphology* (pp. 121–144). Binghampton: University of New York.

Kotef, C., and Pessl, F. J. (1981). Systematic ice retreat in New England. *Geological Survey Professional Paper*, 1179. Washington, D.C.: U.S. Government Printing Office. 19 pages.

Kotilainen, A. T. and Shackleton, N. J. (1995). Rapid climate variability in the North Pacific Ocean during the past 95,000 years. *Nature*, 377(Letters), 323–326.

Kuhle, M. (1987). Subtropical mountain – and highland–glaciation as ice age triggers and the waning of the glacial periods in the Pleistocene. *GeoJournal*, 14(4), 393–421.

Kuhle, M. (1988a). Geomorphological findings on the build-up of Pleistocene glaciation in southern Tibet and on the problem of inland ice – Results of the Shisha Pangma and Mt. Everest Expedition 1984. *GeoJournal,* 17(4), 457–511.

Kuhle, M. (1988b). Topography as a fundamental element of glacial systems: A new approach to ELA calculation and typological classification of paleo–and recent glaciations. *GeoJournal*, 17(4), 545–568.

Kuhle, M. (1988c). The Pleistocene glaciation of Tibet and the onset of ice ages – an autocycle hypothesis. *GeoJournal*, 17(4), 581–595.

Kuhle, M. (1990). The probability of proof in Geomorphology – an example of the application of information theory to a new kind of glacigenetic morphological type, the ice–marginal ramp (Bortensander). *GeoJournal*, 21(3), 195–222.

Kuhle, M. (1991). Observations supporting the Pleistocene inland glaciation of high Asia. *GeoJournal*, 25(2), 133–231.

Kuhle, M. (1994). The problem of historicity in physical geography. *GeoJournal,* 34(4), 339–354.

Kuhle, M.(1995). Glacial isostatic uplift of Tibet as a consequence of a former ice sheet. *GeoJournal*, 37(4), 431–449.

Kuhle, M. (1996). Die Entstehung von Eiszeiten als Folge der Hebung eines subtropischen Hochlandes über die Schneegrenze – dargestellt am Beispiel Tibets. *der Aufschluss*, 47, 145–164.

Kujansuu, R. (1975). Interstadial esker at Marrasjävi, Finnish Lapland. *Geoglogi*, 27(4), 45–50.

Kujansuu, R., and Niemelä, J. (1984). The Quaternary deposits in Finland, 1:100,000 map. Geological Survey of Finland.

Kukla, G. (1987). Loess stratigraphy in Central China. *Quaternary Science Reviews*, 6, 191–219.

Kutzbach, J. E. (1987). Model simulations of the climatic patterns during the deglaciation of North America. In W. F. Ruddiman and H. E. Wright (Eds.), *North America and Adjacent Oceans During the Last Deglaciation* (pp. 424–446). Boulder: The Geological Society of America.

LaBrecque, M. (1990). Clouds and climate: a critical unknown in the global equations. *Mosaic*, 21(3), 2–11.

Lagerbäck, J., and Robertsson, A.–M. (1988). Kettle holes – stratigraphical archives for Weichselian geology and paleoenvironment in northernmost Sweden. *Boreas*, 17, 439–468.

Lagerbäck, R. (1988a). The Veiki moraines in northern Sweden – widespread evidence of an Early Weichselian deglaciation. *Boreas,* 17, 463–486.

Lagerbäck, R. (1988b). Periglacial phenomena in the wooded areas of northern Sweden – relics from the Tärendö interstadial. *Boreas*, 17, 487–500.

Latham, J., and Smith, M. H. (1990). Effect on global warming of wind–dependent aerosol generation at the ocean surface. *Nature*, 347, 372–373.

Laymon, C. A. (1991). Marine episodes in Hudson Strait and Hudson Bay, Canada, during the Wisconsin Glaciation. *Quaternary Research*, 35, 53–62.

Laymon, C. A. (1992). Glacial geology of western Hudson Strait, Canada, with reference to Laurentide Ice Sheet dynamics. *Geological Society of America Bulletin*, 104, 1169–1177.

Lehman, S. J., and Keigwin, L. D. (1992). Sudden changes in North Atlantic circulation during the last deglaciation. *Nature,* 356, 757–762.

Lehman, S. J., Jones, G. A., Keigwin, L. D., Andersen, E. S., Butenko, G., and Ostmo, S.–R. (1991). Initiation of Fennoscandian ice–sheet retreat during the last deglaciation. *Nature*, 349, 513–516.

Letréguilly, A., Huybrechts, P., and Reeh, N. (1991a). Steady state characteristics of the Greenland ice sheet under different climates. *Journal of Glaciology*, 37(125), 149–157.

Letréguilly, A., Reeh, N., and Huybrechts, P. (1991b). The Greenland ice sheet through the last glacial–interglacial cycle. *Palaeogeography, Palaeoclimatology, Palaeoecology (Global and Planetary Change Section)*, 4(4), 385–394.

Licht, K. J., Jennings, A. E., Andres, J. T., and Williams, K. M. (1996). Chronology of late Wisconsin ice retreat from the western Ross Sea, Antarctica. *Geology*, 24(3), 223–226.

Lindstrom, D. R. (1990). The Eurasian Ice Sheet: formation and collapse resulting from natural atmospheric CO_2 concentration variations. *Paleoceanography*, 5(2), 207–227.

Lindstrom, D. R., and Tyler, D. (1984). Preliminary results of Pine Island and Thwaites Glaciers study. *Antarctic Journal of the United States*, 19(5), 53–55.

Lindstrom, D. R., and MacAyeal, D. R. (1986). Paleoclimatic constraints on the maintenance of possible ice–shelf cover in the Norwegian and Greenland seas. *Paleoceanography*, 1, 313–337.

Lindstrom, D. R., and MacAyeal, D. R. (1987). Environmental Constraints on West Antarctic Ice Sheet Formation. *Journal of Glaciology*, 34(115), 128–135.

Lindstrom, D. R., and MacAyeal, D. R. (1989). Scandinavian, Siberian and Arctic Ocean glaciation: effect of Holocene atmospheric CO_2 variations. *Science*, 245, 628–631.

Lindstrom, D. R., and MacAyeal, D. R. (1993). Death of an ice sheet. *Nature*, 365, 214–215.

Lingle, C. S., and Brown, T. J. (1987). A subglacial aquifer bed model and water pressure dependent basal sliding relationship for a West Antarctic ice stream. In C. J. van der Veen and J. Oerlemans (Eds.), *Dynamics of the West Antarctic Ice Sheet* (pp. 249–285). Norwell: D. Reidel.

Lingle, C. S., Hughes, T., and Kollmeyer, R. C. (1981). Tidal flexure of Jakobshavns Glacier, west Greenland. *Journal of Geophysical Research*, 86(B5), 3960–3968.

Lingle, C. S., Schilling, D., Fastook, J. L., and Paterson, W. S. B. (1989). A flowband model of the Ross Ice Shelf, Antarctica: Response to CO_2–nduced climatic warming. *Journal of Geophysical Review–Solid Earth and Planets*, Special Volume 21.

Lingle, C. S., Schilling, D., Fastook, J. L., Paterson, W. S. B., and Brown, T. J. (1991). A flowband model of the Ross Ice Shelf, Antarctica: Response to CO_2–induced climatic warming. *Journal of Geophysical Research*, 96, 6849–6871.

Lliboutry, L. (1968). General theory of subglacial cavitation and sliding of temperate glaciers. *Journal of Glaciology*, 7, 21–58.

Lliboutry, L. (1975). Loi de glissement d'un glacier sans cavitation. *Annales Géophysicae*, 31, 207–226.

Lliboutry, L. (1978). Glissement d'un glacier sur un plan parsemé d'obstacles hémisphériques. *Annales Géophysicae*, 34, 147–162.

Lowell, T. V., Heusser, C. J., Andersen, B. G., Moreno, P. I., Hauser, A., Heusser, L. E., Schlüchter, C., Marchant, D. R., and Denton, G. H. (1995). Interhemispheric correlation of Late Pleistocene glacial events. *Science*, 269, 1541–1549.

Lucchitta, B. K., Rosanova, C. E., and Mullins, K. F. (1995). Velocities of Pine Island Glacier, West Antarctica, from ERS–1 SAR images. *Annals of Glaciology*, 21, 277–283.

Ludwig, K. R., Muhs, D. R., Simmons, K. R., Halley, r. B., and Shinn, E. A. (1996). Sea–level records at ≈ 80 ka from tectonically stable platforms: Florida and Bermuda. *Geology*, 24(3), 211–214.

Lundqvist, J. (1986). Late Weichselian glaciation and deglaciation in Scandinavia. *Quaternary Science Reviews*, 5, 269–292.

Lundqvist, J. (1987). Glaciodynamics of the Younger Dryas marginal zone in Scandinavia. *Geografiska Annaler*, 69(A2), 305–319.

Lundqvist, J. (1992). Glacial stratigraphy in Sweden. In K. Kauranne (Ed.), *Glacial stratigraphy, engineering geology and earth construction* (pp. 43–59). Geological Survey of Finland.

Lundqvist, J., and Saarnisto, M. (1995). Summary of Project IGCP–253. *Quaternary International*, 28, 9–18.

Maasch, K. A. (1992). Ice age dynamics. *Encyclopedia of Earth Systems Science*, 2, 559–569.

MacAyeal, D. R. (1989). Large–scale ice flow over a viscous basal sediment: Theory and application to Ice Stream B, Antarctica. *Journal of Geophysical Research*, 94(B4), 4071–4087.

MacAyeal, D. R. (1992a). A glaciological throttle on fresh water input to the North Atlantic Ocean during glacial climates. In A. F. Spilhaus, S. Cole, and M. C. White (Eds.), *American Geophysical Union Fall Meeting Abstracts*, (p. 158). San Francisco: American Geophysical Union.

MacAyeal, D. R. (1992b). Irregular oscillations of the West Antarctic ice sheet. *Nature*, 359, 29–32.

MacAyeal, D. R. (1993a). A low–order model of the Heinrich–event cycle. *Paleoceanography*, 8(6), 767–773.

MacAyeal, D. R. (1993b). Binge/purge oscillations of the Laurentide ice sheet as a cause of the North Atlantic's Heinrich events. *Paleoceanography*, 8(6), 775–784.

MacAyeal, D. R., Bindschadler, R. A., and Scambos, T. A. (1995). Basal friction of Ice Stream E, West Antarctica. *Journal of Glaciology*, 41(138), 247–262.

Mahrt, L. J., and Schwerdtfeger, W. (1970). Ekman spirals for exponential thermal wind. In *Boundary–Layer Meteorology* (pp. 137–145). Dordrecht: D. Reidel Publishing Company.

Manabe, S., and Broccoli, A. H. (1985). Influence of the continental ice sheets on the climate of an ice age. *Journal of Geophysical Research*, 90, 2167–2190.

Mangerud, J., Svendsen, J. I., Elverhøi, A., Andersen, E., and Solheim, A. (1992). Late Weichselian and Early Holocene Glacial and Climate History of western Spitzbergen, Svalbard. In A. F. Spilhaus Jr., S. Cole, and M. C. White (Eds.), *AGU 1992 Fall Meeting*, (p. 259). San Francisco, CA: American Geophysical Union.

Marshall, S. J., Clarke, G. K. C., Dyke, A. S., and Fisher, D. A. (1996). Geological and topographic controls on fast flow in the Laurentide and Cordilleran Ice Sheets. *Journal of Geophysical Research*, 101, 17827–17839.

Marsiat, I. (1994). Simulation of the Northern Hemisphere continental ice sheets over the last glacial-interglacial cycle: Experiments with a latitude-longitude vertically integrated ice sheet model coupled to a zonally averaged climate model. *Palaeoclimates*, 1(1), 59-98.

Martin, J. H. (1990). Glacial–interglacial CO_2 change: the iron hypothesis. *Paleoceanography*, 5(1), 1–13.

Mayewski, P. A. (1996). *Science and Implementation Plan for the United States Contribution to the International Trans–Antarctic Scientific Expedition (US ITASE)*. Durham, NH: University of New Hampshire, 62 pages.

Mayewski, P. A., Denton, G. H., and Hughes, T. J. (1981). Late Wisconsin ice sheets of North America. In G. H. Denton and T. J. Hughes (Eds.), *The Last Great Ice Sheets* (pp. 67–178). New York: Wiley–Interscience.

Mayewski, P. A., Lyons, W. B., Spencer, M. J., Twickler, M. S., Buck, C. F., and Whitlow, S. (1990). An ice–core record of atmospheric response to anthropogenic sulphate and nitrate. *Nature*, 346, 554–556.

Mayewski, P. A., Meeker, L. D., Whitlow, S., Twickler, M. S., Morrison, M. C., Bloomfield, P., Bond, G. C., Alley, R. B., Gow, A. J., Grootes, P. M., Meese, D. A., Ram, M., Taylor, K. C., and Wumkes, W. (1994). Changes in atmospheric circulation and ocean ice cover over the North Atlantic during the last 41,000 years. *Science*, 263, 1747–1751.

Mayewski, P. A., Meeker, L. D., Twickler, M. S., Whitlow, S., Yang, Q., and Prentice, M. (1997). Major features and forcing of high latitude Northern Hemisphere Atmospheric Circulation Using a 110,000 year long glaciochemical series. *Journal of Geophysical Research*, Special Issue (Oceans/Atmsophere).

McCartney, M. (1997). Is the ocean at the helm? *Nature*, 388, 521–522.

McInnes, B. J., and Budd, W. F. (1984). A cross–sectional model for West Antarctica. *Annals of Glaciology*, 5, 95–99.

McInnes, B., Radok, U., Budd, W. F., and Smith, I. N. (1985). *On the surging potential of polar ice streams, Part I: Sliding and surging of large ice masses – a review* No. DE/ER/60197–2). Cooperative Institute for Research in Environmental Sciences, University of Colorado and Meteorology Department, University of Melbourne.

McInnes, B. J., Budd, W. F., Smith, I. N., and Radok, U. (1986). *On the surging potential of polar ice streams, Part III: Sliding and surging analyses for two West Antarctic Ice Streams* No. DE/ER/60197–4). Meteorology Department, University of Melbourne and Cooperative Institute for Research in Environmental Sciences, University of Colorado.

McWhinnie, M. A. (1973). Physiology and biochemistry. *Antarctic Journal of the United States*, VIII(3), 101–106.

Meeker, L. D., Mayewski, P. A., Twickler, M. S., Whitlow, S. I., and Meese, D. (1997). A 110,000 year history of change in continental biogenic emissions and related atmospheric circulation inferred from the GISP2 Ice Core. *Journal of Geophysical Research*, Special Issue(Oceans/Atmsophere).

Meier, M. F., and Post, A. S. (1962). What are glacier surges? *Canadian Journal of Earth Science*, 6, 807–817.

Melles, M. (Ed.). (1994). *The Expeditions NORLSK/TAYMYR 1993 and BUNGER OASIS 1993/94 of the AWU Research Unit Potsdam*. (Vol. 148/'94). Bremerhaven: Alfred–Wegener Institute for Polar and Marine Research, 80 pages.

Mercer, J. H. (1968a). Antarctic ice and Sangamon sea level.,

International Association of Scientific Hyrdrology Publication Number 79, 217–225.

Mercer, J. H. (1968b). Glacial geology of the Reedy Glacier area, Antarctica. *Geological Society of America Bulletin*, 79, 471–486.

Mercer, J. H. (1969). The Allerod oscillation: A European climatic anomaly. *Arctic and Alpine Research*, 1, 227–234.

Mercer, J. H. (1970). A former ice sheet in the Arctic Ocean. *Palaeogeography, Palaeoclimatology, Palaeoecology*, 8, 19–27.

Mercer, J. H. (1972). Some observations on the glacial geology of the Beardmore Glacier area. In R. J. Adie (Ed.), *Antarctic geology and geophysics*, (pp. 427–433). Oslo: Universitetsforlaget.

Mercer, J. H. (1978). West Antarctic Ice Sheet and CO_2 greenhouse effect: A threat of disaster. *Nature*, 271(5643), 321–325.

Milankovitch, M. M. (1930). *Mathematische Klimalehre und Astronomische Theorie der Klimaschwankungen*. Berlin: Gebruder Borntreger. 176

Milankovitch, M. M. (1941). *Canon of insolation and the ice–age problems*. In Belgrade: Konigllich Serbische Akademie Special Publication 133.

Miller, A., and Thompson, J. C. (1970). Elements of Meteorology. Columbus: Merrill. 402 pages

Miller, G. H., and Kaufman, D. S. (1990). Rapid fluctuations of the Laurentide Ice Sheet at the mouth of Hudson Strait: New evidence for ocean/ice–sheet interactions as a control on the Younger Dryas. *Paleoceanography*, 5(6), 907–919.

Miller, H. G., and de Vernal, A. (1992). Will greenhouse warming lead to Northern Hemisphere ice–sheet growth? *Nature*, 355(Letters to Nature), 244–246.

Mochanov, I. A. (1977). *Drevnétshie Etapy Zaseleniia Chelovekom Severo—Vostochnoi Azii (Ancient Stages of Human Settlement in Northeastern Asia)* Novosibirsk: Izdatrel'stvo "Nauka".

Moran, S. R., Clayton, L., Hooke, R. LeB., Fenton, M. M., and Andriashek, L. D. (1980). Glacier bed landforms of the prairie region of North America. *Journal of Glaciology*, 25(93), 457–476.

Morell, V. (1990). Confusion in earliest America. *Science*, 248, 413–520.

Morlan, R. E., and Cinq–Mars, J. (1989). The peopling of the Americas as seen from northern Yukon Territory. In *First World Summit Conference on the Peopling of the Americas*, . University of Maine:

Mortlock, R. A., Charles, C. D., Froelich, P. N., Zibello, M. A., Saltzman, J., Hays, J. D., and Burckle, L. H. (1991). Evidence for lower productivity in the Antarctic Ocean during the last glaciation. *Nature*, 351, 220–223.

Muhs, D. R., Kennedy, G. L., and Rockwell, T. K. (1994). Uranium–series ages of marine terrace corals from the Pacific coast of North America and implications for last–interglacial sea level history. *Quaternary Research*, 42, 72–87.

Muszynski, I., and Birchfield, G. E. (1987). A coupled marine ice–stream–ice shelf model. *Journal of Glaciology*, 33(113), 3–5.

Nichols, J. (1990). Linguistic diversity and the first settlement of the New World. *Language*, 66, 475–521.

Ninkovitch, D., and Shackleton, N. J. (1975). Distribution, stratigraphic position, and age of ash layer "L" in the Panama Basin region. *Earth and Planetary Science Letters*, 27, 20–34.

Nye, J. F. (1951). The flow of glaciers and ice sheets as a problem in plasticity. *Proceedings of the Royal Society of London*, Series A, 207, 554–572.

Nye, J. F. (1952). A method of calculating the thickness of the ice sheets. *Nature*, 169(4300), 529–530.

Nye, J. F. (1953). The flow law of ice from measurements in glacier tunnels, laboratory experiments and the Jungfraufirn borehole experiment. *Proceedings of the Royal Society of London*, Series A(219), 477–489.

Nye, J. F. (1957). The distribution of stress and velocity in glaciers and ice sheets. *Proceedings of the Royal Society of London*, Series A, 239, 113–133.

Nye, J. F. (1960). *Physical Properties of Crystals*. (Corrected 1st Edition). Oxford: Clarendon Press.

Nye, J. F. (1963a). The response of a glacier to changes in the rate of nourishment and wastage. *Proceedings of the Royal Society of London*, 275(A), 87–112.

Nye, J. F. (1963b). On the theory of the advance and retreat of glaciers. *Royal Astronomical Society Geophysical Journal*, 7, 431–456.

Nye, J. F. (1967). Plasticity solution for a glacier snout. *Journal of Glaciology*, 6, 695–715.

Nye, J. F. (1969). A calculation on the sliding of ice over a wavy surface using a Newtonian viscous approximation. *Proceedings of the Royal Society*, Series A_(311), 445–467.

Nye, J. F. (1970). Glacier sliding without cavitation in a linear viscous approximation. *Proceedings of the Royal Society of London*, Series A (315), 381–403.

Oerlemans, J., and van der Veen, C. J. (1984). *Ice Sheets and Climate* (pp. 61–66). Dordrecht: Reidel.

Oeschger, H., Stauffer, B., Bucher, P., and Moell, M. (1976). Extraction of trace components from large quantities of ice in bore holes. *Journal of Glaciology*, 17, 117–133.

Olafsdottir, T. (1975). Jøkulgardurá sjávaarbotni ut af Breidafirdi. *Náttúrufraedingurinn*, 45(4), 31–36.

Orombelli, G., Baroni, C., and Dentnon, G. H. (1990). Late Cenozoic Glacial History of the Terra Nova Bay Region, Northern Victoria Land, Antarctica. *Geografia Fisica e Dinamica Quaternaria*, 13(2), 139–163.

Oswald, G. K. A., and Robin, G. deQ. (1973). Lakes beneath the Antarctic Ice Sheet. *Nature*, 245, 251–254.

Ota, Y. (1994). *Study on coral reef terraces of the Huon Peninsula, Papua, New Guinea*. Yokohama: Yokohama National University. 429

Parenti, F. (1993) *Le gisement quaternaire de la Toca do Bopqueirão da Pedra Furada (Piaui, Brésil) dans le contexte de la préhistoire américaine*. Fouilles, stratigraphie, chronologie, évolution culturelle thesis, Ecole des Hautes Etudes en Sciences Sociales.

Parish, T. R. (1981). The katabatic winds of Cape Denison and Port Martin. *Polar Record*, 20, 525–532.

Parish, T. R. (1982). Surface airflow over East Antarctica. *Monthly Weather Review*, 110(2), 84–90.

Parish, T. R. (1984). A numerical study of strong katabatic winds over Antarctica. *Monthly Weather Review*, 112, 545–554.

Parish, T. R., and Bromwich, D. H. (1986). The inversion wind pattern over West Antarctica. *Monthly Weather Review*, 114(5), 849–860.

Parish, T. R., and Bromwich, D. H. (1987). The surface windfield over the Antarctic ice sheets. *Nature*, 328(6125), 51–54.

Parish, T. R., and Bromwich, D. H. (1989). Instrumented aircraft observations of the katabatic wind regime near Terra Nova Bay. *Monthly Weather Review*, 117(7), 1570–1585.

Parkinson, C. L., and Washington, W. M. (1979). A large–scale numerical model of sea ice. *Journal of Geophysical Research*, 84, 311–337.

Paterson, W. S. B. (1972). Laurentide Ice Sheet: estimated volumes during late Wisconsin. *Reviews of Geophysics and Space Physics*, 10(4), 885–917.

Paterson, W. S. B. (1981). *The Physics of Glaciers* (Second Edition). Oxford: Pergamon Press. 380 pages.

Paterson, W. S. B. (1991). Why ice–age ice is sometimes "soft". *Cold Regions Science and Technology*, 20, 75–98.

Paterson, W. S. B. (1994). *The Physics of Glaciers* (Third Edition). Oxford, Pergamon Publishing. 480 pages.

Paterson, W. S. B., and Hammer, C. U. (1987). Ice core and other glaciological data. In W. F. Ruddiman and H. E. Wright (Eds.), *North America and Adjacent Oceans During the Last Deglaciation* (pp. 91–109). Boulder: Geological Society of America.

Paterson, W. S. B., Nitsan, U., and Clarke, G. K. C. (1978). An investigation of creep instability as a mechanism for glacier surges. *Materialy Glyatsiologicheskikh Issledovaniy. Khronika Obsuzhdeniya*, 32, 201–209.

Peacock, J. D. (1991). Glacial deposits of the Hebridean region. In J. Ehlers, P. L. Gibbard, and J. Rose (Eds.), *Glacial Deposits in Great Britain and Ireland* (pp. 109–119). Rotterdam: A.A. Balkema.

Peltier, W. R. (1976). Glacial isostatic adjustment: II, The inverse problem. *Journal of the Royal Astronomical Society*, 46, 669–706.

Peltier, W. R. (1983). Glacial geology and glacial isostasy of the Hudson Bay region, *Shorelines and Isostasy*, (pp. 285–319): Institute of British Geographers.

Peltier, W. R. (1987). Glacial isostasy, mantle viscosity, and Pleistocene climatic change. In W. F. Ruddiman and H. E. Wright (Eds.), *North America and Adjacent Oceans During the Last Deglaciaiton* (pp. 155–182). Boulder: Geological Society of North America.

Peltier, W. R. (1994). Ice Age Paleotopography. *Science*, 265, 195–201.

Peltier, W. R. (1995). Response to Paleotopography of glacial–age ice sheets. *Science*, 267(Technical Comments), 536–538.

Peltier, W. R., and Andrews, J. T. (1976). Glacial–isostatic adjustment – I. The forward problem. *Geophysical Journal of the Royal Astronomical Society*, 46, 605–646.

Peltier, W. R., Farrell, W. E., and Clark, J. T. (1978). Glacial isostasy and relative sea level, a global finite element model. *Tectonophysics*, 50, 81–110.

Pelto, M. (1992). Equilibrium line altitude variations with latitude, today and during the Late Wisconsin. *Palaeogeography, Palaeoclimatology, Palaeoecology*, 95, 41–46.

Pelto, M., and Warren, C. R. . (1991). Relationship between tide-water glacier calving velocity and water depth at the calving front. *Annals of Glaciology*, 15, 115–116.

Pelto, M., Higgins, S. M., Hughes, T. J., and Fastook, J. L. (1990). Modeling mass–balance changes during a glaciation cycle. *Annals of Glaciology*, 14, 238–241.

Peng, T. H., and Broecker, W. S. (1991). Dynamical limitations on the Antarctic iron fertilization strategy. *Nature*, 349(227–229).

Perovich, D. K., and Maykut, G. A. (1990). The treatment of shortwave radiation and open water in large–scale models of sea–ice decay. *Annals of Glaciology*, 14.

Pfeffer, W. T., and Bretherton, C. S. (1987). The effect of crevasses on the solar heating of a glacier surface. In *The Physical Basis of Ice–Sheet Modeling* (pp. 1–14). International Association of Hydrological Science.

Pheasant, D. R., and Andrews, J. H. (1972). The Quaternary history of the northern Cumberland Peninsula, Baffin Island, N.W.T. In *24th International Geological Congress, section 12, Part 8, Chronology of Narpaing and Quajon Fiords during the past 120,000 years* (pp. 81–88). Montreal.

Pheasant, D. R., and Andrews, J. H. (1973). Wisconsin glacial chronology and relative sea–level movements, Narpaing Fiord, Broughton Island area, eastern Baffin Island, N.W.T. *Canadian Journal of Earth Sciences*, 10, 1621–1641.

Phillips, R. L., andGrantz, A. (1997). Quaternary history of sea ice and paleoclimate in the Amerasia basin, Arctic Ocean, as recorded in the cyclical strata of Northwind Ridge. *Geological Society of America Bulletin*, 109(9), 1101-1115.

Pisias, N. G., Moore, T. C., Jr., Imbrie, J., and Shackleton, N. J. (1980). The evolution of Pleistocene climate: A time series approach. *Earth and Planetary Science Letters*, 52, 450–453.

Popov, E. P. (1952). *Mechanics of Materials*. Englewood Cliffs: Prentice–Hall., Inc.

Porter, S., Zhisheng, A., and Hongbo, Z. (1992). Cyclic Quaternary alluviation and terracing in a nonglaciated drainage basin on the north flank of the Qinling Shan, Central China. *Quaternary Research*, 38, 157–169.

Prescott, P. (1995, 28 April). *Photogrammetric Examination of the Calving Dynamics of Jakobshavns Isbrae, Greenland*. Unpublished Ph.D. Dissertation, University of Maine.

Prest, V. K. (1969). Retreat of Wisconsin and Recent ice in North America, (Geological Survey of Canada Map 1257A). Ottawa: Geological Survey of Canada.

Punkari, M. (1980). The ice lobes of the Scandinavian ice sheet during the deglaciation in Finland. *Boreas*, 9, 307–310.

Radok, U., Barry, R. G., Jensen, D., Keen, R. A., Kiladis, G. N., and McInnes, B. (1982). *Climatic and Physical Characteristics of the Greenland Ice Sheet, Parts I and II*. Boulder: University of Colorado.

Radok, U., Brown, T. J., Jenssen, D., Smith, I. N., and Budd, W. F. (1986). *On the surging potential of polar ice streams, Part IV: Antarctic ice accumulation basins and their main discharge regions* No. DE/ER/60197–5). Cooperative Institute for Research in Environmental Sciences, University of Colorado and Meteorology Department, University of Melbourne.

Raymond, C. F. (1980). Temperate valley glaciers. In S. C. Colbeck (Ed.), *Dynamics of Snow and Ice Masses* (pp. 79–139). New York: Academic Press Inc.

Raymond, C. F. (1996). Shear margins in glaciers and ice sheets. *Journal of Glaciology*, 42(140), 90–102.

Read, W. T. (1950). Discussion following paper by U. Dehlinger, 1950 – The experimental and theoretical results of plasticity at normal speeds of strain. In *Symposium on Plastic Deformation of Crystalline Solids*, (pp. 111–112). Pittsburgh: Mellon Institute.

Reeh, N. (1968). On the calving of ice from floating glaciers and ice shelves. *Journal of Glaciology*, 7, 215–232.

Reeh, N. (1982). A plasticity theory approach to the steady–state shape of a three–dimensional ice sheet. *Journal of Glaciology*, 28(100), 431–455.

Reeh, N. (1984). Reconstruction of the glacial ice covers of Greenland and the Canadian Arctic islands by three–dimensional, perfectly plastic ice–sheet modeling. *Annals of Glaciology*, 5, 115–121.

Reeh, N. (Ed.). (1994). *Workshop on the calving rate of West Greenland glaciers in response to climate change, 13–15 September 1993*. Danish Polar Center, Copenhagen, Denmark, 171 pages

Richards, D. A., Smart, P. L., and Edwards, R. L. (1994). Maximum sea levels for the last glacial period from U–series ages of submerged speleothems. *Nature*, 367, 357–360.

Rignot, E. J., Gogineni, S. P., Krabill, W. B., and Ekholm, S. (1997). North and Northeast Greenland ice discharge from satellite radar in interferometry. *Science*, 276, 934–937.

Rigsby, G. B. (1958). Effect of hydrostatic pressure on velocity of shear deformation of single ice crystals. *Journal of Glaciology*, 3(24), 271–278.

Ritchie, J. C. (1984). *Past and Present Vegetation of the Far North-west of Canada*. Toronto: University of Toronto Press. 251 pages.

Ritchie, J. C., and Cwynar, L. C. (1982). The Late Quaternary vegetation of the North Yukon. In D. M. Hopkins, J. V. Matthews Jr., C. E. Schweger, and S. B. Young (Eds.), *The Paleoecology of Beringia* (pp. 113–126). New York: Academic Press.

Robin, G. deQ. (1955). Ice movement and temperature distribution in glaciers and ice sheets. *Journal of Glaciology*, 2(18), 523–532.

Robin, G. deQ. (1958). Glaciology III: Seismic shooting and related investigations. *Scientific Results of the Norwegian, British, Swedish Antarctic Expedition, 1949–1952*, 5, 122–125.

Robin, G. deQ. (1967). Surface topography of ice sheets. *Nature*, 215, 1029–1032.

Robin, G. deQ. (1979). Formation, flow, and disintegration of ice shelves. *Journal of Glaciology*, 24, 259–271.

Robin, G. deQ., and Weertman, J. (1973). Cyclic surging of glaciers. *Journal of Glaciology*, 12, 3–18.

Robin, G. deQ., and Millar, D. H. M. (1982). Flow of ice sheets in the vicinity of subglacial peaks. *Annals of Glaciology*, 3, 290–294.

Robin, G. deQ., Swithinbank, C. W. M., and Smith, B. M. E. (1970). Radio–echo exploration of the Antarctic Ice Sheet. *International Association of Scientific Hydrology*, 86, 97–115.

Rohde, L. (1988). Glaciofluvial channels formed prior to the last deglaciation: examples from Swedish Lapland. *Boreas*, 17, 511–516.

Rommelaere, V., and Ritz, C. (1996). A thermomechanical model of ice–shelf flow. *Annals of Glaciology*, 26, 13–20.

Rutter, N. (1995). Problematic ice sheets. *Quaternary International*, 28, 19–37.

Sættem, J., Poole, D. A. R., Ellingsen, K. L., and Sejrup, H. P. (1992). Glacial geology of outer Bjørnøyrenna, southwestern Barents Sea. *Marine Geology*, 103, 15–51.

Salonen, V.–P. (1987). Observations on boulder transport in Finland. *Geological Survey of Finland Special Paper*, 3, 103–110.

Saltzman, B. (1987). Carbon dioxide and the $\delta^{18}0$ record of late–Quaternary climatic change: A global model. *Climate Dynamics*, 1, 77–85.

Saltzman, B., and Maasch, K. A. (1988). Carbon cycle instability as a cause of the late Pleistocene ice age oscillations: Modeling the asymmetric response. *Global Biochemical Cycles*, 2, 177–185.

Saltzman, B., and Maasch, K. A. (1991). A first–order global model of late Cenozoic climatic change II. A simplification of CO_2 dynamics. *Climate Dynamics*, 5, 201–210.

Saltzman, B., and Verbitsky, M. Y. (1992). Asthenospheric ice–load effects in a global dynamical–system model of the Pleistocene climate. *Climate Dynamics*, 8, 1–11.

Saltzman B., and Verbitsky, M. F. (1993). Multiple instabilities and modes of glacial rhythmicity in the Plio–Pleistocene: a general theory of late Cenozoic climatic change. *Climate Dynamics*, 9, 1–15.

Saltzman B., and Verbitsky, M. F. (1994). Late Pleistocene climatic trajectory in the phase space of global ice, ocean state, and CO_2: Observations and theory. Paleoceanography, 9(6), 767–779.

Saltzman B., and Verbitsky, M. F. (1995). Predicting the Vostok CO_2 curve. *Nature*, 377, 690.

Sanderson, T. J. O. (1979). Equilibrium profile of ice shelves. *Journal of Glaciology*, 22, 435–460.

Sanderson, T. J. O., and Doake, C. S. M. (1979). Is vertical shear in an ice shelf negligible? *Journal of Glaciology*, 22, 285–292.

Schmoll, H. R., Szabo, B. J., Rubin, M., and Dobrovolny, E. (1972). Radiometric dating of marine shells from the Bootlegger Cove Clay, Anchorage area, Alaska. *Geological Society of America Bulletin*, 83, 1107–1113.

Schneider, S. H. (1992). Will sea level rise or fall? *Science*, 356, 11–12.

Schnitker, D. (1979). The deep waters of the western North Atlantic during the past 24,000 years, and the reinitiation of the Westerly Boundary Undercurrent. *Marine Micropaleontology*, 4, 265–280.

Schwerdtfeger, W. (1970a). The climate of the Antarctic. In H. E. Landsberg (Ed.), *World Survey of Climatology* (pp. 253–355). Amsterdam: Elsevier Publishing Company.

Schwerdtfeger, W. (1970b). Die Temperaturinversion über dem antarktischen Plateau und die Struktur ihres Windfeldes. *Sonderdruck aus, Meteorologische Rundschau*, 23, 164–171.

Schytt, V., Hoppe, G., Blake, W., Jr., and Grosswald, M. G. (1968). The extent of the Würm glaciation in the European Arctic. *International Association of Sci.entific Hydrology*

I.U.G.G. Publication 79(Comm. of Snow and Ice), 207–216.

Scofield, J. P. (1988) *Flow Charactersitics of an Outlet Glacier: Byrd Glacier, Antarctica*. M.S. Thesis, University of Maine.

Scofield, J. P., Fastook, J. L., and Hughes, T. (1991). Evidence for a frozen bed, Byrd Glacier, Antarctica. *Journal of Geophysical Research*, 96(B7), 11,649–11,655.

Scourse, J., Robinson, E., and Evans, C. (1991). Galciation of the central and southwestern Celtic Sea. In J. Ehlers, P. L. Gibbard, and J. Rose (Eds.), *Glacial deposits in Great Britain and Ireland* (pp. 301–310). Rotterdam: A.A. Balkema.

Shabtaie, S., and Bentley, C. R. (1988). Ice–thickness map of the West Antarctic ice streams by radar sounding. *Annals of Glaciology*, 11, 126–136.

Shabtaie, S., Whillans, I. M., and Bentley, C. R. (1987). The morphology of ice streams A, B, and C, West Antarctica, and their environs. *Journal of Geophysical Research*, 92(B9), 8865–8883.

Shabtaie, S., Bently, C. R., Bindschadler, R. A., and MacAyeal, D. R. (1988). Mass–balance studies of ice streams A, B, and C, West Antarctica, and possible surging behavior of Ice Stream B. *Annals of Glaciology*, 11, 137–149.

Sher, A. V. (1991). Problems of the Last Interglacial in Arctic Siberia. *Quaternary International*, 10–12, 215–222.

Sher, A. V. (1992). Beringian fauna and Early Quaternary mammalian dispersal in Eurasia: Ecological aspects. *Courier Forsc.–Inst. Senckenberg*, 153, 125–133.

Sher, A. V. (1995). Is there any real evidence for a huge shelf ice sheet in East Siberia. *Quaternary International*, 28, 39–40.

Shilts, W. W. (1982). Quaternary evolution of the Hudson/James Bay region. *Naturaliste Canada (Rev. Ecol. Syst.)*, 109, 309.

Simpson, S. (1992). Linguist finds evidence for early peopling of Americas: Diversity of languages indicates long duration. *Mammoth Trumpet*, 7(3), 1, 6–8.

Solheim, A., and Kristoffersen, Y. (1984). The physical environment, western Barents Sea, 1:5000,000: sediment distribution and glacial history of the western Barents Sea. *Norsk Polarinstitutt Skrfter*, 179(B), 1–26.

Spielhagen, R. F. (1992). Timing of deglacial changes in the Arctic Ocean and the deglaciation of NE Greenland. In A. F. Spilhaus, S. Cole, and M. C. White (Ed.), *American Geophysical Union Fall Meeting Abstracts*, (p. 183). San Francisco: American Geophysical Union.

St. Onge, D. A. (1972). Sequence of glacial lakes in north–central Alberta. *Geological Survey of Canada Bulletin, 213*, 16 p.

Starr, D. O. (1987). A cirrus–cloud experiment: Intensive field observations planned for FIRE. *Bulletin of the American Meteorological Society*, 68(2), 119–124.

Stein, R., Nam, S.–I., Schubert, C., Vogt, C., Fütterer, D., and Heinemeier, J. (1994). The last deglaciation event in the Eastern Central Arctic Ocean. *Science*, 264, 692–696.

Stewart, R. W. (1969). The atmosphere and the ocean. *Scientific American*, 221(3), 76–86.

Strahler, A. N. (1973). *Introduction to Physical Geography*. (3rd ed.). New York: Wiley. 468 pages.

Strutt, J. W., Third Baron Rayleigh (1916). On convection currents in a horizontal layer when the higher temperature is on the underside. *Philosophical Magazine (Sixth Series)*, 32(192), 529–546.

Stuiver, M., Denton, G. H., Hughes, T. J., and Fastook, J. L. (1981). History of the marine ice sheet in West Antarctica during the last glaciations: A working hypothesis. In G. H. Denton and T. J. Hughes (Eds.), *The Last Great Ice Sheets* (pp. 319–436). New York: Wiley–Interscience.

Sunder, S. S., and Wu, M. S. (1990). On the constitutive modeling of transient creep polycrystalline ice. *Cold Regions Science and Technology*, 18, 267–294.

Sugden, D. E. (1972). Deglaciation and isostasy in the Sukkertoppen ice cap area, West Greenland. *Arctic and Alpine Research*, 4, 97–117.

Sugden, D. E. (1974). Landscapes of glacial erosion in Greenland and their relationship to ice, topographic and bedrock conditions. *British Geographical Institute Special Publication*, 7, 177–195.

Sugden, D. E. (1977). Reconstruction of the morphology, dynamics, and thermal characteristics of the Laurentide Ice Sheet at its maximum. *Artic and Alpine Research*, 9, 21–47.

Sugden, D. E. (1978). Glacial erosion by the Laurentide Ice Sheet. *Journal of Glaciology*, 20(83), 367–391.

Sugden, D. E., and John, B. (1976). *Glaciers and Landscape – A Geomorphological Approach*. London: Edward Arnold. 320

Sugden, D. E., and Watts, S. H. (1977). Tors, felsenmeer, and glaciation in northern Cumberland Peninsula, Baffin Island. *Canadian Journal of Earth Sciences,* 14, 2817–2823.

Sutherland, D. G. (1991a). Late Devensian glacial deposits and glaciation in Scotland and the adjacent offshore region. In J. Ehlers, P. L. Gibbard, and J. Rose (Eds.), *Glacial Deposits in Great Britain and Ireland* (pp. 53–59). Rotterdam: A.A. Balkema.

Sutherland, D. G. (1991b). The glaciation of the Shetland and Orkney Islands. In J. Ehlers, P. L. Gibbard, and J. Rose (Eds.), *Glacial Deposits in Great Britain and Ireland* (pp. 121–127). Rotterdam: A.A. Balkema.

Sutton, R. T., and Allen, M. R. (1997). Decadal predictability of North Atlantic sea surface temperature and climate. *Nature*, 388, 563–567.

Swithinbank, C. W. M. (1955). Ice shelves. *Geographical Journal*, 121, 64–76.

Swithinbank, C. W. M. (1963). Ice movement of valley glaciers flowing into the Ross Ice Shelf, Antarctica. *Science*, 141(3580), 523–524.

Swithinbank, C. W. M. (1988). Satellite Image Atlas of Glaciers of the World: Antarctica. *United States Geological Survey Professional Paper*, 1386(B), 278 pages.

Sy, A., Rhein, M., Lazier, J. R. N., Koltermann, K. P., Meincke, J., Putzka, A., and Bersch, M. (1997). Surprisingly rapid spreading of newly formed intermediate waters across the North Atlantic Ocean. *Nature*, 386, 675–679.

Szabo, B. J. (1985). Uranium–series dating of fossil corals from marine sediments of southeastern United States Atlantic coastal plain. *Geological Society of America Bulletin*, 96, 398–406.

Szathmary, E. J. E. (1993). Genetics of Aboriginal North Americans. *Evolutionary Anthropology*, 1(6), 202+.

Tarakanov, Y. A., Grosswald, M. G., Kambarov, N. S., and Prixodko, V. A. (1987). New evidence for the connection of the Earth's shape and former glaciation. *Doklady Akad. Nauk*

SSSR, 295(5), 1084–1089.

Taylor, A. (1988). A constraint on the Wisconsinan glacial history, Canadian Arctic Archipelago. *Journal of Quaternary Science*, 3(1), 15–18.

Taylor, K. C., Lamorey, G. W., Doyle, G. A., Alley, R. B., Grootes, P. M., Mayewski, P. A., White, J. W. C., and Barlow, L. K. (1993). The 'flickering switch' of late Pleistocene climate change. *Nature*, 361(letter), 432–435.

Taylor, K. C., Mayewski, P. A., Alley, R. B., Brook, E. J., Gow, A. J., Grootes, P. M., Meese, D. A., Saltzman, E. S., Severinghaus, J. P., Twickler, M. S., White, J. W. C., Whitlow, S., & Zielinski, G. A. (1997). The Holocene-Younger Dryas transition recorded at Summit, Greenland. Science, 278, 825-827.

Teller, J. T. (1995). History and drainage of large ice–dammed lakes along the Laurentide Ice Sheet. *Quaternary International*, 28, 83–92.

Ten Brink, N., and Weideck, A. (1974). Greenland Ice Sheet history since the last glaciation. *Quaternary Research*, 4, 429–440.

Thomas, R. H. (1973a). The creep of ice shelves: theory. *Journal of Glaciology*, 12, 45–53.

Thomas, R. H. (1973b). The creep of ice shelves: interpretation of observed behaviour. *Journal of Glaciology*, 12, 55–70.

Thomas, R. H. (1977). Calving bay dynamics and ice sheet retreat up the St. Lawrence valley system. *Gèographié Physique et Quaternairé*, 31(3–4), 347–356.

Thomas, R. H., and Bentley, C. R. (1978). A model for Holocene retreat of the West Antarctic ice sheet. *Quaternary Research*, 10, 150–170.

Thomas, R. H., and MacAyeal, D. R. (1982). Derived characteristics of the Ross Ice Shelf. *Journal of Glaciology*, 28(100), 397–412.

Thomas, R. H., Sanderson, T. J. O., and Rose, K. E. (1979). Effect of climatic warming on the West Antarctic ice sheet. *Nature*, 277(5695), 355–358.

Thomson, W. (1888). Polar ice–caps and their influence on changing sea levels. *Trans. Geolological Society of Glasgow*, 8, 322–340.

Thorarinsson. (1964). Sudden advance of Vatnajökull outlet glaciers. *Jökull*, 14, 76–89.

Thunell, R. C. and Mortyn, P. G. (1995). Glacial climate instability in the Northeast Pacific Ocean. *Nature*, 376(Letters), 504–506.

Turner, C. G., II (1985). The dental shear for native American origins. In R. Kirk and E. Szathmary (Eds.), *Out of Asia: Peopling the Americas and the Pacific* (pp. 31–78). Canberra: Journal of Pacific History.

Turner, C. G., II (1986). Dentochronological separation estimates for Pacific Rim populations. *Science*, 232, 1140–1142.

Turner, C. G., II (1987). Telltale teeth. *Natural History*, 96(1), 6–10.

Tushingham, A. M., and Peltier, W. R. (1991). Ice–3G: A new global model of Late Pleistocene Deglaciation based upon geophysical predictions of post–glacial relative sea level change. *Journal of Geophysical Research*, 96(B3), 4497–4523.

Van der Veen, C. J. (1983). A note on the equilibrium profile of a free floating ice shelf. In (Vol. 83–15, pp. 15). The Netherlands: Institut voor Meteorologie en Oceanografie, Rijksuniversiteit–Utrecht.

Van der Veen, C. J. (1986). Numerical modelling of ice shelves and ice tongues. *Annales Geophysicae*, 4(B1), 45–54.

Van der Veen, C. J. (1995). Controls on calving rate and basal sliding: Observations from Columbia Glacier, Alaska, prior to and during its rapid retreat, 1976–1993. *Byrd Polar Research Center Report No. 11*, The Ohio State University, 72 pages.

Van der Veen, C. J. (1997). *Calving Glaciers: Report of a Workshop, February 28–March 2, 1997.* BPRC Report No. 15, Byrd Polar Research Center, The Ohio State University, Columbus, Ohio, 194 pages.

Van der Veen, C. J., and Whillans, I. M. (1989). Force budget: I. Theory and numerical methods. *Journal of Glaciology*, 35(119), 53-69.

Van der Veen, C. J., and Whillans, I. M. (1994). Development of fabric in ice. *Cold Regions Science and Technology*, 22, 171–195.

Vartanyan, S. L., Garutt, V. E., and Sher, A. V. (1993). Holocene dwarf mammoths from Wrangel Island in the Siberian Arctic. *Nature*, 362(Letter), 337–340.

Vaughn, D. G. (1993). Relating the occurrence of crevasses to surface strain rates. *Journal of Glaciology*, 39(132), 255–266.

Vaughan, D. G., and Doake, C. S. M. (1996). Recent atmospheric warming and retreat of ice shelves on the Antarctic Peninsula. *Nature*, 379(Letters), 328–330.

Velichko, A. A. (1995). The Pleistocene termination in northern Eurasia. *Quaternary International*, 28, 105–111.

Verbitsky, M. Y., and Oglesby, R. J. (1992). The effect of atmospheric carbon dioxide concentration on continental glaciation of the Northern Hemisphere. *Journal of Geophysical Research*, 97(D5), 5895–5909.

Verbitsky, M. Y., and Saltzman, B. (1994). Heinrich–type glacial surges in a low–order dynamical climate model. *Climate Dynamics*, 10, 39–47.

Verbitsky, M. Y., and Saltzman, B. (1995). A diagnostic analysis of Heinrich glacial surge events. *Paleoceanography*, 10(1), 59–65.

Vernekar, A. D. (1972). Long–period global variations of incoming solar radiation. *Meterological Monographs*, 12(34).

Vincent, J.–S. (1978). Limits of ice advance, glacial lakes, and marine transgressions on Banks Island, District of Franklin: A preliminary interpretation. *Geological Survey of Canada*, 78(13), 53–62.

Vincent, J. T., and Prest, V. K. (1987). The Early Wisconsinan history of the Laurentide Ice Sheet. *Gèographié Physique et Quaternairé*, XLI, 199–213.

Vogt, P. R., Crane, K., and Sundvor, E. (1994). Deep Pleistocene iceberg plowmarks and bottom current effects on the Yermak Plateau, Arctic Ocean: Sidescan and 3.5 k Hz evidence. *Geology*, 22, 403–406.

Vorren, T., and Kristoffersen, Y. (1986). Late Quaternary glaciation in the southwestern Barents Sea. *Boreas*, 15, 51–59.

Vorren, T., Hald, M., and Lebesbye, E. (1988). Late Cenozoic environments in the Barents Sea. *Paleoceanography*, 3, 601–612.

Warren, W. P. (1991). Fenitian (Midlandian) glacial deposits and glaciation in Ireland and the adjacent offshore regions. In J. Ehlers, P. L. Gibbard, and J. Rose (Eds.), *Glacial Deposits in Great Britain and Ireland* (pp. 79–88). Rotterdam: A.A. Balkkema.

Washburn, A. L. (1980). *Geocryology: a survey of periglacial processes and environments.* New York: John Wiley and Sons.

Watkins, N. D. (1973). Marine geology. *Antarctic Journal of the United States*, VIII(3), 69–78.

Weertman, J. (1957a). On the sliding of glaciers. *Journal of Glaciology*, 3(21), 33–38.

Weertman, J. (1957b). Deformation of floating ice shelves. *Journal of Glaciology*, 3(62), 38–42.

Weertman, J. (1961). Stability of ice age ice sheets. *Journal of Geophysical Research*, 66, 3783–3792.

Weertman, J. (1963). Profile and heat balance at the bottom surface of an ice sheet fringed by mountain ranges. *International Association of Scientific Hydrology Publication*, 61, 245–252.

Weertman, J. (1964). The theory of glacier sliding. *Journal of Glaciology*, 5, 287–303.

Weertman, J. (1966). Effect of basal water layer on the dimensions of ice sheets. *Journal of Glaciology*, 6(191–207).

Weertman, J. (1969). Water lubrication mechanism of glacier surges. *Canadian Journal of Earth Sciences*, 6, 929–942.

Weertman, J. (1973). Can a water–filled crevasse reach the bottom surface of a glacier? *International Association of Scientific Hydrology*, Publication 95, 139–145.

Weertman, J. (1974). Stability of the junction of an ice sheet and an ice shelf. *Journal of Glaciology*, 13, 3–11.

Weertman, J. (1976). Glaciology's grand unsolved problem. *Nature*, 260, 284–286.

Weertman, J. (1980). Bottom crevasses. *Journal of Glaciology*, 25(91), 185–188.

Weertman, J. (1986). Basal water and high–pressure basal ice. *Journal of Glaciology*, 32(112), 455–463.

Weertman, J., and Birchfield, G. E. (1982). Subglacial water flow under ice streams and West Antarctic ice sheet stability. *Annals of Glaciology*, 3, 316–320.

Weideck, A. (1968). *Observations on some Holocene glacier fluctuations in West Greenland*, 165. Meddelelser om Grønland, 202 pages.

Weideck, A. (1972). Holocene shorelines and glacial stages in Greenland–An attempt at correlation. *Geological Survey of Greenland Report*, 41, 39 pages.

Weideck, A. (1976). Glaciation of northern Greenland–New evidence. *Polarforschung*, 46, 26–33.

Weinelt, M. S., Sarntheim, M., Vogelsang, E., and Erlenkeuser, H. (1991). Early decay of the Barents Shelf ice sheet – spread of stable isotope signals across the eastern Norwegian Sea. *Norsk Geologisk Tiddskrift*, 71, 137–140.

Weyant, W. S. (1967). The Antarctic atmosphere: Climatology of the surface environment, *Antarctic Map Folio Series 8*. New York: American Geographical Society.

Whillans, I., and Bindschadler. (1988). Mass balance of Ice Stream B, West Antarctica. *Annals of Glaciology*, 11, 187–193.

Whillans, I. M., and Van der Veen, C. J. (1993a). New and improved determination of velocity of Ice Streams B and C, West Antarctica. *Journal of Glaciology*, 39(133), 483–490.

Whillans, I. M., and Van der Veen, C. J. (1993b). Patterns of calculated basal drag on Ice Streams B and C, Antarctica. *Journal of Glaciology*, 39(133), 437–454.

Whillans, I. M., Bolzan, J., and Shabtaie, S. (1987). Velocity of ice streams B, Antarctica, and its mass balance. *Journal of Geophysical Research*, 92(B9), 8895–8902.

Whillans, I. M., Chen, Y. H., van der Veen, C. J., and Hughes, T. (1989). Force Budget III: Application to three–dimensional flow of Byrd Glacier, Antarctica. *Journal of Glaciology*, 35(119), 68–80.

White, J. W. C. (1993). Climate change: Don't touch that dial. *Nature*, 364(6434), 186.

Williams, R. S., Jr., Ferrigno, J. G., Swithinbank, C., Luccitta, B. K., and Seekins, B. A. (1995). Coastal–change and glaciological maps of Antarctica. *Annals of Glaciology*, 21, 284–290.

Wilson, A. T. (1964). Origin of ice ages: an ice shelf theory for Pleistocene glaciation. *Nature*, 201(4915), 147–478.

Wright, H. E. J. (1987). Synthesis: The land south of the ice sheets. In W. F. Ruddiman and H. E. J. Wright (Eds.), *North America and Adjacent Oceans During the Last Deglaciaiton* (pp. 479–487). Boulder: Geological Society of America.

Zhao, Z. (1990) *Measurement, analysis, and modeling of deformation of the shelf–flow, Byrd Glacier*. Doctoral Dissertation Ph.D., University of Maine.

Zilman, J. W., and Dingle, W. R. J. (1973). Meteorology. *Antarctic Journal of the United States*, VIII(3), 111–119.

Zotikov, I. A. (1961). Teplovoi rejim lednika tsentralnoi Antarktidy (Heat regime of the Central Antarctic glacier). *Informatsionny Bulleten Sovetskov Antarktich eskoy Ekspeditsii*, 28.

Zotikov, I. A. (1963). Bottom melting in the central zone of the ice shield on the Antarctic continent and its influence upon the present balance of the ice mass. *Bulletin of the International Association of Scientific Hydrology*, VIIIe(Annee No. 1), 36–44.

Zotikov, I. A. (1986). *The thermophysics of glaciers*. Dordrecht: D. Reidel Publishing Co, 275 pages.

Zumberge, J. H. (1964). Horizontal strain and absolute movement of the Ross ice shelf between Ross Island and Roosevelt Island, Antarctica. *Antarctic Snow and Ice Studies, Antarctic Research Series*, 2, 65–81.

INDEX